# A First Course in Design and Analysis of Experiments

# A First Course in Design and Analysis of Experiments

**Gary W. Oehlert**
**University of Minnesota**

**W. H. Freeman and Company**
**New York**

For Becky
who helped me all the way through
and for Christie and Erica
who put up with a lot while it was getting done

Cover design by Victoria Tomaselli
Cover illustration by Peter Hamlin



Library of Congress Cataloging-in-Publication Data.

Oehlert, Gary W.
A first course in design and analysis of experiments / Gary W. Oehlert.
p. cm.
Includes bibligraphical references and index.
ISBN 0-7167-3510-5
1. Experimental Design I. Title
QA279.O34 2000
519.5—dc21 99-059934



Printed in the United States of America
First printing, 2000

# Contents

# Preface

This text covers the basic topics in experimental design and analysis and is intended for graduate students and advanced undergraduates. Students should have had an introductory statistical methods course at about the level of Moore and McCabe's *Introduction to the Practice of Statistics* (Moore and McCabe 1999) and be familiar with $t$-tests, $p$-values, confidence intervals, and the basics of regression and ANOVA. Most of the text soft-pedals theory and mathematics, but Chapter 19 on response surfaces is a little tougher sledding (eigenvectors and eigenvalues creep in through canonical analysis), and Appendix A is an introduction to the theory of linear models. I use the text in a service course for non-statisticians and in a course for first-year Masters students in statistics. The non-statisticians come from departments scattered all around the university including agronomy, ecology, educational psychology, engineering, food science, pharmacy, sociology, and wildlife.

I wrote this book for the same reason that many textbooks get written: there was no existing book that did things the way I thought was best. I start with single factor, fixed effects, completely randomized designs and cover them thoroughly, including analysis, checking assumptions, and power. I then add factorial treatment structure and random effects to the mix. At this stage, we have a single randomization scheme, a lot of different models for data, and essentially all the analysis techniques we need. I next add blocking designs for reducing variability, covering complete blocks, incomplete blocks, and confounding in factorials. After this I introduce split plots, which can be considered incomplete block designs but really introduce the broader subject of unit structures. Covariate models round out the discussion of variance reduction. I finish with special treatment structures, including fractional factorials and response surface/mixture designs.

This outline is similar in content to a dozen other design texts; how is this book different?

- I include many exercises where the student is required to *choose* an appropriate experimental design for a given situation, or *recognize* the design that was used. Many of the designs in question are from earlier chapters, not the chapter where the question is given. These are important skills that often receive short shrift. See examples on pages 500 and 502.

- I use Hasse diagrams to illustrate models, find test denominators, and compute expected mean squares. I feel that the diagrams provide a much easier and more understandable approach to these problems than the classic approach with tables of subscripts and live and dead indices. I believe that Hasse diagrams should see wider application.
- I spend time trying to sort out the issues with multiple comparisons procedures. These confuse many students, and most texts seem to just present a laundry list of methods and no guidance.
- I try to get students to look beyond saying main effects and/or interactions are significant and to understand the relationships in the data. I want them to learn that understanding what the data have to say is the goal. ANOVA is a tool we use at the beginning of an analysis; it is not the end.
- I describe the difference in philosophy between hierarchical model building and parameter testing in factorials, and discuss how this becomes crucial for unbalanced data. This is important because the different philosophies can lead to different conclusions, and many texts avoid the issue entirely.
- There are three kinds of "problems" in this text, which I have denoted exercises, problems, and questions. Exercises are intended to be simpler than problems, with exercises being more drill on mechanics and problems being more integrative. Not everyone will agree with my classification. Questions are not necessarily more difficult than problems, but they cover more theoretical or mathematical material.

Data files for the examples and problems can be downloaded from the Freeman web site at `http://www.whfreeman.com/`. A second resource is Appendix B, which documents the notation used in the text.

This text contains many formulae, but I try to use formulae only when I think that they will increase a reader's understanding of the ideas. In several settings where closed-form expressions for sums of squares or estimates exist, I do not present them because I do not believe that they help (for example, the Analysis of Covariance). Similarly, presentations of normal equations do not appear. Instead, I approach ANOVA as a comparison of models fit by least squares, and let the computing software take care of the details of fitting. Future statisticians will need to learn the process in more detail, and Appendix A gets them started with the theory behind fixed effects.

Speaking of computing, examples in this text use one of four packages: MacAnova, Minitab, SAS, and S-Plus. MacAnova is a homegrown package that we use here at Minnesota because we can distribute it freely; it runs on Macintosh, Windows, and Unix; and it does everything we need. You can download MacAnova (any version and documentation, even the source) from `http://www.stat.umn.edu/~gary/macanova`. Minitab and SAS are widely used commercial packages. I hadn't used Minitab in twelve years when I started using it for examples; I found it incredibly easy to use. The menu/dialog/spreadsheet interface was very intuitive. In fact, I only opened the manual once, and that was when I was trying to figure out how to do general contrasts (which I was never able to figure out). SAS is far and away the market leader in statistical software. You can do practically every kind of analysis in SAS, but as a novice I spent many hours with the manuals trying to get SAS to do any kind of analysis. In summary, many people swear by SAS, but I found I mostly swore at SAS. I use S-Plus extensively in research; here I've just used it for a couple of graphics.

I need to acknowledge many people who helped me get this job done. First are the students and TA's in the courses where I used preliminary versions. Many of you made suggestions and pointed out mistakes; in particular I thank John Corbett, Alexandre Varbanov, and Jorge de la Vega Gongora. Many others of you contributed data; your footprints are scattered throughout the examples and exercises. Next I have benefited from helpful discussions with my colleagues here in Minnesota, particularly Kit Bingham, Kathryn Chaloner, Sandy Weisberg, and Frank Martin. I thank Sharon Lohr for introducing me to Hasse diagrams, and I received much helpful criticism from reviewers, including Larry Ringer (Texas A&M), Morris Southward (New Mexico State), Robert Price (East Tennessee State), Andrew Schaffner (Cal Poly—San Luis Obispo), Hiroshi Yamauchi (Hawaii—Manoa), and William Notz (Ohio State). My editor Patrick Farace and others at Freeman were a great help. Finally, I thank my family and parents, who supported me in this for years (even if my father did say it looked like a foreign language!).

They say you should never let the camel's nose into the tent, because once the nose is in, there's no stopping the rest of the camel. In a similar vein, student requests for copies of lecture notes lead to student requests for typed lecture notes, which lead to student requests for more complete typed lecture notes, which lead ... well, in my case it leads to a textbook on design and analysis of experiments, which you are reading now. Over the years my students have preferred various more primitive incarnations of this text to other texts; I hope you find this text worthwhile too.

Gary W. Oehlert

# Chapter 1

# Introduction

Researchers use experiments to answer questions. Typical questions might be:

Experiments answer questions

- Is a drug a safe, effective cure for a disease? This could be a test of how AZT affects the progress of AIDS.
- Which combination of protein and carbohydrate sources provides the best nutrition for growing lambs?
- How will long-distance telephone usage change if our company offers a different rate structure to our customers?
- Will an ice cream manufactured with a new kind of stabilizer be as palatable as our current ice cream?
- Does short-term incarceration of spouse abusers deter future assaults?
- Under what conditions should I operate my chemical refinery, given this month's grade of raw material?

This book is meant to help decision makers and researchers design good experiments, analyze them properly, and answer their questions.

## 1.1 Why Experiment?

Consider the spousal assault example mentioned above. Justice officials need to know how they can reduce or delay the recurrence of spousal assault. They are investigating three different actions in response to spousal assaults. The assailant could be warned, sent to counseling but not booked on charges, or arrested for assault. Which of these actions works best? How can they compare the effects of the three actions?

Treatments, experimental units, and responses

This book deals with *comparative experiments*. We wish to compare some *treatments*. For the spousal assault example, the treatments are the three actions by the police. We compare treatments by using them and comparing the outcomes. Specifically, we apply the treatments to *experimental units* and then measure one or more *responses*. In our example, individuals who assault their spouses could be the experimental units, and the response could be the length of time until recurrence of assault. We compare treatments by comparing the responses obtained from the experimental units in the different treatment groups. This could tell us if there are any differences in responses between the treatments, what the estimated sizes of those differences are, which treatment has the greatest estimated delay until recurrence, and so on.

> An experiment is characterized by the treatments and experimental units to be used, the way treatments are assigned to units, and the responses that are measured.

Advantages of experiments

Experiments help us answer questions, but there are also nonexperimental techniques. What is so special about experiments? Consider that:

1. Experiments allow us to set up a direct comparison between the treatments of interest.
2. We can design experiments to minimize any bias in the comparison.
3. We can design experiments so that the error in the comparison is small.
4. Most important, we are in control of experiments, and having that control allows us to make stronger inferences about the nature of differences that we see in the experiment. Specifically, we may make inferences about *causation*.

Control versus observation

This last point distinguishes an experiment from an *observational study*. An observational study also has treatments, units, and responses. However, in the observational study we merely observe which units are in which treatment groups; we don't get to control that assignment.

**Example 1.1** **Does spanking hurt?**

Let's contrast an experiment with an observational study described in Straus, Sugarman, and Giles-Sims (1997). A large survey of women aged 14 to 21 years was begun in 1979; by 1988 these same women had 1239 children between the ages of 6 and 9 years. The women and children were interviewed and tested in 1988 and again in 1990. Two of the items measured were the level of antisocial behavior in the children and the frequency of spanking. Results showed that children who were spanked more frequently in 1988 showed larger increases in antisocial behavior in 1990 than those who were spanked less frequently. Does spanking cause antisocial behavior? Perhaps it does, but there are other possible explanations. Perhaps children who were becoming more troublesome in 1988 may have been spanked more frequently, while children who were becoming less troublesome may have been spanked less frequently in 1988.

The drawback of observational studies is that the grouping into "treatments" is not under the control of the experimenter and its mechanism is usually unknown. Thus observed differences in responses between treatment groups could very well be due to these other hidden mechanisms, rather than the treatments themselves.

Observational studies are useful too

It is important to say that while experiments have some advantages, observational studies are also useful and can produce important results. For example, studies of smoking and human health are observational, but the link that they have established is one of the most important public health issues today. Similarly, observational studies established an association between heart valve disease and the diet drug fen-phen that led to the withdrawal of the drugs fenfluramine and dexfenfluramine from the market (Connolloy et al. 1997 and US FDA 1997).

Causal relationships

Mosteller and Tukey (1977) list three concepts associated with causation and state that two or three are needed to support a causal relationship:

- Consistency
- Responsiveness
- Mechanism.

Consistency means that, all other things being equal, the relationship between two variables is consistent across populations in direction and maybe in amount. Responsiveness means that we can go into a system, change the causal variable, and watch the response variable change accordingly. Mechanism means that we have a step-by-step mechanism leading from cause to effect.

Experiments can demonstrate consistency and responsiveness

In an experiment, we are in control, so we can achieve responsiveness. Thus, if we see a consistent difference in observed response between the various treatments, we can infer that the treatments caused the differences in response. We don't need to know the mechanism—we can demonstrate causation by experiment. (This is not to say that we shouldn't try to learn mechanisms—we should. It's just that we don't need mechanism to infer causation.)

Ethics constrain experimentation

We should note that there are times when experiments are not feasible, even when the knowledge gained would be extremely valuable. For example, we can't perform an experiment proving once and for all that smoking causes cancer in humans. We can observe that smoking is associated with cancer in humans; we have mechanisms for this and can thus infer causation. But we cannot demonstrate responsiveness, since that would involve making some people smoke, and making others not smoke. It is simply unethical.

## 1.2 Components of an Experiment

An experiment has treatments, experimental units, responses, and a method to assign treatments to units.

| Treatments, units, and assignment method specify the *experimental design.* |
|---|

Some authors make a distinction between the selection of treatments to be used, called "treatment design," and the selection of units and assignment of treatments, called "experiment design."

Analysis not part of design, but consider it during planning

Note that there is no mention of a method for analyzing the results. Strictly speaking, the analysis is not part of the design, though a wise experimenter will consider the analysis when planning an experiment. Whereas the design determines the proper analysis to a great extent, we will see that two experiments with similar designs may be analyzed differently, and two experiments with different designs may be analyzed similarly. Proper analysis depends on the design and the kinds of statistical model assumptions we believe are correct and are willing to assume.

Not all experimental designs are created equal. A good experimental design must

- Avoid systematic error
- Be precise
- Allow estimation of error
- Have broad validity.

We consider these in turn.

Comparative experiments estimate differences in response between treatments. If our experiment has systematic error, then our comparisons will be biased, no matter how precise our measurements are or how many experimental units we use. For example, if responses for units receiving treatment one are measured with instrument A, and responses for treatment two are measured with instrument B, then we don't know if any observed differences are due to treatment effects or instrument miscalibrations. Randomization, as will be discussed in Chapter 2, is our main tool to combat systematic error.

Design to avoid systematic error

Even without systematic error, there will be random error in the responses, and this will lead to random error in the treatment comparisons. Experiments are precise when this random error in treatment comparisons is small. Precision depends on the size of the random errors in the responses, the number of units used, and the experimental design used. Several chapters of this book deal with designs to improve precision.

Design to increase precision

Experiments must be designed so that we have an estimate of the size of random error. This permits statistical inference: for example, confidence intervals or tests of significance. We cannot do inference without an estimate of error. Sadly, experiments that cannot estimate error continue to be run.

Design to estimate error

The conclusions we draw from an experiment are applicable to the experimental units we used in the experiment. If the units are actually a statistical sample from some population of units, then the conclusions are also valid for the population. Beyond this, we are extrapolating, and the extrapolation might or might not be successful. For example, suppose we compare two different drugs for treating attention deficit disorder. Our subjects are preadolescent boys from our clinic. We might have a fair case that our results would hold for preadolescent boys elsewhere, but even that might not be true if our clinic's population of subjects is unusual in some way. The results are even less compelling for older boys or for girls. Thus if we wish to have wide validity—for example, broad age range and both genders—then our experimental units should reflect the population about which we wish to draw inference.

Design to widen validity

We need to realize that some compromise will probably be needed between these goals. For example, broadening the scope of validity by using a variety of experimental units may decrease the precision of the responses.

Compromise often needed

## 1.3 Terms and Concepts

Let's define some of the important terms and concepts in design of experiments. We have already seen the terms treatment, experimental unit, and response, but we define them again here for completeness.

**Treatments** are the different procedures we want to compare. These could be different kinds or amounts of fertilizer in agronomy, different long-distance rate structures in marketing, or different temperatures in a reactor vessel in chemical engineering.

**Experimental units** are the things to which we apply the treatments. These could be plots of land receiving fertilizer, groups of customers receiving different rate structures, or batches of feedstock processing at different temperatures.

**Responses** are outcomes that we observe after applying a treatment to an experimental unit. That is, the response is what we measure to judge what happened in the experiment; we often have more than one response. Responses for the above examples might be nitrogen content or biomass of corn plants, profit by customer group, or yield and quality of the product per ton of raw material.

**Randomization** is the use of a known, understood probabilistic mechanism for the assignment of treatments to units. Other aspects of an experiment can also be randomized: for example, the order in which units are evaluated for their responses.

**Experimental Error** is the random variation present in all experimental results. Different experimental units will give different responses to the same treatment, and it is often true that applying the same treatment over and over again to the same unit will result in different responses in different trials. Experimental error does not refer to conducting the wrong experiment or dropping test tubes.

**Measurement units** (or response units) are the actual objects on which the response is measured. These may differ from the experimental units. For example, consider the effect of different fertilizers on the nitrogen content of corn plants. Different field plots are the experimental units, but the measurement units might be a subset of the corn plants on the field plot, or a sample of leaves, stalks, and roots from the field plot.

**Blinding** occurs when the evaluators of a response do not know which treatment was given to which unit. Blinding helps prevent bias in the evaluation, even unconscious bias from well-intentioned evaluators. Double blinding occurs when both the evaluators of the response and the (human subject) experimental units do not know the assignment of treatments to units. Blinding the subjects can also prevent bias, because subject responses can change when subjects have expectations for certain treatments.

**Control** has several different uses in design. First, an experiment is *controlled* because we as experimenters assign treatments to experimental units. Otherwise, we would have an observational study.

Second, a *control* treatment is a "standard" treatment that is used as a baseline or basis of comparison for the other treatments. This control treatment might be the treatment in common use, or it might be a null treatment (no treatment at all). For example, a study of new pain killing drugs could use a standard pain killer as a control treatment, or a study on the efficacy of fertilizer could give some fields no fertilizer at all. This would control for average soil fertility or weather conditions.

**Placebo** is a null treatment that is used when the act of applying a treatment—any treatment—has an effect. Placebos are often used with human subjects, because people often respond to any treatment: for example, reduction in headache pain when given a sugar pill. Blinding is important when placebos are used with human subjects. Placebos are also useful for nonhuman subjects. The apparatus for spraying a field with a pesticide may compact the soil. Thus we drive the apparatus over the field, without actually spraying, as a placebo treatment.

**Factors** combine to form treatments. For example, the baking treatment for a cake involves a given time at a given temperature. The treatment is the combination of time and temperature, but we can vary the time and temperature separately. Thus we speak of a time factor and a temperature factor. Individual settings for each factor are called *levels* of the factor.

**Confounding** occurs when the effect of one factor or treatment cannot be distinguished from that of another factor or treatment. The two factors or treatments are said to be confounded. Except in very special circumstances, confounding should be avoided. Consider planting corn variety A in Minnesota and corn variety B in Iowa. In this experiment, we cannot distinguish location effects from variety effects—the variety factor and the location factor are confounded.

## 1.4 Outline

Here is a road map for this book, so that you can see how it is organized. The remainder of this chapter gives more detail on experimental units and responses. Chapter 2 elaborates on the important concept of randomization. Chapters 3 through 7 introduce the basic experimental design, called the Completely Randomized Design (CRD), and describe its analysis in considerable detail. Chapters 8 through 10 add factorial treatment structure to the CRD, and Chapters 11 and 12 add random effects to the CRD. The idea is that we learn these different treatment structures and analyses in the simplest design setting, the CRD. These structures and analysis techniques can then be used almost without change in the more complicated designs that follow.

We begin learning new experimental designs in Chapter 13, which introduces complete block designs. Chapter 14 introduces general incomplete blocks, and Chapters 15 and 16 deal with incomplete blocks for treatments with factorial structure. Chapter 17 introduces covariates. Chapters 18 and 19 deal with special treatment structures, including fractional factorials and response surfaces. Finally, Chapter 20 provides a framework for planning an experiment.

## 1.5 More About Experimental Units

Experimental and measurement units

Experimentation is so diverse that there are relatively few general statements that can be made about experimental units. A common source of difficulty is the distinction between experimental units and measurement units. Consider an educational study, where six classrooms of 25 first graders each are assigned at random to two different reading programs, with all the first graders evaluated via a common reading exam at the end of the school year. Are there six experimental units (the classrooms) or 150 (the students)?

Experimental unit could get any treatment

One way to determine the experimental unit is via the consideration that an experimental unit should be able to receive any treatment. Thus if students were the experimental units, we could see more than one reading program in each classroom. However, the nature of the experiment makes it clear that all the students in the classroom receive the same program, so the classroom as a whole is the experimental unit. We don't measure how a classroom reads, though; we measure how students read. Thus students are the measurement units for this experiment.

There are many situations where a treatment is applied to group of objects, some of which are later measured for a response. For example,

- Fertilizer is applied to a plot of land containing corn plants, some of which will be harvested and measured. The plot is the experimental unit and the plants are the measurement units.
- Ingots of steel are given different heat treatments, and each ingot is punched in four locations to measure its hardness. Ingots are the experimental units and locations on the ingot are measurement units.

- Mice are caged together, with different cages receiving different nutritional supplements. The cage is the experimental unit, and the mice are the measurement units.

Use a summary of the measurement unit responses as experimental unit response

Treating measurement units as experimental usually leads to overoptimistic analysis more—we will reject null hypotheses more often than we should, and our confidence intervals will be too short and will not have their claimed coverage rates. The usual way around this is to determine a single response for each experimental unit. This single response is typically the average or total of the responses for the measurement units within an experimental unit, but the median, maximum, minimum, variance or some other summary statistic could also be appropriate depending on the goals of the experiment.

Size of units

A second issue with units is determining their "size" or "shape." For agricultural experiments, a unit is generally a plot of land, so size and shape have an obvious meaning. For an animal feeding study, size could be the number of animals per cage. For an ice cream formulation study, size could be the number of liters in a batch of ice cream. For a computer network configuration study, size could be the length of time the network is observed under load conditions.

Edge may be different than center

Not all measurement units in an experimental unit will be equivalent. For the ice cream, samples taken near the edge of a carton (unit) may have more ice crystals than samples taken near the center. Thus it may make sense to plan the units so that the ratio of edge to center is similar to that in the product's intended packaging. Similarly, in agricultural trials, guard rows are often planted to reduce the effect of being on the edge of a plot. You don't want to construct plots that are all edge, and thus all guard row. For experiments that occur over time, such as the computer network study, there may be a transient period at the beginning before the network moves to steady state. You don't want units so small that all you measure is transient.

More experimental units, fewer measurement units usually better

One common situation is that there is a fixed resource available, such as a fixed area, a fixed amount of time, or a fixed number of measurements. This fixed resource needs to be divided into units (and perhaps measurement units). How should the split be made? In general, more experimental units with fewer measurement units per experimental unit works better (see, for example, Fairfield Smith 1938). However, smaller experimental units are inclined to have greater edge effect problems than are larger units, so this recommendation needs to be moderated by consideration of the actual units.

Independence of units

A third important issue is that the response of a given unit should not depend on or be influenced by the treatments given other units or the responses of other units. This is usually ensured through some kind of separation of the units, either in space or time. For example, a forestry experiment would provide separation between units, so that a fast-growing tree does not shade trees in adjacent units and thus make them grow more slowly; and a drug trial giving the same patient different drugs in sequence would include a washout periodbetween treatments, so that a drug would be completely out of a patient's system before the next drug is administered.

When the response of a unit is influenced by the treatment given to other units, we get confounding between the treatments, because we cannot estimate treatment response differences unambiguously. When the response of a unit is influenced by the response of another unit, we get a poor estimate of the precision of our experiment. In particular, we usually overestimate the precision. Failure to achieve this independence can seriously affect the quality of any inferences we might make.

Sample size

A final issue with units is determining how many units are required. We consider this in detail in Chapter 7.

## 1.6 More About Responses

Primary response

We have been discussing "the" response, but it is a rare experiment that measures only a single response. Experiments often address several questions, and we may need a different response for each question. Responses such as these are often called *primary* responses, since they measure the quantity of primary interest for a unit.

Surrogate responses

We cannot always measure the primary response. For example, a drug trial might be used to find drugs that increase life expectancy after initial heart attack: thus the primary response is years of life after heart attack. This response is not likely to be used, however, becauseit may be decades before the patients in the study die, and thus decades before the study is completed. For this reason, experimenters use *surrogate* responses. (It isn't only impatience; it becomes more and more difficult to keep in contact with subjects as time goes on.)

Surrogate responses are responses that are supposed to be related to—and predictive for—the primary response. For example, we might measure the fraction of patients still alive after five years, rather than wait for their actual lifespans. Or we might have an instrumental reading of ice crystals in ice cream, rather than use a human panel and get their subjective assessment of product graininess.

Surrogate responses are common, but not without risks. In particular, we may find that the surrogate response turns out not to be a good predictor of the primary response.

### Cardiac arrhythmias

**Example 1.2**

Acute cardiac arrhythmias can cause death. Encainide and flecanide acetate are two drugs that were known to suppress acute cardiac arrhythmias and stabilize the heartbeat. Chronic arrhythmias are also associated with sudden death, so perhaps these drugs could also work for nonacute cases. The Cardiac Arrhythmia Suppression Trial (CAST) tested these two drugs and a placebo (CAST Investigators 1989). The real response of interest is survival, but regularity of the heartbeat was used as a surrogate response. Both of these drugs were shown to regularize the heartbeat better than the placebo did. Unfortunately, the real response of interest (survival) indicated that the regularized pulse was too often 0. These drugs did improve the surrogate response, but they were actually worse than placebo for the primary response of survival.

By the way, the investigators were originally criticized for including a placebo in this trial. After all, the drugs were *known* to work. It was only the placebo that allowed them to discover that these drugs should not be used for chronic arrhythmias.

Predictive responses

In addition to responses that relate directly to the questions of interest, some experiments collect *predictive* responses. We use predictive responses to model theprimary response. The modeling is done for two reasons. First, such modeling can be used to increase the precision of the experiment and the comparisons of interest. In this case, we call the predictive responses *covariates* (see Chapter 17). Second, the predictive responses may help us understand the mechanism by which the treatment is affecting the primary response. Note, however, that since we observed the predictive responses rather than setting them experimentally, the mechanistic models built using predictive responses are observational.

Audit responses

A final class of responses is *audit* responses. We use audit responses to ensure that treatments were applied as intended and to check that environmental conditions have not changed. Thus in a study looking at nitrogen fertilizers, we might measure soil nitrogen as a check on proper treatment application, and we might monitor soil moisture to check on the uniformity of our irrigation system.

# Chapter 2

# Randomization and Design

We characterize an experiment by the treatments and experimental units to be used, the way we assign the treatments to units, and the responses we measure. An experiment is *randomized* if the method for assigning treatments to units involves a known, well-understood probabilistic scheme. The probabilistic scheme is called a *randomization*. As we will see, an experiment may have several randomized features in addition to the assignment of treatments to units. Randomization is one of the most important elements of a well-designed experiment.

Randomization to assign treatment to units

Let's emphasize first the distinction between a random scheme and a "haphazard" scheme. Consider the following potential mechanisms for assigning treatments to experimental units. In all cases suppose that we have four treatments that need to be assigned to 16 units.

Haphazard is not randomized

- We use sixteen identical slips of paper, four marked with A, four with B, and so on to D. We put the slips of paper into a basket and mix them thoroughly. For each unit, we draw a slip of paper from the basket and use the treatment marked on the slip.
- Treatment A is assigned to the first four units we happen to encounter, treatment B to the next four units, and so on.
- As each unit is encountered, we assign treatments A, B, C, and D based on whether the "seconds" reading on the clock is between 1 and 15, 16 and 30, 31 and 45, or 46 and 60.

The first method clearly uses a precisely-defined probabilistic method. We understand how this method makes it assignments, and we can use this method to obtain statistically equivalent randomizations in replications of the experiment.

The second two methods might be described as "haphazard"; they are not predictable and deterministic, but they do not use a randomization. It is difficult to model and understand the mechanism that is being used. Assignment here depends on the order in which units are encountered, the elapsed time between encountering units, how the treatments were labeled A, B, C, and D, and potentially other factors. I might not be able to replicate your experiment, simply because I tend to encounter units in a different order, or I tend to work a little more slowly. The second two methods are not randomization.

| Haphazard is not randomized. |
|---|

Introducing more randomness into an experiment may seem like a perverse thing to do. After all, we are always battling against random experimental error. However, random assignment of treatments to units has two useful consequences:

Two reasons for randomizing

1. Randomization protects against confounding.
2. Randomization can form the basis for inference.

Randomization is rarely used for inference in practice, primarily due to computational difficulties. Furthermore, some statisticians (Bayesian statisticians in particular) disagree about the usefulness of randomization as a basis for inference.[1] However, the success of randomization in the protection against confounding is so overwhelming that randomization is almost universally recommended.

## 2.1 Randomization Against Confounding

We defined confounding as occurring when the effect of one factor or treatment cannot be distinguished from that of another factor or treatment. How does randomization help prevent confounding? Let's start by looking at the trouble that can happen when we don't randomize.

Consider a new drug treatment for coronary artery disease. We wish to compare this drug treatment with bypass surgery, which is costly and invasive. We have 100 patients in our pool of volunteers that have agreed via

[1]Statisticians don't always agree on philosophy or methodology. This is the first of several ongoing little debates that we will encounter.

informed consent to participate in our study; they need to be assigned to the two treatments. We then measure five-year survival as a response.

What sort of trouble can happen if we fail to randomize? Bypass surgery is a major operation, and patients with severe disease may not be strong enough to survive the operation. It might thus be tempting to assign the stronger patients to surgery and the weaker patients to the drug therapy. This confounds strength of the patient with treatment differences. The drug therapy would likely have a lower survival rate because it is getting the weakest patients, even if the drug therapy is every bit as good as the surgery.

Failure to randomize can cause trouble

Alternatively, perhaps only small quantities of the drug are available early in the experiment, so that we assign more of the early patients to surgery, and more of the later patients to drug therapy. There will be a problem if the early patients are somehow different from the later patients. For example, the earlier patients might be from your own practice, and the later patients might be recruited from other doctors and hospitals. The patients could differ by age, socioeconomic status, and other factors that are known to be associated with survival.

There are several potential randomization schemes for this experiment; here are two:

- Toss a coin for every patient; heads—the patient gets the drug, tails—the patient gets surgery.
- Make up a basket with 50 red balls and 50 white balls well mixed together. Each patient gets a randomly drawn ball; red balls lead to surgery, white balls lead to drug therapy.

Note that for coin tossing the numbers of patients in the two treatment groups are random, while the numbers are fixed for the colored ball scheme.

Here is how randomization has helped us. No matter which features of the population of experimental units are associated with our response, our randomizations put approximately half the patients with these features in each treatment group. Approximately half the men get the drug; approximately half the older patients get the drug; approximately half the stronger patients get the drug; and so on. These are not exactly 50/50 splits, but the deviation from an even split follows rules of probability that we can use when making inference about the treatments.

Randomization balances the population on average

This example is, of course, an oversimplification. A real experimental design would include considerations for age, gender, health status, and so on. The beauty of randomization is that it helps prevent confounding, even for factors that we do not know are important.

Here is another example of randomization. A company is evaluating two different word processing packages for use by its clerical staff. Part of the evaluation is how quickly a test document can be entered correctly using the two programs. We have 20 test secretaries, and each secretary will enter the document twice, using each program once.

As expected, there are potential pitfalls in nonrandomized designs. Suppose that all secretaries did the evaluation in the order A first and B second. Does the second program have an advantage because the secretary will be familiar with the document and thus enter it faster? Or maybe the second program will be at a disadvantage because the secretary will be tired and thus slower.

Two randomized designs that could be considered are:

1. For each secretary, toss a coin: the secretary will use the programs in the orders AB and BA according to whether the coin is a head or a tail, respectively.
2. Choose 10 secretaries at random for the AB order, the rest get the BA order.

Different randomizations are different designs

Both these designs are randomized and will help guard against confounding, but the designs are slightly different and we will see that they should be analyzed differently.

Cochran and Cox (1957) draw the following analogy:

> Randomization is somewhat analogous to insurance, in that it is a precaution against disturbances that may or may not occur and that may or may not be serious if they do occur. It is generally advisable to take the trouble to randomize even when it is not expected that there will be any serious bias from failure to randomize. The experimenter is thus protected against unusual events that upset his expectations.

Randomization generally costs little in time and trouble, but it can save us from disaster.

## 2.2 Randomizing Other Things

We have taken a very simplistic view of experiments; "assign treatments to units and then measure responses" hides a multitude of potential steps and choices that will need to be made. Many of these additional steps can be randomized, as they could also lead to confounding. For example:

- If the experimental units are not used simultaneously, you can randomize the order in which they are used.
- If the experimental units are not used at the same location, you can randomize the locations at which they are used.
- If you use more than one measuring instrument for determining response, you can randomize which units are measured on which instruments.

When we anticipate that one of these might cause a change in the response, we can often design that into the experiment (for example, by using blocking; see Chapter 13). Thus I try to design for the known problems, and randomize everything else.

**Example 2.1**

### One tale of woe

I once evaluated data from a study that was examining cadmium and other metal concentrations in soils around a commercial incinerator. The issue was whether the concentrations were higher in soils near the incinerator. They had eight sites selected (matched for soil type) around the incinerator, and took ten random soil samples at each site.

The samples were all sent to a commercial lab for analysis. The analysis was long and expensive, so they could only do about ten samples a day. Yes indeed, there was almost a perfect match of sites and analysis days. Several elements, including cadmium, were only present in trace concentrations, concentrations that were so low that instrument calibration, which was done daily, was crucial. When the data came back from the lab, we had a very good idea of the variability of their calibrations, and essentially no idea of how the sites differed.

The lab was informed that all the trace analyses, including cadmium, would be redone, all on one day, in a random order that we specified. Fortunately I was not a party to the question of who picked up the $75,000 tab for reanalysis.

## 2.3 Performing a Randomization

Once we decide to use randomization, there is still the problem of actually doing it. Randomizations usually consist of choosing a random order for a set of objects (for example, doing analyses in random order) or choosing random subsets of a set of objects (for example, choosing a subset of units for treatment A). Thus we need methods for putting objects into random orders

Random orders and random subsets

and choosing random subsets. When the sample sizes for the subsets are fixed and known (as they usually are), we will be able to choose random subsets by first choosing random orders.

Randomization methods can be either physical or numerical. Physical randomization is achieved via an actual physical act that is believed to produce random results with known properties. Examples of physical randomization are coin tosses, card draws from shuffled decks, rolls of a die, and tickets in a hat. I say "believed to produce random results with known properties" because cards can be poorly shuffled, tickets in the hat can be poorly mixed, and skilled magicians can toss coins that come up heads every time. Large scale embarrassments due to faulty physical randomization include poor mixing of Selective Service draft induction numbers during World War II (see Mosteller, Rourke, and Thomas 1970). It is important to make sure that any physical randomization that you use is done well.

Physical randomization

Physical generation of random orders is most easily done with cards or tickets in a hat. We must order $N$ objects. We take $N$ cards or tickets, numbered 1 through $N$, and mix them well. The first object is then given the number of the first card or ticket drawn, and so on. The objects are then sorted so that their assigned numbers are in increasing order. With good mixing, all orders of the objects are equally likely.

Physical random order

Once we have a random order, random subsets are easy. Suppose that the $N$ objects are to be broken into $g$ subsets with sizes $n_1, \ldots, n_g$, with $n_1 + \cdots + n_g = N$. For example, eight students are to be grouped into one group of four and two groups of two. First arrange the objects in random order. Once the objects are in random order, assign the first $n_1$ objects to group one, the next $n_2$ objects to group two, and so on. If our eight students were randomly ordered 3, 1, 6, 8, 5, 7, 2, 4, then our three groups would be (3, 1, 6, 8), (5, 7), and (2, 4).

Physical random subsets from random orders

Numerical randomization uses numbers taken from a table of "random" numbers or generated by a "random" number generator in computer software. For example, Appendix Table D.1 contains random digits. We use the table or a generator to produce a random ordering for our $N$ objects, and then proceed as for physical randomization if we need random subsets.

Numerical randomization

We get the random order by obtaining a random number for each object, and then sorting the objects so that the random numbers are in increasing order. Start arbitrarily in the table and read numbers of the required size sequentially from the table. If any number is a repeat of an earlier number, replace the repeat by the next number in the list so that you get $N$ different numbers. For example, suppose that we need 5 numbers and that the random numbers in the table are (4, 3, 7, 4, 6, 7, 2, 1, 9, ...). Then our 5 selected numbers would be (4, 3, 7, 6, 2), the duplicates of 4 and 7 being discarded.

Numerical random order

Now arrange the objects so that their selected numbers are in ascending order. For the sample numbers, the objects, A through E would be reordered E, B, A, D, C. Obviously, you need numbers with more digits as $N$ gets larger.

Getting rid of duplicates makes this procedure a little tedious. You will have fewer duplicates if you use numbers with more digits than are absolutely necessary. For example, for 9 objects, we could use two- or three-digit numbers, and for 30 objects we could use three- or four-digit numbers. The probabilities of 9 random one-, two-, and three-digit numbers having no duplicates are .004, .690, and .965; the probabilities of 30 random two-, three-, and four-digit numbers having no duplicates are .008, .644, and .957 respectively.

Longer random numbers have fewer duplicates

Many computer software packages (and even calculators) can produce "random" numbers. Some produce random integers, others numbers between 0 and 1. In either case, you use these numbers as you would numbers formed by a sequence of digits from a random number table. Suppose that we needed to put 6 units into random order, and that our random number generator produced the following numbers: .52983, .37225, .99139, .48011, .69382, .61181. Associate the 6 units with these random numbers. The second unit has the smallest random number, so the second unit is first in the ordering; the fourth unit has the next smallest random number, so it is second in the ordering; and so on. Thus the random order of the units is B, D, A, F, E, C.

The word *random* is quoted above because these numbers are not truly random. The numbers in the table are the same every time you read it; they don't change unpredictably when you open the book. The numbers produced by the software package are from an algorithm; if you know the algorithm you can predict the numbers perfectly. They are technically *pseudorandom* numbers; that is, numbers that possess many of the attributes of random numbers so that they appear to be random and can usually be used in place of random numbers.

Pseudorandom numbers

## 2.4 Randomization for Inference

Nearly all the analysis that we will do in this book is based on the normal distribution and linear models and will use $t$-tests, F-tests, and the like. As we will see in great detail later, these procedures make assumptions such as "The responses in treatment group A are independent from unit to unit and follow a normal distribution with mean $\mu$ and variance $\sigma^2$." Nowhere in the design of our experiment did we do anything to make this so; all we did was randomize treatments to units and observe responses.

**Table 2.1:** Auxiliary manual times runstitching a collar for 30 workers under standard (S) and ergonomic (E) conditions.

| # | S | E | # | S | E | # | S | E |
|---|---|---|---|---|---|---|---|---|
| 1 | 4.90 | 3.87 | 11 | 4.70 | 4.25 | 21 | 5.06 | 5.54 |
| 2 | 4.50 | 4.54 | 12 | 4.77 | 5.57 | 22 | 4.44 | 5.52 |
| 3 | 4.86 | 4.60 | 13 | 4.75 | 4.36 | 23 | 4.46 | 5.03 |
| 4 | 5.57 | 5.27 | 14 | 4.60 | 4.35 | 24 | 5.43 | 4.33 |
| 5 | 4.62 | 5.59 | 15 | 5.06 | 4.88 | 25 | 4.83 | 4.56 |
| 6 | 4.65 | 4.61 | 16 | 5.51 | 4.56 | 26 | 5.05 | 5.50 |
| 7 | 4.62 | 5.19 | 17 | 4.66 | 4.84 | 27 | 5.78 | 5.16 |
| 8 | 6.39 | 4.64 | 18 | 4.95 | 4.24 | 28 | 5.10 | 4.89 |
| 9 | 4.36 | 4.35 | 19 | 4.75 | 4.33 | 29 | 4.68 | 4.89 |
| 10 | 4.91 | 4.49 | 20 | 4.67 | 4.24 | 30 | 6.06 | 5.24 |

Randomization inference makes few assumptions

In fact, randomization itself can be used as a basis for inference. The advantage of this randomization approach is that it relies only on the randomization that we performed. It does not need independence, normality, and the other assumptions that go with linear models. The disadvantage of the randomization approach is that it can be difficult to implement, even in relatively small problems, though computers make it much easier. Furthermore, the inference that randomization provides is often indistinguishable from that of standard techniques such as ANOVA.

Now that computers are powerful and common, randomization inference procedures can be done with relatively little pain. These ideas of randomization inference are best shown by example. Below we introduce the ideas of randomization inference using two extended examples, one corresponding to a paired $t$-test, and one corresponding to a two sample $t$-test.

### 2.4.1 The paired $t$-test

Bezjak and Knez (1995) provide data on the length of time it takes garment workers to runstitch a collar on a man's shirt, using a standard workplace and a more ergonomic workplace. Table 2.1 gives the "auxiliary manual time" per collar in seconds for 30 workers using both systems.

One question of interest is whether the times are the same on average for the two workplaces. Formally, we test the null hypothesis that the average runstitching time for the standard workplace is the same as the average runstitching time for the ergonomic workplace.

**Table 2.2:** Differences in runstitching times (standard – ergonomic).

| | | | | | | | | | |
|---|---|---|---|---|---|---|---|---|---|
| 1.03 | -.04 | .26 | .30 | -.97 | .04 | -.57 | 1.75 | .01 | .42 |
| .45 | -.80 | .39 | .25 | .18 | .95 | -.18 | .71 | .42 | .43 |
| -.48 | -1.08 | -.57 | 1.10 | .27 | -.45 | .62 | .21 | -.21 | .83 |

A paired $t$-test is the standard procedure for testing this null hypothesis. We use a paired $t$-test because each worker was measured twice, once for each workplace, so the observations on the two workplaces are dependent. Fast workers are probably fast for both workplaces, and slow workers are slow for both. Thus what we do is compute the difference (standard – ergonomic) for each worker, and test the null hypothesis that the average of these differences is zero using a one sample $t$-test on the differences.

Paired $t$-test for paired data

Table 2.2 gives the differences between standard and ergonomic times. Recall the setup for a one sample $t$-test. Let $d_1, d_2, \ldots, d_n$ be the $n$ differences in the sample. We assume that these differences are independent samples from a normal distribution with mean $\mu$ and variance $\sigma^2$, both unknown. Our null hypothesis is that the mean $\mu$ equals prespecified value $\mu_0 = 0$ ($H_0 : \mu = \mu_0 = 0$), and our alternative is $H_1 : \mu > 0$ because we expect the workers to be faster in the ergonomic workplace.

The formula for a one sample $t$-test is

$$t = \frac{\bar{d} - \mu_0}{s/\sqrt{n}} ,$$

where $\bar{d}$ is the mean of the data (here the differences $d_1, d_2, \ldots, d_n$), $n$ is the sample size, and $s$ is the sample standard deviation (of the differences)

The paired $t$-test

$$s = \sqrt{\frac{1}{n-1}\sum_{i=1}^{n}(d_i - \bar{d})^2} .$$

If our null hypothesis is correct and our assumptions are true, then the $t$-statistic follows a $t$-distribution with $n-1$ degrees of freedom.

The $p$-value for a test is the probability, assuming that the null hypothesis is true, of observing a test statistic as extreme or more extreme than the one we did observe. "Extreme" means away from the the null hypothesis towards the alternative hypothesis. Our alternative here is that the true average is larger than the null hypothesis value, so larger values of the test statistic are extreme. Thus the $p$-value is the area under the $t$-curve with $n-1$ degrees of freedom from the observed $t$-value to the right. (If the alternative had been $\mu < \mu_0$, then the $p$-value is the area under the curve to the left of our test

The $p$-value

**Table 2.3:** Paired $t$-tests results for runstitching times (standard – ergonomic) for the last 10 and all 30 workers

| | $n$ | df | $\bar{d}$ | $s$ | $t$ | $p$ |
|---|---|---|---|---|---|---|
| Last 10 | 10 | 9 | .023 | .695 | .10 | .459 |
| All 30 | 30 | 29 | .175 | .645 | 1.49 | .074 |

statistic. For a two sided alternative, the $p$-value is the area under the curve at a distance from 0 as great or greater than our test statistic.)

To illustrate the $t$-test, let's use the data for the last 10 workers and all 30 workers. Table 2.3 shows the results. Looking at the last ten workers, the $p$-value is .46, meaning that we would observe a $t$-statistic this larger or larger in 46% of all tests when the null hypothesis is true. Thus there is little evidence against the null here. When all 30 workers are considered, the $p$-value is .074; this is mild evidence against the null hypothesis. The fact that these two differ probably indicates that the workers are not listed in random order. In fact, Figure 2.1 shows box-plots for the differences by groups of ten workers; the lower numbered differences tend to be greater.

Randomization null hypothesis

Now consider a randomization-based analysis. The randomization null hypothesis is that the two workplaces are completely equivalent and merely act to label the responses that we observed. For example, the first worker had responses of 4.90 and 3.87, which we have labeled as standard and ergonomic. Under the randomization null, the responses would be 4.90 and 3.87 no matter how the random assignment of treatments turned out. The only thing that could change is which of the two is labeled as standard, and which as ergonomic. Thus, under the randomization null hypothesis, we could, with equal probability, have observed 3.87 for standard and 4.90 for ergonomic.

Differences have random signs under randomization null

What does this mean in terms of the differences? We observed a difference of 1.03 for worker 1. Under the randomization null, we could just as easily have observed the difference -1.03, and similarly for all the other differences. Thus in the randomization analogue to a paired $t$-test, the absolute values of the differences are taken to be fixed, and the signs of the differences are random, with each sign independent of the others and having equal probability of positive and negative.

To construct a randomization test, we choose a descriptive statistic for the data and then get the distribution of that statistic under the randomization null hypothesis. The randomization $p$-value is the probability (under this randomization distribution) of getting a descriptive statistic as extreme or more extreme than the one we observed.

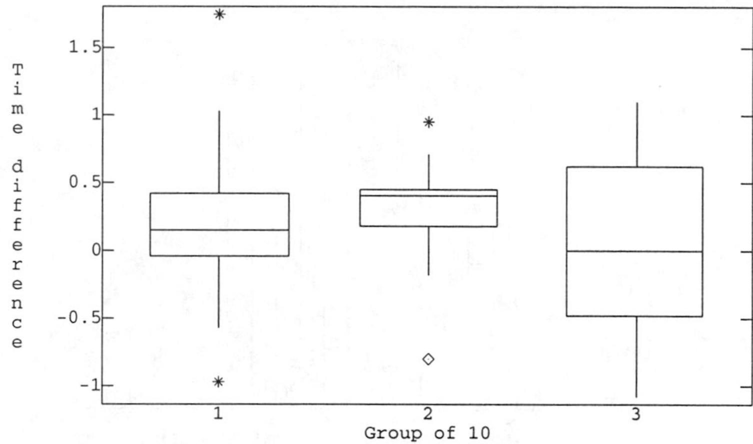


**Figure 2.1:** Box-plots of differences in runstitching times by groups of 10 workers, using MacAnova. Stars and diamonds indicate potential outlier points.

For this problem, we take the sum of the differences as our descriptive statistic. (The average would lead to exactly the same $p$-values, and we could also form tests using the median or other measures of center.) Start with the last 10 workers. The sum of the last 10 observed differences is .23. To get the randomization distribution, we have to get the sum for all possible combinations of signs for the differences. There are two possibilities for each difference, and 10 differences, so there are $2^{10} = 1024$ different equally likely values for the sum in the randomization distribution. We must look at all of them to get the randomization $p$-value.

Randomization statistic and distribution

Figure 2.2 shows a histogram of the randomization distribution for the last 10 workers. The observed value of .23 is clearly in the center of this distribution, so we expect a large $p$-value. In fact, 465 of the 1024 values are .23 or larger, so the randomization $p$-value is 465/1024 = .454, very close to the $t$-test $p$-value.

Randomization $p$-value

We only wanted to do a test on a mean of 10 numbers, and we had to compute 1024 different sums of 10 numbers; you can see one reason why randomization tests have not had a major following. For some data sets, you can compute the randomization $p$-value by hand fairly simply. Consider the last 10 differences in Table 2.2 (reading across rows, rather than columns).

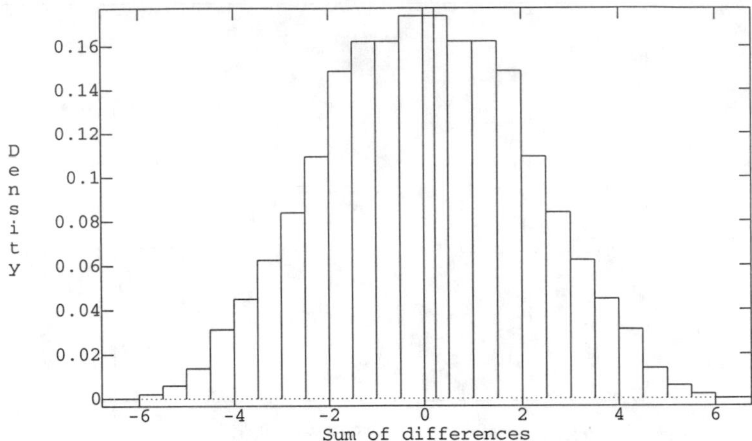


**Figure 2.2:** Histogram of randomization distribution of the sum of the last 10 worker differences for runstitching, with vertical line added at the observed sum.

These differences are

.62 1.75 .71 .21 .01 .42 -.21 .42 .43 .82

Only one of these values is negative (-.21), and seven of the positive differences have absolute value greater than .21. Any change of these seven values can only make the sum less, so we don't have to consider changing their signs, only the signs of .21, .01, and -.21. This is a much smaller problem, and it is fairly easy to work out that four of the 8 possible sign arrangements for testing three differences lead to sums as large or larger than the observed sum. Thus the randomization $p$-value is $4/1024 = .004$, similar to the .007 $p$-value we would get if we used the $t$-test.

Subsample the randomization distribution

Looking at the entire data set, we have $2^{30} = 1,073,741,824$ different sets of signs. That is too many to do comfortably, even on a computer. What is done instead is to have the computer choose a random sample from this complete distribution by choosing random sets of signs, and then use this sample for computing randomization $p$-values as if it were the complete distribution. For a reasonably large sample, say 10,000, the approximation is usually good enough. I took a random sample of size 10,000 and got a $p$-value .069, reasonably close to the $t$-test $p$-value. Two additional samples of 10,000 gave $p$-values of .073 and .068; the binomial distribution suggests

**Table 2.4:** Log whole plant phosphorus ($\ln \mu g$/plant) 15 and 28 days after first harvest.

| 15 Days | | | | 28 Days | | | |
|---|---|---|---|---|---|---|---|
| 4.3 | 4.6 | 4.8 | 5.4 | 5.3 | 5.7 | 6.0 | 6.3 |

that these approximate $p$-values have a standard deviation of about

$$\sqrt{p \times (1-p)/10000} \approx \sqrt{.07 \times .93/10000} = .0026 \ .$$

### 2.4.2 Two-sample $t$-test

Figure 2 of Hunt (1973) provides data from an experiment looking at the absorption of phosphorus by *Rumex acetosa*. Table 2.4 is taken from Figure 2 of Hunt and gives the log phosphorus content of 8 whole plants, 4 each at 15 and 28 days after first harvest. These are 8 plants randomly divided into two groups of 4, with each group getting a different treatment. One natural question is whether the average phosphorus content is the same at the two sampling times. Formally, we test the null hypothesis that the two sampling times have the same average.

A two-sample $t$-test is the standard method for addressing this question. Let $y_{11}, \ldots, y_{14}$ be the responses from the first sample, and let $y_{21}, \ldots, y_{24}$ be the response from the second sample. The usual assumptions for a two-sample $t$-test are that the data $y_{11}, \ldots, y_{14}$ are a sample from a normal distribution with mean $\mu_1$ and variance $\sigma^2$, the data $y_{21}, \ldots, y_{24}$ are a sample from a normal distribution with mean $\mu_2$ and variance $\sigma^2$, and the two samples are independent. Note that while the means may differ, the variances are assumed to be the same. The null hypothesis is $H_0 : \mu_1 = \mu_2$ and our alternative is $H_1 : \mu_1 < \mu_2$ (presumably growing plants will accumulate phosphorus).

Two-sample $t$-test

The two-sample $t$-statistic is

$$t = \frac{\overline{y}_{2\bullet} - \overline{y}_{1\bullet}}{s_p \sqrt{1/n_1 + 1/n_2}} \ ,$$

where $\overline{y}_{1\bullet}$ and $\overline{y}_{2\bullet}$ are the means of the first and second samples, $n_1$ and $n_2$ are the sample sizes, and $s_p^2$ is the pooled estimate of variance defined by

$$s_p = \sqrt{\frac{\sum_{i=1}^{n_1}(y_{1i} - \overline{y}_{1\bullet})^2 + \sum_{i=1}^{n_2}(y_{2i} - \overline{y}_{2\bullet})^2}{n_1 + n_2 - 2}} .$$

If our null hypothesis is correct and our assumptions are true, then the $t$-statistic follows a $t$-distribution with $n_1 + n_2 - 2$ degrees of freedom. The $p$-value for our one-sided alternative is the area under the $t$-distribution curve with $n_1 + n_2 - 2$ degrees of freedom that is to the right of our observed $t$-statistic.

For these data $\overline{y}_{1\bullet} = 4.775$, $\overline{y}_{2\bullet} = 5.825$, $s_p = .446$, and $n_1 = n_2 = 4$. The $t$-statistic is then

$$t = \frac{5.825 - 4.775}{.446\sqrt{1/4 + 1/4}} = 3.33,$$

and the $p$-value is .008, the area under a $t$-curve with 6 degrees of freedom to the right of 3.33. This is strong evidence against the null hypothesis, and we would probably conclude that the null is false.

Randomization null hypothesis

Now consider a randomization analysis. The randomization null hypothesis is that growing time treatments are completely equivalent and serve only as labels. In particular, the responses we observed for the 8 units would be the same no matter which treatments had been applied, and any subset of four units is equally likely to be the 15-day treatment group. For example, under the randomization null wth the 15-day treatment, the responses (4.3, 4.6, 4.8, 5.4), (4.3, 4.6, 5.3, 5.7), and (5.4, 5.7, 6.0, 6.3) are all equally likely.

To construct a randomization test, we choose a descriptive statistic for the data and then get the distribution of that statistic under the randomization null hypothesis. The randomization $p$-value is the probability (under this randomization distribution) of getting a descriptive statistic as extreme or more extreme than the one we observed.

Randomization statistic and distribution

Randomization $p$-value

For this problem, we take the average response at 28 days minus the average response at 15 days as our statistic. The observed value of this statistic is 1.05. There are ${}_8C_4 = 70$ different ways that the 8 plants can be split between the two treatments. Only two of those 70 ways give a difference of averages as large as or larger than the one we observed. Thus the randomization $p$-value is 2/70 = .029. This $p$-value is a bit bigger than that computed from the $t$-test, but both give evidence against the null hypothesis. Note that the smallest possible randomization $p$-value for this experiment is 1/70 = .014.

### 2.4.3 Randomization inference and standard inference

We have seen a couple of examples where the $p$-values for randomization tests were very close to those of $t$-tests, and a couple where the $p$-values differed somewhat. Generally speaking, randomization $p$-values are close to standard $p$-values. The two tend to be very close when the sample size is large and the assumptions of the standard methods are met. For small sample sizes, randomization inference is coarser, in the sense that there are relatively few obtainable $p$-values.

| Randomization $p$-values are usually close to normal theory $p$-values. |
|---|

We will only mention randomization testing in passing in the remainder of this book. Normal theory methods such as ANOVA and $t$-tests are much easier to implement and generalize; furthermore, we get essentially the same inference as the randomization tests, provided we take some care to ensure that the assumptions made by the standard procedures are met. We should consider randomization methods when the assumptions of normal theory cannot be met.

## 2.5 Further Reading and Extensions

Randomization tests, sometimes called permutation tests, were introduced by Fisher (1935) and further developed by Pitman (1937, 1938) and others. Some of the theory behind these tests can be found in Kempthorne (1955) and Lehmann (1959). Fisher's book is undoubtedly a classic and the granddaddy of all modern books on the design of experiments. It is, however, difficult for mere mortals to comprehend and has been debated and discussed since it appeared (see, for example, Kempthorne 1966). Welch (1990) presents a fairly general method for constructing randomization tests.

The randomization distribution for our test statistic is discrete, so there is a nonzero lump of probability on the observed value. We have computed the $p$-value by including all of this probability at the observed value as being in the tail area (as extreme or more extreme than that we observed). One potential variation on the $p$-value is to split the probability at the observed value in half, putting only half in the tail. This can sometimes improve the agreement between randomization and standard methods.

While randomization is traditional in experimental design and its use is generally prescribed, it is only fair to point out that there is an alternative model for statistical inference in which randomization is not *necessary* for valid experimental design, and under which randomization does not form the basis for inference. This is the Bayesian model of statistical inference. The drawback is that the Bayesian analysis must model all the miscellaneous factors which randomization is used to avoid.

The key assumption in many Bayesian analyses is the assumption of *exchangeability*, which is like the assumption of independence in a classical analysis. Many Bayesians will concede that randomization can assist in making exchangeability a reasonable approximation to reality. Thus, some would do randomization to try to get exchangeability. However, Bayesians do not need to randomize and so are free to consider other criteria, such as ethical criteria, much more strongly. Berry (1989) has expounded this view rather forcefully.

Bayesians believe in the likelihood principle, which here implies basing your inference on the data you have instead of the data you might have had. Randomization inference compares the observed results to results that would have been obtained under other randomizations. This is a clear violation of the likelihood principle. Of course, Bayesians don't generally believe in testing or $p$-values to begin with.

A fairly recent cousin of randomization inference is *bootstrapping* (see Efron 1979; Efron and Tibshirani 1993; and many others). Bootstrap inference in the present context does not rerandomize the assignment of treatments to units, rather it randomly reweights the observations in each treatment group in an effort to determine the distribution of statistics of interest.

## 2.6 Problems

**Exercise 2.1** We wish to evaluate a new textbook for a statistics class. There are seven sections; four are chosen at random to receive the new book, three receive the old book. At the end of the semester, student evaluations show the following percentages of students rate the textbook as "very good" or "excellent":

| Section | 1 | 2 | 3 | 4 | 5 | 6 | 7 |
|---|---|---|---|---|---|---|---|
| Book | N | O | O | N | N | O | N |
| Rating | 46 | 37 | 47 | 45 | 32 | 62 | 56 |

Find the one-sided randomization $p$-value for testing the null hypothesis that the two books are equivalent versus the alternative that the new book is better (receives higher scores).

**Exercise 2.2** Dairy cows are bred by selected bulls, but not all cows become pregnant at the first service. A drug is proposed that is hoped to increase the bulls fertility. Each of seven bulls will be bred to 2 herds of 100 cows each (a total of 14 herds). For one herd (selected randomly) the bulls will be given the drug, while no drug will be given for the second herd. Assume the drug has no residual effect. The response we observe for each bull is the number of impregnated cows under drug therapy minus the number of impregnated cows without the drug. The observed differences are -1, 6, 4, 6, 2, -3, 5. Find the $p$-value for the randomization test of the null hypothesis that the drug has no effect versus a one-sided alternative (the drug improves fertility).

**Exercise 2.3**

Suppose we are studying the effect of diet on height of children, and we have two diets to compare: diet A (a well balanced diet with lots of broccoli) and diet B (a diet rich in potato chips and candy bars). We wish to find the diet that helps children grow (in height) fastest. We have decided to use 20 children in the experiment, and we are contemplating the following methods for matching children with diets:

1. Let them choose.
2. Take the first 10 for A, the second 10 for B.
3. Alternate A, B, A, B.
4. Toss a coin for each child in the study: heads $\rightarrow$ A, tails $\rightarrow$ B.
5. Get 20 children; choose 10 at random for A, the rest for B.

Describe the benefits and risks of using these five methods.

**Exercise 2.4**

As part of a larger experiment, Dale (1992) looked at six samples of a wetland soil undergoing a simulated snowmelt. Three were randomly selected for treatment with a neutral pH snowmelt; the other three got a reduced pH snowmelt. The observed response was the number of Copepoda removed from each microcosm during the first 14 days of snowmelt.

| Reduced pH | | | Neutral pH | | |
|---|---|---|---|---|---|
| 256 | 159 | 149 | 54 | 123 | 248 |

Using randomization methods, test the null hypothesis that the two treatments have equal average numbers of Copepoda versus a two-sided alternative.

**Exercise 2.5**

Chu (1970) studied the effect of the insecticide chlordane on the nervous systems of American cockroaches. The coxal muscles from one meso- and one metathoracic leg on opposite sides were surgically extracted from each of six roaches. The roaches were then treated with 50 micrograms of $\alpha$-chlordane, and coxal muscles from the two remaining meso- and metathoracic legs were removed about two hours after treatment. The $Na^+$-$K^+$ATPase activity was measured in each muscle, and the percentage changes for the six roaches are given here:

15.3 -31.8 -35.6 -14.5 3.1 -24.5

Test the null hypothesis that the chlordane treatment has not affected the

$Na^+$-$K^+$ATPas activity. What experimental technique (not mentioned in the description above) must have been used to justify a randomization test?

**Problem 2.1**

McElhoe and Conner (1986) use an instrument called a "Visiplume" to measure ultraviolet light. By comparing absorption in clear air and absorption in polluted air, the concentration of $SO_2$ in the polluted air can be estimated. The EPA has a standard method for measuring $SO_2$, and we wish to compare the two methods across a range of air samples. The recorded response is the ratio of the Visiplume reading to the EPA standard reading. There were six observations on coal plant number 2: .950, .978, .762, .733, .823, and 1.011. If we make the null hypothesis be that the Visiplume and standard measurements are equivalent (and the Visiplume and standard labels are just labels and nothing more), then the ratios could (with equal probability) have been observed as their reciprocals. That is, the ratio of .950 could with equal probability have been 1/.950 = 1.053, since the labels are equivalent and assigned at random. Suppose we take as our summary of the data the sum of the ratios. We observe .95 + ... + 1.011 = 5.257. Test (using randomization methods) the null hypothesis of equivalent measurement procedures against the alternative that Visiplume reads higher than the standard. Report a $p$-value.

**Problem 2.2**

In this problem, a data set of size 5 consists of the numbers 1 through 5; a data set of size 6 consists of the numbers 1 through 6; and so on.

a) For data sets of size 5 and 6, compute the complete randomization distribution for the mean of samples of size 3. (There will be 10 and 20 members respectively in the two distributions.) How normal do these distributions look?

b) For data sets of size 4 and 5, compute the complete randomization distribution for the mean of samples of any size (size 1, size 2, ..., up to all the data in the sample). Again, compare these to normal.

c) Compare the size 5 distributions from parts a) and b). How do they compare for mean, median, variance, and so on.

**Question 2.1**

Let $X_1, X_2, \ldots, X_N$ be independent, uniformly distributed, random $k$-digit integers (that is, less than $10^k$). Find the probability of having no duplicates in $N$ draws.

# Chapter 3

# Completely Randomized Designs

The simplest randomized experiment for comparing several treatments is the Completely Randomized Design, or CRD. We will study CRD's and their analysis in some detail, before considering any other designs, because many of the concepts and methods learned in the CRD context can be transferred with little or no modification to more complicated designs. Here, we define completely randomized designs and describe the initial analysis of results.

## 3.1 Structure of a CRD

We have $g$ treatments to compare and $N$ units to use in our experiment. For a completely randomized design:

All partitions of units with sizes $n_1$ through $n_g$ equally likely in CRD

1. Select sample sizes $n_1, n_2, \ldots, n_g$ with $n_1 + n_2 + \cdots + n_g = N$.
2. Choose $n_1$ units at random to receive treatment 1, $n_2$ units at random from the $N - n_1$ remaining to receive treatment 2, and so on.

This randomization produces a CRD; all possible arrangements of the $N$ units into $g$ groups with sizes $n_1$ though $n_g$ are equally likely. Note that complete randomization only addresses the assignment of treatments to units; selection of treatments, experimental units, and responses is also required.

First consider a CRD

Completely randomized designs are the simplest, most easily understood, most easily analyzed designs. For these reasons, we consider the CRD first when designing an experiment. The CRD may prove to be inadequate for some reason, but I always consider the CRD when developing an experimental design before possibly moving on to a more sophisticated design.

**Example 3.1**

### Acid rain and birch seedlings

Wood and Bormann (1974) studied the effect of acid rain on trees. "Clean" precipitation has a pH in the 5.0 to 5.5 range, but observed precipitation pH in northern New Hampshire is often in the 3.0 to 4.0 range. Is this acid rain harming trees, and if so, does the amount of harm depend on the pH of the rain?

One of their experiments used 240 six-week-old yellow birch seedlings. These seedlings were divided into five groups of 48 *at random*, and the seedlings within each group received an acid mist treatment 6 hours a week for 17 weeks. The five treatments differed by mist pH: 4.7, 4.0, 3.3, 3.0, and 2.3; otherwise, the seedlings were treated identically. After the 17 weeks, the seedlings were weighed, and total plant (dry) weight was taken as response. Thus we have a completely randomized design, with five treatment groups and each $n_i$ fixed at 48. The seedlings were the experimental units, and plant dry weight was the response.

This is a nice, straightforward experiment, but let's look over the steps in planning the experiment and see where some of the choices and compromises were made. It was suspected that damage might vary by pH level, plant developmental stage, and plant species, among other things. This particular experiment only addresses pH level (other experiments were conducted separately). Many factors affect tree growth. The experiment specifically controlled for soil type, seed source, and amounts of light, water, and fertilizer. The desired treatment was real acid rain, but the available treatment was a synthetic acid rain consisting of distilled water and sulfuric acid (rain in northern New Hampshire is basically a weak mixture of sulfuric and nitric acids). There was no placebo *per se*. The experiment used yellow birch seedlings; what about other species or more mature trees? Total plant weight is an important response, but other responses (possibly equally important) are also available. Thus we see that the investigators have narrowed an enormous question down to a workable experiment using artificial acid rain on seedlings of a single species under controlled conditions. A considerable amount of nonstatistical background work and compromise goes into the planning of even the simplest (from a statistical point of view) experiment.

**Example 3.2**

### Resin lifetimes

Mechanical parts such as computer disk drives, light bulbs, and glue bonds eventually fail. Buyers of these parts want to know how long they are likely

**Table 3.1:** $\log_{10}$ times till failure of a resin under stress.

| Temperature (°C) | | | | | | | | | |
|---|---|---|---|---|---|---|---|---|---|
| 175 | | 194 | | 213 | | 231 | | 250 | |
| 2.04 | 1.85 | 1.66 | 1.66 | 1.53 | 1.35 | 1.15 | 1.21 | 1.26 | 1.02 |
| 1.91 | 1.96 | 1.71 | 1.61 | 1.54 | 1.27 | 1.22 | 1.28 | .83 | 1.09 |
| 2.00 | 1.88 | 1.42 | 1.55 | 1.38 | 1.26 | 1.17 | 1.17 | 1.08 | 1.06 |
| 1.92 | 1.90 | 1.76 | 1.66 | 1.31 | 1.38 | 1.16 | | | |

to last, so manufacturers perform tests to determine average lifetime, sometimes expressed as mean time to failure, or mean time between failures for repairable items. The last computer disk drive I bought had a mean time to failure of 800,000 hours (over 90 years). Clearly the manufacturer did not have disks on test for over 90 years; how do they make such claims?

One experimental method for reliability is called an *accelerated life test*. Parts under stress will usually fail sooner than parts that are unstressed. By modeling the lifetimes of parts under various stresses, we can estimate (extrapolate to) the lifetime of parts that are unstressed. That way we get an estimate of the unstressed lifetime without having to wait the complete unstressed lifetime.

Nelson (1990) gave an example where the goal was to estimate the lifetime (in hours) of an encapsulating resin for gold-aluminum bonds in integrated circuits operating at 120°C. Since the lifetimes were expected to be rather long, an accelerated test was used. Thirty-seven units were assigned at random to one of five different temperature stresses, ranging from 175° to 250°. Table 3.1 gives the $\log_{10}$ lifetimes in hours for the test units.

For this experiment, the choice of units was rather clear: integrated circuits with the resin bond of interest. Choice of treatments, however, depended on knowing that temperature stress reduced resin bond lifetime. The actual choice of temperatures probably benefited from knowledge of the results of previous similar experiments. Once again, experimental design is a combination of subject matter knowledge and statistical methods.

## 3.2 Preliminary Exploratory Analysis

It is generally advisable to conduct a preliminary exploratory or graphical analysis of the data prior to any formal modeling, testing, or estimation. Preliminary analysis could include:

- Simple descriptive statistics such as means, medians, standard errors, interquartile ranges;
- Plots, such as stem and leaf diagrams, box-plots, and scatter-plots; and
- The above procedures applied separately to each treatment group.

See, for example, Moore and McCabe (1999) for a description of these exploratory techniques.

Graphical analysis reveals patterns in data

This preliminary analysis presents several possibilities. For example, a set of box-plots with one box for each treatment group can show us the relative sizes of treatment mean differences and experimental error. This often gives us as much understanding of the data as any formal analysis procedure. Preliminary analysis can also be a great help in discovering unusual responses or problems in the data. For example, we might discover an outlying value, perhaps due to data entry error, that was difficult to spot in a table of numbers.

**Example 3.3**

**Resin lifetimes, continued**

We illustrate preliminary analysis by using Minitab to make box-plots of the resin lifetime data of Example 3.2, with a separate box-plot for each treatment; see Figure 3.1. The data in neighboring treatments overlap, but there is a consistent change in the response from treatments one through five, and the change is fairly large relative to the variation within each treatment group. Furthermore, the variation is roughly the same in the different treatment groups (achieving this was a major reason for using log lifetimes).

A second plot shows us something of the challenge we are facing. Figure 3.2 shows the average log lifetimes per treatment group plotted against the stress temperature, with a regression line superimposed. We are trying to extrapolate over to a temperature of $120^{\text{o}}$, well beyond the range of the data. If the relationship is nonlinear (and it looks curved), the linear fit will give a poor prediction and the average log lifetime at $120^{\text{o}}$could be considerably higher than that predicted by the line.

## 3.3 Models and Parameters

A *model* for data is a specification of the statistical distribution for the data. For example, the number of heads in ten tosses of a fair coin would have a Binomial(10,.5) distribution, where .5 gives the probability of a success and 10 is the number of trials. In this instance, the distribution depends on two numbers, called parameters: the success probability and the number of trials.

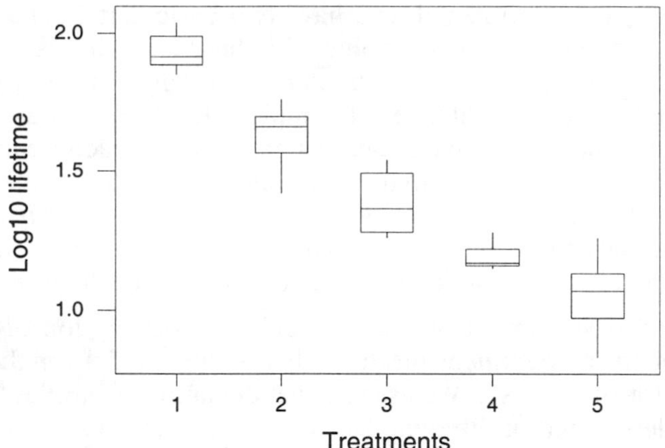


**Figure 3.1:** Box-plots of $\log_{10}$ times till failure of a resin under five different temperature stresses, using Minitab.

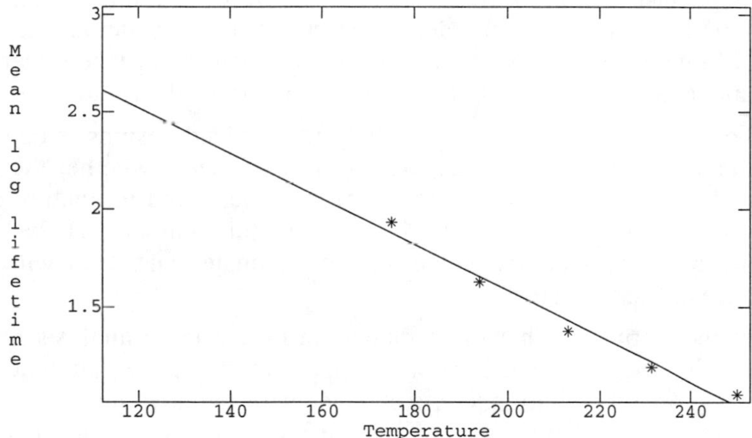


**Figure 3.2:** Average $\log_{10}$ time till failure versus temperature, with linear regression line added, using MacAnova.

For ten tosses of a fair coin, we know both parameters. In the analysis of experimental data, we may posit several different models for the data, all with unknown parameters. The objectives of the experiment can often be described as deciding which model is the best description of the data, and making inferences about the parameters in the models.

Model for the means

Our models for experimental data have two basic parts. The first part describes the average or expected values for the data. This is sometimes called a "model for the means" or "structure for the means." For example, consider the birch tree weights from Example 3.1. We might assume that all the treatments have the same mean response, or that each treatment has its own mean, or that the means in the treatments are a straight line function of the treatment pH. Each one of these models for the means has its own parameters, namely the common mean, the five separate treatment means, and the slope and intercept of the linear relationship, respectively.

Model for the errors

$\sigma^2$

The second basic part of our data models is a description of how the data vary around the treatment means. This is the "model for the errors" or "structure for the errors". We assume that deviations from the treatment means are independent for different data values, have mean zero, and all the deviations have the same variance, denoted by $\sigma^2$.

Normal distribution of errors needed for inference

This description of the model for the errors is incomplete, because we have not described the distribution of the errors. We can actually go a fair way with descriptive statistics using our mean and error models without ever assuming a distribution for the deviations, but we will need to assume a distribution for the deviations in order to do tests, confidence intervals, and other forms of inference. We assume, in addition to independence, zero mean, and constant variance, that the deviations follow a Normal distribution.

Standard analysis explores means

The standard analysis for completely randomized designs is concerned with the structure of the means. We are trying to learn whether the means are all the same, or if some differ from the others, and the nature of any differences that might be present. The error structure is assumed to be known, except for the variance $\sigma^2$, which must be estimated and dealt with but is otherwise of lesser interest.

Standard analysis is not always appropriate

Let me emphasize that these models in the standard analysis may not be the only models of interest; for example, we may have data that do not follow a normal distribution, or we may be interested in variance differences rather than mean differences (see Example 3.4). However, the usual analysis looking at means is a reasonable place to start.

**Example 3.4**

## Luria, Delbrück, and variances

In the 1940s it was known that some strains of bacteria were sensitive to a particular virus and would be killed if exposed. Nonetheless, some members of those strains did not die when exposed to the virus and happily proceeded to reproduce. What caused this phenomenon? Was it spontaneous mutation, or was it an adaptation that occurred after exposure to the virus? These two competing theories for the phenomenon led to the same average numbers of resistant bacteria, but to different variances in the numbers of resistant bacteria—with the mutation theory leading to a much higher variance. Experiments showed that the variances were high, as predicted by the mutation theory. This was an experiment where all the important information was in the variance, not in the mean. It was also the beginning of a research collaboration that eventually led to the 1969 Nobel Prize for Luria and Delbrück.

There are many models for the means; we start with two basic models. We have $g$ treatments and $N$ units. Let $y_{ij}$ be the $j$th response in the $i$th treatment group. Thus $i$ runs between 1 and $g$, and $j$ runs between 1 and $n_i$, in treatment group $i$. The model of separate group means (the full model) assumes that every treatment has its own mean response $\mu_i$. Combined with the error structure, the separate means model implies that all the data are independent and normally distributed with constant variance, but each treatment group may have its own mean:

Separate means model

$$y_{ij} \sim N(\mu_i, \sigma^2) \ .$$

Alternatively, we may write this model as

$$y_{ij} \sim \mu_i + \epsilon_{ij} \ ,$$

where the $\epsilon_{ij}$'s are "errors" or "deviations" that are independent, normally distributed with mean zero and variance $\sigma^2$.

The second basic model for the means is the single mean model (the reduced model). The single mean model assumes that all the treatments have the same mean $\mu$. Combined with the error structure, the single mean model implies that the data are independent and normally distributed with mean $\mu$ and constant variance,

Single mean model

$$y_{ij} \sim N(\mu, \sigma^2) \ .$$

Alternatively, we may write this model as

$$y_{ij} \sim \mu + \epsilon_{ij} \ ,$$

where the $\epsilon_{ij}$'s are independent, normally distributed errors with mean zero and variance $\sigma^2$.

Note that the single mean model is a special case or restriction of the group means model, namely the case when all of the $\mu_i$'s equal each other. Model comparison is easiest when one of the models is a restricted or reduced form of the other.

Compare reduced model to full model

Overall mean $\mu^\star$ and treatment effects $\alpha_i$

We sometimes express the group means $\mu_i$ as $\mu_i = \mu^\star + \alpha_i$. The constant $\mu^\star$ is called the *overall mean*, and $\alpha_i$ is called the $i$th *treatment effect*. In this formulation, the single mean model is the situation where all the $\alpha_i$ values are equal to each other: for example, all zero. This introduction of $\mu^\star$ and $\alpha_i$ seems like a needless complication, and at this stage of the game it really is. However, the treatment effect formulation will be extremely useful later when we look at factorial treatment structures.

Too many parameters

Note that there is something a bit fishy here. There are $g$ means $\mu_i$, one for each of the $g$ treatments, but we are using $g + 1$ parameters ($\mu^\star$ and the $\alpha_i$'s) to describe the $g$ means. This implies that $\mu^\star$ and the $\alpha_i$'s are not uniquely determined. For example, if we add 15 to $\mu^\star$ and subtract 15 from all the $\alpha_i$'s, we get the same treatment means $\mu_i$: the 15's just cancel. However, $\alpha_i - \alpha_j$ will always equal $\mu_i - \mu_j$, so the differences between treatment effects will be the same no matter how we define $\mu^\star$.

Restrictions make treatment effects well defined

We got into this embarrassment by imposing an additional mathematical structure (the overall mean $\mu^\star$) on the set of $g$ group means. We can get out of this embarrassment by deciding what we mean by $\mu^\star$; once we know $\mu^\star$, then we can determine the treatment effects $\alpha_i$ by $\alpha_i = \mu_i - \mu^\star$. Alternatively, we can decide what we mean by $\alpha_i$; then we can get $\mu^\star$ by $\mu^\star = \mu_i - \alpha_i$. These decisions typically take the form of some mathematical restriction on the values for $\mu^\star$ or $\alpha_i$. Restricting $\mu^\star$ or $\alpha_i$ is really two sides of the same coin.

Differences of treatment effects do not depend on restrictions

Mathematically, all choices for defining $\mu^\star$ are equally good. In practice, some choices are more convenient than others. Different statistical software packages use different choices, and different computational formulae use different choices; our major worry is keeping track of which particular choice is in use at any given time. Fortunately, *the important things don't depend on which set of restrictions we use.* Important things are treatment means, differences of treatment means (or equivalently, differences of $\alpha_i$'s), and comparisons of models.

One classical choice is to define $\mu^\star$ as the mean of the treatment means:

$$\mu^\star = \sum_{i=1}^{g} \mu_i/g \ .$$

Sum of treatment effects is zero

For this choice, the sum of the treatment effects is zero:

$$\sum_{i=1}^{g} \alpha_i = 0 \ .$$

An alternative that makes some hand work simpler assumes that $\mu^\star$ is the weighted average of the treatment means, with the sample sizes $n_i$ used as weights:

$$\mu^\star = \sum_{i=1}^{g} n_i\mu_i/N \ .$$

For this choice, the weighted sum of the treatment effects is zero:

Or weighted sum of treatment effects is zero

$$\sum_{i=1}^{g} n_i\alpha_i = 0 \ .$$

When the sample sizes are equal, these two choices coincide. The computational formulae we give in this book will use the restriction that the weighted sum of the $\alpha_i$'s is zero, because it leads to somewhat simpler hand computations. Some of the formulae in later chapters are only valid when the sample sizes are equal.

Degrees of freedom for treatment effects

Our restriction that the treatment effects $\alpha_i$ add to zero (either weighted or not) implies that the treatment effects are not completely free to vary. We can set $g-1$ of them however we wish, but the remaining treatment effect is then determined because it must be whatever value makes the zero sum true. We express this by saying that the treatment effects have $g-1$ degrees of freedom.

## 3.4 Estimating Parameters

Most data analysis these days is done using a computer. Few of us sit down and crunch through the necessary calculations by hand. Nonetheless, knowing the basic formulae and ideas behind our analysis helps us understand and interpret the quantities that come out of the software black box. If we don't understand the quantities printed by the software, we cannot possibly use them to understand the data and answer our questions.

Unbiased estimators correct on average

The parameters of our group means model are the treatment means $\mu_i$ and the variance $\sigma^2$, plus the derived parameters $\mu^\star$ and the $\alpha_i$'s. We will be computing "unbiased" estimates of these parameters. Unbiased means that when you average the values of the estimates across all potential random errors $\epsilon_{ij}$, you get the true parameter values.

It is convenient to introduce a notation to indicate the estimator of a parameter. The usual notation in statistics is to put a "hat" over the parameter to indicate the estimator; thus $\widehat{\mu}$ is an estimator of $\mu$. Because we have parameters that satisfy $\mu_i = \mu^\star + \alpha_i$, we will use estimators that satisfy $\widehat{\mu}_i = \widehat{\mu}^\star + \widehat{\alpha}_i$.

Let's establish some notation for sample averages and the like. The sum of the observations in the $i$th treatment group is

$$y_{i\bullet} = \sum_{j=1}^{n_i} y_{ij} \ .$$

Treatment means

The mean of the observations in the $i$th treatment group is

$$\overline{y}_{i\bullet} = \frac{1}{n_i}\sum_{j=1}^{n_i} y_{ij} = y_{i\bullet}/n_i \ .$$

The overbar indicates averaging, and the dot ($\bullet$) indicates that we have averaged (or summed) over the indicated subscript. The sum of all observations is

$$y_{\bullet\bullet} = \sum_{i=1}^{g}\sum_{j=1}^{n_i} y_{ij} = \sum_{i=1}^{g} y_{i\bullet} \ ,$$

Grand mean

and the grand mean of all observations is

$$\overline{y}_{\bullet\bullet} = \frac{1}{N}\sum_{i=1}^{g}\sum_{j=1}^{n_i} y_{ij} = y_{\bullet\bullet}/N \ .$$

The sum of squared deviations of the data from the group means is

$$SS_E = \sum_{i=1}^{g}\sum_{j=1}^{n_i}(y_{ij} - \overline{y}_{i\bullet})^2 \ .$$

The $SS_E$ measures total variability in the data around the group means.

$\widehat{\mu}_i = \overline{y}_{i\bullet}$

Consider first the separate means model, with each treatment group having its own mean $\mu_i$. The natural estimator of $\mu_i$ is $\overline{y}_{i\bullet}$, the average of the observations in that treatment group. We estimate the expected (or average) response in the $i$th treatment group by the observed average in the $i$th treatment group responses. Thus we have

$$\widehat{\mu}_i = \overline{y}_{i\bullet} \ .$$

The sample average is an unbiased estimator of the population average, so $\widehat{\mu}_i$ is an unbiased estimator of $\mu_i$.

$\widehat{\mu} = \overline{y}_{\bullet\bullet}$

In the single mean model, the only parameter in the model for the means is $\mu$. The natural estimator of $\mu$ is $\overline{y}_{\bullet\bullet}$, the grand mean of all the responses.

That is, if we felt that all the data were responses from the same population, we would estimate the mean of that single population by the grand mean of the data. Thus we have

$$\widehat{\mu} = \overline{y}_{\bullet\bullet} \ .$$

The grand mean is an unbiased estimate of $\mu$ when the data all come from a single population.

We use the restriction that $\mu^\star = \sum_i n_i\mu_i/N$; an unbiased estimate of $\mu^\star$ is

$$\widehat{\mu}^\star = \frac{\sum_{i=1}^{g} n_i\widehat{\mu}_i}{N} = \frac{\sum_{i=1}^{g} n_i\overline{y}_{i\bullet}}{N} = \frac{y_{\bullet\bullet}}{N} = \overline{y}_{\bullet\bullet} \ .$$

This is the same as the estimator we use for $\mu$ in the single mean model. Because $\mu$ and $\mu^\star$ are both estimated by the same value, we will drop the notation $\mu^\star$ and just use the single notation $\mu$ for both roles.

$\mu = \mu^\star$ for weighted sum restriction

The treatment effects $\alpha_i$ are

$$\alpha_i = \mu_i - \mu \ ;$$

these can be estimated by

$\widehat{\alpha}_i = \overline{y}_{i\bullet} - \overline{y}_{\bullet\bullet}$

$$\begin{aligned} \widehat{\alpha}_i &= \widehat{\mu}_i - \widehat{\mu} \\ &= \overline{y}_{i\bullet} - \overline{y}_{\bullet\bullet} \ . \end{aligned}$$

These treatment effects and estimates satisfy the restriction

$$\sum_{i=1}^{g} n_i\alpha_i = \sum_{i=1}^{g} n_i\widehat{\alpha}_i = 0 \ .$$

The only parameter remaining to estimate is $\sigma^2$. Our estimator of $\sigma^2$ is

$$\widehat{\sigma}^2 = MS_E = \frac{SS_E}{N-g} = \frac{\sum_{i=1}^{g}\sum_{j=1}^{n_i}(y_{ij} - \overline{y}_{i\bullet})^2}{N-g} \ .$$

We sometimes use the notation $s$ in place of $\widehat{\sigma}$ in analogy with the sample standard deviation $s$. This estimator $\widehat{\sigma}^2$ is unbiased for $\sigma^2$ in both the separate means and single means models. (Note that $\widehat{\sigma}$ is *not* unbiased for $\sigma$.)

$\widehat{\sigma}^2$ is unbiased for $\sigma^2$

The deviations from the group mean $y_{ij} - \overline{y}_{i\bullet}$ add to zero in any treatment group, so that any $n_i - 1$ of them determine the remaining one. Put another way, there are $n_i - 1$ degrees of freedom for error in each group, or $N - g = \sum_i(n_i - 1)$ degrees of freedom for error for the experiment. There are thus $N - g$ degrees of freedom for our estimate $\widehat{\sigma}^2$. This is analogous to the

Error degrees of freedom

| Model | Parameter | Estimator |
|---|---|---|
| Single mean | $\mu$ | $\overline{y}_{\bullet\bullet}$ |
| | $\sigma^2$ | $\frac{\sum_{i=1}^{g}\sum_{j=1}^{n_i}(y_{ij}-\overline{y}_{i\bullet})^2}{N-g}$ |
| Separate means | $\mu$ | $\overline{y}_{\bullet\bullet}$ |
| | $\mu_i$ | $\overline{y}_{i\bullet}$ |
| | $\alpha_i$ | $\overline{y}_{i\bullet} - \overline{y}_{\bullet\bullet}$ |
| | $\sigma^2$ | $\frac{\sum_{i=1}^{g}\sum_{j=1}^{n_i}(y_{ij}-\overline{y}_{i\bullet})^2}{N-g}$ |

**Display 3.1:** Point estimators in the CRD.

formula $n_1+n_2-2$ for the degrees of freedom in a two-sample $t$-test. Another way to think of $N-g$ is the number of data values minus the number of mean parameters estimated.

The formulae for these estimators are collected in Display 3.1. The next example illustrates their use.

**Example 3.5**

**Resin lifetimes, continued**

Most of the work for computing point estimates is done once we get the average responses overall and in each treatment group. Using the resin lifetime data from Table 3.1, we get the following means and counts:

| Treatment (°C) | 175 | 194 | 213 | 231 | 250 | All data |
|---|---|---|---|---|---|---|
| Average | 1.933 | 1.629 | 1.378 | 1.194 | 1.057 | 1.465 |
| Count | 8 | 8 | 8 | 7 | 6 | 37 |

The estimates $\widehat{\mu}_i$ and $\widehat{\mu}$ can be read from the table:

$$\begin{array}{rclrclrcl} \widehat{\mu}_1 &=& 1.933 & \widehat{\mu}_2 &=& 1.629 & \widehat{\mu}_3 &=& 1.378 \\ \widehat{\mu}_4 &=& 1.194 & \widehat{\mu}_5 &=& 1.057 & \widehat{\mu} &=& 1.465 \end{array}$$

Get the $\widehat{\alpha}_i$ values by subtracting the grand mean from the group means:

$$\begin{array}{rclrcl} \widehat{\alpha}_1 = 1.932 - 1.465 &=& .467 & \widehat{\alpha}_2 = 1.629 - 1.465 &=& .164 \\ \widehat{\alpha}_3 = 1.378 - 1.465 &=& -.088 & \widehat{\alpha}_4 = 1.194 - 1.465 &=& -.271 \\ \widehat{\alpha}_5 = 1.057 - 1.465 &=& -.408 & & & \end{array}$$

Notice that $\sum_{i=1}^{g} n_i\widehat{\alpha}_i = 0$ (except for roundoff error).

The computation for $\widehat{\sigma}^2$ is a bit more work, because we need to compute the $SS_E$. For the resin data, $SS_E$ is

$$\begin{aligned} SS_E &= (2.04 - 1.933)^2 + (1.91 - 1.933)^2 + \cdots + (1.90 - 1.933)^2 + \\ &\quad (1.66 - 1.629)^2 + (1.71 - 1.629)^2 + \cdots + (1.66 - 1.629)^2 + \\ &\quad (1.53 - 1.378)^2 + (1.54 - 1.378)^2 + \cdots + (1.38 - 1.378)^2 + \\ &\quad (1.15 - 1.194)^2 + (1.22 - 1.194)^2 + \cdots + (1.17 - 1.194)^2 + \\ &\quad (1.26 - 1.057)^2 + (.83 - 1.057)^2 + \cdots + (1.06 - 1.057)^2 \\ &= .29369 \end{aligned}$$

Thus we have

$$\widehat{\sigma}^2 = SS_E/(N-g) = .29369/(37-5) = .009178 \ .$$

Confidence intervals for means and effects

A point estimate gives our best guess as to the value of a parameter. A confidence interval gives a plausible range for the parameter, that is, a set of parameter values that are consistent with the data. Confidence intervals for $\mu$ and the $\mu_i$'s are useful and straightforward to compute. Confidence intervals for the $\alpha_i$'s are only slightly more trouble to compute, but are perhaps less useful because there are several potential ways to define the $\alpha$'s. Differences between $\mu_i$'s, or equivalently, differences between $\alpha_i$'s, are extremely useful; these will be considered in depth in Chapter 4. Confidence intervals for the error variance $\sigma^2$ will be considered in Chapter 11.

Generic confidence interval for mean parameter

Confidence intervals for parameters in the mean structure have the general form:

*unbiased estimate* $\pm$ *multiplier* $\times$ *(estimated) standard error of estimate.*

The standard errors for the averages $\overline{y}_{\bullet\bullet}$ and $\overline{y}_{i\bullet}$ are $\sigma/\sqrt{N}$ and $\sigma/\sqrt{n_i}$ respectively. We do not know $\sigma$, so we use $\widehat{\sigma} = s = \sqrt{MS_E}$ as an estimate and obtain $s/\sqrt{N}$ and $s/\sqrt{n_i}$ as estimated standard errors for $\overline{y}_{\bullet\bullet}$ and $\overline{y}_{i\bullet}$.

Use $t$ multiplier when error is estimated

For an interval with coverage $1 - \mathcal{E}$, we use the upper $\mathcal{E}/2$ percent point of the $t$-distribution with $N - g$ degrees of freedom as the multipler. This is denoted $t_{\mathcal{E}/2,N-g}$. We use the $\mathcal{E}/2$ percent point because we are constructing a two-sided confidence interval, and we are allowing error rates of $\mathcal{E}/2$ on both the low and high ends. For example, we use the upper 2.5% point (or 97.5% cumulative point) of $t$ for 95% coverage. The degrees of freedom for the $t$-distribution come from $\widehat{\sigma}^2$, our estimate of the error variance. For the CRD, the degrees of freedom are $N - g$, the number of data points minus the number of treatment groups.

| Parameter | Estimator | Standard Error |
|---|---|---|
| $\mu$ | $\overline{y}_{\bullet\bullet}$ | $s/\sqrt{N}$ |
| $\mu_i$ | $\overline{y}_{i\bullet}$ | $s/\sqrt{n_i}$ |
| $\alpha_i$ | $\overline{y}_{i\bullet} - \overline{y}_{\bullet\bullet}$ | $s\sqrt{1/n_i - 1/N}$ |

**Display 3.2:** Standard errors of point estimators in the CRD.

The standard error of an estimated treatment effect $\widehat{\alpha}_i$ is $\sigma\sqrt{1/n_i - 1/N}$. Again, we must use an estimate of $\sigma$, yielding $s\sqrt{1/n_i - 1/N}$ for the estimated standard error. Keep in mind that the treatment effects $\widehat{\alpha}_i$ are negatively correlated, because they must add to zero.

## 3.5 Comparing Models: The Analysis of Variance

ANOVA compares models

In the standard analysis of a CRD, we are interested in the mean responses of the treatment groups. One obvious place to begin is to decide whether the means are all the same, or if some of them differ. Restating this question in terms of models, we ask whether the data can be adequately described by the model of a single mean, or if we need the model of separate treatment group means. Recall that the single mean model is a special case of the group means model. That is, we can choose the parameters in the group means model so that we actually get the same mean for all groups. The single mean model is said to be a reduced or restricted version of the group means model. Analysis of Variance, usually abbreviated ANOVA, is a method for comparing the fit of two models, one a reduced version of the other.

ANOVA partitions variability

Strictly speaking, ANOVA is an arithmetic procedure for partitioning the variability in a data set into bits associated with different mean structures plus a leftover bit. (It's really just the Pythagorean Theorem, though we've chosen our right triangles pretty carefully in $N$-dimensional space.) When in addition the error structure for the data is independent normal with constant variance, we can use the information provided by an ANOVA to construct statistical tests comparing the different mean structures or models for means that are represented in the ANOVA. The link between the ANOVA decomposition for the variability and tests for models is so tight, however, that we sometimes speak of testing via ANOVA even though the test is not really part of the ANOVA.

Our approach to model comparison is Occam's Razor — we use the simplest model that is consistent with the data. We only move to the more complicated model if the data indicate that the more complicated model is needed.

Use simplest acceptable model

How is this need indicated? The residuals $r_{ij}$ are the differences between the data $y_{ij}$ and the fitted mean model. For the single mean model, the fitted values are all $\overline{y}_{\bullet\bullet}$, so the residuals are $r_{ij} = y_{ij} - \overline{y}_{\bullet\bullet}$; for the separate means model, the fitted values are the group means $\overline{y}_{i\bullet}$, so the residuals are $r_{ij} = y_{ij} - \overline{y}_{i\bullet}$. We measure the closeness of the data to a fitted model by looking at the sum of squared residuals ($SSR$). The point estimators we have chosen for the mean parameters in our models are *least squares* estimators, which implies that they are the parameter estimates that make these sums of squared residuals as small as possible.

Residuals and $SSR$

Least squares

The sum of squared residuals for the separate means model is usually smaller than that for the single mean model; it can never be larger. We will conclude that the more complicated separate means model is needed if its $SSR$ is sufficiently less than that of the single mean model. We still need to construct a criterion for deciding when the $SSR$ has been reduced sufficiently.

One way of constructing a criterion to compare models is via a statistical test, with the null hypothesis that the single mean model is true versus the alternative that the separate means model is true. In common practice, the null and alternative hypotheses are usually expressed in terms of parameters rather than models. Using the $\mu_i = \mu + \alpha_i$ notation for group means, the null hypothesis $H_0$ of a single mean can be expressed as $H_0 : \alpha_i = 0$ for all $i$, and the alternative can be expressed as $H_A : \alpha_i \neq 0$ for some $i$. Note that since we have assumed that $\sum n_i\alpha_i = 0$, one nonzero $\alpha_i$ implies that the $\alpha_i$'s are not all equal to each other. The alternative hypothesis does not mean that all the $\alpha_i$'s are different, just that they are not all the same.

Null and alternative hypotheses

The model comparison point of view opts for the separate means model if that model has sufficiently less residual variation, while the parameter testing view opts for the separate means model if there is sufficiently great variation between the observed group means. These seem like different ideas, but we will see in the ANOVA decomposition that they are really saying the same thing, because less residual variation implies more variation between group means when the total variation is fixed.

## 3.6 Mechanics of ANOVA

ANOVA works by partitioning the total variability in the data into parts that mimic the model. The separate means model says that the data are not all

ANOVA decomposition parallels model

equal to the grand mean because of treatment effects and random error:

$$y_{ij} - \mu = \alpha_i + \epsilon_{ij} .$$

ANOVA decomposes the data similarly into a part that deals with group means, and a part that deals with deviations from group means:

$$\begin{aligned} y_{ij} - \overline{y}_{\bullet\bullet} &= (\overline{y}_{i\bullet} - \overline{y}_{\bullet\bullet}) + (y_{ij} - \overline{y}_{i\bullet}) \\ &= \widehat{\alpha}_i + r_{ij} \ . \end{aligned}$$

$SS_T$

The difference on the left is the deviation of a response from the grand mean. If you square all such differences and add them up you get $SS_T$, the *total sum of squares*.[1]

$SS_{\text{Trt}}$

The first difference on the right is the estimated treatment effect $\widehat{\alpha}_i$. If you squared all these (one for each of the $N$ data values) and added them up, you would get $SS_{\text{Trt}}$, the *treatment sum of squares*:

$$SS_{\text{Trt}} = \sum_{i=1}^{g} \sum_{j=1}^{n_i} (\overline{y}_{i\bullet} - \overline{y}_{\bullet\bullet})^2 = \sum_{i=1}^{g} n_i (\overline{y}_{i\bullet} - \overline{y}_{\bullet\bullet})^2 = \sum_{i=1}^{g} n_i \widehat{\alpha}_i{}^2 \ .$$

I think of this as

1. Square the treatment effect,
2. Multiply by the number of units receiving that effect, and
3. Add over the levels of the effect.

This three-step pattern will appear again frequently.

$SS_E$

The second difference on the right is the $ij$th residual from the model, which gives us some information about $\epsilon_{ij}$. If you squared and added the $r_{ij}$'s you would get $SS_E$, the *error sum of squares*:

$$SS_E = \sum_{i=1}^{g} \sum_{j=1}^{n_i} (y_{ij} - \overline{y}_{i\bullet})^2 \ .$$

This is the same $SS_E$ that we use in estimating $\sigma^2$.

[1] For pedants in the readership, this quantity is the *corrected* total sum of squares. There is also an *uncorrected* total sum of squares. The uncorrected total is the sum of the squared observations; the uncorrected total sum of squares equals $SS_T$ plus $N\overline{y}_{\bullet\bullet}{}^2$. In this book, total sum of squares will mean corrected total sum of squares.

$$
\begin{aligned}
SS_{\text{Trt}} &= \textstyle\sum_{i=1}^{g} n_i \widehat{\alpha}_i^{\,2} \\
SS_E &= \textstyle\sum_{i=1}^{g} \sum_{j=1}^{n_i} (y_{ij} - \overline{y}_{i\bullet})^2/(N-g) \\
SS_T &= SS_{\text{Trt}} + SS_E
\end{aligned}
$$

**Display 3.3:** Sums of squares in the CRD

Recall that

$$y_{ij} - \overline{y}_{\bullet\bullet} = \widehat{\alpha}_i + r_{ij}$$

so that

$$(y_{ij} - \overline{y}_{\bullet\bullet})^2 = \widehat{\alpha}_i^{\,2} + r_{ij}^2 + 2\widehat{\alpha}_i r_{ij} \quad .$$

Adding over $i$ and $j$ we get

$$SS_T = SS_{\text{Trt}} + SS_E + 2\sum_{i=1}^{g}\sum_{j=1}^{n_i} \widehat{\alpha}_i r_{ij} \quad .$$

Total SS

We can show (see Question 3.2) that the sum of the cross-products is zero, so that

$$SS_T = SS_{\text{Trt}} + SS_E \quad .$$

Larger $SS_{\text{Trt}}$ implies smaller $SS_E$

Now we can see the link between testing equality of group means and comparing models via $SSR$. For a given data set (and thus a fixed $SS_T$), more variation between the group means implies a larger $SS_{\text{Trt}}$, which in turn implies that the $SS_E$ must be smaller, which is the $SSR$ for the separate means model.

Display 3.3 summarizes the sums of squares formulae for the CRD. I should mention that there are numerous "calculator" or "shortcut" formulae for computing sums of squares quantities. In my experience, these formulae are more difficult to remember than the ones given here, provide little insight into what the ANOVA is doing, and are in some circumstances more prone to roundoff errors. I do not recommend them.

ANOVA table

ANOVA computations are summarized in a table with columns for source of variation, degrees of freedom, sum of squares, mean squares, and F-statistics. There is a row in the table for every source of variation in the full model. In the CRD, the sources of variation are treatments and errors, sometimes called between- and within-groups variation. Some tables are written with rows for either or both of the grand mean and the total variation, though these rows do not affect the usual model comparisons.

Generic ANOVA table

The following is a generic ANOVA table for a CRD.

| Source | DF | SS | MS | F |
|---|---|---|---|---|
| Treatments | $g-1$ | $SS_{\text{Trt}}$ | $SS_{\text{Trt}}/(g-1)$ | $MS_{\text{Trt}}/MS_E$ |
| Error | $N-g$ | $SS_E$ | $SS_E/(N-g)$ | |

The degrees of freedom are $g-1$ for treatments and $N-g$ for error. We saw the rationale for these in Section 3.4. The formulae for sums of squares were given above, and mean squares are always sums of squares divided by their degrees of freedom. The F-statistic is the ratio of two mean squares, the numerator mean square for a source of variation that we wish to assess, and a denominator (or error) mean square that estimates error variance.

F-test to compare models

We use the F-statistic (or F-ratio) in the ANOVA table to make a test of the null hypothesis that all the treatment means are the same (all the $\alpha_i$ values are zero) versus the alternative that some of the treatment means differ (some of the $\alpha_i$ values are nonzero). When the null hypothesis is true, the F-statistic is about 1, give or take some random variation; when the alternative is true, the F-statistic tends to be bigger than 1. To complete the test, we need to be able to tell how big is too big for the F-statistic. If the null hypothesis is true and our model and distributional assumptions are correct, then the F-statistic follows the F-distribution with $g-1$ and $N-g$ degrees of freedom. Note that the F-distribution has two "degrees of freedom", one from the numerator mean square and one from the denominator mean square.

$p$-value to assess evidence

To do the test, we compute the F-statistic and the degrees of freedom, and then we compute the probability of observing an F-statistic as large or larger than the one we observed, assuming all the $\alpha_i$'s were zero. This probability is called the *$p$-value* or *observed significance level* of the test, and is computed as the area under an F-distribution from the observed F-statistic on to the right, when the F-distribution has degrees of freedom equal to the degrees of freedom for the numerator and denominator mean squares. This $p$-value is usually obtained from a table of the F-distribution (for example, Appendix Table D.5) or via the use of statistical software.

Small values of the $p$-value are evidence that the null may be incorrect: either we have seen a rare event (big F-statistics when the null is actually true, leading to a small $p$-value), or an assumption we used to compute the $p$-value is wrong, namely the assumption that all the $\alpha_i$'s are zero. Given the choice of unlucky or incorrect assumption, most people choose incorrect assumption.

**Table 3.2:** Approximate Type I error probabilities for different $p$-values using the Sellke *et al.* lower bound.

| $p$ | .05 | .01 | .001 | .0001 |
|---|---|---|---|---|
| $\mathcal{P}(p)$ | .29 | .11 | .018 | .0025 |

We have now changed the question from "How big is too big an F?" to "How small is too small a $p$-value?" By tradition, $p$-values less than .05 are termed *statistically significant*, and those less than .01 are termed *highly statistically significant*. These values are reasonable (one chance in 20, one chance in 100), but there is really no reason other than tradition to prefer them over other similar values, say one chance in 30 and one chance in 200. It should also be noted that a person using the traditional values would declare one test with $p$-value of .049 to be significant and another test with a $p$-value of .051 not to be significant, but the two tests are really giving virtually identical results. Thus I prefer to report the $p$-value itself rather than simply report significance or lack thereof.

.05 and .01 significance levels

As with any test, remember that statistical significance is not the same as real world importance. A tiny $p$ value may be obtained with relatively small $\alpha_i$'s if the sample size is large enough or $\sigma^2$ is small enough. Likewise, large important differences between means may not appear significant if the sample size is small or the error variance large.

Practical significance

It is also important not to overinterpret the $p$-value. Reported $p$-values of .05 or .01 carry the magnificent labels of statistically significant or highly statistically significant, but they actually are not terribly strong evidence against the null. What we would really like to know is the probability that rejecting the null is an error; *the p-value does **not** give us that information.* Sellke, Bayarri, and Berger (1999) define an approximate lower bound on this probability. They call their bound a *calibrated p-value*, but I do not like the name because their quantity is not really a $p$-value. Suppose that before seeing any data you thought that the null and alternative each had probability .5 of being true. Then for $p$-values less than $e^{-1} \approx .37$, the Sellke *et al.* approximate error probability is

Approximate error probability

$$\mathcal{P}(p) = \frac{-ep\log(p)}{1 - ep\log(p)} \quad .$$

The interpretation of the approximate error probability $\mathcal{P}(p)$ is that having seen a $p$-value of $p$, the probability that rejecting the null hypothesis is an error is *at least* $\mathcal{P}(p)$. Sellke *et al.* show that this lower bound is pretty good in a wide variety of problems. Table 3.2 shows that the probability that rejection is a Type I error is more than .1, even for a $p$-value of .01.

**Listing 3.1:** Minitab output for resin lifetimes.

```
One-way Analysis of Variance

Analysis of Variance for Lifetime
Source     DF        SS        MS        F        P                      ①
Temp        4   3.53763   0.88441    96.36    0.000
Error      32   0.29369   0.00918
Total      36   3.83132
                                    Individual 95% CIs For Mean          ②
                                    Based on Pooled StDev
Level       N      Mean     StDev   --------+---------+---------+--------
1           8    1.9325    0.0634                              (-*--)
2           8    1.6288    0.1048                    (-*--)
3           8    1.3775    0.1071            (-*-)
4           7    1.1943    0.0458      (--*-)
5           6    1.0567    0.1384   (-*--)
                                    --------+---------+---------+--------
Pooled StDev =   0.0958                    1.20      1.50      1.80
```

**Example 3.6** **Resin lifetimes, continued**

For our resin data, the treatment sum of squares is

$$\begin{aligned} SS_{\text{Trt}} &= \sum_{i=1}^{g} n_i \widehat{\alpha}_i^{\,2} \\ &= 8 \times .467^2 + 8 \times .164^2 + 8 \times (-.088)^2 + \\ &\quad 7 \times (-.271)^2 + 6 \times (-.408)^2 \\ &= 3.5376\ . \end{aligned}$$

We have $g = 5$ treatments so there are $g - 1 = 4$ degrees of freedom between treatments. We computed the $SS_E$ in Example 3.5; it was .29369 with 32 degrees of freedom. The ANOVA table is

ANOVA

| Source | DF | SS | MS | F |
|---|---|---|---|---|
| treatments | 4 | 3.5376 | .88441 | 96.4 |
| error | 32 | .29369 | .0091779 | |
| total | 36 | 3.8313 | | |

The F-statistic is about 96 with 4 and 32 degrees of freedom. There is essentially no probability under the F-curve with 4 and 32 degrees of freedom

**Listing 3.2:** SAS output for resin lifetimes

```
                    Analysis of Variance Procedure

Dependent Variable: LIFETIME
                                     Sum of          Mean
Source                   DF         Squares        Square   F Value      Pr > F   ①

Model                     4      3.53763206    0.88440802     96.36      0.0001

Error                    32      0.29369226    0.00917788

Corrected Total          36      3.83132432

              R-Square            C.V.      Root MSE          LIFETIME Mean

              0.923344        6.538733       0.09580                1.46514

             Level of          -----------LIFETIME----------                      ②
             TEMPER      N         Mean              SD

             1           8       1.93250000        0.06341473
             2           8       1.62875000        0.10480424
             3           8       1.37750000        0.10713810
             4           7       1.19428571        0.04577377
             5           6       1.05666667        0.13837148
```

to the right of 96. (There is only .00001 probability to the right of 11.) Thus the $p$-value for this test is essentially zero, and we would conclude that not all the treatments yield the same mean lifetime. From a practical point of view, the experimenters already knew this; the experiment was run to determine the nature of the dependence of lifetime on temperature, not whether there was any dependence.

Different statistics software packages give slightly different output for the ANOVA of the resin lifetime data. For example, Listing 3.1 gives Minitab ANOVA output. In addition to the ANOVA table ①, the standard Minitab output includes a table of treatment means and a plot of 95% confidence intervals for those means ②. Listing 3.2 gives SAS output (edited to save space) for these data ①. SAS does not automatically print group means, but you can request them as shown here ②.

There is a heuristic for the degrees-of-freedom formulae. Degrees of freedom for a model count the number of additional parameters used for the mean structure when moving from the next simpler model to this model. For example, the degrees of freedom for treatment are $g - 1$. The next simpler

Model df count parameters

model is the model of a single mean for all treatments; the full model has a different mean for each of the $g$ treatments. That is $g - 1$ more parameters. Alternatively, look at the $\alpha_i$'s. Under the null, they are all zero. Under the alternative, they may be nonzero, but only $g - 1$ of them can be set freely, because the last one is then set by the restriction that their weighted sum must be zero. Degrees of freedom for error are the number of data less the number of (mean) parameters estimated.

## 3.7 Why ANOVA Works

$E(MS_E) = \sigma^2$

The mean square for error is a random variable; it depends on the random errors in the data. If we repeated the experiment, we would get different random errors and thus a different mean square for error. However, the expected value of the mean square for error, averaged over all the possible outcomes of the random errors, is the variance of the random errors $\sigma^2$. Thus, the mean square for error estimates the error variance, no matter what the values of the $\alpha_i$'s.

Expected mean square for treatments

The mean square for treatments is also a random variable, but the $MS_{\text{Trt}}$ has expectation:

$$E(MS_{\text{Trt}}) = EMS_{\text{Trt}} = \sigma^2 + \sum_{i=1}^{g} n_i\alpha_i^2/(g-1) \ .$$

The important things to get from this expression are

1. When all of the $\alpha_i$'s are zero, the mean square for treatments also estimates $\sigma^2$.
2. When some of the $\alpha_i$'s are nonzero, the mean square for treatments tends to be bigger than $\sigma^2$.

When the null hypothesis is true, both $MS_{\text{Trt}}$ and $MS_E$ vary around $\sigma^2$, so their ratio (the F-statistic) is about one, give or take some random variation. When the null hypothesis is false, $MS_{\text{Trt}}$ tends to be bigger than $\sigma^2$, and the F-statistic tends to be bigger than one. We thus reject the null hypothesis for sufficiently large values of the F-statistic.

## 3.8 Back to Model Comparison

The preceding section described Analysis of Variance as a test of the null hypothesis that all the $\alpha_i$ values are zero. Another way to look at ANOVA is as a comparison of two models for the data. The reduced model is the model that all treatments have the same expected value (that is, the $\alpha_i$ values are all zero); the full model allows the treatments to have different expected values. From this point of view, we are not testing whether a set of parameters is all zero; we are comparing the adequacy of two different models for the mean structure.

ANOVA compares models

Analysis of Variance uses sums of squared deviations from a model, just as sample standard deviations use squared deviations from a sample mean. For the reduced model (null hypothesis), the estimated model is $\widehat{\mu} = \overline{y}_{\bullet\bullet}$. For the data value $y_{ij}$, the residual is

$$r_{ij} = y_{ij} - \widehat{\mu} = y_{ij} - \overline{y}_{\bullet\bullet}.$$

The residual sum of squares for the reduced model is then

$$SSR_0 = \sum_{ij} r_{ij}^2 = \sum_{ij} (y_{ij} - \overline{y}_{\bullet\bullet})^2.$$

For the full model (alternative hypothesis), the estimated model is $\widehat{\mu}_i = \overline{y}_{i\bullet}$, and the residuals are

$$r_{ij} = y_{ij} - \widehat{\mu}_i = y_{ij} - \overline{y}_{i\bullet}.$$

The residual sum of squares for the full model is then

Model SSR

$$SSR_A = \sum_{ij} r_{ij}^2 = \sum_{ij} (y_{ij} - \overline{y}_{i\bullet})^2.$$

$SSR_A$ can never be bigger than $SSR_0$ and will almost always be smaller. We would prefer the full model if $SSR_A$ is sufficiently smaller than $SSR_0$.

How does this terminology for ANOVA mesh with what we have already seen? The residual sum of squares from the full model, $SSR_A$, is the error sum of squares $SS_E$ in the usual formulation. The residual sum of squares from the reduced model, $SSR_0$, is the total sum of squares $SS_T$ in the usual formulation. The difference $SSR_0 - SSR_A$ is equal to the treatment sum of squares $SS_{\mathrm{Trt}}$. Thus the treatment sum of squares is the additional amount of variation in the data that can be explained by using the more complicated full model instead of the simpler reduced model.

Change in SSR

This idea of comparing models instead of testing hypotheses about parameters is a fairly subtle distinction, and here is why the distinction is important: in our heart of hearts, we almost never believe that the null hypothesis could be true. We usually believe that at some level of precision, there

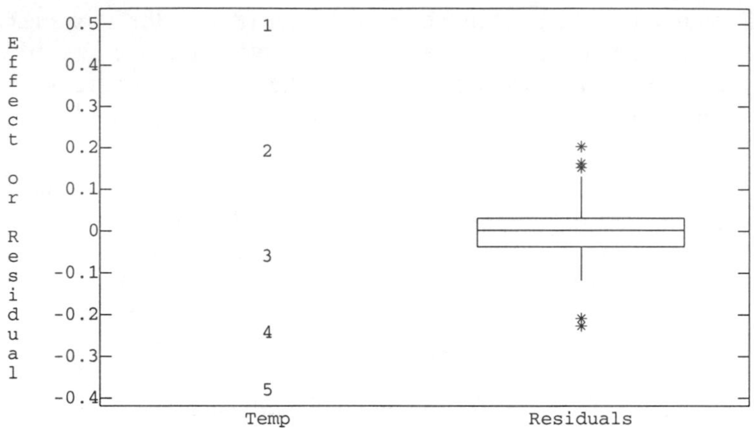


**Figure 3.3:** Side-by-side plot for resin lifetime data, using MacAnova.

is a difference between the mean responses of the treatments. *So why the charade of testing the null hypothesis?*

Choose simplest acceptable model

The answer is that we are choosing a model for the data from a set of potential models. We want a model that is as simple as possible yet still consistent with the data. A more realistic null hypothesis is that the means are so close to being equal that the differences are negligible. When we "reject the null hypothesis" we are making the decision that the data are demonstrably inconsistent with the simpler model, the differences between the means are not negligible, and the more complicated model is required. Thus we use the F-test to guide us in our choice of model. This distinction between testing hypotheses on parameters and selecting models will become more important later.

## 3.9 Side-by-Side Plots

Side-by-side plots show effects and residuals

Hoaglin, Mosteller, and Tukey (1991) introduce the *side-by-side* plot as a method for visualizing treatment effects and residuals. Figure 3.3 shows a side-by-side plot for the resin lifetime data of Example 3.2. We plot the estimated treatment effects $\widehat{\alpha}_i$ in one column and the residuals $r_{ij}$ in a second column. (There will be more columns in more complicated models we will see later.) The vertical scale is in the same units as the response. In this plot, we have used a box-plot for the residuals rather than plot them individually; this will usually be more understandable when there are relatively many points to be put in a single column.

What we see from the side-by-side plot is that the treatment effects are large compared to the size of the residuals. We were also able to see this in the parallel box-plots in the exploratory analysis, but the side-by-side plots will generalize better to more complicated models.

## 3.10 Dose-Response Modeling

Numerical levels or doses

In some experiments, the treatments are associated with numerical levels such as drug dose, baking time, or reaction temperature. We will refer to such levels as *doses,* no matter what they actually are, and the numerical value of the dose for treatment $i$ will be denoted $z_i$. When we have numerical doses, we may reexpress the treatment means as a function of the dose $z_i$:

$$\mu + \alpha_i = f(z_i; \theta) \ ,$$

where $\theta$ is some unknown parameter of the function. For example, we could express the mean weight of yellow birch seedlings as a function of the pH of acid rain.

Polynomial models

The most commonly used functions $f$ are polynomials in the dose $z_i$:

$$\mu + \alpha_i = \theta_0 + \theta_1 z_i + \theta_2 z_i^2 + \cdots + \theta_{g-1} z_i^{g-1} \ .$$

We use the power $g - 1$ because the means at $g$ different doses determine a polynomial of order $g - 1$. Polynomials are used so often because they are simple and easy to understand; they are not always the most appropriate choice.

Polynomials are an alternative to treatment effects

If we know the polynomial coefficients $\theta_0, \theta_1, \ldots, \theta_{g-1}$, then we can determine the treatment means $\mu + \alpha_i$, and vice versa. If we know the polynomial coefficients *except for the constant* $\theta_0$, then we can determine the treatment effects $\alpha_i$, and vice versa. The $g - 1$ parameters $\theta_1$ through $\theta_{g-1}$ in this full polynomial model correspond to the $g - 1$ degrees of freedom between the treatment groups. Thus polynomials in dose are not inherently better or worse than the treatment effects model, just another way to describe the differences between means.

Polynomial modeling is useful in two contexts. First, if only a few of the polynomial coefficients are needed (that is, the others can be set to zero without significantly decreasing the quality of the fit), then this reduced polynomial model represents a reduction in the complexity of our model. For example, learning that the response is linear or quadratic in dose is useful, whereas a polynomial of degree six or seven will be difficult to comprehend (or sell to anyone else). Second, if we wish to estimate the response at some dose other than one used in the experiment, the polynomial model provides a mechanism for generating the estimates. Note that these estimates may be poor if we are extrapolating beyond the range of the doses in our experiment or if the degree of the polynomial is high. High-order polynomials will fit our observed treatment means exactly, but these high-order polynomials can have bizarre behavior away from our data points.

Polynomial models can reduce number of parameters needed and provide interpolation

Consider a sequence of regression models for our data, regressing the responses on dose, dose squared, and so on. The first model just includes the constant $\theta_0$; that is, it fits a single value for all responses. The second model includes the constant $\theta_0$ and a linear term $\theta_1 z_i$; this model fits the responses as a simple linear regression in dose. The third model includes the constant $\theta_0$, a linear term $\theta_1 z_i$, and the quadratic term $\theta_2 z_i^2$; this model fits the responses as a quadratic function (parabola) of dose. Additional models include additional powers of dose up to $g - 1$.

Let $SSR_k$ be the residual sum of squares for the model that includes powers up to $k$, for $k = 0, \ldots, g - 1$. Each successive model will explain a little more of the variability between treatments, so that $SSR_k > SSR_{k+1}$. When we arrive at the full polynomial model, we will have explained all of the between-treatment variability using polynomial terms; that is, $SSR_{g-1} = SS_E$. The "linear sum of squares" is the reduction in residual variability going from the constant model to the model with the linear term:

Polynomial improvement SS for including an additional term

$$SS_{\text{linear}} = SS_1 = SSR_0 - SSR_1 \ .$$

Similarly, the "quadratic sum of squares" is the reduction in residual variability going from the linear model to the quadratic model,

$$SS_{\text{quadratic}} = SS_2 = SSR_1 - SSR_2 \ ,$$

and so on through the remaining orders.

Each of these polynomial sums of squares has 1 degree of freedom, because each is the result of adding one more parameter $\theta_k$ to the model for the means. Thus their mean squares are equal to their sums of squares. In a model with terms up through order $k$, we can test the null hypothesis that $\theta_k = 0$ by forming the F-statistic $SS_k/MS_E$, and comparing it to an F-distribution with 1 and $N - g$ degrees of freedom.

Testing parameters

One method for choosing a polynomial model is to choose the smallest order such that no significant terms are excluded. (More sophisticated model selection methods exist.) It is important to know that the estimated

Model selection

**Listing 3.3:** MacAnova output for resin lifetimes polynomial model.

```
                      DF          SS            MS           F        P-value   ①
CONSTANT               1      79.425        79.425  8653.95365              0
{temperature}          1      3.4593        3.4593   376.91283              0
{(temperature)^2}      1    0.078343      0.078343     8.53610      0.0063378
{(temperature)^3}      1  1.8572e-05    1.8572e-05     0.00202         0.9644
{(temperature)^4}      1  8.2568e-06    8.2568e-06     0.00090        0.97626
ERROR1                32     0.29369     0.0091779

CONSTANT                                                                        ②
(1)        0.96995
{temperature}
(1)       0.075733
{(temperature)^2}
(1)    -0.00076488
{(temperature)^3}
(1)     2.6003e-06
{(temperature)^4}
(1)    -2.9879e-09

                      DF          SS            MS           F        P-value   ③
CONSTANT               1      79.425        79.425  9193.98587              0
{temperature}          1      3.4593        3.4593   400.43330              0
{(temperature)^2}      1    0.078343      0.078343     9.06878      0.0048787
ERROR1                34     0.29372     0.0086388

CONSTANT                                                                        ④
(1)            7.418
{temperature}
(1)      -0.045098
{(temperature)^2}
(1)     7.8604e-05
```

coefficients $\widehat{\theta}_i$ depend on which terms are in the model when the model is estimated. Thus if we decide we only need $\theta_0$, $\theta_1$, and $\theta_2$ when $g$ is 4 or more, we should refit using just those terms to get appropriate parameter estimates.

## Resin lifetimes, continued

**Example 3.7**

The treatments in the resin lifetime data are different temperatures (175, 194, 213, 231, and 250 degrees C), so we can use these temperatures as doses $z_i$ in a dose-response relationship. With $g = 5$ treatments, we can use polynomials up to power 4.

Listing 3.3 shows output for a polynomial dose-response modeling of the resin lifetime data. The first model fits up to temperature to the fourth power. From the ANOVA ① we can see that neither the third nor fourth powers are significant, but the second power is, so a quadratic model seems appropriate. The ANOVA for the reduced model is at ③. The linear and quadratic sums of squares are the same as in ①, but the $SS_E$ in ③ is increased by the cubic and quartic sums of squares in ①. We can also see that the intercept, linear, and quadratic coefficients change dramatically from the full model ② to the reduced model using just those terms ④. We cannot simply take the intercept, linear, and quadratic coefficients from the fourth power model and use them as if they were coefficients in a quadratic model.

Try transforming dose

One additional trick to remember when building a dose-response model is that we can transform or reexpress the dose $z_i$. That is, we can build models using log of dose or square root of dose as simply as we can using dose. For some data it is much simpler to build a model as a function of a transformation of the dose.

## 3.11 Further Reading and Extensions

There is a second randomization that is used occasionally, and unfortunately it also is sometimes called completely randomized.

1. Choose probabilities $p_1$ though $p_g$ with $p_1 + p_2 + \cdots + p_g = 1$.
2. Choose a treatment independently for each unit, choosing treatment $i$ with probability $p_i$.

Now we wind up with $n_i$ units getting treatment i, with $n_1 + n_2 + \cdots + n_g = N$, but the sample sizes $n_i$ are random. This randomization is different than the standard CRD randomization. ANOVA procedures do not distinguish between the fixed and random sample size randomizations, but if we were to do randomization testing, we would use different procedures for the two different randomizations. As a practical matter, we should note that even though we may design for certain fixed sample sizes, we do not always achieve those sample sizes when test tubes get dropped, subjects withdraw from studies, or drunken statistics graduate students drive through experimental fields (you know who you are!).

The estimates we have used for mean parameters are least squares estimates, meaning that they minimize the sum of squared residuals. Least squares estimation goes back to Legendre (1806) and Gauss (1809), who developed the procedure for working with astronomical data. Formal tests based on the $t$-distribution were introduced by Gosset, who wrote under the pseudonym "Student" (Student 1908). Gosset worked at the Guiness Brewery, and he was allowed to publish only under a pseudonym so that the competition would not be alerted to the usefulness of the procedure. What Gosset actually did was posit the $t$-distribution; proof was supplied later by Fisher (1925a).

The Analysis of Variance was introduced by Fisher in the context of population genetics (Fisher 1918); he quickly extended the scope (Fisher 1925b). The 1918 paper actually introduces the terms "variance" and "analysis of variance". Scheffé (1956) describes how models for data essentially the same as those used for ANOVA were in use decades earlier, though analysis methods were different.

From a more theoretical perspective, the $SS_E$ is distributed as $\sigma^2$ times a chi-square random variable with $N - g$ degrees of freedom; $SS_{\text{Trt}}$ is distributed as $\sigma^2$ times a possibly noncentral chi-square random variable with $g - 1$ degrees of freedom; and these two sums of squares are independent. When the null hypothesis is true, $SS_{\text{Trt}}$ is a multiple of an ordinary (central) chi-square; noncentrality arises under the alternative when the expected value of $MS_{\text{Trt}}$ is greater than $\sigma^2$. The ratio of two independent central chi-squares, each divided by their degrees of freedom, is defined to have an F-distribution. Thus the null-hypothesis distribution of the F-statistic is F. Chapter 7 and Appendix A discuss this distribution theory in more detail. Scheffé (1959), Hocking (1985), and others provide book-length expositions of linear models and their related theory.

We have described model selection via testing a null hypothesis. An alternative approach is *prediction*; for example, we can choose the model that we believe will give us the lowest average squared error of prediction. Mallows (1973) defined a quantity called $C_p$

$$C_p = \frac{SSR_p}{MS_E} + 2p - N \quad ,$$

where $SSR_p$ is the residual sum of squares for a means model with $p$ parameters (degrees of freedom including any overall constant), $MS_E$ is the error mean square from the separate means model, and $N$ is the number of observations. We prefer models with small $C_p$.

The separate means model (with $p = g$ parameters) has $C_p = g$. The single mean model, dose-response models, and other models can have $C_p$ values greater or less than $g$. The criterion rewards models with smaller $SSR$ and penalizes models with larger $p$. When comparing two models, one a reduced form of the other, $C_p$ will prefer the larger model if the F-statistic comparing the models is 2 or greater. Thus we see that it generally takes less "evidence" to choose a larger model when using a predictive criterion than when doing testing at the traditional levels.

Quantitative dose-response models as described here are an instance of polynomial regression. Weisberg (1985) is a good general source on regression, including polynomial regression. We have used polynomials because they are simple and traditional, but there are many other sets of functions we could use instead. Some interesting alternatives include sines and cosines, B-splines, and wavelets.

## 3.12 Problems

**Exercise 3.1**

Rats were given one of four different diets at random, and the response measure was liver weight as a percentage of body weight. The responses were

| Treatment | | | |
|---|---|---|---|
| 1 | 2 | 3 | 4 |
| 3.52 | 3.47 | 3.54 | 3.74 |
| 3.36 | 3.73 | 3.52 | 3.83 |
| 3.57 | 3.38 | 3.61 | 3.87 |
| 4.19 | 3.87 | 3.76 | 4.08 |
| 3.88 | 3.69 | 3.65 | 4.31 |
| 3.76 | 3.51 | 3.51 | 3.98 |
| 3.94 | 3.35 | | 3.86 |
| | 3.64 | | 3.71 |

(a) Compute the overall mean and treatment effects.

(b) Compute the Analysis of Variance table for these data. What would you conclude about the four diets?

**Exercise 3.2**

An experimenter randomly allocated 125 male turkeys to five treatment groups: control and treatments A, B, C, and D. There were 25 birds in each group, and the mean results were 2.16, 2.45, 2.91, 3.00, and 2.71, respectively. The sum of squares for experimental error was 153.4. Test the null hypothesis that the five group means are the same against the alternative that one or more of the treatments differs from the control.

**Exercise 3.3**

Twelve orange pulp silage samples were divided at random into four groups of three. One of the groups was left as an untreated control, while the other three groups were treated with formic acid, beet pulp, and sodium chloride, respectively. One of the responses was the moisture content of the silage. The observed moisture contents of the silage are shown below (data from Caro *et al.* 1990):

| | NaCl | Formic acid | Beet pulp | Control |
|---|---|---|---|---|
| | 80.5 | 89.1 | 77.8 | 76.7 |
| | 79.3 | 75.7 | 79.5 | 77.2 |
| | 79.0 | 81.2 | 77.0 | 78.6 |
| Means | 79.6 | 82.0 | 78.1 | 77.5 |
| Grand mean | 79.3 | | | |

Compute an analysis of variance table for these data and test the null hypothesis that all four treatments yield the same average moisture contents.

**Exercise 3.4**

We have five groups and three observations per group. The group means are 6.5, 4.5, 5.7, 5.7, and 5.1, and the mean square for error is .75. Compute an ANOVA table for these data.

**Exercise 3.5**

The leaves of certain plants in the genus *Albizzia* will fold and unfold in various light conditions. We have taken fifteen different leaves and subjected them to red light for 3 minutes. The leaves were divided into three groups of five at random. The leaflet angles were then measured 30, 45, and 60 minutes after light exposure in the three groups. Data from W. Hughes.

| Delay (minutes) | Angle (degrees) | | | | |
|---|---|---|---|---|---|
| 30 | 140 | 138 | 140 | 138 | 142 |
| 45 | 140 | 150 | 120 | 128 | 130 |
| 60 | 118 | 130 | 128 | 118 | 118 |

Analyze these data to test the null hypothesis that delay after exposure does not affect leaflet angle.

**Problem 3.1**

Cardiac pacemakers contain electrical connections that are platinum pins soldered onto a substrate. The question of interest is whether different operators produce solder joints with the same strength. Twelve substrates are randomly assigned to four operators. Each operator solders four pins on each substrate, and then these solder joints are assessed by measuring the shear strength of the pins. Data from T. Kerkow.

| | Strength (lb) | | |
|---|---|---|---|
| Operator | Substrate 1 | Substrate 2 | Substrate 3 |
| 1 | 5.60 6.80 8.32 8.70 | 7.64 7.44 7.48 7.80 | 7.72 8.40 6.98 8.00 |
| 2 | 5.04 7.38 5.56 6.96 | 8.30 6.86 5.62 7.22 | 5.72 6.40 7.54 7.50 |
| 3 | 8.36 7.04 6.92 8.18 | 6.20 6.10 2.75 8.14 | 9.00 8.64 6.60 8.18 |
| 4 | 8.30 8.54 7.68 8.92 | 8.46 7.38 8.08 8.12 | 8.68 8.24 8.09 8.06 |

Analyze these data to determine if there is any evidence that the operators produce different mean shear strengths. (Hint: what are the experimental units?)

**Problem 3.2** Scientists are interested in whether the energy costs involved in reproduction affect longevity. In this experiment, 125 male fruit flies were divided at random into five sets of 25. In one group, the males were kept by themselves. In two groups, the males were supplied with one or eight receptive virgin female fruit flies per day. In the final two groups, the males were supplied with one or eight unreceptive (pregnant) female fruit plies per day. Other than the number and type of companions, the males were treated identically. The longevity of the flies was observed. Data from Hanley and Shapiro (1994).

| Companions | Longevity (days) | | | | | | | | | | | | |
|---|---|---|---|---|---|---|---|---|---|---|---|---|---|
| None | 35 | 37 | 49 | 46 | 63 | 39 | 46 | 56 | 63 | 65 | 56 | 65 | 70 |
| | 63 | 65 | 70 | 77 | 81 | 86 | 70 | 70 | 77 | 77 | 81 | 77 | |
| 1 pregnant | 40 | 37 | 44 | 47 | 47 | 47 | 68 | 47 | 54 | 61 | 71 | 75 | 89 |
| | 58 | 59 | 62 | 79 | 96 | 58 | 62 | 70 | 72 | 75 | 96 | 75 | |
| 1 virgin | 46 | 42 | 65 | 46 | 58 | 42 | 48 | 58 | 50 | 80 | 63 | 65 | 70 |
| | 70 | 72 | 97 | 46 | 56 | 70 | 70 | 72 | 76 | 90 | 76 | 92 | |
| 8 pregnant | 21 | 40 | 44 | 54 | 36 | 40 | 56 | 60 | 48 | 53 | 60 | 60 | 65 |
| | 68 | 60 | 81 | 81 | 48 | 48 | 56 | 68 | 75 | 81 | 48 | 68 | |
| 8 virgin | 16 | 19 | 19 | 32 | 33 | 33 | 30 | 42 | 42 | 33 | 26 | 30 | 40 |
| | 54 | 34 | 34 | 47 | 47 | 42 | 47 | 54 | 54 | 56 | 60 | 44 | |

Analyze these data to test the null hypothesis that reproductive activity does not affect longevity. Write a report on your analysis. Be sure to describe the experiment as well as your results.

**Problem 3.3** Park managers need to know how resistant different vegetative types are to trampling so that the number of visitors can be controlled in sensitive areas. The experiment deals with alpine meadows in the White Mountains of New Hampshire. Twenty lanes were established, each .5 m wide and 1.5 m long. These twenty lanes were randomly assigned to five treatments: 0, 25, 75, 200, or 500 walking passes. Each pass consists of a 70-kg individual wearing lug-soled boots walking in a natural gait down the lane. The response measured is the average height of the vegetation along the lane one year after trampling. Data based on Table 16 of Cole (1993).

| Number of passes | Height (cm) | | | |
|---|---|---|---|---|
| 0 | 20.7 | 15.9 | 17.8 | 17.6 |
| 25 | 12.9 | 13.4 | 12.7 | 9.0 |
| 75 | 11.8 | 12.6 | 11.4 | 12.1 |
| 200 | 7.6 | 9.5 | 9.9 | 9.0 |
| 500 | 7.8 | 9.0 | 8.5 | 6.7 |

Analyze these data to determine if trampling has an effect after one year, and if so, describe that effect.

**Problem 3.4**

Caffeine is a common drug that affects the central nervous system. Among the issues involved with caffeine are how does it get from the blood to the brain, and does the presence of caffeine alter the ability of similar compounds to move across the blood-brain barrier? In this experiment, 43 lab rats were randomly assigned to one of eight treatments. Each treatment consisted of an arterial injection of $C^{14}$-labeled adenine together with a concentration of caffeine (0 to 50 mM). Shortly after injection, the concentration of labeled adenine in the rat brains is measured as the response (data from McCall, Millington, and Wurtman 1982).

| Caffeine (mM) | Adenine | | | | | |
|---|---|---|---|---|---|---|
| 0 | 5.74 | 6.90 | 3.86 | 6.94 | 6.49 | 1.87 |
| 0.1 | 2.91 | 4.14 | 6.29 | 4.40 | 3.77 | |
| 0.5 | 5.80 | 5.84 | 3.18 | 3.18 | | |
| 1 | 3.49 | 2.16 | 7.36 | 1.98 | 5.51 | |
| 5 | 5.92 | 3.66 | 4.62 | 3.47 | 1.33 | |
| 10 | 3.05 | 1.94 | 1.23 | 3.45 | 1.61 | 4.32 |
| 25 | 1.27 | .69 | .85 | .71 | 1.04 | .84 |
| 50 | .93 | 1.47 | 1.27 | 1.13 | 1.25 | .55 |

The main issues in this experiment are whether the amount of caffeine present affects the amount of adenine that can move from the blood to the brain, and if so, what is the dose response relationship. Analyze these data.

**Problem 3.5**

Engineers wish to know the effect of polypropylene fibers on the compressive strength of concrete. Fifteen concrete cubes are produced and randomly assigned to five levels of fiber content (0, .25, .50, .75, and 1%). Data from Figure 2 of Paskova and Meyer (1997).

| Fiber content (%) | Strength (ksi) | | |
|---|---|---|---|
| 0 | 7.8 | 7.4 | 7.2 |
| .25 | 7.9 | 7.5 | 7.3 |
| .50 | 7.4 | 6.9 | 6.3 |
| .75 | 7.0 | 6.7 | 6.4 |
| 1 | 5.9 | 5.8 | 5.6 |

Analyze these data to determine if fiber content has an effect on concrete strength, and if so, describe that effect.

**Question 3.1** Prove that $\mu^\star = \sum_{i=1}^g \mu_i/g$ is equivalent to $\sum_{i=1}^g \alpha_i = 0$.

**Question 3.2** Prove that

$$0 = \sum_{i=1}^{g}\sum_{j=1}^{n_i} \widehat{\alpha}_i r_{ij} \quad .$$

# Chapter 4

# Looking for Specific Differences—Contrasts

An Analysis of Variance can give us an indication that not all the treatment groups have the same mean response, but an ANOVA does not, by itself, tell us which treatments are different or in what ways they differ. To do this, we need to look at the treatment means, or equivalently, at the treatment effects. One method to examine treatment effects is called a *contrast*.

ANOVA is like background lighting that dimly illuminates all of our data, but not giving enough light to see details. Using a contrast is like using a spotlight; it enables us to focus in on a specific, narrow feature of the data. But the contrast has such a narrow focus that it does not give the overall picture. By using several contrasts, we can move our focus around and see more features. Intelligent use of contrasts involves choosing our contrasts so that they highlight interesting features in our data.

Contrasts examine specific differences

## 4.1 Contrast Basics

Contrasts take the form of a difference between means or averages of means. For example, here are two contrasts:

$$(\mu + \alpha_6) - (\mu + \alpha_3)$$

and

$$\frac{\mu + \alpha_2 + \mu + \alpha_4}{2} - \frac{\mu + \alpha_1 + \mu + \alpha_3 + \mu + \alpha_5}{3} \quad .$$

The first compares the means of treatments 6 and 3, while the second compares the mean response in groups 2 and 4 with the mean response in groups 1, 3, and 5.

Contrasts compare averages of means

Formally, a *contrast* is a linear combination of treatment means or effects $\sum_{i=1}^{g} w_i\mu_i = w(\{\mu_i\})$ or $\sum_{i=1}^{g} w_i\alpha_i = w(\{\alpha_i\})$, where the coefficients $w_i$ satisfy $\sum_{i=1}^{g} w_i = 0$.

| Contrast coefficients add to zero. |
|---|

Less formally, we sometimes speak of the set of contrast coefficients $\{w_i\}$ as being a contrast; we will try to avoid ambiguity. Notice that because the sum of the coefficients is zero, we have that

$$\begin{aligned} w(\{\alpha_i\}) &= \sum_{i=1}^{g} w_i\alpha_i = x\sum_{i=1}^{g} w_i + \sum_{i=1}^{g} w_i\alpha_i \\ &= \sum_{i=1}^{g} w_i(x+\alpha_i) = \sum_{i=1}^{g} w_i(\mu+\alpha_i) = w(\{\mu_i\}) \end{aligned}$$

for any fixed constant x (say $\mu$ or $\pi$). We may also make contrasts in the observed data:

$$w(\{\overline{y}_{i\bullet}\}) = \sum_{i=1}^{g} w_i\overline{y}_{i\bullet} = \sum_{i=1}^{g} w_i(\overline{y}_{i\bullet} - \overline{y}_{\bullet\bullet}) = \sum_{i=1}^{g} w_i\widehat{\alpha}_i = w(\{\widehat{\alpha}_i\}) \ .$$

Contrasts do not depend on $\alpha$-restrictions

A contrast depends on the differences between the values being contrasted, but not on the overall level of the values. In particular, a contrast in treatment means depends on the $\alpha_i$'s but not on $\mu$. A contrast in the treatment means or effects will be the same regardless of whether we assume that $\alpha_1 = 0$, or $\sum \alpha_i = 0$, or $\sum n_i\alpha_i = 0$. Recall that with respect to restrictions on the treatment effects, we said that "the important things don't depend on which set of restrictions we use." In particular, contrasts don't depend on the restrictions.

We may use several different kinds of contrasts in any one analysis. The trick is to find or construct contrasts that focus in on interesting features of the data.

Pairwise comparisons

Probably the most common contrasts are *pairwise comparisons*, where we contrast the mean response in one treatment with the mean response in a second treatment. For a pairwise comparison, one contrast coefficient is 1, a second contrast coefficient is -1, and all other contrast coefficients are 0. For example, in an experiment with $g = 4$ treatments, the coefficients (0, 1,

-1, 0) compare the means of treatments 2 and 3, and the coefficients (-1, 0, 1, 0) compare the means of treatments 1 and 3. For $g$ treatments, there are $g(g-1)/2$ different pairwise comparisons. We will consider simultaneous inference for pairwise comparisons in Section 5.4.

A second classic example of contrasts occurs in an experiment with a control and two or more new treatments. Suppose that treatment 1 is a control, and treatments 2 and 3 are new treatments. We might wish to compare the average response in the new treatments to the average response in the control; that is, on average do the new treatments have the same response as the control? Here we could use coefficients (-1, .5, .5), which would subtract the average control response from the average of treatments 2 and 3's average responses. As discussed below, this contrast applied to the observed treatment means $((\overline{y}_{2\bullet} + \overline{y}_{3\bullet})/2 - \overline{y}_{1\bullet})$ would estimate the contrast in the treatment effects $((\alpha_2 + \alpha_3)/2 - \alpha_1)$. Note that we would get the same kind of information from contrasts with coefficients (1, -.5, -.5) or (-6, 3, 3); we've just rescaled the result with no essential loss of information. We might also be interested in the pairwise comparisons, including a comparison of the new treatments to each other (0, 1, -1) and comparisons of each of the new treatments to control (1, -1, 0) and (1, 0, -1).

Control versus other treatments

Consider next an experiment with four treatments examining the growth rate of lambs. The treatments are four different food supplements. Treatment 1 is soy meal and ground corn, treatment 2 is soy meal and ground oats, treatment 3 is fish meal and ground corn, and treatment 4 is fish meal and ground oats. Again, there are many potential contrasts of interest. A contrast with coefficients (.5, .5, -.5, -.5) would take the average response for fish meal treatments and subtract it from the average response for soy meal treatments. This could tell us about how the protein source affects the response. Similarly, a contrast with coefficients (.5, -.5, .5, -.5) would take the average response for ground oats and subtract it from the average response for ground corn, telling us about the effect of the carbohydrate source.

Compare related groups of treatments

Finally, consider an experiment with three treatments examining the effect of development time on the number of defects in computer chips produced using photolithography. The three treatments are 30, 45, and 60 seconds of developing. If we think of the responses as lying on a straight line function of development time, then the contrast with coefficients (-1/30, 0, 1/30) will estimate the slope of the line relating response and time. If instead we think that the responses lie on a quadratic function of development time, then the contrast with coefficients (1/450, -2/450, 1/450) will estimate the quadratic term in the response function. Don't worry for now about where these coefficients come from; they will be discussed in more detail in Section 4.4. For now, consider that the first contrast compares the responses at

Polynomial contrasts for quantitative doses

the ends to get a rate of change, and the second contrast compares the ends to the middle (which yields a 0 comparison for responses on a straight line) to assess curvature.

## 4.2 Inference for Contrasts

We use contrasts in observed treatment means or effects to make inference about the corresponding contrasts in the true treatment means or effects. The kinds of inference we work with here are point estimates, confidence intervals, and tests of significance. The procedures we use for contrasts are similar to the procedures we use when estimating or testing means.

$w(\{\overline{y}_{i\bullet}\})$ estimates $w(\{\mu_i\})$

The observed treatment mean $\overline{y}_{i\bullet}$ is an unbiased estimate of $\mu_i = \mu + \alpha_i$, so a sum or other linear combination of observed treatment means is an unbiased estimate of the corresponding combination of the $\mu_i$'s. In particular, a contrast in the observed treatment means is an unbiased estimate of the corresponding contrast in the true treatment means. Thus we have:

$$E[w(\{\overline{y}_{i\bullet}\})] = E[w(\{\widehat{\alpha}_i\})] = w(\{\mu_i\}) = w(\{\alpha_i\}) \ .$$

The variance of $\overline{y}_{i\bullet}$ is $\sigma^2/n_i$, and the treatment means are independent, so the variance of a contrast in the observed means is

$$\text{Var}\left[w(\{\overline{y}_{i\bullet}\})\right] = \sigma^2 \sum_{i=1}^{g} \frac{w_i^2}{n_i} \ .$$

We will usually not know $\sigma^2$, so we estimate it by the mean square for error from the ANOVA.

Confidence interval for $w(\{\mu_i\})$

We compute a confidence interval for a mean parameter with the general form: unbiased estimate $\pm$ $t$-multiplier $\times$ estimated standard error. Contrasts are linear combinations of mean parameters, so we use the same basic form. We have already seen how to compute an estimate and standard error, so

$$w(\{\overline{y}_{i\bullet}\}) \pm \ t_{\mathcal{E}/2,N-g} \sqrt{MS_E} \sqrt{\sum_{i=1}^{g} \frac{w_i^2}{n_i}}$$

forms a $1 - \mathcal{E}$ confidence interval for $w(\{\mu_i\})$. As usual, the degrees of freedom for our $t$-percent point come from the degrees of freedom for our estimate of error variance, here $N - g$. We use the $\mathcal{E}/2$ percent point because we are forming a two-sided confidence interval, with $\mathcal{E}/2$ error on each side.

The usual $t$-test statistic for a mean parameter takes the form

$$\frac{\textit{unbiased estimate} - \textit{null hypothesis value}}{\textit{estimated standard error of estimate}} .$$

This form also works for contrasts. If we have the null hypothesis $H_0 : w(\{\mu_i\}) = \delta$, then we can do a $t$-test of that null hypothesis by computing the test statistic

$$t = \frac{w(\{\overline{y}_{i\bullet}\}) - \delta}{\sqrt{MS_E}\sqrt{\sum_{i=1}^{g} \frac{w_i^2}{n_i}}} .$$

Under $H_0$, this $t$-statistic will have a $t$-distribution with $N - g$ degrees of freedom. Again, the degrees of freedom come from our estimate of error variance. The $p$-value for this $t$-test is computed by getting the area under the $t$-distribution with $N - g$ degrees of freedom for the appropriate region: either less or greater than the observed $t$-statistic for one-sided alternatives, or twice the tail area for a two-sided alternative.

$t$-test for $w(\{\mu_i\})$

We may also compute a sum of squares for any contrast $w(\{\overline{y}_{i\bullet}\})$:

$$SS_w = \frac{(\sum_{i=1}^{g} w_i \overline{y}_{i\bullet})^2}{\sum_{i=1}^{g} \frac{w_i^2}{n_i}} .$$

This sum of squares has 1 degree of freedom, so its mean square is $MS_w = SS_w/1 = SS_w$. We may use $MS_w$ to test the null hypothesis that $w(\{\mu_i\}) = 0$ by forming the F-statistic $MS_w/MS_E$. If $H_0$ is true, this F-statistic will have an F-distribution with 1 and $N - g$ degrees of freedom ($N - g$ from the $MS_E$). It is not too hard to see that this F is exactly equal to the square of the $t$-statistic computed for same null hypothesis $\delta = 0$. Thus the F-test and two-sided $t$-tests are equivalent for the null hypothesis of zero contrast mean. It is also not too hard to see that if you multiply the contrast coefficients by a nonzero constant (for example, change from (-1, .5, .5) to (2, -1, -1)), then the contrast sum of squares is unchanged. The squared constant cancels from the numerator and denominator of the formula.

SS and F-test for $w(\{\mu_i\})$

## Rat liver weights

**Example 4.1**

Exercise 3.1 provided data on the weight of rat livers as a percentage of body weight for four different diets. Summary statistics from those data follow:

| $i$ | 1 | 2 | 3 | 4 |
|---|---|---|---|---|
| $\overline{y}_{i\bullet}$ | 3.75 | 3.58 | 3.60 | 3.92 |
| $n_i$ | 7 | 8 | 6 | 8 |

$MS_E = .04138$

If diets 1, 2, and 3 are rations made by one manufacturer, and diet 4 is a ration made by a second manufacturer, then it may be of interest to compare the responses from the diets of the two manufacturers to see if there is any difference.

The contrast with coefficients (1/3, 1/3, 1/3, -1) will compare the mean response in the first three diets with the mean response in the last diet. Note that we intend "the mean response in the first three diets" to denote the average of the treatment averages, not the simple average of all the data from those three treatments. The simple average will not be the same as the average of the averages because the sample sizes are different.

Our point estimate of this contrast is

$$w(\{\overline{y}_{i\bullet}\}) = \frac{1}{3}3.75 + \frac{1}{3}3.58 + \frac{1}{3}3.60 + (-1)3.92 = -.277$$

with standard error

$$SE(w(\{\overline{y}_{i\bullet}\})) = \sqrt{.04138}\sqrt{\frac{(\frac{1}{3})^2}{7} + \frac{(\frac{1}{3})^2}{8} + \frac{(\frac{1}{3})^2}{6} + \frac{(-1)^2}{8}} = .0847 \ .$$

The mean square for error has $29 - 4 = 25$ degrees of freedom. To construct a 95% confidence interval for $w(\{\mu_i\})$, we need the upper 2.5% point of a $t$-distribution with 25 degrees of freedom; this is 2.06, as can be found in Appendix Table D.3 or using software. Thus our 95% confidence interval is

$$-.277 \pm 2.06 \times .0847 = -.277 \pm .174 = (-.451, -.103) \ .$$

Suppose that we wish to test the null hypothesis $H_0 : w(\{\mu_i\}) = \delta$. Here we will use the $t$-test and F-test to test $H_0 : w(\{\mu_i\}) = \delta = 0$, but the $t$-test can test other values of $\delta$. Our $t$-test is

$$\frac{-.277 - 0}{.0847} = -3.27 \ ,$$

with 25 degrees of freedom. For a two-sided alternative, we compute the $p$-value by finding the tail area under the $t$-curve and doubling it. Here we get twice .00156 or about .003. This is rather strong evidence against the null hypothesis.

Because our null hypothesis value is zero with a two-sided alternative, we can also test our null hypothesis by computing a mean square for the contrast

**Listing 4.1:** SAS PROC GLM output for the rat liver contrast.

```
Source                   DF      Type I SS   Mean Square   F Value      Pr > F

DIET                      3     0.57820903    0.19273634      4.66      0.0102

Contrast                 DF    Contrast SS   Mean Square   F Value      Pr > F

1,2,3 vs 4                1     0.45617253    0.45617253     11.03      0.0028
```

**Listing 4.2:** MacAnova output for the rat liver contrast.

```
component: estimate
(1)      -0.28115
component: ss
(1)       0.45617
component: se
(1)      0.084674
```

and forming an F-statistic. The sum of squares for our contrast is

$$\frac{(\frac{1}{3}3.75 + \frac{1}{3}3.58 + \frac{1}{3}3.60 + (-1)3.92)^2}{\frac{(1/3)^2}{7} + \frac{(1/3)^2}{8} + \frac{(1/3)^2}{6} + \frac{(-1)^2}{8}} = \frac{(-.277)^2}{.1733} = .443 \ .$$

The mean square is also .443, so the F-statistic is .443/.04138 = 10.7. We compute a $p$-value by finding the area to the right of 10.7 under the F-distribution with 1 and 25 degrees of freedom, getting .003 as for the $t$-test.

Listing 4.1 shows output from SAS for computing the sum of squares for this contrast; Listing 4.2 shows corresponding MacAnova output. The sum of squares in these two listings differs from what we obtained above due to rounding at several steps.

## 4.3 Orthogonal Contrasts

Two contrasts $\{w\}$ and $\{w^\star\}$ are said to be *orthogonal* if

$$\sum_{i=1}^{g} w_i w_i^\star / n_i = 0 \ .$$

$g-1$ orthogonal contrasts

If there are $g$ treatments, you can find a set of $g-1$ contrasts that are mutually orthogonal, that is, each one is orthogonal to all of the others. However, there are infinitely many sets of $g-1$ mutually orthogonal contrasts, and there are no mutually orthogonal sets with more than $g-1$ contrasts. There is an analogy from geometry. In a plane, you can have two lines that are perpendicular (orthogonal), but you *can't* find a third line that is perpendicular to both of the others. On the other hand, there are infinitely many pairs of perpendicular lines.

Orthogonal contrasts are independent and partition variation

The important feature of orthogonal contrasts applied to observed means is that they are independent (as random variables). Thus, the random error of one contrast is not correlated with the random error of an orthogonal contrast. An additional useful fact about orthogonal contrasts is that they partition the between groups sum of squares. That is, if you compute the sums of squares for a full set of orthogonal contrasts ($g-1$ contrasts for $g$ groups), then adding up those $g-1$ sums of squares will give you exactly the between groups sum of squares (which also has $g-1$ degrees of freedom).

**Example 4.2**

## Orthogonal contrast inference

Suppose that we have an experiment with three treatments—a control and two new treatments—with group sizes 10, 5, and 5, and treatment means 6.3, 6.4, and 6.5. The $MS_E$ is .0225 with 17 degrees of freedom. The contrast $w$ with coefficients (1, -.5, -.5) compares the mean response in the control treatment with the average of the mean responses in the new treatments. The contrast with coefficients (0, 1, -1) compares the two new treatments. In our example above, we had a control with 10 units, and two new treatments with 5 units each. These contrasts are orthogonal, because

$$\frac{0 \times 1}{10} + \frac{1 \times -.5}{5} + \frac{-1 \times -.5}{5} = 0 \ .$$

We have three groups so there are 2 degrees of freedom between groups, and we have described above a set of orthogonal contrasts. The sum of squares for the first contrast is

$$\frac{(6.3 - .5 \times 6.4 - .5 \times 6.5)^2}{\frac{1}{10} + \frac{(-.5)^2}{5} + \frac{(-.5)^2}{5}} = .1125 \ ,$$

and the sum of squares for the second contrast is

$$\frac{(0 + 6.4 - 6.5)^2}{\frac{0}{10} + \frac{1^2}{5} + \frac{(-1)^2}{5}} = \frac{.01}{.4} = .025 \ .$$

The between groups sum of squares is

$$10(6.3 - 6.375)^2 + 5(6.4 - 6.375)^2 + 5(6.5 - 6.375)^2 = .1375$$

which equals .1125 + .025.

Contrasts isolate differences

We can see from Example 4.2 one of the advantages of contrasts over the full between groups sum of squares. The control-versus-new contrast has a sum of squares which is 4.5 times larger than the sum of squares for the difference of the new treatments. This indicates that the responses from the new treatments are substantially farther from the control responses than they are from each other. Such indications are not possible using the between groups sum of squares.

The actual contrasts one uses in an analysis arise from the context of the problem. Here we had new versus old and the difference between the two new treatments. In a study on the composition of ice cream, we might compare artificial flavorings with natural flavorings, or expensive flavorings with inexpensive flavorings. It is often difficult to construct a complete set of meaningful orthogonal contrasts, but that should not deter you from using an incomplete set of orthogonal contrasts, or from using contrasts that are nonorthogonal.

| Use contrasts that address the questions you are trying to answer. |
|---|

## 4.4 Polynomial Contrasts

Contrasts yield improvement $SS$ in polynomial dose-response models

Section 3.10 introduced the idea of polynomial modeling of a response when the treatments had a quantitative dose structure. We selected a polynomial model by looking at the improvement sums of squares obtained by adding each polynomial term to the model in sequence. Each of these additional terms in the polynomial has a single degree of freedom, just like a contrast. In fact, each of these improvement sums of squares can be obtained as a contrast sum of squares. We call the contrast that gives us the sum of squares for the linear term the linear contrast, the contrast that gives us the improvement sum of squares for the quadratic term the quadratic contrast, and so on.

Simple contrasts for equally spaced doses with equal $n_i$

When the doses are equally spaced *and* the sample sizes are equal, then the contrast coefficients for polynomial terms are fairly simple and can be found, for example, in Appendix Table D.6; these contrasts are orthogonal and have been scaled to be simple integer values. Equally spaced doses means that the gaps between successive doses are the same, as in 1, 4, 7, 10. Using these tabulated contrast coefficients, we may compute the linear, quadratic, and higher order sums of squares as contrasts without fitting a separate polynomial model. Doses such as 1, 10, 100, 1000 are equally spaced on a logarithmic scale, so we can again use the simple polynomial contrast coefficients, provided we interpret the polynomial as a polynomial in the logarithm of dose.

When the doses are not equally spaced or the sample sizes are not equal, then contrasts for polynomial terms exist, but are rather complicated to derive. In this situation, it is more trouble to derive the coefficients for the polynomial contrasts than it is to fit a polynomial model.

**Example 4.3**

## Leaflet angles

Exercise 3.5 introduced the leaflet angles of plants at 30, 45, and 60 minutes after exposure to red light. Summary information for this experiment is given here:

| | Delay time (min) | | |
|---|---|---|---|
| | 30 | 45 | 60 |
| $\overline{y}_{i\bullet}$ | 139.6 | 133.6 | 122.4 |
| $n_i$ | 5 | 5 | 5 |

$$MS_E = 58.13$$

With three equally spaced groups, the linear and quadratic contrasts are (-1, 0, 1) and (1, -2, 1).

The sum of squares for linear is

$$\frac{((-1)139.6 + (0)133.6 + (1)122.4)^2}{\frac{(-1)^2}{5} + \frac{0}{5} + \frac{1^2}{5}} = 739.6 \quad ,$$

and that for quadratic is

$$\frac{((1)139.6 + (-2)133.6 + (1)122.4)^2}{\frac{1^2}{5} + \frac{(-2)^2}{5} + \frac{1^2}{5}} = 22.53 \quad .$$

Thus the F-tests for linear and quadratic are $739.6/58.13 = 12.7$ and $22.53/58.13 = .39$, both with 1 and 12 degrees of freedom; there is a strong linear trend in the means and almost no nonlinear trend.

## 4.5 Further Reading and Extensions

Contrasts are a special case of *estimable functions*, which are described in some detail in Appendix Section A.6. Treatment means and averages of treatment means are other estimable functions. Estimable functions are those features of the data that do not depend on how we choose to restrict the treatment effects.

## 4.6 Problems

**Exercise 4.1** An experimenter randomly allocated 125 male turkeys to five treatment groups: 0 mg, 20 mg, 40 mg, 60 mg, and 80 mg of estradiol. There were 25 birds in each group, and the mean results were 2.16, 2.45, 2.91, 3.00, and 2.71 respectively. The sum of squares for experimental error was 153.4. Test the null hypothesis that the five group means are the same against the alternative that they are not all the same. Find the linear, quadratic, cubic, and quartic sums of squares (you may lump the cubic and quartic together into a "higher than quadratic" if you like). Test the null hypothesis that the quadratic effect is zero. Be sure to report a $p$-value.

**Exercise 4.2** Use the data from Exercise 3.3. Compute a 99% confidence interval for the difference in response between the average of the three treatment groups (acid, pulp, and salt) and the control group.

**Exercise 4.3** Refer to the data in Problem 3.1. Workers 1 and 2 were experienced, whereas workers 3 and 4 were novices. Find a contrast to compare the experienced and novice workers and test the null hypothesis that experienced and novice works produce the same average shear strength.

**Exercise 4.4** Consider an experiment taste-testing six types of chocolate chip cookies: 1 (brand A, chewy, expensive), 2 (brand A, crispy, expensive), 3 (brand B, chewy, inexpensive), 4 (brand B, crispy, inexpensive), 5 (brand C, chewy, expensive), and 6 (brand D, crispy, inexpensive). We will use twenty different raters randomly assigned to each type (120 total raters).

a) Design contrasts to compare chewy with crispy, and expensive with inexpensive.

b) Are your contrasts in part a) orthogonal? Why or why not?

**Problem 4.1** A consumer testing agency obtains four cars from each of six makes: Ford, Chevrolet, Nissan, Lincoln, Cadillac, and Mercedes. Makes 3 and 6 are imported while the others are domestic; makes 4, 5, and 6 are expensive while 1, 2, and 3 are less expensive; 1 and 4 are Ford products, while 2 and 5 are GM products. We wish to compare the six makes on their oil use per 100,000 miles driven. The mean responses by make of car were 4.6, 4.3, 4.4, 4.7, 4.8, and 6.2, and the sum of squares for error was 2.25.

(a) Compute the Analysis of Variance table for this experiment. What would you conclude?

(b) Design a set of contrasts that seem meaningful. For each contrast, outline its purpose and compute a 95% confidence interval.

**Problem 4.2** Consider the data in Problem 3.2. Design a set of contrasts that seem meaningful. For each contrast, outline its purpose and test the null hypothesis that the contrast has expected value zero.

**Problem 4.3** Consider the data in Problem 3.5. Use polynomial contrasts to choose a quantitative model to describe the effect of fiber proportion on the response.

**Question 4.1** Show that orthogonal contrasts in the observed treatment means are uncorrelated random variables.

# Chapter 5

# Multiple Comparisons

When we make several related tests or interval estimates at the same time, we need to make *multiple comparisons* or do *simultaneous inference*. The issue of multiple comparisons is one of error rates. Each of the individual tests or confidence intervals has a Type I error rate $\mathcal{E}_i$ that can be controlled by the experimenter. If we consider the tests together as a *family*, then we can also compute a combined Type I error rate for the family of tests or intervals. When a family contains more and more true null hypotheses, the probability that one or more of these true null hypotheses is rejected increases, and the probability of any Type I errors in the family can become quite large. Multiple comparisons procedures deal with Type I error rates for families of tests.

Multiple comparisons, simultaneous inference, families of hypotheses

**Example 5.1**

### Carcinogenic mixtures

We are considering a new cleaning solvent that is a mixture of 100 chemicals. Suppose that regulations state that a mixture is safe if all of its constituents are safe (pretending we can ignore chemical interaction). We test the 100 chemicals for causing cancer, running each test at the 5% level. This is the individual error rate that we can control.

What happens if all 100 chemicals are harmless and safe? Because we are testing at the 5% level, we expect 5% of the nulls to be rejected even when all the nulls are true. Thus, on average, 5 of the 100 chemicals will be declared to be carcinogenic, even when all are safe. Moreover, if the tests are independent, then one or more of the chemicals will be declared unsafe in 99.4% of all sets of experiments we run, even if all the chemicals are safe. This 99.4% is a combined Type I error rate; clearly we have a problem.

## 5.1 Error Rates

Determine error rate to control

When we have more than one test or interval to consider, there are several ways to define a combined Type I error rate for the family of tests. This variety of combined Type I error rates is the source of much confusion in the use of multiple comparisons, as different error rates lead to different procedures. People sometimes ask "Which procedure should I use?" when the real question is "Which error rate do I want to control?". As data analyst, you need to decide which error rate is appropriate for your situation and then choose a method of analysis appropriate for that error rate. This choice of error rate is not so much a statistical decision as a scientific decision in the particular area under consideration.

Data snooping performs many implicit tests

*Data snooping* is a practice related to having many tests. Data snooping occurs when we first look over the data and then choose the null hypotheses to be tested based on "interesting" features in the data. What we tend to do is consider many potential features of the data and discard those with uninteresting or null behavior. When we data snoop and then perform a test, we tend to see the smallest $p$-value from the ill-defined family of tests that we considered when we were snooping; we have not really performed just one test. Some multiple comparisons procedures can actually control for data snooping.

> *Simultaneous inference* is deciding which error rate we wish to control, and then using a procedure that controls the desired error rate.

Individual and combined null hypotheses

Let's set up some notation for our problem. We have a set of $K$ null hypotheses $H_{01}$, $H_{02}$, …, $H_{0K}$. We also have the "combined," "overall," or "intersection" null hypotheses $H_0$ which is true if *all* of the $H_{0i}$ are true. In formula,

$$H_0 = H_{01} \cap H_{02} \cap \cdots \cap H_{0K} .$$

The collection $H_{01}$, $H_{02}$, …, $H_{0K}$ is sometimes called a family of null hypotheses. We reject $H_0$ if any of null hypotheses $H_{0i}$ is rejected. In Example 5.1, $K = 100$, $H_{0i}$ is the null hypothesis that chemical $i$ is safe, and $H_0$ is the null hypothesis that all chemicals are safe so that the mixture is safe.

We now define five combined Type I error rates. The definitions of these error rates depend on numbers or fractions of falsely rejected null hypotheses $H_{0i}$, which will never be known in practice. We set up the error rates here and later give procedures that can be shown mathematically to control the error rates.

The *per comparison error rate* or *comparisonwise error rate* is the probability of rejecting a particular $H_{0i}$ in a single test when that $H_{0i}$ is true. Controlling the per comparison error rate at $\mathcal{E}$ means that the expected fraction of individual tests that reject $H_{0i}$ when $H_0$ is true is $\mathcal{E}$. This is just the usual error rate for a $t$-test or F-test; it makes no correction for multiple comparisons. The tests in Example 5.1 controlled the per comparison error rate at 5%.

Comparisonwise error rate

The *per experiment error rate* or *experimentwise error rate* or *familywise error rate* is the probability of rejecting one or more of the $H_{0i}$ (and thus rejecting $H_0$) in a series of tests when all of the $H_{0i}$ are true. Controlling the experimentwise error rate at $\mathcal{E}$ means that the expected fraction of experiments in which we would reject one or more of the $H_{0i}$ when $H_0$ is true is $\mathcal{E}$. In Example 5.1, the per experiment error rate is the fraction of times we would declare one or more of the chemicals unsafe when in fact all were safe. Controlling the experimentwise error rate at $\mathcal{E}$ necessarily controls the comparisonwise error rate at no more than $\mathcal{E}$. The experimentwise error rate considers all individual null hypotheses that were rejected; if any one of them was correctly rejected, then there is no penalty for any false rejections that may have occurred.

Experimentwise error rate

A statistical discovery is the rejection of an $H_{0i}$. The false discovery fraction is 0 if there are no rejections; otherwise it is the number of false discoveries (Type I errors) divided by the total number of discoveries. The *false discovery rate* (FDR) is the expected value of the false discovery fraction. If $H_0$ is true, then all discoveries are false and the FDR is just the experimentwise error rate. Thus controlling the FDR at $\mathcal{E}$ also controls the experimentwise error at $\mathcal{E}$. However, the FDR also controls at $\mathcal{E}$ the average fraction of rejections that are Type I errors when some $H_{0i}$ are true and some are false, a control that the experimentwise error rate does not provide. With the FDR, we are allowed more incorrect rejections as the number of true rejections increases, but the ratio is limited. For example, with FDR at .05, we are allowed just one incorrect rejection with 19 correct rejections.

False discovery rate

The *strong familywise error rate* is the probability of making any false discoveries, that is, the probability that the false discovery fraction is greater than zero. Controlling the strong familywise error rate at $\mathcal{E}$ means that the probability of making any false rejections is $\mathcal{E}$ or less, regardless of how many correct rejections are made. Thus one true rejection cannot make any false rejections more likely. Controlling the strong familywise error rate at $\mathcal{E}$ controls the FDR at no more than $\mathcal{E}$. In Example 5.1, a strong familywise error rate of $\mathcal{E}$ would imply that in a situation where 2 of the chemicals were carcinogenic, the probability of declaring one of the other 98 to be carcinogenic would be no more than $\mathcal{E}$.

Strong familywise error rate

Finally, suppose that each null hypothesis relates to some parameter (for example, a mean), and we put confidence intervals on all these parameters. An error occurs when one of our confidence intervals fails to cover the true parameter value. If this true parameter value is also the null hypothesis value, then an error is a false rejection. The *simultaneous confidence intervals* criterion states that all of our confidence intervals must cover their true parameters simultaneously with confidence $1 - \mathcal{E}$. Simultaneous $1 - \mathcal{E}$ confidence intervals also control the strong familywise error rate at no more than $\mathcal{E}$. (In effect, the strong familywise criterion only requires simultaneous intervals for the null parameters.) In Example 5.1, we could construct simultaneous confidence intervals for the cancer rates of each of the 100 chemicals. Note that a single confidence interval in a collection of intervals with simultaneous coverage $1 - \mathcal{E}$ will have coverage greater than $1 - \mathcal{E}$.

Simultaneous confidence intervals

There is a trade-off between Type I error and Type II error (failing to reject a null when it is false). As we go to more and more stringent Type I error rates, we become more confident in the rejections that we do make, but it also becomes more difficult to make rejections. Thus, when using the more stringent Type I error controls, we are more likely to fail to reject some null hypotheses that should be rejected than when using the less stringent rates. In simultaneous inference, controlling stronger error rates leads to less powerful tests.

More stringent procedures are less powerful

**Example 5.2**

### Functional magnetic resonance imaging

Many functional Magnetic Resonance Imaging (fMRI) studies are interested in determining which areas of the brain are "activated" when a subject is engaged in some task. Any one image slice of the brain may contain 5000 voxels (individual locations to be studied), and one analysis method produces a $t$-test for each of the 5000 voxels. Null hypothesis $H_{0i}$ is that voxel $i$ is not activated. Which error rate should we use?

If we are studying a small, narrowly defined brain region and are unconcerned with other brain regions, then we would want to test individually the voxels in the brain regions of interest. The fact that there are 4999 other voxels is unimportant, so we would use a per comparison method.

Suppose instead that we are interested in determining if there are any activations in the image. We recognize that by making many tests we are likely to find one that is "significant", even when all nulls are true; we want to protect ourselves against that possibility, but otherwise need no stronger control. Here we would use a per experiment error rate.

Suppose that we believe that there will be many activations, so that $H_0$ is not true. We don't want some correct discoveries to open the flood gates for many false discoveries, but we are willing to live with some false discoveries as long as they are a controlled fraction of the total made. This is acceptable because we are going to investigate several subjects; the truly activated rejections should be rejections in most subjects, and the false rejections will be scattered. Here we would use the FDR.

Suppose that in addition to expecting true activations, we are also only looking at a single subject, so that we can't use multiple subjects to determine which activations are real. Here we don't want false activations to cloud our picture, so we use the strong familywise error rate.

Finally, we might want to be able to estimate the amount of activation in every voxel, with simultaneous accuracy for all voxels. Here we would use simultaneous confidence intervals.

> A *multiple comparisons procedure* is a method for controlling a Type I error rate other than the per comparison error rate.

The literature on multiple comparisons is vast, and despite the length of this Chapter, we will only touch the highlights. I have seen quite a bit of nonsense regarding these methods, so I will try to set out rather carefully what the methods are doing. We begin with a discussion of Bonferroni-based methods for combining generic tests. Next we consider the Scheffé procedure, which is useful for contrasts suggested by data (data snooping). Then we turn our attention to pairwise comparisons, for which there are dozens of methods. Finally, we consider comparing treatments to a control or to the best response.

## 5.2 Bonferroni-Based Methods

The Bonferroni technique is the simplest, most widely applicable multiple comparisons procedure. The Bonferroni procedure works for a fixed set of $K$ null hypotheses to test or parameters to estimate. Let $p_i$ be the $p$-value for testing $H_{0i}$. The Bonferroni procedure says to obtain simultaneous $1-\mathcal{E}$ confidence intervals by constructing individual confidence intervals with coverage $1-\mathcal{E}/K$, or reject $H_{0i}$ (and thus $H_0$) if

Ordinary Bonferroni

$$p_i < \mathcal{E}/K \ .$$

That is, simply run each test at level $\mathcal{E}/K$. The testing version controls the strong familywise error rate, and the confidence intervals are simultaneous. The tests and/or intervals need not be independent, of the same type, or related in any way.

| Reject $H_{0(i)}$ if | Method | Control |
|---|---|---|
| $p_{(i)} < \mathcal{E}/K$ | Bonferroni | Simultaneous confidence intervals |
| $p_{(j)} < \mathcal{E}/(K - j + 1)$ for all $j = 1, \ldots, i$ | Holm | Strong familywise error rate |
| $p_{(j)} \leq j\mathcal{E}/K$ for some $j \geq i$ | FDR | False discovery rate; needs independent tests |

**Display 5.1:** Summary of Bonferroni-style methods for $K$ comparisons.

Holm

The *Holm* procedure is a modification of Bonferroni that controls the strong familywise error rate, but does not produce simultaneous confidence intervals (Holm 1979). Let $p_{(1)}, \ldots, p_{(K)}$ be the $p$-values for the $K$ tests sorted into increasing order, and let $H_{0(i)}$ be the null hypotheses sorted along with the $p$-values. Then reject $H_{0(i)}$ if

$$p_{(j)} \leq \mathcal{E}/(K - j + 1) \text{ for all } j = 1, \ldots, i.$$

Thus we start with the smallest $p$-value; if it is rejected we consider the next smallest, and so on. We stop when we reach the first nonsignificant $p$-value. This is a little more complicated, but we gain some power since only the smallest $p$-value is compared to $\mathcal{E}/K$.

FDR modification of Bonferroni requires independent tests

The FDR method of Benjamini and Hochberg (1995) controls the False Discovery Rate. Once again, sort the $p$-values and the hypotheses. For the FDR, start with the largest $p$-value and work down. Reject $H_{0i}$ if

$$p_{(j)} \leq j\mathcal{E}/K \text{ for some } j \geq i.$$

This procedure is correct when the tests are statistically independent. It controls the FDR, but not the strong familywise error rate.

The three Bonferroni methods are summarized in Display 5.1. Example 5.3 illustrates their use.

## Sensory characteristics of cottage cheeses

**Example 5.3**

Table 5.1 shows the results of an experiment comparing the sensory characteristics of nonfat, 2% fat, and 4% fat cottage cheese (Michicich 1995). The table shows the characteristics grouped by type and $p$-values for testing the null hypothesis that there was no difference between the three cheeses in the various sensory characteristics. There are 21 characteristics in three groups of sizes 7, 6, and 8.

How do we do multiple comparisons here? First we need to know:

1. Which error rate is of interest?
2. If we do choose an error rate other than the per comparison error rate, what is the appropriate "family" of tests? Is it all 21 characteristics, or separately within group of characteristic?

There is no automatic answer to either of these questions. The answers depend on the goals of the study, the tolerance of the investigator to Type I error, how the results of the study will be used, whether the investigator views the three groups of characteristics as distinct, and so on.

The last two columns of Table 5.1 give the results of the Bonferroni, Holm, and FDR procedures applied at the 5% level to all 21 comparisons and within each group. The $p$-values are compared to the criteria in Display 5.1 using $K = 21$ for the overall family and $K$ of 7, 6, or 8 for by group comparisons.

Consider the characteristic "cheesy flavor" with a .01 $p$-value. If we use the overall family, this is the tenth smallest $p$-value out of 21 $p$-values. The results are

- *Bonferroni* The critical value is $.05/21 = .0024$—not significant.
- *Holm* The critical value is $.05/(21-10+1) = .0042$—not significant.
- *FDR* The critical value is $10 \times .05/21 = .024$—significant.

If we use the flavor family, this is the fourth smallest $p$-value out of six $p$-values. Now the results are

- *Bonferroni* The critical value is $.05/6 = .008$—not significant.
- *Holm* The critical value is $.05/(6 - 4 + 1) = .017$ (and all smaller $p$-values meet their critical values)—significant.
- *FDR* The critical value is $4 \times .05/6 = .033$—significant.

**Table 5.1:** Sensory attributes of three cottage cheeses: $p$-values and 5% significant results overall and familywise by type of attribute using the Bonferroni (•), Holm (∘), and FDR methods(⋆).

| Characteristic | $p$-value | Overall | By group |
|---|---|---|---|
| **Appearance** | | | |
| White | .004 | ⋆ | •∘⋆ |
| Yellow | .002 | •∘⋆ | •∘⋆ |
| Gray | .13 | | |
| Curd size | .29 | | |
| Size uniformity | .73 | | |
| Shape uniformity | .08 | | |
| Liquid/solid ratio | .02 | ⋆ | ⋆ |
| **Flavor** | | | |
| Sour | .40 | | |
| Sweet | .24 | | |
| Cheesy | .01 | ⋆ | ∘⋆ |
| Rancid | .0001 | •∘⋆ | •∘⋆ |
| Cardboard | .0001 | •∘⋆ | •∘⋆ |
| Storage | .001 | •∘⋆ | •∘⋆ |
| **Texture** | | | |
| Breakdown rate | .001 | •∘⋆ | •∘⋆ |
| Firm | .0001 | •∘⋆ | •∘⋆ |
| Sticky | .41 | | |
| Slippery | .07 | | |
| Heavy | .15 | | |
| Particle size | .42 | | |
| Runny | .002 | •∘⋆ | •∘⋆ |
| Rubbery | .006 | ⋆ | •∘⋆ |

These results illustrate that more null hypotheses are rejected considering each group of characteristics to be a family of tests rather than overall (the $K$ is smaller for the individual groups), and fewer rejections are made using the more stringent error rates. Again, the choices of error rate and family of tests are not purely statistical, and controlling an error rate within a group of tests does not control that error rate for all tests.

## 5.3 The Scheffé Method for *All* Contrasts

The Scheffé method is a multiple comparisons technique for contrasts that produces simultaneous confidence intervals for *any* and *all* contrasts, *including contrasts suggested by the data.* Thus Scheffé is the appropriate technique for assessing contrasts that result from data snooping. This sounds like the ultimate in error rate control—arbitrarily many comparisons, even ones suggested from the data! The downside of this amazing protection is low power. Thus we only use the Scheffé method in those situations where we have a contrast suggested by the data, or many, many contrasts that cannot be handled by other techniques. In addition, pairwise comparison contrasts $\overline{y}_{i\bullet} - \overline{y}_{j\bullet}$, even pairwise comparisons suggested by the data, are better handled by methods specifically designed for pairwise comparisons.

Scheffé protects against data snooping, but has low power

We begin with the Scheffé test of the null hypothesis $H_0 : w(\{\alpha_i\}) = 0$ against a two-sided alternative. The Scheffé test statistic is the ratio

$$\frac{SS_w/(g-1)}{MS_E} ;$$

we get a $p$-value as the area under an F-distribution with $g-1$ and $\nu$ degrees of freedom to the right of the test statistic. The degrees of freedom $\nu$ are from our denominator $MS_E$; $\nu = N - g$ for the completely randomized designs we have been considering so far. Reject the null hypothesis if this $p$-value is less than our Type I error rate $\mathcal{E}$. In effect, the Scheffé procedure treats the mean square for any single contrast as if it were the full $g-1$ degrees of freedom between groups mean square.

Scheffé F-test

There is also a Scheffé $t$-test for contrasts. Suppose that we are testing the null hypothesis $H_0 : w(\{\alpha_i\}) = \delta$ against a two-sided alternative. The Scheffé $t$-test controls the Type I error rate at $\mathcal{E}$ by rejecting the null hypothesis when

Scheffé $t$-test

$$\frac{|w(\{\overline{y}_{i\bullet}\}) - \delta|}{\sqrt{MS_E \sum_{i=1}^{g} \frac{w_i^2}{n_i}}} > \sqrt{(g-1)F_{\mathcal{E},g-1,\nu}} ,$$

where $F_{\mathcal{E},g-1,\nu}$ is the upper $\mathcal{E}$ percent point of an F-distribution with $g-1$ and $\nu$ degrees of freedom. Again, $\nu$ is the degrees of freedom for $MS_E$. For the usual null hypothesis value $\delta = 0$, this is equivalent to the ratio-of-mean-squares version given above.

We may also use the Scheffé approach to form simultaneous confidence intervals for any $w(\{\alpha_i\})$:

Scheffé confidence interval

$$w(\{\overline{y}_{i\bullet}\}) \pm \sqrt{(g-1)F_{\mathcal{E},g-1,\nu}} \times \sqrt{MS_E \sum_{i=1}^{g} \frac{w_i^2}{n_i}} .$$

These Scheffé intervals have simultaneous $1 - \mathcal{E}$ coverage over any set of contrasts, including contrasts suggested by the data.

**Example 5.4**

### Acid rain and birch seedlings, continued

Example 3.1 introduced an experiment in which birch seedlings were exposed to various levels of artificial acid rain. The following table gives some summaries for the data:

| pH | 4.7 | 4.0 | 3.3 | 3.0 | 2.3 |
|---|---|---|---|---|---|
| weight | .337 | .296 | .320 | .298 | .177 |
| n | 48 | 48 | 48 | 48 | 48 |

The $MS_E$ was .0119 with 235 degrees of freedom.

Inspection of the means shows that most of the response means are about .3, but the response for the pH 2.3 treatment is much lower. This suggests that a contrast comparing the pH 2.3 treatment with the mean of the other treatments would have a large value. The coefficients for this contrast are (.25, .25, .25, .25, -1). This contrast has value

$$\frac{.337 + .296 + .320 + .298}{4} - .177 = .1357$$

and standard error

$$\sqrt{.0119\left(\frac{.0625}{48} + \frac{.0625}{48} + \frac{.0625}{48} + \frac{.0625}{48} + \frac{1}{48}\right)} = .0176 \ .$$

We must use the Scheffé procedure to construct a confidence interval or assess the significance of this contrast, because the contrast was suggested by the data. For a 99% confidence interval, the Scheffé multiplier is

$$\sqrt{4\ F_{.01,4,235}} = 3.688 \ .$$

Thus the 99% confidence interval for this contrast is $.1357 - 3.688 \times .0176$ up to $.1357 + 3.688 \times .0176$, or (.0708, .2006). Alternatively, the $t$-statistic for testing the null hypothesis that the mean response in the last group is equal to the average of the mean responses in the other four groups is $.1357/.0176 = 7.71$. The Scheffé critical value for testing the null hypothesis at the $\mathcal{E} = .001$ level is

$$\sqrt{(g-1)F_{\mathcal{E},g-1,N-g}} = \sqrt{4\ F_{.001,4,235}} = \sqrt{4 \times 5.876} = 4.85 \ ,$$

so we can reject the null at the .001 level.

Remember, it is not fair to hunt around through the data for a big contrast, test it, and think that you've only done one comparison.

## 5.4 Pairwise Comparisons

A *pairwise comparison* is a contrast that examines the difference between two treatment means $\overline{y}_{i\bullet} - \overline{y}_{j\bullet}$. For $g$ treatment groups, there are

$$\binom{g}{2} = \frac{g(g-1)}{2}$$

different pairwise comparisons. Pairwise comparisons procedures control a Type I error rate at $\mathcal{E}$ for all pairwise comparisons. If we data snoop, choose the biggest and smallest $\overline{y}_{i\bullet}$'s and take the difference, we have not made just one comparison; rather we have made all $g(g-1)/2$ pairwise comparisons, and selected the largest. Controlling a Type I error rate for this greatest difference is one way to control the error rate for all differences.

Tests or confidence intervals

As with many other inference problems, pairwise comparisons can be approached using confidence intervals or tests. That is, we may compute confidence intervals for the differences $\mu_i - \mu_j$ or $\alpha_i - \alpha_j$ or test the null hypotheses $H_{0ij} : \mu_i = \mu_j$ or $H_{0ij} : \alpha_i = \alpha_j$. Confidence regions for the differences of means are generally more informative than tests.

A pairwise comparisons procedure can generally be viewed as a critical value (or set of values) for the $t$-tests of the pairwise comparison contrasts. Thus we would reject the null hypothesis that $\alpha_i - \alpha_j = 0$ if

$$\frac{|\overline{y}_{i\bullet} - \overline{y}_{j\bullet}|}{\sqrt{MS_E}\sqrt{1/n_i + 1/n_j}} > u ,$$

Critical values $u$ for $t$-tests

where $u$ is a critical value. Various pairwise comparisons procedures differ in how they define the critical value $u$, and $u$ may depend on several things, including $\mathcal{E}$, the degrees of freedom for $MS_E$, the number of treatments, the number of treatments with means between $\overline{y}_{i\bullet}$ and $\overline{y}_{j\bullet}$, and the number of treatment comparisons with larger $t$-statistics.

An equivalent form of the test will reject if

$$|\overline{y}_{i\bullet} - \overline{y}_{j\bullet}| > u\ \sqrt{MS_E}\sqrt{1/n_i + 1/n_j} = D_{ij} \quad .$$

Significant differences $D_{ij}$

If all sample sizes are equal and the critical value $u$ is constant, then $D_{ij}$ will be the same for all $i, j$ pairs and we would reject the null if any pair of treatments had mean responses that differed by $D$ or more. This quantity $D$ is called a *significant difference*; for example, using a Bonferroni adjustment to the $g(g-1)/2$ pairwise comparisons tests leads to a Bonferroni significant difference (BSD).

Confidence intervals for pairwise differences $\mu_i - \mu_j$ can be formed from the pairwise tests via

$$(\overline{y}_{i\bullet} - \overline{y}_{j\bullet}) \pm u\sqrt{MS_E}\sqrt{1/n_i + 1/n_j} \; .$$

The remainder of this section presents methods for displaying the results of pairwise comparisons, introduces the Studentized range, discusses several pairwise comparisons methods, and then illustrates the methods with an example.

### 5.4.1 Displaying the results

Underline diagram summarizes pairwise comparisons

Pairwise comparisons generate a lot of tests, so we need convenient and compact ways to present the results. An *underline diagram* is a graphical presentation of pairwise comparison results; construct the underline diagram in the following steps.

1. Sort the treatment means into increasing order and write out treatment labels (numbers or names) along a horizontal axis. The $\overline{y}_{i\bullet}$ values may be added if desired.
2. Draw a line segment under a group of treatments if no pair of treatments in that group is significantly different. Do not include short lines that are implied by long lines. That is, if treatments 4, 5, and 6 are not significantly different, only use one line under all of them—not a line under 4 and 5, and a line under 5 and 6, and a line under 4, 5, and 6.

Here is a sample diagram for three treatments that we label A, B, and C:

```
C   A   B
―――――――
    ―――――――
```

This diagram includes treatment labels, but not treatment means. From this summary we can see that C can be distinguished from B (there is no underline that covers both B and C), but A cannot be distinguished from either B or C (there are underlines under A and C, and under A and B).

Note that there can be some confusion after pairwise comparisons. You must not confuse "is not significantly different from" or "cannot be distinguished from" with "is equal to." Treatment mean A cannot be equal to treatment means B and C and still have treatment means B and C not equal each other. Such a pattern can hold for results of significance tests.

Insignificant difference does not imply equality

There are also several nongraphical methods for displaying pairwise comparisons results. In one method, we sort the treatments into order of increasing means and print the treatment labels. Each treatment label is followed by one or more numbers (letters are sometimes used instead). Any treatments sharing a number (or letter) are not significantly different. Thus treatments sharing common numbers or letters are analogous to treatments being connected by an underline. The grouping letters are often put in parentheses or set as sub- or superscripts. The results in our sample underline diagram might thus be presented as one of the following:

Letter or number tags

| | | | | | |
|---|---|---|---|---|---|
| C (1) | A (12) | B (2) | C (a) | A (ab) | B (b) |
| $C^1$ | $A^{12}$ | $B^2$ | $C^a$ | $A^{ab}$ | $B^b$ |

There are several other variations on this theme.

A third way to present pairwise comparisons is as a table, with treatments labeling both rows and columns. Table elements can flag significant differences or contain confidence intervals for the differences. Only entries above or below the diagonal of the table are needed.

Table of CI's or significant differences

### 5.4.2 The Studentized range

The range of a set is the maximum value minus the minimum value, and *Studentization* means dividing a statistic by an estimate of its standard error. Thus the *Studentized range* for a set of treatment means is

Range, Studentization, and Studentized range

$$\max_i \frac{\overline{y}_{i\bullet}}{\sqrt{MS_E/n}} - \min_j \frac{\overline{y}_{j\bullet}}{\sqrt{MS_E/n}} .$$

Note that we have implicitly assumed that all the sample sizes $n_i$ are the same.

If all the treatments have the same mean, that is, if $H_0$ is true, then the Studentized range statistic follows the Studentized range distribution. Large values of the Studentized range are less likely under $H_0$ and more likely under the alternative when the means are not all equal, so we may use the Studentized range as a test statistic for $H_0$, rejecting $H_0$ when the Studentized

Studentized range distribution

range statistic is sufficiently large. This Studentized range test is a legitimate alternative to the ANOVA F-test.

The Studentized range distribution is important for pairwise comparisons because it is the distribution of the biggest (scaled) difference between treatment means when the null hypothesis is true. We will use it as a building block in several pairwise comparisons methods.

Percent points $q_{\mathcal{E}}(g, \nu)$

The Studentized range distribution depends only on $g$ and $\nu$, the number of groups and the degrees of freedom for the error estimate $MS_E$. The quantity $q_{\mathcal{E}}(g, \nu)$ is the upper $\mathcal{E}$ percent point of the Studentized range distribution for $g$ groups and $\nu$ error degrees of freedom; it is tabulated in Appendix Table D.8.

### 5.4.3 Simultaneous confidence intervals

Tukey HSD or honest significant difference

The Tukey honest significant difference (HSD) is a pairwise comparisons technique that uses the Studentized range distribution to construct simultaneous confidence intervals for differences of all pairs of means. If we reject the null hypothesis $H_{0ij}$ when the (simultaneous) confidence interval for $\mu_i - \mu_j$ does not include 0, then the HSD also controls the strong familywise error rate.

The HSD uses the critical value

$$u(\mathcal{E}, \nu, g) = \frac{q_{\mathcal{E}}(g, \nu)}{\sqrt{2}} ,$$

The HSD

leading to

$$HSD = \frac{q_{\mathcal{E}}(g, \nu)}{\sqrt{2}} \sqrt{MS_E} \sqrt{\frac{1}{n} + \frac{1}{n}} = \frac{q_{\mathcal{E}}(g, \nu)\sqrt{MS_E}}{\sqrt{n}} .$$

Form simultaneous $1 - \mathcal{E}$ confidence intervals via

$$\overline{y}_{i\bullet} - \overline{y}_{j\bullet} \pm \frac{q_{\mathcal{E}}(g, \nu)}{\sqrt{2}} \sqrt{MS_E} \sqrt{\frac{1}{n} + \frac{1}{n}} .$$

The degrees of freedom $\nu$ are the degrees of freedom for the error estimate $MS_E$.

Strictly speaking, the HSD is only applicable to the equal sample size situation. For the unequal sample size case, the approximate HSD is

$$HSD_{ij} = q_{\mathcal{E}}(g, \nu)\sqrt{MS_E}\sqrt{\frac{1}{2n_i n_j/(n_i + n_j)}}$$

**Table 5.2:** Total free amino acids in cheeses after 168 days of ripening.

| Strain added | | | |
|---|---|---|---|
| None | A | B | A&B |
| 4.195 | 4.125 | 4.865 | 6.155 |
| 4.175 | 4.735 | 5.745 | 6.488 |

or, equivalently,

Tukey-Kramer form for unequal sample sizes

$$HSD_{ij} = \frac{q_{\mathcal{E}}(g,\nu)}{\sqrt{2}}\sqrt{MS_E}\sqrt{(\frac{1}{n_i}+\frac{1}{n_j})} \ .$$

This approximate HSD, often called the Tukey-Kramer form, tends to be slightly conservative (that is, the true error rate is slightly less than $\mathcal{E}$).

Bonferroni significant difference or BSD

The Bonferroni significant difference (BSD) is simply the application of the Bonferroni technique to the pairwise comparisons problem to obtain

$$\begin{aligned} u = u(\mathcal{E},\nu,K) &= t_{\mathcal{E}/(2K),\nu} \ , \\ BSD_{ij} &= t_{\mathcal{E}/(2K),\nu}\sqrt{MS_E}\sqrt{1/n_i + 1/n_j} \ , \end{aligned}$$

where $K$ is the number of pairwise comparisons. We have $K = g(g-1)/2$ for all pairwise comparisons between $g$ groups. BSD produces simultaneous confidence intervals and controls the strong familywise error rate.

Use HSD when making all pairwise comparisons

When making all pairwise comparisons, the HSD is less than the BSD. Thus we prefer the HSD to the BSD for all pairwise comparisons, because the HSD will produce shorter confidence intervals that are still simultaneous. When only a preplanned subset of all the pairs is being considered, the BSD may be less than and thus preferable to the HSD.

**Example 5.5**

### Free amino acids in cheese

Cheese is produced by bacterial fermentation of milk. Some bacteria in cheese are added by the cheese producer. Other bacteria are present but were not added deliberately; these are called nonstarter bacteria. Nonstarter bacteria vary from facility to facility and are believed to influence the quality of cheese.

Two strains (A and B) of nonstarter bacteria were isolated at a premium cheese facility. These strains will be added experimentally to cheese to determine their effects. Eight cheeses are made. These cheeses all get a standard starter bacteria. In addition, two cheeses will be randomly selected for each of the following four treatments: control, add strain A, add strain B, or add both strains A and B. Table 5.2 gives the total free amino acids in the cheeses after 168 days of ripening. (Free amino acids are thought to contribute to flavor.)

Listing 5.1 gives Minitab output showing an Analysis of Variance for these data ①, as well as HSD comparisons (called Tukey's pairwise comparisons) using $\mathcal{E} = .1$ ②; we use the $MS_E$ from this ANOVA in constructing the HSD. HSD is appropriate if we want simultaneous confidence intervals on the pairwise differences. The HSD is

$$\begin{aligned} \frac{q_{\mathcal{E}}(g,\nu)}{\sqrt{2}} \sqrt{MS_E} \sqrt{\frac{1}{n_i} + \frac{1}{n_j}} &= \frac{q_{.1}(4,4)}{\sqrt{2}} \sqrt{.1572} \sqrt{\frac{1}{2} + \frac{1}{2}} \\ &= 4.586 \times .3965/1.414 = 1.286 \quad . \end{aligned}$$

We form confidence intervals as the observed difference in treatment means, plus or minus 1.286; so for A&B minus control, we have

$$6.322 - 4.185 \pm 1.286 \text{ or } (.851, 3.423) \quad .$$

In fact, only two confidence intervals for pairwise differences do not include zero (see Listing 5.1 ②). The underline diagram is:

| C | A | B | A&B |
|---|---|---|---|
| 4.19 | 4.43 | 5.31 | 6.32 |

(Underlines: C, A, B; B, A&B.)

Note in Listing 5.1 ② that Minitab displays pairwise comparisons as a table of confidence intervals for differences.

### 5.4.4 Strong familywise error rate

Step-down methods work inward from the outside comparisons

A *step-down method* is a procedure for organizing pairwise comparisons starting with the most extreme pair and then working in. Relabel the groups so that the sample means are in increasing order with $\overline{y}_{(1)\bullet}$ smallest and $\overline{y}_{(g)\bullet}$ largest. (The relabeled estimated effects $\widehat{\alpha}_{(i)}$ will also be in increasing order, but the relabeled true effects $\alpha_{[i]}$ may or may not be in increasing order.) With this ordering, $\overline{y}_{(1)\bullet}$ to $\overline{y}_{(g)\bullet}$ is a stretch of $g$ means, $\overline{y}_{(1)\bullet}$ to $\overline{y}_{(g-1)\bullet}$ is a stretch of $g-1$ means, and $\overline{y}_{(i)\bullet}$ to $\overline{y}_{(j)\bullet}$ is a stretch of $j-i+1$ means. In a step-down procedure, all comparisons for stretches of $k$ means use the same critical value, but we may use different critical values for different $k$. This

**Listing 5.1:** Minitab output for free amino acids in cheese.

```
Source      DF        SS         MS         F        P            ①
Trt          3     5.628      1.876     11.93    0.018
Error        4     0.629      0.157
Total        7     6.257
                                    Individual 95% CIs For Mean
                                    Based on Pooled StDev
Level        N      Mean      StDev  ------+---------+---------+---------+
A            2    4.4300     0.4313     (------*-------)
A+B          2    6.3215     0.2355                      (-------*-------)
B            2    5.3050     0.6223              (-------*-------)
control      2    4.1850     0.0141  (-------*-------)
                                     ------+---------+---------+---------+
Pooled StDev =    0.3965                  4.0       5.0       6.0       7.0

Tukey's pairwise comparisons                                          ②
    Family error rate = 0.100
Individual error rate = 0.0315

Critical value = 4.59

Intervals for (column level mean) - (row level mean)

           A           A+B          B

 A+B        -3.1784
            -0.6046

 B          -2.1619     -0.2704
             0.4119      2.3034

 control    -1.0419      0.8496     -0.1669
             1.5319      3.4234      2.4069

Fisher's pairwise comparisons                                         ③
    Family error rate = 0.283
Individual error rate = 0.100

Critical value = 2.132

Intervals for (column level mean) - (row level mean)

                   A           A+B           B

 A+B        -2.7369
            -1.0461

 B          -1.7204      0.1711
            -0.0296      1.8619

 control    -0.6004      1.2911       0.2746
             1.0904      2.9819       1.9654
```

has the advantage that we can use larger critical values for long stretches and smaller critical values for short stretches.

$(i)$ and $(j)$ are different if their stretch and all containing stretches reject

Begin with the most extreme pair (1) and $(g)$. Test the null hypothesis that all the means for (1) up through $(g)$ are equal. If you fail to reject, declare all means equal and stop. If you reject, declare (1) different from $(g)$ and go on to the next step. At the next step, we consider the stretches (1) through $(g-1)$ and (2) through $(g)$. If one of these rejects, we declare its ends to be different and then look at shorter stretches within it. If we fail to reject for a stretch, we do not consider any substretches within the stretch. We repeat this subdivision till there are no more rejections. In other words, we declare that means $(i)$ and $(j)$ are different if the stretch from $(i)$ to $(j)$ rejects its null hypothesis and all stretches containing $(i)$ to $(j)$ also reject their null hypotheses.

REGWR is step-down with Studentized range based critical values

The REGWR procedure is a step-down range method that controls the strong familywise error rate without producing simultaneous confidence intervals. The awkward name REGWR abbreviates the Ryan-Einot-Gabriel-Welsch range test, named for the authors who worked on it. The REGWR critical value for testing a stretch of length $k$ depends on $\mathcal{E}$, $\nu$, $k$, and $g$. Specifically, we use

$$u = u(\mathcal{E}, \nu, k, g) = q_{\mathcal{E}}(k, \nu)/\sqrt{2} \quad k = g, g-1,$$

and

$$u = u(\mathcal{E}, \nu, k, g) = q_{k\mathcal{E}/g}(k, \nu)/\sqrt{2} \quad k = g-2, g-1, \ldots, 2.$$

This critical value derives from a Studentized range with $k$ groups, and we use percent points with smaller tail areas as we move in to smaller stretches.

As with the HSD, REGWR error rate control is approximate when the sample sizes are not equal.

**Example 5.6** **Free amino acids in cheese, continued**

Suppose that we only wished to control the strong familywise error rate instead of producing simultaneous confidence intervals. Then we could use REGWR instead of HSD and could potentially see additional significant differences. Listing 5.2 ② gives SAS output for REGWR (called REGWQ in SAS) for the amino acid data.

REGWR is a step-down method that begins like the HSD. Comparing C and A&B, we conclude as in the HSD that they are different. We may now compare C with B and A with A&B. These are comparisons that involve

**Listing 5.2:** SAS output for free amino acids in cheese.

```
             Student-Newman-Keuls test for variable: FAA                ①

                    Alpha= 0.1   df= 4   MSE= 0.157224

            Number of Means          2          3          4
            Critical Range     0.84531 1.1146718 1.2859073

      Means with the same letter are not significantly different.

              SNK Grouping                  Mean       N   TRT

                          A               6.3215       2   4

                          B               5.3050       2   3

                          C               4.4300       2   2
                          C
                          C               4.1850       2   1


   Ryan-Einot-Gabriel-Welsch Multiple Range Test for variable: FAA     ②

                    Alpha= 0.1   df= 4   MSE= 0.157224

            Number of Means          2          3          4
            Critical Range   1.0908529 1.1146718 1.2859073

      Means with the same letter are not significantly different.

             REGWQ Grouping                 Mean       N   TRT

                          A               6.3215       2   4
                          A
                    B     A               5.3050       2   3
                    B
                    B     C               4.4300       2   2
                          C
                          C               4.1850       2   1
```

stretches of $k = 3$ means; since $k = g - 1$, we still use $\mathcal{E}$ as the error rate. The significant difference for these comparisons is

$$\frac{q_{\mathcal{E}}(k,\nu)}{\sqrt{2}}\sqrt{MS_E}\sqrt{\frac{1}{n_i}+\frac{1}{n_j}} = \frac{q_{.1}(3,4)}{\sqrt{2}}\sqrt{.1572}\sqrt{\frac{1}{2}+\frac{1}{2}} = 1.115 \ .$$

Both the B-C and A&B-A differences (1.12 and 1.89) exceed this cutoff, so REGWR concludes that B differs from C, and A differs from A&B. Recall that the HSD did not distinguish C from B.

Having concluded that there are B-C and A&B-A differences, we can now compare stretches of means within them, namely C to A, A to B, and B to A&B. These are stretches of $k = 2$ means, so for REGWR we use the error rate $k\mathcal{E}/g = .05$. The significant difference for these comparisons is

$$\frac{q_{\mathcal{E}/2}(k,\nu)}{\sqrt{2}}\sqrt{MS_E}\sqrt{\frac{1}{n_i}+\frac{1}{n_j}} = \frac{q_{.05}(2,4)}{\sqrt{2}}\sqrt{.1572}\sqrt{\frac{1}{2}+\frac{1}{2}} = 1.101 \ .$$

None of the three differences exceeds this cutoff, so we fail to conclude that those treatments differ and finish. The underline diagram is:

| C | A | B | A&B |
|---|---|---|---|
| 4.19 | 4.43 | 5.31 | 6.32 |

(Underlines: C–A; A–B; B–A&B.)

Note in Listing 5.2 ② that SAS displays pairwise comparisons using what amounts to an underline diagram turned on its side, with vertical lines formed by letters.

### 5.4.5 False discovery rate

SNK

The Student-Newman-Keuls (SNK) procedure is a step-down method that uses the Studentized range test with critical value

$$u = u(\mathcal{E}, \nu, k, g) = q_{\mathcal{E}}(k,\nu)/\sqrt{2}$$

for a stretch of $k$ means. This is similar to REGWR, except that we keep the percent point of the Studentized range constant as we go to shorter stretches. The SNK controls the false discovery rate, but not the strong familywise error rate. As with the HSD, SNK error rate control is approximate when the sample sizes are not equal.

**Example 5.7** **Free amino acids in cheese, continued**

Suppose that we only wished to control the false discovery rate; now we would use SNK instead of the more stringent HSD or REGWR. Listing 5.2 ① gives SAS output for SNK for the amino acid data.

SNK is identical to REGWR in the first two stages, so SNK will also get to the point of making the comparisons of the three pairs C to A, A to B, and

B to A&B. However, the SNK significant difference for these pairs is less than that used in REGWR:

$$\frac{q_{\mathcal{E}}(k,\nu)}{\sqrt{2}}\sqrt{MS_E}\sqrt{\frac{1}{n_i}+\frac{1}{n_j}}=\frac{q_{.1}(2,4)}{\sqrt{2}}\sqrt{.1572}\sqrt{\frac{1}{2}+\frac{1}{2}}=.845\ .$$

Both the B-A and A&B-B differences (1.02 and .98) exceed the cutoff, but the A-C difference (.14) does not. The underline diagram for SNK is:

| C | A | B | A&B |
|---|---|---|---|
| 4.19 | 4.43 | 5.31 | 6.32 |

(C and A are joined by an underline.)

### 5.4.6 Experimentwise error rate

Protected LSD uses F-test to control experimentwise error rate

The Analysis of Variance F-test for equality of means controls the experimentwise error rate. Thus investigating pairwise differences only when the F-test has a $p$-value less than $\mathcal{E}$ will control the experimentwise error rate. This is the basis for the Protected least significant difference, or Protected LSD. If the F-test rejects at level $\mathcal{E}$, then do simple $t$-tests at level $\mathcal{E}$ among the different treatments.

The critical values are from a $t$-distribution:

$$u(\mathcal{E},\nu)=t_{\mathcal{E}/2,\nu}\ ,$$

leading to the significant difference

$$LSD=t_{\mathcal{E}/2,\nu}\sqrt{MS_E}\sqrt{1/n_i+1/n_j}\ .$$

As usual, $\nu$ is the degrees of freedom for $MS_E$, and $t_{\mathcal{E}/2,\nu}$ is the upper $\mathcal{E}/2$ percent point of a $t$-curve with $\nu$ degrees of freedom.

Confidence intervals produced from the protected LSD do not have the anticipated $1-\mathcal{E}$ coverage rate, either individually or simultaneously. See Section 5.7.

**Example 5.8**

**Free amino acids in cheese, continued**

Finally, suppose that we only wish to control the experimentwise error rate. Protected LSD will work here. Listing 5.1 ① shows that the ANOVA F-test is significant at level $\mathcal{E}$, so we may proceed with pairwise comparisons.

Listing 5.1 ③ shows Minitab output for the LSD (called Fisher's pairwise comparisons) as confidence intervals.

LSD uses the same significant difference for all pairs:

$$t_{\mathcal{E}/2,\nu}\sqrt{MS_E}\sqrt{\frac{1}{n_i}+\frac{1}{n_j}} = t_{.05,4}\sqrt{.1572}\sqrt{\frac{1}{2}+\frac{1}{2}} = .845 \ .$$

This is the same as the SNK comparison for a stretch of length 2. All differences except A-C exceed the cutoff, so the underline diagram for LSD is:

| C | A | B | A&B |
|---|---|---|---|
| 4.19 | 4.43 | 5.31 | 6.32 |

(C and A are joined by a single underline.)

### 5.4.7 Comparisonwise error rate

LSD

Ordinary $t$-tests and confidence intervals without any adjustment control the comparisonwise error rate. In the context of pairwise comparisons, this is called the least significant difference (LSD) method.

The critical values are the same as for the protected LSD:

$$u(\mathcal{E},\nu) = t_{\mathcal{E}/2,\nu},$$

and

$$LSD = t_{\mathcal{E}/2,\nu}\sqrt{MS_E}\sqrt{1/n_i + 1/n_j} \ .$$

### 5.4.8 Pairwise testing reprise

Choose your error rate, not your method

It is easy to get overwhelmed by the abundance of methods, and there are still more that we haven't discussed. Your anchor in all this is your error rate. Once you have determined your error rate, the choice of method is reasonably automatic, as summarized in Display 5.2. Your choice of error rate is determined by the needs of your study, bearing in mind that the more stringent error rates have fewer false rejections, and also fewer correct rejections.

### 5.4.9 Pairwise comparisons methods that do *not* control combined Type I error rates

There are many other pairwise comparisons methods beyond those already mentioned. In this Section we discuss two methods that are motivated by

| Error rate | Method |
|---|---|
| Simultaneous confidence intervals | BSD or HSD |
| Strong familywise | REGWR |
| False discovery rate | SNK |
| Experimentwise | Protected LSD |
| Comparisonwise | LSD |

**Display 5.2:** Summary of pairwise comparison methods.

completely different criteria than controlling a combined Type I error rate. These two techniques do *not* control the experimentwise error rate or any of the more stringent error rates, and you should not use them with the expectation that they do. You should only use them when the situation and assumptions under which they were developed are appropriate for your experimental analysis.

Duncan's multiple range if there is a cost per error or you believe $H_0$ less likely as $g$ increases

Suppose that you believe *a priori* that the overall null hypothesis $H_0$ is less and less likely to be true as the number of treatments increases. Then the strength of evidence required to reject $H_0$ should decrease as the number of groups increases. Alternatively, suppose that there is a quantifiable penalty for each incorrect (pairwise comparison) decision we make, and that the total loss for the overall test is the sum of the losses from the individual decisions. Under either of these assumptions, the Duncan multiple range (given below) or something like it is appropriate. Note by comparison that the procedures that control combined Type I error rates require more evidence to reject $H_0$ as the number of groups increases, while Duncan's method requires less. Also, a procedure that controls the experimentwise error rate has a penalty of 1 if there are any rejections when $H_0$ is true and a penalty of 0 otherwise; this is very different from the summed loss that leads to Duncan's multiple range.

Duncan's Multiple Range

Duncan's multiple range (sometimes called Duncan's test or Duncan's new multiple range) is a step-down Studentized range method. You specify a "protection level" $\mathcal{E}$ and proceed in step-down fashion using

$$u = u(\mathcal{E}, \nu, k, g) = q_{1-(1-\mathcal{E})^{k-1}}(k, \nu)/\sqrt{2}$$

for the critical values. Notice that $\mathcal{E}$ is the comparisonwise error rate for testing a stretch of length 2, and the experimentwise error rate will be $1 - (1 - \mathcal{E})^{g-1}$, which can be considerably more than $\mathcal{E}$. Thus *fixing Duncan's protection level at $\mathcal{E}$ does **not** control the experimentwise error rate or any more stringent rate.* Do not use Duncan's procedure if you are interested in controlling any of the combined Type I error rates.

Experimentwise error rate very large for Duncan

As a second alternative to combined Type I error rates, suppose that our interest is in predicting future observations from the treatment groups, and that we would like to have a prediction method that makes the average squared prediction error small. One way to do this prediction is to first partition the $g$ treatments into $p$ classes, $1 \leq p \leq g$; second, find the average response in each of these $p$ classes; and third, predict a future observation from a treatment by the observed mean response of the class for the treatment. We thus look for partitions that will lead to good predictions.

Minimize prediction error instead of testing

One way to choose among the partitions is to use Mallows' $C_p$ statistic:

$$C_p = \frac{SSR_p}{MS_E} + 2p - N ,$$

where $SSR_p$ is the sum of squared errors for the Analysis of Variance, partitioning the data into $p$ groups. Partitions with low values of $C_p$ should give better predictions.

Predictive Pairwise Comparisons

This predictive approach makes no attempt to control any Type I error rate; in fact, the Type I error rate is .15 or greater even for $g = 2$ groups! This approach is useful when prediction is the goal, but can be quite misleading if interpreted as a test of $H_0$.

### 5.4.10 Confident directions

In our heart of hearts, we often believe that all treatment means differ when examined sufficiently precisely. Thus our concern with null hypotheses $H_{0ij}$ is misplaced. As an alternative, we can make statements of *direction*. After having collected data, we consider $\mu_i$ and $\mu_j$; assume $\mu_i < \mu_j$. We could decide from the data that $\mu_i < \mu_j$, or that $\mu_i > \mu_j$, or that we don't know—that is, we don't have enough information to decide. These decisions correspond in the testing paradigm to rejecting $H_{0ij}$ in favor of $\mu_i < \mu_j$, rejecting $H_{0ij}$ in favor of $\mu_j < \mu_i$, and failing to reject $H_{0ij}$. In the confident directions framework, only the decision $\mu_i > \mu_j$ is an error. See Tukey (1991).

All means differ, but their order is uncertain

Can only make an error in one direction

*Confident directions* procedures are pairwise comparisons testing procedures, but with results interpreted in a directional context. Confident directions procedures bound error rates when making statements about direction.

If a testing procedure bounds an error rate at $\mathcal{E}$, then the corresponding confident directions procedure bounds a confident directions error rate at $\mathcal{E}/2$, the factor of 2 arising because we cannot falsely reject in the correct direction.

Pairwise comparisons can be used for confident directions

Let us reinterpret our usual error rates in terms of directions. Suppose that we use a pairwise comparisons procedure with error rate bounded at $\mathcal{E}$. In a confident directions setting, we have the following:

| | |
|---|---|
| Strong familywise | The probability of making any incorrect statements of direction is bounded by $\mathcal{E}/2$. |
| FDR | Incorrect statements of direction will on average be no more than a fraction $\mathcal{E}/2$ of the total number of statements of direction. |
| Experimentwise | The probability of making any incorrect statements of direction when all the means are very nearly equal is bounded by $\mathcal{E}/2$. |
| Comparisonwise | The probability of making an incorrect statement of direction for a given comparison is bounded by $\mathcal{E}/2$. |

There is no directional analog of simultaneous confidence intervals, so procedures that produce simultaneous intervals should be considered procedures that control the strong familywise error rate (which they do).

## 5.5 Comparison with Control or the Best

Comparison with control does not do all tests

There are some situations where we do not do all pairwise comparisons, but rather make comparisons between a control and the other treatments, or the best responding treatment (highest or lowest average) and the other treatments. For example, you may be producing new standardized mathematics tests for elementary school children, and you need to compare the new tests with the current test to assure comparability of the results. The procedures for comparing to a control or the best are similar.

### 5.5.1 Comparison with a control

Two-sided Dunnett

Suppose that there is a special treatment, say treatment $g$, with which we wish to compare the other $g-1$ treatments. Typically, treatment $g$ is a control treatment. The Dunnett procedure allows us to construct simultaneous $1-\mathcal{E}$ confidence intervals on $\mu_i - \mu_g$, for $i = 1, \ldots, g-1$ when all sample sizes are equal via

$$\overline{y}_i - \overline{y}_g \pm d_{\mathcal{E}}(g-1,\nu)\, \sqrt{MS_E}\, \sqrt{\frac{1}{n_i} + \frac{1}{n_g}}\ ,$$

where $\nu$ is the degrees of freedom for $MS_E$. The value $d_{\mathcal{E}}(g-1,\nu)$ is tabulated in Appendix Table D.9. These table values are exact when all sample sizes are equal and only approximate when the sizes are not equal.

For testing, we can use

$$u(\mathcal{E},i,j) = d_{\mathcal{E}}(g-1,\nu)\ ,$$

DSD, the Dunnett significant difference

which controls the strong familywise error rate and leads to

$$DSD = d_{\mathcal{E}}(g-1,\nu)\sqrt{MS_E}\, \sqrt{\frac{1}{n_i} + \frac{1}{n_g}}\ ,$$

the Dunnett significant difference. There is also a step-down modification that still controls the strong familywise error rate and is slightly more powerful. We have $g-1$ $t$-statistics. Compare the largest (in absolute value) to $d_{\mathcal{E}}(g-1,\nu)$. If the test fails to reject the null, stop; otherwise compare the second largest to $d_{\mathcal{E}}(g-2,\nu)$ and so on.

One-sided Dunnett

There are also one-sided versions of the confidence and testing procedures. For example, you might reject the null hypothesis of equality only if the noncontrol treatments provide a higher response than the control treatments. For these, test using the critical value

$$u(\mathcal{E},i,j) = d'_{\mathcal{E}}(g-1,\nu)\ ,$$

tabulated in Appendix Table D.9, or form simultaneous one-sided confidence intervals on $\mu_i - \mu_g$ with

$$\overline{y}_i - \overline{y}_g \geq d'_{\mathcal{E}}(g-1,\nu)\, \sqrt{MS_E}\, \sqrt{\frac{1}{n_i} + \frac{1}{n_g}}.$$

For $t$-critical values, a one-sided cutoff is equal to a two-sided cutoff with a doubled $\mathcal{E}$. The same is not true for Dunnett critical values, so that $d'_{\mathcal{E}}(g-1,\nu) \neq d_{2\mathcal{E}}(g-1,\nu)$.

### Alfalfa meal and turkeys

**Example 5.9**

An experiment is conducted to study the effect of alfalfa meal in the diet of male turkey poults (chicks). There are nine treatments. Treatment 1 is a control treatment; treatments 2 through 9 contain alfalfa meal of two different types in differing proportions. Units consist of 72 pens of eight birds each, so there are eight pens per treatment. One response of interest is average daily weight gains per bird for birds aged 7 to 14 days. We would like to know which alfalfa treatments are significantly different from the control in weight gain, and which are not.

Here are the average weight gains (g/day) for the nine treatments:

|  |  |  |  |  |
|---|---|---|---|---|
| 22.668 | 21.542 | 20.001 | 19.964 | 20.893 |
| 21.946 | 19.965 | 20.062 | 21.450 |  |

The $MS_E$ is 2.487 with 55 degrees of freedom. (The observant student will find this degrees of freedom curious; more on this data set later.) Two-sided, 95% confidence intervals for the differences between control and the other treatments are computed using

$$\begin{aligned} d_{\mathcal{E}}(g-1,\nu)\,\sqrt{MS_E}\,\sqrt{\frac{1}{n_i}+\frac{1}{n_g}} &= d_{.05}(8,55)\,\sqrt{2.487}\,\sqrt{\frac{1}{8}+\frac{1}{8}} \\ &= 2.74 \times 1.577/2 \\ &= 2.16 \ . \end{aligned}$$

Any treatment with mean less than 2.16 from the control mean of 22.668 is not significantly different from the control. These are treatments 2, 5, 6, and 9.

Give the control more replication

It is a good idea to give the control (treatment $g$) greater replication than the other treatments. The control is involved in every comparison, so it makes sense to estimate its mean more precisely. More specifically, if you had a fixed number of units to spread among the treatments, and you wished to minimize the average variance of the differences $\overline{y}_{g\bullet} - \overline{y}_{i\bullet}$, then you would do best when the ratio $n_g/n_i$ is about equal to $\sqrt{g-1}$.

Personally, I rarely use the Dunnett procedure, because I nearly always get the itch to compare the noncontrol treatments with each other as well as with the control.

### 5.5.2 Comparison with the best

Suppose that the goal of our experiment is to screen a number of treatments and determine those that give the best response—to pick the winner. The multiple comparisons with best (MCB) procedure produces two results:

Use MCB to choose best subset of treatments

- It produces a subset of treatments that cannot be distinguished from the best; the treatment having the true largest mean response will be in this subset with probability $1 - \mathcal{E}$.
- It produces simultaneous $1-\mathcal{E}$ confidence intervals on $\mu_i - \max_{j \neq i} \mu_j$, the difference between a treatment mean and the best of the other treatment means.

The subset selection procedure is the more useful product, so we only discuss the selection procedure.

The best subset consists of all treatments $i$ such that

$$\overline{y}_{i\bullet} > \overline{y}_{j\bullet} - d'_{\mathcal{E}}(g-1,\nu)\ \sqrt{MS_E}\ \sqrt{\frac{1}{n_i} + \frac{1}{n_j}} \quad \text{for all } j \neq i$$

In words, treatment $i$ is in the best subset if its mean response is greater than the largest treatment mean less a one-sided Dunnett allowance. When small responses are good, a treatment $i$ is in the best subset if its mean response is less than the smallest treatment mean plus a one-sided Dunnett allowance.

**Example 5.10** **Weed control in soybeans**

Weeds reduce crop yields, so farmers are always looking for better ways to control weeds. Fourteen weed control treatments were randomized to 56 experimental plots that were planted in soybeans. The plots were later visually assessed for weed control, the fraction of the plot without weeds. The percent responses are given in Table 5.3. We are interested in finding a subset of treatments that contains the treatment giving the best weed control (largest response) with confidence 99%.

For reasons that will be explained in Chapter 6, we will analyze as our response the square root of percent weeds (that is, 100 minus the percent weed control). Because we have subtracted weed control, small values of the transformed response are good. On this scale, the fourteen treatment means are

| | | | | | | |
|---|---|---|---|---|---|---|
| 1.000 | 2.616 | 2.680 | 2.543 | 2.941 | 1.413 | 1.618 |
| 2.519 | 2.847 | 1.618 | 1.000 | 4.115 | 4.988 | 5.755 |

**Table 5.3:** Percent weed control in soybeans under 14 treatments.

| 1 | 2 | 3 | 4 | 5 | 6 | 7 |
|---|---|---|---|---|---|---|
| 99 | 95 | 92 | 95 | 85 | 98 | 99 |
| 99 | 92 | 95 | 88 | 92 | 99 | 95 |
| 99 | 95 | 92 | 95 | 92 | 95 | 99 |
| 99 | 90 | 92 | 95 | 95 | 99 | 95 |

| 8 | 9 | 10 | 11 | 12 | 13 | 14 |
|---|---|---|---|---|---|---|
| 95 | 92 | 99 | 99 | 88 | 65 | 75 |
| 85 | 90 | 95 | 99 | 88 | 65 | 50 |
| 95 | 95 | 99 | 99 | 85 | 92 | 72 |
| 97 | 90 | 95 | 99 | 68 | 72 | 68 |

and the $MS_E$ is .547 with 42 degrees of freedom. The smallest treatment mean is 1.000, and the Dunnett allowance is

$$\begin{aligned} d'_{\mathcal{E}}(g-1,\nu)\sqrt{MS_E}\sqrt{\frac{1}{n_i}+\frac{1}{n_j}} &= d'_{.01}(13,42)\sqrt{.547}\sqrt{\frac{1}{4}+\frac{1}{4}} \\ &= 3.29 \times .740 \times .707 \\ &= 1.72. \end{aligned}$$

So, any treatment with a mean of $1 + 1.72 = 2.72$ or less is included in the 99% grouping. These are treatments 1, 2, 3, 4, 6, 7, 8, 10, and 11.

## 5.6 Reality Check on Coverage Rates

We already pointed out that the error rate control for some multiple comparisons procedures is only approximate if the sample sizes are not equal or the tests are dependent. However, even in the "exact" situations, these procedures depend on assumptions about the distribution of the data for the coverage rates to hold: for example normality or constant error variance. These assumptions are often violated—data are frequently nonnormal and error variances are often nonconstant.

Violation of distributional assumptions usually leads to true error rates that are not equal to the nominal $\mathcal{E}$. The amount of discrepancy depends on the nature of the violation. Unequal sample sizes or dependent tests are just another variable to consider.

The point is that we need to get some idea of what the true error is, and not get worked up about the fact that it is not *exactly* equal to $\mathcal{E}$.

| In the real world, coverage and error rates are always approximate. |
|---|

## 5.7 A Warning About Conditioning

Except for the protected LSD, the multiple comparisons procedures discussed above do not require the ANOVA F-test to be significant for protection of the experimentwise error rate. They stand apart from the F-test, protecting the experimentwise error rate by other means. In fact, requiring that the ANOVA F-test be significant will alter their error rates.

Requiring the F-test to be significant alters the error rates of pairwise procedures

Bernhardson (1975) reported on how conditioning on the ANOVA F-test being significant affected the per comparison and per experiment error rates of pairwise comparisons, including LSD, HSD, SNK, Duncan's procedure, and Scheffé. Requiring the F to be significant lowered the per comparison error rate of the LSD from 5% to about 1% and lowered the per experiment error rate for HSD from 5% to about 3%, both for 6 to 10 groups. Looking just at those null cases where the F-test rejected, the LSD had a per comparison error rate of 20 to 30% and the HSD per experiment error rate was about 65%—both for 6 to 10 groups. Again looking at just the null cases where the F was significant, even the Scheffé procedure's per experiment error rate increased to 49% for 4 groups, 22% for 6 groups, and down to about 6% for 10 groups.

The problem is that when the ANOVA F-test is significant in the null case, one cause might be an unusually low estimate of the error variance. This unusually low variance estimate gets used in the multiple comparisons procedures leading to smaller than normal HSD's, and so on.

## 5.8 Some Controversy

Simultaneous inference is deciding which error rate to control and then using an appropriate technique for that error rate. Controversy arises because

- Users cannot always agree on the appropriate error rate. In particular, some statisticians (including Bayesian statisticians) argue strongly that the only relevant error rate is the per comparison error rate.

- Users cannot always agree on what constitutes the appropriate family of tests. Different groupings of the tests lead to different results.
- Standard statistical practice seems to be inconsistent in its application of multiple comparisons ideas. For example, multiple comparisons are fairly common when comparing treatment means, but almost unheard of when examining multiple factors in factorial designs (see Chapter 8).

You as experimenter and data analyst must decide what is the proper approach for inference. See Carmer and Walker (1982) for an amusing allegory on this topic.

## 5.9 Further Reading and Extensions

There is much more to the subject of multiple comparisons than what we have discussed here. For example, many procedures for contrasts can be adapted to other linear combinations of parameters, and many of the pairwise comparisons techniques can be adapted to contrasts. A good place to start is Miller (1981), an instant classic when it appeared and still an excellent and readable reference; much of the discussion here follows Miller. Hochberg and Tamhane (1987) contains some of the more recent developments.

The first multiple comparisons technique appears to be the LSD suggested by Fisher (1935). Curiously, the next proposal was the SNK (though not so labeled) by Newman (1939). Multiple comparisons then lay dormant till around 1950, when there was an explosion of ideas: Duncan's multiple range procedure (Duncan 1955), Tukey's HSD (Tukey 1952), Scheffé's all contrasts method (Scheffé 1953), Dunnett's method (Dunnett 1955), and another proposal for SNK (Keuls 1952). The pace of introduction then slowed again. The REGW procedures appeared in 1960 and evolved through the 1970's (Ryan 1960; Einot and Gabriel 1975; Welsch 1977). Improvements in the Bonferroni inequality lead to the modified Bonferroni procedures in the 1970's and later (Holm 1979; Simes 1986; Hochberg 1988; Benjamini and Hochberg 1995).

Curiously, procedures sometimes predate a careful understanding of the error rates they control. For example, SNK has often been advocated as a less conservative alternative to the HSD, but the false discovery rate was only defined recently (Benjamini and Hochberg 1995). Furthermore, many textbook introductions to multiple comparisons procedures do not discuss the different error rates, thus leading to considerable confusion over the choice of procedure.

One historical feature of multiple comparisons is the heavy reliance on tables of critical values and the limitations imposed by having tables only for selected percent points or equal sample sizes. Computers and software remove many of these limitations. For example, the software in Lund and Lund (1983) can be used to compute percent points of the Studentized range for $\mathcal{E}$'s not usually tabulated, while the software in Dunnett (1989) can compute critical values for the Dunnett test with unequal sample sizes. When no software for exact computation is available (for example, Studentized range for unequal sample sizes), percent points can be approximated through simulation (see, for example, Ripley 1987).

Hayter (1984) has shown that the Tukey-Kramer adjustment to the HSD procedure is conservative when the sample sizes are not equal.

## 5.10 Problems

**Exercise 5.1** We have five groups and three observations per group. The group means are 6.5, 4.5, 5.7, 5.6, and 5.1, and the mean square for error is .75. Compute simultaneous confidence intervals (95% level) for the differences of all treatment pairs.

**Exercise 5.2** Consider a completely randomized design with five treatments, four units per treatment, and treatment means

$$3.2892 \quad 10.256 \quad 8.1157 \quad 8.1825 \quad 7.5622 \ .$$

The MSE is 4.012.

a) Construct an ANOVA table for this experiment and test the null hypothesis that all treatments have the same mean.

b) Test the null hypothesis that the average response in treatments 1 and 2 is the same as the average response in treatments 3, 4, and 5.

c) Use the HSD procedure to compare the means of the five treatments.

**Exercise 5.3** Refer to the data in Problem 3.1. Test the null hypothesis that all pairs of workers produce solder joints with the same average strength against the alternative that some workers produce different average strengths. Control the strong familywise error rate at .05.

**Exercise 5.4** Refer to the data in Exercise 3.1. Test the null hypothesis that all pairs of diets produce the same average weight liver against the alternative that some diets produce different average weights. Control the FDR at .05.

**Exercise 5.5** Use the data from Exercise 3.3. Compute 95% simultaneous confidence intervals for the differences in response between the the three treatment groups (acid, pulp, and salt) and the control group.

**Problem 5.1** Use the data from Problem 3.2. Use the Tukey procedure to make all pairwise comparisons between the treatment groups. Summarize your results with an underline diagram.

**Problem 5.2** In an experiment with four groups, each with five observations, the group means are 12, 16, 21, and 19, and the MSE is 20. A colleague points out that the contrast with coefficients -4, -2, 3, 3 has a rather large sum of squares. No one knows to begin with why this contrast has a large sum of squares, but after some detective work, you discover that the contrast coefficients are roughly the same (except for the overall mean) as the time the samples had to wait in the lab before being analyzed (3, 5, 10, and 10 days). What is the significance of this contrast?

**Problem 5.3** Consider an experiment taste-testing six types of chocolate chip cookies: 1 (brand A, chewy, expensive), 2 (brand A, crispy, expensive), 3 (brand B, chewy, inexpensive), 4 (brand B, crispy, inexpensive), 5 (brand C, chewy, expensive), 6 (brand D, crispy, inexpensive). We will use twenty different raters randomly assigned to each type (120 total raters). I have constructed five preplanned contrasts for these treatments, and I obtain $p$-values of .03, .04, .23, .47, and .68 for these contrasts. Discuss how you would assess the statistical significance of these contrasts, including what issues need to be resolved.

**Question 5.1** In an experiment with five groups and 25 degrees of freedom for error, for what numbers of contrasts is the Bonferroni procedure more powerful than the Scheffé procedure?

# Chapter 6

# Checking Assumptions

We analyze experimental results by comparing the average responses in different treatment groups using an overall test based on ANOVA or more focussed procedures based on contrasts and pairwise comparisons. All of these procedures are based on the *assumption* that our data follow the model

$$y_{ij} = \mu + \alpha_i + \epsilon_{ij},$$

where the $\alpha_i$'s are fixed but unknown numbers and the $\epsilon_{ij}$'s are independent normals with constant variance. We have done nothing to ensure that these assumptions are reasonably accurate.

What we did was random assignment of treatments to units, followed by measurement of the response. As discussed briefly in Chapter 2, randomization methods permit us to make inferences based solely on the randomization, but these methods tend to be computationally tedious and difficult to extend. Model-based methods with distributional assumptions usually yield good approximations to the randomization inferences, provided that the model assumptions are themselves reasonably accurate. If we apply the model-based methods in situations where the model assumptions do not hold, the inferences we obtain may be misleading. We thus need to look to the accuracy of the model assumptions.

Accuracy of inference depends on assumptions being true

## 6.1 Assumptions

The three basic assumptions we need to check are that the errors are 1) independent, 2) normally distributed, and 3) have constant variance. Independence is the most important of these assumptions, and also the most difficult to accommodate when it fails. We will not discuss accommodating dependent errors in this book. For the kinds of models we have been using, normality is the least important assumption, particularly for large sample sizes; see Chapter 11 for a different kind of model that is extremely dependent on normality. Constant variance is intermediate, in that nonconstant variance can have a substantial effect on our inferences, but nonconstant variance can also be accommodated in many situations.

Independence, constant variance, normality

Note that the quality of our inference depends on how well the errors $\epsilon_{ij}$ conform to our assumptions, but that we do not observe the errors $\epsilon_{ij}$. The closest we can get to the errors are $r_{ij}$, the residuals from the full model. Thus we must make decisions about how well the errors meet our assumptions based not on the errors themselves, but instead on residual quantities that we can observe. This unobservable nature of the errors can make diagnosis difficult in some situations.

In any real-world data set, we are almost sure to have one or more of the three assumptions be false. For example, real-world data are never exactly normally distributed. Thus there is no profit in formal testing of our assumptions; we already know that they are not true. The good news is that our procedures can still give reasonable inferences when the departures from our assumptions are not too large. This is called *robustness of validity*, which means that our inferences are reasonably valid across a range of departures from our assumptions. Thus the real question is whether the deviations from our assumptions are sufficiently great to cause us to mistrust our inference. At a minimum, we would like to know in what way to mistrust the inference (for example, our confidence intervals are shorter than they should be), and ideally we would like to be able to correct any problems.

Robustness of validity

The remaining sections of this chapter consider diagnostics and remedies for failed model assumptions. To some extent, we are falling prey to the syndrome of "When all you have is a hammer, the whole world looks like a nail," because we will go through a variety of maneuvers to make our linear models with normally distributed errors applicable to many kinds of data. There are other models and methods that we could use instead, including generalized linear models, robust methods, randomization methods, and nonparametric rank-based methods. For certain kinds of data, some of these alternative methods can be considerably more efficient (for example, produce shorter confidence intervals with the same coverage) than the linear models/normal distribution based methods used here, even when the normal based methods are still reasonably valid. However, these alternative methods are each another book in themselves, so we just mention them here and in Section 6.7.

Many other methods exist

## 6.2 Transformations

The primary tool for dealing with violations of assumptions is a transformation, or reexpression, of the response. For example, we might analyze the logarithm of the response. The idea is that the responses on the transformed scale match our assumptions more closely, so that we can use standard methods on the transformed data. There are several schemes for choosing transformations, some of which will be discussed below. For now, we note that transformations often help, and discuss the effect that transformations have on inference. The alternative to transformations is to develop specialized methods that deal with the violated assumptions. These alternative methods exist, but we will discuss only some of them. There is a tendency for these alternative methods to proliferate as various more complicated designs and analyses are considered.

Transformed data may meet assumptions

The null hypothesis tested by an F-test is that all the treatment means are equal. Together with the other assumptions we have about the responses, the null hypothesis implies that the distributions of the responses in all the treatment groups are exactly the same. Because these distributions are the same before transformation, they will be the same after transformation, provided that we used the same transformation for all the data. Thus we may test the null hypothesis of equal treatment means on any transformation scale that makes our assumptions tenable. By the same argument, we may test pairwise comparisons null hypotheses on any transformation scale.

Transformations don't affect the null

Confidence intervals are more problematic. We construct confidence intervals for means or linear combinations of means, such as contrasts. However, the center described by a mean depends on the scale in which the mean was computed. For example, the average of a data set is not equal to the square of the average of the square roots of the data set. This implies that confidence intervals for means or contrasts of means computed on a transformed scale do not back-transform into confidence intervals for the analogous means or contrasts of means on the original scale.

Transformations affect means

A confidence interval for an individual treatment *median* can be obtained by back-transforming a confidence interval for the corresponding mean from the scale where the data satisfy our assumptions. This works because medians are preserved through monotone transformations. If we truly need confidence intervals for differences of means on the original scale, then there is little choice but to do the intervals on the original scale (perhaps using some alternative procedure) and accept whatever inaccuracy results from violated assumptions. Large-sample, approximate confidence intervals on the original scale can sometimes be constructed from data on the transformed scale by using the delta method (Oehlert 1992).

Medians follow transformations

Special rules for logs

The logarithm is something of a special case. Exponentiating a confidence interval for the *difference* of two means on the log scale leads to a confidence interval for the *ratio* of the means on the original scale. We can also construct an approximate confidence interval for a mean on the original scale using data on the log scale. Land (1972) suggests the following: let $\hat{\mu}$ and $\hat{\sigma}^2$ be estimates of the mean and variance on the log scale, and let $\hat{\eta}^2 = \hat{\sigma}^2/n + \hat{\sigma}^4/[2(n+1)]$ where $n$ is the sample size. Then form a $1-\mathcal{E}$ confidence interval for the mean on the original scale by computing

Land's method

$$\exp(\hat{\mu} + \hat{\sigma}^2/2 \pm z_{\mathcal{E}/2}\,\hat{\eta}) ,$$

where $z_{\mathcal{E}/2}$ is the upper $\mathcal{E}/2$ percent point of the standard normal.

## 6.3 Assessing Violations of Assumptions

Our assumptions of independent, normally distributed errors with constant variance are not true for real-world data. However, our procedures may still give us reasonably good inferences, provided that the departures from our assumptions are not too great. Therefore we *assess* the nature and degree to which the assumptions are violated and take corrective measures if they are needed. The $p$-value of a formal test of some assumption does not by itself tell us the nature and degree of violations, so formal *testing* is of limited utility. Graphical and numerical assessments are the way to go.

Assess — don't test

Our assessments of assumptions about the errors are based on residuals. The raw residuals $r_{ij}$ are simply the differences between the data $y_{ij}$ and the treatment means $\overline{y}_{i\bullet}$. In later chapters there will be more complicated structures for the means, but the raw residuals are always the differences between the data and the fitted value.

Assessments based on residuals

We sometimes modify the raw residuals to make them more interpretable (see Cook and Weisberg 1982). For example, the variance of a raw residual is $\sigma^2(1-H_{ij})$, so we might divide raw residuals by an estimate of their standard error to put all the residuals on an equal footing. (See below for $H_{ij}$.) This is the *internally Studentized* residual $s_{ij}$, defined by

Internally Studentized residual

$$s_{ij} = \frac{r_{ij}}{\sqrt{MS_E(1-H_{ij})}} .$$

Internally Studentized residuals have a variance of approximately 1.

Alternatively, we might wish to get a sense of how far a data value is from what would be predicted for it from all the other data. This is the *externally*

*Studentized* residual $t_{ij}$, defined by

$$t_{ij} = s_{ij} \left( \frac{N-g-1}{N-g-s_{ij}^2} \right)^{1/2} ,$$

where $s_{ij}$ in this formula is the internally Studentized residual. The externally Studentized residual helps us determine whether a data point follows the pattern of the other data. When the data actually come from our assumed model, the externally Studentized residuals $t_{ij}$ follow a $t$-distribution with $N-g-1$ degrees of freedom.

Externally Studentized residual

The quantity $H_{ij}$ used in computing $s_{ij}$ (and thus $t_{ij}$) is called the *leverage* and depends on the model being fit to the data and sample sizes; $H_{ij}$ is $1/n_i$ for the separate treatment means model we are using now. Most statistical software will produce leverages and various kinds of residuals.

Leverage

### 6.3.1 Assessing nonnormality

The normal probability plot (NPP), sometimes called a rankit plot, is a graph ical procedure for assessing normality. We plot the ordered data on the vertical axis against the ordered normal scores on the horizontal axis. For assessing the normality of residuals, we plot the ordered residuals on the vertical axis. If you make an NPP of normally distributed data, you get a more or less straight line. It won't be perfectly straight due to sampling variability. If you make an NPP of nonnormal data, the plot will tend to be curved, and the shape of curvature tells you how the data depart from normality.

Normal probability plot (NPP)

Normal scores are the expected values for the smallest, second smallest, and so on, up to the largest data point in a sample that really came from a normal distribution with mean 0 and variance 1. The *rankit* is a simple approximation to the normal score. The $i$th rankit from a sample of size $n$ is the $(i - 3/8)/(n + 1/4)$ percent point of a standard normal.

Normal scores and rankits

In our diagnostic setting, we make a normal probability plot of the residuals from fitting the full model; it generally matters little whether we use raw or Studentized residuals. We then examine this plot for systematic deviation from linearity, which would indicate nonnormality. Figure 6.1 shows prototype normal probability plots for long and short tailed data and data skewed to the left and right. All sample sizes are 50.

It takes some practice to be able to look at an NPP and tell whether the deviation from linearity is due to nonnormality or sampling variability, and even with practice there is considerable room for error. If you have software that can produce NPP's for data from different distributions and sample sizes,

Practice!

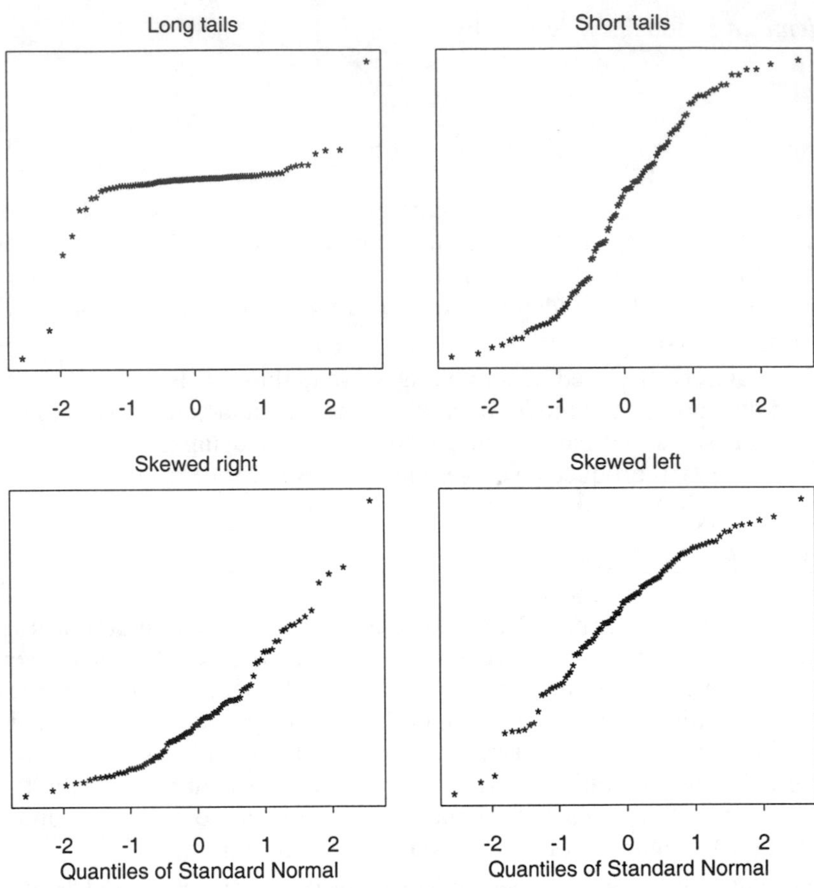


**Figure 6.1:** Rankit plots of nonnormal data, using S-Plus.

it is well worth your time to look at a bunch of plots to get a feel for how they may vary.

Outliers

*Outliers* are an extreme form of nonnormality. Roughly speaking, an outlier is an observation "different" from the bulk of the data, where different is usually taken to mean far away from or not following the pattern of the bulk of the data. Outliers can show up on an NPP as isolated points in the corners that lie off the pattern shown by the rest of the data.

We can use externally Studentized residuals to construct a formal outlier test. Each externally Studentized residual is a test statistic for the null hypothesis that the corresponding data value follows the pattern of the rest of

**Table 6.1:** Rainfall in acre feet from 52 clouds.

| Seeded | | | Unseeded | | |
|---|---|---|---|---|---|
| 1202.6 | 87.0 | 26.1 | 2745.6 | 274.7 | 115.3 |
| 830.1 | 81.2 | 24.4 | 1697.8 | 274.7 | 92.4 |
| 372.4 | 68.5 | 21.7 | 1656.0 | 255.0 | 40.6 |
| 345.5 | 47.3 | 17.3 | 978.0 | 242.5 | 32.7 |
| 321.2 | 41.1 | 11.5 | 703.4 | 200.7 | 31.4 |
| 244.3 | 36.6 | 4.9 | 489.1 | 198.6 | 17.5 |
| 163.0 | 29.0 | 4.9 | 430.0 | 129.6 | 7.7 |
| 147.8 | 28.6 | 1.0 | 334.1 | 119.0 | 4.1 |
| 95.0 | 26.3 | | 302.8 | 118.3 | |

the data, against an alternative that it has a different mean. Large absolute values of the Studentized residual are compatible with the alternative, so we reject the null and declare a given point to be an outlier if that point's Studentized residual exceeds in absolute value the upper $\mathcal{E}/2$ percent point of a $t$-distribution with $N - g - 1$ degrees of freedom. To test all data values (or equivalently, to test the maximum Studentized residual), make a Bonferroni correction and test the maximum Studentized residual against the upper $\mathcal{E}/(2N)$ percent point of a $t$-distribution with $N - g - 1$ degrees of freedom. This test can be fooled if there is more than one outlier.

**Example 6.1**

### Cloud seeding

Simpson, Olsen, and Eden (1975) provide data giving the rainfall in acre feet of 52 clouds, 26 of which were chosen at random for seeding with silver oxide. The problem is to determine if seeding has an effect and what size the effect is (if present). Data are given in Table 6.1.

An analysis of variance yields an F of 3.99 with 1 and 50 degrees of freedom.

| Source | DF | SS | MS | F |
|---|---|---|---|---|
| Seeding | 1 | 1.0003e+06 | 1.0003e+06 | 3.99 |
| Error | 50 | 1.2526e+07 | 2.5052e+05 | |

This has a $p$-value of about .05, giving moderate evidence of a difference between the treatments.

Figure 6.2 shows an NPP for the cloud seeding data residuals. The plot is angled with the bend in the lower right corner, indicating that the residuals are skewed to the right. This skewness is pretty evident if you make box-plots of the data, or simply look at the data in Table 6.1.

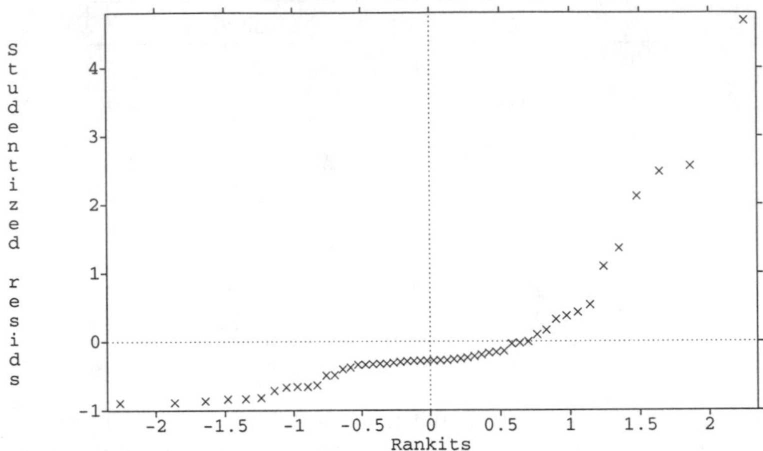


**Figure 6.2:** Normal probability plot for cloud seeding data, using MacAnova.

Now compute the externally Studentized residuals. The largest (corresponding to 2745.6) is 6.21, and is well beyond any reasonable cutoff for being an outlier. The next largest studentized residual is 2.71. If we remove the outlier from the data set and reanalyze, we now find that the largest studentized residual is 4.21, corresponding to 1697.5. This has a Bonferroni $p$-value of about .003 for the outlier test. This is an example of *masking*, where one apparently outlying value can hide a second. If we remove this second outlier and repeat the analysis, we now find that 1656 has a Studentized residual of 5.35, again an "outlier". Still more data values will be indicated as outliers as we pick them off one by one. The problem we have here is not so much that the data are mostly normal with a few outliers, but that the data do not follow a normal distribution at all. The outlier test is based on normality, and doesn't work well for nonnormal data.

### 6.3.2 Assessing nonconstant variance

Don't test equality of variances

There are formal tests for equality of variance—*do not use them!* This is for two reasons. First, $p$-values from such tests do not tell us what we need to know: the amount of nonconstant variance that is present and how it affects our inferences. Second, classical tests of constant variance (such as Bartlett's test or Hartley's test) are *so incredibly sensitive* to nonnormality that their inferences are worthless in practice.

We will look for nonconstant variance that occurs when the responses within a treatment group all have the same variance $\sigma_i^2$, but the variances differ between groups. We cannot distinguish nonconstant variance within a treatment group from nonnormality of the errors.

Does variance differ by treatment?

We assess nonconstant variance by making a plot of the residuals $r_{ij}$ (or $s_{ij}$ or $t_{ij}$) on the vertical axis against the fitted values $y_{ij} - r_{ij} = \overline{y}_{i\bullet}$ on the horizontal axis. This plot will look like several vertical stripes of points, one stripe for each treatment group. If the variance is constant, the vertical spread in the stripes will be about the same. Nonconstant variance is revealed as a pattern in the spread of the residuals. Note that groups with larger sample sizes will tend to have some residuals with slightly larger absolute values, simply because the sample size is bigger. It is the overall pattern that we are looking for.

Residual plots reveal nonconstant variance

The most common deviations from constant variance are those where the residual variation depends on the mean. Usually we see variances increasing as the mean increases, but other patterns can occur. When the variance increases with the mean, the residual plot has what is called a right-opening megaphone shape; it's wider on the right than on the left. When the variance decreases with the mean, the megaphone opens to the left. A third possible shape arises when the responses are proportions; proportions around .5 tend to have more variability than proportions near 0 or 1. Other shapes are possible, but these are the most common.

Right-opening megaphone is most common nonconstant variance

If you absolutely must test equality of variances—for example if change of variance is the treatment effect of interest—Conover, Johnson, and Johnson (1981) suggest a modified Levene test. Let $y_{ij}$ be the data. First compute $\tilde{y}_i$, the median of the data in group $i$; then compute $d_{ij} = |y_{ij} - \tilde{y}_i|$, the absolute deviations from the group medians. Now treat the $d_{ij}$ as data, and use the ANOVA F-test to test the null hypothesis that the groups have the same average value of $d_{ij}$. This test for means of the $d_{ij}$ is equivalent to a test for the equality of standard deviations of the original data $y_{ij}$. The Levene test as described here is a general test and is not tuned to look for specific kinds of nonconstant variance, such as right-opening megaphones. Just as contrasts and polynomial models are more focused than ANOVA, corresponding variants of ANOVA in the Levene test may be more sensitive to specific ways in which constant variance can be violated.

Levene test

## Resin lifetimes, continued

**Example 6.2**

In Example 3.2 we analyzed the $\log_{10}$ lifetimes of an encapsulating resin under different temperature stresses. What happens if we look at the lifetimes on the original scale rather than the log scale? Figure 6.3 shows a residual plot for these data on the original scale. A right-opening megaphone shape is

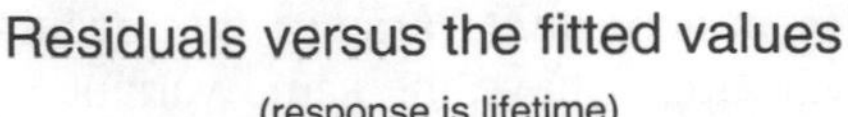


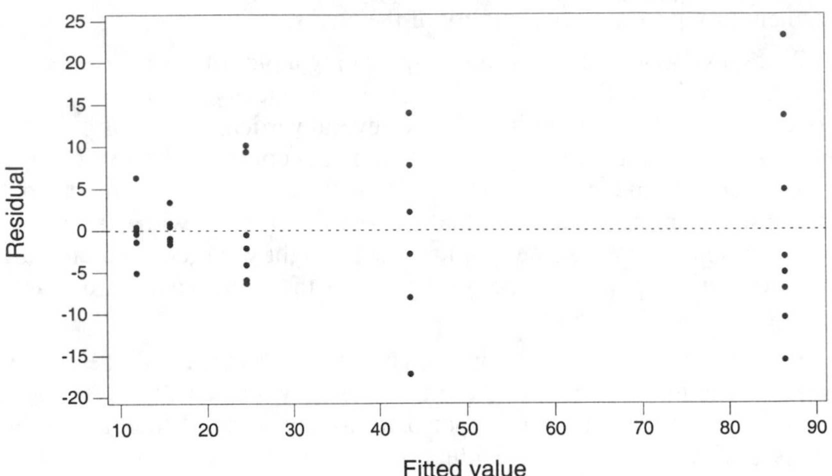


**Figure 6.3:** Residuals versus predicted plot for resin lifetime data, using Minitab.

clear, showing that the variability of the residuals increases with the response mean. The Levene test for the null hypothesis of constant variance has a $p$-value of about .07.

### 6.3.3 Assessing dependence

Serial dependence

*Serial dependence* or *autocorrelation* is one of the more common ways that independence can fail. Serial dependence arises when results close in time tend to be too similar (*positive* dependence) or too dissimilar (*negative* dependence). Positive dependence is far more common. Serial dependence could result from a "drift" in the measuring instruments, a change in skill of the experimenter, changing environmental conditions, and so on. If there is no idea of time order for the units, then there can be no serial dependence.

Index plot to detect serial dependence

A graphical method for detecting serial dependence is to plot the residuals on the vertical axis versus time sequence on the horizontal axis. The plot is sometimes called an *index plot* (that is, residuals-against-time index). Index plots give a visual impression of whether neighbors are too close to-

**Table 6.2:** Temperature differences in degrees Celsius between two thermocouples for 64 consecutive readings, time order along rows.

| | | | | | | | |
|---|---|---|---|---|---|---|---|
| 3.19 | 3.15 | 3.13 | 3.14 | 3.14 | 3.13 | 3.13 | 3.11 |
| 3.16 | 3.17 | 3.17 | 3.14 | 3.14 | 3.14 | 3.15 | 3.15 |
| 3.14 | 3.15 | 3.12 | 3.05 | 3.12 | 3.16 | 3.15 | 3.17 |
| 3.15 | 3.16 | 3.15 | 3.16 | 3.15 | 3.15 | 3.14 | 3.14 |
| 3.14 | 3.15 | 3.13 | 3.12 | 3.15 | 3.17 | 3.16 | 3.15 |
| 3.13 | 3.13 | 3.15 | 3.15 | 3.05 | 3.16 | 3.15 | 3.18 |
| 3.15 | 3.15 | 3.17 | 3.17 | 3.14 | 3.13 | 3.10 | 3.14 |
| 3.07 | 3.13 | 3.13 | 3.12 | 3.14 | 3.15 | 3.14 | 3.14 |

gether (positive dependence), or too far apart (negative dependence). Positive dependence appears as drifting patterns across the plot, while negatively dependent data have residuals that center at zero and rapidly alternate positive and negative.

The Durbin-Watson statistic is a simple numerical method for checking serial dependence. Let $r_k$ be the residuals sorted into time order. Then the Durbin-Watson statistic is:

Durbin-Watson statistic to detect serial dependence

$$DW = \frac{\sum_{k=1}^{n-1}(r_k - r_{k+1})^2}{\sum_{k=1}^{n} r_k^2} .$$

If there is no serial correlation, the DW should be about 2, give or take sampling variation. Positive serial correlation will make DW less than 2, and negative serial correlation will make DW more than 2. As a rough rule, serial correlations corresponding to DW outside the range 1.5 to 2.5 are large enough to have a noticeable effect on our inference techniques. Note that DW itself is random and may be outside the range 1.5 to 2.5, even if the errors are uncorrelated. For data sets with long runs of units from the same treatment, the variance of DW is a bit less than $4/N$.

### Temperature differences

**Example 6.3**

Christensen and Blackwood (1993) provide data from five thermocouples that were inserted into a high-temperature furnace to ascertain their relative bias. Sixty-four temperature readings were taken using each thermocouple, with the readings taken simultaneously from the five devices. Table 6.2 gives the differences between thermocouples 3 and 5.

We can estimate the relative bias by the average of the observed differences. Figure 6.4 shows the residuals (deviations from the mean) plotted in

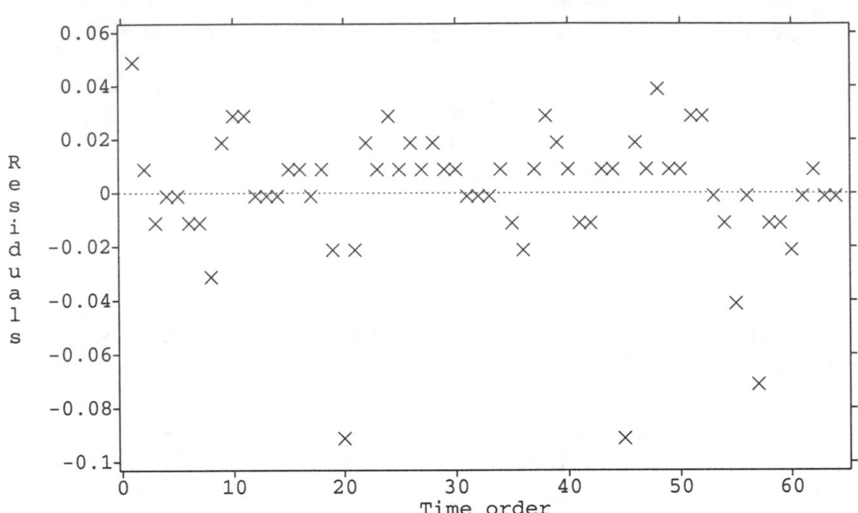


**Figure 6.4:** Deviations from the mean for paired differences of 64 readings from two thermocouples, using MacAnova.

time order. There is a tendency for positive and negative residuals to cluster in time, indicating positive autocorrelation. The Durbin-Watson statistic for these data is 1.5, indicating that the autocorrelation may be strong enough to affect our inferences.

Spatial association

*Spatial association,* another common form of dependence, arises when units are distributed in space and neighboring units have responses more similar than distant units. For example, spatial association might occur in an agronomy experiment when neighboring plots tend to have similar fertility, but distant plots could have differing fertilities.

Variogram to detect spatial association

One method for diagnosing spatial association is the *variogram.* We make a plot with a point for every pair of units. The plotting coordinates for a pair are the distance between the pair (horizontal axis) and the squared difference between their residuals (vertical axis). If there is a pattern in this figure—for example, the points in the variogram tend to increase with increasing distance—then we have spatial association.

Plot binned averages in variogram

This plot can look pretty messy, so we usually do some averaging. Let $D_{max}$ be the maximum distance between a pair of units. Choose some number of bins $K$, say 10 or 15, and then divide the distance values into $K$ groups: those from 0 to $D_{max}/K$, $D_{max}/K$ up to $2D_{max}/K$, and so on.

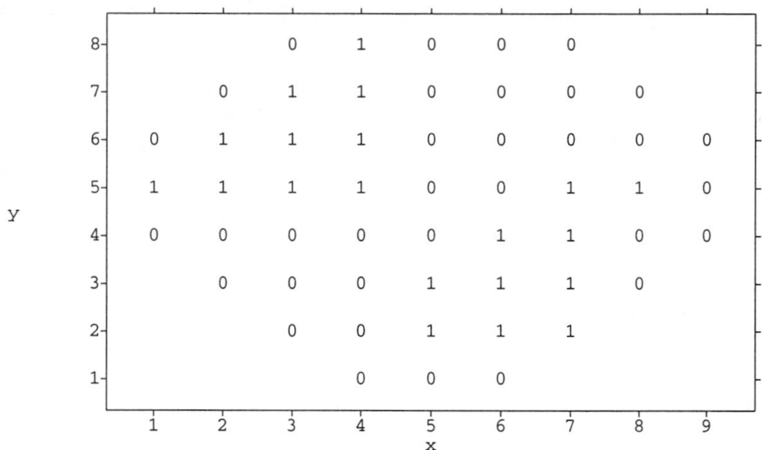


**Figure 6.5:** Horizontal (x) and vertical (y) locations of good (1) and bad (0) integrated circuits on a wafer

Now plot the average of the squared difference in residuals for each group of pairs. This plot should be roughly flat for data with no spatial association; it will usually have small average squared differences for small distances when there is spatial association.

**Example 6.4**

## Defective integrated circuits on a wafer

Taam and Hamada (1993) provide an example from the manufacture of integrated circuit chips. Many IC chips are made on a single silicon wafer, from which the individual ICs are cut after manufacture. Figure 6.5 (Taam and Hamada's Figure 1) shows the location of good (1) and bad (0) chips on a single wafer.

Describe the location of each chip by its $x$ (1 to 9) and $y$ (1 to 8) coordinates, and compute distances between pairs of chips using the usual Euclidean distance. Bin the pairs into those with distances from 1 to 2, 2 to 3, and so on. Figure 6.6 shows the variogram with this binning. We see that chips close together, and also chips far apart, tend to be more similar than those at intermediate distances. The similarity close together arises because the good chips are clustered together on the wafer. The similarity at large distances arises because almost all the edge chips are bad, and the only way to get a pair with a large distance is for them to cross the chip completely.

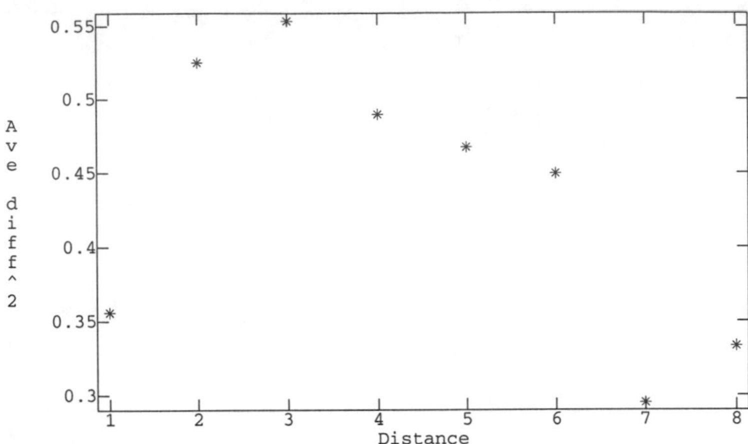


**Figure 6.6:** Variogram for chips on a wafer.

## 6.4 Fixing Problems

When our assessments indicate that our data do not meet our assumptions, we must either modify the data so that they do meet the assumptions, or modify our methods so that the assumptions are less important. We will give examples of both strategies.

### 6.4.1 Accommodating nonnormality

Transformations to improve normality

Nonnormality, particularly asymmetry, can sometimes be lessened by transforming the response to a different scale. Skewness to the right is lessened by a square root, logarithm, or other transformation to a power less than one, while skewness to the left is lessened by a square, cube, or other transformation to a power greater than one. Symmetric long tails do not easily yield to a transformation. Robust and rank-based methods can also be used in cases of nonnormality.

Try analysis with and without outliers

Individual outliers can affect our analysis. It is often useful to perform the analysis both with the full data set and with outliers excluded. If your conclusions change when the outliers are excluded, then you must be fairly careful in interpreting the results, because the results depend rather delicately on a few outlier data values. Some outliers are truly "bad" data, and their extremity draws our attention to them. For example, we may have miscopied

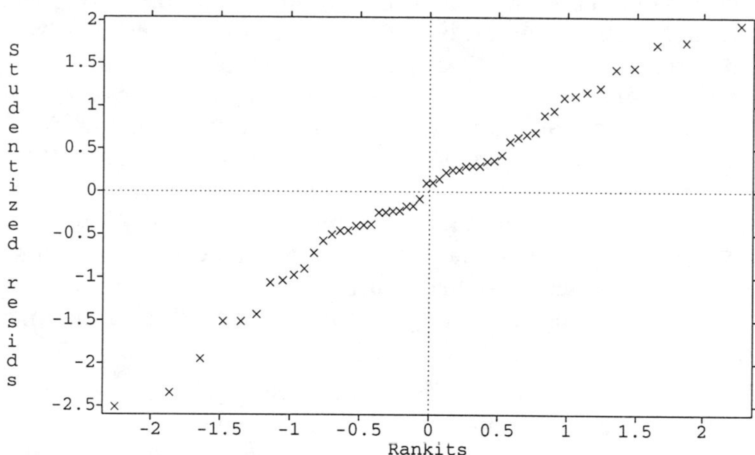


**Figure 6.7:** Normal probability plot for log-transformed cloud seeding data, using MacAnova.

the data so that 17.4 becomes 71.4, an outlier; or perhaps Joe sneezed in a test tube, and the yield on that run was less than satisfactory. However, outliers need not be bad data points; in fact, they may be the most interesting and informative data points in the whole data set. They just don't fit the model, which probably means that the model is wrong.

Outliers can be interesting data

## Cloud seeding, continued

**Example 6.5**

The cloud seeding data introduced in Example 6.1 showed considerable skewness to the right. Thus a square root or logarithm should help make things look more normal. Here is an Analysis of Variance for the data on the logarithmic scale.

| Source | DF | SS | MS | F |
|---|---|---|---|---|
| Seeding | 1 | 17.007 | 17.007 | 6.47382 |
| Error | 50 | 131.35 | 2.6271 | |

Figure 6.7 shows an NPP for the logged cloudseeding data residuals. This plot is much straighter than the NPP for the natural scale residuals, indicating that the error distribution is more nearly normal. The $p$-value for the test on the log scale is .014; the change is due more to stabilizing variance (see Section 6.5.2) than improved normality.

Since the cloud seeding data arose from a randomized experiment, we could use a randomization test on the difference of the means of the seeded and unseeded cloud rainfalls. There are almost $5 \times 10^{14}$ different possible randomizations, so it is necessary to take a random subsample of them when computing the randomization $p$-value. The two-sided randomization $p$-values using data on the original and log scales are .047 and .014 respectively. Comparing these with the corresponding $p$-values from the ANOVAs (.051 and .014), we see that they agree pretty well, but are closer on the log scale. We also note that the randomization inferences depend on scale as well. We used the same test statistic (difference of means) on both scales, but the difference of means on the log scale is the ratio of geometric means on the original scale.

We also wish to estimate the effect of seeding. On the log scale, a 95% confidence interval for the difference between seeded and unseeded is (.24, 2.05). This converts to a confidence interval on the ratio of the means of (1.27, 7.76) by back-exponentiating. A 95% confidence interval for the mean of the seeded cloud rainfalls, based on the original data and using a $t$-interval, is (179.1, 704.8); this interval is symmetric around the sample mean 442.0. Using Land's method for log-normal data, we get (247.2, 1612.2); this interval is not symmetric around the sample mean and reflects the asymmetry in log-normal data.

### 6.4.2 Accommodating nonconstant variance

The usual way to fix nonconstant error variances is by transformation of the response. For some distributions, there are standard transformations that equalize or stabilize the variance. In other distributions, we use a more ad hoc approach. We can also use some alternative methods instead of the usual ANOVA.

#### Transformations of the response

There is a general theory of variance-stabilizing transformations that applies to distributions where the variance depends on the mean. For example, Binomial(1, $p$) data have a mean of $p$ and a variance of $p(1-p)$. This method uses the relationship between the mean and the variance to construct a transformation such that the variance of the data after transformation is constant and no longer depends on the mean. (See Bishop, Fienberg, and Holland 1975.) These transformations generally work better when the sample size is large

Variance-stabilizing transformations

**Table 6.3:** Variance-stabilizing transformations.

| Distribution | Transformation | New variance |
|---|---|---|
| Binomial proportions<br>$X \sim \text{Bin}(n, p)$<br>$\hat{p} = X/n$<br>$\text{Var}(\hat{p}) = p(1-p)/n$ | $\arcsin(\sqrt{\hat{p}})$ | $1/(4n)$ |
| Poisson<br>$X \sim \text{Poisson}(\lambda)$<br>$\text{Var}(X) = E(X) = \lambda$ | $\sqrt{X}$ | $\frac{1}{4}$ |
| Correlation coefficient<br>$(u_i, v_i), i = 1, \ldots, n$ are independent, bivariate normal pairs with correlation $\rho$ and sample correlation $\hat{\rho}$ | $\frac{1}{2}\log\left(\frac{1+\hat{\rho}}{1-\hat{\rho}}\right)$ | 1 |

(or the mean is large relative to the standard deviation); modifications may be needed otherwise.

Table 6.3 lists a few distributions with their variance-stabilizing transformations. Binomial proportions model the fraction of success in some number of trials. If all proportions are between about .2 and .8, then the variance is fairly constant and the transformation gives little improvement. The Poisson distribution is often used to model counts; for example, the number of bacteria in a volume of solution or the number of asbestos particles in a volume of air.

## Artificial insemination in chickens

**Example 6.6**

Tajima (1987) describes an experiment examining the effect of a freeze-thaw cycle on the potency of semen used for artificial insemination in chickens. Four semen mixtures are prepared. Each mixture consists of equal volumes of semen from Rhode Island Red and White Leghorn roosters. Mixture 1 has both varieties fresh, mixture 4 has both varieties frozen, and mixtures 2 and 3 each have one variety fresh and the other frozen. Sixteen batches of Rhode Island Red hens are inseminated with the mixtures, using a balanced completely randomized design. The response is the fraction of chicks from each batch that have white feathers (white feathers indicate a White Leghorn father).

It is natural to model these fractions as binomial proportions. Each chick in a given treatment group has the same probability of having a White Leghorn father, though this probability may vary between groups due to the freeze-thaw treatments. Thus the total number of chicks with white feathers in a given batch should have a binomial distribution, and the fraction of chicks is a binomial proportion. The observed proportions ranged from .19 to .95, so the arcsine square root transformation is a good bet to stabilize the variability.

Power family transformations

When we don't have a distribution with a known variance-stabilizing transformation (and we generally don't), then we usually try a *power family* transformation. The power family of transformations includes

$$y \to \text{sign}(\lambda)y^\lambda$$

and

$$y \to \log(y) \ ,$$

where sign($\lambda$) is +1 for positive $\lambda$ and –1 for negative $\lambda$. The log function corresponds to $\lambda$ equal to zero. We multiply by the sign of $\lambda$ so that the order of the responses is preserved when $\lambda$ is negative.

Need positive data with max/min fairly large

Power family transformations are not likely to have much effect unless the ratio of the largest to smallest value is bigger than 4 or so. Furthermore, power family transformations only make sense when the data are all positive. When we have data with both signs, we can add a constant to all the data to make them positive before transforming. Different constants added lead to different transformations.

Regression method for choosing $\lambda$

Here is a simple method for finding an approximate variance-stabilizing transformation power $\lambda$. Compute the mean and standard deviation for the data in each treatment group. Regress the logarithms of the standard deviations on the logarithms of the group means; let $\hat{\beta}$ be the estimated regression slope. Then the estimated variance stabilizing power transformation is $\lambda = 1 - \hat{\beta}$. If there is no relationship between mean and standard deviation ($\hat{\beta} = 0$), then the estimated transformation is the power 1, which doesn't change the data. If the standard deviation increases proportionally to the mean ($\hat{\beta} = 1$), then the log transformation (power 0) is appropriate for variance stabilization.

Box-Cox transformations

The Box-Cox method for determining a transformation power is somewhat more complicated than the simple regression-based estimate, but it tends to find a better power and also yields a confidence interval for $\lambda$. Furthermore, Box-Cox can be used on more complicated designs where the simple method is difficult to adapt. Box-Cox transformations rescale the power family transformation to make the different powers easier to compare. Let $\dot{y}$

denote the geometric mean of all the responses, where the geometric mean is the product of all the responses raised to the 1/N power:

$$\dot{y} = \left( \prod_{i=1}^{g} \prod_{j=1}^{n_i} y_{ij} \right)^{1/N} .$$

The Box-Cox transformations are then

$$y^{(\lambda)} = \begin{cases} \dfrac{y^\lambda - 1}{\lambda \dot{y}^{\lambda-1}} & \lambda \neq 0 \\ \dot{y} \log(y) & \lambda = 0 \end{cases} .$$

In the Box-Cox technique, we transform the data using a range of $\lambda$ values from, say, -2 to 3, and do the ANOVA for each of these transformations. From these we can get $SS_E(\lambda)$, the sum of squared errors as a function of the transformation power $\lambda$. The best transformation power $\lambda^\star$ is the power that minimizes $SS_E(\lambda)$. We generally use a convenient transformation power $\lambda$ close to $\lambda^\star$, where by convenient I mean a "pretty" power, like .5 or 0, rather than the actual minimizing power which might be something like .427.

Use best convenient power

The Box-Cox minimizing power $\lambda^\star$ will rarely be exactly 1; when should you actually use a transformation? A graphical answer is obtained by making the suggested transformation and seeing if the residual plot looks better. If there was little change in the variances or the group variances were not that different to start with, then there is little to be gained by making the transformation. A more formal answer can be obtained by computing an approximate $1 - \mathcal{E}$ confidence interval for the transformation power $\lambda$. This confidence interval consists of all powers $\lambda$ such that

Confidence interval for $\lambda$

$$SS_E(\lambda) \leq SS_E(\lambda^\star)(1 + \frac{F_{\mathcal{E},1,\nu}}{\nu}) ,$$

where $\nu$ is the degrees of freedom for error. Very crudely, if the transformation doesn't decrease the error sum of squares by a factor of at least $\nu/(\nu+4)$, then $\lambda = 1$ is in the confidence interval, and a transformation may not be needed. When I decide whether a transformation is indicated, I tend to rely mostly on a visual judgement of whether the residuals improve after transformation, and secondarily on the confidence interval.

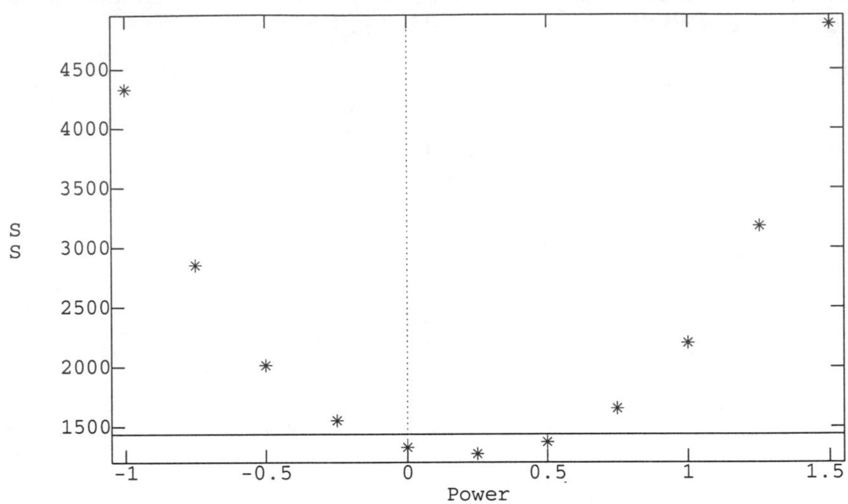


**Figure 6.8:** Box-Cox error SS versus transformation power for resin lifetime data.

**Example 6.7**

**Resin lifetimes, continued**

The resin lifetime data on the original scale show considerable nonconstant variance. The treatment means and variances are

| | 1 | 2 | 3 | 4 | 5 |
|---|---|---|---|---|---|
| Mean | 86.42 | 43.56 | 24.52 | 15.72 | 11.87 |
| Variance | 169.75 | 91.45 | 41.07 | 3.00 | 13.69 |

If we regress the log standard deviations on the log means, we get a slope of .86 for an estimated transformation power of .14; we would probably use a log (power 0) or quarter power since they are near the estimated power.

We can use Box-Cox to suggest an appropriate transformation. Figure 6.8 shows $SS_E(\lambda)$ plotted against transformation power for powers between $-1$ and 1.5; the minimum appears to be about 1270 near a power of .25. The logarithm does nearly as well as the quarter power ($SS_E(0)$ is nearly as small as $SS_E(.25)$), and the log is easier to work with, so we will use the log transformation. As a check, the 95% confidence interval for the transformation power includes all powers with Box-Cox error SS less than $1270(1+F_{.05,1,32}/32) = 1436$. The horizontal line on the plot is at this level;

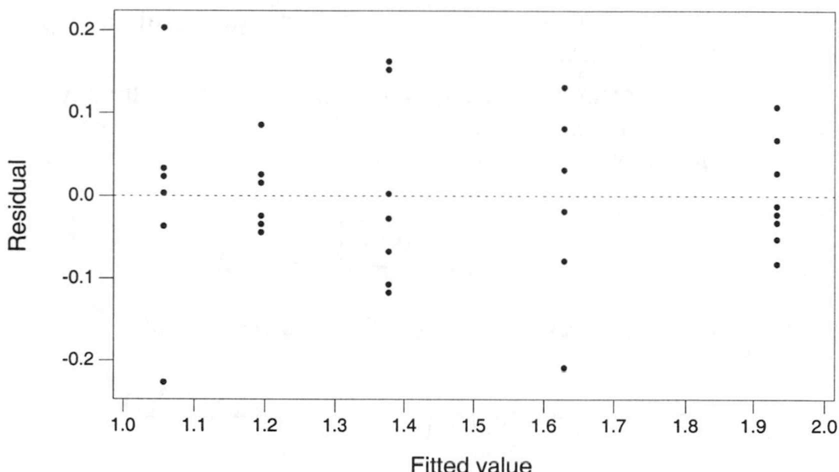


**Figure 6.9:** Residuals versus predicted plot for resin log lifetime data, using Minitab.

the log has an $SS_E$ well below the line, and the original scale has an $SS_E$ well above the line, suggesting that the logarithm is the way to go. Figure 6.9 shows the improvement in residuals versus fitted values after transformation. There is no longer as strong a tendency for the residuals to be larger when the mean is larger.

### Alternative methods

Dealing with nonconstant variance has provided gainful employment to statisticians for many years, so there are a number of alternative methods to consider. The simplest situation may be when the ratio of the variances in the different groups is known. For example, suppose that the response for each unit in treatments 1 and 2 is the average from five measurement units, and the response for each unit in treatments 3 and 4 is the average from seven measurement units. If the variance among measurement units is the same, then the variance between experimental units in treatments 3 and 4 would be 5/7 the size of the variance between experimental units in treatments 1

Weighted ANOVA when ratio of variances is known

and 2 (assuming no other sources of variation), simply due to different numbers of values in each average. Situations such as this can be handled using *weighted ANOVA*, where each unit receives a weight proportional to the number of measurement units used in its average. Most statistical packages can handle weighted ANOVA.

Welch's $t$ for pairwise comparisons with unequal variance

For pairwise comparisons, the Welch procedure is quite attractive. This procedure is sometimes called the "unpooled" $t$-test. Let $s_i^2$ denote the sample variance in treatment $i$. Then the Welch test statistic for testing $\mu_i = \mu_j$ is

$$t_{ij} = \frac{\overline{y}_{i\bullet} - \overline{y}_{j\bullet}}{\sqrt{s_i^2/n_i + s_j^2/n_j}} .$$

This test statistic is compared to a Student's $t$ distribution with

$$\nu = (s_i^2/n_i + s_j^2/n_j)^2 / \left( \frac{1}{n_i - 1} \frac{s_i^4}{n_i^2} + \frac{1}{n_j - 1} \frac{s_j^4}{n_j^2} \right)$$

degrees of freedom. For a confidence interval, we compute

$$t_{ij} = \overline{y}_{i\bullet} - \overline{y}_{j\bullet} \pm t_{\mathcal{E}/2,\nu} \sqrt{s_i^2/n_i + s_j^2/n_j} \ ,$$

with $\nu$ computed in the same way. More generally, for a contrast we use

$$t = \frac{\sum_i^g w_i \overline{y}_{i\bullet}}{\sqrt{\sum_i^g w_i^2 s_i^2/n_i}}$$

with approximate degrees of freedom

$$\nu = (\sum_{i=1}^g w_i^2 s_i^2/n_i)^2 / \left( \sum_{i=1}^g \frac{1}{n_i - 1} \frac{w_i^4 s_i^4}{n_i^2} \right) .$$

Confidence intervals are computed in an analogous way.

Welch's $t$ works well

The Welch procedure generally gives observed error rates close to the nominal error rates. Furthermore, the accuracy improves quickly as the sample sizes increase, something that cannot be said for the $t$ and F-tests under nonconstant variance. Better still, there is almost no loss in power for using the Welch procedure, even when the variances are equal. For simple comparisons, the Welch procedure can be used routinely. The problem arises in generalizing it to more complicated situations.

The next most complicated procedure is an ANOVA alternative for nonconstant variance. The Brown-Forsythe method is much less sensitive to nonconstant variance than is the usual ANOVA F test. Again let $s_i^2$ denote the sample variance in treatment $i$, and let $d_i = s_i^2(1 - n_i/N)$. The Brown-Forsythe modified F-test is

Brown-Forsythe modified F

$$BF = \frac{\sum_{i=1}^{g} n_i(\overline{y}_{i\bullet} - \overline{y}_{\bullet\bullet})^2}{\sum_{i=1}^{g} s_i^2(1 - n_i/N)} \ .$$

Under the null hypothesis of equal treatment means, BF is approximately distributed as F with $g - 1$ and $\nu$ degrees of freedom, where

$$\nu = \frac{(\sum_i d_i)^2}{\sum_i d_i^2/(n_i - 1)} \ .$$

**Example 6.8**

### Resin lifetimes, continued

Suppose that we needed confidence intervals for the difference in means between the pairs of temperatures on the original scale for the resin lifetime data. If we use the usual method and ignore the nonconstant variance, then pairwise differences have an estimated standard deviation of

$$\sqrt{68.82(1/n_i + 1/n_j)} \ ;$$

these range from 4.14 to 4.61, depending on sample sizes, and all would use 35 degrees of freedom. Using the Welch procedure, we get standard deviations for pairwise differences ranging from 5.71 (treatments 1 and 2) to 1.65 (treatments 4 and 5), with degrees of freedom ranging from 6.8 to 12.8. Thus the comparisons using the usual method are much too short for pairs such as 1 and 2, and much too long for pairs such as 4 and 5.

Consider now testing the null hypothesis that all groups have the same mean on the original scale. The F ratio from ANOVA is 101.8, with 4 and 32 degrees of freedom. The Brown-Forsythe F is 111.7, with 4 and 18.3 degrees of freedom. Both clearly reject the null hypothesis.

### 6.4.3 Accommodating dependence

There are no simple methods for dealing with dependence in data. Time series analysis and spatial statistics can be used to model data with dependence, but these methods are considerably beyond the scope of this book.

## 6.5 Effects of Incorrect Assumptions

Our methods work as advertised when the data meet our assumptions. Some violations of the assumptions have little effect on the quality of our inference, but others can cause almost catastrophic failure. This section gives an overview of how failed assumptions affect inference.

### 6.5.1 Effects of nonnormality

Before describing the effects of nonnormality, we need some way to quantify the degree to which a distribution is nonnormal. For this we will use the *skewness* and *kurtosis*, which measure asymmetry and tail length respectively. The skewness $\gamma_1$ and kurtosis $\gamma_2$ deal with third and fourth powers of the data:

$$\gamma_1 = \frac{E[(X-\mu)^3]}{\sigma^3} \quad \text{and} \quad \gamma_2 = \frac{E[(X-\mu)^4]}{\sigma^4} - 3.$$

Skewness measures asymmetry

Kurtosis measures tail length

For a normal distribution, both the skewness and kurtosis are 0. Distributions with a longer right tail have positive skewness, while distributions with a longer left tail have negative skewness. Symmetric distributions, like the normal, have zero skewness. Distributions with longer tails than the normal (more outlier prone) have positive kurtosis, and those with shorter tails than the normal (less outlier prone) have negative kurtosis. The "-3" in the definition of kurtosis is there to make the normal distribution have zero kurtosis. Note that neither skewness nor kurtosis depends on location or scale.

Table 6.4 lists the skewness and kurtosis for several distributions, giving you an idea of some plausible values. We could estimate the skewness and kurtosis for the residuals in our analysis, but these values are of limited diagnostic value, as sample estimates of skewness and kurtosis are notoriously variable.

For our discussion of nonnormal data, we will assume that the distribution of responses in each treatment group is the same apart from different means, but we will allow this common distribution to be nonnormal instead of requiring it to be normal. Our usual point estimates of group means and the common variance ($\overline{y}_{i\bullet}$ and $MS_E$ respectively) are still unbiased.

Long tails conservative for balanced data

The nominal $p$-values for F-tests are only slightly affected by moderate nonnormality of the errors. For balanced data sets (where all treatment groups have the same sample size), long tails tend to make the F-tests conservative; that is, the nominal $p$-value is usually a bit larger than it should be; so we reject the null too rarely. Again for balanced data, short tails will tend to

**Table 6.4:** Skewness and kurtosis for selected distributions.

| Distribution | $\gamma_1$ | $\gamma_2$ |
|---|---|---|
| Normal | 0 | 0 |
| Uniform | 0 | −1.2 |
| Normal truncated at | | |
| ±1 | 0 | −1.06 |
| ±2 | 0 | −0.63 |
| Student's t (df) | | |
| 5 | 0 | 6 |
| 6 | 0 | 3 |
| 8 | 0 | 1.5 |
| 20 | 0 | .38 |
| Chi-square (df) | | |
| 1 | 2.83 | 12 |
| 2 | 2 | 6 |
| 4 | 1.41 | 3 |
| 8 | 1 | 1.5 |

make the F-tests liberal; that is, the nominal $p$-value is usually a bit smaller than it should be, so that we reject the null too frequently. Asymmetry generally has a smaller effect than tail length on $p$-values. Unbalanced data sets are less predictable and can be less affected by nonnormality than balanced data sets, or even affected in the opposite direction. The effect of nonnormality decreases quickly with sample size. Table 6.5 gives the true Type I error rate of a nominal 5% F-test for various combinations of sample size, skewness, and kurtosis.

Short tails liberal for balanced data

The situation is not quite so good for confidence intervals, with skewness generally having a larger effect than kurtosis. When the data are normal, two-sided $t$-confidence intervals have the correct coverage, and the errors are evenly split high and low. When the data are from a distribution with nonzero skewness, two-sided $t$-confidence intervals still have approximately the correct coverage, but the errors tend to be to one side or the other, rather than split evenly high and low. One-sided confidence intervals for a mean can be seriously in error. The skewness for a contrast is less than that for a single mean, so the errors will be more evenly split. In fact, for a pairwise comparison when the sample sizes are equal, skewness essentially cancels out, and confidence intervals behave much as for normal data.

Skewness affects confidence intervals

**Table 6.5:** Actual Type I error rates for ANOVA F-test with nominal 5% error rate for various sample sizes and values of $\gamma_1$ and $\gamma_2$ using the methods of Gayen (1950).

| Four Samples of Size 5 | | | | | | | |
|---|---|---|---|---|---|---|---|
| | | | | $\gamma_2$ | | | |
| $\gamma_1$ | -1 | -.5 | 0 | .5 | 1 | 1.5 | 2 |
| 0 | .0527 | .0514 | .0500 | .0486 | .0473 | .0459 | .0446 |
| .5 | .0530 | .0516 | .0503 | .0489 | .0476 | .0462 | .0448 |
| 1 | .0538 | .0524 | .0511 | .0497 | .0484 | .0470 | .0457 |
| 1.5 | .0552 | .0538 | .0525 | .0511 | .0497 | .0484 | .0470 |

| $\gamma_1 = 0$ and $\gamma_2 = 1.5$ | | | | | |
|---|---|---|---|---|---|
| 4 groups of $k$ | | $k$ groups of 5 | | $(k_1, k_1, k_2, k_2)$ | |
| $k$ | Error | $k$ | Error | $k_1, k_2$ | Error |
| 2 | .0427 | 4 | .0459 | 10,10 | .0480 |
| 10 | .0480 | 8 | .0474 | 8,12 | .0483 |
| 20 | .0490 | 16 | .0485 | 5,15 | .0500 |
| 40 | .0495 | 32 | .0492 | 2,18 | .0588 |

Outliers, robustness, resistance

Individual outliers can so influence both treatment means and the mean square for error that the entire inference can change if repeated excluding the outlier. It may be useful here to distinguish between robustness (of validity) and resistance (to outliers). Robustness of validity means that our procedures give us inferences that are still approximately correct, even when some of our assumptions (such as normality) are incorrect. Thus we say that the ANOVA F-test is robust, because a nominal 5% F-test still rejects the null in about 5% of all samples when the null is true, even when the data are somewhat nonnormal. A procedure is resistant when it is not overwhelmed by one or a few individual data values. Our linear models methods are somewhat robust, but they are not resistant to outliers.

### 6.5.2 Effects of nonconstant variance

Nonconstant variance affects F-test $p$-values

When there are $g = 2$ groups and the sample sizes are equal, the Type I error rate of the F-test is very insensitive to nonconstant variance. indexAssumptions!constant variance When there are more than two groups or the sample sizes are not equal, the deviation from nominal Type I error rate is noticeable and can in fact be quite large. The basic facts are as follows:

**Table 6.6:** Approximate Type I error rate $\mathcal{E}$ for nominal 5% ANOVA F-test when the error variance is not constant.

| $g$ | $\sigma_i^2$ | $n_i$ | $\mathcal{E}$ |
|---|---|---|---|
| 3 | 1, 1, 1 | 5, 5, 5 | .05 |
| | 1, 2, 3 | 5, 5, 5 | .0579 |
| | 1, 2, 5 | 5, 5, 5 | .0685 |
| | 1, 2, 10 | 5, 5, 5 | .0864 |
| | 1, 1, 10 | 5, 5, 5 | .0954 |
| | 1, 1, 10 | 50, 50, 50 | .0748 |
| 3 | 1, 2, 5 | 2, 5, 8 | .0202 |
| | 1, 2, 5 | 8, 5, 2 | .1833 |
| | 1, 2, 10 | 2, 5, 8 | .0178 |
| | 1, 2, 10 | 8, 5, 2 | .2831 |
| | 1, 2, 10 | 20, 50, 80 | .0116 |
| | 1, 2, 10 | 80, 50, 20 | .2384 |
| 5 | 1, 2, 2, 2, 5 | 5, 5, 5, 5, 5 | .0682 |
| | 1, 2, 2, 2, 5 | 2, 2, 5, 8, 8 | .0292 |
| | 1, 2, 2, 2, 5 | 8, 8, 5, 2, 2 | .1453 |
| | 1, 1, 1, 1, 5 | 5, 5, 5, 5, 5 | .0908 |
| | 1, 1, 1, 1, 5 | 2, 2, 5, 8, 8 | .0347 |
| | 1, 1, 1, 1, 5 | 8, 8, 5, 2, 2 | .2029 |

1. If all the $n_i$'s are equal, then the effect of unequal variances on the $p$-value of the F-test is relatively small.

2. If big $n_i$'s go with big variances, then the nominal $p$-value will be bigger than the true $p$-value (we overestimate the variance and get a conservative test).

3. If big $n_i$'s go with small variances, then the nominal $p$-value will be less than the true $p$-value (we underestimate the variance and get a liberal test).

We can be more quantitative by using an approximation given in Box (1954). Table 6.6 gives the approximate Type I error rates for the usual F test when error variance is not constant. Clearly, nonconstant variance can dramatically affect our inference. These examples show (approximate) true type I error rates ranging from under .02 to almost .3; these are deviations from the nominal .05 that cannot be ignored.

Our usual form of confidence intervals uses the $MS_E$ as an estimate of error. When the error variance is not constant, the $MS_E$ will overestimate

Nonconstant variance affects confidence intervals

the error for contrasts between groups with small errors and underestimate the error for contrasts between groups with large errors. Thus our confidence intervals will be too long when comparing groups with small errors and too short when comparing groups with large errors. The intervals that are too long will have coverage greater than the nominal $1 - \mathcal{E}$, and vice versa for the intervals that are too short. The degree to which these intervals are too long or short can be arbitrarily large depending on sample sizes, the number of groups, and the group error variances.

### 6.5.3 Effects of dependence

Variance of average not $\sigma^2/n$ for dependent data

When the errors are dependent but otherwise meet our assumptions, indexAssumptions!independenceour estimates of treatment effects are still unbiased, and the $MS_E$ is nearly unbiased for $\sigma^2$ when the sample size is large. The big change is that the variance of an average is no longer just $\sigma^2$ divided by the sample size. This means that our estimates of standard errors for treatment means and contrasts are biased (whether too large or small depends on the pattern of dependence), so that confidence intervals and tests will not have their claimed error rates. The usual ANOVA F-test will be affected for similar reasons.

F robust to dependence averaged across randomizations

Let's be a little more careful. The ANOVA F-test is robust to dependence when considered as a randomization test. This means that averaged across all possible randomizations, the F-test will reject the null hypothesis about the correct fraction of times when the null is true. However, when the original data arise with a dependence structure, certain outcomes of the randomization will tend to have too many rejections, while other outcomes of the randomization will have too few.

More severe problems can arise when there was no randomization across the dependence. For example, treatments may have been assigned to units at random; but when responses were measured, all treatment 1 units were measured, followed by all treatment 2 units, and so on. Random assignment of treatment to units will not help us, even on average, if there is a strong correlation across time in the measurement errors.

**Example 6.9** **Correlated errors**

Consider a situation with two treatments and large, equal sample sizes. Suppose that the units have a time order, and that there is a correlation of $\rho$ between the errors $\epsilon_{ij}$ for time-adjacent units and a correlation of 0 between

**Table 6.7:** Error rates $\times 100$ of nominal 95% confidence intervals for $\mu_1 - \mu_2$, when neighboring data values have correlation $\rho$ and data patterns are consecutive or alternate.

| | $\rho$ | | | | | | | |
|---|---|---|---|---|---|---|---|---|
| | –.3 | –.2 | –.1 | 0 | .1 | .2 | .3 | .4 |
| Con. | .19 | 1.1 | 2.8 | 5 | 7.4 | 9.8 | 12 | 14 |
| Alt. | 12 | 9.8 | 7.4 | 5 | 2.8 | 1.1 | .19 | .001 |

the errors of other pairs. As a basis for comparison, Durbin-Watson values of 1.5 and 2.5 correspond to $\rho$ of $\pm.125$. For two treatments, the F-test is equivalent to a $t$-test. The $t$-test assumes that the difference of the treatment means has variance $2\sigma^2/n$. The actual variance of the difference depends on the correlation $\rho$ and the temporal pattern of the two treatments.

Consider first two temporal patterns for the treatments; call them consecutive and alternate. In the consecutive pattern, all of one treatment occurs, followed by all of the second treatment. In the alternate pattern, the treatments alternate every other unit. For the consecutive pattern, the actual variance of the difference of treatment means is $2(1 + 2\rho)\sigma^2/n$, while for the alternate pattern the variance is $2(1 - 2\rho)\sigma^2/n$. For the usual situation of $\rho > 0$, the alternate pattern gives a more precise comparison than the consecutive pattern, but the estimated variance in the $t$-test $(2\sigma^2/n)$ is the same for both patterns and correct for neither. So for $\rho > 0$, confidence intervals in the consecutive case are too short by a factor of $1/\sqrt{1+2\rho}$, and the intervals will not cover the difference of means as often as they claim, whereas confidence intervals in the alternate case are too long by a factor of $1/\sqrt{1-2\rho}$ and will cover the difference of means more often than they claim.

Table 6.7 gives the true error rates for a nominal 95% confidence interval under the type of serial correlation described above and the consecutive and alternate treatment patterns. These will also be the true error rates for the two-group F-test, and the consecutive results will be the true error rates for a confidence interval for a single treatment mean when the data for that treatment are consecutive.

In contrast, consider randomized assignment of treatments for the same kind of units. We could get consecutive or alternate patterns by chance, but that is very unlikely. Under the randomization, each unit has on average one neighbor with the same treatment and one neighbor with the other treatment, tending to make the effects of serial correlation cancel out. Table 6.8 shows median, upper, and lower quartiles of error rates for $\rho = .4$ and sample sizes

**Table 6.8:** Median, upper and lower quartiles of error rates × 100 of nominal 95% confidence intervals for $\mu_1 - \mu_2$ when neighboring data values have correlation .4 and treatments are assigned randomly, based on 10,000 simulations.

| | n | | | | |
|---|---|---|---|---|---|
| | 10 | 20 | 30 | 50 | 100 |
| Lower quartile | 3.7 | 3.9 | 4.0 | 4.2 | 4.5 |
| Median | 4.5 | 4.8 | 4.8 | 4.9 | 5.0 |
| Upper quartile | 6.5 | 5.7 | 5.8 | 5.5 | 5.4 |

from 10 to 100 based on 10,000 simulations. The best and worst case error rates are those from Table 6.7; but we can see in Table 6.8 that most randomizations lead to reasonable error rates, and the deviation from the nominal error rate gets smaller as the sample size increases.

Positive serial correlation has a smaller effective sample size

Here is another way of thinking about the effect of serial correlation when treatments are in a consecutive pattern. Positive serial correlation leads to variances for treatment means that are larger than $\sigma^2/n$, say $\sigma^2/(En)$, for $E < 1$. The effective sample size $En$ is less than our actual sample size $n$, because an additional measurement correlated with other measurements doesn't give us a full unit's worth of new information. Thus if we use the nominal sample size, we are being overly optimistic about how much precision we have for estimation and testing.

The effects of spatial association are similar to those of serial correlation, because the effects are due to correlation itself, not spatial correlation as opposed to temporal correlation.

## 6.6 Implications for Design

Use balanced designs

The major implication for design is that balanced data sets are usually a good idea. Balanced data are less susceptible to the effects of nonnormality and nonconstant variance. Furthermore, when there is nonconstant variance, we can usually determine the direction in which we err for balanced data.

When we know that our measurements will be subject to temporal or spatial correlation, we should take care to block and randomize carefully. We can, in principle, use the correlation in our design and analysis to increase precision, but these methods are beyond this text.

## 6.7 Further Reading and Extensions

Statisticians started worrying about what would happen to their $t$-tests and F-tests on real data almost immediately after they started using the tests. See, for example, Pearson (1931). Scheffé (1959) provides a more mathematical introduction to the effects of violated assumptions than we have given here. Ito (1980) also reviews the subject.

Transformations have long been used in Analysis of Variance. Tukey (1957a) puts the power transformations together as a family, and Box and Cox (1964) introduce the scaling required to make the $SS_E$'s comparable. Atkinson (1985) and Hoaglin, Mosteller, and Tukey (1983) give more extensive treatments of transformations for several goals, including symmetry and equalization of spread.

The Type I error rates for nonnormal data were computed using the methods of Gayen (1950). Gayen assumed that the data followed an Edgeworth distribution, which is specified by its first four moments, and then computed the distribution of the F-ratio (after several pages of awe-inspiring calculus). Our Table 6.5 is computed with his formula (2.30), though note that there are typos in his paper.

Box and Andersen (1955) approached the same problem from a different tack. They computed the mean and expectation of a transformation of the F-ratio under the permutation distribution when the data come from nonnormal distributions. From these moments they compute adjusted degrees of freedom for the F-ratio. They concluded that multiplying the numerator and denominator degrees of freedom by $(1+\gamma_2/N)$ gave $p$-values that more closely matched the permutation distribution.

There are two enormous, parallel areas of literature that deal with outliers. One direction is outlier identification, which deals with finding outliers, and to some extent with estimating and testing after outliers are found and removed. Major references include Hawkins (1980), Beckman and Cook (1983), and Barnett and Lewis (1994). The second direction is robustness, which deals with procedures that are valid and efficient for nonnormal data (particularly outlier-prone data). Major references include Andrews *et al.* (1972), Huber (1981), and Hampel *et al.* (1986). Hoaglin, Mosteller, and Tukey (1983) and Rey (1983) provide gentler introductions.

Rank-based, nonparametric methods are a classical alternative to linear methods for nonnormal data. In the simplest situation, the numerical values of the responses are replaced by their ranks, and we then do randomization analysis on the ranks. This is feasible because the randomization distribution of a rank test can often be computed analytically. Rank-based methods have sometimes been advertised as assumption-free; this is not true. Rank methods have their own strengths and weakness. For example, the power of two-sample rank tests for equality of medians can be very low when the two samples have different spreads. Conover (1980) is a standard introduction to nonparametric statistics.

We have been modifying the data to make them fit the assumptions of our linear analysis. Where possible, a better approach is to use an analysis that is appropriate for the data. Generalized Linear Models (GLM's) permit the kinds of mean structures we have been using to be combined with a variety of error structures, including Poisson, binomial, gamma, and other distributions. GLM's allow direct modeling of many forms of nonnormality and nonconstant variance. On the down side, GLM's are more difficult to compute, and most of their inference is asymptotic. McCullagh and Nelder (1989) is the standard reference for GLM's.

We computed approximate test sizes for F under nonconstant variance using a method given in Box (1954). When our distributional assumptions and the null hypothesis are true, then our observed F-statistic $F_{\text{obs}}$ is distributed as F with $g-1$ and $N-g$ degrees of freedom, and

$$P(F_{\text{obs}} > F_{\mathcal{E},g-1,N-g}) = \mathcal{E}.$$

If the null is true but we have different variances in the different groups, then $F_{\text{obs}}/b$ is distributed approximately as $F(\nu_1, \nu_2)$, where

$$b = \frac{N-g}{N(g-1)} \frac{\sum_i (N-n_i)\sigma_i^2}{\sum_i (n_i-1)\sigma_i^2} ,$$

$$\nu_1 = \frac{[\sum_i (N-n_i)\sigma_i^2]^2}{[\sum_i n_i\sigma_i^2]^2 + N\sum_i (N-2n_i)\sigma_i^4} ,$$

$$\nu_2 = \frac{[\sum_i (n_i-1)\sigma_i^2]^2}{\sum_i (n_i-1)\sigma_i^4} .$$

Thus the actual Type I error rate of the usual F test under nonconstant variance is approximately the probability that an F with $\nu_1$ and $\nu_2$ degrees of freedom is greater than $F_{\mathcal{E},g-1,N-g}/b$.

The Durbin-Watson statistic was developed in a series of papers (Durbin and Watson 1950, Durbin and Watson 1951, and Durbin and Watson 1971). The distribution of DW is complicated in even simple situations. Ali (1984) gives a (relatively) simple approximation to the distribution of DW.

There are many more methods to test for serial correlation. Several fairly simple related tests are called runs tests. These tests are based on the idea that if the residuals are arranged in time order, then positive serial correlation will lead to "runs" in the residuals. Different procedures measure runs differently. For example, Geary's test is the total number of consecutive pairs of residuals that have the same sign (Geary 1970). Other runs include maximum number of consecutive residuals of the same sign, the number of runs up (residuals increasing) and down (residuals decreasing), and so on.

In some instances we might believe that we know the correlation structure of the errors. For example, in some genetics studies we might believe that correlation can be deduced from pedigree information. If the correlation is known, it can be handled simply and directly by using generalized least squares (Weisberg 1985).

We usually have to use advanced methods from times series or spatial statistics to deal with correlation. Anderson (1954), Durbin (1960), Pierce (1971), and Tsay (1984) all deal with the problem of regression when the residuals are temporally correlated. Kriging is a class of methods for dealing with spatially correlated data that has become widely used, particularly in geology and environmental sciences. Cressie (1991) is a standard reference for spatial statistics. Grondona and Cressie (1991) describe using spatial statistics in the analysis of designed experiments.

## 6.8 Problems

**Exercise 6.1**

As part of a larger experiment, 32 male hamsters were assigned to four treatments in a completely randomized fashion, eight hamsters per treatment. The treatments were 0, 1, 10, and 100 nmole of melatonin daily, 1 hour prior to lights out for 12 weeks. The response was paired testes weight (in mg). Below are the means and standard deviations for each treatment group (data from Rollag 1982). What is the problem with these data and what needs to be done to fix it?

| Melatonin | Mean | SD |
|---:|---:|---:|
| 0 nmole | 3296 | 90 |
| 1 nmole | 2574 | 153 |
| 10 nmole | 1466 | 207 |
| 100 nmole | 692 | 332 |

**Exercise 6.2**

Bacteria in solution are often counted by a method known as serial dilution plating. Petri dishes with a nutrient agar are inoculated with a measured amount of solution. After 3 days of growth, an individual bacterium will have grown into a small colony that can be seen with the naked eye. Counting original bacteria in the inoculum is then done by counting the colonies on the plate. Trouble arises because we don't know how much solution to add. If we get too many bacteria in the inoculum, the petri dish will be covered with a lawn of bacterial growth and we won't be able to identify the colonies. If we get too few bacteria in the inoculum, there may be no colonies to count. The resolution is to make several dilutions of the original solution (1:1, 10:1, 100:1, and so on) and make a plate for each of these dilutions. One of the dilutions should produce a plate with 10 to 100 colonies on it, and that is the one we use. The count in the original sample is obtained by multiplying by the dilution factor.

Suppose that we are trying to compare three different Pasteurization treatments for milk. Fifteen samples of milk are randomly assigned to the three treatments, and we determine the bacterial load in each sample after treatment via serial dilution plating. The following table gives the counts.

| | | | | | |
|---|---|---|---|---|---|
| Treatment 1 | $26 \times 10^2$ | $29 \times 10^2$ | $20 \times 10^2$ | $22 \times 10^2$ | $32 \times 10^2$ |
| Treatment 2 | $35 \times 10^3$ | $23 \times 10^3$ | $20 \times 10^3$ | $30 \times 10^3$ | $27 \times 10^3$ |
| Treatment 3 | $29 \times 10^5$ | $23 \times 10^5$ | $17 \times 10^5$ | $29 \times 10^5$ | $20 \times 10^5$ |

Test the null hypothesis that the three treatments have the same effect on bacterial concentration.

**Exercise 6.3**

In order to determine the efficacy and lethal dosage of cardiac relaxants, anesthetized guinea pigs are infused with a drug (the treatment) till death occurs. The total dosage required for death is the response; smaller lethal doses are considered more effective. There are four drugs, and ten guinea pigs are chosen at random for each drug. Lethal dosages follow.

| | | | | | | | | | | |
|---|---|---|---|---|---|---|---|---|---|---|
| 1 | 18.2 | 16.4 | 10.0 | 13.5 | 13.5 | 6.7 | 12.2 | 18.2 | 13.5 | 16.4 |
| 2 | 5.5 | 12.2 | 11.0 | 6.7 | 16.4 | 8.2 | 7.4 | 12.2 | 6.7 | 11.0 |
| 3 | 5.5 | 5.0 | 8.2 | 9.0 | 10.0 | 6.0 | 7.4 | 5.5 | 12.2 | 8.2 |
| 4 | 6.0 | 7.4 | 12.2 | 11.0 | 5.0 | 7.4 | 7.4 | 5.5 | 6.7 | 5.5 |

Determine which drugs are equivalent, which are more effective, and which less effective.

**Exercise 6.4**

Four overnight delivery services are tested for "gentleness" by shipping fragile items. The breakage rates observed are given below:

| | | | | | |
|---|---|---|---|---|---|
| A | 17 | 20 | 15 | 21 | 28 |
| B | 7 | 11 | 15 | 10 | 10 |
| C | 11 | 9 | 5 | 12 | 6 |
| D | 5 | 4 | 3 | 7 | 6 |

You immediately realize that the variance is not stable. Find an approximate 95% confidence interval for the transformation power using the Box-Cox method.

**Exercise 6.5**

Consider the following four plots. Describe what each plot tells you about the assumptions of normality, independence, and constant variance. (Some plots may tell you nothing about assumptions.)

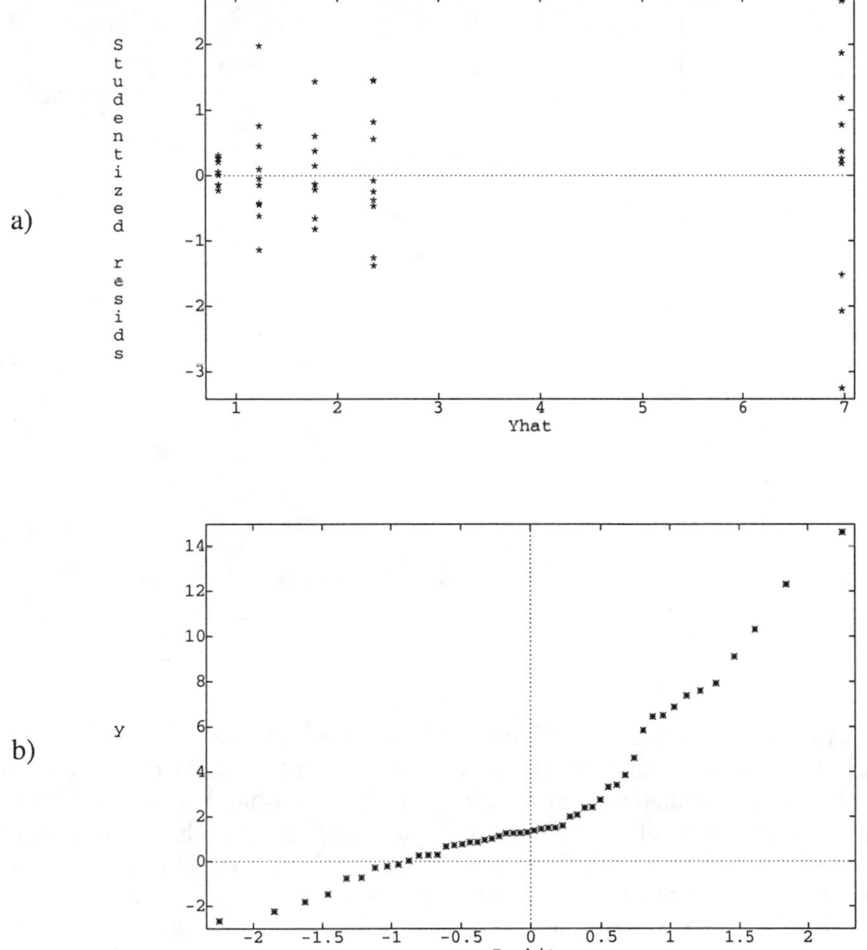

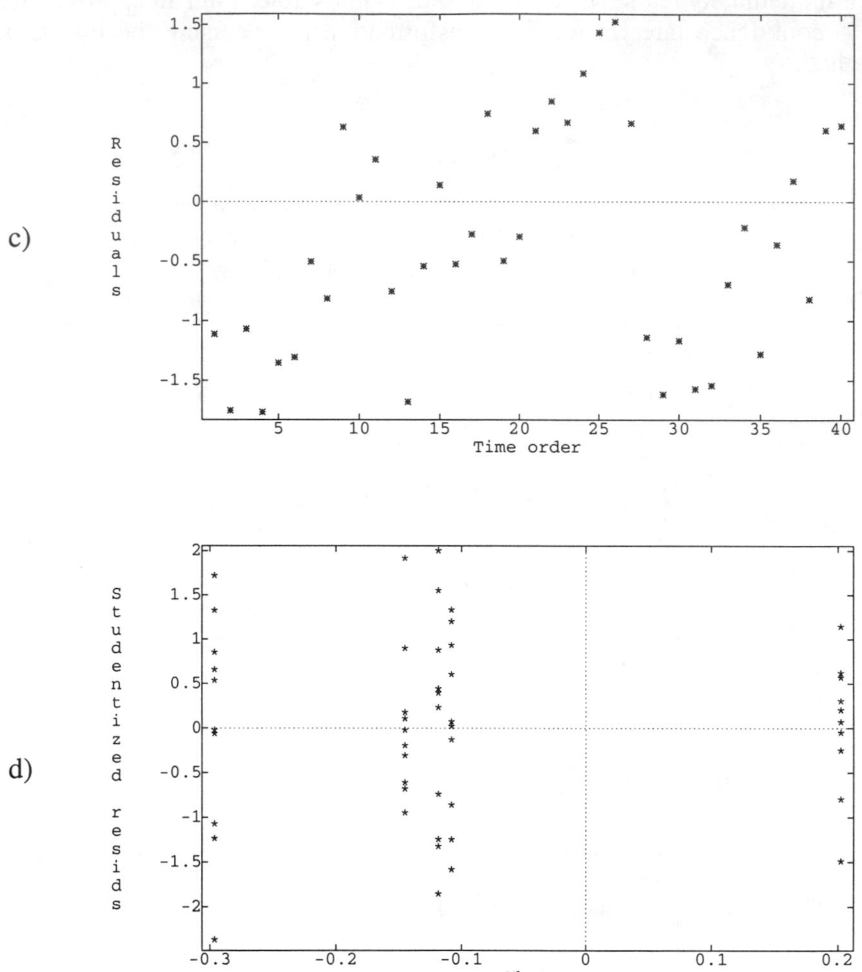


**Exercise 6.6** An instrument called a "Visiplume" measures ultraviolet light. By comparing absorption in clear air and absorption in polluted air, the concentration of $SO_2$ in the polluted air can be estimated. The EPA has a standard method for measuring $SO_2$, and we wish to compare the two methods across a range of air samples. The recorded response is the ratio of the Visiplume reading to the EPA standard reading. The four experimental conditions are: measurements of $SO_2$ in an inflated bag (n = 9), measurements of a smoke generator with $SO_2$ injected (n = 11), measurements at two coal-fired plants (n = 5 and 6). We are interested in whether the Visiplume instrument performs the same

relative to the standard method across all experimental conditions, between the coal-fired plants, and between the generated smoke and the real coal-fired smoke. The data follow (McElhoe and Conner 1986):

| | | | | | | | |
|---|---|---|---|---|---|---|---|
| Bag | 1.055 | 1.272 | .824 | 1.019 | 1.069 | .983 | 1.025 |
| | 1.076 | 1.100 | | | | | |
| Smoke | 1.131 | 1.236 | 1.161 | 1.219 | 1.169 | 1.238 | 1.197 |
| | 1.252 | 1.435 | .827 | 3.188 | | | |
| Plant no. 1 | .798 | .971 | .923 | 1.079 | 1.065 | | |
| Plant no. 2 | .950 | .978 | .762 | .733 | .823 | 1.011 | |

**Problem 6.1**

We wish to study the competition of grass species: in particular, big bluestem (from the tall grass prairie) versus quack grass (a weed). We set up an experimental garden with 24 plots. These plots were randomly allocated to the six treatments: nitrogen level 1 (200 mg N/kg soil) and no irrigation; nitrogen level 1 and 1cm/week irrigation; nitrogen level 2 (400 mg N/kg soil) and no irrigation; nitrogen level 3 (600 mg N/kg soil) no irrigation; nitrogen level 4 (800 mg N/kg soil) and no irrigation; and nitrogen level 4 and 1 cm/week irrigation. Big bluestem was seeded in these plots and allowed to establish itself. After one year, we added a measured amount of quack grass seed to each plot. After another year, we harvest the grass and measure the fraction of living material in each plot that is big bluestem. We wish to determine the effects (if any) of nitrogen and/or irrigation on the ability of quack grass to invade big bluestem. (Based on Wedin 1990.)

| N level<br>Irrigation | 1<br>N | 1<br>Y | 2<br>N | 3<br>N | 4<br>N | 4<br>Y |
|---|---|---|---|---|---|---|
| | 97 | 83 | 85 | 64 | 52 | 48 |
| | 96 | 87 | 84 | 72 | 56 | 58 |
| | 92 | 78 | 78 | 63 | 44 | 49 |
| | 95 | 81 | 79 | 74 | 50 | 53 |

(a) Do the data need a transformation? If so, which transformation?

(b) Provide an Analysis of Variance for these data. Are all the treatments equivalent?

(c) Are there significant quadratic effects of nitrogen under nonirrigated conditions?

(d) Is there a significant effect of irrigation?

(e) Under which conditions is big bluestem best able to prevent the invasion by quack grass? Is the response at this set of conditions significantly different from the other conditions?

**Question 6.1** What happens to the $t$-statistic as one of the values becomes extremely large? Look at the data set consisting of the five numbers 0, 0, 0, 0, K, and compute the $t$-test for testing the null hypothesis that these numbers come from a population with mean 0. What happens to the $t$-statistic as K goes to infinity?

**Question 6.2** Why would we expect the log transformation to be the variance-stabilizing transformation for the data in Exercise 6.2?

# Chapter 7

# Power and Sample Size

The last four chapters have dealt with analyzing experimental results. In this chapter we return to design and consider the issues of choosing and assessing sample sizes. As we know, an experimental design is determined by the units, the treatments, and the assignment mechanism. Once we have chosen a pool of experimental units, decided which treatments to use, and settled on a completely randomized design, the major thing left to decide is the sample sizes for the various treatments. Choice of sample size is important because we want our experiment to be as small as possible to save time and money, but big enough to get the job done. What we need is a way to figure out how large an experiment needs to be to meet our goals; a bigger experiment would be wasteful, and a smaller experiment won't meet our needs.

Decide how large an experiment is needed

## 7.1 Approaches to Sample Size Selection

There are two approaches to specifying our needs from an experiment, and both require that we know something about the system under test to do effective sample size planning. First, we can require that confidence intervals for means or contrasts should be no wider than a specified length. For example, we might require that a confidence interval for the difference in average weight loss under two diets should be no wider than 1 kg. The width of a confidence interval depends on the desired coverage, the error variance, and the sample size, so we must know the error variance at least roughly before we can compute the required sample size. If we have no idea about the size of the error variance, then we cannot say how wide our intervals will be, and we cannot plan an appropriate sample size.

Specify maximum CI width

The second approach to sample size selection involves error rates for the fixed level ANOVA F-test. While we prefer to use $p$-values for analysis, fixed level testing turns out to be a convenient framework for choosing sample size. In a fixed level test, we either reject the null hypothesis or we fail to reject the null hypothesis. If we reject a true null hypothesis, we have made a Type I error, and if we fail to reject a false null hypothesis, we have made a Type II error. The probability of making a Type I error is $\mathcal{E}_I$; $\mathcal{E}_I$ is under our control. We choose a Type I error rate $\mathcal{E}_I$ (5%, 1%, etc.), and reject $H_0$ if the $p$-value is less than $\mathcal{E}_I$. The probability of making a Type II error is $\mathcal{E}_{II}$; the probability of rejecting $H_0$ when $H_0$ is false is $1 - \mathcal{E}_{II}$ and is called *power*. The Type II error rate $\mathcal{E}_{II}$ depends on virtually everything: $\mathcal{E}_I$, $g$, $\sigma^2$, and the $\alpha_i$'s and $n_i$'s. Most books use the symbols $\alpha$ and $\beta$ for the Type I and II error rates. We use $\mathcal{E}$ for error rates, and use subscripts here to distinguish types of errors.

Power is probability of rejecting a false null hypothesis

It is more or less true that we can fix all but one of the interrelated parameters and solve for the missing one. For example, we may choose $\mathcal{E}_I$, $g$, $\sigma^2$, and the $\alpha_i$'s and $n_i$ and then solve for $1 - \mathcal{E}_{II}$. This is called a power analysis, because we are determining the power of the experiment for the alternative specified by the particular $\alpha_i$'s. We may also choose $\mathcal{E}_I$, $g$, $1 - \mathcal{E}_{II}$, $\sigma^2$ and the $\alpha_i$'s and then solve for the sample sizes. This, of course, is called a sample size analysis, because we have specified a required power and now find a sample size that achieves that power. For example, consider a situation with three diets, and $\mathcal{E}_I$ is .05. How large should $N$ be (assuming equal $n_i$'s) to have a 90% chance of rejecting $H_0$ when $\sigma^2$ is 9 and the treatment mean responses are -7, -5, 3 ($\alpha_i$'s are -4, -2, and 6)?

Find minimum sample size that gives desired power

The use of power or sample size analysis begins by deciding on interesting values of the treatment effects and likely ranges for the error variance. "Interesting" values of treatment effects could be anticipated effects, or they could be effects that are of a size to be scientifically significant; in either case, we want to be able to detect interesting effects. For each combination of treatment effects, error variance, sample sizes, and Type I error rate, we may compute the power of the experiment. Sample size computation amounts to repeating this exercise again and again until we find the smallest sample sizes that give us at least as much power as required. Thus what we do is set up a set of circumstances that we would like to detect with a given probability, and then design for those circumstances.

Use prior knowledge of system

**Example 7.1** **VOR in ataxia patients**

Spinocerebellar ataxias (SCA's) are inherited, degenerative, neurological diseases. Clinical evidence suggests that eye movements and posture are affected by SCA. There are several distinct types of SCA's, and we would like to determine if the types differ in observable ways that could be used to classify patients and measure the progress of the disease.

We have some preliminary data. One response is the "amplitude of the vestibulo-ocular reflex for 20 deg/s$^2$ velocity ramps"; let's just call it VOR. VOR deals with how your eyes move when trying to focus on a fixed target while you are seated on a chair on a turntable that is rotating increasingly quickly. We have preliminary observations on a total of seventeen patients from SCA groups 1, 5, and 6, with sample sizes 5, 11, and 1. The response appears to have stable variance on the log scale, on which scale the group means of VOR are 2.82, 3.89, and 3.04, and the variance is .075. Thus it looks like the average response (on the original scale) in SCA 5 is about three times that of SCA 1, while the average response of SCA 6 is only about 25% higher than that of SCA 1.

We would like to know the required sample sizes for three criteria. First, 95% confidence intervals for pairwise differences (on the log scale) should be no wider than .5. Second, power should be .99 when testing at the .01 level for two null hypotheses: the null hypothesis that all three SCAs have the same mean VOR, and the null hypothesis that SCA 1 and SCA 6 have the same mean VOR. We must specify the means and error variance to compute power, so we use those from the preliminary data. Note that there is only one subject in SCA 6, so our knowledge there is pretty slim and our computed sample sizes involving SCA 6 will not have a very firm foundation.

## 7.2 Sample Size for Confidence Intervals

We can compute confidence intervals for means of treatment groups and contrasts between treatment groups. One sample size criterion is to choose the sample sizes so that confidence intervals of interest are no wider than a maximum allowable width $W$. For the mean of group $i$, a $1 - \mathcal{E}_I$ confidence interval has width

Width of confidence interval

$$2\, t_{\mathcal{E}_I/2,N-g}\sqrt{MS_E/n_i}\ ;$$

for a contrast, the confidence interval has width

$$2\, t_{\mathcal{E}_I/2,N-g}\sqrt{MS_E}\ \sqrt{\sum_i \frac{w_i^2}{n_i}}\ .$$

In principle, the required sample size can be found by equating either of these widths with $W$ and solving for the sample sizes. In practice, we don't know $MS_E$ until the experiment has been performed, so we must anticipate a reasonable value for $MS_E$ when planning the experiment.

Calculating sample size

Assuming that we use equal sample sizes $n_i = n$, we find that

$$n \approx \frac{4\, t^2_{\mathcal{E}_I/2,g(n-1)}\, MS_E \sum w_i^2}{W^2} \quad .$$

This is an approximation because $n$ must be a whole number and the quantity on the right can have a fractional part; what we want is the smallest $n$ such that the left-hand side is at least as big as the right-hand side. The sample size $n$ appears in the degrees of freedom for $t$ on the right-hand side, so we don't have a simple formula for $n$. We can compute a reasonable lower bound for $n$ by substituting the upper $\mathcal{E}_I/2$ percent point of a normal for $t^2_{\mathcal{E}_I/2,g(n-1)}$. Then increase $n$ from the lower bound until the criterion is met.

If in doubt, design for largest plausible $MS_E$

Often the best we can do is provide a plausible range of values for $MS_E$. Larger values of $MS_E$ lead to larger sample sizes to meet maximum confidence interval width requires. To play it safe, choose your sample size so that you will meet your goals, even if you encounter the largest plausible $MS_E$.

**Example 7.2** **VOR in ataxia patients, continued**

Example 7.1 gave a requirement that 95% confidence intervals for pairwise differences should be no wider than .5. The preliminary data had an $MS_E$ of .075, so that is a plausible value for future data. The starting approximation is then

$$n \approx \frac{4 \times 4 \times .075 \times (1^2 + (-1)^2)}{.5^2} = 9.6 \quad ,$$

so we round up to 10 and start there. With a sample size of 10, there are 27 degrees of freedom for error, so we now use $t_{.025,27} = 2.052$. Feeding in this sample size, we get

$$n \approx \frac{4 \times 2.052^2 \times .075 \times (1+1)}{.5^2} = 10.1 \quad ,$$

and we round up to 11. There are now 30 degrees of freedom for error, and $t_{.025,30} = 2.042$, and

$$n \approx \frac{4 \times 2.042^2 \times .075 \times (1+1)}{.5^2} = 10.01 \quad ,$$

so $n = 11$ is the required sample size.

Taking a more conservative approach, we might feel that the $MS_E$ in a future experiment could be as large as .15 (we will see in Chapter 11 that this is not unlikely). Repeating our sample size calculation with the new $MS_E$ value we get

$$n \approx \frac{4 \times 4 \times .15 \times (1+1)}{.5^2} = 20.2 ,$$

or 21 for the first approximation. Because $t_{.025,60} = 2.0003$, the first approximation is the correct sample size.

On the other hand, we might be feeling extremely lucky and think that the $MS_E$ will only be .0375 in the experiment. Repeat the calculation again, and we get

$$n \approx \frac{4 \times 4 \times .0375 \times (1+1)}{.5^2} = 4.8 ,$$

or 5 for the first approximation; $t_{.025,12} = 2.18$, so the second guess is

$$n \approx \frac{4 \times 2.18^2 \times .0375 \times (1+1)}{.5^2} = 5.7 ,$$

and $n = 6$ works out to be the required sample size.

Note from the example that doubling the assumed $MS_E$ does not quite double the required sample size. This is because changing the sample size also changes the degrees of freedom and thus the percent point of $t$ that we use. This effect is strongest for small sample sizes.

Sample size affects df and $t$-percent point

## 7.3 Power and Sample Size for ANOVA

The ANOVA F-statistic is the ratio of the mean square for treatments to the mean square for error. When the null hypothesis is true, the F-statistic follows an F-distribution with degrees of freedom from the two mean squares. We reject the null when the observed F-statistic is larger than the upper $\mathcal{E}_I$ percent point of the F-distribution. When the null hypothesis is false, the F-statistic follows a *noncentral F-distribution*. Power, the probability of rejecting the null when the null is false, is the probability that the F-statistic (which follows a noncentral F-distribution when the alternative is true) exceeds a cutoff based on the usual (central) F distribution.

F-statistic follows noncentral F-distribution when null is false

This is illustrated in Figure 7.1. The thin line gives a typical null distribution for the F-test. The vertical line is at the 5% cutoff point; 5% of the area under the null curve is to the right, and 95% is to the left. This 5% is the Type I error rate, or $\mathcal{E}_I$. The thick curve is the distribution of the F-ratio for one alternative. We would reject the null at the 5% level if our F-statistic is greater than the cutoff. The probability of this happening is the area under

Power computed with noncentral F

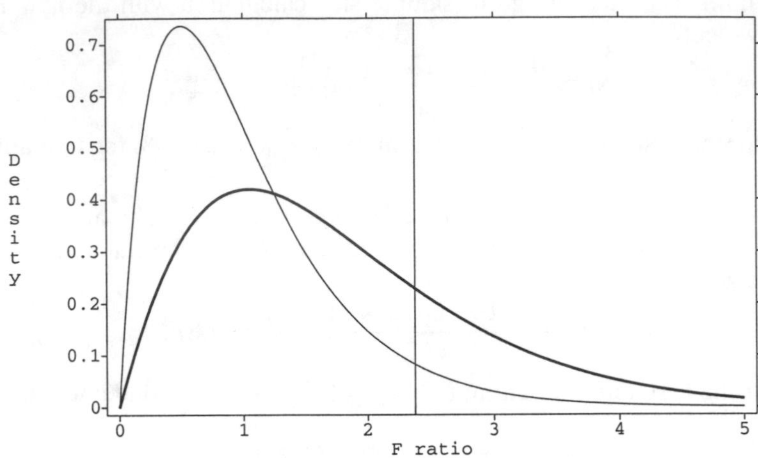


**Figure 7.1:** Null distribution (thin line) and alternative distribution (thick line) for an F test, with the 5% cutoff marked.

the alternative distribution curve to the right of the cutoff (the power); the area under the alternative curve to the left of the cutoff is the Type II error rate $\mathcal{E}_{II}$.

Noncentrality parameter measures distance from null

The noncentral F-distribution has numerator and denominator degrees of freedom the same as the ordinary (central) F, and it also has a *noncentrality parameter* $\zeta$ defined by

$$\zeta = \frac{\sum_i n_i \alpha_i^2}{\sigma^2} \ .$$

The noncentrality parameter measures how far the treatment means are from being equal ($\alpha_i^2$) relative to the variation of $\overline{y}_{i\bullet}$ ($\sigma^2/n_i$). The ordinary central F-distribution has $\zeta = 0$, and the bigger the value of $\zeta$, the more likely we are to reject $H_0$.

We must use the noncentral F-distribution when computing power or $\mathcal{E}_{II}$. This wouldn't be too bad, except that there is a different noncentral F-distribution for every noncentrality parameter. Thus there is a different alternative distribution for each value of the noncentrality parameter, and we will only be able to tabulate power for a selection of parameters.

Power curves

There are two methods available to compute power. The first is to use power tables—figures really—such as Appendix Table D.10, part of which is reproduced here as Figure 7.2. There is a separate figure for each numerator degrees of freedom, with power on the vertical axis and noncentrality param-

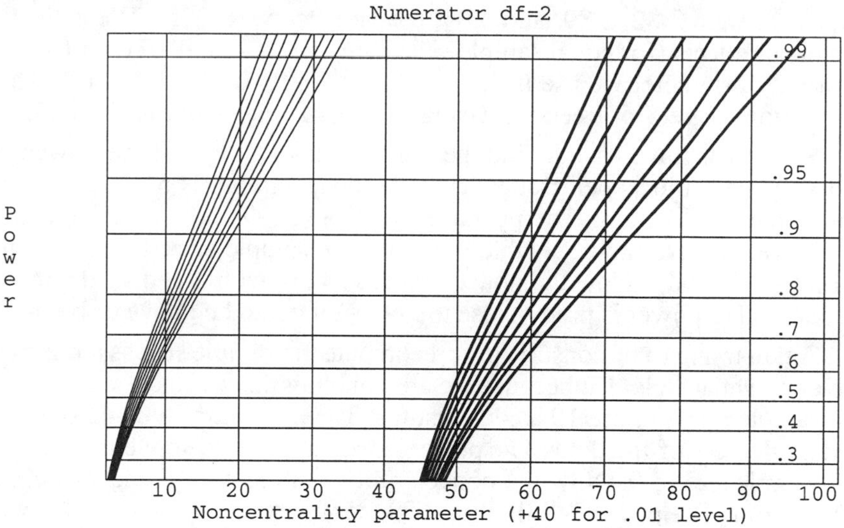


**Figure 7.2:** Sample power curves for 2 numerator degrees of freedom, .05 (thin) and .01 (thick) Type I error rates, and 8, 9, 10, 12, 15, 20, 30, and 60 denominator degrees of freedom (right to left within each group).

eter on the horizontal axis. Within a figure, each curve shows the power for a particular denominator degrees of freedom (8, 9, 10, 12, 15, 20, 30, 60) and Type I error rate (5% or 1%). The power curves for level .01 are shifted to the right by 40 units to prevent overlap with the .05 curves.

Find required sample sizes iteratively

To compute power, you first get the correct figure (according to numerator degrees of freedom); then find the correct horizontal position on the figure (according to the noncentrality parameter, shifted right for .01 tests); then move up to the curve corresponding to the correct denominator degrees of freedom (you may need to interpolate between the values shown); and then read across to get power. Computing minimum sample sizes for a required power is a trial-and-error procedure. We investigate a collection of sample sizes until we find the smallest sample size that yields our required power.

## VOR in ataxia patients, continued

**Example 7.3**

We wish to compute the power for a test of the null hypothesis that the mean VOR of the three SCA's are all equal against the alternative that the means are as observed in the preliminary data, when we have four subjects per group and test at the .01 level. On the log scale, the group means in the preliminary data were 2.82, 3.89, and 3.04; the variance was .075. The estimated treatment effects (for equal sample sizes) are -.43, .64, and -.21, so the noncentrality parameter we use is $4(.43^2 + .64^2 + .21^2)/.075 = 34.06$. There are 2 and 9 degrees of freedom. Using Figure 7.2, the power is about .92.

Suppose that we wish to find the sample size required to have power .99. Let's try six subjects per group. Then the noncentrality is 51.1, with 2 and 15 degrees of freedom. The power is now above .99 and well off the chart in Figure 7.2. We might be able to reduce the sample size, so let's try five subjects per group. Now the noncentrality is 42.6, with 2 and 12 degrees of freedom. The power is pretty close to .99, but it could be above or below.

Again trying to be conservative, recompute the sample size assuming that the error variance is .15; because we are doubling the variance, we'll double the sample size and use 10 as our first try. The noncentrality is 42.6, with 2 and 27 degrees of freedom. The power is well above .99, so we try reducing the sample size to 9. Now the noncentrality is 38.3, with 2 and 24 degrees of freedom. The power is still above .99, so we try sample size 8. Now the noncentrality is 34.06 with 2 and 21 degrees of freedom. It is difficult to tell from the graph, but the power seems to be less than .99; thus 9 is the required sample size.

Power curves are difficult to use

This example illustrates the major problems with using power curves. Often there is not a curve for the denominator degrees of freedom that we need, and even when there is, reading power off the curves is not very accurate. These power curves are usable, but tedious and somewhat crude, and certain to lead to eyestrain and frustration.

Power software

A better way to compute power or sample size is to use computer software designed for that task. Unfortunately, many statistical systems don't provide power or sample size computations. Thomas and Krebs (1997) review power analysis software available in late 1996. As of summer 1999, they also maintain a Web pagelisting power analysis capabilities and sources for extensions for several dozen packages.[1] Minitab and MacAnova can both compute power and minimum sample size for several situations, including ANOVA problems with equal replication. The user interfaces for power software computations differ dramatically; for example, in Minitab one enters the means, and in MacAnova one enters the noncentrality parameter.

**Example 7.4** **VOR in ataxia patients, continued**

Let's redo the power and sample size computations using Minitab. Listing 7.1 shows Minitab output for the first two computations of Example 7.3. First we

[1] `http://sustain.forestry.ubc.ca/cacb/power/review/powrev.html`

**Listing 7.1:** Minitab output for power and sample size computation.

```
Power and Sample Size                        ①

One-way ANOVA

Sigma = 0.2739  Alpha = 0.01  Number of Levels = 3
Corrected Sum of Squares of Means = 0.6386
Means = 2.82, 3.89, 3.04

Sample
  Size   Power
     4  0.9297


Power and Sample Size                        ②

One-way ANOVA

Sigma = 0.2739  Alpha = 0.01  Number of Levels = 3
Corrected Sum of Squares of Means = 0.6386
Means = 2.82, 3.89, 3.04

Sample   Target  Actual
  Size    Power   Power
     5   0.9900  0.9903
```

find the power when we have four subjects per group; this is shown in section ① of the listing. The computed power is almost .93; we read about .92 from the curves. Second, we can find minimum the sample size to get power .99; this is shown in section ② of the listing. The minimum sample size for .99 power is 5, as we had guessed but were not sure about from the tables. The exact power is .9903, so in this case we were actually pretty close using the tables.

Specify minimum difference

Here is a useful trick for choosing sample size. Sometimes it is difficult to specify an interesting alternative completely; that is, we can't specify all the means or effects $\alpha_i$, but we can say that any configuration of means that has two means that differ by an amount D or more would be interesting. The smallest possible value for the noncentrality parameter when this condition is met is $nD^2/(2\sigma^2)$, corresponding to two means D units apart and all the other means in the middle (with zero $\alpha_i$'s). If we design for this alternative, then we will have at least as much power for any other alternative with two treatments D units apart.

## 7.4 Power and Sample Size for a Contrast

The Analysis of Variance F-test is sensitive to all departures from the null hypothesis of equal treatment means. A contrast is sensitive to particular departures from the null. In some situations, we may be particularly interested in one or two contrasts, and less interested in other contrasts. In that case, we might wish to design our experiment so that the contrasts of particular interest had adequate power.

Noncentrality parameter for a contrast

Suppose that we have a contrast with coefficients $\{w_i\}$. Test the null hypothesis that the contrast has expected value zero by using an F-test (the sum of squares for the contrast divided by the $MS_E$). The F-test has 1 and $N - g$ degrees of freedom and noncentrality parameter

$$\frac{(\sum_{i=1}^{g} w_i\alpha_i)^2}{\sigma^2 \sum_{i=1}^{g} w_i^2/n_i} .$$

We now use power curves or software for 1 numerator degree of freedom to compute power.

**Example 7.5**

**VOR in ataxia patients, continued**

Suppose that we are particularly interested in comparing the VOR for SCA 1 to the average VOR for SCA 5 and 6 using a contrast with coefficients (1, -.5, -.5). On the basis of the observed means and $MS_E$ and equal sample sizes, the noncentrality parameter is

$$\frac{(2.82 - .5(3.89 + 3.04))^2}{.075(1/n + .25/n + .25/n)} = 3.698n \ .$$

The noncentrality parameter for $n = 5$ is 18.49; this would have 1 and 12 degrees of freedom. The power from the tables is about .86; the exact power is .867.

## 7.5 More about Units and Measurement Units

Thinking about sample size, cost, and power brings us back to some issues involved in choosing experimental units and measurement units. The basic problems are those of dividing fixed resources (there is never enough money, time, material, etc.) and trying to get the most bang for the buck.

Consider first the situation where there is a fixed amount of experimental material that can be divided into experimental units. In agronomy, the limited resource might be an agricultural field of a fixed size. In textiles, the limited resource might be a bolt of cloth of fixed size. The problem is choosing into how many units the field or bolt should be divided. Larger units have the advantage that their responses tend to have smaller variance, since these responses are computed from more material. Their disadvantage is that you end up with fewer units to average across. Smaller units have the opposite properties; there are more of them, but they have higher variance.

Subdividing spatial units

There is usually some positive spatial association between neighboring areas of experimental material. Because of that, the variance of the average of $k$ adjacent spatial units is greater than the variance of the average of $k$ randomly chosen units. (How much greater is very experiment specific.) This greater variance for contiguous blocks implies that randomizing treatments across more little units will lead to smaller variances for treatment averages and comparisons than using fewer big units.

More little units generally better

There are limits to this splitting, of course. For example, there may be an expensive or time-consuming analytical measurement that must be made on each unit. An upper bound on time or cost thus limits the number of units that can be considered. A second limit comes from edge guard wastage. When units are treated and analyzed *in situ* rather then being physically separated, it is common to exclude from analysis the edge of each unit. This is done because treatments may spill over and have effects on neighboring units; excluding the edge reduces this spillover. The limit arises because as the units become smaller and smaller, more and more of the unit becomes edge, and we eventually we have little analyzable center left.

A second situation occurs when we have experimental units and measurement units. Are we better off taking more measurements on fewer units or fewer measurement on more units? In general, we have more power and shorter confidence intervals if we take fewer measurements on more units. However, this approach may have a higher cost per unit of information.

More units or measurement units?

For example, consider an experiment where we wish to study the possible effects of heated animal pens on winter weight gain. Each animal will be a measurement unit, and each pen is an experimental unit. We have $g$ treatments with $n$ pens per treatment ($N = gn$ total pens) and $r$ animals per pen. The cost of the experiment might well be represented as $C_1 + gnC_2 + gnrC_3$. That is, there is a fixed cost, a cost per pen, and a cost per animal. The cost per pen is no doubt very high. Let $\sigma_1^2$ be the variation from pen to pen, and let $\sigma_2^2$ be the variation from animal to animal. Then the variance of a treatment average is

Costs may vary by unit type

$$\frac{\sigma_1^2}{n} + \frac{\sigma_2^2}{nr}.$$

The question is now, "What values of $n$ and $r$ give us minimal variance of a treatment average for fixed total cost?" We need to know a great deal about the costs and sources of variation before we can complete the exercise.

## 7.6 Allocation of Units for Two Special Cases

We have considered computing power and sample size for balanced allocations of units to treatments. Indeed, Chapter 6 gave some compelling reasons for favoring balanced designs. However, there are some situations where unequal sample sizes could increase the power for alternatives of interest. We examine two of these.

Comparison with control

Suppose that one of the $g$ treatments is a control treatment, say treatment 1, and we are only interested in determining whether the other treatments differ from treatment 1. That is, we wish to compare treatment 2 to control, treatment 3 to control, ..., treatment $g$ to control, but we don't compare noncontrol treatments. This is the standard setup where Dunnett's test is applied. For such an experiment, the control plays a special role (it appears in all contrasts), so it makes sense that we should estimate the control response more precisely by putting more units on the control. In fact, we can show that we should choose group sizes so that the noncontrol treatments sizes ($n_t$) are equal and the control treatment size ($n_c$) is about $n_c = n_t\sqrt{g-1}$.

Allocation for polynomial contrasts

A second special case occurs when the $g$ treatments correspond to numerical levels or doses. For example, the treatments could correspond to four different temperatures of a reaction vessel, and we can view the differences in responses at the four treatments as linear, quadratic, and cubic temperature effects. If one of these effects is of particular interest, we can allocate units to treatments in such a way to make the standard error for that selected effect small.

Suppose that we believe that the temperature effect, if it is nonzero, is essentially linear with only small nonlinearities. Thus we would be most interested in estimating the linear effect and less interested in estimating the quadratic and cubic effects. In such a situation, we could put more units at the lowest and highest temperatures, thereby decreasing the variance for the linear effect contrast. We would still need to keep some observations in the intermediate groups to estimate quadratic and cubic effects, though we wouldn't need as many as in the high and low groups since determining curvature is assumed to be of less importance than determining the presence of a linear effect.

Note that we need to exercise some caution. If our assumptions about shape of the response and importance of different contrasts are incorrect, we could wind up with an experiment that is much less informative than the equal sample size design. For example, suppose we are near the peak of a quadratic response instead of on an essentially linear response. Then the linear contrast (on which we spent all our units to lower its variance) is estimating zero, and the quadratic contrast, which in this case is the one with all the interesting information, has a high variance.

Sample sizes based on incorrect assumptions can lower power

## 7.7 Further Reading and Extensions

When the null hypothesis is true, the treatment and error sums of squares are distributed as $\sigma^2$ times chi-square distributions. Mathematically, the ratio of two independent chi-squares, each divided by their degrees of freedom, has an F-distribution; thus the F-ratio has an F-distribution when the null is true. When the null hypothesis is false, the error sum of squares still has its chi-square distribution, but the treatment sum of squares has a *noncentral chi-square* distribution. Here we briefly describe the noncentral chi-square.

If $Z_1, Z_2, \cdots, Z_n$ are independent normal random variables with mean 0 and variance 1, then $Z_1^2 + Z_2^2 + \cdots + Z_n^2$ (a sum of squares) has a chi-square distribution with $n$ degrees of freedom, denoted by $\chi_n^2$. If the $Z_i$'s have variance $\sigma^2$, then their sum of squares is distributed as $\sigma^2$ times a $\chi_n^2$. Now suppose that the $Z_i$'s are independent with means $\delta_i$ and variance $\sigma^2$. Then the sum of squares $Z_1^2 + Z_2^2 + \cdots + Z_n^2$ has a distribution which is $\sigma^2$ times a *noncentral* chi-square distribution with n degrees of freedom and noncentrality parameter $\sum_{i=1}^n \delta_i^2/\sigma^2$. Let $\chi_n^2(\zeta)$ denote a noncentral chi-square with n degrees of freedom and noncentrality parameter $\zeta$. If the noncentrality parameter is zero, we just have an ordinary chi-square.

In Analysis of Variance, the treatment sum of squares has a distribution that is $\sigma^2$ times a noncentral chi-square distribution with $g-1$ degrees of freedom and noncentrality parameter $\sum_{i=1}^g n_i\alpha_i^2/\sigma^2$. See Appendix A. The mean square for treatments thus has a distribution

$$MS_{\text{trt}} \sim \frac{\sigma^2}{g-1}\chi_{g-1}^2\left(\frac{\sum_{i=1}^g n_i\alpha_i^2}{\sigma^2}\right) .$$

The expected value of a noncentral chi-square is the sum of its degrees of freedom and noncentrality parameter, so the expected value of the mean square for treatments is $\sigma^2 + \sum_{i=1}^g n_i\alpha_i^2/(g-1)$. When the null is false, the F-ratio is a noncentral chi-square divided by a central chi-square (each divided by its degrees of freedom); this is a noncentral F-distribution, with the noncentrality of the F coming from the noncentrality of the numerator chi-square.

## 7.8 Problems

**Exercise 7.1** Find the smallest sample size giving power of at least .7 when testing equality of six groups at the .05 level when $\zeta = 4n$.

**Exercise 7.2** We are planning an experiment comparing three fertilizers. We will have six experimental units per fertilizer and will do our test at the 5% level. One of the fertilizers is the standard and the other two are new; the standard fertilizer has an average yield of 10, and we would like to be able to detect the situation when the new fertilizers have average yield 11 each. We expect the error variance to be about 4. What sample size would we need if we want power .9?

**Exercise 7.3** What is the probability of rejecting the null hypothesis when there are four groups, the sum of the squared treatment effects is 6, the error variance is 3, the group sample sizes are 4, and $\mathcal{E}$ is .01?

**Exercise 7.4** I conduct an experiment doing fixed-level testing with $\mathcal{E}$ = .05; I know that for a given set of alternatives my power will be .85. True or False?

1. The probability of rejecting the null hypothesis when the null hypothesis is false is .15.
2. The probability of failing to reject the null hypothesis when the null hypothesis is true is .05.

**Exercise 7.5** We are planning an experiment on the quality of video tape and have purchased 24 tapes, four tapes from each of six types. The six types of tape were 1) brand A high cost, 2) brand A low cost, 3) brand B high cost, 4) brand B low cost, 5) brand C high cost, 6) brand D high cost. Each tape will be recorded with a series of standard test patterns, replayed 10 times, and then replayed an eleventh time into a device that measures the distortion on the tape. The distortion measure is the response, and the tapes will be recorded and replayed in random order. Previous similar tests had an error variance of about .25.

a) What is the power when testing at the .01 level if the high cost tapes have an average one unit different from the low cost tapes?

b) How large should the sample size have been to have a 95% brand A versus brand B confidence interval of no wider than 2?

**Problem 7.1** We are interested in the effects of soy additives to diets on the blood concentration of estradiol in premenopausal women. We have historical data on six subjects, each of whose estradiol concentration was measured at the same stage of the menstrual cycle over two consecutive cycles. On the log scale, the error variance is about .109. In our experiment, we will have a pretreatment measurement, followed by a treatment, followed by a posttreatment measurement. Our response is the difference (post − pre), so the variance of our response should be about .218. Half the women will receive the soy treatment, and the other half will receive a control treatment.

How large should the sample size be if we want power .9 when testing at the .05 level for the alternative that the soy treatment raises the estradiol concentration 25% (about .22 log units)?

**Problem 7.2**

Nondigestible carbohydrates can be used in diet foods, but they may have effects on colonic hydrogen production in humans. We want to test to see if inulin, fructooligosaccharide, and lactulose are equivalent in their hydrogen production. Preliminary data suggest that the treatment means could be about 45, 32, and 60 respectively, with the error variance conservatively estimated at 35. How many subjects do we need to have power .95 for this situation when testing at the $\mathcal{E}_I = .01$ level?

**Problem 7.3**

Consider the situation of Exercise 3.5. The data we have appear to depend linearly on delay with no quadratic component. Suppose that the true expected value for the contrast with coefficients (1,-2,1) is 1 (representing a slight amount of curvature) and that the error variance is 60. What sample size would be needed to have power .9 when testing at the .01 level?

# Chapter 8

# Factorial Treatment Structure

We have been working with completely randomized designs, where $g$ treatments are assigned at random to $N$ units. Up till now, the treatments have had no structure; they were just $g$ treatments. *Factorial treatment structure* exists when the $g$ treatments are the combinations of the levels of two or more factors. We call these combination treatments *factor-level combinations* or *factorial combinations* to emphasize that each treatment is a combination of one level of each of the factors. We have not changed the randomization; we still have a completely randomized design. It is just that now we are considering treatments that have a factorial structure. We will learn that there are compelling reasons for preferring a factorial experiment to a sequence of experiments investigating the factors separately.

Factorials combine the levels of two or more factors to create treatments

## 8.1 Factorial Structure

It is best to start with some examples of factorial treatment structure. Lynch and Strain (1990) performed an experiment with six treatments studying how milk-based diets and copper supplements affect trace element levels in rat livers. The six treatments were the combinations of three milk-based diets (skim milk protein, whey, or casein) and two copper supplements (low and high levels). Whey itself was not a treatment, and low copper was not a treatment, but a low copper/whey diet was a treatment. Nelson, Kriby, and Johnson (1990) studied the effects of six dietary supplements on the occurrence of leg abnormalities in young chickens. The six treatments were the combinations of two levels of phosphorus supplement and three levels of calcium supplement. Finally, Hunt and Larson (1990) studied the effects of

Table 8.1: Barley sprouting data.

| ml $H_2O$ | Age of Seeds (weeks) 1 | 3 | 6 | 9 | 12 |
|---|---|---|---|---|---|
| 4 | 11 | 7 | 9 | 13 | 20 |
| | 9 | 16 | 19 | 35 | 37 |
| | 6 | 17 | 35 | 28 | 45 |
| 8 | 8 | 1 | 5 | 1 | 11 |
| | 3 | 7 | 9 | 10 | 15 |
| | 3 | 3 | 9 | 9 | 25 |

sixteen treatments on zinc retention in the bodies of rats. The treatments were the combinations of two levels of zinc in the usual diet, two levels of zinc in the final meal, and four levels of protein in the final meal. Again, it is the combination of factor levels that makes a factorial treatment.

Two-factor designs

We begin our study of factorial treatment structure by looking at two-factor designs. We may present the responses of a two-way factorial as a table with rows corresponding to the levels of one factor (which we call factor A) and columns corresponding to the levels of the second factor (factor B). For example, Table 8.1 shows the results of an experiment on sprouting barley (these data reappear in Problem 8.1). Barley seeds are divided into 30 lots of 100 seeds each. The 30 lots are divided at random into ten groups of three lots each, with each group receiving a different treatment. The ten treatments are the factorial combinations of amount of water used for sprouting (factor A) with two levels, and age of the seeds (factor B) with five levels. The response measured is the number of seeds sprouting.

Multiple subscripts denote factor levels and replication

We use the notation $y_{ijk}$ to indicate responses in the two-way factorial. In this notation, $y_{ijk}$ is the $k$th response in the treatment formed from the $i$th level of factor A and the $j$th level of factor B. Thus in Table 8.1, $y_{2,5,3} = 25$. For a four by three factorial design (factor A has four levels, factor B has three levels), we could tabulate the responses as in Table 8.2. This table is just a convenient representation that emphasizes the factorial structure; treatments were still assigned to units at random.

Balanced data have equal replication

Notice in both Tables 8.1 and 8.2 that we have the same number of responses in every factor-level combination. This is called *balance*. Balance turns out to be important for the standard analysis of factorial responses. We will assume for now that our data are balanced with $n$ responses in every factor-level combination. Chapter 10 will consider analysis of unbalanced factorials.

**Table 8.2:** A two-way factorial treatment structure.

| | B1 | B2 | B3 |
|---|---|---|---|
| A1 | $y_{111}$ ⋮ $y_{11n}$ | $y_{121}$ ⋮ $y_{12n}$ | $y_{131}$ ⋮ $y_{13n}$ |
| A2 | $y_{211}$ ⋮ $y_{21n}$ | $y_{221}$ ⋮ $y_{22n}$ | $y_{231}$ ⋮ $y_{23n}$ |
| A3 | $y_{311}$ ⋮ $y_{31n}$ | $y_{321}$ ⋮ $y_{32n}$ | $y_{331}$ ⋮ $y_{33n}$ |
| A4 | $y_{411}$ ⋮ $y_{41n}$ | $y_{421}$ ⋮ $y_{42n}$ | $y_{431}$ ⋮ $y_{43n}$ |

## 8.2 Factorial Analysis: Main Effect and Interaction

When our treatments have a factorial structure, we may also use a factorial analysis of the data. The major concepts of this factorial analysis are main effect and interaction.

Consider a two-way factorial where factor A has four levels and factor B has three levels, as in Table 8.2. There are $g = 12$ treatments, with 11 degrees of freedom between the treatments. We use $i$ and $j$ to index the levels of factors A and B. The expected values in the twelve treatments may be denoted $\mu_{ij}$, coefficients for a contrast in the twelve means may be denoted $w_{ij}$ (where as usual $\sum_{ij} w_{ij} = 0$), and the contrast sum is $\sum_{ij} w_{ij}\mu_{ij}$. Similarly, $\overline{y}_{ij\bullet}$ is the observed mean in the $ij$ treatment group, and $\overline{y}_{i\bullet\bullet}$ and $\overline{y}_{\bullet j\bullet}$ are the observed means for all responses having level $i$ of factor A or level $j$ of B, respectively. It is often convenient to visualize the expected values, means, and contrast coefficients in matrix form, as in Table 8.3.

Treatment, row, and column means

For the moment, forget about factor B and consider the experiment to be a completely randomized design just in factor A (it *is* completely randomized in factor A). Analyzing this design with four "treatments," we may compute a sum of squares with 3 degrees of freedom. The variation summarized by this sum of squares is denoted $SS_A$ and depends on just the level of factor A. The expected value for the mean of the responses in row $i$ is $\mu + \alpha_i$, where we assume that $\sum_i \alpha_i = 0$.

Factor A ignoring factor B

**Table 8.3:** Matrix arrangement of (a) expected values, (b) means, and (c) contrast coefficients in a four by three factorial.

(a)

| | | |
|---|---|---|
| $\mu_{11}$ | $\mu_{12}$ | $\mu_{13}$ |
| $\mu_{21}$ | $\mu_{22}$ | $\mu_{23}$ |
| $\mu_{31}$ | $\mu_{32}$ | $\mu_{33}$ |
| $\mu_{41}$ | $\mu_{42}$ | $\mu_{43}$ |

(b)

| | | |
|---|---|---|
| $\overline{y}_{11\bullet}$ | $\overline{y}_{12\bullet}$ | $\overline{y}_{13\bullet}$ |
| $\overline{y}_{21\bullet}$ | $\overline{y}_{22\bullet}$ | $\overline{y}_{23\bullet}$ |
| $\overline{y}_{31\bullet}$ | $\overline{y}_{32\bullet}$ | $\overline{y}_{33\bullet}$ |
| $\overline{y}_{41\bullet}$ | $\overline{y}_{42\bullet}$ | $\overline{y}_{43\bullet}$ |

(c)

| | | |
|---|---|---|
| $w_{11}$ | $w_{12}$ | $w_{13}$ |
| $w_{21}$ | $w_{22}$ | $w_{23}$ |
| $w_{31}$ | $w_{32}$ | $w_{33}$ |
| $w_{41}$ | $w_{42}$ | $w_{43}$ |

Factor B ignoring factor A

Now, reverse the roles of A and B. Ignore factor A and consider the experiment to be a completely randomized design in factor B. We have an experiment with three "treatments" and treatment sum of squares $SS_B$ with 2 degrees of freedom. The expected value for the mean of the responses in column $j$ is $\mu + \beta_j$, where we assume that $\sum_j \beta_j = 0$.

A main effect describes variation due to a single factor

The effects $\alpha_i$ and $\beta_j$ are called the *main effects* of factors A and B, respectively. The main effect of factor A describes variation due solely to the level of factor A (row of the response matrix), and the main effect of factor B describes variation due solely to the level of factor B (column of the response matrix). We have analogously that $SS_A$ and $SS_B$ are main-effects sums of squares.

Interaction is variation not described by main effects

The variation described by the main effects is variation that occurs from row to row or column to column of the data matrix. The example has twelve treatments and 11 degrees of freedom between treatments. We have described 5 degrees of freedom using the A and B main effects, so there must be 6 more degrees of freedom left to model. These 6 remaining degrees of freedom describe variation that arises from changing rows and columns simultaneously. We call such variation *interaction* between factors A and B, or between the rows and columns, and denote it by $SS_{AB}$.

Here is another way to think about main effect and interaction. The main effect of rows tells us how the response changes when we move from one row to another, averaged across all columns. The main effect of columns tells us how the response changes when we move from one column to another, averaged across all rows. The interaction tells us how the change in response depends on columns when moving between rows, or how the change in response depends on rows when moving between columns. Interaction between factors A and B means that the change in mean response going from level $i_1$ of factor A to level $i_2$ of factor A depends on the level of factor B under consideration. We can't simply say that changing the level of factor A changes the response by a given amount; we may need a different amount of change for each level of factor B.

**Table 8.4:** Sample main-effects and interaction contrast coefficients for a four by three factorial design.

A

| | | |
|---|---|---|
| -3 | -3 | -3 |
| -1 | -1 | -1 |
| 1 | 1 | 1 |
| 3 | 3 | 3 |

| | | |
|---|---|---|
| 1 | 1 | 1 |
| -1 | -1 | -1 |
| -1 | -1 | -1 |
| 1 | 1 | 1 |

| | | |
|---|---|---|
| -1 | -1 | -1 |
| 3 | 3 | 3 |
| -3 | -3 | -3 |
| 1 | 1 | 1 |

B

| | | |
|---|---|---|
| -1 | 0 | 1 |
| -1 | 0 | 1 |
| -1 | 0 | 1 |
| -1 | 0 | 1 |

| | | |
|---|---|---|
| 1 | -2 | 1 |
| 1 | -2 | 1 |
| 1 | -2 | 1 |
| 1 | -2 | 1 |

AB

| | | |
|---|---|---|
| 3 | 0 | -3 |
| 1 | 0 | -1 |
| -1 | 0 | 1 |
| -3 | 0 | 3 |

| | | |
|---|---|---|
| -1 | 0 | 1 |
| 1 | 0 | -1 |
| 1 | 0 | -1 |
| -1 | 0 | 1 |

| | | |
|---|---|---|
| 1 | 0 | -1 |
| -3 | 0 | 3 |
| 3 | 0 | -3 |
| -1 | 0 | 1 |

| | | |
|---|---|---|
| -3 | 6 | -3 |
| -1 | 2 | -1 |
| 1 | -2 | 1 |
| 3 | -6 | 3 |

| | | |
|---|---|---|
| 1 | -2 | 1 |
| -1 | 2 | -1 |
| -1 | 2 | -1 |
| 1 | -2 | 1 |

| | | |
|---|---|---|
| -1 | 2 | -1 |
| 3 | -6 | 3 |
| -3 | 6 | -3 |
| 1 | -2 | 1 |

We can make our description of main-effect and interaction variation more precise by using contrasts. Any contrast in factor A (ignoring B) has four coefficients $w_i^\star$ and observed value $w^\star(\{\overline{y}_{i\bullet\bullet}\})$. This is a contrast in the four row means. We can make an equivalent contrast in the twelve treatment means by using the coefficients $w_{ij} = w_i^\star/3$. This contrast just repeats $w_i^\star$ across each row and then divides by the number of columns to match up with the division used when computing row means. Factor A has four levels, so three orthogonal contrasts partition $SS_A$. There are three analogous orthogonal $w_{ij}$ contrasts that partition the same variation. (See Question 8.1.) Table 8.4 shows one set of three orthogonal contrasts describing the factor A variation; many other sets would do as well.

Main-effects contrasts

The variation in $SS_B$ can be described by two orthogonal contrasts between the three levels of factor B. Equivalently, we can describe $SS_B$ with orthogonal contrasts in the twelve treatment means, using a matrix of contrast coefficients that is constant on columns (that is, $w_{1j} = w_{2j} = w_{3j} = w_{4j}$ for all columns $j$). Table 8.4 also shows one set of orthogonal contrasts for factor B.

A contrasts orthogonal to B contrasts for balanced data

Inspection of Table 8.4 shows that not only are the factor A contrasts orthogonal to each other, and the factor B contrasts orthogonal to each other, but the factor A contrasts are also orthogonal to the factor B contrasts. This orthogonality depends on balanced data and is the key reason why balanced data are easier to analyze.

Interaction contrasts

There are 11 degrees of freedom between the twelve treatments, and the A and B contrasts describe 5 of those 11 degrees of freedom. The 6 additional degrees of freedom are interaction degrees of freedom; sample interaction contrasts are also shown in Table 8.4. Again, inspection shows that the interaction contrasts are orthogonal to both sets of main-effects contrasts. Thus the 11 degrees of freedom between-treatment sum of squares can be partitioned using contrasts into $SS_A$, $SS_B$, and $SS_{AB}$.

Contrast coefficients satisfy zero-sum restrictions

Look once again at the form of the contrast coefficients in Table 8.4. Row-main-effects contrast coefficients are constant along each row, and add to zero down each column. Column-main-effects contrasts are constant down each column and add to zero along each row. Interaction contrasts add to zero down columns and along rows. This pattern of zero sums will occur again when we look at parameters in factorial models.

## 8.3 Advantages of Factorials

Factorial structure versus analysis

Before discussing advantages, let us first recall the difference between factorial treatment structure and factorial analysis. Factorial analysis is an option we have when the treatments have factorial structure; we can always ignore main effects and interaction and just analyze the $g$ treatment groups.

One-at-a-time designs

It is easiest to see the advantages of factorial treatment structure by comparing it to a design wherein we only vary the levels of a single factor. This second design is sometimes referred to as "one-at-a-time." The sprouting data in Table 8.1 were from a factorial experiment where the levels of sprouting water and seed age were varied. We might instead use two one-at-a-time designs. In the first, we fix the sprouting water at the lower level and vary the seed age across the five levels. In the second experiment, we fix the seed age at the middle level, and vary the sprouting water across two levels.

Factorial treatment structure has two advantages:

1. When the factors interact, factorial experiments can estimate the interaction. One-at-at-time experiments cannot estimate interaction. Use of one-at-a-time experiments in the presence of interaction can lead to serious misunderstanding of how the response varies as a function of the factors.

2. When the factors do not interact, factorial experiments are more efficient than one-at-a-time experiments, in that the units can be used to assess the (main) effects for both factors. Units in a one-at-a-time experiment can only be used to assess the effects of one factor.

There are thus two times when you should use factorial treatment structure—when your factors interact, and when your factors do not interact. Factorial structure is a win, whether or not we have interaction.

Use factorials!

The argument for factorial analysis is somewhat less compelling. We usually wish to have a model for the data that is as simple as possible. When there is no interaction, then main effects alone are sufficient to describe the means of the responses. Such a model (or data) is said to be *additive*. An additive model is simpler (in particular, uses fewer degrees of freedom) than a model with a mean for every treatment. When interaction is moderate compared to main effects, the factorial analysis is still useful. However, in some experiments the interactions are so large that the idea of main effects as the primary actors and interaction as fine tuning becomes untenable. For such experiments it may be better to revert to an analysis of $g$ treatment groups, ignoring factorial structure.

Additive model has only main effects

**Example 8.1**

### Pure interactive response

Consider a chemistry experiment involving two catalysts where, unknown to us, both catalysts must be present for the reaction to proceed. The response is one or zero depending on whether or not the reaction occurs. The four treatments are the factorial combinations of Catalyst A present or absent, and Catalyst B present or absent. We will have a response of one for the combination of both catalysts, but the other three responses will be zero. While it is possible to break this down as main effect and interaction, it is clearly more comprehensible to say that the response is one when both catalysts are present and zero otherwise. Note here that the factorial treatment structure was still a good idea, just not the main-effects/interactions analysis.

## 8.4 Visualizing Interaction

An *interaction plot*, also called a *profile plot*, is a graphic for assessing the relative size of main effects and interaction; an example is shown in Figure 8.1. Consider first a two-factor factorial design. We construct an interaction plot in a "connect-the-dots" fashion. Choose a factor, say A, to put on the horizontal axis. For each factor level combination, plot the pair $(i, \overline{y}_{ij\bullet})$. Then "connect-the-dots" corresponding to the points with the same level of factor

Interaction plots connect-the-dots between treatment means

**Table 8.5:** Iron levels in liver tissue, mg/g dry weight.

| Diet | Control | Cu deficient |
|---|---|---|
| Skim milk protein | .70 | 1.28 |
| Whey | .93 | 1.87 |
| Casein | 2.11 | 2.53 |

B; that is, connect $(1, \overline{y}_{1j\bullet})$, $(2, \overline{y}_{2j\bullet})$, up to $(a, \overline{y}_{aj\bullet})$. In our four by three prototype factorial, the level of factor A will be a number between one and four; there will be three points plotted above one, three points plotted above two, and so on; and there will be three "connect-the-dots" lines, one for each level of factor B.

Interaction plot shows relative size of main effects and interaction

For additive data, the change in response moving between levels of factor A does not depend on the level of factor B. In an interaction plot, that similarity in change of level shows up as parallel line segments. Thus interaction is small compared to the main effects when the connect-the-dots lines are parallel, or nearly so. Even with visible interaction, the degree of interaction may be sufficiently small that the main-effects-plus-interaction description is still useful. It is worth noting that we sometimes get visually different impressions of the interaction by reversing the roles of factors A and B.

**Example 8.2**

**Rat liver iron**

Table 8.5 gives the treatment means for liver tissue iron in the Lynch and Strain (1990) experiment. Figure 8.1 shows an interaction plot with milk diet factor on the horizontal axis and the copper treatments indicated by different lines. The lines seem fairly parallel, indicating little interaction.

Interpret "parallel" in light of variability

Figure 8.1 points out a deficiency in the interaction plot as we have defined it. The observed means that we plot are subject to error, so the line segments will not be exactly parallel—even if the true means are additive. The degree to which the lines are not parallel must be interpreted in light of the likely size of the variation in the observed means. As the data become more variable, greater departures from parallel line segments become more likely, even for truly additive data.

**Example 8.3**

**Rat liver iron, continued**

The line segments are fairly parallel, so there is not much evidence of interaction, though it appears that the effect of copper may be somewhat larger for milk diet 2. The mean square for error in the Lynch and Strain experiment was approximately .26, and each treatment had replication $n = 5$. Thus the standard errors of a treatment mean, the difference of two treatment means,

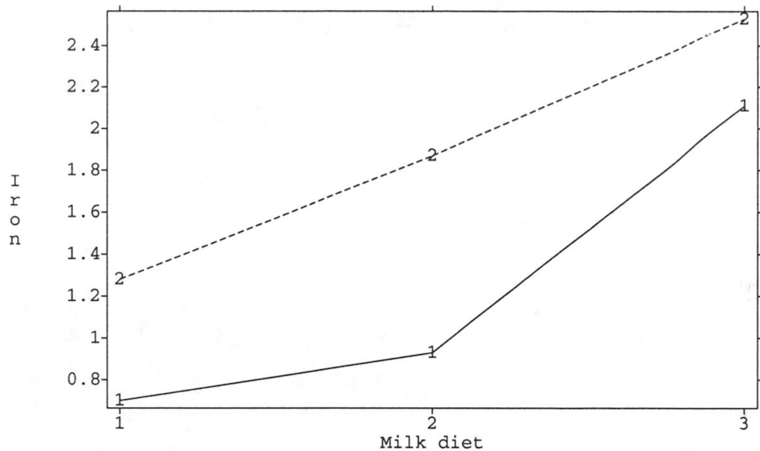


**Figure 8.1:** Interaction plot of liver iron data with diet factor on the horizontal axis, using MacAnova.

and the difference of two such differences are about .23, .32, and .46 respectively. The slope of a line segment in the interaction plot is the difference of two treatment means. The slopes from milk diet 1 to 2 are .23 and .59, and the slopes from milk diets 2 to 3 are 1.18 and .66; each of these slopes was calculated as the difference of two treatment means. The differences of the slopes (which have standard error .46 because they are differences of differences of means) are .36 and .48. Neither of these differences is large compared to its standard error, so there is still no evidence for interaction.

We finish this section with interaction plots for the other two nutrition experiments described in the first section.

### Chick body weights

**Example 8.4**

Figure 8.2 is an interaction plot of the chick body weights from the Nelson, Kriby, and Johnson (1990) data with the calcium factor on the horizontal axis and a separate line for each level of phosphorus. Here, interaction is clear. At the upper level of phosphorus, chick weight does not depend on calcium. At the lower level of phosphorus, weight decreases with increasing calcium. Thus the effect of changing calcium levels depends on the level of phosphorus.

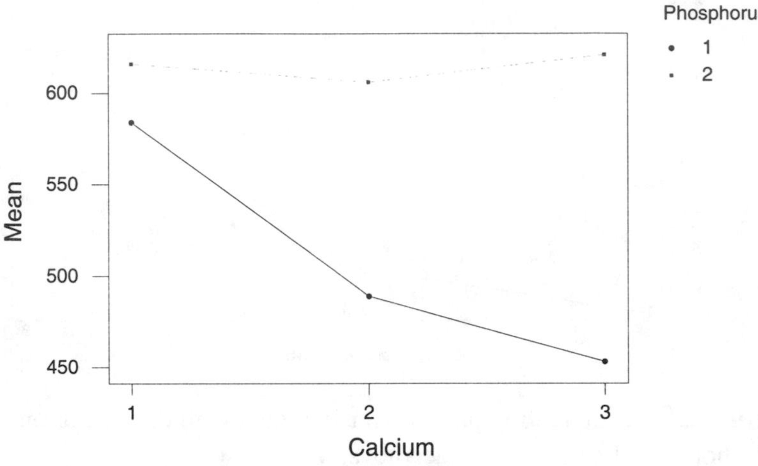


**Figure 8.2:** Interaction plot of chick body weights data with calcium on the horizontal axis, using Minitab.

**Example 8.5**

**Zinc retention**

Finally, let's look at the zinc retention data of Hunt and Larson (1990). This is a three-factor factorial design (four by two by two), so we need to modify our approach a bit. Figure 8.3 is an interaction plot of percent zinc retention with final meal protein on the horizontal axis. The other four factor-level combinations are coded 1 (low meal zinc, low diet zinc), 2 (low meal zinc, high diet zinc), 3 (high meal zinc, low diet zinc), and 4 (high meal zinc, high diet zinc). Lines 1 and 2 are low meal zinc, and lines 3 and 4 are high meal zinc. The 1,2 pattern across protein is rather different from the 3,4 pattern across protein, so we conclude that meal zinc and meal protein interact.

On the other hand, the 1,3 pair of lines (low diet zinc) has the same basic pattern as the 2,4 pair of lines (high diet zinc), so the average of the 1,3 lines should look like the average of the 2,4 lines. This means that diet zinc and meal protein appear to be additive.

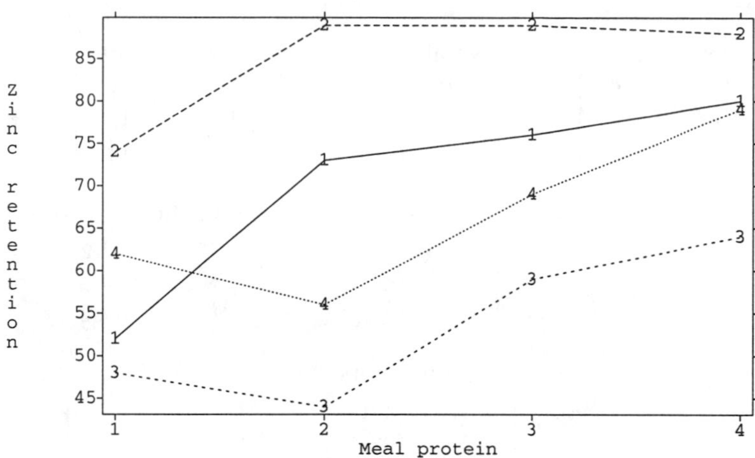


**Figure 8.3:** Interaction plot of percent zinc retention data with meal protein on the horizontal axis, using MacAnova.

## 8.5 Models with Parameters

Let us now look at the factorial analysis model for a two-way factorial treatment structure. Factor A has $a$ levels, factor B has $b$ levels, and there are $n$ experimental units assigned to each factor-level combination. The $k$th response at the $i$th level of A and $j$th level of B is $y_{ijk}$. The model is

A has $a$ levels, B has $b$ levels, $n$ replications

$$y_{ijk} = \mu + \alpha_i + \beta_j + \alpha\beta_{ij} + \epsilon_{ijk} ,$$

where $i$ runs from 1 to $a$, $j$ runs from 1 to $b$, $k$ runs from 1 to $n$, and the $\epsilon_{ijk}$'s are independent and normally distributed with mean zero and variance $\sigma^2$. The $\alpha_i$, $\beta_j$, and $\alpha\beta_{ij}$ parameters in this model are fixed, unknown constants. There is a total of $N = nab$ experimental units.

Factorial model

Another way of viewing the model is that the table of responses is broken down into a set of tables which, when summed element by element, give the response. Display 8.1 is an example of this breakdown for a three by two factorial with $n = 1$.

The term $\mu$ is called the overall mean; it is the expected value for the responses averaged across all treatments. The term $\alpha_i$ is called the main effect of A at level $i$. It is the average effect (averaged over levels of B) for level $i$ of factor A. Since the average of all the row averages must be the overall average, these row effects $\alpha_i$ must sum to zero. The same is true for

Main effects

$$
\begin{array}{c} \text{responses} \\ \begin{bmatrix} y_{111} & y_{121} \\ y_{211} & y_{221} \\ y_{311} & y_{321} \end{bmatrix} \end{array} =
\begin{array}{c} \text{overall mean} \\ \begin{bmatrix} \mu & \mu \\ \mu & \mu \\ \mu & \mu \end{bmatrix} \end{array} +
\begin{array}{c} \text{row effects} \\ \begin{bmatrix} \alpha_1 & \alpha_1 \\ \alpha_2 & \alpha_2 \\ \alpha_3 & \alpha_3 \end{bmatrix} \end{array} +
$$

$$
\begin{array}{c} \text{column effects} \\ \begin{bmatrix} \beta_1 & \beta_2 \\ \beta_1 & \beta_2 \\ \beta_1 & \beta_2 \end{bmatrix} \end{array} +
\begin{array}{c} \text{interaction effects} \\ \begin{bmatrix} \alpha\beta_{11} & \alpha\beta_{12} \\ \alpha\beta_{21} & \alpha\beta_{22} \\ \alpha\beta_{31} & \alpha\beta_{32} \end{bmatrix} \end{array} +
$$

$$
\begin{array}{c} \text{random errors} \\ \begin{bmatrix} \epsilon_{111} & \epsilon_{121} \\ \epsilon_{211} & \epsilon_{221} \\ \epsilon_{311} & \epsilon_{321} \end{bmatrix} \end{array}
$$

**Display 8.1:** Breakdown of a three by two table into factorial effects.

Interaction effects

$\beta_j$, which is the main effect of factor B at level $j$. The term $\alpha\beta_{ij}$ is called the interaction effect of A and B in the $ij$ treatment. Do not confuse $\alpha\beta_{ij}$ with the product of $\alpha_i$ and $\beta_j$; they are different ideas. The interaction effect is a measure of how far the treatment means differ from additivity. Because the average effect in the $i$th row must be $\alpha_i$, the sum of the interaction effects in the $i$th row must be zero. Similarly, the sum of the interaction effects in the $j$th column must be zero.

The expected value of the response for treatment $ij$ is

$$E\, y_{ijk} = \mu + \alpha_i + \beta_j + \alpha\beta_{ij} \ .$$

Expected value

There are $ab$ different treatment means, but we have $1 + a + b + ab$ parameters, so we have vastly overparameterized. Recall that in Chapter 3 we had to choose a set of restrictions to make treatment effects well defined; we must again choose some restrictions for factorial models. We will use the following set of restrictions on the parameters:

Zero-sum restrictions on parameters

$$0 = \sum_{i=1}^{a} \alpha_i = \sum_{j=1}^{b} \beta_j = \sum_{i=1}^{a} \alpha\beta_{ij} = \sum_{j=1}^{b} \alpha\beta_{ij} \ .$$

This set of restrictions is standard and matches the description of the parameters in the preceding paragraph. The $\alpha_i$ values must sum to 0, so at most

$$\begin{aligned}
\widehat{\mu} &= \overline{y}_{\bullet\bullet\bullet} \\
\widehat{\alpha}_i &= \overline{y}_{i\bullet\bullet} - \widehat{\mu} = \overline{y}_{i\bullet\bullet} - \overline{y}_{\bullet\bullet\bullet} \\
\widehat{\beta}_j &= \overline{y}_{\bullet j\bullet} - \widehat{\mu} = \overline{y}_{\bullet j\bullet} - \overline{y}_{\bullet\bullet\bullet} \\
\widehat{\alpha\beta}_{ij} &= \overline{y}_{ij\bullet} - \widehat{\mu} - \widehat{\alpha}_i - \widehat{\beta}_j \\
&= \overline{y}_{ij\bullet} - \overline{y}_{i\bullet\bullet} - \overline{y}_{\bullet j\bullet} + \overline{y}_{\bullet\bullet\bullet}
\end{aligned}$$

**Display 8.2:** Estimators for main effects and interactions in a two-way factorial.

$a - 1$ of them can vary freely; there are $a - 1$ degrees of freedom for factor A. Similarly, the $\beta_j$ values must sum to 0, so at most $b - 1$ of them can vary freely, giving $b - 1$ degrees of freedom for factor B. For the interaction, we have $ab$ effects, but they must add to 0 when summed over $i$ or $j$. We can show that this leads to $(a-1)(b-1)$ degrees of freedom for the interaction. Note that the parameters obey the same restrictions as the corresponding contrasts: main-effects contrasts and effects add to zero across the subscript, and interaction contrasts and effects add to zero across rows or columns.

Main-effect and interaction degrees of freedom

When we add the degrees of freedom for A, B, and AB, we get $a - 1 + b - 1 + (a-1)(b-1) = ab - 1 = g - 1$. That is, the $ab - 1$ degrees of freedom between the means of the $ab$ factor level combinations have been partitioned into three sets: A, B, and the AB interaction. Within each factor-level combination there are $n - 1$ degrees of freedom about the treatment mean. The error degrees of freedom are $N - g = N - ab = (n-1)ab$, exactly as we would get ignoring factorial structure.

Main effects and interactions partition between treatments variability

The Lynch and Strain data had a three by two factorial structure with $n = 5$. Thus there are 2 degrees of freedom for factor A, 1 degree of freedom for factor B, 2 degrees of freedom for the AB interaction, and 24 degrees of freedom for error.

Display 8.2 gives the formulae for estimating the effects in a two-way factorial. Estimate $\mu$ by the mean of all the data $\overline{y}_{\bullet\bullet\bullet}$. Estimate $\mu + \alpha_i$ by the mean of all responses that had treatment A at level $i$, $\overline{y}_{i\bullet\bullet}$. To get an estimate of $\alpha_i$ itself, subtract our estimate of $\mu$ from our estimate of $\mu + \alpha_i$. Do similarly for factor B, using $\overline{y}_{\bullet j\bullet}$ as an estimate of $\mu + \beta_j$. We can extend this basic idea to estimate the interaction terms $\alpha\beta_{ij}$. The expected value in treatment $ij$ is $\mu+\alpha_i+\beta_j+\alpha\beta_{ij}$, which we can estimate by $\overline{y}_{ij\bullet}$, the observed treatment mean. To get an estimate of $\alpha\beta_{ij}$, simply subtract the estimates of

Estimating factorial effects

**Table 8.6:** Total free amino acids in cheddar cheese after 56 days of ripening.

| Control | R50#10 | R21#2 | blend |
|---|---|---|---|
| 1.697 | 2.032 | 2.211 | 2.091 |
| 1.601 | 2.017 | 1.673 | 2.255 |
| 1.830 | 2.409 | 1.973 | 2.987 |

the lower order parameters (parameters that contain no additional subscripts beyond those found in this term) from the estimate of the treatment mean.

We examine the estimated effects to determine which treatment levels lead to large or small responses, and where factors interact (that is, which combinations of levels have large interaction effects).

**Example 8.6**

**Nonstarter bacteria in cheddar cheese**

Cheese is made by bacterial fermentation of Pasteurized milk. Most of the bacteria are purposefully added; these are the starter cultures. Some "wild" bacteria are also present in cheese; these are nonstarter bacteria. This experiment explores how intentionally-added nonstarter bacteria affect cheese quality. We use two strains of nonstarter bacteria: R50#10 and R21#2. Our four treatments will be control, addition of R50, addition of R21, and addition of a blend of R50 and R21. Twelve cheeses are made, three for each of the four treatments, with the treatments being randomized to the cheeses. After 56 days of ripening, each cheese is measured for total free amino acids (a measure of bacterial activity related to cheese quality). Responses are given in Table 8.6 (data from Peggy Swearingen).

Let's estimate the effects in these data. The four treatment means are

$$\begin{aligned}
\overline{y}_{11\bullet} &= (1.697 + 1.601 + 1.830)/3 = 1.709 \ \text{Control} \\
\overline{y}_{21\bullet} &= (2.032 + 2.017 + 2.409)/3 = 2.153 \ \text{R50} \\
\overline{y}_{12\bullet} &= (2.211 + 1.673 + 1.973)/3 = 1.952 \ \text{R21} \\
\overline{y}_{22\bullet} &= (2.091 + 2.255 + 2.987)/3 = 2.444 \ \text{Blend.}
\end{aligned}$$

The grand mean is the total of all the data divided by 12,

$$\overline{y}_{\bullet\bullet\bullet} = 24.776/12 = 2.065 \ ;$$

the R50 (row or first factor) means are

$$\begin{aligned}
\overline{y}_{1\bullet\bullet} &= (1.709 + 1.952)/2 = 1.831 \\
\overline{y}_{2\bullet\bullet} &= (2.153 + 2.444)/2 = 2.299 \ ;
\end{aligned}$$

and the R21 (column or second factor) means are

$$\begin{aligned} \overline{y}_{\bullet 1 \bullet} &= (1.709 + 2.153)/2 = 1.931 \\ \overline{y}_{\bullet 2 \bullet} &= (1.952 + 2.444)/2 = 2.198 \ . \end{aligned}$$

Using the formulae in Display 8.2 we have the estimates

$$\begin{aligned} \widehat{\mu} &= \overline{y}_{\bullet\bullet\bullet} &&= 2.065 \\ \\ \widehat{\alpha}_1 &= 1.831 - 2.065 &&= -.234 \\ \widehat{\alpha}_2 &= 2.299 - 2.065 &&= \phantom{-}.234 \\ \\ \widehat{\beta}_1 &= 1.931 - 2.065 &&= -.134 \\ \widehat{\beta}_2 &= 2.198 - 2.065 &&= \phantom{-}.134 \ . \end{aligned}$$

Finally, use the treatment means and the previously estimated effects to get the estimated interaction effects:

$$\begin{aligned} \widehat{\alpha\beta}_{11} &= 1.709 - (2.065 + -.234 + -.134) &&= \phantom{-}.012 \\ \widehat{\alpha\beta}_{21} &= 2.153 - (2.065 + -.234 + \phantom{-}.134) &&= -.012 \\ \widehat{\alpha\beta}_{12} &= 1.952 - (2.065 + \phantom{-}.234 + -.134) &&= -.012 \\ \widehat{\alpha\beta}_{22} &= 2.444 - (2.065 + \phantom{-}.234 + \phantom{-}.134) &&= \phantom{-}.012 \ . \end{aligned}$$

## 8.6 The Analysis of Variance for Balanced Factorials

We have described the Analysis of Variance as an algorithm for partitioning variability in data, a method for testing null hypotheses, and a method for comparing models for data. The same roles hold in factorial analysis, but we now have more null hypotheses to test and/or models to compare.

ANOVA source for every term in model

Sum of squares

We partition the variability in the data by using ANOVA. There is a source of variability for every term in our model; for a two-factor analysis, these are factor A, factor B, the AB interaction, and error. In a one-factor ANOVA, we obtained the sum of squares for treatments by first squaring an estimated effect (for example, $\widehat{\alpha}_i{}^2$), then multiplying by the number of units receiving that effect ($n_i$), and finally adding over the index of the effect (for example, add over $i$ for $\alpha_i$). The total sum of squares was found by summing the squared deviations of the data from the overall mean, and the error sum of squares was found by summing the squared deviations of the data

| Term | Sum of Squares | Degrees of Freedom |
|---|---|---|
| A | $\sum_{i=1}^{a} bn(\widehat{\alpha}_i)^2$ | $a-1$ |
| B | $\sum_{j=1}^{b} an(\widehat{\beta}_j)^2$ | $b-1$ |
| AB | $\sum_{i=1,j=1}^{a,b} n(\widehat{\alpha\beta}_{ij})^2$ | $(a-1)(b-1)$ |
| Error | $\sum_{i=1,j=1,k=1}^{a,b,n} (y_{ijk} - \overline{y}_{ij\bullet})^2$ | $ab(n-1)$ |
| Total | $\sum_{i=1,j=1,k=1}^{a,b,n} (y_{ijk} - \overline{y}_{\bullet\bullet\bullet})^2$ | $abn-1$ |

**Display 8.3:** Sums of squares in a balanced two-way factorial.

from the treatment means. We follow exactly the same program for balanced factorials, obtaining the formulae in Display 8.3.

The sums of squares must add up in various ways. For example

$$SS_T = SS_A + SS_B + SS_{AB} + SS_E \ .$$

SS partitions

Also recall that $SS_A$, $SS_B$, and $SS_{AB}$ must add up to the sum of squares between treatments, when considering the experiment to have $g = ab$ treatments, so that

$$\sum_{i=1,j=1}^{a,b} n(\overline{y}_{ij\bullet} - \overline{y}_{\bullet\bullet\bullet})^2 = SS_A + SS_B + SS_{AB} \ .$$

These identities can provide useful checks on ANOVA computations.

Two-factor ANOVA table

We display the results of an ANOVA decomposition in an Analysis of Variance table. As before, the ANOVA table has columns for source, degrees of freedom, sum of squares, mean square, and F. For the two-way factorial, the sources of variation are factor A, factor B, the AB interaction, and error, so the table looks like this:

| Source | DF | SS | MS | F |
|---|---|---|---|---|
| A | a-1 | $SS_A$ | $SS_A/(a-1)$ | $MS_A/MS_E$ |
| B | b-1 | $SS_B$ | $SS_B/(b-1)$ | $MS_B/MS_E$ |
| AB | (a-1)(b-1) | $SS_{AB}$ | $SS_{AB}/[(a-1)(b-1)]$ | $MS_{AB}/MS_E$ |
| Error | (n-1)ab | $SS_E$ | $SS_E/[(n-1)ab]$ | |

Normality needed for testing

Tests or model comparisons require assumptions on the errors. We have assumed that the errors $\epsilon_{ijk}$ are independent and normally distributed with constant variance. When the assumptions are true, the sums of squares as random variables are independent of each other and the tests discussed below are valid.

F-tests for factorial null hypotheses

To test the null hypothesis $H_0 : \alpha_1 = \alpha_2 = \ldots = \alpha_a = 0$ against the alternative that some $\alpha_i$'s are not zero, we use the F-statistic $MS_A/MS_E$ with $a-1$ and $ab(n-1)$ degrees of freedom. This is a test of the main effect of A. The $p$-value is calculated as before. To test $H_0 : \beta_1 = \beta_2 = \ldots = \beta_b = 0$ against the null hypothesis that at least one $\beta$ is nonzero, use the F-statistic $MS_B/MS_E$, with $b-1$ and $ab(n-1)$ degrees of freedom. Similarly, the test statistic for the null hypothesis that the $\alpha\beta$ interaction terms are all zero is $MS_{AB}/MS_E$, with $(a-1)(b-1)$ and $ab(n-1)$ degrees of freedom. Alternatively, these tests may be viewed as comparisons between models that include and exclude the terms under consideration.

**Example 8.7**

### Nonstarter bacteria, continued

We compute sums of squares using the effects of Example 8.6 and the formulae of Display 8.3.

$$\begin{aligned} SS_{R50} &= 6 \times ((-.234)^2 + .234^2) = .656 \\ SS_{R21} &= 6 \times ((-.134)^2 + .134^2) = .214 \\ SS_{R50.R21} &= 3 \times (.012^2 + (-.012)^2 + (-.012)^2 + .012^2) = .002 \end{aligned}$$

Computing $SS_E$ is more work:

$$\begin{aligned} SS_E &= (1.697 - 1.709)^2 + (2.032 - 2.153)^2 + (2.211 - 1.952)^2 \\ &\quad + (2.091 - 2.444)^2 + \cdots + (2.987 - 2.444)^2 = .726 \ . \end{aligned}$$

We have $a = 2$ and $b = 2$, so the main effects and the two-factor interaction have 1 degree of freedom each; there are $12-4 = 8$ error degrees of freedom. Combining, we get the ANOVA table:

**Listing 8.1:** SAS output for nonstarter bacteria.

```
                   General Linear Models Procedure

Dependent Variable: TFAA
                                      Sum of          Mean
Source                    DF         Squares        Square   F Value      Pr > F

Model                      3      0.87231400    0.29077133      3.21      0.0834

Error                      8      0.72566267    0.09070783

Corrected Total           11      1.59797667

                   General Linear Models Procedure

Dependent Variable: TFAA

Source                    DF       Type I SS   Mean Square   F Value      Pr > F

R50                        1      0.65613633    0.65613633      7.23      0.0275
R21                        1      0.21440133    0.21440133      2.36      0.1627
R50*R21                    1      0.00177633    0.00177633      0.02      0.8922
```

| Source | DF | SS | MS | F | $p$-value |
|---|---|---|---|---|---|
| R50 | 1 | .656 | .656 | 7.23 | .028 |
| R21 | 1 | .214 | .214 | 2.36 | .16 |
| R50.R21 | 1 | .002 | .002 | .02 | .89 |
| Error | 8 | .726 | .091 | | |

The large $p$-values indicate that we have no evidence that R21 interacts with R50 or causes a change in total free amino acids. The $p$-value of .028 indicates moderate evidence that R50 may affect total free amino acids.

Listing 8.1 shows SAS output for these data. Note that SAS gives the ANOVA table in two parts. In the first, all model degrees of freedom are combined into a single 3 degree-of-freedom term. In the second, the main effects and interactions are broken out individually.

## 8.7 General Factorial Models

The model and analysis of a multi-way factorial are similar to those of a two-way factorial. Consider a four-way factorial with factors A, B, C and D, which match with the letters $\alpha$, $\beta$, $\gamma$, and $\delta$. The model is

$$\begin{aligned} y_{ijklm} &= \mu + \alpha_i + \beta_j + \gamma_k + \delta_l \\ &\quad + \alpha\beta_{ij} + \alpha\gamma_{ik} + \alpha\delta_{il} + \beta\gamma_{jk} + \beta\delta_{jl} + \gamma\delta_{kl} \\ &\quad + \alpha\beta\gamma_{ijk} + \alpha\beta\delta_{ijl} + \alpha\gamma\delta_{ikl} + \beta\gamma\delta_{jkl} \\ &\quad + \alpha\beta\gamma\delta_{ijkl} \\ &\quad + \epsilon_{ijklm} \ . \end{aligned}$$

The first line contains the overall mean and main effects for the four factors; the second line has all six two-factor interactions; the third line has three-factor interactions; the fourth line has the four-factor interaction; and the last line has the error. Just as a two-factor interaction describes how a main effect changes depending on the level of a second factor, a three-factor interaction like $\alpha\beta\gamma_{ijk}$ describes how a two-factor interaction changes depending on the level of a third factor. Similarly, four-factor interactions describe how three-factor interactions depend on a fourth factor, and so on for higher order interactions.

Multi-factor interactions

We still have the assumption that the $\epsilon$'s are independent normals with mean 0 and variance $\sigma^2$. Analogous with the two-factor case, we restrict our effects so that they will add to zero when summed over any subscript. For example,

Zero-sum restrictions on parameters

$$0 = \sum_l \delta_l = \sum_k \beta\gamma_{jk} = \sum_j \alpha\beta\delta_{ijl} = \sum_i \alpha\beta\gamma\delta_{ijkl} \ .$$

These zero-sum restrictions make the model parameters unique. The *abcd* − 1 degrees of freedom between the *abcd* treatments are assorted among the terms as follows. Each term contains some number of factors—one, two, three, or four—and each factor has some number of levels—*a*, *b*, *c*, or *d*. To get the degrees of freedom for a term, subtract one from the number of levels for each factor in the term and take the product. Thus, for the ABD term, we have $(a-1)(b-1)(d-1)$ degrees of freedom.

Degrees of freedom for general factorials

Effects in the model are estimated analogously with how we estimated effects for a two-way factorial, building up from overall mean, to main effects, to two-factor interactions, to three-factor interactions, and so on. The estimate of the overall mean is $\widehat{\mu} = \sum_{ijklm} y_{ijklm}/N = \overline{y}_{\bullet\bullet\bullet\bullet\bullet}$. Main-effect and two-factor interaction estimates are just like for two-factor factorials, ignoring all factors but the two of interest. For example, to estimate a main effect, say the $k$th level of factor C, we take the mean of all responses that received the $k$th level of factor C, and subtract out the lower order estimated effects, here just $\widehat{\mu}$:

Main effects and two-factor estimates as before

$$\widehat{\gamma}_k = \overline{y}_{\bullet\bullet k\bullet\bullet} - \widehat{\mu} \ .$$

Multi-way effects for general factorials

For a three-way interaction, say the $ijk$th level of factors A, B, and C, we take the mean response at the $ijk$ combination of factors A, B, and C, and then subtract out the lower order terms—the overall mean; main effects of A, B, and C; and two-factor interactions in A, B, and C:

$$\widehat{\alpha\beta\gamma}_{ijk} = \overline{y}_{ijk\bullet\bullet} - (\widehat{\mu} + \widehat{\alpha}_i + \widehat{\beta}_j + \widehat{\gamma}_k + \widehat{\alpha\beta}_{ij} + \widehat{\alpha\gamma}_{ik} + \widehat{\beta\gamma}_{jk}) \quad .$$

Simply continue this general rule for higher order interactions.

Sums of squares for general factorials

The rules for computing sums of squares follow the usual pattern: square each effect, multiply by the number of units that receive that effect, and add over the levels. Thus,

$$SS_{ABD} = \sum_{ijl} nc(\widehat{\alpha\beta\delta}_{ijl})^2 \quad ,$$

and so on.

ANOVA and F-tests for multi-way factorial

As with the two-factor factorial, the results of the Analysis of Variance are summarized in a table with the usual columns and a row for each term in the model. We test the null hypothesis that the effects in a given term are all zeroes by taking the ratio of the mean square for that term to the mean square for error and comparing this observed F to the F-distribution with the corresponding numerator and denominator degrees of freedom. Alternatively, we can consider these F-tests to be tests of whether a given term is needed in a model for the data.

Alternate computational algorithm

It is clear by now that the computations for a multi-way factorial are tedious at best and should be performed on a computer using statistical software. However, you might be stranded on a desert island (or in an exam room) and need to do a factorial analysis by hand. Here is a technique for multi-way factorials that reorganizes the computations required for computing factorial effects; some find this easier for hand work. The general approach is to compute an effect, subtract it from the data, and then compute the next effect on the differences from the preceding step. This way we only need to subtract out lower order terms once, and it is easier to keep track of things.

Estimate marginal means and subtract

First compute the overall mean $\widehat{\mu}$ and subtract it from the all data values. Now, compute the mean of the differences at each level of factor A. Because we have already subtracted out the overall mean, these means are the estimated effects for factor A. Now subtract these factor A effects from their corresponding entries in the differences. Proceed similarly with the other main effects, estimating and then sweeping the effects out of the differences. To get a two-factor interaction, get the two-way table of difference means. Because we have already subtracted out the grand mean and main effects, these means are the two-factor interaction effects. Continue by computing two-way means and sweeping the effects out of the differences. Proceed up through higher order interactions. As long as we proceed in a hierarchical fashion, we will obtain the desired estimated effects.

## 8.8 Assumptions and Transformations

The validity of our inference procedures still depends on the accuracy of our assumptions. We still need to check for normality, constant variance, and independence and take corrective action as required, just as we did in single-factor models.

Check assumptions

One new wrinkle that occurs for factorial data is that violations of assumptions may sometimes follow the factorial structure. For example, we may find that error variance is constant within a given level of factor B, but differs among levels of factor B.

A second wrinkle with factorials is that the appropriate model for the mean structure depends on the scale in which we are analyzing the data. Specifically, interaction terms may appear to be needed on one scale but not on another. This is easily seen in the following example. Suppose that the means for the factor level combinations follow the model

Transformation affects mean structure

$$\mu_{ij} = M \exp \alpha_i \exp \beta_j \ .$$

This model is multiplicative in the sense that changing levels of factor A or B rescales the response by multiplying rather than adding to the response. If we fit the usual factorial model to such data, we will need the interaction term, because an additive model won't fit multiplicative data well. For log-transformed data the mean structure is

$$\log(\mu_{ij}) = \log(M) + \alpha_i + \beta_j \ .$$

Multiplicative data look additive after log transformation; no interaction term is needed. Serendipitously, log transformations often fix nonconstant variance at the same time.

Some people find this confusing at first, and it begs the question of what do we mean by interaction. How can the data have interaction on one scale but not on another? Data are interactive *when analyzed on a particular scale* if the main-effects-only model is inadequate and one or more interaction terms are required. Whether or not interaction terms are needed depends on the scale of the response.

Interaction depends on scale

## 8.9 Single Replicates

Some factorial experiments are run with only one unit at each factor-level combination ($n = 1$). Clearly, this will lead to trouble, because we have no degrees of freedom for estimating error. What can we do? At this point, analysis of factorials becomes art as well as science, because you must choose among several approaches and variations on the approaches. None of these approaches is guaranteed to work, because none provides the estimate of pure experimental error that we can get from replication. If we use an approach that has an error estimate that is biased upwards, then we will have a conservative procedure. Conservative in this context means that the $p$-value that we compute is generally larger than the true $p$-value; thus we reject null hypotheses less often than we should and wind up with models with fewer terms than might be appropriate. On the other hand, if we use a procedure with an error estimate that is biased downwards, then we will have a liberal procedure. Liberal means that the computed $p$-value is generally smaller than the true $p$-value; thus we reject null hypotheses too often and wind up with models with too many terms.

No estimate of pure error in single replicates

Biased estimates of error lead to biased tests

The most common approach is to combine one or more high-order interaction mean squares into an estimate of error; that is, select one or more interaction terms and add their sums of squares and degrees of freedom to get a surrogate error sum of squares and degrees of freedom. If the underlying true interactions are null (zero), then the surrogate error mean square is an unbiased estimate of error. If any of these interactions is nonnull, then the surrogate error mean square tends on average to be a little bigger than error. Thus, if we use a surrogate error mean square as an estimate of error and make tests on other effects, we will have tests that range from valid (when interaction is absent) to conservative (when interaction is present).

High-order interactions can estimate error

This valid to conservative range for surrogate errors assumes that you haven't peeked at the data. It is very tempting to look at interaction mean squares, decide that the small ones must be error and the large ones must be genuine effects. However, this approach tends to give you error estimates that are too small, leading to a liberal test. It is generally safer to choose the mean squares to use as error before looking at the data.

Data snooping makes $MS_E$ too small

A second approach to single replicates is to use an external estimate of error. That is, we may have run similar experiments before, and we know what the size of the random errors was in those experiments. Thus we might use an $MS_E$ from a similar experiment in place of an $MS_E$ from this experiment. This *might* work, but it is a risky way of proceeding. The reason it is risky is that we need to be sure that the external estimate of error is really estimating the error that we incurred during this experiment. If the size of the

External estimates of error are possible but risky

**Table 8.7:** Page faults for a CPU experiment.

| | | | Allocation | | |
|---|---|---|---|---|---|
| Algorithm | Sequence | Size | 1 | 2 | 3 |
| 1 | 1 | 1 | 32 | 48 | 538 |
| | | 2 | 53 | 81 | 1901 |
| | | 3 | 142 | 197 | 5689 |
| | 2 | 1 | 52 | 244 | 998 |
| | | 2 | 112 | 776 | 3621 |
| | | 3 | 262 | 2625 | 10012 |
| | 3 | 1 | 59 | 536 | 1348 |
| | | 2 | 121 | 1879 | 4637 |
| | | 3 | 980 | 5698 | 12880 |
| 2 | 1 | 1 | 49 | 67 | 789 |
| | | 2 | 100 | 134 | 3152 |
| | | 3 | 233 | 350 | 9100 |
| | 2 | 1 | 79 | 390 | 1373 |
| | | 2 | 164 | 1255 | 4912 |
| | | 3 | 458 | 3688 | 13531 |
| | 3 | 1 | 85 | 814 | 1693 |
| | | 2 | 206 | 3394 | 5838 |
| | | 3 | 1633 | 10022 | 17117 |

random errors is not stable, that is, if the size of the random errors changes from experiment to experiment or depends on the conditions under which the experiment is run, then an external estimate of error will likely be estimating something other than the error of this experiment.

Model interaction

A final approach is to use one of the models for interaction described in the next chapter. These interaction models often allow us to fit the bulk of an interaction with relatively few degrees of freedom, leaving the other degrees of freedom for interaction available as potential estimates of error.

**Example 8.8**

## CPU page faults

Some computers divide memory into pages. When a program runs, it is allocated a certain number of pages of RAM. The program itself may require more pages than were allocated. When this is the case, currently unused pages are stored on disk. From time to time, a page stored on disk is needed; this is called a *page fault*. When a page fault occurs, one of the currently active pages must be moved to disk in order to make room for the page that must be brought in from disk. The trick is to choose a "good" page to send out to disk, where "good" means a page that will not be used soon.

**Listing 8.2:** SAS output for log page faults.

```
                    General Linear Models Procedure

Dependent Variable: LFAULTS
                                    Sum of            Mean
Source                   DF        Squares          Square    F Value      Pr > F

Model                    45     173.570364        3.857119    1353.60      0.0001

Error                     8       0.022796        0.002850

Corrected Total          53     173.593160

Source                   DF      Type I SS    Mean Square    F Value      Pr > F

SEQ                       2     24.6392528     12.3196264    4323.41      0.0001
SIZE                      2     41.6916546     20.8458273    7315.56      0.0001
ALLOC                     2     92.6972988     46.3486494   16265.43      0.0001
ALG                       1      2.5018372      2.5018372     877.99      0.0001
SEQ*SIZE                  4      0.8289576      0.2072394      72.73      0.0001
SEQ*ALLOC                 4      9.5104719      2.3776180     834.39      0.0001
SEQ*ALG                   2      0.0176369      0.0088184       3.09      0.1010
SIZE*ALLOC                4      0.5043045      0.1260761      44.24      0.0001
SIZE*ALG                  2      0.0222145      0.0111073       3.90      0.0658
ALLOC*ALG                 2      0.0600396      0.0300198      10.54      0.0057
SEQ*SIZE*ALLOC            8      1.0521223      0.1315153      46.15      0.0001
SEQ*ALLOC*ALG             4      0.0260076      0.0065019       2.28      0.1491
SEQ*SIZE*ALG              4      0.0145640      0.0036410       1.28      0.3548
SIZE*ALLOC*ALG            4      0.0040015      0.0010004       0.35      0.8365
```

The experiment consists of running different programs on a computer under different configurations and counting the number of page faults. There were two paging algorithms to study, and this is the factor of primary interest. A second factor with three levels was the sequence in which system routines were initialized. Factor three was the size of the program (small, medium, or large memory requirements), and factor four was the amount of RAM memory allocated (large, medium, or small). Table 8.7 shows the number of page faults that occurred for each of the 54 combinations.

Before computing any ANOVA's, look at the data. There is no replication, so there is no estimate of error. We will need to use some of the interactions as experimental error. The obvious choice is the four-way interaction with 8 degrees of freedom. Eight is on the low end of acceptable; I'd like to have 15 or 20, but I don't know which other interactions I should use—

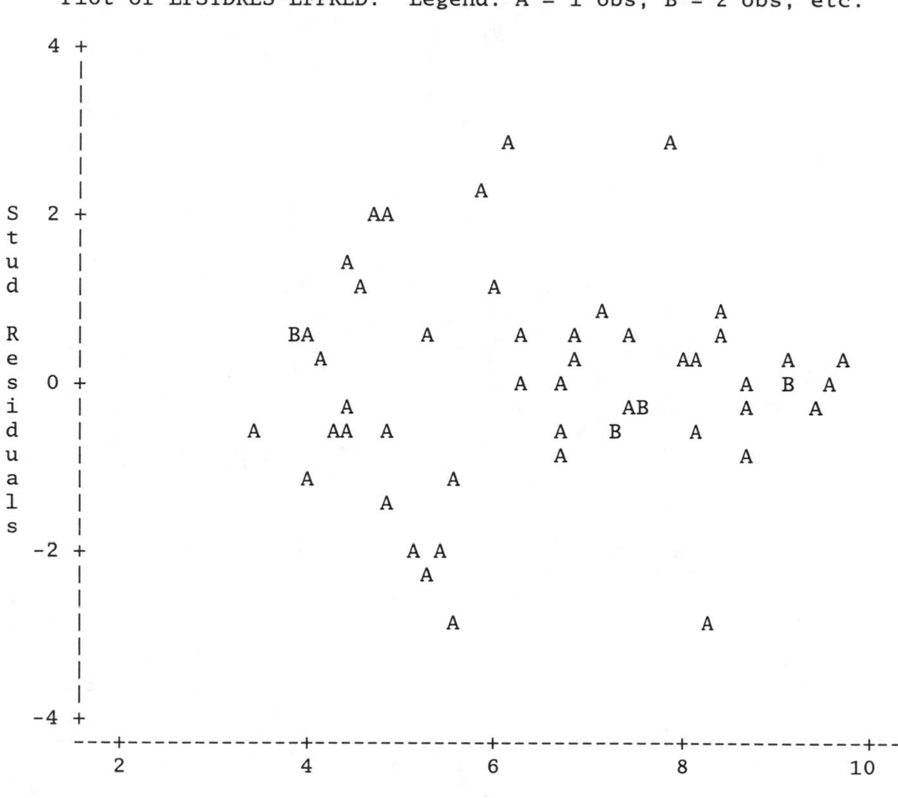


**Figure 8.4:** Studentized residuals versus predicted values for log page fault data, using SAS.

all three- and four-way interactions, perhaps? I will stay with the four-way interaction as a proxy error term.

The second thing to notice is that the data range over several orders of magnitude and just look multiplicative. Increasing the program size or changing the allocation seems to double or triple the number of page faults, rather than just adding a constant number. This suggests that a log transform of the response is advisable, and we begin by analyzing the log number of page faults.

Listing 8.2 gives the ANOVA for log page faults. All main effects are significant, and all interactions involving just allocation, program size, and load

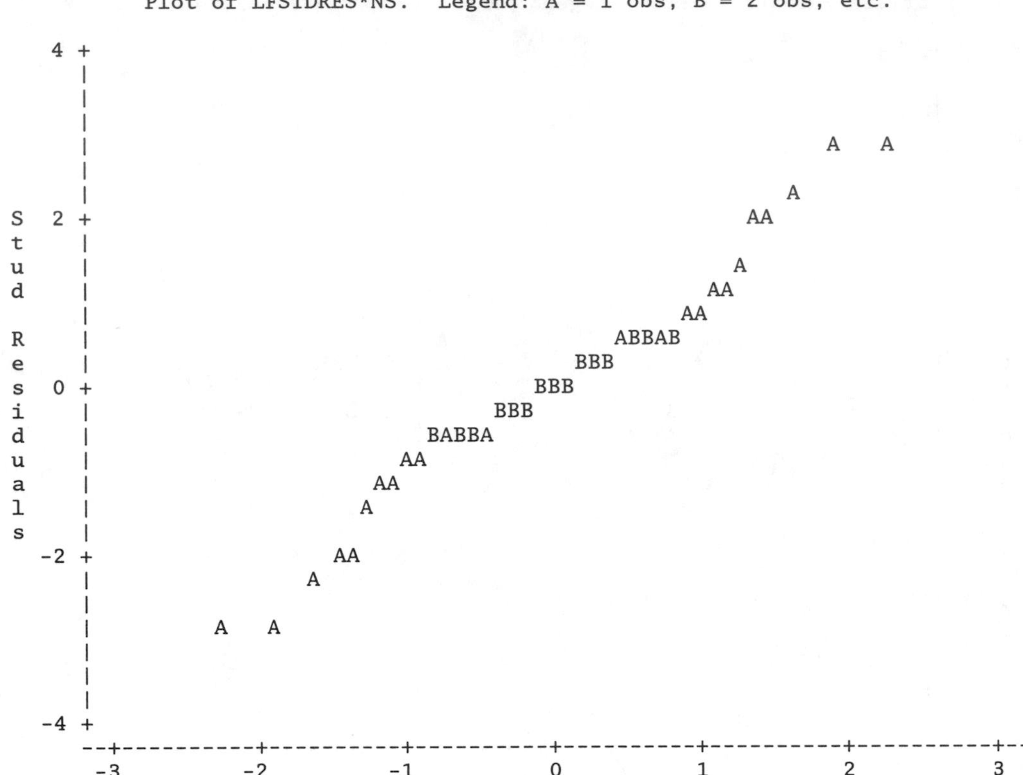


**Figure 8.5:** Normal probability plot of studentized residuals for log page fault data, using SAS.

sequence are significant. There is fairly strong evidence for an allocation by algorithm interaction ($p$-value .006), but interactions that include sequence and algorithm or size and algorithm are not highly significant.

The variance is fairly stable on this scale (see Figure 8.4), and normality looks good too (Figure 8.5). Thus we believe that our inferences are fairly sound.

The full model explains 173.6 SS; of that, 170.9 is explained by allocation, size, load sequence, and their interactions. Thus while algorithm and some of its interactions may be significant, their effects are tiny compared to the other effects. This is clear in the side-by-side plot (Figure 8.6).

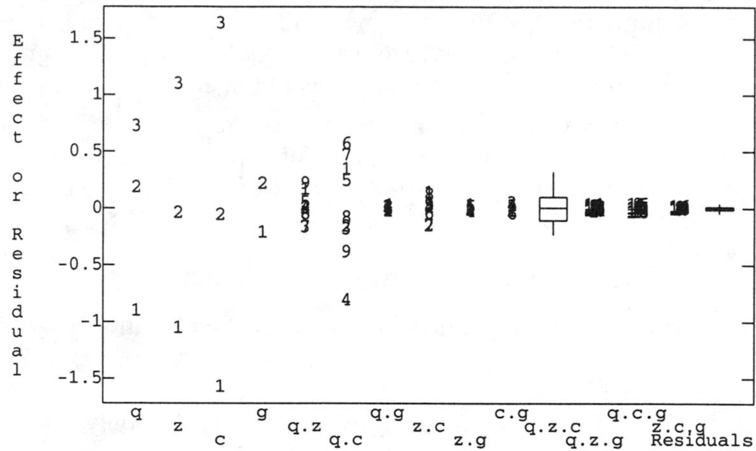


**Figure 8.6:** Side-by-side plot for log page fault data, using MacAnova. Factor labels size-z, sequence-q, allocation-c, algorithm-g.

Since algorithm is the factor of interest, we examine it more closely. The effects for algorithm are -.215 and .215. Recalling that the data are on the log scale, the difference from algorithm 1 to 2 is about a factor of $\exp(2 \times .215) = 1.54$, so algorithm 2 produces about 1.54 times as many page faults as does algorithm 1. This is worth knowing, since page faults take a lot of time on a computer. Looking at the algorithm by allocation interaction, we find the effects

| | |
|---|---|
| -.0249 | .0249 |
| -.0223 | .0223 |
| .0471 | -.0471 |

Thus while algorithm 1 is considerably better overall, its comparative advantage over algorithm 2 is a few percent less on small allocations.

## 8.10 Pooling Terms into Error

Pooling is the practice of adding sums of squares and degrees of freedom for nonsignificant model terms with those of error to form a new (pooled

Pooling leads to biased estimates of error

together) error term for further testing. In statistical software, this is usually done by computing the ANOVA for a model that does not include the terms to be pooled into error. I do *not* recommend pooling as standard practice, because pooling may lead to biased estimates of the error.

Pooling may be advantageous if there are very few error degrees of freedom. In that case, the loss of power from possible overestimation of the error may be offset by the increase in error degrees of freedom. Only consider pooling a term into error if

Rules for pooling

1. There are 10 or fewer error degrees of freedom, and
2. The term under consideration for pooling has an F-ratio less than 2.

Otherwise, do not pool.

For unbalanced factorials, refitting with a model that only includes required terms has other uses. See Chapter 10.

## 8.11 Hierarchy

Hierarchical models don't skip terms

A factorial model for data is called *hierarchical* if the presence of any term in the model implies the presence of all lower order terms. For example, a hierarchical model including the AB interaction must include the A and B main effects, and a hierarchical model including the BCD interaction must include the B, C, and D main effects and the BC, BD, and CD interactions. One potential source of confusion is that lower-order terms occur earlier in a model and thus appear above higher-order terms in the ANOVA table; lower-order terms are above.

Choose among hierarchical models

One view of data analysis for factorial treatment structure is the selection of an appropriate model for the data; that is, determining which terms are needed, and which terms can be eliminated without loss of explanatory ability. Use hierarchical models when modeling factorial data. Do not automatically test terms above (that is, lower-order to) a needed interaction. If factors A and B interact, conclude that A and B act jointly to influence the response; there is no need to test the A and B main effects.

Building a model versus testing hypotheses

The F-test allows us to test whether any term is needed, even the main effect of A when the AB interaction is needed. Why should we not test these lower-order terms, and possibly break hierarchy, when we have the ability to do so? The distinction is one between generic modeling of how the response depends on factors and interactions, and testing specific hypotheses about the treatment means. Tests of main effects are tests that certain very specific contrasts are zero. If those specific contrasts are genuinely of interest, then

**Table 8.8:** Number of rats that died after exposure to three strains of bacteria and treatment with one of two antibiotics, and factorial decompositions using equal weighting and 1,2,1 weighting of rows.

| Means | | Equal Weights | | | Row Weighted | | |
|---|---|---|---|---|---|---|---|
| 120 | 168 | -24 | 24 | -8 | -21 | 21 | -9 |
| 144 | 168 | -12 | 12 | 4 | -9 | 9 | 3 |
| 192 | 120 | 36 | -36 | 4 | 39 | -39 | 3 |
| | | 0 | 0 | 152 | -3 | 3 | 153 |

testing main effects is appropriate, even if interactions exist. Thus I only consider nonhierarchical models when I know that the main-effects contrasts, and thus the nonhierarchical model, make sense in the experimental context.

The problem with breaking hierarchy is that we have chosen to define main effects and interactions using equally weighted averages of treatment means, but we could instead define main effects and interactions using unequally weighted averages. This new set of main effects and interactions is just as valid mathematically as our usual set, but one set may have zero main effects and the other set have nonzero main effects. Which do we want to test? We need to know the appropriate set of weights, or equivalently, the appropriate contrast coefficients, for the problem at hand.

Are equally weighted averages appropriate?

**Example 8.9**

## Unequal weights

Suppose that we have a three by two factorial design testing two antibiotics against three strains of bacteria. The response is the number of rats (out of 500) that die from the given infection when treated with the given antibiotic. Our goal is to find the antibiotic with the lower death rate. Table 8.8 gives hypothetical data and two ways to decompose the means into grand mean, row effects, column effects, and interaction effects.

The first decomposition in Table 8.8 (labeled equal weights) is our usual factorial decomposition. The row effects and column effects add to zero, and the interaction effects add to zero across any row or column. With this standard factorial decomposition, the column (antibiotic) effects are zero, so there is no average difference between the antibiotics.

On the other hand, suppose that we knew that strain 2 of bacteria was twice as prevalent as the other two strains. Then we would probably want to weight row 2 twice as heavily as the other rows in all averages that we make. The second decomposition uses 1,2,1 row weights; all these factorial effects are different from the equally weighted effects. In particular, the antibiotic

**Table 8.9:** Amylase specific activity (IU), for two varieties of sprouted maize under different growth and analysis temperatures (degrees C).

| | | Analysis Temperature | | | | | | | |
|---|---|---|---|---|---|---|---|---|---|
| GT | Var. | 40 | 35 | 30 | 25 | 20 | 15 | 13 | 10 |
| 25 | B73 | 391.8 | 427.7 | 486.6 | 469.2 | 383.1 | 338.9 | 283.7 | 269.3 |
| | | 311.8 | 388.1 | 426.6 | 436.8 | 408.8 | 355.5 | 309.4 | 278.7 |
| | | 367.4 | 468.1 | 499.8 | 444.0 | 429.0 | 304.5 | 309.9 | 313.0 |
| | O43 | 301.3 | 352.9 | 376.3 | 373.6 | 377.5 | 308.8 | 234.3 | 197.1 |
| | | 271.4 | 296.4 | 393.0 | 364.8 | 364.3 | 279.0 | 255.4 | 198.3 |
| | | 300.3 | 346.7 | 334.7 | 386.6 | 329.2 | 261.3 | 239.4 | 216.7 |
| 13 | B73 | 292.7 | 422.6 | 443.5 | 438.5 | 350.6 | 305.9 | 319.9 | 286.7 |
| | | 283.3 | 359.5 | 431.2 | 398.9 | 383.9 | 342.8 | 283.2 | 266.5 |
| | | 348.1 | 381.9 | 388.3 | 413.7 | 408.4 | 332.2 | 287.9 | 259.8 |
| | O43 | 269.7 | 380.9 | 389.4 | 400.3 | 340.5 | 288.6 | 260.9 | 221.9 |
| | | 284.0 | 357.1 | 420.2 | 412.8 | 309.5 | 271.8 | 253.6 | 254.4 |
| | | 235.3 | 339.0 | 453.4 | 371.9 | 313.0 | 333.7 | 289.5 | 246.7 |

effects change, and with this weighting antibiotic 1 has a mean response 6 units lower on average than antibiotic 2 and is thus preferred to antibiotic 2.

Analogous examples have zero column effects for weighted averages and nonzero column effects in the usual decomposition. Note in the weighted decomposition that column effects add to zero and the interactions add to zero across columns, but row effects and interaction effects down columns only add to zero with 1,2,1 weights.

Weighting matters due to interaction

If factors A and B do not interact, then the A and B main effects are the same regardless of how we weight the means. In the absence of AB interaction, testing the main effects of A and B computed using our equally weighted averages gives the same results as for any other weighting. Similarly, if there is no ABC interaction, then testing AB, AC, or BC using the standard ANOVA gives the same results as for any weighting.

Use correct weighting

Factorial effects are only defined in the context of a particular weighting scheme for averages. As long as we are comparing hierarchical models, we know that the parameter tests make sense for any weighting. When we test lower-order terms in the presence of an including interaction, we must use the correct weighting.

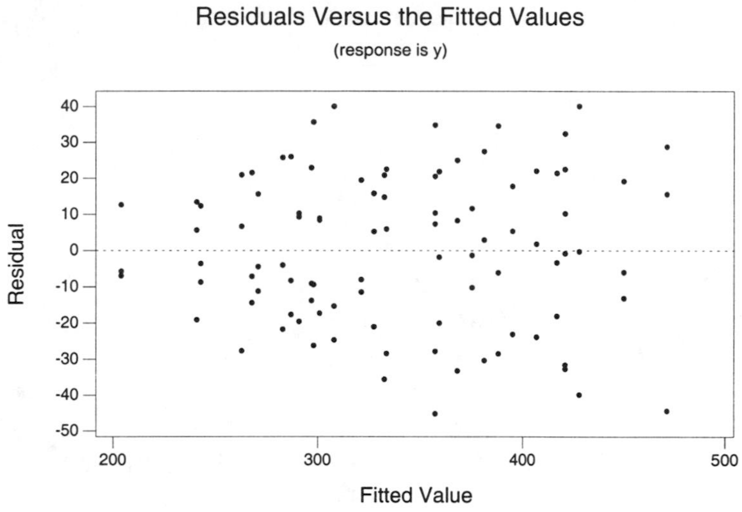

**Figure 8.7:** Residuals versus predicted values for amylase activity data, using Minitab.

## Amylase activity

**Example 8.10**

Orman (1986) studied germinating maize. One of his experiments looked at the amylase specific activity of sprouted maize under 32 different treatment conditions. These treatment conditions were the factorial combinations of analysis temperature (eight levels, 40, 35, 30, 25, 20, 15, 13, and 10 degrees C), growth temperature of the sprouts (25 or 13 degrees C), and variety of maize (B73 or Oh43). There were 96 units assigned at random to these 32 treatments. Table 8.9 gives the amylase specific activities in International Units.

This is an eight by two by two factorial with replication, so we fit the full factorial model. Figure 8.7 shows that the variability of the residuals increases slightly with mean. The best Box-Cox transformation is the log (power 0), and power 1 is slightly outside a 95% confidence interval for the transformation power. After transformation to the log scale, the constant variance assumption is somewhat more plausible (Figure 8.8), but the improvement is fairly small. The normal probability plot shows that the residuals are slightly short-tailed.

We will analyze on the log scale. Listing 8.3 shows an ANOVA for the log scale data (`at` is analysis temperature, `gt` is growth temperature,

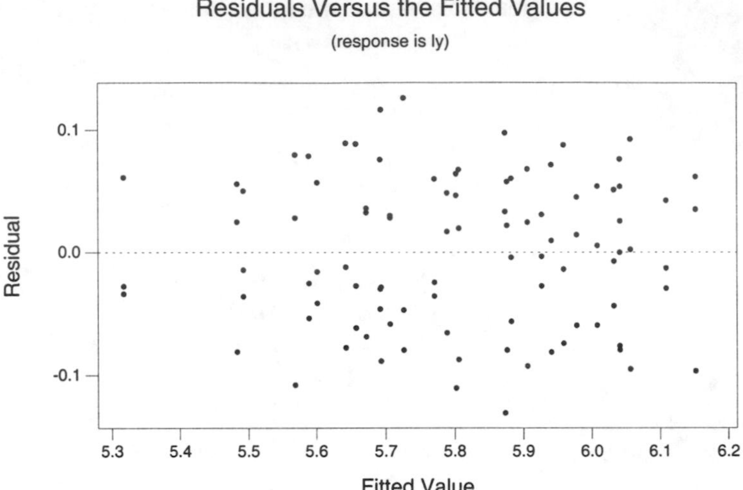


**Figure 8.8:** Residuals versus predicted values for log amylase activity data, using Minitab.

and v is variety). Analysis temperature, variety, and the growth temperature by variety interaction are all highly significant; the analysis temperature by growth temperature interaction is marginally significant. I include in any final model the main effect of growth temperature (even though it has a fairly large $p$-value), because growth temperature interacts with variety, and I wish to maintain hierarchy.

Note that the analysis is not finished. We should look more closely at the actual effects and interactions to describe them in more detail. We will continue this example in Chapter 9, but for now we examine the side-by-side plot of all the effects and residuals, shown in Figure 8.9. Analysis temperature and variety have the largest effects. Some of the analysis temperature by growth temperature and analysis temperature by variety interaction effects (neither terribly significant) are as large or larger than the growth temperature by variety interactions. Occasional large effects in nonsignificant terms can occur because the F-test averages across all the degrees of freedom in a term, and many small effects can mask one large one.

**Listing 8.3:** ANOVA for log amylase activity, using Minitab.

```
Analysis of Variance for ly

Source      DF        SS        MS       F      P
at           7   3.01613   0.43088   78.86  0.000
gt           1   0.00438   0.00438    0.80  0.374
v            1   0.58957   0.58957  107.91  0.000
at*gt        7   0.08106   0.01158    2.12  0.054
at*v         7   0.02758   0.00394    0.72  0.654
gt*v         1   0.08599   0.08599   15.74  0.000
at*gt*v      7   0.04764   0.00681    1.25  0.292
Error       64   0.34967   0.00546
Total       95   4.20202
```

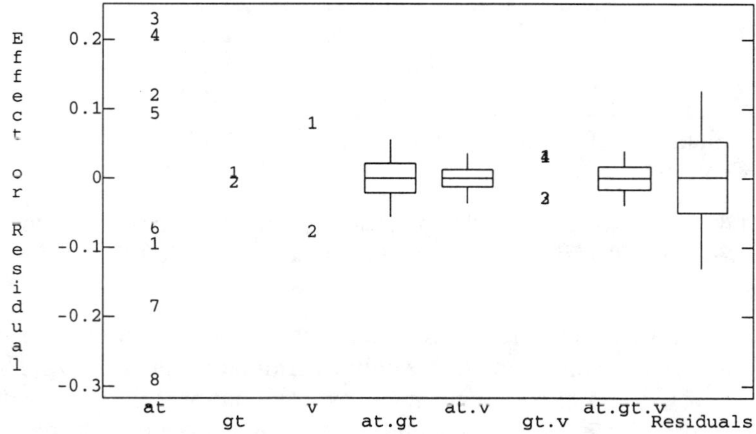


**Figure 8.9:** Side-by-side plot for effects in analysis of log amylase activity data.

## 8.12 Problems

**Exercise 8.1**

Diet affects weight gain. We wish to compare nine diets; these diets are the factor-level combinations of protein source (beef, pork, and grain) and number of calories (low, medium, and high). There are eighteen test animals that were randomly assigned to the nine diets, two animals per diet. The mean responses (weight gain) are:

| | | Calories | | |
|---|---|---|---|---|
| | | Low | Medium | High |
| Source | Beef | 76.0 | 86.8 | 101.8 |
| | Pork | 83.3 | 89.5 | 98.2 |
| | Grain | 83.8 | 83.5 | 86.2 |

The mean square for error was 8.75. Analyze these data to determine an appropriate model.

**Exercise 8.2** An experiment was conducted to determine the effect of germination time (in days) and temperature (degrees C) on the free alpha amino nitrogen (FAN) content of rice malt. The values shown in the following are the treatment means of FAN with $n = 2$ (data from Aniche and Okafor 1989).

| | Temperature | | | | |
|---|---|---|---|---|---|
| Days | 22 | 24 | 26 | 28 | Row Means |
| 1 | 39.4 | 49.9 | 55.1 | 59.5 | 50.98 |
| 2 | 56.4 | 68.0 | 76.4 | 88.8 | 72.40 |
| 3 | 70.2 | 81.5 | 95.6 | 99.6 | 86.72 |
| Column Means | 55.33 | 66.47 | 75.70 | 82.63 | |
| Grand Mean | 70.03 | | | | |

The total sum of squares was 8097. Draw an interaction plot for these data. Compute an ANOVA table and determine which terms are needed to describe the means.

**Problem 8.1** Brewer's malt is produced from germinating barley, so brewers like to know under what conditions they should germinate their barley. The following is part of an experiment on barley germination. Barley seeds were divided into 30 lots of 100 seeds, and each lot of 100 seeds was germinated under one of ten conditions chosen at random. The conditions are the ten combinations of weeks after harvest (1, 3, 6, 9, or 12 weeks) and amount of water used in germination (4 ml or 8 ml). The response is the number of seeds germinating. We are interested in whether timing and/or amount of water affect germination. The data for this problem are in Table 8.1 (Hareland and Madson 1989). Analyze these data to determine how the germination rate depends on the treatments.

**Problem 8.2** Particleboard is made from wood chips and resins. An experiment is conducted to study the effect of using slash chips (waste wood chips) along with standard chips. The researchers make eighteen boards by varying the target density (42 or 48 lb/ft$^3$), the amount of resin (6, 9, or 12%), and the fraction of slash (0, 25, or 50%). The response is the actual density of the boards produced (lb/ft$^3$, data from Boehner 1975). Analyze these data to determine the effects of the factors on particleboard density and how the density differs from target.

| | 42 Target | | | 48 Target | | |
|---|---|---|---|---|---|---|
| | 0% | 25% | 50% | 0% | 25% | 50% |
| 6 | 40.9 | 41.9 | 42.0 | 44.4 | 46.2 | 48.4 |
| 9 | 42.8 | 43.9 | 44.8 | 48.2 | 48.6 | 50.7 |
| 12 | 45.4 | 46.0 | 46.2 | 49.9 | 50.8 | 50.3 |

**Problem 8.3**

We have data from a four by three factorial with 24 units. Below are ANOVA tables and residual versus predicted plots for the data and the log-transformed data. What would you conclude about interaction in the data?

Original data:

| | DF | SS | MS |
|---|---|---|---|
| r | 3 | 7.8416e+06 | 2.6139e+06 |
| c | 2 | 2.7756e+06 | 1.3878e+06 |
| r.c | 6 | 4.7148e+06 | 7.858e+05 |
| Error | 12 | 1.7453e+06 | 1.4544e+05 |

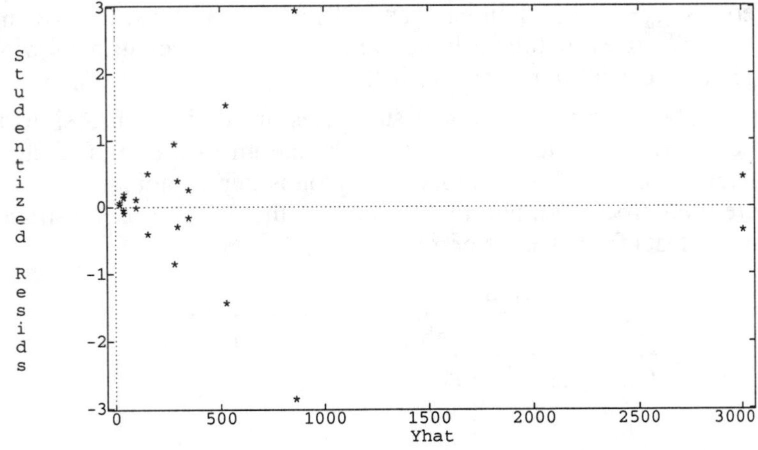


Log data:

| | DF | SS | MS |
|---|---|---|---|
| r | 3 | 27.185 | 9.0617 |
| c | 2 | 17.803 | 8.9015 |
| r.c | 6 | 7.5461 | 1.2577 |
| Error | 12 | 20.77 | 1.7308 |

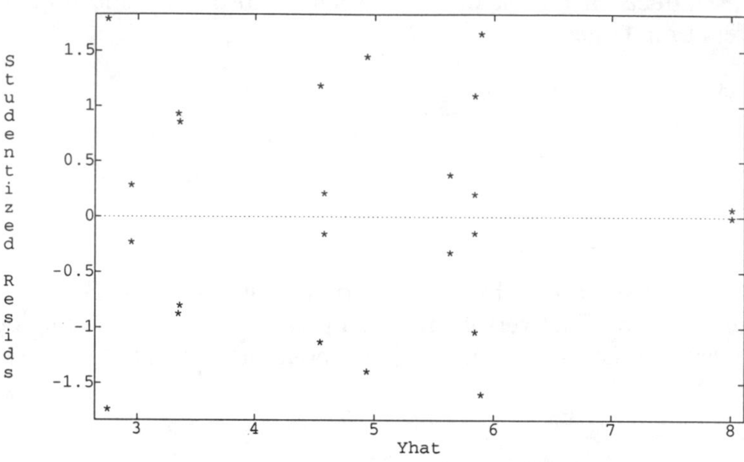


**Problem 8.4**

Implantable heart pacemakers contain small circuit boards called substrates. These substrates are assembled, cut to shape, and fired. Some of the substrates will separate, or delaminate, making them useless. The purpose of this experiment was to study the effects of three factors on the rate of delamination. The factors were A: firing profile time, 8 versus 13 hours with the theory suggesting 13 hours is better; B: furnace airflow, low versus high, with theory suggesting high is better; and C: laser, old versus new, with theory suggesting new cutting lasers are better.

A large number of raw, assembled substrates are divided into sixteen groups. These sixteen groups are assigned at random to the eight factor-level combinations of the three factors, two groups per combination. The substrates are then processed, and the response is the fraction of substrates that delaminate. Data from Todd Kerkow.

| | 8 hrs | | 13 hrs | |
|---|---|---|---|---|
| | Low | High | Low | High |
| Old | .83 | .68 | .18 | .25 |
| | .78 | .90 | .16 | .20 |
| New | .86 | .72 | .30 | .10 |
| | .67 | .81 | .23 | .14 |

Analyze these data to determine how the treatments affect delamination.

**Problem 8.5**

Pine oleoresin is obtained by tapping the trunks of pine trees. Tapping is done by cutting a hole in the bark and collecting the resin that oozes out. This experiment compares four shapes for the holes and the efficacy of acid treating the holes. Twenty-four pine trees are selected at random from a plantation, and the 24 trees are assigned at random to the eight combinations of whole shape (circular, diagonal slash, check, rectangular) and acid treatment (yes or no). The response is total grams of resin collected from the hole (data from Low and Bin Mohd. Ali 1985).

| | Circular | Diagonal | Check | Rect. |
|---|---|---|---|---|
| Control | 9 | 43 | 60 | 77 |
| | 13 | 48 | 65 | 70 |
| | 12 | 57 | 70 | 91 |
| Acid | 15 | 66 | 75 | 97 |
| | 13 | 58 | 78 | 108 |
| | 20 | 73 | 90 | 99 |

Analyze these data to determine how the treatments affect resin yield.

**Problem 8.6**

A study looked into the management of various tropical grasses for improved production, measured as dry matter yield in hundreds of pounds per acre over a 54-week study period. The management variables were height of cut (1, 3, or 6 inches), the cutting interval (1, 3, 6, or 9 weeks), and amount of nitrogen fertilizer (0, 8, 16, or 32 hundred pounds of ammonium sulfate per acre per year). Forty-eight plots were assigned in completely randomized fashion to the 48 factor-level combinations. Dry matter yields for the plots are shown in the table below (data from Richards 1965). Analyze these data and write your conclusions in a report of at most two pages.

| | | Cutting Interval | | | |
|---|---|---|---|---|---|
| | | 1 wks. | 3 wks. | 6 wks. | 9 wks. |
| Ht 1 | F 0 | 74.1 | 65.4 | 96.7 | 147.1 |
| | F 8 | 87.4 | 117.7 | 190.2 | 188.6 |
| | F 16 | 96.5 | 122.2 | 197.9 | 232.0 |
| | F 32 | 107.6 | 140.5 | 241.3 | 192.0 |
| Ht 3 | F 0 | 61.7 | 83.7 | 88.8 | 155.6 |
| | F 8 | 112.5 | 129.4 | 145.0 | 208.1 |
| | F 16 | 102.3 | 137.8 | 173.6 | 203.2 |
| | F 32 | 115.3 | 154.3 | 211.2 | 245.2 |
| Ht 6 | F 0 | 49.9 | 72.7 | 113.9 | 143.4 |
| | F 8 | 92.9 | 126.4 | 175.5 | 207.5 |
| | F 16 | 100.8 | 153.5 | 184.5 | 194.2 |
| | F 32 | 115.8 | 160.0 | 224.8 | 197.5 |

**Problem 8.7** Big sagebrush is often planted in range restoration projects. An experiment is performed to determine the effects of storage length and relative humidity on the viability of seeds. Sixty-three batches of 300 seeds each are randomly divided into 21 groups of three. These 21 groups each receive a different treatment, namely the combinations of storage length (0, 60, 120, 180, 240, 300, or 360 days) and storage relative humidity (0, 32, or 45%). After the storage time, the seeds are planted, and the response is the percentage of seeds that sprout (data from Welch 1996). Analyze these data for the effects of the factors on viability.

| | Days | | | | | | |
|---|---|---|---|---|---|---|---|
| | 0 | 60 | 120 | 180 | 240 | 300 | 360 |
| 0% | 82.1 | 78.6 | 79.8 | 82.3 | 81.7 | 85.0 | 82.7 |
| | 79.0 | 80.8 | 79.1 | 75.5 | 80.1 | 87.9 | 84.6 |
| | 81.9 | 80.5 | 78.2 | 79.1 | 81.1 | 82.1 | 81.7 |
| 32% | 83.1 | 78.1 | 80.4 | 77.8 | 83.8 | 82.0 | 81.0 |
| | 80.5 | 83.6 | 81.8 | 80.4 | 83.7 | 77.6 | 78.9 |
| | 82.4 | 78.3 | 83.8 | 78.8 | 81.5 | 80.3 | 83.1 |
| 45% | 83.1 | 66.5 | 52.9 | 52.9 | 52.2 | 38.6 | 25.2 |
| | 78.9 | 61.4 | 58.9 | 54.3 | 51.9 | 37.9 | 25.8 |
| | 81.0 | 61.2 | 59.3 | 48.7 | 48.8 | 40.6 | 21.0 |

**Question 8.1** Consider a balanced four by three factorial. Show that orthogonal contrasts in row means (ignoring factor B) are also orthogonal contrasts for all twelve treatments when the contrast coefficients have been repeated across rows ($w_{ij} = w_i$). Show that a contrast in the row means and the analogous contrast in all twelve treatment means have the same sums of squares.

**Question 8.2** In a two-way factorial, we have defined $\widehat{\mu}$ as the grand mean of the data, $\widehat{\mu}+\widehat{\alpha}_i$ as the mean of the responses for the $i$th level of factor A, $\widehat{\mu}+\widehat{\beta}_j$ as the mean of the responses for the $j$th level of factor B, and $\widehat{\mu} + \widehat{\alpha}_i + \widehat{\beta}_j + \widehat{\alpha\beta}_{ij}$ as the mean of the $ij$th factor-level combination. Show that this implies our zero-sum restrictions on the estimated effects.

Suppose that we use the same idea, but instead of ordinary averages we use weighted averages with $v_{ij}$ as the weight for the $ij$th factor-level combination. Derive the new zero-sum restrictions for these weighted averages.

# Chapter 9

# A Closer Look at Factorial Data

Analysis of factorially structured data should be more than just an enumeration of which main effects and interactions are significant. We should look closely at the data to try to determine what the data are telling us by understanding the main effects and interactions in the data. For example, reporting that factor B only affects the response at the high level of factor A is more informative than reporting that factors A and B have significant main effects and interactions. One of my pet peeves is an analysis that just reports significant terms. This chapter explores a few techniques for exploring factorial data more closely.

Look at more than just significance of main effects and interactions

## 9.1 Contrasts for Factorial Data

Contrasts allow us to examine particular ways in which treatments differ. With factorial data, we can use contrasts to look at how specific main effects differ and to see patterns in interactions. Indeed, we have seen that the usual factorial ANOVA can be built from sets of contrasts. Chapters 4 and 5 discussed contrasts and multiple comparisons in the context of single factor analysis. These procedures carry over to factorial treatment structures with little or no modification.

Use contrasts to explore the response

In this section we will discuss contrasts in the context of a three-way factorial; generalization to other numbers of factors is straightforward. The factors in our example experiment are drug (one standard drug and two new

| | |
|---|---|
| Expected value | $\sum_{ijk} w_{ijk}\, \mu_{ijk}$ |
| Variance | $\sigma^2 \sum_{ijk} \frac{w_{ijk}^2}{n_{ijk}}$ |
| Sum of squares | $\frac{(\sum_{ijk} w_{ijk}\, \overline{y}_{ijk\bullet})^2}{\sum_{ijk} w_{ijk}^2/n_{ijk}}$ |
| Confidence interval | $\sum_{ijk} w_{ijk}\, \overline{y}_{ijk\bullet} \pm t_{\mathcal{E}/2,N-abc} \times \sqrt{MS_E \sum_{ijk} w_{ijk}^2/n_{ijk}}$ |
| F-test | $\frac{(\sum_{ijk} w_{ijk}\, \overline{y}_{ijk\bullet})^2}{MS_E \sum_{ijk} w_{ijk}^2/n_{ijk}}$ |

**Display 9.1:** Contrast formulae for a three-way factorial.

drugs), dose (four levels, equally spaced), and administration time (morning or evening). We will usually assume balanced data, because contrasts for balanced factorial data have simpler orthogonality relationships.

Inference for contrasts remains the same

We saw in one-way analysis that the arithmetic of contrasts is not too hard; the big issue was finding contrast coefficients that address an interesting question. The same is true for factorials. Suppose that we have a set of contrast coefficients $w_{ijk}$. We can work with this contrast for a factorial just as we did with contrasts in the one-way case using the formulae in Display 9.1. These formulae are nothing new, merely the application of our usual contrast formulae to the design with $g = abc$ treatments. We still need to find meaningful contrast coefficients.

Pairwise comparisons

Pairwise comparisons are differences between two treatments, ignoring the factorial structure. We might compare the standard drug at the lowest dose with morning administration to the first new drug at the lowest dose with evening administration. As we have seen previously with pairwise comparisons, there may be a multiple testing issue to consider, and our pairwise multiple comparisons procedures (for example, HSD) carry over directly to the factorial setting.

A *simple effect* is a particular kind of pairwise comparison. A simple effect is a difference between two treatments that have the same levels of all factors but one. A comparison between the standard drug at the lowest dose with morning administration and the standard drug at the lowest dose with evening administration is a simple effect. Differences of main effects are averages of simple effects.

Simple effects are pairwise differences that vary just one factor

The structure of a factorial design suggests that we should also consider contrasts that reflect the design, namely main-effect contrasts and interaction contrasts. In general, we use contrasts with coefficient patterns that mimic those of factorial effects. A *main-effect contrast* is one where the coefficients $w_{ijk}$ depend only on a single index; for example, $k$ for a factor C contrast. That is, two contrast coefficients are equal if they have the same $k$ index. These coefficients will add to zero across $k$ for any $i$ and $j$. For *interaction* contrasts, the coefficients depend only on the indices of factors in the interaction in question and satisfy the same zero-sum restrictions as their corresponding model terms. Thus a BC interaction contrast has coefficients $w_{ijk}$ that depend only on $j$ and $k$ and add to zero across $j$ or $k$ when the other subscript is kept constant. For an ABC contrast, the coefficients $w_{ijk}$ must add to zero across any subscript.

Main-effect and interaction contrasts examine factorial components

We can use pairwise multiple comparisons procedures such as HSD for marginal means. Thus to compare all levels of factor B using HSD, we treat the means $\overline{y}_{\bullet j \bullet \bullet}$ as $b$ treatment means each with sample size $acn$ and do multiple comparisons with $abc(n-1)$ degrees of freedom for error. The same approach works for two-way and higher marginal tables of means. For example, treat $\overline{y}_{\bullet jk \bullet}$ as $bc$ treatment means each with sample size $an$ and $abc(n-1)$ degrees of freedom for error. Pairwise multiple comparisons procedures also work when applied to main effects—for example, $\widehat{\beta}_j$—but most do not work for interaction effects due to the additional zero sum restrictions. (Bonferroni does work.)

Pairwise multiple comparisons work for marginal means

Please note: simple-effects, main-effects, and interaction contrasts are examples of contrasts that are frequently useful in analysis of factorial data; there are many other kinds of contrasts. Use contrasts that address your questions. Don't be put off if a contrast that makes sense to you does not fit into one of these neat categories.

**Example 9.1**

### Factorial contrasts

Let's look at some factorial contrasts for our three-way drug test example. Coefficients $w_{ijk}$ for these contrasts are shown in Table 9.1. Suppose that we want to compare morning and evening administration times averaged across all drugs and doses. The first contrast in Table 9.1 has coefficients -1 for

**Table 9.1:** Example contrasts.

| Morning versus Evening | | | | | | | | | |
|---|---|---|---|---|---|---|---|---|---|
| | Dose | | | | | Dose | | | |
| Drug | 1 | 2 | 3 | 4 | Drug | 1 | 2 | 3 | 4 |
| 1 | 1 | 1 | 1 | 1 | 1 | -1 | -1 | -1 | -1 |
| 2 | 1 | 1 | 1 | 1 | 2 | -1 | -1 | -1 | -1 |
| 3 | 1 | 1 | 1 | 1 | 3 | -1 | -1 | -1 | -1 |

| Linear in Dose | | | | | | | | | |
|---|---|---|---|---|---|---|---|---|---|
| | Dose | | | | | Dose | | | |
| Drug | 1 | 2 | 3 | 4 | Drug | 1 | 2 | 3 | 4 |
| 1 | -3 | -1 | 1 | 3 | 1 | -3 | -1 | 1 | 3 |
| 2 | -3 | -1 | 1 | 3 | 2 | -3 | -1 | 1 | 3 |
| 3 | -3 | -1 | 1 | 3 | 3 | -3 | -1 | 1 | 3 |

| Linear in Dose by Morning versus Evening | | | | | | | | | |
|---|---|---|---|---|---|---|---|---|---|
| | Dose | | | | | Dose | | | |
| Drug | 1 | 2 | 3 | 4 | Drug | 1 | 2 | 3 | 4 |
| 1 | -3 | -1 | 1 | 3 | 1 | 3 | 1 | -1 | -3 |
| 2 | -3 | -1 | 1 | 3 | 2 | 3 | 1 | -1 | -3 |
| 3 | -3 | -1 | 1 | 3 | 3 | 3 | 1 | -1 | -3 |

| Linear in Dose by Morning versus Evening by Drug 2 versus Drug 3 | | | | | | | | | |
|---|---|---|---|---|---|---|---|---|---|
| | Dose | | | | | Dose | | | |
| Drug | 1 | 2 | 3 | 4 | Drug | 1 | 2 | 3 | 4 |
| 1 | 0 | 0 | 0 | 0 | 1 | 0 | 0 | 0 | 0 |
| 2 | -3 | -1 | 1 | 3 | 2 | 3 | 1 | -1 | -3 |
| 3 | 3 | 1 | -1 | -3 | 3 | -3 | -1 | 1 | 3 |

| Linear in Dose for Drug 1 | | | | | | | | | |
|---|---|---|---|---|---|---|---|---|---|
| | Dose | | | | | Dose | | | |
| Drug | 1 | 2 | 3 | 4 | Drug | 1 | 2 | 3 | 4 |
| 1 | -3 | -1 | 1 | 3 | 1 | -3 | -1 | 1 | 3 |
| 2 | 0 | 0 | 0 | 0 | 2 | 0 | 0 | 0 | 0 |
| 3 | 0 | 0 | 0 | 0 | 2 | 0 | 0 | 0 | 0 |

evening and 1 for morning and thus makes the desired comparison. This is a main-effect contrast (coefficients only depend on administration time, factor C). We can get the same information by using a contrast with coefficients (1, -1) and the means $\overline{y}_{\bullet\bullet k\bullet}$ or effects $\widehat{\gamma}_k$.

The response presumably changes with drug dose (factor B), so it makes sense to examine dose as a quantitative effect. To determine the linear effect of dose, use a main-effect contrast with coefficients -3, -1, 1, and 3 for doses 1 through 4 (Appendix Table D.6); this is the second contrast in Table 9.1. As with the first example, we could again get the same information from a contrast in the means $\overline{y}_{\bullet j\bullet\bullet}$ or effects $\widehat{\beta}_j$ using the same coefficients. The simple coefficients -3, -1, 1, and 3 are applicable here because the doses are equally spaced and balance gives equal sample sizes.

A somewhat more complex question is whether the linear effect of dose is the same for the two administration times. To determine this, we compute the linear effect of dose from the morning data, and then subtract the linear effect of dose from the evening data. This is the third contrast in Table 9.1. This is a two-factor interaction contrast; the coefficients add to zero across dose or administration time. Note that this contrast is literally the elementwise product of the two corresponding main-effects contrasts.

A still more complex question is whether the dependence of the linear effect of dose on administration times is the same for drugs 2 and 3. To determine this, we compute the linear in dose by administration time interaction contrast for drug 2, and then subtract the corresponding contrast for drug 3. This three-factor interaction contrast is the fourth contrast in Table 9.1. It is formed as the elementwise product of the linear in dose by administration time two-way contrast and a main-effect contrast between drugs 2 and 3.

Finally, the last contrast in Table 9.1 is an example of a useful contrast that is not a simple effect, main effect, or interaction contrast. This contrast examines the linear effect of dose for drug one, averaged across time.

Products of main-effect contrasts

The interaction contrasts in Example 9.1 illustrate an important special case of interaction contrasts, namely, products of main-effect contrasts. These products allow us to determine if an interesting contrast in one main effect varies systematically according to an interesting contrast in a second main effect.

We can reexpress a main-effect contrast in the individual treatment means $\overline{y}_{ijk\bullet}$ in terms of a contrast in the factor main effects or the factor marginal means. For example, a contrast in factor C can be reexpressed as

$$\begin{aligned}\sum_{ijk} w_{ijk}\,\overline{y}_{ijk\bullet} &= \sum_k \left[ w_{11k} \sum_{ij} \overline{y}_{ijk\bullet} \right] \\ &= \sum_k w_k\,\overline{y}_{\bullet\bullet k\bullet} \\ &= \sum_k w_k\,\widehat{\gamma}_k \ ,\end{aligned}$$

Contrasts for treatment means or marginal means

where $w_k = abw_{11k}$. Because scale is somewhat arbitrary for contrast coefficients, we could also use $w_k = w_{11k}$ and still get the same kind of information. For balanced data, two main-effect contrasts for the same factor with coefficients $w_k$ and $w_k^\star$ are orthogonal if

$$\sum_k w_k\,w_k^\star = 0 \ .$$

Interaction contrasts of means or effects

We can also express an interaction contrast in the individual treatment means as a contrast in marginal means or interaction effects. For example, suppose $w_{ijk}$ is a set of contrast coefficients for a BC interaction contrast. Then we can rewrite the contrast in terms of marginal means or interaction effects:

$$\begin{aligned}\sum_{ijk} w_{ijk}\,\overline{y}_{ijk\bullet} &= \sum_{jk} w_{jk}\,\overline{y}_{\bullet jk\bullet} \\ &= \sum_{jk} w_{jk}\,\widehat{\beta\gamma}_{jk}\end{aligned}$$

where $aw_{1jk} = w_{jk}$. Two interaction contrasts for the same interaction with coefficients $w_{jk}$ and $w_{jk}^\star$ are orthogonal if

$$\sum_{jk} w_{jk}\,w_{jk}^\star = 0 \ .$$

Simplied formulae for main-effect and interaction contrasts

For balanced data, the formulae in Display 9.1 can be simplified by replacing the sample size $n_{ijk}$ by the common sample size $n$. The formulae can be simplified even further for main-effect and interaction contrasts, because they can be rewritten in terms of the effects or marginal means of interest instead of using all treatment means. Consider a main-effect contrast in factor

C with coefficients $w_k$; the number of observations at the $k$th level of factor C is $abn$. We have for the contrast $\sum_k w_k \overline{y}_{\bullet\bullet k\bullet}$:

| | |
|---|---|
| Expected value | $\sum_k w_k \gamma_k$ |
| Variance | $\sum_k w_k^2 \sigma^2/(abn)$ |
| Sum of squares | $\dfrac{(\sum_k w_k \overline{y}_{\bullet\bullet k\bullet})^2}{\sum_k w_k^2/(abn)}$ |
| Confidence interval | $\sum_k w_k \overline{y}_{\bullet\bullet k\bullet} \pm t_{\mathcal{E}/2,N-abc}\sqrt{MS_E \sum_k w_k^2/(abn)}$ |
| F-test | $\dfrac{(\sum_k w_k \overline{y}_{\bullet\bullet k\bullet})^2}{MS_E \sum_k w_k^2/(abn)}$ |

The simplification is similar for interaction contrasts. For example, the BC interaction contrast $\sum_{jk} w_{jk} \overline{y}_{\bullet jk\bullet}$ has sum of squares

$$\frac{(\sum_{jk} w_{jk} \overline{y}_{\bullet jk\bullet})^2}{\sum_{jk} w_{jk}^2/(an)}$$

($an$ is the "sample size" at each $jk$ combination).

## 9.2 Modeling Interaction

An interaction is a deviation from additivity. If the effect of going from dose 1 to dose 2 changes from drug 2 to drug 3, then there is an interaction between drug and dose. Similarly, if the interaction of drug and dose is different in morning and evening applications, then there is a three-factor interaction between drug, dose, and time. Try to understand and model any interaction that may be present in your data. This is not always easy, but when it can be done it leads to much greater insight into what the data have to say. This section discusses three specific models for interaction; there are many others.

Models for interaction help to understand data

### 9.2.1 Interaction plots

We introduced interaction plots in Section 8.4 as a method for visualizing interaction. These plots continue to be important tools, but there are a few variations on interaction plots that can make them more useful in multi-way factorials. The first variation is to plot marginal means. If, for example, we are exploring the AB interaction, then we can make an interaction plot using the means $\overline{y}_{ij\bullet\bullet}$. Thus we do not plot every treatment mean individually but instead average across any other factors. This makes for a cleaner picture of the AB interaction, because it hides all other interactions.

Interaction plots of marginal means

A second variation is to plot interaction effects rather than marginal means. Marginal means such as $\overline{y}_{ij\bullet\bullet}$ satisfy

$$\overline{y}_{ij\bullet\bullet} = \widehat{\mu} + \widehat{\alpha}_i + \widehat{\beta}_j + \widehat{\alpha\beta}_{ij} \ ,$$

so they contain main effects as well as interaction. By making the interaction plot using $\widehat{\alpha\beta}_{ij}$ instead of $\overline{y}_{ij\bullet\bullet}$, we eliminate the main effects information and concentrate on the interaction. This is good for understanding the nature of the interaction once we are reasonably certain that interaction is there, but it works poorly for diagnosing the presence of interaction because interaction plots of interaction effects will always show interaction. So first decide whether interaction is present by looking at means or by using ANOVA. If interaction is present, a plot of interaction effects can be useful in understanding the interaction.

Interaction plots of interaction effects

### 9.2.2 One-cell interaction

A *one-cell interaction* is a common type of interaction where most of the experiment is additive, but one treatment deviates from the additive structure. The name "cell" comes from the idea that one cell in the table of treatment means does not follow the additive model. More generally, there may be one or a few cells that deviate from a relatively simple model. If the deviation from the simple model in these few cells is great enough, all the usual factorial interaction effects can be large and statistically significant.

A single unusual treatment can make all interactions significant

Understanding one-cell interaction is easy: the data follow a simple model except for a single cell or a few cells. Finding a one-cell interaction is harder. It requires a careful study of the interaction effects or plots or a more sophisticated estimation technique than the least squares we have been using (see Daniel 1976 or Oehlert 1994). Be warned, large one-cell interactions can be masked or hidden by other large one-cell interactions.

One-cell interactions can sometimes be detected by examination of interaction effects. A table of interaction effects adds to zero across rows or columns. A one-cell interaction shows up in the effects as an entry with a large absolute value. The other entries in the same row and column are moderate and of the opposite sign, and the remaining entries are small and of

**Table 9.2:** Data from a replicated four-factor experiment. All factors have two levels, labeled low and high.

| A | B | C | D | | | |
|---|---|---|---|---|---|---|
| | | | Low | | High | |
| low | low | low | 26.1 | 27.5 | 23.5 | 21.1 |
| low | low | high | 22.8 | 23.8 | 30.6 | 32.5 |
| low | high | low | 22.0 | 20.2 | 28.1 | 29.9 |
| low | high | high | 30.0 | 29.3 | 38.3 | 38.5 |
| high | low | low | 11.4 | 11.0 | 20.4 | 22.0 |
| high | low | high | 22.3 | 20.2 | 28.7 | 28.8 |
| high | high | low | 18.9 | 16.4 | 26.6 | 26.5 |
| high | high | high | 29.6 | 29.8 | 34.5 | 34.9 |

the same sign as the interacting cell. For example, a three by four factorial with all responses 0 except for 12 in the (2,2) cell has interaction effects as follows:

Characteristic pattern of effects for a one-cell interaction

$$\begin{array}{rrrr} 1 & -3 & 1 & 1 \\ -2 & 6 & -2 & -2 \\ 1 & -3 & 1 & 1 \end{array}$$

Rearranging the rows and columns to put the one-cell interaction in a corner emphasizes the pattern:

$$\begin{array}{rrrr} 6 & -2 & -2 & -2 \\ -3 & 1 & 1 & 1 \\ -3 & 1 & 1 & 1 \end{array}$$

**Example 9.2**

## One-cell interaction

Consider the data in Table 9.2 (Table 1 of Oehlert 1994). These data are responses from an experiment with four factors, each at two levels labeled low and high, and replicated twice. A standard factorial ANOVA of these data shows that all main effects and interactions are highly significant, and analysis of the residuals reveals no problems. In fact, these data follow an additive model, except for one unusual treatment. Thus all interaction in these data is one-cell interaction.

The interacting cell is the treatment combination with all factors low (it is about 12.5 units higher than the additive model predicts); casual inspection of the data would probably suggest the treatment with mean 11.2, but that is incorrect. We can see the one-cell interaction in Figure 9.1, which shows an

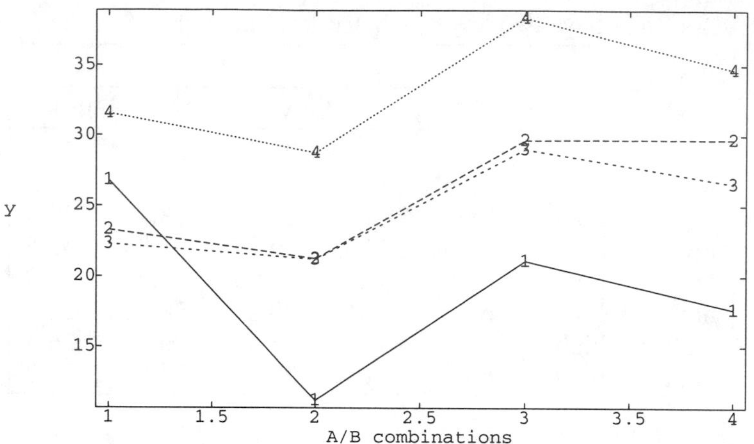


**Figure 9.1:** Interaction plot for data in Table 9.2, using MacAnova. Horizontal locations 1 through 4 correspond to (A low, B low), (A high, B low), (A low, B high), and (A high, B high). Curves 1 through 4 correspond to (C low, D low), (C high, D low), (C low, D high), and (C high, D high).

interaction plot of the treatment means. The first mean in the line labeled 1 is too high, but the other segments are basically parallel.

### 9.2.3 Quantitative factors

Polynomial models for quantitative factors

A second type of interaction that can be easily modeled occurs when one or more of the factors have quantitative levels (doses). First consider the situation when the interacting factors are all quantitative. Suppose that the doses for factor A are $z_{Ai}$, and those for factor B are $z_{Bj}$. We can build a polynomial regression model for cell means as

$$\mu_{ij} = \theta_0 + \sum_{r=1}^{a-1} \theta_{Ar} z_{Ai}^r + \sum_{s=1}^{b-1} \theta_{Bs} z_{Bj}^s + \sum_{r=1}^{a-1} \sum_{s=1}^{b-1} \theta_{ArBs} z_{Ai}^r z_{Bj}^s \quad .$$

Polynomial terms in $z_{Ai}$ model the main effects of factor A, polynomial terms in $z_{Bj}$ model the main effects of factor B, and cross product terms model the AB interaction. Models of this sort are most useful when relatively few of the polynomial terms are needed to provide an adequate description of the response.

A polynomial term $z_{Ai}^{r} z_{Bj}^{s}$ is characterized by its exponents $(r, s)$. A term with exponents $(r, s)$ is "above" a term with exponents $(u, v)$ if $r \leq u$ *and* $s \leq v$; we also say that $(u, v)$ is below $(r, s)$. The mnemonic here is that in an ANOVA table, simpler terms (such as main effects) are above more complicated terms (such as interactions). This is a little confusing, because we also use the phrase *higher order* for the more complicated terms, but higher order terms appear below the simpler terms.

Lower powers are above higher powers

A term in this polynomial model is needed if its own sum of squares is large, or if it is above a term with a large sum of squares. This preserves a polynomial hierarchy. We compute the sum of squares for a term by looking at the difference in error sums of squares for two models: subtract the error sum of squares for the model that contains the term of interest, and all terms that are above it from the error sum of squares for the model that contains only the terms above the term of interest. Thus, the sum of squares for the term $z_{Ai}^{2} z_{Bi}^{1}$ is the error sum of squares for the model with terms $z_{Ai}$, $z_{Ai}^{2}$, $z_{Bi}$ and $z_{Ai}z_{Bi}$, less the error sum of squares for the model with terms $z_{Ai}$, $z_{Ai}^{2}$, $z_{Bi}$, $z_{Ai}z_{Bi}$, and $z_{Ai}^{2} z_{Bi}^{1}$.

Use hierarchical polynomial models

Computing polynomial sums of squares

Computation of the polynomial sums of squares can usually be accomplished in statistical software with one command. Recall, however, that the polynomial coefficients $\theta$ depend on what other polynomial terms are in a given regression model. Thus if we determine that only linear and quadratic terms are needed, we must refit the model with just those terms to find their coefficients when the higher order terms are omitted. In particular, you should not use coefficients from the full model when predicting with a model with fewer terms. Use the full model $MS_E$ for determining which terms to include, but use coefficients computed for a model including just your selected terms.

Compute polynomial coefficients for final model including only selected terms

For single-factor models, we were able to compute polynomial sums of squares using polynomial contrasts when the sample sizes are equal and the doses are equally spaced. The same is true for balanced factorials with equally spaced doses. Polynomial main-effect contrast coefficients are the same as the polynomial contrast coefficients for single-factor models, and polynomial interaction contrast coefficients are the elementwise products of the polynomial main-effect contrasts.

Polynomial contrasts

### Amylase activity, continued

**Example 9.3**

Recall the amylase specific activity data of Example 8.10. The three factors are analysis temperature, growth temperature, and variety. On the log scale,

**Listing 9.1:** MacAnova output for polynomial effects in the log amylase activity data.

```
              DF         SS            MS          P-value
at^1          1     0.87537       0.87537                0
at^2          1      2.0897        2.0897                0
at^3          1    0.041993      0.041993        0.0072804
at^4          1   0.0028388     0.0028388          0.47364
at^5          1  1.3373e-06    1.3373e-06          0.98757
at^6          1   0.0034234     0.0034234          0.43154
at^7          1    0.002784      0.002784          0.47792
gt            1   0.0043795     0.0043795          0.37398
gt*at^1       1    0.035429      0.035429         0.013298
gt*at^2       1  8.9037e-05    8.9037e-05          0.89882
gt*at^3       1    0.029112      0.029112         0.024224
gt*at^4       1   0.0062113     0.0062113          0.29033
gt*at^5       1   0.0068862     0.0068862          0.26577
gt*at^6       1   0.0009846     0.0009846          0.67262
gt*at^7       1   0.0023474     0.0023474          0.51452
```

the analysis temperature by growth temperature interaction (both quantitative variables) was marginally significant. Let us explore the main effects and interactions using quantitative variables. We cannot use the tabulated contrast coefficients here because the levels of analysis temperature are not equally spaced.

Listing 9.1 gives the ANOVA for the polynomial main effects and interactions of analysis temperature (`at`) and growth temperature (`gt`). The $MS_E$ for this experiment was .00546 with 64 degrees of freedom. We see that linear, quadratic, and cubic terms in analysis temperature are significant, but no higher order terms. Also the cross products of linear in growth temperature and linear and cubic analysis temperature are significant. Thus a succinct model would include the three lowest order terms for analysis temperature, growth temperature, and their cross products. We need to refit with just those terms to get coefficients.

This example also illustrates a bothersome phenomenon—the averaging involved in multi-degree-of-freedom mean squares can obscure some interesting effects in a cloud of uninteresting effects. The 7 degree-of-freedom growth temperature by analysis temperature interaction is marginally significant with a $p$-value of .054, but some individual degrees of freedom in that 7 degree-of-freedom bundle are rather more significant.

There can also be interaction between a quantitative factor and a non-quantitative factor. Here are a couple of ways to proceed. First, we can use interaction contrasts that are products of a polynomial contrast in the quantitative factor and an interesting contrast in the qualitative factor. For example, we might have three drugs at four doses, with one control drug and two new drugs. An interesting contrast with coefficients (1, -.5, -.5) compares the control drug to the mean of the new drugs. The interaction contrast formed by the product of this contrast and linear in dose would compare the linear effect of dose in the new drugs with the linear effect of dose in the control drug.

Interaction of quantitative and qualitative factors

Second, we can make polynomial models of the response (as a function of the quantitative factor) separately for each level of the qualitative factor. Let $\mu_{ij}$ be the expected response at level $i$ of a quantitative factor with dose $z_{Ai}$ and level $j$ of a qualitative factor. We have a choice of several equivalent models, including:

Separate polynomial models

$$\mu_{ij} = \theta_j + \sum_{r=1}^{a-1} \theta_{Arj} z_{Ai}^r$$

and

$$\mu_{ij} = \theta_0 + \beta_j + \sum_{r=1}^{a-1} \theta_{Ar0} z_{Ai}^r + \sum_{r=1}^{a-1} \theta\beta_{Arj} z_{Ai}^r \ ,$$

where $\theta_j = \theta_0 + \beta_j$, $\theta_{Arj} = \theta_{Ar0} + \theta\beta_{Arj}$, and the parameters have the zero sum restrictions $\sum_j \beta_j = 0$ and $\sum_j \theta\beta_{Arj} = 0$.

In both forms there is a separate polynomial of degree $a - 1$ in $z_{Ai}$ for each level of factor B. The only difference between these models is how the regression coefficients are expressed. In the first version the constant terms of the model are expressed as $\theta_j$; in the second version the constant terms are expressed as an overall constant $\theta_0$ plus deviations $\beta_j$ that depend on the qualitative factor. In the first version the coefficients for power $r$ are expressed as $\theta_{Arj}$; in the second version the coefficients for power $r$ are expressed as an overall coefficient $\theta_{Ar0}$ plus deviations $\theta\beta_{Arj}$ that depend on the qualitative factor. These are analogous to having treatment means $\mu_i$ written as $\mu + \alpha_i$, an overall mean plus treatment effects.

Alternate forms for regression coefficients

Suppose again that we have three drugs at four doses; do we need separate cubic coefficients for the different drugs, or will one overall coefficient suffice? To answer this we can test the null hypothesis that all the $\theta_{A3j}$'s equal each other, or equivalently, that all the $\theta\beta_{A3j}$'s are zero. In many statistics packages it is easier to do the tests using the overall-plus-deviation form of the model.

Overall plus deviation form can be easier for testing

### Seed viability

**Example 9.4**

Let's examine the interaction in the data from Problem 8.7. The interaction plot in Figure 9.2 shows the interaction very clearly: there is almost

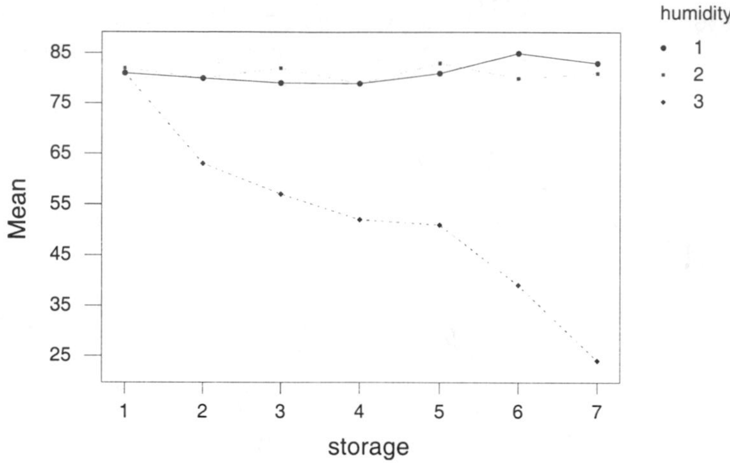


**Figure 9.2:** Interaction plot for seed viability data, using Minitab.

no dependence on storage time at the two lowest humidities, and considerable dependence on storage time at the highest humidity. Thus even though humidity is a quantitative variable, it is descriptive to treat it as qualitative.

Listing 9.2 shows MacAnova output for the viability data. This model begins with an overall constant and polynomial terms in storage, and then adds the deviations from the overall terms that allow separate polynomial coefficients for each level of humidity. Terms up to cubic in storage time are significant. There is modest evidence for some terms higher order than cubic, but their effects are small compared to the included terms and so will be ignored. To get the coefficients for the needed terms, refit using only those terms; the estimated values for the coefficients will change dramatically.

The overall storage by humidity interaction has 12 degrees of freedom and 4154.2 sum of squares. It appears from the interaction plot that most of the interaction is a difference in slope (coefficient of the linear term) between the highest level of humidity and the lower two levels. We can address that observation with an interaction contrast with coefficients

**Listing 9.2:** MacAnova output for polynomial effects in the viability activity data.

```
              DF           SS           MS      P-value
CONSTANT       1   3.2226e+05   3.2226e+05            0
{s}            1         1562         1562            0
{(s)^2}        1       5.3842       5.3842      0.29892
{(s)^3}        1       191.16       191.16   1.6402e-07
{(s)^4}        1     0.001039     0.001039      0.98841
{(s)^5}        1      0.22354      0.22354      0.83134
{(s)^6}        1       29.942       29.942     0.017221
h              2        11476       5738.2            0
{s}.h          2       3900.5       1950.2            0
{(s)^2}.h      2       17.672       8.8359      0.17532
{(s)^3}.h      2       185.81       92.906   1.2687e-06
{(s)^4}.h      2       25.719        12.86     0.083028
{(s)^5}.h      2       5.6293       2.8147      0.56527
{(s)^6}.h      2       18.881       9.4405      0.15643
ERROR1        42       204.43       4.8673
```

$$\begin{array}{ccccccc} -3 & -2 & -1 & 0 & 1 & 2 & 3 \\ -3 & -2 & -1 & 0 & 1 & 2 & 3 \\ 6 & 4 & 2 & 0 & -2 & -4 & -6 \end{array} .$$

This contrast has sum of squares 3878.9, which is over 93% of the total interaction sum of squares.

### 9.2.4 Tukey one-degree-of-freedom for nonadditivity

Transformable nonadditivity is reduced on the correct scale

The *Tukey one-degree-of-freedom* model for interaction is also called *transformable nonadditivity,* because interaction of this kind can usually be reduced or even eliminated by transforming the response by an appropriate power. (Some care needs to be taken when using this kind of transformation, because the transformation to reduce interaction could introduce nonconstant variance.) The form of a Tukey interaction is similar to that of a linear by linear interaction, but the Tukey model can be used with nonquantitative factors.

The Tukey model can be particularly useful in single replicates, where we have no estimate of pure error and generally must use high-order interactions as surrogate error. If we can transform to a scale that removes much of the interaction, then using high-order interactions as surrogate error is much more palatable.

In a two-factor model, Tukey interaction has the form $\alpha\beta_{ij} = \eta\alpha_i\beta_j/\mu$, for some multiplier $\eta$. If interaction is of this form, then transforming the

Tukey interaction is a scaled product of main effects

responses with a power $1 - \eta$ will approximately remove the interaction. You may recall our earlier admonition that an interaction effect $\alpha\beta_{ij}$ was not the product of the main effects; well, the Tukey model of interaction for the two-factor model is a multiple of just that product. The Tukey model adds one additional parameter $\eta$, so it is a one-degree-of-freedom model for nonadditivity. The form of the Tukey interaction for more general models is discussed in Section 9.3, but it is always a single degree of freedom scale factor times a combination of other model parameters.

Algorithm to fit a Tukey one-degree-of-freedom interaction

There are several algorithms for fitting a Tukey interaction and testing its significance. The following algorithm is fairly general, though somewhat obscure.

1. Fit a preliminary model; this will usually be an additive model.
2. Get the predicted values from the preliminary model; square them and divide their squares by twice the mean of the data.
3. Fit the data with a model that includes the preliminary model and the rescaled squared predicted values as explanatory variables.
4. The improvement sum of squares going from the preliminary model to the model including the rescaled squared predicted values is the single degree of freedom sum of squares for the Tukey model.
5. Test for significance of a Tukey type interaction by dividing the Tukey sum of squares by the error mean square from the model including squared predicted terms.
6. The coefficient for the rescaled squared predicted values is $\widehat{\eta}$, an estimate of $\eta$. If Tukey interaction is present, transform the data to the power $1 - \widehat{\eta}$ to remove the Tukey interaction.

The transforming power $1-\eta$ found in this way is approximate and can often be improved slightly.

**Example 9.5**

## CPU page faults, continued

Recall the CPU page fault data from Example 8.8. We originally analyzed those data on the log scale because they simply looked multiplicative. Would we have reached the same conclusion via a Tukey interaction analysis?

Listing 9.3 ① shows the ANOVA for the four main effects and rescaled, squared predicted values from the additive model on the raw data. The Tukey interaction is highly significant, with an F-statistic of 241. The coefficient for the rescaled, squared predicted values is .899 with a standard error of about .06 ②, so the estimated power transformation is $1 - .899 = .101$ with the

**Listing 9.3:** SAS output for Tukey one-degree-of-freedom interaction in the page faults data.

```
                        General Linear Models Procedure

Dependent Variable: FAULTS
                                        Sum of           Mean
Source                    DF           Squares         Square   F Value       Pr > F

Model                      8         764997314       95624664    107.38       0.0001

Error                     45          40074226         890538

Corrected Total           53         805071540

                    R-Square              C.V.       Root MSE              FAULTS Mean

                    0.950223          37.42933        943.683                  2521.24

                        General Linear Models Procedure

Dependent Variable: FAULTS

Source                    DF       Type I SS    Mean Square   F Value       Pr > F

SEQ                        2        59565822       29782911     33.44       0.0001
SIZE                       2       216880816      108440408    121.77       0.0001
ALLOC                      2       261546317      130773159    146.85       0.0001
ALG                        1        11671500       11671500     13.11       0.0007
RSPV                       1       215332859      215332859    241.80       0.0001   ①

                        General Linear Models Procedure

Dependent Variable: FAULTS

                                             T for H0:    Pr > |T|   Std Error of
Parameter                    Estimate     Parameter=0                  Estimate

Tukey eta                  0.89877776           15.55      0.0001     0.05779942     ②
```

same standard error, or approximately a log transformation. Thus a Tukey interaction analysis confirms our choice of the log transformation.

The main effects account for about 68% of the total sum of squares before transformation, and about 93% after transformation. As we saw, some interactions are still significant, but they are smaller compared to the main effects after transformation.

## 9.3 Further Reading and Extensions

One way of understanding Tukey models is to suppose that we have a simple structure for values $\mu_{ij} = \mu + \alpha_i + \beta_j$. Let's divide through by $\mu$ and assume that row and column effects are relatively small compared to the mean. We now have $\mu_{ij} = \mu(1 + \alpha_i/\mu + \beta_j/\mu)$. But instead of working with data on this scale, suppose that we have these data raised to the $1/\lambda$ power. Then the observed mean structure looks like

$$\begin{aligned}
(1+\frac{\alpha_i}{\mu}+\frac{\beta_j}{\mu})^{1/\lambda} &\approx 1+\frac{\alpha_i}{\mu}+\frac{\beta_j}{\mu}+\frac{1-\lambda}{2\mu^2\lambda^2}(\alpha_i^2+2\alpha_i\beta_j+\beta_j^2) \\
&= 1+\frac{\alpha_i}{\mu}+\frac{1-\lambda}{2\mu^2\lambda^2}\alpha_i^2+\frac{\beta_j}{\mu}+\frac{1-\lambda}{2\mu^2\lambda^2}\beta_j^2+\frac{1-\lambda}{\mu^2\lambda^2}\alpha_i\beta_j \\
&\approx 1+\frac{\alpha_i}{\mu}+\frac{1-\lambda}{2\mu^2\lambda^2}\alpha_i^2+\frac{\beta_j}{\mu}+\frac{1-\lambda}{2\mu^2\lambda^2}\beta_j^2+ \\
&\qquad (1-\lambda)(\frac{\alpha_i}{\mu}+\frac{1-\lambda}{2\mu^2\lambda^2}\alpha_i^2)(\frac{\beta_j}{\mu}+\frac{1-\lambda}{2\mu^2\lambda^2}\beta_j^2) \\
&= (\mu+r_i+c_j+(1-\lambda)\frac{r_ic_j}{\mu})\frac{1}{\mu} \quad ,
\end{aligned}$$

where the first approximation is via a Taylor series and

$$\begin{aligned}
r_i &= \frac{\alpha_i}{\mu}+\frac{1-\lambda}{2\mu^2\lambda^2}\alpha_i^2 \\
c_j &= \frac{\beta_j}{\mu}+\frac{1-\lambda}{2\mu^2\lambda^2}\beta_j^2 \quad .
\end{aligned}$$

Thus when we see mean structure of the form $\mu + r_i + c_j + (1-\lambda)r_ic_j/\mu$, we should be able to recover an additive structure by taking the data to the power $\lambda$. That is, the transformation power is one minus the coefficient of the cross product term. The cross products $r_ic_j/\mu$ are called the comparison values, because we can compare the residuals from the additive model to these comparison values to see if Tukey style interaction is present.

Here is why our algorithm works for assessing Tukey interaction. We are computing the improvement sum of squares for adding a single degree of freedom term $X$ to a model $M$. In any ANOVA or regression, the improvement sum of squares obtained by adding the $X$ to $M$ is the same as the sum of squares for the single degree of freedom model consisting of the residuals of $X$ fit to $M$. For the Tukey interaction procedure in a two-way factorial, the predicted values have the form $\widehat{\mu} + \widehat{\alpha}_i + \widehat{\beta}_j$, so the rescaled squared predicted values equal

$$\frac{\widehat{\mu}}{2} + (\widehat{\alpha}_i + \frac{\widehat{\alpha}_i^{\,2}}{2\widehat{\mu}}) + (\widehat{\beta}_j + \frac{\widehat{\beta}_j^{\,2}}{2\widehat{\mu}}) + \frac{\widehat{\alpha}_i\widehat{\beta}_j}{\widehat{\mu}} \ .$$

If we fit the additive model to these rescaled squared predicted values, the residuals will be $\widehat{\alpha}_i\widehat{\beta}_j/\widehat{\mu}$. These residuals are exactly the comparison values, so the sum of squares for the squared predicted values entered last will be equal to the sum of squares for the comparison values.

What do we do for comparison values in more complicated models; for example, three factors instead of two? For two factors, the comparison values are the product of the row and column effects divided by the mean. The comparison values for other models are the sums of the cross products of all the terms in the simple model divided by the mean. For example:

| Simple Model | Tukey Interaction |
|---|---|
| $\mu + \alpha_i + \beta_i + \gamma_k$ | $\eta(\frac{\alpha_i\beta_j}{\mu} + \frac{\alpha_i\gamma_k}{\mu} + \frac{\beta_i\gamma_k}{\mu})$ |
| $\mu + \alpha_i + \beta_i + \gamma_k + \delta_l$ | $\eta(\frac{\alpha_i\beta_i}{\mu} + \frac{\alpha_i\gamma_k}{\mu} + \frac{\alpha_i\delta_i}{\mu} + \frac{\beta_i\gamma_k}{\mu} + \frac{\beta_i\delta_l}{\mu} + \frac{\gamma_k\delta_l}{\mu})$ |
| $\mu + \alpha_i + \beta_i + \alpha\beta_{ij} + \gamma_k$ | $\eta(\frac{\alpha_i\beta_i}{\mu} + \frac{\alpha_i\gamma_k}{\mu} + \frac{\alpha_i\alpha\beta_{ij}}{\mu} + \frac{\beta_i\gamma_k}{\mu} + \frac{\beta_i\alpha\beta_{ij}}{\mu} + \frac{\gamma_k\alpha\beta_{ij}}{\mu})$ |

Once we have the comparison values, we can get their coefficient and the Tukey sum of squares by adding the comparison values to our ANOVA model. In all cases, using the rescaled squared predicted values from the base model accomplishes the same task.

There are several further models of interaction that can be useful, particularly for designs with only one data value per treatment. (See Cook and Weisberg 1982, section 2.5, for a fuller discussion.) Mandel (1961) introduced the *row-model, column-model,* and *slopes-model.* These are generalizations of the Tukey model of interaction, and take the forms

| | |
|---|---|
| Row-model: | $\mu_{ij} = \mu + \alpha_i + \beta_j + \zeta_j\alpha_i$ |
| Column-model: | $\mu_{ij} = \mu + \alpha_i + \beta_j + \xi_i\beta_j$ |
| Slopes-model: | $\mu_{ij} = \mu + \alpha_i + \beta_j + \zeta_j\alpha_i + \xi_i\beta_j$ . |

Clearly, the slopes-model is just the union of the row- and column-models. These models have the restrictions that

$$\sum_j \zeta_j = \sum_i \xi_i = 0 \ ,$$

so they represent $b-1$, $a-1$, and $a+b-2$ degrees of freedom respectively in the $(a-1)(b-1)$ degree of freedom interaction. The Tukey model is the special case where $\zeta_j = \eta\beta_j$ or $\xi_i = \eta\alpha_i$. It is not difficult to verify that the row- and column-models of interaction are orthogonal to the main effects and each other (though not to the Tukey model, which they include, or the slopes-model, which includes both of them).

The interpretation of these models is not too hard. The row-model states that mean value of each treatment is a linear function of the row effects, but the slope $(1+\zeta_j)$ and intercept $(\mu+\beta_j)$ differ from column to column. Similarly, the column-model states that the mean value of each treatment is a linear function of the column effects, but the slope $(1+\xi_i)$ and intercept $(\mu+\alpha_i)$ differ from row to row.

Johnson and Graybill (1972) proposed a model of interaction that does not depend on the main effects:

$$\alpha\beta_{ij} = \delta v_i u_j \ ,$$

with the restrictions that $\sum_i v_i = \sum_j u_j = 0$, and $\sum_i v_i^2 = \sum_j u_j^2 = 1$. This more general structure can model several forms of nonadditivity, including one cell interactions and breakdown of the table into separate additive parts. The components $\delta$, $v_i$, and $u_j$ are computed from the singular value decomposition of the residuals from the additive model. See Cook and Weisberg for a detailed discussion of this procedure.

## 9.4 Problems

**Problem 9.1** Fat acidity is a measure of flour quality that depends on the kind of flour, how the flour has been treated, and how long the flour is stored. In this experiment there are two types of flour (Patent or First Clear); the flour treatment factor (extraction) has eleven levels, and the flour has been stored for one of six periods (0, 3, 6, 9, 15, or 21 weeks). We observe only one unit for each factor-level combination. The response is fat acidity in mg KOH/100 g flour (data from Nelson 1961). Analyze these data. Of particular interest are the effect of storage time and how that might depend on the other factors.

| T | W | Extraction 1 | 2 | 3 | 4 | 5 | 6 | 7 | 8 | 9 | 10 | 11 |
|---|---|---|---|---|---|---|---|---|---|---|---|---|
| P | 0 | 12.7 | 12.3 | 15.4 | 13.3 | 13.9 | 30.3 | 123.9 | 53.4 | 29.4 | 11.4 | 19.0 |
| | 3 | 11.3 | 16.4 | 18.1 | 14.6 | 10.5 | 27.5 | 112.3 | 48.9 | 31.4 | 11.6 | 29.1 |
| | 6 | 16.5 | 24.3 | 27.2 | 10.9 | 11.6 | 34.1 | 117.5 | 52.9 | 38.3 | 15.8 | 17.1 |
| | 9 | 10.9 | 30.8 | 24.5 | 13.5 | 13.2 | 33.2 | 107.4 | 49.6 | 42.9 | 17.8 | 15.9 |
| | 15 | 12.5 | 30.6 | 26.5 | 15.8 | 13.3 | 36.2 | 109.5 | 51.0 | 15.2 | 18.2 | 13.5 |
| | 21 | 15.2 | 36.3 | 36.8 | 14.4 | 13.1 | 43.2 | 98.6 | 48.2 | 58.6 | 22.2 | 17.6 |
| FC | 0 | 36.5 | 38.5 | 38.4 | 27.1 | 35.0 | 38.3 | 274.6 | 241.4 | 21.8 | 34.2 | 34.2 |
| | 3 | 35.4 | 68.5 | 63.6 | 41.4 | 34.5 | 76.8 | 282.8 | 231.8 | 47.9 | 33.9 | 33.2 |
| | 6 | 35.7 | 93.2 | 76.7 | 50.2 | 34.0 | 96.4 | 270.8 | 223.2 | 65.2 | 38.9 | 35.2 |
| | 9 | 33.8 | 95.0 | 113.0 | 44.9 | 36.1 | 94.5 | 271.6 | 200.1 | 75.0 | 39.0 | 34.7 |
| | 15 | 43.0 | 156.7 | 160.0 | 30.2 | 33.0 | 75.8 | 269.5 | 213.6 | 88.9 | 37.9 | 33.0 |
| | 21 | 53.0 | 189.3 | 199.3 | 41.0 | 45.5 | 143.9 | 136.1 | 198.9 | 104.0 | 39.2 | 37.1 |

**Problem 9.2**

Artificial insemination is an important tool in agriculture, but freezing semen for later use can reduce its potency (ability to produce offspring). Here we are trying to understand the effect of freezing on the potency of chicken semen. Four semen mixtures are prepared, consisting of equal parts of either fresh or frozen Rhode Island Red semen, and either fresh or frozen White Leghorn semen. Sixteen batches of Rhode Island Red hens are assigned at random, four to each of the four treatments. Each batch of hens is inseminated with the appropriate mixture, and the response measured is the fraction of the hatching eggs that have white feathers and thus White Leghorn fathers (data from Tajima 1987). Analyze these data to determine how freezing affects potency of chicken semen.

| RIR | WL | | | | |
|---|---|---|---|---|---|
| Fresh | Fresh | .435 | .625 | .643 | .615 |
| Frozen | Frozen | .500 | .600 | .750 | .750 |
| Fresh | Frozen | .250 | .267 | .188 | .200 |
| Frozen | Fresh | .867 | .850 | .846 | .950 |

**Problem 9.3**

Explore the interaction in the pacemaker delamination data introduced in Problem 8.4.

**Problem 9.4**

Explore the interaction in the tropical grass production data introduced in Problem 8.6.

**Problem 9.5**

One measure of the effectiveness of cancer drugs is their ability to reduce the number of viable cancer cells in laboratory settings. In this experiment, the A549 line of malignant cells is plated onto petri dishes with various concentrations of the drug cisplatin. After 7 days of incubation, half the petri dishes at each dose are treated with a dye, and the number of viable cell colonies per 500 mm$^2$ is determined as a response for all petri dishes (after Figure 1 of Alley, Uhl, and Lieber 1982). The dye is supposed to make the counting machinery more specific to the cancer cells.

| | Cisplatin (ng/ml) | | | | | |
|---|---|---|---|---|---|---|
| | 0 | .15 | 1.5 | 15 | 150 | 1500 |
| Conventional | 200 | 178 | 158 | 132 | 63 | 40 |
| Dye added | 56 | 50 | 45 | 63 | 18 | 14 |

Analyze these data for the effects of concentration and dye. What can you say about interaction?

**Problem 9.6** An experiment studied the effects of starch source, starch concentration, and temperature on the strength of gels. This experiment was completely randomized with sixteen units. There are four starch sources (adzuki bean, corn, wheat, and potato), two starch percentages (5% and 7%), and two temperatures (22°C and 4°C). The response is gel strength in grams (data from Tjahjadi 1983).

| Temperature | Percent | Bean | Corn | Wheat | Potato |
|---|---|---|---|---|---|
| 22 | 5 | 62.9 | 44.0 | 43.8 | 34.4 |
| | 7 | 110.3 | 115.6 | 123.4 | 53.6 |
| 4 | 5 | 60.1 | 57.9 | 58.2 | 63.0 |
| | 7 | 147.6 | 180.7 | 163.8 | 92.0 |

Analyze these data to determine the effects of the factors on gel strength.

**Question 9.1** Show how to construct simultaneous confidence intervals for all pairwise differences of interaction effects $\widehat{\alpha\beta}_{ij}$ using Bonferroni. Hint: first find the variances of the differences.

**Question 9.2** Determine the condition for orthogonality of two main-effects contrasts for the same factor when the data are unbalanced.

**Question 9.3** Show that an interaction contrast $w_{ij}$ in the means $\overline{y}_{ij\bullet\bullet}$ equals the corresponding contrast in the interaction effects $\widehat{\alpha\beta}_{ij}$.

# Chapter 10

# Further Topics in Factorials

There are many more things to learn about factorials; this chapter covers just a few, including dealing with unbalanced data, power and sample size for factorials, and special methods for two-series designs.

## 10.1 Unbalanced Data

Our discussion of factorials to this point has assumed *balance;* that is, that all factor-level combinations have the same amount of replication. When this is not true, the data are said to be *unbalanced*. The analysis of unbalanced data is more complicated, in part because there are no simple formulae for the quantities of interest. Thus we will need to rely on statistical software for all of our computation, and we will need to know just exactly what the software is computing, because there are several variations on the basic computations.

Balanced versus unbalanced data

The root cause of these complications has to do with orthogonality, or rather the lack of it. When the data are balanced, a contrast for one main effect or interaction is orthogonal to a contrast for any other main effect or interaction. One consequence of this orthogonality is that we can estimate effects and compute sums of squares one term at a time, and the results for that term do not depend on what other terms are in the model. When the data are unbalanced, the results we get for one term depend on what other terms are in the model, so we must to some extent do all the computations simultaneously.

Imbalance destroys orthogonality

The questions we want to answer do not change because the data are unbalanced. We still want to determine which terms are required to model

Build models and/or test hypotheses

the response adequately, and we may wish to test specific null hypotheses about model parameters. We made this distinction for balanced data in Section 8.11, even though the test statistics for comparing models or testing hypotheses are the same. For unbalanced data, this distinction actually leads to different tests.

Use exact methods

Our discussion will be divided into two parts: building models and testing hypotheses about parameters. We will consider only exact approaches for computing sums of squares and doing tests. There are approximate methods for unbalanced factorials that were popular before the easy availability of computers for doing all the hard computations. But when you have the computational horsepower, you might as well use it to get exact results.

### 10.1.1 Sums of squares in unbalanced data

SS adjusted for terms in reduced model

Terms in model affect SS

We have formulated the sum of squares for a term in a balanced ANOVA model as the difference in error sum of squares for a reduced model that excludes the term of interest, and that same model with the term of interest included. The term of interest is said to have been "adjusted for" the terms in the reduced model. We also presented simple formulae for these sums of squares. When the data are unbalanced, we still compute the sum of squares for a term as a difference in error sums of squares for two models, but there are no simple formulae to accomplish that task. Furthermore, precisely which two models are used doesn't matter in balanced data so long as they only differ by the term of interest, but which models are used *does* matter for unbalanced data.

$SS(B|1, A)$ is SS of B adjusted for 1 and A

Models are usually specified as a sequence of terms. For example, in a three-factor design we might specify (1, A, B, C) for main effects, or (1, A, B, AB, C) for main effects and the AB interaction. The "1" denotes the overall grand mean $\mu$ that is included in all models. The sum of squares for a term is the difference in error sums of squares for two models that differ only by that term. For example, if we look at the the two models (1, A, C) and (1, A, B, C), then the difference in error sums of squares will be the sum of squares for B adjusted for 1, A, and C. We write this as $SS(B|1, A, C)$.

**Example 10.1** **Unbalanced amylase data**

Recall the amylase data of Example 8.10, where we explore how amylase activity depends on analysis temperature (A), variety (B), and growth temperature (C). Suppose that the first observation in growth temperature 25, analysis temperature 40, and variety B73 were missing, making the data unbalanced. The sum of squares for factor C is computed as the difference in error sums of squares for a pair of models differing only in the term C.

Here are five such model pairs: (1), (1, C); (1, A), (1, A, C); (1, B), (1, B, C); (1, A, B), (1, A, B, C); (1, A, B, AB), (1, A, B, AB, C). The sums of squares for C computed using these five model pairs are denoted $SS(C|1)$, $SS(C|1, A)$, $SS(C|1, B)$, $SS(C|1, A, B)$ and $SS(C|1, A, B, AB)$, and are shown in following table (sum of squares $\times 10^6$, data on log scale):

| | |
|---|---|
| $SS(C\|1)$ | 2444.1 |
| $SS(C\|1, A)$ | 1396.0 |
| $SS(C\|1, B)$ | 3303.0 |
| $SS(C\|1, A, B)$ | 2107.4 |
| $SS(C\|1, A, B, AB)$ | 2069.4 |

All five of these sums of squares differ, some rather substantially. There is no single sum of squares for C, so we must explicitly state which one we are using at any give time.

The simplest choice for a sum of squares is *sequential* sums of squares. This is called Type I in SAS. For sequential sums of squares, we specify a model and the sum of squares for any term is adjusted for those terms that precede it in the model. If the model is (1, A, B, AB, C), then the sequential sums of squares are $SS(A, 1)$, $SS(B|1, A)$, $SS(AB|1, A, B)$, and $SS(C|1, A, B, AB)$. Notice that if you specify the terms in a different order, you get different sums of squares; the sequential sums of squares for (1, A, B, C, AB) are $SS(A, 1)$, $SS(B|1, A)$, $SS(C|1, A, B)$, and $SS(AB|1, A, B, C)$.

Type I SS is sequential

Type I SS depends on order of terms

Two models that include the same terms in different order will have the same estimated treatment effects and interactions. However, models that include different terms may have different estimated effects for the terms they have in common. Thus (1, A, B, AB, C) and (1, A, B, C, AB) will have the same $\widehat{\alpha}_i$'s, but (1, A, B, AB, C) and (1, A, B, C) may have different $\widehat{\alpha}_i$'s.

Estimated effects don't depend on order of terms

### 10.1.2 Building models

*Building models* means deciding which main effects and interactions are needed to describe the data adequately. I build hierarchical models. In a hierarchical model, the inclusion of any interaction in a model implies the inclusion of any term that is "above" it, where we say that a factorial term U is above a factorial term V if every factor in term U is also in term V. The goal is to find the hierarchical model that includes all terms that must be included, but does not include any unnecessary terms.

Compare hierarchical models

Our approach to computing sums of squares when model-building is to use as the reduced model for term U the largest hierarchical model M that

Type II SS or Yates' fitting constants

does not contain U. This is called Type II in the SAS statistical program. In two-factor models, this might be called "Yates' fitting constants" or "each adjusted for the other."

Type II adjusts for largest hierarchal model not including term

Consider computing Type II sums of squares for all the terms in a three-factor model. The largest hierarchical models not including ABC, BC, and C are (1, A, B, C, AB, AC, BC), (1, A, B, C, AC, AB), and (1, A, B, AB), respectively. Thus for Type II sums of squares, the three-factor interaction is adjusted for all main effects and two-factor interactions, a two-factor interaction is adjusted for all main effects and the other two-factor interactions, and a main effect is adjusted for the other main effects and their interactions, or $SS(ABC|1, A, B, C, AB, AC, BC)$, $SS(BC|1, A, B, C, AB, AC)$, and $SS(C|1, A, B, AB)$. In Example 10.1, the Type II sum of squares for growth temperature (factor C) is $2069 \times 10^{-6}$.

Use $MS_E$ from full model

It is important to point out that the denominator mean square used for testing is $MS_E$ from the full model. We do not pool "unused" terms into error. Thus, the Type II SS for C is $SS(C|1, A, B, AB)$, but the error mean square is from the model (1, A, B, C, AB, AC, BC, ABC).

**Example 10.2**

### Unbalanced amylase data, continued

Listing 10.1 ① shows SAS output giving the Type II analysis for the unbalanced amylase data of Example 10.1. Choose the hierarchical model by starting at the three-factor interaction. The three-factor interaction is not significant ($p$-value .21) and so will not be retained in the model. Because it is not needed, we can now test to see if any of the two-factor interactions are needed. Growth temperature by variety is highly significant; therefore, that interaction and the main effects of growth temperature and variety will be in our final model. Neither the analysis temperature by growth temperature interaction nor the analysis temperature by variety interaction is significant, so they will not be retained. We may now test analysis temperature, which is significant. We do not test the other main effects because they are implied by the significant two-factor interaction. The final model is all three main effects and the growth temperature by variety interaction.

Get Type II SS from Type I SS

If your software does not compute Type II sums of squares directly, you can determine them from Type I sums of squares for a sequence of models with the terms arranged in different orders. For example, suppose we have the Type I sums of squares for the model (1, A, B, AB, C, AC, BC, ABC). Then the Type I sums of squares for ABC, BC, and C are also Type II sums of squares. Type I sums of squares for (1, B, C, BC, A, AB, AC, ABC) allow us to get Type II sums of squares for A, AC, ABC, and so on.

**Listing 10.1:** SAS output for unbalanced amylase data.

```
                    General Linear Models Procedure

Dependent Variable: LY
                                  Sum of           Mean
Source                 DF        Squares         Square    F Value      Pr > F

Model                  31     3.83918760     0.12384476      23.26      0.0001

Error                  63     0.33537806     0.00532346

Source                 DF     Type II SS    Mean Square    F Value      Pr > F

ATEMP                   7     3.03750534     0.43392933      81.51      0.0001
GTEMP                   1     0.00206944     0.00206944       0.39      0.5352
ATEMP*GTEMP             7     0.06715614     0.00959373       1.80      0.1024
VAR                     1     0.55989306     0.55989306     105.17      0.0001
ATEMP*VAR               7     0.02602887     0.00371841       0.70      0.6731
GTEMP*VAR               1     0.07863197     0.07863197      14.77      0.0003
ATEMP*GTEMP*VAR         7     0.05355441     0.00765063       1.44      0.2065  ①

Source                 DF    Type III SS    Mean Square    F Value      Pr > F

ATEMP                   7     3.03041604     0.43291658      81.32      0.0001
GTEMP                   1     0.00258454     0.00258454       0.49      0.4885
ATEMP*GTEMP             7     0.06351586     0.00907369       1.70      0.1241
VAR                     1     0.55812333     0.55812333     104.84      0.0001
ATEMP*VAR               7     0.02580103     0.00369872       0.69      0.6701
GTEMP*VAR               1     0.07625999     0.07625999      14.33      0.0003
ATEMP*GTEMP*VAR         7     0.05355441     0.00765063       1.44      0.2065  ②

Contrast               DF    Contrast SS    Mean Square    F Value      Pr > F

gtemp low vs high       1     0.00258454     0.00258454       0.49      0.4885  ③
```

Type I sums of squares for the terms in a model will sum to the overall model sum of squares with $g - 1$ degrees of freedom. This is not true for Type II sums of squares, as can be seen in Listing 10.1; the model sum of squares is 3.8392, but the Type II sums of squares add to 3.8248.

The Type II approach to model building is not foolproof. The following example shows that in some situations the overall model can be highly significant, even though none of the individual terms in the model is significant.

### Unbalanced data puzzle

**Example 10.3**

Consider the data in Table 10.1. These data are highly unbalanced. Listing 10.2 gives SAS output for these data, including Type I and II sums of

**Table 10.1:** A highly unbalanced two by two factorial.

| A | B | |
|---|---|---|
| | 1 | 2 |
| 1 | 2.7 7.9 26.3 -1.9 30.6<br>3.8 27.2 20.9 20.6 14.6 | 21.5 |
| 2 | 26.1 | 41.1 46.7 57.8 38 39.3 |

squares at ② and ③. Note that the Type I and II sums of squares for B and AB are the same, because B enters the model after A and so is adjusted for A in Type I; similarly, AB enters after A and B and is adjusted for them in the Type I analysis. A enters first, so its Type I sum of squares $SS(A|1)$ is not Type II.

Also shown at ① is the sum of squares with 3 degrees of freedom for the overall model, ignoring the factorial structure. The overall model is significant with a $p$-value of about .002. However, neither the interaction nor either main effect has a Type II $p$-value less than .058. Thus the overall model is highly significant, but none of the individual terms is significant.

What has actually happened in these data is that either A or B alone explains a large amount of variation (see the sum of squares for A in ②), but they are in some sense explaining the same variation. Thus B is not needed if A is already present, A is not needed if B is already present, and the interaction is never needed.

### 10.1.3 Testing hypotheses

Standard tests are for equally weighted factorial parameters

In some situations we may wish to test specific hypotheses about treatment means rather than building a model to describe the means. Many of these hypotheses can be expressed in terms of the factorial parameters, but recall that the parameters we use in our factorial decomposition carry a certain amount of arbitrariness in that they assume equally weighted averages. When the hypotheses of interest correspond to our usual, equally weighted factorial parameters, testing is reasonably straightforward; otherwise, special purpose contrasts must be used.

Let's review how means and parameters correspond in the two-factor situation. Let $\mu_{ij}$ be the mean of the $ij$th treatment:

$$\mu_{ij} = \mu + \alpha_i + \beta_j + \alpha\beta_{ij}$$

**Listing 10.2:** SAS output for data in Table 10.1.

```
                  General Linear Models Procedure

Dependent Variable: Y
                                     Sum of           Mean
Source                   DF         Squares         Square   F Value      Pr > F

Model                     3      2876.88041      958.96014      8.53      0.0022  ①

Error                    13      1460.78900      112.36838

Corrected Total          16      4337.66941

Source                   DF       Type I SS    Mean Square   F Value      Pr > F

A                         1      2557.00396     2557.00396     22.76      0.0004  ②
B                         1       254.63189      254.63189      2.27      0.1561
A*B                       1        65.24457       65.24457      0.58      0.4597

Source                   DF      Type II SS    Mean Square   F Value      Pr > F

A                         1      485.287041     485.287041      4.32      0.0581  ③
B                         1      254.631889     254.631889      2.27      0.1561
A*B                       1       65.244565      65.244565      0.58      0.4597

Source                   DF     Type III SS    Mean Square   F Value      Pr > F

A                         1      499.951348     499.951348      4.45      0.0549  ④
B                         1      265.471348     265.471348      2.36      0.1483
A*B                       1       65.244565      65.244565      0.58      0.4597
```

with

$$0 = \sum_i \alpha_i = \sum_j \beta_j = \sum_i \alpha\beta_{ij} = \sum_j \alpha\beta_{ij} \ .$$

Let $n_{ij}$ be the number of observations in the $ij$th treatment. Form row and column averages of treatment means using equal weights for the treatment means:

Row and column averages of treatment expected values

$$\begin{aligned}
\mu_{i\bullet} &= \sum_{j=1}^{b} \mu_{ij}/b \\
&= \mu + \alpha_i \ , \\
\mu_{\bullet j} &= \sum_{i=1}^{a} \mu_{ij}/a \\
&= \mu + \beta_j \ .
\end{aligned}$$

The null hypothesis that the main effects of factor A are all zero ($\alpha_i \equiv 0$) is the same as the null hypothesis that all the row averages of the treatment means are equal ($\mu_{1\bullet} = \mu_{2\bullet} = \cdots = \mu_{a\bullet}$). This is also the same as the null hypothesis that all factor A main-effects contrasts evaluate to zero.

Test equally weighted hypotheses using Type III SS or standard parametric

Testing the null hypothesis that the main effects of factor A are all zero ($\alpha_i \equiv 0$) is accomplished with an F-test. We compute the sum of squares for this hypothesis by taking the difference in error sum of squares for two models: the full model with all factors and interactions, and that model with the main effect of factor A deleted, or $SS(A|1, B, C, AB, AC, BC, ABC)$ in a three-factor model. This reduced model is not hierarchical; it includes interactions with A but not the main effect of A. Similarly, we compute a sum of squares for any other hypothesis that a set of factorial effects is all zero by comparing the sum of squares for the full model with the sum of squares for the model with that effect removed. This may be called "standard parametric," "Yates' weighted squares of means," or "fully adjusted"; in SAS it is called Type III.

**Example 10.4**

### Unbalanced data puzzle, continued

Let us continue Example 10.3. If we wish to test the null hypothesis that $\alpha_i \equiv 0$ or $\beta_j \equiv 0$, we need to use Type III sums of squares. This is shown at ④ of Listing 10.2. None of the null hypotheses about main effects or interaction is anywhere near as significant as the overall model; all have *p*-values greater than .05.

How can this be so when we know that there are large differences between treatment means in the data? Consider for a moment the test for factor A main effects. The null hypothesis is that the factor A main effects are zero, but no constraint is placed on factor B main effects or interactions. We can fit the data fairly well with the $\alpha_i$'s equal to zero, so long as we can manipulate the $\beta_j$'s and $\alpha\beta_{ij}$'s to take up the slack. Similarly, when testing factor B, no constraint is placed on factor A main effects or AB interactions. These three tests of A, B, and AB do not test that all three null hypotheses are true simultaneously. For that we need to test the overall model with 3 degrees of freedom.

Contrast SS are Type III

When we test the null hypothesis that a contrast in treatment effects is zero, we are testing the null hypothesis that a particular linear combination of treatment means is zero with no other restrictions on the cell means. This is equivalent to testing that the single degree of freedom represented by the contrast can be removed from the full model, so the contrast has been adjusted for all other effects in the model. Thus the sum of squares for any contrast is a Type III sum of squares.

### Unbalanced amylase data, continued

Example 10.5

Continuing Example 10.1, the Type III ANOVA can be found at ② in Listing 10.1. The Type III sum of squares for growth temperature is .0025845, different from both Types I and II. If you compute the main-effect contrast in growth temperature with coefficients 1 and -1, you get the results shown at ③ in Listing 10.1, including the same sum of squares as the Type III analysis. This equivalence of the effect sum of squares and the contrast sum of squares is due to the fact that the effect has only a single degree of freedom, and thus the contrast describes the entire effect.

The only factorial null hypotheses that would be rejected are those for the main effects of analysis temperature and variety and the interaction of growth temperature and variety. Thus while growth temperature and variety jointly act to influence the response, there is no evidence that the average responses for the two growth temperatures differ (equally weighted averages across all analysis temperatures and varieties).

## 10.1.4 Empty cells

The problems of unbalanced data are increased when one or more of the cells are empty, that is, when there are no data for some factor-level combinations. The model building/Type II approach to analysis doesn't really change. We can just keep comparing hierarchical models. The hypothesis testing approach becomes very problematic, however, because the parameters about which we are making hypotheses are no longer uniquely defined, even when we are sure we want to work with equal weighting.

Empty cells make factorial effects ambiguous

When there are empty cells, there are infinitely many different sets of factorial effects that fit the observed treatment means exactly; these different sets of effects disagree on what they fit for the empty cells. Consider the following three by two table of means with one empty value, and two different factorial decompositions of the means into grand mean, row, column, and interaction effects.

Multiple sets of parameters with different fits for empty cells

| | | | | | | | | |
|---|---|---|---|---|---|---|---|---|
| | | 156.0 | -23.0 | 23.0 | | 133.0 | .0 | .0 |
| 196 | 124 | 4.0 | 59.0 | -59.0 | | 27.0 | 36.0 | -36.0 |
| 156 | 309 | 76.5 | -53.5 | 53.5 | | 99.5 | -76.5 | 76.5 |
| 47 | | -80.5 | -5.5 | 5.5 | | -126.5 | 40.5 | -40.5 |

Both of these factorial decompositions meet the usual zero-sum requirements, and both add together to match the table of means exactly. The first is what would be obtained if the empty cell had mean 104, and the second if the empty cell had mean -34.

Use contrasts to analyze data with empty cells

Because the factorial effects are ambiguous, it makes no sense to test hypotheses about the factorial model parameters. For example, are the column effects above zero or nonzero? What does make sense is to look at simple effects and to set up contrasts that make factorial-like comparisons where possible. For example, levels 1 and 2 of factor A are complete, so we can compare those two levels with a contrast. Note that the difference of row means is 72.5, and $\alpha_2 - \alpha_1$ is 72.5 in both decompositions. We might also want to compare level 1 of factor B with level 2 of factor B for the two levels of factor A that are complete. There are many potential ways to choose interesting contrasts for designs with empty cells.

## 10.2 Multiple Comparisons

F-tests in factorial ANOVA not usually adjusted for multiple comparisons

The perceptive reader may have noticed that we can do a lot of F-tests in the analysis of a factorial, but we haven't been talking about multiple comparisons adjustments. Why this resounding silence, when we were so careful to describe and account for multiple testing for pairwise comparisons? I have no good answer; common statistical practice seems inconsistent in this regard. What common practice does is treat each main effect and interaction as a separate "family" of hypotheses and make multiple comparisons adjustments within a family (Section 9.1) but not between families.

Be wary of isolated significant interactions

We sometimes use an informal multiple comparisons correction when building hierarchical models. Suppose that we have a three-way factorial, and only the three-way interaction is significant, with a $p$-value of .04; the main-effects and two-factor interactions are not near significance. I would probably conclude that the low $p$-value for the three-way interaction is due to chance rather than interaction effects. I conclude this because I usually expect main effects to be bigger than two-factor interactions, and two-factor interactions to be bigger than three-factor interactions. I thus interpret an isolated, marginally significant three-way interaction as a null result. I know that isolated three-way interaction can occur, but it seems less likely to me than chance occurrence of a moderately low $p$-value.

We could also adopt a predictive approach to model selection (as in Section 5.4.9) and choose that hierarchical model that has lowest Mallows' $C_p$. Models chosen by predictive criteria can include more terms than those chosen via tests, because the $C_p$ criterion corresponds to including terms with F-tests greater than 2.

## 10.3 Power and Sample Size

Chapter 7 described the computation of power and sample size for completely randomized designs. If we ignore the factorial structure and consider our treatments simply as $g$ treatments, then we can use the methods of Chapter 7 to compute power and sample size for the overall null hypothesis of no model effects. Power depends on the Type I error rate $\mathcal{E}_I$, numerator and denominator degrees of freedom, and the effects, sample sizes, and error variance through the noncentrality parameter.

Compute power for main effects and interactions separately

For factorial data, we usually test null hypotheses about main effects or interactions in addition to the overall null hypothesis of no model effects. Power for these tests again depends on the Type I error rate $\mathcal{E}_I$, numerator and denominator degrees of freedom, and the effects, sample sizes, and error variance through the noncentrality parameter, so we can do the same kinds of power and sample size computations for factorial effects once we identify the degrees of freedom and noncentrality parameters.

Power for balanced data

We will address power and sample size only for balanced data, because most factorial experiments are designed to be balanced, and simple formulae for noncentrality parameters exist only for balanced data. For concreteness, we present the formulae in terms of a three-factor design; the generalization to more factors is straightforward. In a factorial, main effects and interactions are tested separately, so we can perform a separate power analysis for each main effect and interaction. The numerator degrees of freedom are simply the degrees of freedom for the factorial effect: for example, $(b-1)(c-1)$ for the BC interaction. Error degrees of freedom $(N-abc)$ are the denominator degrees of freedom.

The noncentrality parameter depends on the factorial parameters, sample size, and error variance. The algorithm for a noncentrality parameter in a balanced design is

Noncentrality parameter

1. Square the factorial effects and sum them,
2. Multiply this sum by the total number of data in the design divided by the number of levels in the effect, and
3. Divide that product by the error variance.

For the AB interaction, this noncentrality parameter is

$$\frac{\frac{N}{ab}\sum_{ij}\alpha\beta_{ij}^2}{\sigma^2} = \frac{nc\sum_{ij}\alpha\beta_{ij}^2}{\sigma^2} .$$

The factor in step 2 equals the number of data values observed at each level of the given effect. For the AB interaction, there are $n$ values in each treatment, and $c$ treatments with the same $ij$ levels, for a total of $nc$ observations in each $ij$ combination.

As in Chapter 7, minimum sample sizes to achieve a given power are found iteratively, literally by trying different sample sizes and finding the smallest one that does the job.

## 10.4 Two-Series Factorials

All factors have exactly two levels in two-series factorials

A *two-series* factorial design is one in which all the factors have just two levels. For $k$ factors, we call this a $2^k$ design, because there are $2^k$ different factor-level combinations. Similarly, a design with $k$ factors, each with three levels, is a three-series design and denoted by $3^k$. Two-series designs are somewhat special, because they are the smallest designs with $k$ factors. They are often used when screening many factors.

Levels called low and high

Because two-series designs are so common, there are special notations and techniques associated with them. The two levels for each factor are generally called *low* and *high*. These terms have clear meanings if the factors are quantitative, but they are often used as labels even when the factors are not quantitative. Note that "off" and "on" would work just as well, but low and high are the usual terms.

Lower-case letters denote factors at high levels

Do not confuse treatments like $bc$ with effects like BC

There are two methods for denoting a factor-level combination in a two-series design. The first uses letters and is probably the more common. Denote a factor-level combination by a string of lower-case letters: for example, *bcd*. We have been using these lower-case letters to denote the number of levels in different factors, but all factors in a two-series design have two levels, so there should be no confusion. Letters that are present correspond to factors at their high levels, and letters that are absent correspond to factors at their low levels. Thus *ac* is the combination where factors A and C are at their high levels and all other factors are at their low levels. Use the symbol (1) to denote the combination where all factors are at their low levels. Denote the mean response at a given factor-level combination by $\overline{y}$ with a subscript, for example $\overline{y}_{ab}$. Do not confuse the factor-level combination $bc$ with the interaction BC; the former is a single treatment, and the latter is a contrast among treatments.

Binary digits, 1 for high, 0 for low

The second method uses numbers and generalizes to three-series and higher-order factorials as well. A factor-level combination is denoted by $k$ binary digits, with one digit giving the level of each factor: a zero denotes a factor at its low level, and a one denotes a factor at its high level. Thus 000 is all factors at low level, the same as (1), and 011 is factors B and C at high level, the same as *bc*. This generalizes to other factorials by using more

**Table 10.2:** Pluses and minuses for a $2^3$ design.

| | A | B | C |
|---|---|---|---|
| (1) | $-$ | $-$ | $-$ |
| a | $+$ | $-$ | $-$ |
| b | $-$ | $+$ | $-$ |
| ab | $+$ | $+$ | $-$ |
| c | $-$ | $-$ | $+$ |
| ac | $+$ | $-$ | $+$ |
| bc | $-$ | $+$ | $+$ |
| abc | $+$ | $+$ | $+$ |

digits. For example, we use the digits 0, 1, and 2 to denote the three levels of a three-series.

Standard order prescribes a pattern for listing factor-level combinations

It is customary to arrange the factor-level combinations of a two-series factorial in *standard order*. Standard order will help us keep track of factor-level combinations when we later modify two-series designs. Historically, standard order was useful for Yates' algorithm (see next section). Standard order for a two-series design begins with (1). Then proceed through the remainder of the factor-level combinations with factor A varying fastest, then factor B, and so on. In standard order, factor A will repeat the pattern low, high; factor B will repeat the pattern low, low, high, high; factor C will repeat the pattern low, low, low, low, high, high, high, high; and so on though other factors. In general, the $j$th factor will repeat a pattern of $2^{j-1}$ lows followed by $2^{j-1}$ highs. For a $2^4$, standard order is (1), $a$, $b$, $ab$, $c$, $ac$, $bc$, $abc$, $d$, $ad$, $bd$, $abd$, $cd$, $acd$, $bcd$, and $abcd$.

Table of + and –

Two-series factorials form the basis of several designs we will consider later, and one of the tools we will use is a table of pluses and minuses. For a $2^k$ design, build a table with $2^k$ rows and $k$ columns. The rows are labeled with factor-level combinations in standard order, and the columns are labeled with the $k$ factors. In principle, the body of the table contains $+1$'s and $-1$'s, with $+1$ indicating a factor at a high level, and $-1$ indicating a factor at a low level. In practice, we use just plus and minus signs to denote the factor levels. Table 10.2 shows this table for a $2^3$ design.

### 10.4.1 Contrasts

One nice thing about a two-series design is that every main effect and interaction is just a single degree of freedom, so we may represent any main effect or interaction by a single contrast. For example, the main effect of factor A in a $2^3$ can be expressed as

$$\begin{aligned}
\widehat{\alpha}_2 &= -\widehat{\alpha}_1 \\
&= \overline{y}_{2\bullet\bullet} - \overline{y}_{\bullet\bullet\bullet} \\
&= \frac{1}{8}(\overline{y}_a + \overline{y}_{ab} + \overline{y}_{ac} + \overline{y}_{abc} - \overline{y}_{(1)} - \overline{y}_b - \overline{y}_c - \overline{y}_{bc}) \\
&= \frac{1}{8}(-\overline{y}_{(1)} + \overline{y}_a - \overline{y}_b + \overline{y}_{ab} - \overline{y}_c + \overline{y}_{ac} - \overline{y}_{bc} + \overline{y}_{abc}) \ ,
\end{aligned}$$

Two-series effects are contrasts

which is a contrast in the eight treatment means with plus signs where A is high and minus signs where A is low. Similarly, the sum of squares for A can be written

$$\begin{aligned}
SS_A &= 4n\widehat{\alpha}_1{}^2 + 4n\widehat{\alpha}_2{}^2 \\
&= \frac{n}{8}(\overline{y}_a + \overline{y}_{ab} + \overline{y}_{ac} + \overline{y}_{abc} - \overline{y}_{(1)} - \overline{y}_b - \overline{y}_c - \overline{y}_{bc})^2 \\
&= \frac{n}{8}(-\overline{y}_{(1)} + \overline{y}_a - \overline{y}_b + \overline{y}_{ab} - \overline{y}_c + \overline{y}_{ac} - \overline{y}_{bc} + \overline{y}_{abc})^2 \ ,
\end{aligned}$$

Effect contrasts same as columns of pluses and minuses

which is the sum of squares for the contrast $w_A$ with coefficients $+1$ where A is high and $-1$ where A is low (or .25 and $-.25$, or $-17.321$ and 17.321, as the sum of squares is unaffected by a nonzero multiplier for the contrast coefficients). Note that this contrast $w_A$ has exactly the same pattern of pluses and minuses as the column for factor A in Table 10.2.

The difference

$$\overline{y}_{2\bullet\bullet\bullet} - \overline{y}_{1\bullet\bullet\bullet} = \widehat{\alpha}_2 - \widehat{\alpha}_1 = 2\widehat{\alpha}_2$$

Total effect

is the *total effect* of factor A. The total effect is the average response where A is high, minus the average response where A is low, so we can also obtain the total effect of factor A by rescaling the contrast $w_A$

$$\overline{y}_{2\bullet\bullet\bullet} - \overline{y}_{1\bullet\bullet\bullet} = \frac{1}{4}\sum_{ijk} w_{Aijk}\,\overline{y}_{ijk\bullet} \ ,$$

where the divisor of 4 is replaced by $2^{k-1}$ for a $2^k$ design.

Interaction contrasts are products of main-effects contrasts

The columns of Table 10.2 give us contrasts for the main effects. Interactions in the two-series are also single degrees of freedom, so there must be contrasts for them as well. We obtain these interaction contrasts by taking elementwise products of main-effects contrasts. For example, the coefficients in the contrast for the BC interaction are the products of the coefficients for the B and C contrasts. A three-way interaction contrast is the product of the

**Table 10.3:** All contrasts for a $2^3$ design.

| | A | B | C | AB | AC | BC | ABC |
|---|---|---|---|---|---|---|---|
| (1) | − | − | − | + | + | + | − |
| a | + | − | − | − | − | + | + |
| b | − | + | − | − | + | − | + |
| ab | + | + | − | + | − | − | − |
| c | − | − | + | + | − | − | + |
| ac | + | − | + | − | + | − | − |
| bc | − | + | + | − | − | + | − |
| abc | + | + | + | + | + | + | + |

three main-effects contrasts, and so on. This is most easily done by referring to the columns of Table 10.2, with $+$ and $-$ interpreted as $+1$ and $-1$. We show these contrasts for a $2^3$ design in Table 10.3.

Yates' algorithm efficiently computes effects in two-series

Yates' algorithm is a method for efficient computation of the effects in a two-series factorial. It can be modified to work in three-series and general factorials, but we will only discuss it for the two-series. Yates' algorithm begins with the treatment means in standard order and produces the grand mean and factorial effects in standard order with a minimum of computation. Looking at Table 10.3, we see that there are $2^k$ effect columns (adding a column of all ones for the overall mean) each involving $2^k$ additions, subtractions, or multiplications for a total of $2^{2k}$ operations. Yates' algorithm allows us to get the same results with $k2^k$ operations, a substantial savings for hand computation and worth consideration in computer software as well.

Each column is sums and differences of preceding column

Arrange the treatment means of a $2^k$ in standard order in a column; call it column 0. Yates' algorithm computes the effects in $k$ passes through the data, each pass producing a new column. We perform an operation on column 0 to get column 1; then we perform the same operation on column 1 to get column 2; and so on. The operation is sums and differences of successive pairs. To make a new column, the first half of the elements are found as sums of successive pairs in the preceding column. The last half of the elements are found as differences of successive pairs in the preceding column.

For example, in a $2^3$, the elements of column 0 (the data) are $\overline{y}_{(1)}$, $\overline{y}_a$, $\overline{y}_b$, $\overline{y}_{ab}$, $\overline{y}_c$, $\overline{y}_{ac}$, $\overline{y}_{bc}$, $\overline{y}_{abc}$. The elements in column 1 are: $\overline{y}_{(1)} + \overline{y}_a$, $\overline{y}_b + \overline{y}_{ab}$, $\overline{y}_c + \overline{y}_{ac}$, $\overline{y}_{bc} + \overline{y}_{abc}$, $\overline{y}_a - \overline{y}_{(1)}$, $\overline{y}_{ab} - \overline{y}_b$, $\overline{y}_{ac} - \overline{y}_c$, and $\overline{y}_{abc} - \overline{y}_{bc}$. We repeat the same operation on column 1 to get column 2: $\overline{y}_{(1)} + \overline{y}_a + \overline{y}_b + \overline{y}_{ab}$, $\overline{y}_c + \overline{y}_{ac} + \overline{y}_{bc} + \overline{y}_{abc}$, $\overline{y}_a - \overline{y}_{(1)} + \overline{y}_{ab} - \overline{y}_b$, $\overline{y}_{ac} - \overline{y}_c + \overline{y}_{abc} - \overline{y}_{bc}$, $\overline{y}_b + \overline{y}_{ab} - \overline{y}_{(1)} - \overline{y}_a$, $\overline{y}_{bc} + \overline{y}_{abc} - \overline{y}_c - \overline{y}_{ac}$, $\overline{y}_{ab} - \overline{y}_b - \overline{y}_a + \overline{y}_{(1)}$, and $\overline{y}_{abc} - \overline{y}_{bc} - \overline{y}_{ac} + \overline{y}_c$. This procedure continues through the remaining columns.

**Table 10.4:** Yates' algorithm for the pacemaker substrate data.

| | Data | 1 | 2 | 3 | Effects | |
|---|---|---|---|---|---|---|
| (1) | 4.388 | 7.219 | 14.686 | 29.090 | 3.636 | Mean |
| a | 2.831 | 7.467 | 14.404 | -5.735 | -.717 | A |
| b | 4.360 | 7.598 | -2.809 | -.544 | -.068 | B |
| ab | 3.107 | 6.806 | -2.926 | -.500 | -.062 | AB |
| c | 4.330 | -1.556 | .248 | -.282 | -.035 | C |
| ac | 3.268 | -1.252 | -.791 | -.117 | -.015 | AC |
| bc | 4.336 | -1.061 | .304 | -1.039 | -.130 | BC |
| abc | 2.471 | -1.865 | -.804 | -1.108 | -.138 | ABC |

After $k$ passes, the $k$th column contains the total of the treatment means and the effect contrasts with $\pm 1$ coefficients applied to the treatment means. These results are in standard order (total, A effect, B effect, AB effect, and so on). To get the grand mean and effects, divide column $k$ by $2^k$.

**Example 10.6** **Pacemaker substrates**

We use the data of Problem 8.4. This was a $2^3$ experiment with two replications; factors A—profile time, B—airflow, and C—laser; and response the fraction of substrates delaminating. The column labeled *Data* in Table 10.4 shows the treatment means for the log scale data. Columns labeled *1, 2,* and *3* are the three steps of Yates' algorithm, and the final column is the grand mean followed by the seven factorial effects in standard order. Profile time (A) clearly has the largest effect (in absolute value).

### 10.4.2 Single replicates

Single replicates need an estimate of error

As with all factorials, a single replication in a two-series design means that we have no degrees of freedom for error. We can apply any of the usual methods for single replicates to a two-series design, but there are also methods developed especially for single replicate two-series. We describe two of these methods. The first is graphically based and is subjective; it does not provide $p$-values. The second is just slightly more complicated, but it does allow at least approximate testing.

Effects are independent with constant variance

Both methods are based on the idea that if our original data are independent and normally distributed with constant variance, then use of the effects contrasts in Table 10.3 gives us results that are also independent and normally distributed with constant variance. The expected value of any of these contrasts is zero if the corresponding null hypothesis of no main effect or interaction is correct. If that null hypothesis is not correct, then the expected value of the contrast is not zero. So, when we look at the results, contrasts corresponding to null effects should look like a sample from a normal distribution with mean zero and fixed variance, and contrasts corresponding to non-null effects will have different means and should look like outliers. We now need a technique to identify outliers.

Significant effects are outliers

We implicitly make an assumption here. We assume that we will have mostly null results, with a few non-null results that should look like outliers. This is called *effect sparsity*. These techniques will work poorly if there are many non-null effects, because we won't have a good basis for deciding what null behavior is.

We assume effect sparsity

The first method is graphical and is usually attributed to Daniel (1959). Simply make a normal probability plot of the contrasts and look for outliers. Alternatively, we can use a *half-normal* probability plot, because we don't care about the signs of the effects when determining which ones are outliers. A half-normal probability plot plots the sorted absolute values on the vertical axis against the sorted expected scores from a half-normal distribution (that is, the expected value of $i$th smallest absolute value from a sample of size $2^k - 1$ from a normal distribution). I usually find the half-normal plots easier to interpret.

Half-normal plot of effects

The second method computes a *pseudo-standard error* (PSE) for the contrasts, allowing us to do $t$-tests. Lenth (1989) computes the PSE in two steps. First, let $s_0$ be 1.5 times the median of the absolute values of the contrast results. Second, delete any contrasts results whose absolute values are greater than $2.5s_0$, and let the PSE be 1.5 times the median of the remaining absolute contrast results. Treat the PSE as a standard error for the contrasts with $(2^k - 1)/3$ degrees of freedom, and do $t$-tests. These can be individual tests, or you can do simultaneous tests using a Bonferroni correction.

Lenth's pseudo-standard error

## Pacemaker substrates, continued

**Example 10.7**

We illustrate both methods using the pacemaker substrate data from Table 10.4. The column labeled *Effects* gives the grand mean and effects. Removing the grand mean, we make a half-normal plot of the remaining seven effects, as shown in Figure 10.1. Effect 1, the main effect of A, appears as a clear outlier, and the rest appear to follow a nice line. Thus we would conclude subjectively that A is significant, but no other effects are significant.

To use Lenth's method, we first need the median of the absolute factorial effects, .068 for these data. We next delete any absolute effects greater than $2.5 \times .068 = .17$; only the the main effect of A meets this cutoff. The median of the remaining absolute effects is .065, so the PSE is $1.5 \times .065 = .098$.

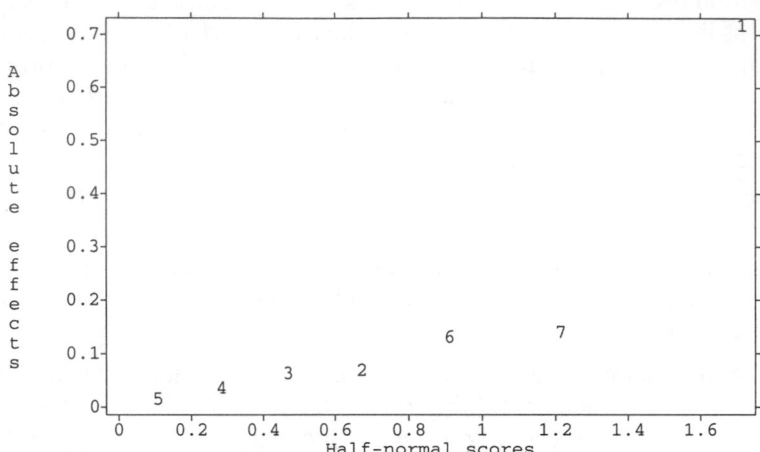


**Figure 10.1:** Half-normal plot of factorial effects for the log pacemaker substrate data, using MacAnova. Numbers indicate standard order: 1 is A, 7 is ABC, and so on.

We treat this PSE as having 7/3 degrees of freedom. With this criterion, the main effect of A has a two-sided $p$-value of about .01, in agreement with our subjective conclusion.

A single nonzero response yields effects equal in absolute value

An interesting feature of two-series factorials can be seen if you look at a data set consisting of all zeroes except for a single nonzero value. All factorial effects for such a data set are equal in absolute value, but some will be positive and some negative, depending on which data value is nonzero and the pattern of pluses and minuses. For example, suppose that $c$ has a positive value and all other responses are zero. Looking at the row for $c$ in Table 10.3, the effects for C, AB, and ABC should be positive, and the effects for A, B, AC, and BC should be negative. Similarly, if $bc$ had a negative value and all other responses were zero, then the row for $bc$ shows us that A, AB, AC, and ABC would be positive, and B, C, and BC would be negative. The patterns of positive and negative effects are unique for all combinations of which response is nonzero and whether the response is positive or negative.

Flat spots in half normal plot may mean one-cell interaction

When a two-series design contains a large one-cell interaction, many of what should be null effects will have about the same absolute value, and we will see an approximate horizontal line in the half-normal plot. By matching the signs of the seemingly constant effects (or their inverses) to rows of tables of pluses and minuses, we can determine which cell is interacting.

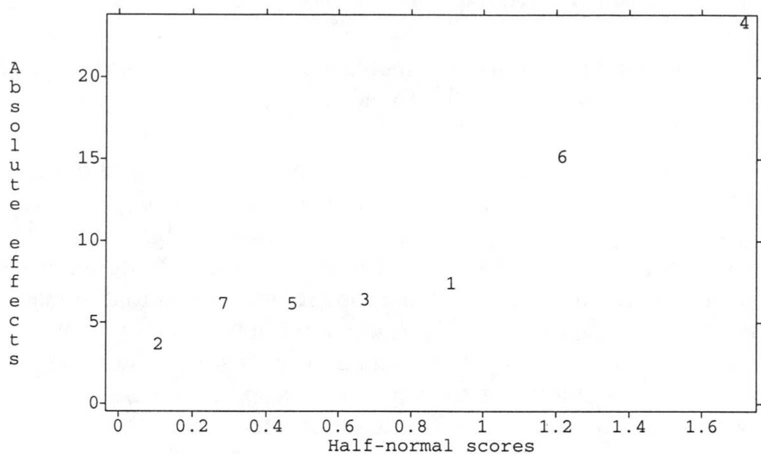


**Figure 10.2:** Half-normal plot of factorial effects for seed maturation data, using MacAnova.

**Example 10.8**

## Seed maturation on cut stems

Sixteen heliopsis (sunflower) blooms were cut with 15 cm stems and the stems were randomly placed in eight water solutions with the combinations of the following three factors: preservative at one-quarter or one-half strength, MG or MS preservative, 1% or 2% sucrose. After the blooms had dried, the total number of seeds for the two blooms was determined as response (data from David Zlesak). In standard order, the responses were:

| (1) | a | b | ab | c | ac | bc | abc |
|---|---|---|---|---|---|---|---|
| 12 | 10 | 60 | 8 | 89 | 87 | 52 | 49 |

Figure 10.2 shows a half-normal plot of the factorial effects. Effects 1, 2, 3, 5, and 7 (A, B, AB, AC, and ABC) seem roughly constant. Examination of the effects (not shown) reveals that A, B, and AB have negative effects, and AC and ABC have positive effects. Looking at Table 10.3, we can see that the only factor-level combination where the A, B, and AB contrasts have the same sign—and the AC and ABC contrasts have the same sign and opposite that of A, B, and AB—is the *ab* combination. Examining the data, the response of 8 for *ab* indeed looks like a one-cell interaction.

## 10.5 Further Reading and Extensions

A good expository discussion of unbalance can be found in Herr (1986); more advanced treatments can be found in texts on linear models, such as Hocking (1985).

The computational woes of unbalance are less for *proportional balance.* In a two-factor design, we have proportional balance if $n_{ij}/N = n_{i\bullet}/N \times n_{\bullet j}/N$. For example, treatments at level 1 of factor A might have replication 4, and all other treatments have replication 2. Under proportional balance, contrasts in one main effect or interaction are orthogonal to contrasts in any other main effect or interaction. Thus the order in which terms enter a model does not matter, and ordinary, Type II, and Type III sums of squares all agree. Balanced data are obviously a special case of proportional balance. For more than two factors, the rule for proportional balance is that the fraction of the data in one cell should be the product of the fractions in the different margins.

When we have specific hypotheses that we would like to test, but they do not correspond to standard factorial terms, then we must address them with special-purpose contrasts. This is reasonably easy for a single degree of freedom. For hypotheses with several degrees of freedom, we can form multidegree of freedom sums of squares for a set of contrasts using methods described in Hocking (1985) and implemented in many software packages. Alternatively, we may use Bonferroni to combine the tests of individual degrees of freedom.

It is somewhat instructive to see the hypotheses tested by approaches other than Type III. Form row and column averages of treatment means using weights proportional to cell counts:

$$
\begin{aligned}
\mu_{i\star} &= \sum_{j=1}^{b} n_{ij}\mu_{ij}/n_{i\bullet} \\
\mu_{\star j} &= \sum_{i=1}^{a} n_{ij}\mu_{ij}/n_{\bullet j} \ ;
\end{aligned}
$$

and form averages for each row of the column weighted averages, and weighted averages for each column of the row weighted averages:

$$
\begin{aligned}
(\mu_{\star j})_{i\star} &= \sum_{j=1}^{b} n_{ij}\mu_{\star j}/n_{i\bullet} \\
(\mu_{i\star})_{\star j} &= \sum_{i=1}^{a} n_{ij}\mu_{i\star}/n_{\bullet j} \ .
\end{aligned}
$$

Thus there is a $(\mu_{\star j})_{i\star}$ value for each row $i$, formed by taking a weighted average of the column weighted averages $\mu_{\star j}$. The values may differ between rows because the counts $n_{ij}$ may differ between rows, leading to different weighted averages.

Consider two methods for computing a sum of squares for factor A. We can calculate the sum of squares for factor A ignoring all other factors; this is SAS Type I for factor A first in the model, and is also called "weighted means." This sum of squares is the change in error sum of squares in going from a model with just a grand mean to a model with row effects and is appropriate for testing the null hypothesis

$$\mu_{1\star} = \mu_{2\star} = \cdots = \mu_{a\star} \ .$$

Alternatively, calculate the sum of squares for factor A adjusted for factor B; this is a Type II sum of squares for a two-way model and is appropriate for testing the null hypothesis

$$\mu_{1\star} = (\mu_{\star j})_{1\star}; \quad \mu_{2\star} = (\mu_{\star j})_{2\star}; \quad \ldots; \ \mu_{a\star} = (\mu_{\star j})_{a\star} \ .$$

That is, the Type II null hypothesis for factor A allows the row weighted means to differ, but only because they are different weighted averages of the column weighted means.

Daniel (1976) is an excellent source for the analysis of two-series designs, including unreplicated two-series designs. Much data-analytic wisdom can be found there.

## 10.6 Problems

**Exercise 10.1** Three ANOVA tables are given for the results of a single experiment. These tables give sequential (Type I) sums of squares. Construct a Type II ANOVA table. What would you conclude about which effects and interactions are needed?

| | DF | SS | MS |
|---|---|---|---|
| a | 1 | 1.9242 | 1.9242 |
| b | 2 | 1584.2 | 792.11 |
| a.b | 2 | 19.519 | 9.7595 |
| c | 1 | 1476.7 | 1476.7 |
| a.c | 1 | 17.527 | 17.527 |
| b.c | 2 | 191.84 | 95.922 |
| a.b.c | 2 | 28.567 | 14.283 |
| ERROR | 11 | 166.71 | 15.156 |

| | DF | SS | MS |
|---|---|---|---|
| b | 2 | 1585.4 | 792.7 |
| c | 1 | 1447.2 | 1447.2 |
| b.c | 2 | 221.12 | 110.56 |
| a | 1 | 2.656 | 2.656 |
| b.a | 2 | 26.876 | 13.438 |
| c.a | 1 | 8.4376 | 8.4376 |
| b.c.a | 2 | 28.567 | 14.283 |
| ERROR | 11 | 166.71 | 15.156 |

| | DF | SS | MS |
|---|---|---|---|
| c | 1 | 1275.8 | 1275.8 |
| b | 2 | 1756.8 | 878.42 |
| c.b | 2 | 221.12 | 110.56 |
| a | 1 | 2.656 | 2.656 |
| c.a | 1 | 6.569 | 6.569 |
| b.a | 2 | 28.744 | 14.372 |
| c.b.a | 2 | 28.567 | 14.283 |
| ERROR | 11 | 166.71 | 15.156 |

**Exercise 10.2** A single replicate of a $2^4$ factorial is run. The results in standard order are 1.106, 2.295, 7.074, 6.931, 4.132, 2.148, 10.2, 10.12, 3.337, 1.827, 8.698, 6.255, 3.755, 2.789, 10.99, and 11.85. Analyze the data to determine the important factors and find which factor-level combination should be used to maximize the response.

**Exercise 10.3** Here are two sequential (Type I) ANOVA tables for the same data. Complete the second table. What do you conclude about the significance of row effects, column effects, and interactions?

| | DF | SS | MS |
|---|---|---|---|
| r | 3 | 3.3255 | 1.1085 |
| c | 3 | 112.95 | 37.65 |
| r.c | 9 | 0.48787 | 0.054207 |
| ERROR | 14 | 0.8223 | 0.058736 |

| | DF | SS | MS |
|---|---|---|---|
| c | 3 | 116.25 | 38.749 |
| r | 3 | | |
| c.r | 9 | | |
| ERROR | 14 | | |

**Exercise 10.4** Consider the following two plots, which show normal and half-normal plots of the effects from an unreplicated $2^5$ factorial design. The effects are numbered starting with A as 1 and are in standard order. What would you conclude?

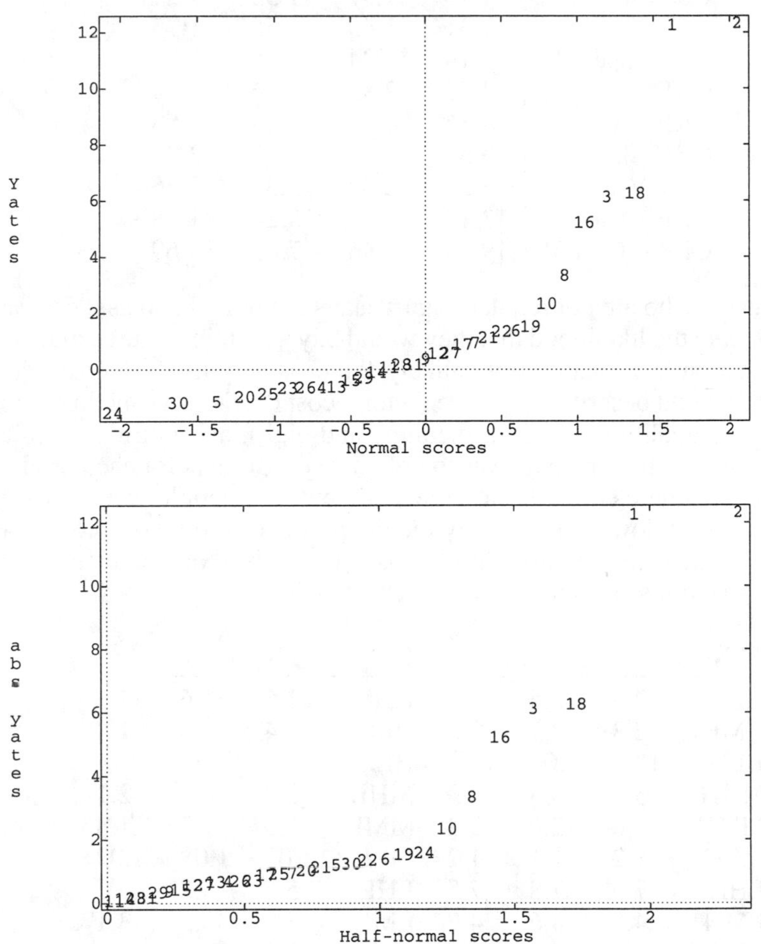


**Problem 10.1**

An experiment investigated the release of the hormone ACTH from rat pituitary glands under eight treatments: the factorial combinations of CRF (0 or 100 nM; CRF is believed to increase ACTH release), calcium (0 or 2 mM of $CaCl_2$), and Verapamil (0 or 50 $\mu$M; Verapamil is thought to block the effect of calcium). Thirty-six rat pituitary cell cultures are assigned at random to the factor-level combinations, with control (all treatments 0) getting 8 units, and other combinations getting 4. The data follow (Giguere, Lefevre, and Labrie 1982). Analyze these data and report your conclusions.

| | | | | |
|---|---|---|---|---|
| Control | 1.73 | 1.57 | 1.53 | 2.1 |
| | 1.31 | 1.45 | 1.55 | 1.75 |
| V (Verapamil) | 2.14 | 2.24 | 2.15 | 1.87 |
| CRF | 4.72 | 2.82 | 2.76 | 4.44 |
| CRF + V | 4.36 | 4.05 | 6.08 | 4.58 |
| Ca (Calcium) | 3.53 | 3.13 | 3.47 | 2.99 |
| Ca + V | 3.22 | 2.89 | 3.32 | 3.56 |
| CRF + Ca | 13.18 | 14.26 | 15.24 | 11.18 |
| CRF + Ca + V | 19.53 | 16.46 | 17.89 | 14.69 |

**Problem 10.2** Consumers who are not regular yogurt eaters are polled and asked to rate on a 1 to 9 scale the likelihood that they would buy a certain yogurt product at least once a month; 1 means very unlikely, 9 means very likely. The product is hypothetical and described by three factors: cost ("C"—low, medium, and high), sensory quality ("S"—low, medium, and high), and nutritional value ("N"—low and high). The plan was to poll three consumers for each product type, but it became clear early in the experiment that people were unlikely to buy a high-cost, low-nutrition, low-quality product, so only one consumer was polled for that combination. Each consumer received one of the eighteen product descriptions chosen at random. The data follow:

| CSN | Scores | | | CSN | Scores | | |
|---|---|---|---|---|---|---|---|
| HHH | 2.6 | 2.5 | 2.9 | HHL | 1.5 | 1.6 | 1.5 |
| HMH | 2.3 | 2.1 | 2.3 | HML | 1.4 | 1.5 | 1.4 |
| HLH | 1.05 | 1.06 | 1.05 | HLL | 1.01 | | |
| MHH | 3.3 | 3.5 | 3.3 | MHL | 2.2 | 2.0 | 2.1 |
| MMH | 2.6 | 2.6 | 2.3 | MML | 1.8 | 1.7 | 1.8 |
| MLH | 1.2 | 1.1 | 1.2 | MLL | 1.07 | 1.08 | 1.07 |
| LHH | 7.9 | 7.8 | 7.5 | LHL | 5.5 | 5.7 | 5.7 |
| LMH | 4.5 | 4.6 | 4.0 | LML | 3.8 | 3.3 | 3.1 |
| LLH | 1.7 | 1.8 | 1.8 | LLL | 1.5 | 1.6 | 1.5 |

Analyze these data for the effects of cost, quality, and nutrition on likelihood of purchase.

**Problem 10.3** Modern ice creams are not simple recipes. Many use some type of gum to enhance texture, and a non-cream protein source (for example, whey protein solids). A food scientist is trying to determine how types of gum and protein added change a sensory rating of the ice cream. She runs a five by five factorial with two replications using five gum types and five protein sources. Unfortunately, six of the units did not freeze properly, and these units were not rated. Ratings for the other units are given below (higher numbers are better).

| Gum | Protein 1 | 2 | 3 | 4 | 5 |
|---|---|---|---|---|---|
| 1 | 3.5 | 3.6 | 2.1 | 4.0 | 3.1 |
| | 3.0 | 2.9 | 4.5 | | |
| 2 | 7.2 | 6.8 | 6.7 | 7.5 | 6.8 |
| | | | 4.8 | 6.9 | 9.3 |
| 3 | 4.1 | 5.8 | 4.5 | 5.3 | 4.1 |
| | 5.6 | 4.8 | 4.6 | 7.3 | 5.3 |
| 4 | 5.3 | 4.8 | 5.0 | 6.7 | 5.2 |
| | | 3.2 | 7.2 | 6.7 | 4.2 |
| 5 | 4.5 | 5.1 | 5.0 | 4.9 | 4.5 |
| | 2.7 | 3.7 | 4.5 | 4.7 | |

Analyze these data to determine if protein and/or gum have any effect on the sensory rating. Determine which, if any, proteins and/or gums differ in their sensory ratings.

**Problem 10.4**

Gums are used to alter the texture and other properties of foods, in part by binding water. An experiment studied the water-binding of various carrageenan gums in gel systems under various conditions. The experiment had factorial treatment structure with four factors. Factor 1 was the type of gum (kappa, mostly kappa with some lambda, and iota). Factor 2 was the concentration of the gum in the gel in g/100g $H_20$ (level 1 is .1; level 2 is .5; and level 3 is 2 for gums 1 and 2, and 1 for gum 3). The third factor was type of solute (NaCl, $Na_2SO_4$, sucrose). The fourth factor was solute concentration (ku/kg $H_20$). For sucrose, the three levels were .05, .1, and .25; for NaCl and $Na_2SO_4$, the levels were .1, .25, and 1. The response is the water-binding for the gel in mOsm (data from Rey 1981). This experiment was completely randomized. There were two units at each factor-level combination except solute concentration 3, where all but one combination had four units.

Analyze these data to determine the effects and interactions of the factors. Summarize your analysis and conclusions in a report.

| S. | S. conc. | G. conc. 1 G. 1 | G. conc. 1 G. 2 | G. conc. 1 G. 3 | G. conc. 2 G. 1 | G. conc. 2 G. 2 | G. conc. 2 G. 3 | G. conc. 3 G. 1 | G. conc. 3 G. 2 | G. conc. 3 G. 3 |
|---|---|---|---|---|---|---|---|---|---|---|
| 1 | 1 | 99.7 | 97.6 | 99.0 | 100.0 | 104.7 | 107.3 | 123.0 | 125.7 | 117.3 |
| | | 98.3 | 103.7 | 98.0 | 104.3 | 105.7 | 106.7 | 116.3 | 121.7 | 117.3 |
| 1 | 2 | 239.0 | 239.7 | 237.0 | 249.7 | 244.7 | 243.7 | 277.0 | 266.3 | 268.0 |
| | | 236.0 | 246.7 | 237.7 | 255.7 | 245.7 | 247.7 | 262.3 | 276.3 | 266.7 |
| 1 | 3 | 928.7 | 940.0 | 899.3 | 937.0 | 942.7 | 953.3 | 968.0 | 992.7 | 1183.7 |
| | | 930.0 | 961.3 | 941.0 | 938.7 | 988.0 | 991.0 | 975.7 | 1019.0 | 1242.0 |
| | | 929.0 | 939.7 | 944.3 | 939.7 | 945.7 | 988.7 | 972.7 | 1018.7 | 1133.0 |
| | | 930.0 | 931.3 | 919.0 | 924.3 | 933.0 | 965.7 | 968.0 | 1021.0 | 1157.0 |
| 2 | 1 | 87.3 | 80.0 | 88.0 | 92.3 | 94.5 | 86.7 | 104.3 | 115.7 | 101.0 |
| | | 89.0 | 89.3 | 89.0 | 97.7 | 94.3 | 95.3 | 104.0 | 118.0 | 104.3 |
| 2 | 2 | 203.7 | 204.0 | 203.0 | 209.0 | 210.7 | 203.7 | 218.0 | 241.0 | 214.7 |
| | | 204.0 | 206.3 | 201.7 | 209.3 | 210.0 | 209.0 | 221.5 | 232.7 | 222.7 |
| 2 | 3 | 695.0 | 653.0 | 668.7 | 688.7 | 697.7 | 726.7 | 726.0 | 731.0 | 747.7 |
| | | 679.7 | 642.7 | 686.7 | 701.3 | 701.7 | 744.7 | 747.7 | 790.3 | 897.0 |
| | | 692.7 | 686.0 | 665.0 | 698.0 | 698.0 | 741.0 | 736.7 | 799.7 | 812.7 |
| | | 688.0 | 646.0 | 688.3 | 711.7 | 698.7 | 708.7 | 743.7 | 806.0 | 885.0 |
| 3 | 1 | 55.0 | 56.7 | 54.7 | 61.7 | 62.7 | 63.7 | 90.7 | 99.0 | 72.7 |
| | | 55.3 | 56.0 | 56.3 | 62.0 | 64.0 | 65.0 | 99.3 | 102.3 | 75.0 |
| | 2 | 123.7 | 109.7 | 105.0 | 113.3 | 115.0 | 114.3 | 229.3 | 213.4 | 123.7 |
| | | 106.0 | 111.0 | 105.7 | 115.0 | 115.7 | 116.7 | 193.7 | 196.3 | 132.7 |
| 3 | 3 | 283.3 | 271.7 | 258.3 | 277.3 | 279.3 | 282.0 | 426.5 | 399.7 | 291.7 |
| | | 276.0 | 275.3 | 268.0 | 277.0 | 283.0 | 279.3 | 389.3 | 410.3 | 308.0 |
| | | 266.0 | 267.3 | 273.3 | 281.3 | | 282.7 | 420.0 | 360.0 | 310.0 |
| | | 263.0 | 268.7 | 272.7 | 279.0 | | 281.0 | 421.7 | 409.3 | 303.3 |

**Problem 10.5** Expanded/extruded wheat flours have air cells that vary in size, and the size may depend on the variety of wheat used to make the flour, the location where the wheat was grown, and the temperature at which the flour was extruded. An experiment has been conducted to assess these factors. The first factor is the variety of wheat used (Butte 86, 2371, or Grandin). The second factor is the growth location (MN or ND). The third factor is the temperature of the extrusion (120°C or 180°C). The response is the area in $mm^2$ of the air cells (data from Sutheerawattananonda 1994).

Analyze these data and report your conclusions; variety and temperature effects are of particular interest.

| Temp. | Loc. | Var. | Response | | |
|---|---|---|---|---|---|
| 1 | 1 | 1 | 4.63 | 10.37 | 7.53 |
| 1 | 1 | 2 | 6.83 | 7.43 | 2.99 |
| 1 | 1 | 3 | 11.02 | 13.87 | 2.47 |
| 1 | 2 | 1 | 3.44 | 5.88 | |
| 1 | 2 | 2 | 2.60 | 4.48 | |
| 1 | 2 | 3 | 4.29 | 2.67 | |
| 2 | 1 | 1 | 2.80 | 3.32 | |
| 2 | 1 | 2 | 3.01 | 4.51 | |
| 2 | 1 | 3 | 5.30 | 3.58 | |
| 2 | 2 | 1 | 3.12 | 2.58 | 2.97 |
| 2 | 2 | 2 | 2.15 | 2.62 | 3.00 |
| 2 | 2 | 3 | 2.24 | 2.80 | 3.18 |

**Problem 10.6**

Anticonvulsant drugs may be effective because they encourage the effect of the neurotransmitter GABA ($\gamma$-aminobutyric acid). Calcium transport may also be involved. The present experiment randomly assigned 48 rats to eight experimental conditions. These eight conditions are the factor-level combinations of three factors, each at two levels. The factors are the anticonvulsant Trifluoperazine (brand name Stelazine) present or absent, the anticonvulsant Diazepam (brand name Valium) present or absent, and the calcium-binding protein calmodulin present or absent. The response is the amount of GABA released when brain tissues are treated with 33 mM $K^+$ (data based on Table I of de Belleroche, Dick, and Wyrley-Birch 1982).

| Tri | Dia | Cal | | | | | | | | |
|---|---|---|---|---|---|---|---|---|---|---|
| A | A | A | 1.19 | 1.33 | 1.34 | 1.23 | 1.24 | 1.23 | 1.28 | 1.32 |
| | | P | 1.07 | 1.44 | 1.14 | .87 | 1.35 | 1.19 | 1.17 | .89 |
| | P | A | .58 | .54 | .63 | .81 | | | | |
| | | P | .61 | .60 | .51 | .88 | | | | |
| P | A | A | .89 | .40 | .89 | .80 | .65 | .85 | .45 | .37 |
| | | P | 1.21 | 1.20 | 1.40 | .70 | 1.10 | 1.09 | .90 | 1.28 |
| | P | A | .19 | .34 | .61 | .30 | | | | |
| | | P | .34 | .41 | .29 | .52 | | | | |

Analyze these data and report your findings. We are interested in whether the drugs affect the GABA release, by how much, and if the calmodulin changes the drug effects.

**Problem 10.7**

In a study of patient confidentiality, a large number of pediatricians was surveyed. Each pediatrician was given a "fable" about a female patient less than 18 years old. There were sixteen different fables, the combinations of the factors complaint (C: 1—drug problem, 2—venereal disease), age (A:

1—14 years, 2—17 years), the length of time the pediatrician had known the family (L: 1—less than 1 year, 2—more than 5 years), and the maturity of patient (M: 1—immature for age, 2—mature for age). The response at each combination of factor levels is the fraction of doctors who would keep confidentiality and not inform the patient's parents (data modeled on Moses 1987). Analyze these data to determine which factors influence the pediatrician's decision.

| C | A | L | M | Response | C | A | L | M | Response |
|---|---|---|---|---|---|---|---|---|---|
| 1 | 1 | 1 | 1 | .445 | 2 | 1 | 1 | 1 | .578 |
| 1 | 1 | 1 | 2 | .624 | 2 | 1 | 1 | 2 | .786 |
| 1 | 1 | 2 | 1 | .360 | 2 | 1 | 2 | 1 | .622 |
| 1 | 1 | 2 | 2 | .493 | 2 | 1 | 2 | 2 | .755 |
| 1 | 2 | 1 | 1 | .513 | 2 | 2 | 1 | 1 | .814 |
| 1 | 2 | 1 | 2 | .693 | 2 | 2 | 1 | 2 | .902 |
| 1 | 2 | 2 | 1 | .534 | 2 | 2 | 2 | 1 | .869 |
| 1 | 2 | 2 | 2 | .675 | 2 | 2 | 2 | 2 | .902 |

**Problem 10.8** An animal nutrition experiment was conducted to study the effects of protein in the diet on the level of leucine in the plasma of pigs. Pigs were randomly assigned to one of twelve treatments. These treatments are the combinations of protein source (fish meal, soybean meal, and dried skim milk) and protein concentration in the diet (9, 12, 15, or 18 percent). The response is the free plasma leucine level in mcg/ml (data from Windels 1964)

| Meal | 9% | 12% | 15% | 18% |
|---|---|---|---|---|
| Fish | 27.8 | 31.5 | 34.0 | 30.6 |
| | 23.7 | 28.5 | 28.7 | 32.7 |
| | | 32.8 | 28.3 | 33.7 |
| Soy | 39.3 | 39.8 | 38.5 | 42.9 |
| | 34.8 | 40.0 | 39.2 | 49.0 |
| | 29.8 | 39.1 | 40.0 | 44.4 |
| Milk | 40.6 | 42.9 | 59.5 | 72.1 |
| | 31.0 | 50.1 | 48.9 | 59.8 |
| | 34.6 | 37.4 | 41.4 | 67.6 |

Analyze these data to determine the effects of the factors on leucine level.

# Chapter 11

# Random Effects

*Random effects* are another approach to designing experiments and modeling data. Random effects are appropriate when the treatments are random samples from a population of potential treatments. They are also useful for random subsampling from populations. Random-effects models make the same kinds of decompositions into overall mean, treatment effects, and random error that we have been using, but random-effects models assume that the treatment effects are random variables. Also, the focus of inference is on the population, not the individual treatment effects. This chapter introduces random-effects models.

Random effects for randomly chosen treatments and subsamples

## 11.1 Models for Random Effects

A company has 50 machines that make cardboard cartons for canned goods, and they want to understand the variation in strength of the cartons. They choose ten machines at random from the 50 and make 40 cartons on each machine, assigning 400 lots of feedstock cardboard at random to the ten chosen machines. The resulting cartons are tested for strength. This is a completely randomized design, with ten treatments and 400 units; we will refer to this as carton experiment one.

Carton experiment one, a single random factor

We have been using models for data that take the form

$$y_{ij} = \mu_i + \epsilon_{ij} = \mu + \alpha_i + \epsilon_{ij} \; .$$

The parameters of the mean structure ($\mu_i$, $\mu$, and $\alpha_i$) have been treated as fixed, unknown numbers with the treatment effects summing to zero, and

Fixed effects

the primary thrust of our inference has been learning about these mean parameters. These sorts of models are called *fixed-effects* models, because the treatment effects are fixed numbers.

Random-effects designs study populations of treatments

These fixed-effects models are not appropriate for our carton strength data. It still makes sense to decompose the data into an overall mean, treatment effects, and random error, but the fixed-effects assumptions don't make much sense here for a couple of reasons. First, we are trying to learn about and make inferences about the whole population of machines, not just these ten machines that we tested in the experiment, so we need to be able to make statements for the whole population, not just the random sample that we used in the experiment. Second, we can learn all we want about these ten machines, but a replication of the experiment will give us an entirely different set of machines. Learning about $\alpha_1$ in the first experiment tells us nothing about $\alpha_1$ in the second experiment—they are probably different machines. We need a new kind of model.

The basic random effects model begins with the usual decomposition:

$$y_{ij} = \mu + \alpha_i + \epsilon_{ij} \quad .$$

Treatment effects are random in random-effects models

We assume that the $\epsilon_{ij}$ are independent normal with mean 0 and variance $\sigma^2$, as we did in fixed effects. For random effects, we also assume that the treatment effects $\alpha_i$ are independent normal with mean 0 and variance $\sigma^2_\alpha$, and that the $\alpha_i$'s and the $\epsilon_{ij}$'s are independent of each other. Random effects models do not require that the sum of the $\alpha_i$'s be zero.

Variance components

The variance of $y_{ij}$ is $\sigma^2_\alpha + \sigma^2$. The terms $\sigma^2_\alpha$ and $\sigma^2$ are called *components of variance* or *variance components*. Thus the random-effects model is sometimes called a components of variance model. The correlation between $y_{ij}$ and $y_{kl}$ is

$$\mathrm{Cor}(y_{ij}, y_{kl}) = \begin{cases} 0 & i \neq k \\ \sigma^2_\alpha/(\sigma^2_\alpha + \sigma^2) \quad \text{for} & i = k \text{ and } j \neq l \\ 1 & i = k \text{ and } j = l \end{cases} \quad .$$

Intraclass correlation

Random effects can be specified by correlation structure

The correlation is nonzero when $i = k$ because the two responses share a common value of the random variable $\alpha_i$. The correlation between two responses in the same treatment group is called the *intraclass* correlation. Another way of thinking about responses in a random-effects model is that they all have mean $\mu$, variance $\sigma^2_\alpha + \sigma^2$, and a correlation structure determined by the variance components. The additive random-effects model and the correlation structure approach are nearly equivalent (the additive random-effects model can only induce positive correlations, but the general correlation structure model allows negative correlations).

The parameters of the random effects model are the overall mean $\mu$, the error variance $\sigma^2$, and the variance of the treatment effects $\sigma_\alpha^2$; the treatment effects $\alpha_i$ are random variables, not parameters. We want to make inferences about these parameters; we are not so interested in making inferences about the $\alpha_i$'s and $\epsilon_{ij}$'s, which will be different in the next experiment anyway. Typical inferences would be point estimates or confidence intervals for the variance components, or a test of the null hypothesis that the treatment variance $\sigma_\alpha^2$ is 0.

Tests and confidence intervals for parameters

Now extend carton experiment one. Suppose that machine operators may also influence the strength of the cartons. In addition to the ten machines chosen at random, the manufacturer also chooses ten operators at random. Each operator will produce four cartons on each machine, with the cardboard feedstock assigned at random to the machine-operator combinations. We now have a two-way factorial treatment structure with both factors random effects and completely randomized assignment of treatments to units. This is carton experiment two.

Carton experiment two, two random factors

The model for two-way random effects is

$$y_{ijk} = \mu + \alpha_i + \beta_j + \alpha\beta_{ij} + \epsilon_{ijk} \ ,$$

where $\alpha_i$ is a main effect for factor A, $\beta_j$ is a main effect for factor B, $\alpha\beta_{ij}$ is an AB interaction, and $\epsilon_{ijk}$ is random error. The model assumptions are that all the random effects $\alpha_i$, $\beta_j$, $\alpha\beta_{ij}$, and $\epsilon_{ijk}$ are independent, normally distributed, with mean 0. Each effect has its own variance: Var($\alpha_i$) = $\sigma_\alpha^2$, Var($\beta_j$) = $\sigma_\beta^2$, Var($\alpha\beta_{ij}$) = $\sigma_{\alpha\beta}^2$, and Var($\epsilon_{ijk}$) = $\sigma^2$. The variance of $y_{ijk}$ is $\sigma_\alpha^2 + \sigma_\beta^2 + \sigma_{\alpha\beta}^2 + \sigma^2$, and the correlation of two responses is the sum of the variances of the random components that they share, divided by their common variance $\sigma_\alpha^2 + \sigma_\beta^2 + \sigma_{\alpha\beta}^2 + \sigma^2$.

Two-factor model

This brings us to another way that random effects differ from fixed effects. In fixed effects, we have a table of means onto which we impose a structure of equally weighted main effects and interactions. There are other plausible structures based on unequal weightings that can have different main effects and interactions, so testing main effects when interactions are present in fixed effects makes sense only when we are truly interested in the specific, equally-weighted null hypothesis corresponding to the main effect. Random effects set up a correlation structure among the responses, with autonomous contributions from the different variance components. It is reasonable to ask if a main-effect contribution to correlation is absent even if interaction contribution to correlation is present. Similarly, equal weighting is about the only weighting that makes sense in random effects; after all, the row effects and column effects are chosen randomly and exchangeably. Why weight one

Hierarchy less important in random-effects models

row or column more than any other? So for random effects, we more or less automatically test for main effects, even if interactions are present.

Carton experiment three, three random factors

We can, of course, have random effects models with more than two factors. Suppose that there are many batches of glue, and we choose two of them at random. Now each operator makes two cartons on each machine with each batch of glue. We now have 200 factor-level combinations assigned at random to the 400 units. This is carton experiment three.

The model for three-way random effects is

$$y_{ijkl} = \mu + \alpha_i + \beta_j + \alpha\beta_{ij} + \gamma_k + \alpha\gamma_{ik} + \beta\gamma_{jk} + \alpha\beta\gamma_{ijk} + \epsilon_{ijkl} ,$$

Three-factor model

where $\alpha_i$, $\beta_j$, and $\gamma_k$ are main effects; $\alpha\beta_{ij}$, $\alpha\gamma_{ik}$, $\beta\gamma_{ik}$, and $\alpha\beta\gamma_{ijk}$ are interactions; and $\epsilon_{ijkl}$ is random error. The model assumptions remain that all the random effects are independent and normally distributed with mean 0. Each effect has its own variance: $\text{Var}(\alpha_i) = \sigma^2_\alpha$, $\text{Var}(\beta_j) = \sigma^2_\beta$, $\text{Var}(\gamma_k) = \sigma^2_\gamma$, $\text{Var}(\alpha\beta_{ij}) = \sigma^2_{\alpha\beta}$, $\text{Var}(\alpha\gamma_{ik}) = \sigma^2_{\alpha\gamma}$, $\text{Var}(\beta\gamma_{jk}) = \sigma^2_{\beta\gamma}$, $\text{Var}(\alpha\beta\gamma_{ijk}) = \sigma^2_{\alpha\beta\gamma}$, and $\text{Var}(\epsilon_{ijkl}) = \sigma^2$. Generalization to more factors is straightforward, and Chapter 12 describes some additional variations that can occur for factorials with random effects.

## 11.2 Why Use Random Effects?

The carton experiments described above are all completely randomized designs: the units are assigned at random to the treatments. The difference from what we have seen before is that the treatments have been randomly sampled from a population. Why should anyone design an experiment that uses randomly chosen treatments?

Random effects study variances in populations

The answer is that we are trying to draw inferences about the population from which the treatments were sampled. Specifically, we are trying to learn about variation in the treatment effects. Thus we want to design an experiment that looks at variation in a population by looking at the variability that arises when we sample from the population. When you want to study variances and variability, think random effects.

Use random effects when subsampling

Random-effects models are also used in subsampling situations. Revise carton experiment one. The manufacturer still chooses ten machines at random, but instead of making new cartons, she simply goes to the warehouse and collects 40 cartons at random from those made by each machine. It still makes sense to model the carton strengths with a random effect for the randomly chosen machine and a random error for the randomly chosen cartons from each machine's stock; that is precisely the random effects model.

| Source | DF | EMS |
|---|---|---|
| Treatments | g-1 | $\sigma^2 + n\sigma_\alpha^2$ |
| Error | N-g | $\sigma^2$ |

**Display 11.1:** Generic skeleton ANOVA for a one-factor model.

In the subsampling version of the carton example, we have done no experimentation in the sense of applying randomly assigned treatments to units. Instead, the stochastic nature of the data arises because we have sampled from a population. The items we have sampled are not exactly alike, so the responses differ. Furthermore, the sampling was done in a structured way (in the example, first choose machines, then cartons for each machine) that produces some correlation between the responses. For example, we expect cartons from the same machine to be a bit similar, but cartons from different machines should be unrelated. The pattern of correlation for subsampling is the same as the pattern of correlation for randomly chosen treatments applied to units, so we can use the same models for both.

Subsampling induces random variation

## 11.3 ANOVA for Random Effects

An analysis of variance for random effects is computed *exactly* the same as for fixed effects. (And yes, this implies that unbalanced data give us difficulties in random effects factorials too; see Section 12.8.) The ANOVA table has rows for every term in the model and columns for source, sums of squares, degrees of freedom, mean squares, and F-statistics.

No changes in SS or df

A random-effects ANOVA table usually includes an additional column for expected mean squares (EMS's). The EMS for a term is literally the expected value of its mean square. We saw EMS's briefly for fixed effects, but their utility there was limited to their relationship with noncentrality parameters and power. The EMS is much more useful for random effects. Chapter 12 will give general rules for computing EMS's in balanced factorials. For now, we will produce them magically and see how they are used.

ANOVA table includes column for EMS

The EMS for error is $\sigma^2$, exactly the same as in fixed effects. For balanced single-factor data, the EMS for treatments is $\sigma^2 + n\sigma_\alpha^2$. Display 11.1 gives the general form for a one-factor skeleton ANOVA (just sources, degrees of freedom, and EMS). For carton experiment one, the EMS for machines is $\sigma^2 + 40\sigma_\alpha^2$.

One-factor EMS

| Source | DF | EMS |
|---|---|---|
| A | $a-1$ | $\sigma^2 + n\sigma_{\alpha\beta}^2 + nb\sigma_\alpha^2$ |
| B | $b-1$ | $\sigma^2 + n\sigma_{\alpha\beta}^2 + na\sigma_\beta^2$ |
| AB | $(a-1)(b-1)$ | $\sigma^2 + n\sigma_{\alpha\beta}^2$ |
| Error | $N-ab=ab(n-1)$ | $\sigma^2$ |

**Display 11.2:** Generic skeleton ANOVA for a two-factor model.

To test the null hypothesis that $\sigma_\alpha^2 = 0$, we use the F-ratio $MS_{\text{Trt}}/MS_E$ and compare it to an F-distribution with $g-1$ and $N-g$ degrees of freedom to get a $p$-value. Let's start looking for the pattern now. To test the null hypothesis that $\sigma_\alpha^2 = 0$, we try to find two expected mean squares that would be the same if the null hypothesis were true and would differ otherwise. Put the mean square with the larger EMS in the numerator. If the null hypothesis is true, then the ratio of these mean squares should be about 1 (give or take some random variation). If the null hypothesis is false, then the ratio tends to be larger than 1, and we reject the null for large values of the ratio. In a one-factor ANOVA such as carton experiment one, there are only two mean squares to choose from, and we use $MS_{\text{Trt}}/MS_E$ to test the null hypothesis of no treatment variation.

Construct tests by examining EMS

It's a bit puzzling at first that fixed- and random-effects models, which have such different assumptions about parameters, should have the same test for the standard null hypothesis. However, think about the effects when the null hypotheses are true. For fixed effects, the $\alpha_i$ are fixed and all zero; for random effects, the $\alpha_i$ are random and all zero. Either way, they're all zero. It is this commonality under the null hypothesis that makes the two tests the same.

Two-factor EMS

Now look at a two-factor experiment such as carton experiment two. The sources in a two-factor ANOVA are A, B, the AB interaction, and error; Display 11.2 gives the general two-factor skeleton ANOVA. For carton experiment 2, this table is

| Source | DF | EMS |
|---|---|---|
| Machine | 9 | $\sigma^2 + 4\sigma_{\alpha\beta}^2 + 40\sigma_\alpha^2$ |
| Operator | 9 | $\sigma^2 + 4\sigma_{\alpha\beta}^2 + 40\sigma_\beta^2$ |
| Machine.operator | 81 | $\sigma^2 + 4\sigma_{\alpha\beta}^2$ |
| Error | 300 | $\sigma^2$ |

| Source | $EMS$ |
|---|---|
| A | $\sigma^2 + n\sigma^2_{\alpha\beta\gamma} + nc\sigma^2_{\alpha\beta} + nb\sigma^2_{\alpha\gamma} + nbc\sigma^2_{\alpha}$ |
| B | $\sigma^2 + n\sigma^2_{\alpha\beta\gamma} + nc\sigma^2_{\alpha\beta} + na\sigma^2_{\beta\gamma} + nac\sigma^2_{\beta}$ |
| C | $\sigma^2 + n\sigma^2_{\alpha\beta\gamma} + nb\sigma^2_{\alpha\gamma} + na\sigma^2_{\beta\gamma} + nab\sigma^2_{\gamma}$ |
| AB | $\sigma^2 + n\sigma^2_{\alpha\beta\gamma} + nc\sigma^2_{\alpha\beta}$ |
| AC | $\sigma^2 + n\sigma^2_{\alpha\beta\gamma} + nb\sigma^2_{\alpha\gamma}$ |
| BC | $\sigma^2 + n\sigma^2_{\alpha\beta\gamma} + na\sigma^2_{\beta\gamma}$ |
| ABC | $\sigma^2 + n\sigma^2_{\alpha\beta\gamma}$ |
| Error | $\sigma^2$ |

**Display 11.3:** Expected mean squares for a three-factor model.

Suppose that we want to test the null hypothesis that $\sigma^2_{\alpha\beta} = 0$. The EMS for the AB interaction is $\sigma^2 + n\sigma^2_{\alpha\beta}$, and the EMS for error is $\sigma^2$. These differ only by the variance component of interest, so we can test this null hypothesis using the ratio $MS_{AB}/MS_E$, with $(a-1)(b-1)$ and $ab(n-1)$ degrees of freedom.

That was pretty familiar; how about testing the null hypothesis that $\sigma^2_\alpha = 0$? The only two lines that have EMS's that differ by a multiple of $\sigma^2_\alpha$ are A and the AB interaction. Thus we use the F-ratio $MS_A/MS_{AB}$ with $a-1$ and $(a-1)(b-1)$ degrees of freedom to test $\sigma^2_\alpha = 0$. Similarly, the test for $\sigma^2_\beta = 0$ is $MS_B/MS_{AB}$ with $b-1$ and $(a-1)(b-1)$ degrees of freedom. Not having $MS_E$ in the denominator is a major change from fixed effects, and figuring out appropriate denominators is one of the main uses of EMS.

The denominator mean square for F-tests in random effects models will not always be $MS_E$!

Three-factor model

Let's press on to three random factors. The sources in a three-factor ANOVA are A, B, and C; the AB, AC, BC, and ABC interactions; and error. Display 11.3 gives the generic expected mean squares. For carton experiment 3, with m, o, and g indicating machine, operator, and glue, this table is

| Source | DF | EMS |
|---|---|---|
| m | 9 | $\sigma^2 + 2\sigma^2_{\alpha\beta\gamma} + 4\sigma^2_{\alpha\beta} + 20\sigma^2_{\alpha\gamma} + 40\sigma^2_{\alpha}$ |
| o | 9 | $\sigma^2 + 2\sigma^2_{\alpha\beta\gamma} + 4\sigma^2_{\alpha\beta} + 20\sigma^2_{\beta\gamma} + 40\sigma^2_{\beta}$ |
| g | 1 | $\sigma^2 + 2\sigma^2_{\alpha\beta\gamma} + 20\sigma^2_{\alpha\gamma} + 20\sigma^2_{\beta\gamma} + 200\sigma^2_{\gamma}$ |
| m.o | 81 | $\sigma^2 + 2\sigma^2_{\alpha\beta\gamma} + 4\sigma^2_{\alpha\beta}$ |
| m.g | 9 | $\sigma^2 + 2\sigma^2_{\alpha\beta\gamma} + 20\sigma^2_{\alpha\gamma}$ |
| o.g | 9 | $\sigma^2 + 2\sigma^2_{\alpha\beta\gamma} + 20\sigma^2_{\beta\gamma}$ |
| m.o.g | 81 | $\sigma^2 + 2\sigma^2_{\alpha\beta\gamma}$ |
| Error | 200 | $\sigma^2$ |

Testing for interactions is straightforward using our rule for finding two terms with EMS's that differ only by the variance component of interest. Thus error is the denominator for ABC, and ABC is the denominator for AB, AC, and BC. What do we do about main effects? Suppose we want to test the main effect of A, that is, test whether $\sigma^2_\alpha = 0$. If we set $\sigma^2_\alpha$ to 0 in the EMS for A, then we get $\sigma^2 + 2\sigma^2_{\alpha\beta\gamma} + 4\sigma^2_{\alpha\beta} + 20\sigma^2_{\alpha\gamma}$. A quick scan of the table of EMS's shows that *no* term has $\sigma^2 + 2\sigma^2_{\alpha\beta\gamma} + 4\sigma^2_{\alpha\beta} + 20\sigma^2_{\alpha\gamma}$ for its EMS. What we have seen is that there is no exact F-test for the null hypothesis that a main effect is zero in a three-way random-effects model. The lack of an exact F-test turns out to be not so unusual in models with many random effects. The next section describes how we handle this.

No exact F-tests for some hypotheses

## 11.4 Approximate Tests

Some null hypotheses have no exact F-tests in models with random effects. For example, there is no exact F-test for a main effect in a model with three random factors. This Section describes how to construct approximate tests for such hypotheses.

An exact F-test is the ratio of two positive, independently distributed random quantities (mean squares). The denominator is distributed as a multiple $\tau_d$ of a chi-square random variable divided by its degrees of freedom (the denominator degrees of freedom), and the numerator is distributed as a multiple $\tau_n$ of a chi-square random variable divided by its degrees of freedom (the numerator degrees of freedom). The multipliers $\tau_d$ and $\tau_n$ are the expected mean squares; $\tau_n = \tau_d$ when the null hypothesis is true, and $\tau_n > \tau_d$ when the null hypothesis is false. Putting these together gives us a test statistic that has an F-distribution when the null hypothesis is true and tends to be bigger when the null is false.

Mean squares are multiples of chi-squares divided by their degrees of freedom

1. Find a mean square to start the numerator. This mean square should have an EMS that includes the variance component of interest.
2. Find the EMS of the numerator when the variance component of interest is zero, that is, under the null hypothesis.
3. Find a sum of mean squares for the denominator. The sum of the EMS for these mean squares must include every variance component in the null hypothesis EMS of the numerator, include only those variance components in the null hypothesis EMS of the numerator, and be at least as big as the null hypothesis EMS of the numerator. The mean squares in the denominator should not appear in the numerator.
4. Add mean squares to the numerator as needed to make its expectation at least as big as that of the denominator but not larger than necessary. The mean squares added to the numerator should not appear in the denominator and should contain no variance components that have not already appeared.
5. If the numerator and denominator expectations are not the same, repeat the last two steps until they are.

**Display 11.4:** Steps to find mean squares for approximate F-tests.

Approximate tests mimic exact tests

We want the approximate test to mimic the exact test as much as possible. The approximate F-test should be the ratio of two positive, independently distributed random quantities. When the null hypothesis is true, both quantities should have the same expected value. For exact tests, the numerator and denominator are each a single mean square. For approximate tests, the numerator and denominator are sums of mean squares. Because the numerator and denominator should be independent, we need to use different mean squares for the two sums.

The key to the approximate test is to find sums for the numerator and denominator that have the same expectation when the null hypothesis is true. We do this by inspection of the table of EMS's using the steps given in Display 11.4; there is also a graphical technique we will discuss in the next chapter. One helpful comment: you always have the same number of mean squares in the numerator and denominator.

**Example 11.1** **Finding mean squares for an approximate test**

Consider testing for no factor A effect ($H_0 : \sigma^2_\alpha = 0$) in a three-way model with all random factors. Referring to the expected mean squares in Display 11.3 and the steps in Display 11.4, we construct the approximate test as follows:

1. The only mean square with an EMS that involves $\sigma^2_\alpha$ is $MS_A$, so it must be in the numerator.
2. The EMS for A under the null hypothesis $\sigma^2_\alpha = 0$ is $\sigma^2 + n\sigma^2_{\alpha\beta\gamma} + nc\sigma^2_{\alpha\beta} + nb\sigma^2_{\alpha\gamma}$.
3. We need to find a term or terms that will include $nc\sigma^2_{\alpha\beta}$ and $nb\sigma^2_{\alpha\gamma}$ without extraneous variance components. We can get $nc\sigma^2_{\alpha\beta}$ from $MS_{AB}$, and we can get $nb\sigma^2_{\alpha\gamma}$ from $MS_{AC}$. Our provisional denominator is now $MS_{AB} + MS_{AC}$; its expected value is $2\sigma^2 + 2n\sigma^2_{\alpha\beta\gamma} + nc\sigma^2_{\alpha\beta} + nb\sigma^2_{\alpha\gamma}$, which meets our criteria.
4. The denominator now has an expected value that is $\sigma^2 + n\sigma^2_{\alpha\beta\gamma}$ larger than that of the numerator. We can make them equal in expectation by adding $MS_{ABC}$ to the numerator.
5. The numerator $MS_A + MS_{ABC}$ and denominator $MS_{AB} + MS_{AC}$ have the same expectations under the null hypothesis, so we can stop and use them in our test.

Get approximate $p$-value using F-distribution

Now that we have the numerator and denominator, the test statistic is their ratio. To compute a $p$-value, we have to know the distribution of the ratio, and this is where the approximation comes in. We don't know the distribution of the ratio exactly; we approximate it. Exact F-tests follow the F-distribution, and we are going to compute $p$-values assuming that our approximate F-test also follows an F-distribution, even though it doesn't really. The degrees of freedom for our approximating F-distribution come from Satterthwaite formula (Satterthwaite 1946) shown below. These degrees of freedom will almost never be integers, but that is not a problem for most software. If you only have a table, rounding the degrees of freedom down gives a conservative result.

The simplest situation is when we have the sum of several mean squares, say $MS_1$, $MS_2$, and $MS_3$, with degrees of freedom $\nu_1$, $\nu_2$, and $\nu_3$. The approximate degrees of freedom are calculated as

$$\nu^\star = \frac{(MS_1 + MS_2 + MS_3)^2}{MS_1^2/\nu_1 + MS_2^2/\nu_2 + MS_3^2/\nu_3} .$$

Satterthwaite approximate degrees of freedom

In more complicated situations, we may have a general linear combination of mean squares $\sum_k g_k MS_k$. This linear combination has approximate degrees of freedom

$$\nu^\star = \frac{(\sum_k g_k MS_k)^2}{\sum_k g_k^2 MS_k^2/\nu_k} .$$

Unbalanced data will lead to these more complicated forms. The approximation tends to work better when all the coefficients $g_k$ are positive.

### Carton experiment three (F-tests)

**Example 11.2**

Suppose that we obtain the following ANOVA table for carton experiment 3 (data not shown):

| | DF | SS | MS | EMS |
|---|---|---|---|---|
| m | 9 | 2706 | 300.7 | $\sigma^2 + 2\sigma^2_{\alpha\beta\gamma} + 4\sigma^2_{\alpha\beta} + 20\sigma^2_{\alpha\gamma} + 40\sigma^2_{\alpha}$ |
| o | 9 | 8887 | 987.5 | $\sigma^2 + 2\sigma^2_{\alpha\beta\gamma} + 4\sigma^2_{\alpha\beta} + 20\sigma^2_{\beta\gamma} + 40\sigma^2_{\beta}$ |
| g | 1 | 2376 | 2376 | $\sigma^2 + 2\sigma^2_{\alpha\beta\gamma} + 20\sigma^2_{\alpha\gamma} + 20\sigma^2_{\beta\gamma} + 200\sigma^2_{\gamma}$ |
| m.o | 81 | 1683 | 20.78 | $\sigma^2 + 2\sigma^2_{\alpha\beta\gamma} + 4\sigma^2_{\alpha\beta}$ |
| m.g | 9 | 420.4 | 46.71 | $\sigma^2 + 2\sigma^2_{\alpha\beta\gamma} + 20\sigma^2_{\alpha\gamma}$ |
| o.g | 9 | 145.3 | 16.14 | $\sigma^2 + 2\sigma^2_{\alpha\beta\gamma} + 20\sigma^2_{\beta\gamma}$ |
| m.o.g | 81 | 1650 | 20.37 | $\sigma^2 + 2\sigma^2_{\alpha\beta\gamma}$ |
| error | 200 | 4646 | 23.23 | $\sigma^2$ |

The test for the three-way interaction uses error as the denominator; the F is $20.368/23.231 = .88$ with 81 and 200 degrees of freedom and $p$-value .75. The tests for the two-way interactions use the three-way interaction as denominator. Of these, only the machine by glue interaction has an F much larger than 1. Its F is 2.29 with 9 and 81 degrees of freedom and a $p$-value of .024, moderately significant.

We illustrate approximate tests with a test for machine. We have already discovered that the numerator should be the sum of the mean squares for machine and the three-way interaction; these are 300.7 and 20.37 with 9 and 81 degrees of freedom. Our numerator is 321.07, and the approximate degrees of freedom are:

$$\nu_n^\star = \frac{321.07^2}{300.7^2/9 + 20.37^2/81} \approx 10.3 .$$

The denominator is the sum of the mean squares for the machine by operator and the machine by glue interactions; these are 20.78 and 46.71 with 81 and 9 degrees of freedom. The denominator is 67.49, and the approximate degrees of freedom are

$$\nu_d^\star = \frac{67.49^2}{20.78^2/81 + 46.71^2/9} \approx 18.4 \ .$$

The F test is $321.07/67.49 = 4.76$ with 10.3 and 18.4 approximate degrees of freedom and an approximate $p$-value of .0018; this is strong evidence against the null hypothesis of no machine to machine variation.

## 11.5 Point Estimates of Variance Components

The parameters of a random-effects model are the variance components, and we would like to get estimates of them. Specifically, we would like both point estimates and confidence intervals. There are many point estimators for variance components; we will describe only the easiest method. There is an $MS$ and $EMS$ for each term in the model. Choose estimates of the variance components so that the observed mean squares equal their expectations when we use the estimated variance components in the EMS formulae. Operationally, we get the estimates by equating the observed mean squares with their expectations and solving the resulting set of equations for the variance components. These are called the ANOVA estimates of the variance components. ANOVA estimates are unbiased, but they can take negative values.

ANOVA estimates of variance components are unbiased but may be negative

In a one-factor design, the mean squares are $MS_A$ and $MS_E$ with expectations $\sigma^2 + n\sigma_\alpha^2$ and $\sigma^2$, so we get the equations:

$$\begin{aligned} MS_A &= \widehat{\sigma}^2 + n\widehat{\sigma}_\alpha^2 \\ MS_E &= \widehat{\sigma}^2 \end{aligned}$$

with solutions

$$\begin{aligned} \widehat{\sigma}_\alpha^2 &= \frac{MS_A - MS_E}{n} \\ \widehat{\sigma}^2 &= MS_E \ . \end{aligned}$$

It is clear that $\widehat{\sigma}_\alpha^2$ will be negative whenever $MS_A < MS_E$.

We follow the same pattern in bigger designs, but things are more complicated. For a three-way random-effects model, we get the equations:

$$MS_A = \widehat{\sigma}^2 + n\widehat{\sigma}_{\alpha\beta\gamma}^2 + nc\widehat{\sigma}_{\alpha\beta}^2 + nb\widehat{\sigma}_{\alpha\gamma}^2 + nbc\widehat{\sigma}_\alpha^2$$

$$\begin{aligned}
MS_B &= \widehat{\sigma}^2 + n\widehat{\sigma}^2_{\alpha\beta\gamma} + nc\widehat{\sigma}^2_{\alpha\beta} + na\widehat{\sigma}^2_{\beta\gamma} + nac\widehat{\sigma}^2_{\beta} \\
MS_C &= \widehat{\sigma}^2 + n\widehat{\sigma}^2_{\alpha\beta\gamma} + nb\widehat{\sigma}^2_{\alpha\gamma} + na\widehat{\sigma}^2_{\beta\gamma} + nab\widehat{\sigma}^2_{\gamma} \\
MS_{AB} &= \widehat{\sigma}^2 + n\widehat{\sigma}^2_{\alpha\beta\gamma} + nc\widehat{\sigma}^2_{\alpha\beta} \\
MS_{AC} &= \widehat{\sigma}^2 + n\widehat{\sigma}^2_{\alpha\beta\gamma} + nb\widehat{\sigma}^2_{\alpha\gamma} \\
MS_{BC} &= \widehat{\sigma}^2 + n\widehat{\sigma}^2_{\alpha\beta\gamma} + na\widehat{\sigma}^2_{\beta\gamma} \\
MS_{ABC} &= \widehat{\sigma}^2 + n\widehat{\sigma}^2_{\alpha\beta\gamma} \\
MS_E &= \widehat{\sigma}^2 .
\end{aligned}$$

It's usually easiest to solve these from the bottom up. The solutions are

$$\begin{aligned}
\widehat{\sigma}^2 &= MS_E \\
\widehat{\sigma}^2_{\alpha\beta\gamma} &= \frac{MS_{ABC} - MS_E}{n} \\
\widehat{\sigma}^2_{\beta\gamma} &= \frac{MS_{BC} - MS_{ABC}}{na} \\
\widehat{\sigma}^2_{\alpha\gamma} &= \frac{MS_{AC} - MS_{ABC}}{nb} \\
\widehat{\sigma}^2_{\alpha\beta} &= \frac{MS_{AB} - MS_{ABC}}{nc} \\
\widehat{\sigma}^2_{\gamma} &= \frac{MS_C - MS_{AC} - MS_{BC} + MS_{ABC}}{nab} \\
\widehat{\sigma}^2_{\beta} &= \frac{MS_B - MS_{AB} - MS_{BC} + MS_{ABC}}{nac} \\
\widehat{\sigma}^2_{\alpha} &= \frac{MS_A - MS_{AB} - MS_{AC} + MS_{ABC}}{nbc}
\end{aligned}$$

You can see a relationship between the formulae for variance component estimates and test numerators and denominators: mean squares in the test numerator are added in estimates, and mean squares in the test denominator are subtracted. Thus a variance component with an exact test will have an estimate that is just a difference of two mean squares.

Numerator MS's are added, denominator MS's are subtracted in estimates

Each ANOVA estimate of a variance component is a linear combination of mean squares, so we can again use the Satterthwaite formula to compute an approximate degrees of freedom for each estimated variance component.

**Example 11.3**

## Carton experiment three (estimates of variance components)

Let's compute ANOVA estimates of variance components and their approximate degrees of freedom for the data from carton experiment 3.

| Effect | Estimate | Calculation | DF |
|---|---|---|---|
| $\widehat{\sigma}^2$ | 23.231 | | 200 |
| $\widehat{\sigma}^2_{\alpha\beta\gamma}$ | $-1.43$ | $(20.368 - 23.231)/2$ | 1.05 |
| $\widehat{\sigma}^2_{\beta\gamma}$ | $-.21$ | $(16.15 - 20.368)/20$ | .52 |
| $\widehat{\sigma}^2_{\alpha\gamma}$ | 1.317 | $(46.71 - 20.368)/20$ | 2.80 |
| $\widehat{\sigma}^2_{\alpha\beta}$ | .10 | $(20.775 - 20.368)/20$ | 2.80 |
| $\widehat{\sigma}^2_{\gamma}$ | 11.67 | $(2375.8 - 46.71 - 16.15 + 20.368)/200$ | .96 |
| $\widehat{\sigma}^2_{\beta}$ | 24.27 | $(987.47 - 20.775 - 16.15 + 20.368)/40$ | 8.70 |
| $\widehat{\sigma}^2_{\alpha}$ | 6.34 | $(300.71 - 20.775 - 46.71 + 20.368)/40$ | 6.24 |

We can see several things from this example. First, negative estimates for variance components are not just a theoretical anomaly; they happen regularly in practice. Second, the four terms that were significant (the three main effects and the machine by glue interaction) have estimated variance components that are positive and reasonably far from zero in some cases. Third, the approximate degrees of freedom for a variance component estimate can be much less than the degrees of freedom for the corresponding term. For example, AB is an 81 degree of freedom term, but its estimated variance component has fewer than 3 degrees of freedom.

Negative estimates of variance components can cause problems and may indicate model inadequacy

We know that variance components are nonnegative, but ANOVA estimates of variance components can be negative. What should we do if we get negative estimates? The three possibilities are to ignore the issue, to get a new estimator, or to get a new model for the data. Ignoring the issue is certainly easiest, but this may lead to problems in a subsequent analysis that uses estimated variance components. The simplest new estimator is to replace the negative estimate by zero, though this revised estimator is no longer unbiased. Section 11.9 mentions some other estimation approaches that do not give negative results. Finally, negative variance estimates may indicate that our variance component model is inadequate. For example, consider an animal feeding study where each pen gets a fixed amount of food. If some animals get more food so that others get less food, then the weight gains of these animals will be negatively correlated. Our variance component models handle positive correlations nicely but are more likely to give negative estimates of variance when there is negative correlation.

## 11.6 Confidence Intervals for Variance Components

Degrees of freedom tell us something about how precisely we know a positive quantity—the larger the degrees of freedom, the smaller the standard deviation is as a fraction of the mean. Variances are difficult quantities to estimate, in the sense that you need lots of data to get a firm handle on a variance. The standard deviation of a mean square with $\nu$ degrees of freedom is $\sqrt{2/\nu}$ times the expected value, so if you want the standard deviation to be about 10% of the mean, you need 200 degrees of freedom! We rarely get that kind of precision.

Precise estimates of variances need lots of data

We can compute a standard error for estimates of variance components, but it is of limited use unless the degrees of freedom are fairly high. The usual interpretation for a standard error is something like "plus or minus 2 standard errors is approximately a 95% confidence interval." That works for normally distributed estimates, but it only works for variance estimates with many degrees of freedom. Estimates with few or moderate degrees of freedom have so much asymmetry that the symmetric-plus-or-minus idea is more misleading than helpful. Nevertheless, we can estimate the standard error of a linear combination of mean squares $\sum_k g_k MS_k$ via

SE of a variance estimate only useful with many degrees of freedom

$$\sqrt{2 \sum_k (g_k^2 MS_k^2 / \nu_k)} \ ,$$

where $MS_k$ has $\nu_k$ degrees of freedom. This looks like the approximate degrees-of-freedom formula because the variance is used in computing approximate degrees of freedom.

### Carton experiment three (standard errors)

**Example 11.4**

Let's compute standard errors for the estimates of the error, machine by glue, and machine variance components in carton experiment three. We estimate the error variance by $MS_E$ with 200 degrees of freedom, so its standard deviation is estimated to be

$$\sqrt{2 \times 23.231^2/200} = 2.3231 \ .$$

The machine by glue variance component estimate $\widehat{\sigma}^2_{\alpha\gamma}$ is $(MS_{AC} - MS_{ABC})/20$, so the coefficients $g_k^2 = 1/400$, and the standard deviation is

$$\sqrt{\frac{2}{400}(46.71^2/9 + 20.368^2/81)} = 1.11 \ .$$

$$\frac{\nu MS}{\chi^2_{\mathcal{E}/2,\nu}} \leq EMS \leq \frac{\nu MS}{\chi^2_{1-\mathcal{E}/2,\nu}}$$

**Display 11.5:** $1 - \mathcal{E}$ confidence interval for an EMS based on its MS with $\nu$ degrees of freedom.

Finally, the machine variance component estimate $\widehat{\sigma}^2_\alpha$ is $(MS_A - MS_{AB} - MS_{AC} + MS_{ABC})/40$, so the coefficients $g_k^2 = 1/1600$, and the standard deviation is

$$\sqrt{\frac{2}{1600}(300.71^2/9 + 20.775^2/81 + 46.71^2/9 + 20.368^2/81)} = 3.588 \quad .$$

Recall from Examples 11.2 and 11.3 that the $p$-values for testing the null hypotheses of no machine variation and no machine by glue variation were .0018 and .024, and that the corresponding variance component estimates were 6.34 and 1.32. We have just estimated their standard errors to be 3.588 and 1.11, so the estimates are only 1.8 and 1.2 standard errors from their null hypothesis values of zero, even though the individual terms are rather significant. The usual plus or minus two standard errors interpretation simply doesn't work for variance components with few degrees of freedom.

Confidence interval for $\sigma^2$

We can construct confidence intervals that account for the asymmetry of variance estimates, but these intervals are exact in only a few situations. One easy situation is a confidence interval for the expected value of a mean square. If we let $\chi^2_{\mathcal{E},\nu}$ be the upper $\mathcal{E}$ percent point of a chi-square distribution with $\nu$ degrees of freedom, then a $1 - \mathcal{E}$ confidence interval for the EMS of an MS can be formed as shown in Display 11.5. The typical use for this is an interval estimate for $\sigma^2$ based on $MS_E$:

$$\frac{\nu MS_E}{\chi^2_{\mathcal{E}/2,\nu}} \leq \sigma^2 \leq \frac{\nu MS_E}{\chi^2_{1-\mathcal{E}/2,\nu}} \quad .$$

**Example 11.5**

**Carton experiment three (confidence interval for $\sigma^2$)**

Use the method of Display 11.5 to compute a confidence interval for $\sigma^2$. The error mean square was 23.231 with 200 degrees of freedom. For a 95% interval, we need the upper and lower 2.5% points of $\chi^2$ with 200 degrees of freedom; these are 162.73 and 241.06. Our interval is

$$\frac{F}{F_{\mathcal{E}/2,\nu_1,\nu_2}} \leq \frac{EMS_1}{EMS_2} \leq \frac{F}{F_{(1-\mathcal{E})/2,\nu_1,\nu_2}}$$

**Display 11.6:** $1 - \mathcal{E}$ confidence interval for the ratio $EMS_1/EMS_2$ based on $F = MS_1/MS_2$ with $\nu_1$ and $\nu_2$ degrees of freedom.

$$\frac{200 \times 23.231}{241.06} = 19.27 \leq \sigma^2 \leq 28.55 = \frac{200 \times 23.231}{162.73} \quad .$$

Even with 200 degrees of freedom, this interval is not symmetric around the estimated component. The length of the interval is about 4 standard errors, however.

Confidence intervals for ratios of EMS's

We can also construct confidence intervals for ratios of EMS's from ratios of the corresponding mean squares. Let $MS_1$ and $MS_2$ have $EMS_1$ and $EMS_2$ as their expectations. Then a $1 - \mathcal{E}$ confidence interval for $EMS_1/EMS_2$ is shown in Display 11.6. This confidence interval is rarely used as is; instead, it is used as a building block for other confidence intervals. Consider a one-way random effects model; the EMS's are shown in Display 11.1. Using the confidence interval in Display 11.6, we get

$$\frac{MS_{\text{Trt}}/MS_E}{F_{\mathcal{E}/2,\nu_1,\nu_2}} \leq \frac{\sigma^2 + n\sigma_\alpha^2}{\sigma^2} \leq \frac{MS_{\text{Trt}}/MS_E}{F_{(1-\mathcal{E})/2,\nu_1,\nu_2}} \quad .$$

Subtracting 1 and dividing by $n$, we get a confidence interval for $\sigma_\alpha^2/\sigma^2$:

$$L = \frac{1}{n}\left(\frac{MS_{\text{Trt}}/MS_E}{F_{\mathcal{E}/2,\nu_1,\nu_2}} - 1\right) \leq \frac{\sigma_\alpha^2}{\sigma^2} \leq \frac{1}{n}\left(\frac{MS_{\text{Trt}}/MS_E}{F_{(1-\mathcal{E})/2,\nu_1,\nu_2}} - 1\right) = U \quad .$$

Confidence interval for intraclass correlation

Continuing, we can get a confidence interval for the intraclass correlation via

$$\frac{L}{1+L} \leq \frac{\sigma_\alpha^2}{\sigma^2 + \sigma_\alpha^2} \leq \frac{U}{1+U} \quad .$$

This same approach works for any pair of mean squares with $EMS_2 = \tau$ and $EMS_1 = \tau + n\sigma_\eta^2$ to get confidence intervals for $\sigma_\eta^2/\tau$ and $\tau/(\tau + \sigma_\eta^2)$.

Example 11.6

**Carton experiment three (confidence interval for $\sigma^2_{\alpha\gamma}/(\sigma^2 + 2\sigma^2_{\alpha\beta\gamma})$)**

The machine by glue interaction was moderately significant in Example 11.2, so we would like to look more closely at the machine by glue interaction variance component. The mean square for machine by glue was 46.706 with 9 degrees of freedom and EMS $\sigma^2 + 2\sigma^2_{\alpha\beta\gamma} + 20\sigma^2_{\alpha\gamma}$. The mean square for the three-way interaction was 20.368 with 81 degrees of freedom and EMS $\sigma^2 + 2\sigma^2_{\alpha\beta\gamma}$. For a 90% confidence interval, we need the upper and lower 5% points of F with 9 and 81 degrees of freedom; these are .361 and 1.998.

The confidence interval is

$$\frac{1}{20}\left(\frac{46.706/20.368}{1.998} - 1\right) \leq \frac{\sigma^2_{\alpha\gamma}}{\sigma^2 + 2\sigma^2_{\alpha\beta\gamma}} \leq \frac{1}{20}\left(\frac{46.706/20.368}{.361} - 1\right)$$

$$.0074 \leq \frac{\sigma^2_{\alpha\gamma}}{\sigma^2 + 2\sigma^2_{\alpha\beta\gamma}} \leq .268 .$$

Confidence intervals for ratios of variances often cover more than one order of magnitude

Example 11.6 illustrates that even for a significant term ($p$-value = .024) with reasonably large degrees of freedom (9, 81), a confidence interval for a ratio of variances with a reasonable coverage rate can cover an order of magnitude. Here we saw the upper endpoint of a 90% confidence interval for a variance ratio to be 36 times as large as the lower endpoint. The problem gets worse with higher coverage and lower degrees of freedom. Variance ratios are even harder to estimate than variances.

Williams' approximate confidence interval for a variance component with an exact test

There are no simple, exact confidence intervals for any variance components other than $\sigma^2$, but a couple of approximate methods are available. In one, Williams (1962) provided a conservative confidence interval for variance components that have exact F-tests. Suppose that we wish to construct a confidence interval for a component $\sigma^2_\eta$, and that we have two mean squares with expectations $\text{EMS}_1 = \tau + k\sigma^2_\eta$ and $\text{EMS}_2 = \tau$ and degrees of freedom $\nu_1$ and $\nu_2$. The test for $\sigma^2_\eta$ has an observed F-ratio of $F_O = MS_1/MS_2$. We construct a confidence interval for $\sigma^2_\eta$ with coverage at least $1-\mathcal{E}$ as follows:

$$\frac{\nu_1 MS_1(1 - F_{\mathcal{E}/4,\nu_1,\nu_2}/F_O)}{k\chi^2_{\mathcal{E}/4,\nu_1}} \leq \sigma^2_\eta \leq \frac{\nu_1 MS_1(1 - F_{1-\mathcal{E}/4,\nu_1,\nu_2}/F_O)}{k\chi^2_{1-\mathcal{E}/4,\nu_1}} .$$

The use of $\mathcal{E}/4$ arises because we are combining two exact $1-\mathcal{E}/2$ confidence intervals (on $\tau + k\sigma^2_\eta$ and $\sigma^2_\eta/\tau$) to get a $1 - \mathcal{E}$ interval on $\sigma^2_\eta$. In fact, we can use $F_{\mathcal{E}_F/2,\nu_1,\nu_2}$, $F_{1-\mathcal{E}_F/2,\nu_1,\nu_2}$, $\chi^2_{\mathcal{E}_\chi/2,\nu_1}$, and $\chi^2_{1-\mathcal{E}_\chi/2,\nu_1}$ for any $\mathcal{E}_F$ and $\mathcal{E}_\chi$ that add to $\mathcal{E}$.

The other method is simple and works for any variance component estimated with the ANOVA method, but it is also very approximate. Each estimated variance component has an approximate degrees of freedom from Satterthwaite; use the formula in Display 11.5, treating our estimate and its approximate degrees of freedom as if they were a mean square and a true degrees of freedom.

Approximate CI by treating as a single mean square

**Example 11.7**

### Carton experiment three (confidence interval for $\sigma^2_{\alpha\gamma}$)

Consider $\sigma^2_{\alpha\gamma}$ in carton experiment three. Example 11.3 gave a point estimate of 1.32 with 2.8 approximate degrees of freedom. For a 95% confidence interval the approximate method gives us:

$$\frac{2.8 \times 1.32}{.174} \leq \sigma^2_{\alpha\gamma} \leq \frac{2.8 \times 1.32}{8.97}$$
$$.412 \leq \sigma^2_{\alpha\gamma} \leq 21.2 \ .$$

This more than an order of magnitude from top to bottom is fairly typical for estimates with few degrees of freedom.

We can also use the Williams' method. The mean squares we use are $MS_{AC}$ (46.706 with expectation $\sigma^2 + 2\sigma^2_{\alpha\beta\gamma} + 20\sigma^2_{\alpha\gamma}$ and 9 degrees of freedom) and $MS_{ABC}$ (20.368 with expectation $\sigma^2 + 2\sigma^2_{\alpha\beta\gamma}$ and 81 degrees of freedom); the observed F is $F_O$ = 2.29. The required percent points are $F_{.0125,9,81} = 2.55$, $F_{.9875,9,81} = .240$, $\chi_{.0125,9} = 21.0$, and $\chi_{.9875,9} = 2.22$. Computing, we get

$$\frac{9 \times 46.71(1 - 2.55/2.29)}{20 \times 21.0} \leq \sigma^2_{\eta} \leq \frac{9 \times 46.71(1 - .240/2.293)}{20 \times 2.22}$$
$$-.114 \leq \sigma^2_{\eta} \leq 8.48$$

This interval is considerably shorter than the interval computed via the other approximation, but it does include zero. If we use $\mathcal{E}_F = .0495$ and $\mathcal{E}_\chi = .0005$, then we get the interval (.0031, 22.32), which is much more similar to the approximate interval.

## 11.7 Assumptions

We have discussed tests of null hypotheses that variance components are zero, point estimates for variance components, and interval estimates for variance components. Nonnormality and nonconstant variance affect the tests in

Random effects tests affected similarly to fixed effects tests

random-effects models in much the same way as they do tests of fixed effects. This is because the fixed and random tests are essentially the same under the null hypothesis, though the notion of "error" changes from test to test when we have different denominators. Transformation of the response can improve the quality of inference for random effects, just as it does for fixed effects.

Point estimates of variance components remain unbiased when the distributions of the random effects are nonnormal.

Confidence intervals depend strongly on normality

But now the bad news: the validity of the confidence intervals we have constructed for variance components is horribly, horribly dependent on normality. Only a little bit of nonnormality is needed before the coverage rate diverges greatly from $1 - \mathcal{E}$. Furthermore, not just the errors $\epsilon_{ijk}$ need to be normal; other random effects must be normal as well, depending on which confidence intervals we are computing. While we often have enough data to make a reasonable check on the normality of the residuals, we rarely have enough levels of treatments to make any kind of check on the normality of treatment effects. Only the most blatant outliers seem likely to be identified.

To give you some idea of how bad things are, suppose that we have a 25 degree of freedom estimate for error, and we want a 95% confidence interval for $\sigma^2$. If one in 20 of the data values has a standard deviation 3 times that of the other 24, then a 95% confidence interval will have only about 80% coverage.

> Confidence intervals for variance components of real-world data are quite likely to miss their stated coverage rather badly, and we should consider them approximate at best.

## 11.8 Power

Power for random effects uses central F

Power is one of the few places where random effects are simpler than fixed effects, because there are no noncentrality parameters to deal with in random effects. Suppose that we wish to compute the power for testing the null hypothesis that $\sigma_\eta^2 = 0$, and that we have two mean squares with expectations $\text{EMS}_1 = \tau + k\sigma_\eta^2$ and $\text{EMS}_2 = \tau$ and degrees of freedom $\nu_1$ and $\nu_2$. The test for $\sigma_\eta^2$ is the F-ratio $MS_1/MS_2$.

When the null hypothesis is true, the F-ratio has an F-distribution with $\nu_1$ and $\nu_2$ degrees of freedom. We reject the null when the observed F-statistic is greater than $F_{\mathcal{E},\nu_1,\nu_2}$. When the null hypothesis is false, the observed F-statistic is distributed as $(\tau + k\sigma_\eta^2)/\tau$ times an F with $\nu_1$ and $\nu_2$ degrees of

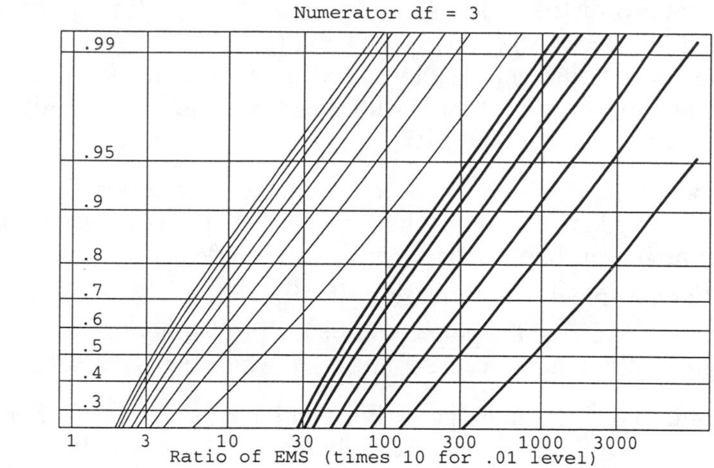


**Figure 11.1:** Power for random effects F-tests with 3 numerator degrees of freedom, testing at the .05 and .01 levels, and 2, 3, 4, 6, 8, 16, 32, or 256 denominator degrees of freedom. Curves for .01 have been shifted right by a factor of 10.

freedom. Thus the power is the probability than an F with $\nu_1$ and $\nu_2$ degrees of freedom exceeds $\tau/(\tau + k\sigma_\eta^2)F_{\mathcal{E},\nu_1,\nu_2}$. This probability can be computed with any software that can compute $p$-values and critical points for the F-distribution.

Alternatively, power curves are available in the Appendix Tables for random effects tests with small numerator degrees of freedom. The curves for three numerator degrees of freedom are reproduced in Figure 11.1. Looking at these curves, we see that the ratio of expected mean squares must be greater than 10 before power is .9 or above.

You may need to change number of levels $a$ instead of replications $n$

Changing the sample size $n$ or the number of levels $a$, $b$, or $c$ can affect $\tau$, $k$, $\nu_1$, or $\nu_2$, depending on the mean squares in use. However, there is a major difference between fixed-effects power and random-effects power that must be stressed. In fixed effects, power can be made as high as desired by increasing the replication $n$. That is *not* necessarily true for random effects; in random effects, you may need to increase $a$, $b$, or $c$ instead.

**Example 11.8**

## Carton experiment three (power)

Consider the power for testing the null hypothesis that $\sigma_{\alpha\gamma}^2$ is zero when $\sigma_{\alpha\gamma}^2 = 1$, $\sigma^2 + 2\sigma_{\alpha\beta\gamma}^2 = 20$, and $\mathcal{E}_I = .01$. The F-ratio is $MS_{AC}/MS_{ABC}$.

This F-ratio is distributed as $(\sigma^2 + n\sigma^2_{\alpha\beta\gamma} + nb\sigma^2_{\alpha\gamma})/(\sigma^2 + n\sigma^2_{\alpha\beta\gamma})$ times an F-distribution with $(a-1)(c-1)$ and $(a-1)(b-1)(c-1)$ degrees of freedom, here 2 times an F with 9 and 81 degrees of freedom. Power for this test is the probability that an F with 9 and 81 degrees of freedom exceeds $F_{.01,9,81}/2 = 1.32$, or about 24%.

Suppose that we want 95% power. Increasing $n$ does not change the degrees of freedom, but it does change the multiplier. However, the multiplier can get no bigger than $1 + b\sigma^2_{\alpha\gamma}/\sigma^2_{\alpha\beta\gamma} = 1 + 10\sigma^2_{\alpha\gamma}/\sigma^2_{\alpha\beta\gamma} = 1 + 10/\sigma^2_{\alpha\beta\gamma}$ no matter how much you increase $n$. If $\sigma^2_{\alpha\beta\gamma} = 2$, then the largest multiplier is $1 + 10/2 = 6$, and the power will be the probability that an F with 9 and 81 degrees of freedom exceeds $F_{.01,9,81}/6$, which is only 91%.

To make this test more powerful, you have to increase $b$. For example, $b = 62$ and $n = 2$ has the F-test distributed as 7.2 times an F with 9 and 549 degrees of freedom (assuming still that $\sigma^2_{\alpha\gamma} = 1$ and $\sigma^2_{\alpha\beta\gamma} = 2$). This gives the required power.

## 11.9 Further Reading and Extensions

We have only scratched the surface of the subject of random effects. Searle (1971) provides a review, and Searle, Casella, and McCulloch (1992) provide book-length coverage.

In the single-factor situation, there is a simple formula for the EMS for treatments when the data are unbalanced: $\sigma^2 + n'\sigma^2_\alpha$, where

$$n' = \frac{1}{a-1}[N - \frac{1}{N}\sum_{i=1}^{a} n_i^2] \ .$$

The formula for $n'$ reduces to $n$ for balanced data.

Expected mean squares do not depend on normality, though the chi-square distribution for mean square and F-distribution for test statistics do depend on normality. Tukey (1956) and Tukey (1957b) work out variances for variance components, though the notation and algebra are rather heavy going.

The Satterthwaite formula is based on matching the mean and variance of an unknown distribution to that of an approximating distribution. There are quite a few other possibilities; Johnson and Kotz (1970) describe the major ones.

We have discussed the ANOVA method for estimating variance components. There are several others, including maximum likelihood estimates, restricted maximum likelihood estimates (REML), and minimum norm quadratic unbiased estimates (MINQUE). All of these have the advantage of providing estimates that will be nonnegative, but they are all much more complicated to compute. See Searle, Casella, and McCulloch (1992) or Hocking (1985).

## 11.10 Problems

**Exercise 11.1**

The following ANOVA table is from an experiment where four identically equipped cars were chosen at random from a car dealership, and each car was tested 3 times for gas mileage on a dynamometer.

| Source | DF | SS | MS |
|---|---|---|---|
| Cars | 3 | 15 | 5 |
| Error | 8 | 16 | 2 |

Find estimates of the variance components and a 95% confidence interval for the intraclass correlation of the mileage measurements.

**Exercise 11.2**

We wish to examine the average daily weight gain by calves sired by four bulls selected at random from a population of bulls. Bulls denoted A through D were mated with randomly selected cows. Average daily weight gain by the calves is given below.

| A | B | C | D |
|---|---|---|---|
| 1.46 | 1.17 | .98 | .95 |
| 1.23 | 1.08 | 1.06 | 1.10 |
| 1.12 | 1.20 | 1.15 | 1.07 |
| 1.23 | 1.08 | 1.11 | 1.11 |
| 1.02 | 1.01 | .83 | .89 |
| 1.15 | .86 | .86 | 1.12 |

a) Test the null hypothesis that there is no sire to sire variability in the response.

b) Find 90% confidence intervals for the error variance and the sire to sire variance.

**Exercise 11.3** Five tire types (brand/model combinations like Goodyear/Arriva) in the size 175/80R-13 are chosen at random from those available in a metropolitan area, and six tires of each type are taken at random from warehouses. The tires are placed (in random order) on a machine that will test tread durability and report a response in thousands of miles. The data follow:

| Brand | Miles | | | | | |
|---|---|---|---|---|---|---|
| 1 | 55 | 56 | 59 | 55 | 60 | 57 |
| 2 | 39 | 42 | 43 | 41 | 41 | 42 |
| 3 | 39 | 41 | 43 | 40 | 43 | 43 |
| 4 | 44 | 44 | 42 | 39 | 40 | 43 |
| 5 | 46 | 42 | 45 | 42 | 42 | 44 |

Compute a 99% confidence interval for the ratio of type to type variability to tire within type variability ($\sigma_\alpha^2/\sigma^2$). Do you believe that this interval actually has 99% coverage? Explain.

**Exercise 11.4** A 24-head machine fills bottles with vegetable oil. Five of the heads are chosen at random, and several consecutive bottles from these heads were taken from the line. The net weight of oil in these bottles is given in the following table (data from Swallow and Searle 1978):

| Group | | | | |
|---|---|---|---|---|
| 1 | 2 | 3 | 4 | 5 |
| 15.70 | 15.69 | 15.75 | 15.68 | 15.65 |
| 15.68 | 15.71 | 15.82 | 15.66 | 15.60 |
| 15.64 | | 15.75 | 15.59 | |
| 15.60 | | 15.71 | | |
| | | 15.84 | | |

Is there any evidence for head to head variability? Estimate the head to head and error variabilities.

**Exercise 11.5** The burrowing mayfly *Hexagenia* can be used as an indicator of water quality (it likes clean water). Before starting a monitoring program using *Hexagenia* we take three samples from each of ten randomly chosen locations along the upper Mississippi between Lake Peppin and the St. Anthony Lock and Dam. We use these data to estimate the within location and between location variability in *Hexagenia* abundance. An ANOVA follows; the data are in hundreds of insects per square meter.

```
                    DF          SS          MS
Location             9       11.59       1.288
Error               20       1.842      0.0921
```

a) Give a point estimate for the between location variance in *Hexagenia* abundance.

b) Give a 95% confidence interval for the within location variance in *Hexagenia* abundance.

**Exercise 11.6**

Anecdotal evidence suggests that some individuals can tolerate alcohol better than others. As part of a traffic safety study, you are planning an experiment to test for the presence of individual to individual variation. Volunteers will be recruited who have given their informed consent for participation after having been informed of the risks of the study. Each individual will participate in two sessions one week apart. In each session, the individual will arrive not having eaten for at least 4 hours. They will take a hand-eye coordination test, drink 12 ounces of beer, wait 15 minutes, and then take a second hand-eye coordination test. The score for a session is the change in hand-eye coordination. There are two sessions, so $n = 2$. We believe that the individual to individual variation $\sigma_\alpha^2$ will be about the same size as the error $\sigma^2$. If we are testing at the 1% level, how many individuals should be tested to have power .9 for this setup?

**Problem 11.1**

Suppose that you are interested in estimating the variation in serum cholesterol in a student population; in particular, you are interested in the ratio $\sigma_\alpha^2/\sigma^2$. Resources limit you to 100 cholesterol measurements. Are you better off taking ten measurements on each of ten students, or two measurements on each of 50 students? (Hint: which one should give you a shorter interval?)

**Problem 11.2**

Milk is tested after Pasteurization to assure that Pasteurization was effective. This experiment was conducted to determine variability in test results between laboratories, and to determine if the interlaboratory differences depend on the concentration of bacteria.

Five contract laboratories are selected at random from those available in a large metropolitan area. Four levels of contamination are chosen at random by choosing four samples of milk from a collection of samples at various stages of spoilage. A batch of fresh milk from a dairy was obtained and split into 40 units. These 40 units are assigned at random to the twenty combinations of laboratory and contamination sample. Each unit is contaminated with 5 ml from its selected sample, marked with a numeric code, and sent to the selected laboratory. The laboratories count the bacteria in each sample by serial dilution plate counts without knowing that they received four pairs, rather than eight separate samples. Data follow (colony forming units per $\mu$l):

| Lab | Sample 1 | 2 | 3 | 4 |
|---|---|---|---|---|
| 1 | 2200 | 3000 | 210 | 270 |
| | 2200 | 2900 | 200 | 260 |
| 2 | 2600 | 3600 | 290 | 360 |
| | 2500 | 3500 | 240 | 380 |
| 3 | 1900 | 2500 | 160 | 230 |
| | 2100 | 2200 | 200 | 230 |
| 4 | 2600 | 2800 | 330 | 350 |
| | 4300 | 1800 | 340 | 290 |
| 5 | 4000 | 4800 | 370 | 500 |
| | 3900 | 4800 | 340 | 480 |

Analyze these data to determine if the effects of interest are present. If so, estimate them.

**Problem 11.3** Composite materials used in the manufacture of aircraft components must be tested to determine tensile strength. A manufacturer tests five random specimens from five randomly selected batches, obtaining the following coded strengths (data from Vangel 1992).

| Batch | | | | | |
|---|---|---|---|---|---|
| 1 | 379 | 357 | 390 | 376 | 376 |
| 2 | 363 | 367 | 382 | 381 | 359 |
| 3 | 401 | 402 | 407 | 402 | 396 |
| 4 | 402 | 387 | 392 | 395 | 394 |
| 5 | 415 | 405 | 396 | 390 | 395 |

Compute point estimates for the between batch and within batch variance components, and compute a 95% confidence interval for $\sigma_\alpha^2/\sigma^2$.

**Question 11.1** Why do you always wind up with the same number of numerator and denominator terms in approximate tests?

**Question 11.2** Derive the confidence interval formula given in Display 11.5.

**Question 11.3** Derive the Satterthwaite approximate degrees of freedom for a sum of mean squares by matching the first two moments of the sum of mean squares to a multiple of a chi-square.

# Chapter 12

# Nesting, Mixed Effects, and Expected Mean Squares

We have seen fixed effects and random effects in the factorial context of forming treatments by combining levels of factors, and we have seen how sampling from a population can introduce structure for which random effects are appropriate. This chapter introduces new ways in which factors can be combined, discusses models that contain both fixed and random effects, and describes the rules for deriving expected mean squares.

## 12.1 Nesting Versus Crossing

The vitamin A content of baby food carrots may not be consistent. To evaluate this possibility, we go to the grocery store and select four jars of carrots at random from each of the three brands of baby food that are sold in our region. We then take two samples from each jar and measure the vitamin A in every sample for a total of 24 responses.

Multiple sources of variation

It makes sense to consider decomposing the variation in the 24 responses into various sources. There is variation between the brands, variation between individual jars for each brand, and variation between samples for every jar.

No jar effect across brands

It does *not* make sense to consider jar main effects and brand by jar interaction. Jar one for brand A has absolutely nothing to do with jar one for brand B. They might both have lots of vitamin A by chance, but it would just be chance. They are not linked, so there should be no jar main effect across the brands. If the main effect of jar doesn't make sense, then neither does a jar by brand interaction, because that two-factor interaction can be interpreted as how the main effect of jar must be altered at each level of brand to obtain treatment means.

Crossed factors form treatments with their combinations

Main effects and interaction are appropriate when the treatment factors are *crossed*. Two factors are crossed when treatments are formed as the combinations of levels of the two factors, and we use the same levels of the first factor for every level of the second factor, and vice versa. All factors we have considered until the baby carrots have been crossed factors. The jar and brand factors are not crossed, because we have different jars (levels of the jar factor) for every brand.

Factor B nested in A has different levels for every level of A

The alternative to crossed factors is *nested* factors. Factor B is nested in factor A if there is a completely different set of levels of B for every level of A. Thus the jars are nested in the brands and not crossed with the brands, because we have a completely new set of jars for every brand. We write nested models using parentheses in the subscripts to indicate the nesting. If brand is factor A and jar (nested in brand) is factor B, then the model is written

$$y_{ijk} = \mu + \alpha_i + \beta_{j(i)} + \epsilon_{k(ij)} \ .$$

The $j(i)$ indicates that the factor corresponding to $j$ (factor B) is nested in the factor corresponding to $i$ (factor A). Thus there is a different $\beta_j$ for each level $i$ of A.

Errors are nested

Note that we wrote $\epsilon_{k(ij)}$, nesting the random errors in the brand-jar combinations. This means that we get a different, unrelated set of random errors for each brand-jar combination. In the crossed factorials we have used until now, the random error is nested in the all-way interaction, so that for a three-way factorial the error $\epsilon_{ijkl}$ could more properly have been written $\epsilon_{l(ijk)}$. Random errors are always nested in some model term; we've just not needed to deal with it before now.

Nested factors are usually random

Nested factors can be random or fixed, though they are usually random and often arise from some kind of subsampling. As an example of a factor that is fixed and nested, consider a company with work crews, each crew consisting of four members. Members are nested in crews, and we get the same four crew members whenever we look at a given crew, making member a fixed effect.

Fully nested design

When we have a chain of factors, each nested in its predecessor, we say that the design is fully nested. The baby carrots example is fully nested, with jars nested in brand, and sample nested in jar. Another example comes from genetics. There are three subspecies. We randomly choose five males from each subspecies (a total of fifteen males); each male is mated with four

| Source | DF | EMS |
|---|---|---|
| A | $a-1$ | $\sigma^2 + n\sigma_\delta^2 + nd\sigma_\gamma^2 + ncd\sigma_\beta^2 + nbcd\sigma_\alpha^2$ |
| B(A) | $a(b-1)$ | $\sigma^2 + n\sigma_\delta^2 + nd\sigma_\gamma^2 + ncd\sigma_\beta^2$ |
| C(AB) | $ab(c-1)$ | $\sigma^2 + n\sigma_\delta^2 + nd\sigma_\gamma^2$ |
| D(ABC) | $abc(d-1)$ | $\sigma^2 + n\sigma_\delta^2$ |
| Error | $abcd(n-1)$ | $\sigma^2$ |

**Display 12.1:** Skeleton ANOVA and EMS for a generic fully-nested four-factor design.

females (of the same subspecies, a total of 60 females); we observe three offspring per mating (a total of 180 offspring); and we make two measurements on each offspring (a total of 360 measurements). Offspring are nested in females, which are nested in males, which are nested in subspecies.

EMS for fully-nested model

The expected mean squares for a balanced, fully nested design with random terms are simple; Display 12.1 shows a skeleton ANOVA and EMS for a four-factor fully-nested design. Note that in parallel to the subscript notation, factor B nested in A can be denoted B(A). Rules for deriving the EMS will be given in Section 12.6. The degrees of freedom for any term are the total number of effects for that term minus the number of degrees of freedom above the term, counting 1 for the constant. For example, B(A) has $ab$ effects ($b$ for each of the $a$ levels of A), so $ab - (a-1) - 1 = a(b-1)$ degrees of freedom for B(A). The denominator for any term is the term immediately below it.

For the fully-nested genetics example we have:

| Source | DF | EMS |
|---|---|---|
| s | 2 | $\sigma^2 + 2\sigma_\delta^2 + 6\sigma_\gamma^2 + 24\sigma_\beta^2 + 120\sigma_\alpha^2$ |
| m(s) | 12 | $\sigma^2 + 2\sigma_\delta^2 + 6\sigma_\gamma^2 + 24\sigma_\beta^2$ |
| f(ms) | 45 | $\sigma^2 + 2\sigma_\delta^2 + 6\sigma_\gamma^2$ |
| o(fms) | 120 | $\sigma^2 + 2\sigma_\delta^2$ |
| Error | 180 | $\sigma^2$ |

where s, m, f, and o indicate subspecies, males, females, and offspring. To test the null hypothesis $\sigma_\beta^2 = 0$, that is, no male to male variation, we would use the F-statistic $MS_m/MS_f$ with 12 and 45 degrees of freedom.

| Component | Estimate |
|---|---|
| $\sigma_\alpha^2$ | $(MS_A - MS_B)/(nbcd)$ |
| $\sigma_\beta^2$ | $(MS_B - MS_C)/(ncd)$ |
| $\sigma_\gamma^2$ | $(MS_C - MS_D)/(nd)$ |
| $\sigma_\delta^2$ | $(MS_D - MS_E)/n$ |
| $\sigma^2$ | $MS_E$ |

**Display 12.2:** ANOVA estimates for variance components in a fully-nested four-factor design.

Most df at bottom

One potential problem with fully-nested designs is that the degrees of freedom tend to pile up at the bottom. That is, the effects that are nested more and more deeply tend to have more degrees of freedom. This can be a problem if we are as interested in the variance components at the top of the hierarchy as we are those at the bottom. We return to this issue in Section 12.9.

ANOVA estimates of variance components

The ANOVA estimates of variance components are again found by equating observed mean squares with their expectations and solving for the parameters. Display 12.2 shows that each variance component is estimated by a rescaled difference of two mean squares. As before, these simple estimates of variance components can be negative. Confidence intervals for these variance components can be found using the methods of Section 11.6.

Sums of squares for fully nested designs

Here are two approaches to computing sums of squares for completely nested designs. In the first, obtain the sum of squares for factor A as usual. There are $ab$ different $j(i)$ combinations for B(A). Get the sum of squares treating these $ab$ different $j(i)$ combinations as $ab$ different treatments. Note that the sum of squares for factor A is included in what we just calculated for the $j(i)$ groups. Therefore, subtract the sum of squares for factor A from that for the $j(i)$ groups to get the improvement from adding B(A) to the model. For C(AB), there are $abc$ different $k(ij)$ combinations. Again, get the sum of squares between these different groups, but subtract from this the sums of squares of the terms that are above C, namely A and B(A). The same is done for later terms in the model.

The second method begins with a fully-crossed factorial decomposition with main effects and interactions and then combines these factorial pieces (some of which do not make sense by themselves in a nested design) to get the results we need. The sum of squares, degrees of freedom, and estimated effects for A can be taken straight from this factorial decomposition. The sum of squares and degrees of freedom for B(A) are the totals of those quantities for B and AB from the factorial. Similarly, the estimated effects are found by addition:

SS and effects by recombination of factorial terms

$$\widehat{\beta}_{j(i)} = \widehat{\beta}_j + \widehat{\alpha\beta}_{ij}$$

In general, the sum of squares, degrees of freedom, and estimated effects for a term X nested in a term Y are the sums of the corresponding quantities for term X and term X crossed with any subset of factors from term Y in the full factorial. Thus for D nested in ABC, the sums will be over D, AD, BD, ABD, CD, ACD, BCD, and ABCD; and for CD nested in AB, the sums will be over CD, ACD, BCD, and ABCD.

## 12.2 Why Nesting?

We may design an experiment with nested treatment structure for several reasons. Subsampling produces small units by one or more layers of selection from larger bundles of units. For the baby carrots we went from brands to jars to samples, with each layer being a group of units from the layer beneath it. Subsampling can be used to select treatments as well as units. In some experiments crossing is theoretically possible, but logistically impractical. There may be two or three clinics scattered around the country that can perform a new diagnostic technique. We could in principle send our patients to all three clinics to cross clinics with patients, but it is more realistic to send each patient to just one clinic. In other experiments, crossing simply cannot be done. For example, consider a genetics experiment with females nested in males. We need to be able to identify the father of the offspring, so we can only breed each female to one male at a time. However, if females of the species under study only live through one breeding, we must have different females for every male.

Unit generation, logistics, and constraints may lead to nesting

We do not simply choose to use a nested model for an experiment. We use a nested model because the treatment structure of the experiment was nested, and we must build our models to match our treatment structure.

Models must match designs

## 12.3 Crossed and Nested Factors

Designs can have both crossed and nested factors. One common source of this situation is that "units" are produced in some sense through a nesting structure. In addition to the nesting structure, there are treatment factors, the combinations of which are assigned at random to the units in such a way

Units with nesting crossed with treatments

that all the combinations of nesting factors and treatment factors get an equal number of units.

**Example 12.1**

### Gum arabic

Gum arabic is used to lengthen the shelf life of emulsions, including soft drinks, and we wish to see how different gums and gum preparations affect emulsion shelf life. Raw gums are ground, dissolved, treated (possible treatments include Pasteurization, demineralization, and acidification), and then dried; the resulting dry powder is used as an emulsifier in food products.

Gum arabic comes from acacia trees; we obtain four raw gum samples from each of two varieties of acacia tree (a total of eight samples). Each sample is split into two subsamples. One of the subsamples (chosen at random) will be demineralized during treatment, the other will not. The sixteen subsamples are now dried, and we make five emulsions from each subsample and measure as the response the time until the ingredients in the emulsion begin to separate.

This design includes both crossed and nested factors. The samples of raw gum are nested in variety of acacia tree; we have completely different samples for each variety. The subsamples are nested in the samples. Subsample is now a unit to which we apply one of the two levels of the demineralization factor. Because one subsample from each sample will be demineralized and the other won't be, each sample occurs with both levels of the demineralization treatment factor. Thus sample and treatment factor are crossed. Similarly, each variety of acacia occurs with both levels of demineralization so that variety and treatment factor are crossed. The five individual emulsions from a single subsample are nested in that subsample, or equivalently, in the variety-sample-treatment combinations. They are measurement units.

If we let variety, sample, and demineralization be factors A, B, and C, then an appropriate model for the responses is

$$y_{ijkl} = \mu + \alpha_i + \beta_{j(i)} + \gamma_k + \alpha\gamma_{ik} + \beta\gamma_{jk(i)} + \epsilon_{l(ijk)} \ .$$

Treatments and units not always clear

Not all designs with crossed and nested factors have such a clear idea of unit. For some designs, we can identify the sources of variation among responses as factors crossed or nested, but identifying "treatments" randomly assigned to "units" takes some mental gymnastics.

**Example 12.2**

### Cheese tasting

Food scientists wish to study how urban and rural consumers rate cheddar cheeses for bitterness. Four 50-pound blocks of cheddar cheese of different types are obtained. Each block of cheese represents one of the segments of the market (for example, a sharp New York style cheese). The raters are students from a large introductory food science class. Ten students from rural backgrounds and ten students from urban backgrounds are selected at random from the pool of possible raters. Each rater will taste eight bites of cheese presented in random order. The eight bites are two each from the four different cheeses, but the raters don't know that. Each rater rates each bite for bitterness.

The factors in this experiment are background, rater, and type of cheese. The raters are nested in the backgrounds, but both background and rater are crossed with cheese type, because all background-cheese type combinations and all rater/cheese type combinations occur. This is an experiment with both crossed and nested factors. Perhaps the most sensible formulation of this as treatments and units is to say that bites of cheese are units (nested in type of cheese) and that raters nested in background are treatments applied to bites of cheese.

If we let background, rater, and type be factors A, B, and C, then an appropriate model for the responses is

$$y_{ijkl} = \mu + \alpha_i + \beta_{j(i)} + \gamma_k + \alpha\gamma_{ik} + \beta\gamma_{jk(i)} + \epsilon_{l(ijk)} \ .$$

This is the same model as Example 12.1, even though the structure of units and treatments is very different!

These two examples illustrate some of the issues of working with designs having both crossed and nested factors. You need to

Steps to build a model

1. Determine the sources of variation,
2. Decide which cross and which nest,
3. Decide which factors are fixed and which are random, and
4. Decide which interactions should be in the model.

Identifying the appropriate model is the hard part of working with fixed-random-crossed-nested designs; it takes a lot of practice. We will return to model choice in Section 12.5.

## 12.4 Mixed Effects

In addition to having both crossed and nested factors, Example 12.1 has both fixed (variety and demineralization) and random (sample) factors; Example 12.2 also has fixed (background and cheese type) and random (rater)

Mixed effects models have fixed and random factors

factors. An experiment with both fixed and random effects is said to have *mixed* effects. The interaction of a fixed effect and a random effect must be random, because a new random sample of factor levels will also lead to a new sample of interactions.

Two standards for analysis of mixed effects

Analysis of mixed-effects models reminds me of the joke in the computer business about standards: "The wonderful thing about standards is that there are so many to choose from." For mixed effects, there are two sets of assumptions that have a reasonable claim to being standard. Unfortunately, the two sets of assumptions lead to different analyses, and potentially different answers.

Two mechanisms to generate mixed data

Before stating the mathematical assumptions, let's visualize two mechanisms for producing the data in a mixed-effects model; each mechanism leads to a different set of assumptions. By thinking about the mechanisms behind the assumptions, we should be able to choose the appropriate assumptions in any particular experiment. Let's consider a two-factor model, with factor A fixed and factor B random, and a very small error variance so that the data are really just the sums of the row, column, and interaction effects.

Mechanism 1: sampling columns from a table

Here is one way to get the data. Imagine a table with $a$ rows and a very large number of columns. Our random factor B corresponds to selecting $b$ of the columns from the table at random, and the data we observe are the items in the table for the columns that we select.

Restricted model has interaction effects that add to zero across the fixed levels

This construction implies that if we repeated the experiment and we happened to get the same column twice, then the column totals of the data for the repeated column would be the same in the two experiments. Put another way, once we know the column we choose, we know the total for that column; we don't need to wait and see what particular interaction effects are chosen before we see the column total. Thus column differences are determined by the main effects of column; we can assume that the interaction effects in a given column add to zero. This approach leads to the *restricted* model, since it restricts the interaction effects to add to zero when summed across a fixed effect.

Mechanism 2: independent sampling from effects populations

The second approach treats the main effects and interactions independently. Now we have two populations of effects; one population contains random column main effects $\beta_j$, and the other population contains random interaction effects $\alpha\beta_{ij}$. In this second approach, we have fixed row effects, we choose column effects randomly and independently from the column main effects population, and we choose interaction effects randomly and independently from the interaction effects population; the column and interaction effects are also independent.

When we look at column totals in these data, the column total of the interaction effects can change the column total of the data. Another sample with the same column will have a different column total, because we will have a different set of interaction effects. This second approach leads to the *unrestricted* model, because it has no zero-sum restrictions.

No zero sums when unrestricted

Choose between these models by answering the following question: if you reran the experiment and got a column twice, would you have the same interaction effects or an independent set of interaction effects for that repeated column? If you have the same set of interaction effects, use the restricted model. If you have new interaction effects, use the unrestricted model. I tend to use the restricted model by default and switch to the unrestricted model when appropriate.

Restricted model if repeated main effect implies repeated interaction

### Cheese tasting, continued

**Example 12.3**

In the cheese tasting example, one of our raters is Mary; Mary likes sharp cheddar cheese and dislikes mild cheese. Any time we happen to get Mary in our sample, she will rate the sharp cheese higher and the mild cheese lower. We get the same rater by cheese interaction effects every time we choose Mary, so the restricted model is appropriate.

### Particle sampling

**Example 12.4**

To monitor air pollution, a fixed volume of air is drawn through disk-shaped filters, and particulates deposit on the filters. Unfortunately, the particulate deposition is not uniform across the filter. Cadmium particulates on a filter are measured by X-ray fluorescence. The filter is placed in an instrument that chooses a random location on the filter, irradiates that location twice, measures the resulting fluorescence spectra, and converts them to cadmium concentrations. We compare three instruments by choosing ten filters at random and running each filter through all three instruments, for a total of 60 cadmium measurements.

In this experiment we believe that the primary interaction between filter and instrument arises because of the randomly chosen locations on that filter that are scanned and the nonuniformity of the particulate on the filter. Each time the filter is run through an instrument, we get a different location and thus a different "interaction" effect, so the unrestricted model is appropriate.

Unfortunately, the choice between restricted and unrestricted models is not always clear.

### Gum arabic, continued

**Example 12.5**

Gum sample is random (nested in variety) and crosses with the fixed demineralization factor. Should we use the restricted or unrestricted model? If a gum sample is fairly heterogeneous, then at least some of any interaction that we observe is probably due to the random split of the sample into two subsamples. The next time we do the experiment, we will get different subsamples and probably different responses. In this case, the demineralization by sample interaction should be treated as unrestricted, because we would get a new set of effects every time we redid a sample.

On the other hand, how a sample reacts to demineralization may be a shared property of the complete sample. In this case, we would get the same interaction effects each time we redid a sample, so the restricted model would be appropriate.

We need to know more about the gum samples before we can make a reasoned decision on the appropriate model.

Unrestricted model assumptions

Here are the technical assumptions for mixed effects. For the unrestricted model, all random effects are independent and have normal distributions with mean 0. Random effects corresponding to the same term have the same variance: $\sigma^2_\beta$, $\sigma^2_{\alpha\beta}$, and so on. Any purely fixed effect or interaction must add to zero across any subscript.

Restricted model assumptions

The assumptions for the restricted model are the same, except for interactions that include both fixed and random factors. Random effects in a mixed-interaction term have the same variance, which is written as a factor times the usual variance component: for example, $r_{ab}\,\sigma^2_{\alpha\beta}$. These effects must sum to zero across any subscript corresponding to a fixed factor, but are independent if the random subscripts are not the same. The zero sum requirement induces negative correlation among the random effects with the same random subscripts.

Scale factors in restricted model variances

The scaling factors like $r_{ab}$ are found as follows. Get the number of levels for all fixed factors involved in the interaction. Let $r_1$ be the product of these levels, and let $r_2$ be the product of the levels each reduced by 1. Then the multiplier is $r_2/r_1$. For an AB interaction with A fixed and B random, this is $(a-1)/a$; for an ABC interaction with A and B fixed and C random, the multiplier is $(a-1)(b-1)/(ab)$.

## 12.5 Choosing a Model

Analysis depends on model

A table of data alone does not tell us the correct model. Before we can analyze data, we have to have a model on which to build the analysis. This model reflects both the structure of the experiment (nesting and/or crossing of effects), how broadly we are trying to make inference (just these treatments or a whole population of treatments), and whether mixed effects should be restricted or unrestricted. Once we have answered these questions, we can build a model. Parameters are only defined within a model, so we need the model to make tests, compute confidence intervals, and so on.

We must decide whether each factor is fixed or random. This decision is usually straightforward but can actually vary depending upon the goals of an experiment. Suppose that we have an animal breeding experiment with four sires. Now we know that the four sires we used are the four sires that were available; we did no random sampling from a population. If we are trying to make inferences about just these four sires, we treat sire as a fixed effect. On the other hand, if we are trying to make inferences about the population of potential sires, we would treat sires as a random effect. This is reasonable, provided that we can consider the four sires at hand to be a random sample from the population, even though we did no actual sampling. If these four sires are systematically different from the population, trying to use them to make inferences about the population will not work well.

Fixed or random factors?

We must decide whether each factor is nested in some other factor or interaction. The answer is determined by examining the construction of an experiment. Do all the levels of the factor appear with all the levels of another effect (crossing), or do some levels of the factor appear with some levels of the effect and other levels of the factor appear with other levels of the effect? For the cheese raters example, we see a different set of raters for rural and urban backgrounds, so rater must be nested in background. Conversely, all the raters taste all the different kinds of cheese, so rater is crossed with cheese type.

Nesting or crossing?

My model generally includes interactions for all effects that could interact, but we will see in some designs later on (for example, split plots) that not all possible interactions are always included in models. To some degree the decision as to which interactions to include is based on knowledge of the treatments and experimental materials in use, but there is also a degree of tradition in the choice of certain models.

Which interactions?

Finally, we must decide between restricted and unrestricted model assumptions. I generally use the restricted model as a default, but we must think carefully in any given situation about whether the zero-sum restrictions are appropriate.

Restricted or unrestricted?

## 12.6 Hasse Diagrams and Expected Mean Squares

One of the major issues in random and mixed effects is finding expected mean squares and appropriate denominators for tests. The tool that we use to address these issues for balanced data is the Hasse diagram (Lohr 1995).

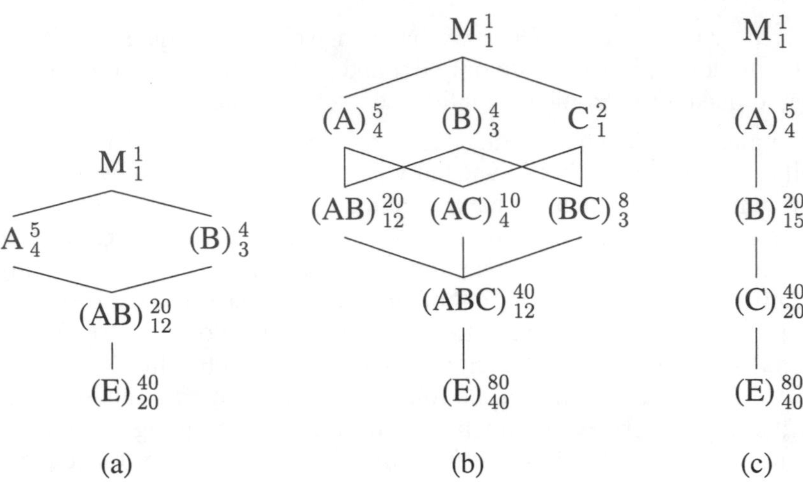


**Figure 12.1:** Hasse diagrams: (a) two-way factorial with A fixed and B random, A and B crossed; (b) three-way factorial with A and B random, C fixed, all factors crossed; (c) fully nested, with B fixed, A and C random. In all cases, A has 5 levels, B has 4 levels, and C has 2 levels.

A Hasse diagram is a graphical representation of a model showing the nesting/crossing and random/fixed structure. We can go back and forth between models and Hasse diagrams. I find Hasse diagrams to be useful when I am trying to build my model, as I find the graphic easier to work with and comprehend than a cryptic set of parameters and subscripts.

Nodes for terms, joined by lines for above/below

Figure 12.1 shows three Hasse diagrams that we will use for illustration. First, every term in a model has a *node* on the Hasse diagram. A node consists of a label to identify the term (for example, AB), a subscript giving the degrees of freedom for the term, and a superscript giving the number of different effects in a given term (for example, *ab* for $\beta_{j(i)}$). Some nodes are joined by line segments. Term U is above term V (or term V is below term U) if you can go from U to V by moving *down* line segments. For example, in Figure 12.1(b), AC is below A, but BC is not. The label for a random factor or any term below a random factor is enclosed in parentheses to indicate that it is random.

Random terms in parentheses

### 12.6.1 Test denominators

Hasse diagrams look the same whether you use the restricted model or the unrestricted model, but the models are different and we must therefore use

1. The denominator for testing a term U is the leading eligible random term below U in the Hasse diagram.
2. An eligible random term V below U is leading if there is no eligible random term that is above V and below U.
3. If there are two or more leading eligible random terms, then we must use an approximate test.
4. In the unrestricted model, all random terms below U are eligible.
5. In the restricted model, all random terms below U are eligible except those that contain a fixed factor not found in U.

**Display 12.3:** Rules for finding test denominators in balanced factorials using the Hasse diagram.

the Hasse diagram slightly differently for restricted and unrestricted models. Display 12.3 gives the steps for finding test denominators using the Hasse diagram. In general, you find the leading random term below the term to be tested, but only random terms without additional fixed factors are eligible in the restricted model. If there is more than one leading random term, we have an approximate test.

Finding test denominators

## Test denominators in the restricted model

**Example 12.6**

Consider the Hasse diagram in Figure 12.1(a). The next random term below A is the AB interaction. The only fixed factor in AB is A, so AB is the denominator for A. The next random term below B is also the AB interaction. However, AB contains A, an additional fixed factor not found in B, so AB is ineligible to be the denominator for B. Proceeding down, we get to error, which is random and does not contain any additional fixed factors. Therefore, error is the denominator for B. Similarly, error is the denominator for AB.

Figure 12.1(b) is a Hasse diagram for a three-way factorial with factors A and B random, and factor C fixed. The denominator for ABC is error. Immediately below AB is the random interaction ABC. However, ABC is not an eligible denominator for AB because it includes the additional fixed factor C. Therefore, the denominator for AB is error. For AC and BC, the denominator will be ABC, because it is random, immediately below, and contains no additional fixed factor. Next consider main effects. We see two random terms immediately below A, the AB and AC interactions. However, AC is not an eligible denominator for A, because it includes the additional fixed factor C. Therefore, the denominator for A is AB. Similarly, the denominator for B is AB. Finally consider C. There are two random terms immediately below C (AC and BC), and both of these are eligible to be denominators for C because neither includes an additional fixed factor. Thus we have an approximate test for C (C and ABC in the numerator, AC and BC in the denominator, as we will see when we get to expected mean squares).

Figure 12.1(c) is a Hasse diagram for a three-factor, fully-nested model, with A and C random and B fixed. Nesting structure appears as a vertical chain, with one factor below another. Note that the B nested in A *term* is a random term, even though B is a fixed factor. This seems odd, but consider that there is a different set of B effects for every level of A; we have a random set of A levels, so we must have a random set of B levels, so B nested in A is a random term. The denominator for C is E, and the denominator for B is C. The next random term below A is B, but B contains the fixed factor B not found in A, so B is not an eligible denominator. The closest eligible random term below A is C, which is the denominator for A.

When all the nested effects are random, the denominator for any term is simply the term below it. A fixed factor nested in a random factor is something of an oddity—it is a random term consisting only of a fixed factor. It will never be an eligible denominator in the restricted model.

**Example 12.7**

### Test denominators in the unrestricted model

Figure 12.1(a) shows a two-factor mixed-effects design. Using the unrestricted model, error is the denominator for AB, and AB is the denominator for both A and B. This is a change from the restricted model, which had error as the denominator for B.

Using the unrestricted model in the three-way mixed effects design shown in Figure 12.1(b), we find that error is the denominator for ABC, and ABC is the denominator for AB, BC, and AC; error was the denominator for AB in the restricted model. All three main effects have approximate tests, because there are two leading eligible random two-factor interactions below every main effect.

In the three-way nested design shown in Figure 12.1(c), the denominator for every term is the term immediately below it. This is again different from the restricted model, which used C as the denominator for A.

One side effect of using the unrestricted model is that there are more approximate tests, because there are more eligible denominators.

1. The representative element for a random term is its variance component.
2. The representative element for a fixed term is a function Q equal to the sum of the squared effects for the term divided by the degrees of freedom.
3. The contribution of a term is the number of data values N, divided by the number of effects for that term (the superscript for the term in the Hasse diagram), times the representative element for the term.
4. The expected mean square for a term U is the sum of the contributions for U and all eligible random terms below U in the Hasse diagram.
5. In the unrestricted model, all random terms below U are eligible.
6. In the restricted model, all random terms below U are eligible except those that contains a fixed factor not found in U.

**Display 12.4:** Rules for computing expected mean squares in balanced factorials using the Hasse diagram.

### 12.6.2 Expected mean squares

The rules for computing expected mean squares are given in Display 12.4. The description of the representative element for a fixed term seems a little arcane, but we have seen this Q before in expected mean squares. For a fixed main effect A, the representative element is $\sum_i \alpha_i^2/(a-1) = Q(\alpha)$. For a fixed interaction AB, the representative element is $\sum_{ij}(\alpha\beta_{ij})^2/[(a-1)(b-1)] = Q(\alpha\beta)$. These are the same forms we saw in Chapters 3 and 10 when discussing EMS, noncentrality parameters, and power.

Representative elements appear in noncentrality parameters

**Example 12.8**

#### Expected mean squares in the restricted model

Consider the term A in Figure 12.1(b). In the restricted model, the eligible random terms below A are AB and E; AC and ABC are ineligible due to the inclusion of the additional fixed factor C. Thus the expected mean square for A is

$$\sigma^2 + \frac{80}{20}\sigma^2_{\alpha\beta} + \frac{80}{5}\sigma^2_\alpha = \sigma^2 + 4\sigma^2_{\alpha\beta} + 16\sigma^2_\alpha \ .$$

For term C in Figure 12.1(b), all random terms below C are eligible, so the EMS for C is

$$\sigma^2 + \frac{80}{40}\sigma^2_{\alpha\beta\gamma} + \frac{80}{8}\sigma^2_{\beta\gamma} + \frac{80}{10}\sigma^2_{\alpha\gamma} + \frac{80}{2}Q(\gamma) =$$

$$\sigma^2 + 2\sigma^2_{\alpha\beta\gamma} + 10\sigma^2_{\beta\gamma} + 8\sigma^2_{\alpha\gamma} + 40Q(\gamma) \ .$$

For term A in Figure 12.1(c), the eligible random terms are C and E; B is ineligible. Thus the expected mean square for A is

$$\sigma^2 + \frac{80}{40}\sigma^2_\gamma + \frac{80}{5}\sigma^2_\alpha = \sigma^2 + 2\sigma^2_\gamma + 16\sigma^2_\alpha \ .$$

**Example 12.9** **Expected mean squares in the unrestricted model**

We now recompute two of the expected mean squares from Example 12.8 using the unrestricted model. There are four random terms below A in Figure 12.1(b); all of these are eligible in the unrestricted model, so the expected mean square for A is

$$\sigma^2 + \frac{80}{40}\sigma^2_{\alpha\beta\gamma} + \frac{80}{20}\sigma^2_{\alpha\beta} + \frac{80}{10}\sigma^2_{\alpha\gamma} + \frac{80}{5}\sigma^2_\alpha =$$

$$\sigma^2 + 2\sigma^2_{\alpha\beta\gamma} + 4\sigma^2_{\alpha\beta} + 8\sigma^2_{\alpha\gamma} + 16\sigma^2_\alpha \ .$$

This includes two additional contributions that were not present in the restricted model.

For term A in Figure 12.1(c), B, C, and E are all eligible random terms. Thus the expected mean square for A is

$$\sigma^2 + \frac{80}{40}\sigma^2_\gamma + + \frac{80}{20}\sigma^2_\beta + \frac{80}{5}\sigma^2_\alpha = \sigma^2 + 2\sigma^2_\gamma + 4\sigma^2_\beta + 16\sigma^2_\alpha \ .$$

Term B contributes to the expected mean square of A in the unrestricted model.

We can figure out approximate tests by using the rules for expected mean squares and the Hasse diagram. Consider testing C in Figure 12.1(b). AC and BC are both eligible random terms below C, so both of their expected

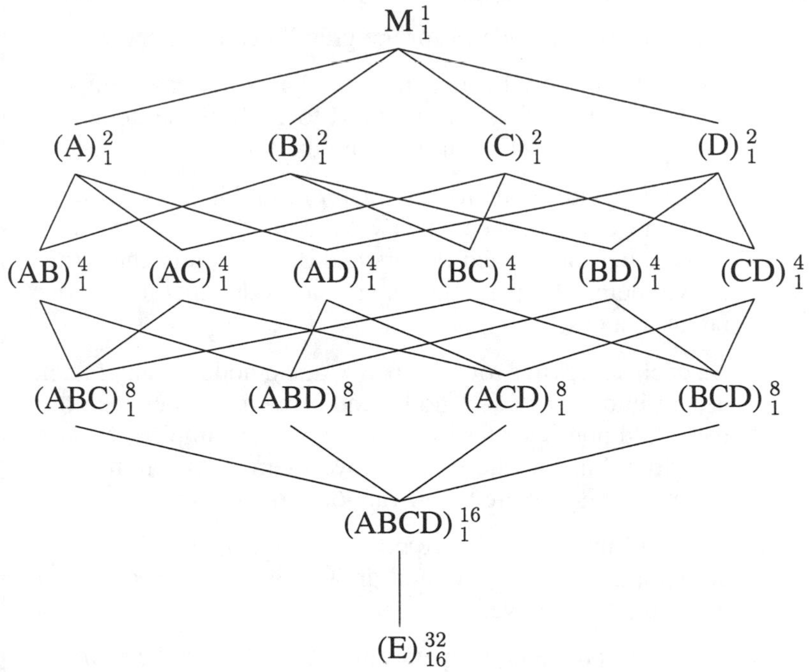


**Figure 12.2:** Hasse diagram for a four-way factorial with all random effects.

mean squares will appear in the EMS for C; thus both AC and BC need to be in the denominator for C. However, putting both AC and BC in the denominator double-counts the terms below AC and BC, namely ABC and error. Therefore, we add ABC to the numerator to match the double-counting.

Use Hasse diagrams to find approximate tests

Here is a more complicated example: testing a main effect in a four-factor model with all factors random. Figure 12.2 shows the Hasse diagram. Suppose that we wanted to test A. Terms AB, AC, and AD are all eligible random terms below A, so all would appear in the EMS for A, and all must appear in the denominator for A. If we put AB, AC, and AD in the denominator, then the expectations of ABC, ABD, and ACD will be double-counted there. Thus we must add them to the numerator to compensate. With A, ABC, ABD, and ACD in the numerator, ABCD and error are quadruple-counted in the numerator but only triple-counted in the denominator, so we must add ABCD to the denominator. We now have a numerator (A + ABC + ABD + ACD) and a denominator (AB + AC + AD + ABCD) with expectations that differ only by a multiple of $\sigma_\alpha^2$.

1. Start row 0 with node M for the grand mean.
2. Put a node on row 1 for each factor that is not nested in any term. Add lines from the node M to each of the nodes on row 1. Put parentheses around random factors.
3. On row 2, add a node for any factor nested in a row 1 node, and draw a line between the two. Add nodes for terms with two explicit or implied factors and draw lines to the terms above them. Put parentheses around nodes that are below random nodes.
4. On each successive row, say row $i$, add a node for any factor nested into a row $i-1$ node, and draw a line between the two. Add nodes for terms with $i$ explicit or implied factors and draw lines to the terms above them. Put parentheses around nodes that are below random nodes.
5. When all interactions have been exhausted, add a node for error on the bottom line, and draw a line from error to the dangling node above it.
6. For each node, add a superscript that indicates the number of effects in the term.
7. For each node, add a subscript that indicates the degrees of freedom for the term. Degrees of freedom for a term U are found by starting with the superscript for U and subtracting out the degrees of freedom for all terms above U.

**Display 12.5:** Steps for constructing a Hasse diagram.

### 12.6.3 Constructing a Hasse diagram

Build from top down

A Hasse diagram always has a node M at the top for the grand mean, a node (E) at the bottom for random error, and nodes for each factorial term in between. I build Hasse diagrams from the top down, but to do that I need to know which terms go above other terms. Hasse diagrams have the same above/below relationships as ANOVA tables.

Nested factors include implicit factors

A term U is above a term V in an ANOVA table if all of the factors in term U are in term V. Sometimes these factors are explicit; for example, factors A, B, and C are in the ABC interaction. When nesting is present, some of the factors may be implicit or implied in a term. For example, factors A, B, and

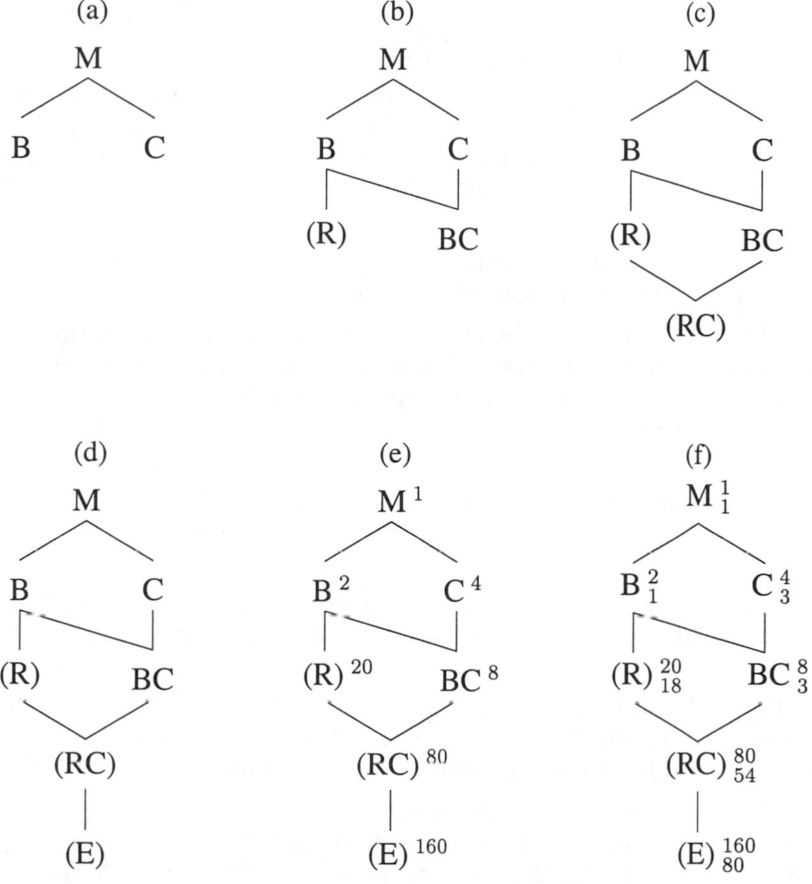


**Figure 12.3:** Stages in the construction of Hasse diagram for the cheese rating example.

C are all in the term C nested in the AB interaction. When we write the term as C, A and B are there implicitly. We will say that term U is above term V if all of the factors in term U are present or implied in term V.

Before we start the Hasse diagram, we must determine the factors in the model, which are random and which are fixed, and which nest and which cross. Once these have been determined, we can construct the diagram using the steps in Display 12.5.

**Example 12.10** **Cheese tasting Hasse diagram**

The cheese tasting experiment of Example 12.2 had three factors: the fixed factor for background (two levels, labeled B), the fixed factor cheese type (four levels, labeled C), and the random factor for rater (ten levels, random, nested in background, labeled R). Cheese type crosses with both background and rater.

Figure 12.3(a) shows the first stage of the diagram, with the M node for the mean and nodes for each factor that is not nested.

Figure 12.3(b) shows the next step. We have added rater nested in background. It is in parentheses to denote that it is random, and we have a line up to background to show the nesting. Also in this row is the BC two-factor interaction, with lines up to B and C.

Figure 12.3(c) shows the third stage, with the rater by cheese RC interaction. This is random (in parentheses) because it is below rater. It is also below BC; B is present implicitly in any term containing R, because R nests in B.

Figure 12.3(d) adds the node for random error. You can determine the appropriate denominators for tests at this stage without completing the Hasse diagram.

Figure 12.3(e) adds the superscripts for each term. The superscript is the number of different effects in the term and equals the product of the number of levels of all the implied or explicit factors in a term.

Finally, Figure 12.3(f) adds the subscripts, which give the degrees of freedom. Compute the degrees of freedom by starting with the superscript and subtracting out the degrees of freedom for all terms above the given term. It is easiest to get degrees of freedom by starting with terms at the top and working down.

## 12.7 Variances of Means and Contrasts

Distinct means can be correlated in mixed effects models

Variances of treatment means are easy to calculate in a fixed-effects models—simply divide $\sigma^2$ by the number of responses in the average. Furthermore, distinct means are independent. Things are more complicated for mixed-effects models, because there are multiple random terms that can all contribute to the variance of a mean, and some of these random terms can cause nonzero covariances as well. In this section we give a set of rules for calculating the variance and covariance of treatment means. We can use the

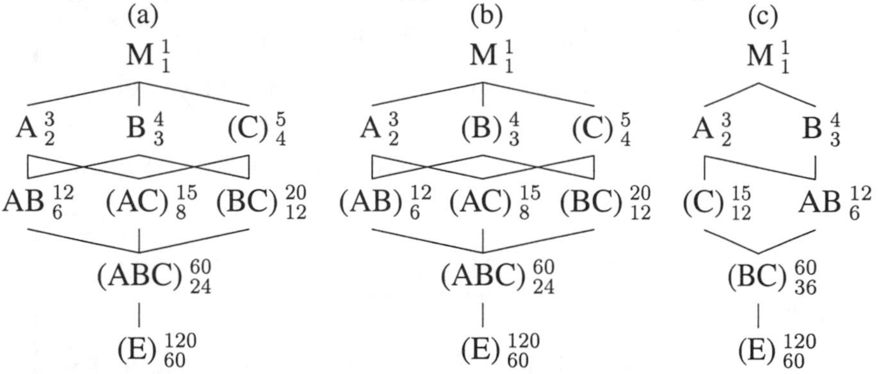


**Figure 12.4:** Hasse diagrams for three three-way factorials. (a) C random; (b) B and C random; (c) C random and nested in A.

covariance to determine the variance of pairwise comparisons and other contrasts.

Treatment means make sense for combinations of fixed factors, but are generally less interesting for random effects. Consider the Hasse diagrams in Figure 12.4. All are three-way factorials with $a = 3$, $b = 4$, $c = 5$, and $n = 2$. In panels (a) and (c), factors A and B are fixed. Thus it makes sense to consider means for levels of factor A ($\overline{y}_{i\bullet\bullet\bullet}$), for levels of factor B ($\overline{y}_{\bullet j\bullet\bullet}$), and for AB combinations ($\overline{y}_{ij\bullet\bullet}$). In panel (b), only factor A is fixed, so only means $\overline{y}_{i\bullet\bullet\bullet}$ are usually of interest.

Look at treatment means for fixed factors

It is tempting to use the denominator mean square for A as the variance for means $\overline{y}_{i\bullet\bullet\bullet}$. *This does not work!* We must go through the steps given in Display 12.6 to compute variances for means. We can use the denominator mean square for A when computing the variance for a *contrast* in factor A means; simply substitute the denominator mean square as an estimate of variance into the usual formula for the variance of a contrast. Similarly, we can use the denominator mean square for the AB interaction when we compute the variance of an AB interaction contrast, but this will not work for means $\overline{y}_{ij\bullet\bullet}$ or paired differences or other combinations that are not interaction contrasts.

Do not use denominator mean squares as variances for means

Display 12.6 gives the steps required to compute the variance of a mean. For a mean $\overline{y}_{i\bullet\bullet\bullet}$, the base term is A and the base factor is A; for a mean $\overline{y}_{ij\bullet\bullet}$, the base term is AB and the base factors are A and B.

1. Make a Hasse diagram for the model.
2. Identify the base term and base factors for the mean of interest.
3. The variance of the mean of interest will be the sum over all contributing terms T of

$$\sigma_T^2 \frac{\text{product of superscripts of all base factors above T}}{\text{superscript of term T}}$$

4. In the unrestricted model, all random terms contribute to the variance of the mean of interest.
5. In the restricted model, all random terms contribute to the variance of the mean of interest except those that contain a fixed factor not found in the base term.

**Display 12.6:** Steps for determining the variance of a marginal mean.

**Example 12.11** **Variances of means**

Let's compute variances for some means in the models of Figure 12.4 using restricted model assumptions. Consider first the mean $\overline{y}_{i\bullet\bullet\bullet}$. The base term is A, and the base factor is A. In panel (a), there will be contributions from C, AC, and E (but not BC or ABC because they contain the additional fixed factor B). The variance is

$$\sigma_\gamma^2 \frac{1}{5} + \sigma_{\alpha\gamma}^2 \frac{3}{15} + \sigma^2 \frac{3}{120} \ .$$

In panel (b), there will be contributions from all random terms (A is the only fixed term). Thus the variance is

$$\sigma_\beta^2 \frac{1}{4} + \sigma_\gamma^2 \frac{1}{5} + \sigma_{\alpha\beta}^2 \frac{3}{12} + \sigma_{\alpha\gamma}^2 \frac{3}{15} + \sigma_{\beta\gamma}^2 \frac{1}{20} + \sigma_{\alpha\beta\gamma}^2 \frac{3}{60} + \sigma^2 \frac{3}{120} \ .$$

Finally, in panel (c), there will be contributions from C and E (but not BC). The variance is

$$\sigma_\gamma^2 \frac{3}{15} + \sigma^2 \frac{3}{120} \ .$$

Now consider a mean $\overline{y}_{\bullet j\bullet\bullet}$ in model (c). The contributing terms will be C, BC, and E, and the variance is

1. Identify the base term and base factors for the means of interest.
2. Determine whether the subscripts agree or disagree for each base factor.
3. The covariance of the means will be the sum over all contributing terms T of

$$\sigma_T^2 \frac{\text{product of superscripts of all base factors above T}}{\text{superscript of term T}}$$

4. In the unrestricted model, all random terms contribute to the covariance *except* those that are below a base factor with disagreeing subscripts.
5. In the restricted model, all random terms contribute to the covariance *except* those that contain a fixed factor not found in the base term and those that are below a base factor with disagreeing subscripts.

**Display 12.7:** Steps for determining the covariance between two marginal means.

$$\sigma_\gamma^2 \frac{1}{15} + \sigma_{\beta\gamma}^2 \frac{4}{60} + \sigma^2 \frac{4}{120} \ .$$

Finally, consider the variance of $\overline{y}_{ij\bullet\bullet}$; this mean does not make sense in panel (b). In panel (a), all random terms contribute to the variance, which is

$$\sigma_\gamma^2 \frac{1}{5} + \sigma_{\alpha\gamma}^2 \frac{3}{15} + \sigma_{\beta\gamma}^2 \frac{4}{20} + \sigma_{\alpha\beta\gamma}^2 \frac{3 \times 4}{60} + \sigma^2 \frac{3 \times 4}{120} \ .$$

In panel (c), all random terms contribute, but the variance here is

$$\sigma_\gamma^2 \frac{3}{15} + \sigma_{\beta\gamma}^2 \frac{3 \times 4}{60} + \sigma^2 \frac{3 \times 4}{120} \ .$$

The variance of a difference is the sum of the individual variances minus twice the covariance. We thus need to compute covariances of means in order to get variances of differences of means. Display 12.7 gives the steps for computing the covariance between two means, which are similar to those

Need covariances to get variance of a difference

for variances, with the additional twist that we need to know which of the subscripts in the means agree and which disagree. For example, the factor A subscripts in $\overline{y}_{i\bullet\bullet\bullet} - \overline{y}_{i'\bullet\bullet\bullet}$ disagree, but in $\overline{y}_{ij\bullet\bullet} - \overline{y}_{ij'\bullet\bullet}, j \neq j'$, the factor A subscripts agree while the factor B subscripts disagree.

**Example 12.12** **Covariances of means**

Now compute covariances for some means in the models of Figure 12.4 using restricted model assumptions. Consider the means $\overline{y}_{i\bullet\bullet\bullet}$ and $\overline{y}_{i'\bullet\bullet\bullet}$. The base term is A, the base factor is A, and the factor A subscripts disagree. In model (a), only term C contributes to the covariance, which is

$$\sigma^2_\gamma \frac{1}{5}$$

Using the variance for $\overline{y}_{i\bullet\bullet\bullet}$ computed in Example 12.11, we find

$$\begin{aligned}
\text{Var}(\overline{y}_{i\bullet\bullet\bullet} - \overline{y}_{i'\bullet\bullet\bullet}) &= \text{Var}(\overline{y}_{i\bullet\bullet\bullet}) + \text{Var}(\overline{y}_{i'\bullet\bullet\bullet}) - 2 \times \text{Cov}(\overline{y}_{i\bullet\bullet\bullet}, \overline{y}_{i'\bullet\bullet\bullet}) \\
&= 2 \times (\sigma^2_\gamma \frac{1}{5} + \sigma^2_{\alpha\gamma} \frac{1}{5} + \sigma^2 \frac{1}{40}) - 2 \times \sigma^2_\gamma \frac{1}{5} \\
&= 2 \times (\sigma^2_{\alpha\gamma} \frac{1}{5} + \sigma^2 \frac{1}{40}) \\
&= EMS_{AC}(\frac{1}{40} + \frac{1}{40}) \ .
\end{aligned}$$

The last line is what we would get by using the denominator for A and applying the usual contrast formulae with a sample size of 40 in each mean.

In model (b), B, C, and BC contribute to the covariance, which is

$$\sigma^2_\beta \frac{1}{4} + \sigma^2_\gamma \frac{1}{5} + \sigma^2_{\beta\gamma} \frac{1}{20}$$

and leads to

$$\begin{aligned}
\text{Var}(\overline{y}_{i\bullet\bullet\bullet} - \overline{y}_{i'\bullet\bullet\bullet}) &= \text{Var}(\overline{y}_{i\bullet\bullet\bullet}) + \text{Var}(\overline{y}_{i'\bullet\bullet\bullet}) - 2 \times \text{Cov}(\overline{y}_{i\bullet\bullet\bullet}, \overline{y}_{i'\bullet\bullet\bullet}) \\
&= 2 \times (\sigma^2_{\alpha\beta} \frac{1}{4} + \sigma^2_{\alpha\gamma} \frac{1}{5} + \sigma^2_{\alpha\beta\gamma} \frac{1}{20} + \sigma^2 \frac{1}{40})
\end{aligned}$$

In panel (c), all the random terms are below A, so none can contribute to the covariance, which is thus 0.

Consider now $\overline{y}_{\bullet j\bullet\bullet} - \overline{y}_{\bullet j'\bullet\bullet}$ in model (c). Only the term C contributes to the covariance, which is

**Table 12.1:** Covariances and variances of differences of two-factor means $\overline{y}_{ij\bullet\bullet}$ for models (a) and (c) of Figure 12.4 as a function of which subscripts disagree.

| | | Covariance | Variance of difference |
|---|---|---|---|
| (a) | A | $\frac{1}{5}\sigma^2_\gamma + \frac{1}{5}\sigma^2_{\beta\gamma}$ | $2 \times (\frac{1}{5}\sigma^2_{\alpha\gamma} + \frac{1}{5}\sigma^2_{\alpha\beta\gamma} + \frac{1}{10}\sigma^2)$ |
| (a) | B | $\frac{1}{5}\sigma^2_\gamma + \frac{1}{5}\sigma^2_{\alpha\gamma}$ | $2 \times (\frac{1}{5}\sigma^2_{\beta\gamma} + \frac{1}{5}\sigma^2_{\alpha\beta\gamma} + \frac{1}{10}\sigma^2)$ |
| (a) | A and B | $\frac{1}{5}\sigma^2_\gamma$ | $2 \times (\frac{1}{5}\sigma^2_{\alpha\gamma} + \frac{1}{5}\sigma^2_{\beta\gamma} + \frac{1}{5}\sigma^2_{\alpha\beta\gamma} + \frac{1}{10}\sigma^2)$ |
| (c) | A | 0 | $2 \times (\frac{1}{5}\sigma^2_\gamma + \frac{1}{5}\sigma^2_{\beta\gamma} + \frac{1}{10}\sigma^2)$ |
| (c) | B | $\frac{1}{5}\sigma^2_\gamma$ | $2 \times (\frac{1}{5}\sigma^2_{\beta\gamma} + \frac{1}{10}\sigma^2)$ |
| (c) | A and B | 0 | $2 \times (\frac{1}{5}\sigma^2_\gamma + \frac{1}{5}\sigma^2_{\beta\gamma} + \frac{1}{10}\sigma^2)$ |

$$\sigma^2_\gamma \frac{1}{15} \ ;$$

and leads to

$$\begin{aligned}
\text{Var}(\overline{y}_{\bullet j\bullet\bullet} - \overline{y}_{\bullet j'\bullet\bullet}) &= \text{Var}(\overline{y}_{\bullet j\bullet\bullet}) + \text{Var}(\overline{y}_{\bullet j'\bullet\bullet}) - 2 \times \text{Cov}(\overline{y}_{\bullet j\bullet\bullet}, \overline{y}_{\bullet j'\bullet\bullet}) \\
&= 2 \times (\sigma^2_{\beta\gamma}\frac{1}{15} + \sigma^2\frac{1}{30}) \\
&= \frac{2}{30} EMS_{BC} \ ;
\end{aligned}$$

which is what would be obtained by using the denominator for B in the standard contrast formulae for means with sample size 30.

Things get a little more interesting with two-factor means, because we can have the first, the second, or both subscripts disagreeing, and we can get different covariances for each. Of course there are even more possibilities with three-factor means. Consider covariances for AB means in panel (a) of Figure 12.4. If the A subscripts differ, then only C and BC can contribute to the covariance; if the B subscripts differ, then C and AC contribute to the covariance; if both differ, then only C contributes to the covariance. In panel (c), if the A subscripts differ, then no terms contribute to covariance; if the B subscripts differ, then only C contributes to covariance. Table 12.1 summarizes the covariances and variances of differences of means for these cases.

**Listing 12.1:** MacAnova output for restricted Type III EMS.

```
EMS(a) = V(ERROR1) + 1.9753V(a.b.c) + 7.8752V(a.c) + 9.8424V(a.b) + 39.516Q(a)
EMS(b) = V(ERROR1) + 5.9048V(b.c) + 29.524V(b)
EMS(a.b) = V(ERROR1) + 1.9758V(a.b.c) + 9.8469V(a.b)
EMS(c) = V(ERROR1) + 5.9062V(b.c) + 23.625V(c)
EMS(a.c) = V(ERROR1) + 1.976V(a.b.c) + 7.8803V(a.c)
EMS(b.c) = V(ERROR1) + 5.9167V(b.c)
EMS(a.b.c) = V(ERROR1) + 1.9774V(a.b.c)
EMS(ERROR1) = V(ERROR1)
```

## 12.8 Unbalanced Data and Random Effects

EMS for Types I, II, and III, and restricted or unrestricted models by computer

Unbalanced data or random effects make data analysis more complicated; life gets very interesting with unbalanced data and random effects. Mean squares change depending on how they are computed (Type I, II, or III), so there are also Type I, II, and III expected mean squares to go along with them. Type III mean squares are generally more usable in unbalanced mixed-effects models than those of Types I or II, because they have simpler expected mean squares. As with balanced data, expected mean squares for unbalanced data depend on whether we are using the restricted or unrestricted model assumptions. Expected mean squares cannot usually be determined by hand; in particular, the Hasse diagram method for finding denominators and expected mean squares is for balanced data and does not work for unbalanced data.

Do not use Hasse diagram with unbalanced data

Many statistical software packages can compute expected mean squares for unbalanced data, but most do not compute all the possibilities. For example, SAS PROC GLM can compute Type I, II, or III expected mean squares, but only for the unrestricted model. Similarly, Minitab computes sequential (Type I) and "adjusted" (Type III) expected mean squares for the unrestricted model. MacAnova can compute sequential and "marginal" (Type III) expected mean squares for both restricted and unrestricted assumptions.

**Example 12.13** **Unbalanced expected mean squares**

Suppose we make the three-way factorial of Figure 12.4(b) unbalanced by having only one response when all factors are at their low levels. Listings 12.1, 12.2, and 12.3 show the EMS's for Type III restricted, Type III unrestricted, and Type II unrestricted, computed respectively using MacAnova, Minitab, and SAS. All three tables of expected mean squares differ, indicating that the different sums of squares and assumptions lead to different tests and possibly different inferences.

**Listing 12.2:** Minitab output for unrestricted Type III EMS.

```
Expected Mean Squares, using Adjusted SS

Source        Expected Mean Square for Each Term
 1 A          (8) +  1.9677(7) +  7.8710(5) +  9.8387(4) + Q[1]
 2 B          (8) +  1.9683(7) +  5.9048(6) +  9.8413(4) + 29.5238(2)
 3 C          (8) +  1.9688(7) +  5.9063(6) +  7.8750(5) + 23.6250(3)
 4 A*B        (8) +  1.9697(7) +  9.8485(4)
 5 A*C        (8) +  1.9706(7) +  7.8824(5)
 6 B*C        (8) +  1.9722(7) +  5.9167(6)
 7 A*B*C      (8) +  1.9762(7)
 8 Error      (8)
```

**Listing 12.3:** SAS output for unrestricted Type II EMS.

```
Source       Type II Expected Mean Square

A            Var(Error) + 1.9878 Var(A*B*C) + 7.9265 Var(A*C)
             + 9.9061 Var(A*B) + Q(A)

B            Var(Error) + 1.9888 Var(A*B*C) + 5.9496 Var(B*C)
             + 9.9104 Var(A*B) + 29.714 Var(B)

A*B          Var(Error) + 1.9841 Var(A*B*C) + 9.8889 Var(A*B)

C            Var(Error) + 1.9893 Var(A*B*C) + 5.9509 Var(B*C)
             + 7.9316 Var(A*C) + 23.778 Var(C)

A*C          Var(Error) + 1.9845 Var(A*B*C) + 7.913 Var(A*C)

B*C          Var(Error) + 1.9851 Var(A*B*C) + 5.9375 Var(B*C)

A*B*C        Var(Error) + 1.9762 Var(A*B*C)
```

For unbalanced data, almost all tests are approximate tests. For example, consider testing $\sigma^2_\gamma = 0$ using the Type III unrestricted analysis in Listing 12.2. The expected mean square for C is

Use general linear combinations of MS to get denominators

$$\sigma^2 + 1.9688\sigma^2_{\alpha\beta\gamma} + 5.9063\sigma^2_{\beta\gamma} + 7.8750\sigma^2_{\alpha\gamma} + 23.625\sigma^2_\gamma \ ,$$

so we need to find a linear combination of mean squares with expectation

$$\sigma^2 + 1.9688\sigma^2_{\alpha\beta\gamma} + 5.9063\sigma^2_{\beta\gamma} + 7.8750\sigma^2_{\alpha\gamma}$$

to use as a denominator. The combination

$$.9991MS_{AC} + .9982MS_{BC} - .9962MS_{ABC} - .0011MS_E$$

has the correct expectation, so we could use this as our denominator for $MS_C$ with approximate degrees of freedom computed with Satterthwaite's formula.

Rearrange so that all MS's are added

Alternatively, we could use $MS_C + .9962MS_{ABC} + .0011MS_E$ as the numerator and $.9991MS_{AC} + .9982MS_{BC}$ as the denominator, computing approximate degrees of freedom for both the numerator and denominator. This second form avoids subtracting mean squares and generally leads to larger approximate degrees of freedom. It does move the F-ratio towards one, however.

ANOVA estimates of variance components

We can compute point estimates and confidence intervals for variance components in unbalanced problems using exactly the same methods we used in the balanced case. To get point estimates, equate the observed mean squares with their expectations and solve for the variance components (the ANOVA method). Confidence intervals are approximate, based on the Satterthwaite degrees of freedom for the point estimate, and of dubious coverage.

## 12.9 Staggered Nested Designs

Ordinary nesting has more degrees of freedom for nested terms

One feature of standard fully-nested designs is that we have few degrees of freedom for the top-level mean squares and many for the low-level mean squares. For example, in Figure 12.1(c), we have a fully-nested design with 4, 15, 20, and 40 degrees of freedom for A, B, C, and error. This difference in degrees of freedom implies that our estimates for the top-level variance components will be more variable than those for the lower-level components. If we are equally interested in all the variance components, then some other experimental design might be preferred.

Staggered nested designs nest in an unbalanced way

*Staggered nested designs* can be used to distribute the degrees of freedom more evenly (Smith and Beverly 1981). There are several variants on these designs; we will only discuss the simplest. Factor A has $a$ levels, where we'd like $a$ as large as feasible. A has $(a-1)$ degrees of freedom. Factor B has two levels and is nested in factor A; B appears at two levels for every level of A. B has $a(2-1) = a$ degrees of freedom. Factor C has two levels and is nested in B, but in an unbalanced way. Only level 2 of factor B will have two levels of factor C; level 1 of factor B will have just one level of factor C. Factor D is nested in factor C, but in the same unbalanced way. Only level 2 of factor C will have two levels of factor D; level 1 of factor C will have just one level of factor D. Any subsequent factors are nested in the same unbalanced fashion. Figure 12.5 illustrates the idea for a four-factor model.

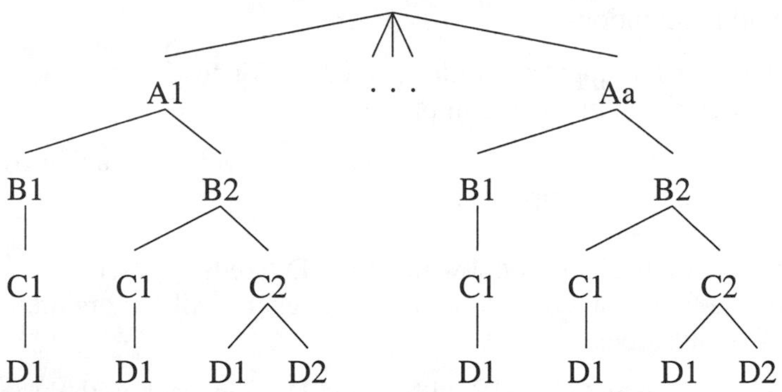


**Figure 12.5:** Example of staggered nested design.

**Listing 12.4:** SAS output for Type I EMS in a staggered nested design.

```
Source      Type I Expected Mean Square

A           Var(Error) + 1.5 Var(C(A*B)) + 2.5 Var(B(A)) + 4 Var(A)

B(A)        Var(Error) + 1.1667 Var(C(A*B)) + 1.5 Var(B(A))

C(A*B)      Var(Error) + 1.3333 Var(C(A*B))
```

For a staggered nested design with $h$ factors (counting error), there are $ha$ units. There is 1 degree of freedom for the overall mean, $a-1$ degrees of freedom for A, and $a$ degrees of freedom for each nested factor below A. The expected mean squares will generally be determined using software. For example, Listing 12.4 gives the Type I expected mean squares for a staggered nested design with $h=4$ factors counting error and $a=10$ levels for factor A; the degrees of freedom are 9 for A and 10 for B, C, and error.

Staggered nested designs spread degrees of freedom evenly

## 12.10 Problems

Many of the problems in this Chapter will ask *the standard five questions*:

(a) Draw the Hasse diagram for this model.

(b) Determine the appropriate denominators for testing each term using the restricted model assumptions.

(c) Determine the expected mean squares for each term using the restricted model assumptions.

(d) Determine the appropriate denominators for testing each term using the unrestricted model assumptions.

(e) Determine the expected mean squares for each term using the unrestricted model assumptions.

**Exercise 12.1** Consider a four-factor model with A and D fixed, each with three levels. Factors B and C are random with two levels each. All factors are crossed. Standard five questions.

**Exercise 12.2** Consider a four-factor model with A and D fixed, each with three levels. Factors B and C are random with two levels each. B nests in A, C nests in B, and D crosses with the others. There is a total of 72 observations. Standard five questions.

**Exercise 12.3** Consider a four-factor model with A and D fixed, each with three levels. Factors B and C are random with two levels each. B nests in A, C nests in D, and all other combinations cross. There is a total of 72 observations. Standard five questions.

**Exercise 12.4** Briefly describe the treatment structure you would choose for each of the following situations. Describe the factors, the number of levels for each, whether they are fixed or random, and which are crossed.

(a) One of the expenses in animal experiments is feeding the animals. A company salesperson has made the claim that their new rat chow (35% less expensive) is equivalent to the two standard chows on the market. You wish to test this claim by measuring weight gain of rat pups on the three chows. You have a population of 30 inbred, basically exchangeable female rat pups to work with, each with her own cage.

(b) Different gallons of premixed house paints with the same label color do not always turn out the same. A manufacturer of paint believes that color variability is due to three sources: supplier of tint materials, miscalibration of the devices that add the tint to the base paint, and uncontrollable random variation between gallon cans. The manufacturer wishes to assess the sizes of these sources of variation and is willing to use 60 gallons of paint in the process. There are three suppliers of tint and 100 tint-mixing machines at the plant.

(c) Insect infestations in croplands are not uniform; that is, the number of insects present in meter-square plots can vary considerably. Our interest is in determining the variability at different geographic scales.

That is, how much do insect counts vary from meter square to meter square within a hectare field, from hectare to hectare within a county, and from county to county? We have resources for at most 10 counties in southwestern Minnesota, and at most 100 total meter-square insect counts.

(d) The disposable diaper business is very competitive, with all manufacturers trying to get a leg up, as it were. You are a consumer testing agency comparing the absorbency of two brands of "newborn" size diapers. The test is to put a diaper on a female doll and pump body-temperature water through the doll into the diaper at a fixed rate until the diaper leaks. The response is the amount of liquid pumped before leakage. We are primarily interested in brand differences, but we are also interested in variability between individual diapers and between batches of diapers (which we can only measure as between boxes of diapers, since we do not know the actual manufacturing time or place of the diapers). We can afford to buy 32 boxes of diapers and test 64 diapers.

**Problem 12.1**

Answer the standard five questions for each of the following experiments.

(a) We are interested in the relationship between atmospheric sulfate aerosol concentration and visibility. As a preliminary to this study, we examine how we will measure sulfate aerosol. Sulfate aerosol is measured by drawing a fixed volume of air through a filter and then chemically analyzing the filter for sulfate. There are four brands of filter available and two methods to analyze the filters chemically. We randomly select eight filters for each brand-method combination. These 64 filters are then used (by drawing a volume of air with a known concentration of sulfate through the filter), split in half, and both halves are chemically analyzed with whatever method was assigned to the filter, for a total of 128 responses.

(b) A research group often uses six contract analytical laboratories to determine total nitrogen in plant tissues. However, there is a possibility that some labs are biased with respect to the others. Forty-two tissue samples are taken at random from the freezer and split at random into six groups of seven, one group for each lab. Each lab then makes two measurements on each of the seven samples they receive, for a total of 84 measurements.

(c) A research group often uses six contract analytical laboratories to determine total nitrogen in plant tissues. However, there is a possibility that some labs are biased with respect to the others. Seven tissue samples are taken at random from the freezer and each is split into six parts, one part for each lab. We expect some variation among the subsamples of a given sample. Each lab then makes two measurements on each of the seven samples they receive, for a total of 84 measurements.

**Problem 12.2** Dental fillings made with gold can vary in hardness depending on how the metal is treated prior to its placement in the tooth. Two factors are thought to influence the hardness: the gold alloy and the condensation method. In addition, some dentists doing the work are better at some types of fillings than others.

Five dentists were selected at random. Each dentist prepares 24 fillings (in random order), one for each of the combinations of method (three levels) and alloy (eight levels). The fillings were then measured for hardness using the Diamond Pyramid Hardness Number (big scores are better). The data follow (from Xhonga 1971 via Brown 1975):

| | | Alloy | | | | | | | |
|---|---|---|---|---|---|---|---|---|---|
| Dentist | Method | 1 | 2 | 3 | 4 | 5 | 6 | 7 | 8 |
| 1 | 1 | 792 | 824 | 813 | 792 | 792 | 907 | 792 | 835 |
| | 2 | 772 | 772 | 782 | 698 | 665 | 1115 | 835 | 870 |
| | 3 | 782 | 803 | 752 | 620 | 835 | 847 | 560 | 585 |
| 2 | 1 | 803 | 803 | 715 | 803 | 813 | 858 | 907 | 882 |
| | 2 | 752 | 772 | 772 | 782 | 743 | 933 | 792 | 824 |
| | 3 | 715 | 707 | 835 | 715 | 673 | 698 | 734 | 681 |
| 3 | 1 | 715 | 724 | 743 | 627 | 752 | 858 | 762 | 724 |
| | 2 | 792 | 715 | 813 | 743 | 613 | 824 | 847 | 782 |
| | 3 | 762 | 606 | 743 | 681 | 743 | 715 | 824 | 681 |
| 4 | 1 | 673 | 946 | 792 | 743 | 762 | 894 | 792 | 649 |
| | 2 | 657 | 743 | 690 | 882 | 772 | 813 | 870 | 858 |
| | 3 | 690 | 245 | 493 | 707 | 289 | 715 | 813 | 312 |
| 5 | 1 | 634 | 715 | 707 | 698 | 715 | 772 | 1048 | 870 |
| | 2 | 649 | 724 | 803 | 665 | 752 | 824 | 933 | 835 |
| | 3 | 724 | 627 | 421 | 483 | 405 | 536 | 405 | 312 |

Analyze these data to determine which factors influence the response and how they influence the response. (Hint: the dentist by method interaction can use close inspection.)

**Problem 12.3** An investigative group at a television station wishes to determine if doctors treat patients on public assistance differently from those with private insurance. They measure this by how long the doctor spends with the patient. There are four large clinics in the city, and the station chooses three pediatricians at random from each of the four clinics. Ninety-six families on public assistance are located and divided into four groups of 24 at random. All 96 families have a one-year-old child and a child just entering school. Half the families will request a one-year checkup, and the others will request a preschool checkup. Half the families will be given temporary private insurance for the study, and the others will use public assistance. The four groupings of families are the factorial combinations of checkup type and insurance type. Each group of 24 is now divided at random into twelve sets of two, with each set of two assigned to one of the twelve selected doctors. Thus each doctor will see eight patients from the investigation. Recap: 96 units (families); the response is how long the doctor spends with each family; and treatments are clinic, doctor, checkup type, and insurance type. Standard five questions.

**Problem 12.4**

Eurasian water milfoil is an exotic water plant that is infesting North American waters. Some weevils will eat milfoil, so we conduct an experiment to see what may influence weevils' preferences for Eurasian milfoil over the native northern milfoil. We may obtain weevils that were raised on Eurasian milfoil or northern milfoil. From each source, we take ten randomly chosen males (a total of twenty males). Each male is mated with three randomly chosen females raised on the same kind of milfoil (a total of 60 females). Each female produces many eggs. Eight eggs are chosen at random from the eggs of each female (a total of 480 eggs). The eight eggs for each female are split at random into four groups of two, with each set of two assigned to one of the factor-level combinations of hatching species and growth species (an egg may be hatched on either northern or Eurasian milfoil, and after hatching grows to maturity on either northern or Eurasian milfoil). After the hatched weevils have grown to maturity, they are given ten opportunities to swim to a plant. The response is the number of times they swim to Eurasian. Standard five questions.

**Problem 12.5**

City hall wishes to learn about the rate of parking meter use. They choose eight downtown blocks at random (these are *city* blocks, not *statistical* blocks!), and on each block they choose five meters at random. Six weeks are chosen randomly from the year, and the usage (money collected) on each meter is measured every day (Monday through Sunday) for all the meters on those weeks. Standard five questions.

**Problem 12.6**

Eight 1-gallon containers of raw milk are obtained from a dairy and are assigned at random to four abuse treatments, two containers per treatment. Abuse consists of keeping the milk at 25$^o$C for a period of time; the four abuse treatments are four randomly selected durations between 1 and 18 hours. After abuse, each gallon is split into five equal portions and frozen.

We have selected five contract laboratories at random from those available in the state. For each gallon, the five portions are randomly assigned to the five laboratories. The eight portions for a given laboratory are then placed in an insulated shipping container cooled with dry ice and shipped. Each laboratory is asked to provide duplicate counts of bacteria in each milk portion. Data follow (bacteria counts per $\mu$l).

| | Abuse | | | | | | | |
|---|---|---|---|---|---|---|---|---|
| Lab | 1 | | 2 | | 3 | | 4 | |
| 1 | 7800 | 7000 | 870 | 490 | 1300 | 1000 | 31000 | 36000 |
| | 7500 | 7200 | 690 | 530 | 1200 | 980 | 35000 | 34000 |
| 2 | 8300 | 9700 | 900 | 930 | 2500 | 2300 | 27000 | 28000 |
| | 8200 | 10000 | 940 | 840 | 1900 | 2300 | 34000 | 32000 |
| 3 | 7300 | 7300 | 760 | 840 | 2100 | 2300 | 34000 | 34000 |
| | 7600 | 7900 | 790 | 780 | 2000 | 2200 | 34000 | 33000 |
| 4 | 5400 | 5500 | 520 | 750 | 1400 | 1100 | 16000 | 16000 |
| | 5700 | 5600 | 770 | 620 | 1300 | 1400 | 16000 | 15000 |
| 5 | 15000 | 12000 | 1200 | 800 | 4600 | 3500 | 41000 | 39000 |
| | 14000 | 12000 | 1100 | 600 | 4000 | 3600 | 40000 | 39000 |

Analyze these data. The main issues are the sources and sizes of variation, with an eye toward reliability of future measurements.

**Problem 12.7** Cheese is made by bacterial fermentation of Pasteurized milk. Most of the bacteria are purposefully added to do the fermentation; these are the starter cultures. Some "wild" bacteria are also present in cheese; these are the nonstarter bacteria. One hypothesis is that nonstarter bacteria may affect the quality of a cheese, so that otherwise identical cheese making facilities produce different cheeses due to their different indigenous nonstarter bacteria.

Two strains of nonstarter bacteria were isolated at a premium cheese facility: R50#10 and R21#2. We will add these nonstarter bacteria to cheese to see if they affect quality. Our four treatments will be control, addition of R50, addition of R21, and addition of a blend of R50 and R21. Twelve cheeses are made, three for each of the four treatments, with the treatments being randomized to the cheeses. Each cheese is then divided into four portions, and the four portions for each cheese are randomly assigned to one of four aging times: 1 day, 28 days, 56 days, and 84 days. Each portion is measured for total free amino acids (a measure of bacterial activity) after it has aged for its specified number of days (data from Peggy Swearingen).

| | | Days | | | |
|---|---|---|---|---|---|
| Treatment | Cheese | 1 | 28 | 56 | 84 |
| Control | 1 | .637 | 1.250 | 1.697 | 2.892 |
| | 2 | .549 | .794 | 1.601 | 2.922 |
| | 3 | .604 | .871 | 1.830 | 3.198 |
| R50 | 1 | .678 | 1.062 | 2.032 | 2.567 |
| | 2 | .736 | .817 | 2.017 | 3.000 |
| | 3 | .659 | .968 | 2.409 | 3.022 |
| R21 | 1 | .607 | 1.228 | 2.211 | 3.705 |
| | 2 | .661 | .944 | 1.673 | 2.905 |
| | 3 | .755 | .924 | 1.973 | 2.478 |
| R50+R21 | 1 | .643 | 1.100 | 2.091 | 3.757 |
| | 2 | .581 | 1.245 | 2.255 | 3.891 |
| | 3 | .754 | .968 | 2.987 | 3.322 |

We are particularly interested in the bacterial treatment effects and interactions, and less interested in the main effect of time.

**Problem 12.8**

As part of a larger experiment, researchers are looking at the amount of beer that remains in the mouth after expectoration. Ten subjects will repeat the experiment on two separate days. Each subject will place 10 ml or 20 ml of beer in his or her mouth for five seconds, and then expectorate the beer. The beer has a dye, so the amount of expectorated beer can be determined, and thus the amount of beer retained in the mouth (in ml, data from Bréfort, Guinard, and Lewis 1989)

| | 10 ml | | 20 ml | |
|---|---|---|---|---|
| Subject | Day 1 | Day 2 | Day 1 | Day 2 |
| 1 | 1.86 | 2.18 | 2.49 | 3.75 |
| 2 | 2.08 | 2.19 | 3.15 | 2.67 |
| 3 | 1.76 | 1.68 | 1.76 | 2.57 |
| 4 | 2.02 | 3.87 | 2.99 | 4.51 |
| 5 | 2.60 | 1.85 | 3.25 | 2.42 |
| 6 | 2.26 | 2.71 | 2.86 | 3.60 |
| 7 | 2.03 | 2.63 | 2.37 | 4.12 |
| 8 | 2.39 | 2.58 | 2.19 | 2.84 |
| 9 | 2.40 | 1.91 | 3.25 | 2.52 |
| 10 | 1.63 | 2.43 | 2.00 | 2.70 |

Compute confidence intervals for the amount of beer retained in the mouth for both volumes.

**Problem 12.9** An experiment is performed to determine the effects of different Pasteurization methods on bacterial survival. We work with whole milk, 2% milk, and skim milk. We obtain four gallons of each kind of milk from a grocery store. These gallons are assumed to be a random sample from all potential gallons. Each gallon is then dosed with an equal number of bacteria. (We assume that this dosing is really equal so that dosing is not a factor of interest in the model.) Each gallon is then subdivided into two parts, with the two Pasteurization methods assigned at random to the two parts. Our observations are 24 bacterial concentrations after Pasteurization. Standard five questions.

**Question 12.1** Start with a four by three table of independent normals with mean 0 and variance 1. Compute the row means and then subtract out these row means. Find the distribution of the resulting differences and relate this to the restricted model for mixed effects.

**Question 12.2** Consider a three-factor model with A and B fixed and C random. Show that the variance for the difference $\overline{y}_{ij\bullet} - \overline{y}_{i'j\bullet} - \overline{y}_{ij'\bullet} + \overline{y}_{i'j'\bullet}$ can be computed using the usual formula for contrast variance with the "denominator" expected mean square as error variance.

# Chapter 13

# Complete Block Designs

We now begin the study of *variance reduction design*. Experimental error makes inference difficult. As the variance of experimental error ($\sigma^2$) increases, confidence intervals get longer and test power decreases. All other things being equal, we would thus prefer to conduct our experiments with units that are homogeneous so that $\sigma^2$ will be small. Unfortunately, all other things are rarely equal. For example, there may be few units available, and we must simply take what we can get. Or we might be able to find homogeneous units, but using the homogeneous units would restrict our inference to a subset of the population of interest. Variance reduction designs can give us many of the benefits of small $\sigma^2$, without necessarily restricting us to a subset of the population of units.

Variance reduction design

## 13.1 Blocking

Variance reduction design deals almost exclusively with a technique called *blocking*. A *block* of units is a set of units that are homogeneous in some sense. Perhaps they are field plots located in the same general area, or are samples analyzed at about the same time, or are units that came from a single supplier. These similarities in the units themselves lead us to anticipate that units within a block may also have similar responses. So when constructing blocks, we try to achieve homogeneity of the units within blocks, but units in different blocks may be dissimilar.

A block is a set of homogeneous units

Blocking designs are not completely randomized designs. The Randomized Complete Block design described in the next section is the first design we study that uses some kind of restricted randomization. When we design

Blocking restricts randomization

an experiment, we know the design we choose to use and thus the randomization that is used. When we look at an experiment designed by someone else, we can determine the design from the way the randomization was done, that is, from the kinds of restrictions that were placed on the randomization, not on the actual outcome of which units got which treatments.

Complete blocks include every treatment

There are many, many blocking designs, and we will only cover some of the more widely used designs. This chapter deals with *complete block designs* in which every treatment is used in every block; later chapters deal with *incomplete block designs* (not every treatment is used in every block) and some special block designs for treatments with factorial structure.

## 13.2 The Randomized Complete Block Design

RCB has $r$ blocks of $g$ units each

Block for homogeneity

The Randomized Complete Block design (RCB) is the basic blocking design. There are $g$ treatments, and each treatment will be assigned to $r$ units for a total of $N = gr$ units. We partition the $N$ units into $r$ groups of $g$ units each; these $r$ groups are our blocks. We make this partition into blocks in such a way that the units within a block are somehow alike; we anticipate that these alike units will have similar responses. In the first block, we randomly assign the $g$ treatments to the $g$ units; we do an independent randomization, assigning treatments to units in each of the other blocks. This is the RCB design.

Blocks exist at the time of the randomization of treatments to units. We cannot impose blocking structure on a completely randomized design after the fact; either the randomization was blocked or it was not.

**Example 13.1** **Mealybugs on cycads**

Modern zoos try to reproduce natural habitats in their exhibits as much as possible. They therefore use appropriate plants, but these plants can be infested with inappropriate insects. Zoos need to take great care with pesticides, because the variety of species in a zoo makes it more likely that a sensitive species is present.

Cycads (plants that look vaguely like palms) can be infested with mealybug, and the zoo wishes to test three treatments: water (a control), horticultural oil (a standard no-mammalian-toxicity pesticide), and fungal spores in water (*Beauveria bassiana*, a fungus that grows exclusively on insects). Five infested cycads are removed to a testing area. Three branches are randomly chosen on each cycad, and two 3 cm by 3 cm patches are marked on each branch; the number of mealybugs in these patches is noted. The three

**Table 13.1:** Changes in mealybug counts on cycads after treatment. Treatments are water, *Beauveria bassiana* spores, and horticultural oil.

| | Plant | | | | |
|---|---|---|---|---|---|
| | 1 | 2 | 3 | 4 | 5 |
| Water | -9 | 18 | 10 | 9 | -6 |
| | -6 | 5 | 9 | 0 | 13 |
| Spores | -4 | 29 | 4 | -2 | 11 |
| | 7 | 10 | -1 | 6 | -1 |
| Oil | 4 | 29 | 14 | 14 | 7 |
| | 11 | 36 | 16 | 18 | 15 |

branches on each cycad are randomly assigned to the three treatments. After three days, the patches are counted again, and the response is the change in the number of mealybugs (before – after). Data for this experiment are given in Table 13.1 (data from Scott Smith).

How can we decode the experimental design from the description just given? *Follow the randomization!* Looking at the randomization, we see that the treatments were applied to the branches (or pairs of patches). Thus the branches (or pairs) must be experimental units. Furthermore, the randomization was done so that each treatment was applied once on each cycad. There was no possibility of two branches from the same plant receiving the same treatment. This is a restriction on the randomization, with cycads acting as blocks. The patches are measurement units. When we analyze these data, we can take the average or sum of the two patches on each branch as the response for the branch. To recap, there were $g = 3$ treatments applied to $N = 15$ units arranged in $r = 5$ blocks of size 3 according to an RCB design; there were two measurement units per experimental unit.

Why did the experimenter block? Experience and intuition lead the experimenter to believe that branches on the same cycad will tend to be more alike than branches on different cycads—genetically, environmentally, and perhaps in other ways. Thus blocking by plant may be advantageous.

It is important to realize that tables like Table 13.1 hide the randomization that has occurred. The table makes it appear as though the first unit in every block received the water treatment, the second unit the spores, and so on. This is not true. The table ignores the randomization for the convenience of a readable display. The water treatment may have been applied to any of the three units in the block, chosen at random.

You cannot determine the design used in an experiment just by looking at a table of results, you have to know the randomization. There may be many

Follow the randomization to determine design

different designs that could produce the same data, and you will not know the correct analysis for those data without knowing the design. *Follow the randomization to determine the design.*

General treatment structure

An important feature to note about the RCB is that we have placed no restrictions on the treatments. The treatments could simply be $g$ treatments, or they could be the factor-level combinations of two or more factors. These factors could be fixed or random, crossed or nested. All of these treatment structures can be incorporated when we use blocking designs to achieve variance reduction.

**Example 13.2**

**Protein/amino acid effects on growing rats**

Male albino laboratory rats (Sprague-Dawley strain) are used routinely in many kinds of experiments. Proper nutrition for the rats is important. This experiment was conducted to determine the requirements for protein and the amino acid threonine. Specifically, this experiment will examine the factorial combinations of the amount of protein in diet and the amount of threonine in diet. The general protein in the diet is threonine deficient. There are eight levels of threonine (.2 through .9% of diet) and five levels of protein (8.68, 12, 15, 18, and 21% of diet), for a total of 40 treatments.

Two-hundred weanling rats were acclimated to cages. On the second day after arrival, all rats were weighed, and the rats were separated into five groups of 40 to provide groupings of approximately uniform weight. The 40 rats in each group were randomly assigned to the 40 treatments. Body weight and food consumption were measured twice weekly, and the response we consider is average daily weight gain over 21 days.

This is a randomized complete block design. Initial body weight is a good predictor of body weight in 3 weeks, so the rats were blocked by initial weight in an attempt to find homogeneous groups of units. There are 40 treatments, which have an eight by five factorial structure.

### 13.2.1 Why and when to use the RCB

We use an RCB to increase the power and precision of an experiment by decreasing the error variance. This decrease in error variance is achieved by finding groups of units that are homogeneous (blocks) and, in effect, repeating the experiment independently in the different blocks. The RCB is an effective design when there is a single source of extraneous variation in the responses that we can identify ahead of time and use to partition the units into blocks. Blocking is done at the time of randomization; you can't construct blocks after the experiment has been run.

Block when you can identify a source of variation

There is an almost infinite number of ways in which units can be grouped into blocks, but a few examples may suffice to get the ideas across. We would like to group into blocks on the basis of homogeneity of the responses, but that is not possible. Instead, we must group into blocks on the basis of other similarities that we think may be associated with responses.

Some blocking is fairly obvious. For example, you need milk to make cheese, and you get a new milk supply every day. Each batch of milk makes slightly different cheese. If your batches are such that you can make several types of cheese per batch, then blocking on batch of raw material is a natural.

Block on batch

Units may be grouped spatially. For example, some units may be located in one city, and other units in a second city. Or, some units may be in cages on the top shelf, and others in cages on the bottom shelf. It is common for units close in space to have more similar responses, so spatial blocking is also common.

Block spatially

Units may be grouped temporally. That is, some units may be treated or measured at one time, and other units at another time. For example, you may only be able to make four measurements a day, and the instrument may need to be recalibrated every day. As with spatial grouping, units close in time may tend to have similar responses, so temporal blocking is common.

Block temporally

Age and gender blocking are common for animal subjects. Sometimes units have a "history." The number of previous pregnancies could be a blocking factor. In general, any source of variation that you think may influence the response and which can be identified prior to the experiment is a candidate for blocking.

Age, gender, and history blocks

### 13.2.2 Analysis for the RCB

Now all the hard work in the earlier chapters studying analysis methods pays off. The design of an RCB is new, but there is nothing new in the analysis of an RCB. Once we have the correct model, we do point estimates, confidence intervals, multiple comparisons, testing, residual analysis, and so on, in the same way as for the CRD.

Nothing new in analysis of RCB

Let $y_{ij}$ be the response for the $i$th treatment in the $j$th block. The standard model for an RCB has a grand mean, a treatment effect, a block effect, and experimental error, as in

Blocks usually assumed additive

$$y_{ij} = \mu + \alpha_i + \beta_j + \epsilon_{ij} .$$

This standard model says that treatments and blocks are additive, so that treatments have the same effect in every block and blocks only serve to shift

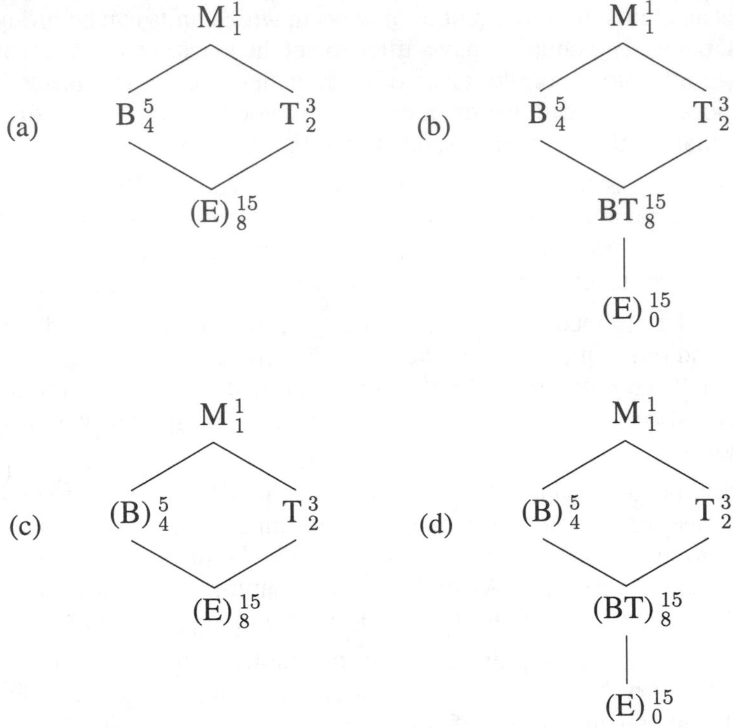


**Figure 13.1:** Models for a Randomized Complete Block.

the mean response up or down. Hasse diagrams (a) or (c) in Figure 13.1 correspond to this standard model.

All reasonable models for RCB use the same analysis

To complete the model, we must decide which terms are random and which are fixed; we must also decide whether to use the standard additive model given above or to allow for the possibility that treatments and blocks interact. Fortunately, all variations lead to the same operational analysis procedure for the RCB design. Figure 13.1 shows Hasse diagrams for four different sets of assumptions for the RCB. Panels (a) and (b) assume the blocks are fixed, and panels (c) and (d) assume the blocks are random. Panels (a) and (c) assume that blocks do not interact with treatments (as in the standard model above), and panels (b) and (d) include an interaction between blocks and treatments. In all four cases, we will use the $(r-1)(g-1)$ degree of freedom term below treatments as the denominator for treatments. This is true whether we think that the treatments are fixed or random; what differs is how this denominator term is interpreted.

In panels (a) and (c), where we assume that blocks and treatments are additive, the $(r-1)(g-1)$ degree of freedom term is the usual error and the only random term below treatments. In panel (d), this term is the block by treatment interaction and is again the natural denominator for treatments. In panel (b), the correct denominator for treatments is "error," but "error" cannot be estimated because we have 0 degrees of freedom for error (only one observation for each treatment in each block). Instead, we must use the block by treatment interaction as a surrogate for error and recognize that this surrogate error may be too large if interaction is indeed present. Thus we will arrive at the same inferences regardless of our assumptions on randomness of blocks and interaction between treatments and blocks.

Denominator for treatments is $(r-1)(g-1)$ degree of freedom interaction or error

The computation of estimated effects, sums of squares, contrasts, and so on is done exactly as for a two-way factorial. In this the *model* we are using to analyze an RCB is just the same as a two-way factorial with replication $n = 1$, even though the *design* of an RCB is not the same.

One difference between an RCB and a factorial is that we do not try to make inferences about blocks, even though the machinery of our model allows us to do so. The reason for this goes back to thinking of F-tests as approximations to randomization tests. Under the RCB randomization, units are assigned at random to treatments, but units always stay in the same block. Thus the block effects and sums of squares are not random, and there is no test for blocks; blocks simply exist. More pragmatically, we blocked because we believed that the units within blocks were more similar, so finding a block effect is not a major revelation.

Do not test blocks—they were not randomized

## Mealybugs, continued

**Example 13.3**

We take as our response the mean of the two measurements for each branch from Table 13.1. The ANOVA table follows:

| | DF | SS | MS | F-stat | $p$-value |
|---|---|---|---|---|---|
| Blocks | 4 | 686.4 | 171.60 | | |
| Treatments | 2 | 432.03 | 216.02 | 12.2 | .0037 |
| Error | 8 | 141.8 | 17.725 | | |

There is fairly strong evidence for differences in mealybugs between the treatments, and there is no evidence that assumptions were violated.

Looking more closely, we can use pairwise comparisons to examine the differences. We compute the pairwise comparisons (HSD's or LSD's or whatever) exactly as for ordinary factorial data. The underline diagram below shows the HSD at the 5% level:

| Water | Spores | Oil |
|---|---|---|
| -4.57 | -2.97 | 7.53 |

Here we see that spores treatment cannot be distinguished from the control (water) treatment, but both can be distinguished from the oil treatment.

Standard model has blocks additive

The usual assumption made for an RCB model is that blocks and treatments do not interact. To some degree this assumption is forced on us, because as we saw from the Hasse diagrams, there is little we can do besides assume additivity. When the treatments have a factorial structure, we could have a model with blocks random and interacting with the various factors. In such a model, the error for factor A would be the A by block interaction, the error for factor B would be the B by block interaction, and so on. However, the standard model allows treatment factors to interact, whereas blocks are still additive.

Transform for additivity

Assuming that blocks and treatments are additive does not make them so. One thing we can do with potential interaction in the RCB is investigate transformable nonadditivity using Tukey one-degree-of-freedom procedures. When there is transformable nonadditivity, reexpressing the data on the appropriate scale can make the data more additive. When the data are more additive, the term that we use as error contains less interaction and is a better surrogate for error.

### 13.2.3 How well did the blocking work?

Gain in variance, lose in degrees of freedom

The gain from using an RCB instead of a CRD is a decrease in error variance, and the loss is a decrease in error degrees of freedom by $(r-1)$. This loss is only severe for small experiments. How can we quantify our gain or loss from an RCB? As discussed above, the "F-test" for blocks does not correspond to a valid randomization test for blocks. Even if it did, knowing simply that the blocks are not all the same does not tell us what we need to know: how much have we saved by using blocks? We need something other than the F-test to measure that gain.

Relative efficiency measures sample size savings

Suppose that we have an RCB and a CRD to test the same treatments; both designs have the same total size N, and both use the same population of units. The efficiency of the RCB relative to the CRD is the factor by which the sample size of the CRD would need to be increased to have the same information as the RCB. (Information is a technical term; think of two designs with the same information as having approximately the same power or yielding approximately the same length of confidence intervals.) For example, if an RCB with fifteen units has relative efficiency 2, then a CRD using the same population of units would need 30 units to obtain the same information. Units almost always translate to time or money, so reducing $N$ by blocking is one good way to save money.

Efficiency is denoted by E with a subscript to identify the designs being compared. The relative efficiency of an RCB to a CRD is given in the following formula:

Relative efficiency is the ratio of variances times a degrees of freedom adjustment

$$E_{\text{RCB:CRD}} = \frac{(\nu_{rcb}+1)(\nu_{crd}+3)}{(\nu_{rcb}+3)(\nu_{crd}+1)} \frac{\sigma^2_{crd}}{\sigma^2_{rcb}} ,$$

where $\sigma^2_{crd}$ and $\sigma^2_{rcb}$ are the error variances for the CRD and RCB, $\nu_{rcb} = (r-1)(g-1)$ is the error degrees of freedom for the RCB design, and $\nu_{crd} = (r-1)g$ is the error degrees of freedom for the CRD of the same size. The first part is a degrees of freedom adjustment; variances must be estimated and we get better estimates with more degrees of freedom. The second part is the ratio of the error variances for the two different designs. The efficiency is determined primarily by this ratio of variances; the degrees of freedom adjustment is usually a smaller effect.

We will never know the actual variances $\sigma^2_{crd}$ or $\sigma^2_{rcb}$; we must estimate them. Suppose that we have conducted an RCB experiment. We can estimate $\sigma^2_{rcb}$ using $MS_E$ for the RCB design. We estimate $\sigma^2_{crd}$ via

Estimate $\sigma^2_{crd}$ with a weighted average of $MS_E$ and $MS_{\text{Blocks}}$

$$\widehat{\sigma}^2_{crd} = \frac{(r-1)MS_{\text{Blocks}} + ((g-1)+(r-1)(g-1))MS_E}{(r-1)+(g-1)+(r-1)(g-1)}$$

This is the weighted average of $MS_{\text{Blocks}}$ and $MS_E$ with $MS_{\text{Blocks}}$ having weight equal to the degrees of freedom for blocks and $MS_E$ having weight equal to the sum of the degrees of freedom for treatment and error. This is *not* the result of simply pooling sums of squares and degrees of freedom for blocks and error in the RCB.

## Mealybugs, continued

**Example 13.4**

For the mealybug experiment, we have $g = 3$, $r = 5$, $\nu_{rcb} = (r-1)(g-1) = 8$, $\nu_{crd} = r(g-1) = 12$, $MS_{\text{Blocks}} = 171.6$, and $MS_E = 17.725$, so we get

$$\begin{aligned}
\widehat{\sigma}^2_{crd} &= \frac{4 \times 171.6 + (2+8) \times 17.725}{4+2+8} = 61.69 , \\
\frac{(\nu_{rcb}+1)(\nu_{crd}+3)}{(\nu_{rcb}+3)(\nu_{crd}+1)} &= \frac{9 \times 15}{11 \times 13} = .944 , \\
\widehat{E}_{\text{RCB:CRD}} &= \frac{(\nu_{rcb}+1)(\nu_{crd}+3)}{(\nu_{rcb}+3)(\nu_{crd}+1)} \frac{\widehat{\sigma}^2_{crd}}{MS_E} ,
\end{aligned}$$

$$= .944 \times \frac{61.69}{17.725} = 3.29 \ .$$

We had five units for each treatment, so an equivalent CRD would have needed $5 \times 3.29 = 16.45$, call it seventeen units per treatment. This blocking was rather successful. Observe that even in this fairly small experiment, the loss from degrees of freedom was rather minor.

### 13.2.4 Balance and missing data

Balance makes inference easier

The standard RCB is balanced, in the sense that each treatment occurs once in each block. Balance was helpful in factorials, and it is helpful in randomized complete blocks for the same reason: it makes the calculations and inference easier. When the data are balanced, simple formulae can be used, exactly as for balanced factorials. When the data are balanced, adding 1 million to all the responses in a given block does not change any contrast between treatment means.

Treatments adjusted for blocks

Missing data in an RCB destroy balance. The approach to inference is to look at treatment effects adjusted for blocks. If the treatments are themselves factorial, we can compute whatever type of sum of squares we feel is appropriate, but we always adjust for blocks prior to treatments. The reason is that we believed, before any experimentation, that blocks affected the response. We thus allow blocks to account for any variability they can before examining any additional variability that can be explained by treatments. This "ordering" for sums of squares and testing does not affect the final estimated effects for either treatments or blocks.

## 13.3 Latin Squares and Related Row/Column Designs

Randomized Complete Block designs allow us to block on a single source of variation in the responses. There are experimental situations with more than one source of extraneous variation, and we need designs for these situations.

**Example 13.5** **Addled goose eggs**

The Canada goose (*Branta canadensis*) is a magnificent bird, but it can be a nuisance in urban areas when present in large numbers. One population control method is to addle eggs in nests to prevent them from hatching. This method may be harmful to the adult females, because the females fast while incubating and tend to incubate as long as they can if the eggs are unhatched.

Would the removal of addled eggs at the usual hatch date prevent these potential side effects?

An experiment is proposed to compare egg removal and no egg removal treatments. The birds in the study will be banded and observed in the future so that survival can be estimated for the two treatments. It is suspected that geese nesting together at a site may be similar due to both environmental and interbreeding effects. Furthermore, we know older females tend to nest earlier, and they may be more fit.

We need to block on both site and age. We would like each treatment to be used equally often at all sites (to block on populations), and we would like each treatment to be used equally often with young and old birds (to block on age).

A Latin Square (LS) is a design that blocks for two sources of variation. A Latin Square design for $g$ treatments uses $g^2$ units and is thus a little restrictive on experiment size. Latin Squares are usually presented pictorially. Here are examples of LS designs for $g = 2, 3,$ and 4 treatments:

LS has $g^2$ units for $g$ treatments and blocks two ways

| B | A |
|---|---|
| A | B |

| A | B | C |
|---|---|---|
| B | C | A |
| C | A | B |

| A | B | C | D |
|---|---|---|---|
| B | A | D | C |
| C | D | A | B |
| D | C | B | A |

The $g^2$ units are represented as a square (what a surprise!). By convention, the letters A, B, and so on represent the $g$ different treatments. There are two blocking factors in a Latin Square, and these are represented by the rows and columns of the square. Each treatment occurs once in each row and once in each column. Thus in the goose egg example, we might have rows one and two be different nesting sites, with column one being young birds and column two being older birds. This square uses four units, one young and one old bird from each of two sites. Using the two by two square above, treatment A is given to the site 1 old female and the site 2 young female, and treatment B is given to the site 1 young female and the site 2 old female.

Each treatment once in each row and column

Look a little closer at what the LS design is accomplishing. If you ignore the row blocking factor, the LS design is an RCB for the column blocking factor (each treatment appears once in each column). If you ignore the column blocking factor, the LS design is an RCB for the row blocking factor (each treatment appears once in each row). The rows and columns are also balanced because of the square arrangement of units. A Latin Square blocks on both rows and columns *simultaneously*.

Rows and columns of LS form RCBs

We use Latin Squares because they allow blocking on two sources of variation, but Latin Squares do have drawbacks. First, a single Latin Square has exactly $g^2$ units. This may be too few or even too many units. Second, Latin Squares generally have relatively few degrees of freedom for estimating error; this problem is particularly serious for small designs. Third, it may be difficult to obtain units that block nicely on both sources of variation. For example, we may two sources of variation, but one source of variation may only have $g - 1$ units per block.

### 13.3.1 The crossover design

Crossover design has subject and time period blocks

One of the more common uses for a Latin Square arises when a sequence of treatments is given to a subject over several time periods. We need to block on subjects, because each subject tends to respond differently, and we need to block on time period, because there may consistent differences over time due to growth, aging, disease progression, or other factors. A *crossover* design has each treatment given once to each subject, and has each treatment occurring an equal number of times in each time period. With $g$ treatments given to $g$ subjects over $g$ time periods, the crossover design is a Latin Square. (We will also consider a more sophisticated view of and analysis for the crossover design in Chapter 16.)

**Example 13.6** **Bioequivalence of drug delivery**

Consider the blood concentration of a drug after the drug has been administered. The concentration will typically start at zero, increase to some maximum level as the drug gets into the bloodstream, and then decrease back to zero as the drug is metabolized or excreted. These time-concentration curves may differ if the drug is delivered in a different form, say a tablet versus a capsule. Bioequivalence studies seek to determine if different drug delivery systems have similar biological effects. One variable to compare is the area under the time-concentration curve. This area is proportional to the average concentration of the drug.

We wish to compare three methods for delivering a drug: a solution, a tablet, and a capsule. Our response will be the area under the time-concentration curve. We anticipate large subject to subject differences, so we block on subject. There are three subjects, and each subject will be given the drug three times, once with each of the three methods. Because the body may adapt to the drug in some way, each drug will be used once in the first period, once in the second period, and once in the third period. Table 13.2 gives the assignment of treatments and the responses (data from Selwyn and Hall 1984). This Latin Square is a crossover design.

**Table 13.2:** Area under the curve for administering a drug via A—solution, B—tablet, and C—capsule. Table entries are treatments and responses.

| | | Subject | | | | | |
|---|---|---|---|---|---|---|---|
| | | 1 | | 2 | | 3 | |
| Period | 1 | A | 1799 | C | 2075 | B | 1396 |
| | 2 | C | 1846 | B | 1156 | A | 868 |
| | 3 | B | 2147 | A | 1777 | C | 2291 |

### 13.3.2 Randomizing the LS design

It is trivial to produce an LS for any number of treatments $g$. Assign the treatments in the first row in order. In the remaining rows, shift left all the treatments in the row above, bringing the first element of the row above around to the end of this row. The three by three square on page 325 was produced in this fashion. It is much less trivial to choose a square randomly. In principle, you assign treatments to units randomly, subject to the restrictions that each treatment occurs once in each row and once in each column, but effecting that randomization is harder than it sounds.

*One* LS is easy, random LS is harder

The recommended randomization is described in Fisher and Yates (1963). This randomization starts with *standard squares*, which are squares with the letters in the first row and first column in order. The three by three and four by four squares on page 325 are standard squares. For $g$ of 2, 3, 4, 5, and 6, there are 1, 1, 4, 56, and 9408 standard squares. Appendix C contains several standard Latin Square plans.

Standard squares

The Fisher and Yates randomization goes as follows. For $g$ of 3, 4, or 5, first choose a standard square at random. Then randomly permute all rows except the first, randomly permute all columns, and randomly assign the treatments to the letters. For $g$ of 6, select a standard square at random, randomly permute all rows and columns, and randomly assign the treatments to the letters. For $g$ of 7 or greater, choose any square, randomly permute the rows and columns, and randomly assign treatments to the letters.

Fisher-Yates randomization

### 13.3.3 Analysis for the LS design

The standard model for a Latin Square has a grand mean, effects for row and column blocks and treatments, and experimental error. Let $y_{ijk}$ be the response from the unit given the $i$th treatment in the $j$th row block and $k$th column block. The standard model is

Additive treatment, row, and column effects

$$y_{ijk} = \mu + \alpha_i + \beta_j + \gamma_k + \epsilon_{ijk} \;,$$

where $\alpha_i$ is the effect of the $i$th treatment, $\beta_j$ is the effect of the $j$ row block, and $\gamma_k$ is the effect of the $k$th column block. As with the RCB, block effects are assumed to be additive.

Usual formulae still work for LS

Here is something new: we do not observe all $g^3$ of the $i, j, k$ combinations in an LS; we only observe $g^2$ of them. However, the LS is constructed so that we have balance when we look at rows and columns, rows and treatments, or columns and treatments. This balance implies that contrasts between rows, contrasts between columns, and contrasts between treatments are all orthogonal, and the standard calculations for effects, sums of squares, contrasts, and so on work for the LS. Thus, for example,

$$\begin{aligned} \widehat{\alpha}_i &= \overline{y}_{i\bullet\bullet} - \overline{y}_{\bullet\bullet\bullet} \\ SS_{\text{Trt}} &= \sum_{i=1}^{g} g\widehat{\alpha}_i{}^2 \;. \end{aligned}$$

Note that $\overline{y}_{\bullet\bullet\bullet}$ and $\overline{y}_{i\bullet\bullet}$ are means over $g^2$ and $g$ units respectively. The sum of squares for error is usually found by subtracting the sums of squares for treatments, rows, and columns from the total sum of squares.

The Analysis of Variance table for a Latin Square design has sources for rows, columns, treatments, and error. We test the null hypothesis of no treatment effects via the F-ratio formed by mean square for treatments over mean square for error. As in the RCB, we do not test row or column blocking. Here is a schematic ANOVA table for a Latin Square:

| Source | SS | DF | MS | F |
|---|---|---|---|---|
| Rows | $SS_{\text{Rows}}$ | $g-1$ | $SS_{\text{Rows}}/(g-1)$ | |
| Columns | $SS_{\text{Cols}}$ | $g-1$ | $SS_{\text{Cols}}/(g-1)$ | |
| Treatments | $SS_{\text{Trt}}$ | $g-1$ | $SS_{\text{Trt}}/(g-1)$ | $MS_{Trt}/MS_E$ |
| Error | $SS_E$ | $(g-2)(g-1)$ | $SS_E/[(g-2)(g-1)]$ | |

Few degrees of freedom for error

There is no intuitive rule for the degrees of freedom for error $(g-2)(g-1)$; we just have to do our sums. Start with the total degrees of freedom $g^2$ and subtract one for the constant and all the degrees of freedom in the model, $3(g-1)$. The difference is $(g-2)(g-1)$. Latin Squares can have few degrees of freedom for error.

**Listing 13.1:** SAS output for bioequivalence Latin Square.

```
                    General Linear Models Procedure

Dependent Variable: AREA
                                  Sum of             Mean
Source                  DF       Squares           Square    F Value       Pr > F

Model                    6    1798011.33        299668.56      66.67       0.0149

Error                    2       8989.56          4494.78

Source                  DF      Type I SS     Mean Square    F Value       Pr > F

PERIOD                   2    928005.556      464002.778      103.23       0.0096
SUBJECT                  2    261114.889      130557.444       29.05       0.0333
TRT                      2    608890.889      304445.444       67.73       0.0145 ①

           Tukey's Studentized Range (HSD) Test for variable: AREA

                   Alpha= 0.05  df= 2  MSE= 4494.778
               Critical Value of Studentized Range= 8.331
                 Minimum Significant Difference= 322.46

        Means with the same letter are not significantly different.

                Tukey Grouping              Mean      N  TRT

                             A           2070.67      3  3                    ②

                             B           1566.33      3  2
                             B
                             B           1481.33      3  1
```

## Bioequivalence, continued

**Example 13.7**

Listing 13.1 shows the ANOVA for the bioequivalence data from Table 13.2. There is reasonable evidence against the null hypothesis that all three methods have the same area under the curve, $p$-value .0145 ①. Looking at the Tukey HSD output ②, it appears that treatment 3, the capsule, gives a higher area under the curve than the other two treatments.

Note that this three by three Latin Square has only 2 degrees of freedom for error.

The output in Listing 13.1 shows F-tests for both period and subject. We should ignore these, because period and subject are unrandomized blocking factors. The software does not know this and simply computes F-tests for all model terms.

### 13.3.4 Replicating Latin Squares

Replicate for better precision and error estimates

Increased replication gives us better estimates of error and increased power through averaging. We often need better estimates of error in LS designs, because a single Latin Square has relatively few degrees of freedom for error (for example, Listing 13.1). Thus using multiple Latin Squares in a single experiment is common practice.

Some blocks can be reused

When we replicate a Latin Square, we may be able to "reuse" row or column blocks. For example, we may believe that the period effects in a crossover design will be the same in all squares; this reuses the period blocks across the squares. Replicated Latin Squares can reuse both row and column blocks, reuse neither row nor column blocks, or reuse one of the row or column blocks. Whether we reuse any or all of the blocks when replicating an LS depends on the experimental and logistical constraints. Some blocks may represent small batches of material or time periods when weather is fairly constant; these blocks may be unavailable or have been consumed prior to the second replication. Other blocks may represent equipment that could be reused in principle, but we might want to use several pieces of equipment at once to conclude the experiment sooner rather than later.

Reusability depends on experiment and logistics

From an analysis point of view, the advantage of reusing a block factor is that we will have more degrees of freedom for error. The risk when reusing a block factor is that the block effects will actually change, so that the assumption of constant block effects across the squares is invalid.

**Example 13.8** **Carbon monoxide emissions**

Carbon monoxide (CO) emissions from automobiles can be influenced by the formulation of the gasoline that is used. In Minnesota, we use "oxygenated fuels" in the winter to decrease CO emissions. We have four gasoline blends, the combinations of factors A and B, each at two levels, and we wish to test the effects of these blends on CO emissions in nonlaboratory conditions, that is, in real cars driven over city streets. We know that there are car to car differences in CO emissions, and we suspect that there are route to route differences in the city (stop and go versus freeway, for example). With two blocking factors, a Latin Square seems appropriate. We will use three squares to get enough replication.

If we have only four cars and four routes, and these will be used in all three replications, then we are reusing the row and column blocking factors across squares. Alternatively, we might be using only four cars, but we have twelve different routes. Then we are reusing the row blocks (cars), but not the column blocks (routes). Finally, we could have twelve cars and twelve routes, which we divide into three sets of four each to create squares. For this design, neither rows nor columns is reused.

Models depend on which blocks are reused

The analysis of a replicated Latin Square varies slightly depending on which blocks are reused. Let $y_{ijkl}$ be the response for treatment $i$ in row $j$ and column $k$ of square $l$. There are $g$ treatments (and rows and columns in each block) and $m$ squares. Consider the provisional model

$$y_{ijkl} = \mu + \alpha_i + \beta_{j(l)} + \gamma_{k(l)} + \delta_l + \epsilon_{ijkl} \ .$$

This model has an overall mean $\mu$, the treatment effects $\alpha_i$, square effects $\delta_l$, and row and column block effects $\beta_{j(l)}$ and $\gamma_{k(l)}$. As usual in block designs, block effects are additive.

Df when neither rows nor columns reused

This model has row and column effects nested in square, so that each square will have its own set of row and column effects. This model is appropriate when neither row nor column blocks are reused. The degrees of freedom for this model are one for the grand mean, $g-1$ between treatments, $m-1$ between squares, $m(g-1)$ for each of rows and columns, and $(mg-m-1)(g-1)$ for error.

The model terms and degrees of freedom for the row and column block effects depend on whether we are reusing the row and/or column blocks. Suppose that we reuse row blocks, but not column blocks; reusing columns but not rows can be handled similarly. The model is now

$$y_{ijkl} = \mu + \alpha_i + \beta_j + \gamma_{k(l)} + \delta_l + \epsilon_{ijkl} \ ,$$

Df when rows reused

and the degrees of freedom are one for the grand mean, $g-1$ between treatments, $m-1$ between squares, $g-1$ between rows, $m(g-1)$ between columns, and $(mg-2)(g-1)$ for error. Finally, consider reusing both row and column blocks. Then the model is

$$y_{ijkl} = \mu + \alpha_i + \beta_j + \gamma_k + \delta_l + \epsilon_{ijkl} \ ,$$

Df when rows and columns reused

and the degrees of freedom are one for the grand mean, $g-1$ between treatments, rows and columns, $m-1$ between squares, and $(mg+m-3)(g-1)$ for error.

### Example 13.9

### CO emissions, continued

Consider again the three versions of the CO emissions example given above. The degrees of freedom for the sources of variation are

| Source | 4 cars, 4 routes DF | 4 cars, 12 routes DF | 12 cars, 12 routes DF |
|---|---|---|---|
| Squares | $(m-1)=2$ | $(m-1)=2$ | $(m-1)=2$ |
| Cars | $(g-1)=3$ | $(g-1)=3$ | $m(g-1)=9$ |
| Routes | $(g-1)=3$ | $m(g-1)=9$ | $m(g-1)=9$ |
| Fuels | $(g-1)=3$ | $(g-1)=3$ | $(g-1)=3$ |
| or A | 1 | 1 | 1 |
| B | 1 | 1 | 1 |
| AB | 1 | 1 | 1 |
| Error | $(mg+m-3)(g-1)$ $=12\times 3=36$ | $(mg-2)(g-1)$ $=10\times 3=30$ | $(mg-m-1)(g-1)$ $=8\times 3=24$ |
| or | | | |
| Error | $47-11=36$ | $47-17=30$ | $47-23=24$ |

Note that we have computed error degrees of freedom twice, once by applying the formulae, and once by subtracting model degrees of freedom from total degrees of freedom. I usually obtain error degrees of freedom by subtraction.

Estimated effects and sums of squares follow the usual patterns

Estimated effects follow the usual patterns, because even though we do not see all the $ijkl$ combinations, the combinations we do see are such that treatment, row, and column effects are orthogonal. So, for example,

$$\begin{aligned}\widehat{\alpha}_i &= \overline{y}_{i\bullet\bullet\bullet} - \overline{y}_{\bullet\bullet\bullet\bullet} \\ \widehat{\delta}_l &= \overline{y}_{\bullet\bullet\bullet l} - \overline{y}_{\bullet\bullet\bullet\bullet} \ .\end{aligned}$$

If row blocks are reused, we have

$$\widehat{\beta}_j = \overline{y}_{\bullet j\bullet\bullet} - \overline{y}_{\bullet\bullet\bullet\bullet} \ ,$$

and if row blocks are not reused we have

$$\begin{aligned}\widehat{\beta}_{j(l)} &= \overline{y}_{\bullet j\bullet l} - \widehat{\delta}_l - \widehat{\mu} \\ &= \overline{y}_{\bullet j\bullet l} - \overline{y}_{\bullet\bullet\bullet l} \ .\end{aligned}$$

The rules for column block effects are analogous. In all cases, the sum of squares for a source of variation is found by squaring an effect, multiplying that by the number of responses that received that effect, and adding across all levels of the effect.

Can combine between squares with columns

When only one of the blocking factors (rows, for example) is reused, it is fairly common to combine the terms for "between squares" ($m-1$ degrees of freedom) and "between columns within squares" ($m(g-1)$ degrees of freedom) into an overall between columns factor with $gm-1$ degrees of freedom.

**Table 13.3:** Area under the curve for administering a drug via A—solution, B—tablet, and C—capsule. Table entries are treatments and responses.

| | Period | | | | | |
|---|---|---|---|---|---|---|
| Subject | 1 | | 2 | | 3 | |
| 1 | A | 1799 | C | 1846 | B | 2147 |
| 2 | C | 2075 | B | 1156 | A | 1777 |
| 3 | B | 1396 | A | 868 | C | 2291 |
| 4 | B | 3100 | A | 3065 | C | 4077 |
| 5 | C | 1451 | B | 1217 | A | 1288 |
| 6 | A | 3174 | C | 1714 | B | 2919 |
| 7 | C | 1430 | A | 836 | B | 1063 |
| 8 | A | 1186 | B | 642 | C | 1183 |
| 9 | B | 1135 | C | 1305 | A | 984 |
| 10 | C | 873 | A | 1426 | B | 1540 |
| 11 | A | 2061 | B | 2433 | C | 1337 |
| 12 | B | 1053 | C | 1534 | A | 1583 |

This is not necessary, but it sometimes makes the software commands easier. Note that when neither rows nor columns is reused, you cannot get combined $m(g-1)$ degrees of freedom terms for both rows and columns at the same time. The "between squares" sums of squares and degrees of freedom comes from contrasts between the means of the different squares and can be considered as either a row or column difference, but it cannot be combined into *both* rows and columns in the same analysis.

## Bioequivalence (continued)

**Example 13.10**

Example 13.6 introduced a three by three Latin Square for comparing delivery of a drug via solution, tablet, and capsule. In fact, this crossover design included $m = 4$ Latin Squares. These squares involve twelve different subjects, but the same three time periods. Data are given in Table 13.3.

Listing 13.2 ① gives an Analysis of Variance for the complete bioequivalence data. The residuals show some signs of nonconstant variance, but the power 1 is reasonably within a confidence interval for the Box-Cox transformation and the residuals do not look much better on the log or quarter power scale, so we will stick with the original data.

**Listing 13.2:** SAS output for bioequivalence replicated Latin Square.

```
Dependent Variable: AREA
                                   Sum of           Mean
Source                 DF         Squares         Square   F Value      Pr > F

Error                  20       4106499.6       205325.0

SQ                      3      8636113.56     2878704.52     14.02      0.0001
PERIOD                  2       737750.72      368875.36      1.80      0.1916
SUBJECT                 8      7748946.67      968618.33      4.72      0.0023
TRT                     2        81458.39       40729.19      0.20      0.8217   ①

                                   Sum of           Mean
Source                 DF         Squares         Square   F Value      Pr > F

Error                  14       2957837.9       211274.1

SQ                      3      8636113.56     2878704.52     13.63      0.0002
PERIOD                  2       737750.72      368875.36      1.75      0.2104
SUBJECT                 8      7748946.67      968618.33      4.58      0.0065
TRT                     2        81458.39       40729.19      0.19      0.8268
SQ*TRT                  6      1148661.61      191443.60      0.91      0.5179   ②

          Level of   Level of          -------------AREA------------               ③
          SQ         TRT         N      Mean                SD

          1          1           3      1481.33333           531.27614
          1          2           3      1566.33333           516.99162
          1          3           3      2070.66667           222.53165
          2          1           3      2509.00000          1058.82057
          2          2           3      2412.00000          1038.84984
          2          3           3      2414.00000          1446.19120
          3          1           3      1002.00000           175.69291
          3          2           3       946.66667           266.29370
          3          3           3      1306.00000           123.50304
          4          1           3      1690.00000           330.74613
          4          2           3      1675.33333           699.88309
          4          3           3      1248.00000           339.36853

                                   Sum of           Mean
Source                 DF         Squares         Square   F Value      Pr > F   ④
----------------------------------- SQ=1 -----------------------------------
Error                   2         8989.56        4494.78
TRT                     2      608890.889     304445.444     67.73      0.0145
----------------------------------- SQ=2 -----------------------------------
Error                   2       937992.67      468996.33
TRT                     2        18438.00        9219.00      0.02      0.9807
----------------------------------- SQ=3 -----------------------------------
Error                   2       46400.889      23200.444
TRT                     2      224598.222     112299.111      4.84      0.1712
----------------------------------- SQ=4 -----------------------------------
Error                   2       327956.22      163978.11
TRT                     2      378192.889     189096.444      1.15      0.4644
```

Note that the complete data set is compatible with the null hypothesis of no treatment effects. Those of you keeping score may recall from Example 13.7 that the data from just the first square seemed to indicate that there were differences between the treatments. Also the $MS_E$ in the complete data is about 45 times bigger than for the first square. What has happened?

Here are three possibilities. First, the subjects may not have been numbered in a random order, so the early subjects could be systematically different from the later subjects. This can lead to some dramatic differences between analysis of subsets and complete sets of data, though we have no real evidence of that here.

Second, there could be subject by treatment interaction giving rise to different treatment effects for different subsets of the data. Our Latin Square blocking model is based on the assumption of additivity, but interaction could be present. The error term in our ANOVA contains any effects not explicitly modeled, so it would be inflated in the presence of subject by treatment interaction, and interaction could obviously lead to different treatment effects being estimated in different squares.

We explore this somewhat at ② of Listing 13.2, which shows a second ANOVA that includes a square by treatment interaction. This term explains a reasonable sum of squares, but is not significant as a 6 degree of freedom mean square. Listing 13.2 ③ shows the response means separately by square and treatment. Means by square for treatments 1 and 2 are generally not too far apart. The mean for treatment 3 is higher than the other two in squares 1 and 3, about the same in square 2, and lower in square 4. The interaction contrast making this comparison has a large sum of squares, but it is not significant after making a Scheffé adjustment for having data snooped. This is suggestive that the effect of treatment 3 depends on subject, but certainly not conclusive; a follow up experiment may be in order.

Third, we may simply have been unlucky. Listing 13.2 ④ shows error and treatment sums of squares for each square separately. The $MS_E$ in the first square is unusually low, and the $MS_{\text{Trt}}$ is somewhat high. It seems most likely that the results in the first square appear significant due to an unusually small error mean square.

### 13.3.5 Efficiency of Latin Squares

We approach the efficiency of Latin Squares much as we did the efficiency of RCB designs. That is, we try to estimate by what factor the sample sizes would need to be increased in order for a simpler design to have as much

Efficiency of LS relative to RCB or CRD

information as the LS design. We can compare an LS design to an RCB by considering the elimination of either row or column blocks, or we can compare an LS design to a CRD by considering the elimination of both row and column blocks.

Error degrees of freedom

As with RCB's, our estimate of efficiency is the product of two factors, the first a correction for degrees of freedom for error and the second an estimate of the ratio of the error variances for the two designs. With $g^2$ units in a Latin Square, there are $\nu_{ls} = (g-1)(g-2)$ degrees of freedom for error; if either row or column blocks are eliminated, there are $\nu_{rcb} = (g-1)(g-1)$ degrees of freedom for error; and if both row and column blocks are eliminated, there are $\nu_{crd} = (g-1)g$ degrees of freedom for error.

$E_{\text{LS:RCB}}$

The efficiency of a Latin Square relative to an RCB is

$$E_{\text{LS:RCB}} = \frac{(\nu_{ls}+1)(\nu_{rcb}+3)}{(\nu_{ls}+3)(\nu_{rcb}+1)} \frac{\sigma^2_{rcb}}{\sigma^2_{ls}} ,$$

$E_{\text{LS:CRD}}$

and the efficiency of a Latin Square relative to a CRD is

$$E_{\text{LS:CRD}} = \frac{(\nu_{ls}+1)(\nu_{crd}+3)}{(\nu_{ls}+3)(\nu_{crd}+1)} \frac{\sigma^2_{crd}}{\sigma^2_{ls}} .$$

We have already computed the degrees of freedom, so all that remains is the estimates of variance for the three designs.

The estimated variance for the LS design is simply $MS_E$ from the LS design. For the RCB and CRD we estimate the error variance in the simpler design with a weighted average of the $MS_E$ from the LS and the mean squares from the blocking factors to be eliminated. The weight for $MS_E$ is $(g-1)^2$, the sum of treatment and error degrees of freedom, and the weights for blocking factors are their degrees of freedom $(g-1)$. In formulae:

$$\begin{aligned} \widehat{\sigma}^2_{rcb} &= \frac{(g-1)MS_{\text{Rows}} + ((g-1)+(g-1)(g-2))MS_E}{2(g-1)+(g-1)(g-2)} \\ &= \frac{MS_{\text{Rows}} + (g-1)MS_E}{g} \qquad \text{(row blocks eliminated),} \end{aligned}$$

or

$$\begin{aligned} \widehat{\sigma}^2_{rcb} &= \frac{(g-1)MS_{\text{Cols}} + ((g-1)+(g-1)(g-2))MS_E}{2(g-1)+(g-1)(g-2)} \\ &= \frac{MS_{\text{Cols}} + (g-1)MS_E}{g} \qquad \text{(column blocks eliminated),} \end{aligned}$$

or

$$\begin{aligned}\widehat{\sigma}^2_{crd} &= \frac{(g-1)(MS_{\text{Rows}} + MS_{col} + MS_E) + (g-1)(g-2)MS_E}{3(g-1)+(g-1)(g-2)} \\ &= \frac{MS_{\text{Rows}} + MS_{\text{Cols}} + (g-1)MS_E}{g+1} \qquad \text{(both eliminated).}\end{aligned}$$

The two versions of $\widehat{\sigma}^2_{rcb}$ are for eliminating row and column blocking, respectively.

**Example 13.11**

## Bioequivalence, continued

Example 13.7 gave the ANOVA table for the first square of the bioequivalence data. The mean squares for subject, period, and error were 130,557; 464,003; and 4494.8 respectively. All three of these and treatments had 2 degrees of freedom each. Thus we have $\nu_{ls} = 2$, $\nu_{rcb} = 4$, and $\nu_{crd} = 6$. The estimated variances are

Blocking removed

$$\begin{aligned}
&\text{Neither} & \widehat{\sigma}^2_{ls} &= 4494.8 \\
&\text{Subjects} & \widehat{\sigma}^2_{rcb} &= \frac{130,557 + 2 \times 4494.8}{3} = 46516 \\
&\text{Periods} & \widehat{\sigma}^2_{rcb} &= \frac{464,003 + 2 \times 4494.8}{3} = 157664 \\
&\text{Both} & \widehat{\sigma}^2_{crd} &= \frac{130557 + 464,003 + 2 \times 4494.8}{4} = 150887 \ .
\end{aligned}$$

The estimated efficiencies are

$$\begin{aligned}
&\text{Subjects} & E &= \frac{(2+1)(4+3)}{(2+3)(4+1)}\frac{46516}{4494.8} = 8.69 \\
&\text{Periods} & E &= \frac{(2+1)(4+3)}{(2+3)(4+1)}\frac{157664}{4494.8} = 29.46 \\
&\text{Both} & E &= \frac{(2+1)(6+3)}{(2+3)(6+1)}\frac{150887}{4494.8} = 25.90 \ .
\end{aligned}$$

Both subject and period blocking were effective, particularly the period blocking.

### 13.3.6 Designs balanced for residual effects

Crossover designs give all treatments to all subjects and use subjects and periods as blocking factors. The standard analysis includes terms for subject, period, and treatment. There is an implicit assumption that the response in a given time period depends on the treatment for that period, and not at all on treatments from prior periods. This is not always true. For example, a drug that is toxic and has terrible side effects may alter the responses for a subject, even after the drug is no longer being given. These effects that linger after treatment are called *residual effects* or *carryover effects.*

Residual effects affect subsequent treatment periods

There are experimental considerations when treatments may have residual effects. A *washout period* is a time delay inserted between successive treatments for a subject. The idea is that residual effects will decrease or perhaps even disappear given some time, so that if we can design this time into the experiment between treatments, we won't need to worry about the residual effects. Washout periods are not always practical or completely effective, so alternative designs and models have been developed.

A washout period may reduce residual effects

In an experiment with no residual effects, only the treatment from the current period affects the response. The simplest form of residual effect occurs when only the current treatment and the immediately preceding treatment affect the response. A design balanced for residual effects, or carryover design, is a crossover design with the additional constraint that each treatment follows every other treatment an equal number of times.

Balance for residual effects of preceding treatment

Look at these two Latin Squares with rows as periods and columns as subjects.

| | | | |
|---|---|---|---|
| A | B | C | D |
| B | A | D | C |
| C | D | A | B |
| D | C | B | A |

| | | | |
|---|---|---|---|
| A | B | C | D |
| B | D | A | C |
| C | A | D | B |
| D | C | B | A |

In the first square, A occurs first once, follows B twice, and follows D once. Other treatments have a similar pattern. The first square is a crossover design, but it is not balanced for residual effects. In the second square, A occurs first once, and follows B, C, and D once each. A similar pattern occurs for the other treatments, so the second square is balanced for residual effects. When $g$ is even, we can find a design balanced for residual effects using $g$ subjects; when $g$ is odd, we need $2g$ subjects (two squares) to balance for residuals effects. A design that includes all possible orders for the treatments an equal number of times will be balanced for residual effects.

**Table 13.4:** Milk production (pounds per 6 weeks) for eighteen cows fed A—roughage, B—limited grain, and C—full grain.

| | Cow | | | | | |
|---|---|---|---|---|---|---|
| Period | 1 | 2 | 3 | 4 | 5 | 6 |
| 1 | A 1376 | B 2088 | C 2238 | A 1863 | B 1748 | C 2012 |
| 2 | B 1246 | C 1864 | A 1724 | C 1755 | A 1353 | B 1626 |
| 3 | C 1151 | A 1392 | B 1272 | B 1462 | C 1339 | A 1010 |
| Period | 7 | 8 | 9 | 10 | 11 | 12 |
| 1 | A 1655 | B 1938 | C 1855 | A 1384 | B 1640 | C 1677 |
| 2 | B 1517 | C 1804 | A 1298 | C 1535 | A 1284 | B 1497 |
| 3 | C 1366 | A 969 | B 1233 | B 1289 | C 1370 | A 1059 |
| Period | 13 | 14 | 15 | 16 | 17 | 18 |
| 1 | A 1342 | B 1344 | C 1627 | A 1180 | B 1287 | C 1547 |
| 2 | B 1294 | C 1312 | A 1186 | C 1245 | A 1000 | B 1297 |
| 3 | C 1371 | A 903 | B 1066 | B 1082 | C 1078 | A 887 |

The model for a residual-effects design has terms for subject, period, direct effect of a treatment, residual effect of a treatment, and error. Specifically, let $y_{ijkl}$ be the response for the $k$th subject in the $l$th time period; the subject received treatment $i$ in period $l$ and treatment $j$ in period $l-1$. The indices $i$ and $l$ run from 1 to $g$, and $k$ runs across the number of subjects. Use $j = 0$ to indicate that there was no earlier treatment (that is, when $l = 1$ and we are in the first period); $j$ then runs from 0 to $g$. Our model is

Residual-effects model has subject, period, direct treatment, and residual treatment effects

$$y_{ijkl} = \mu + \alpha_i + \beta_j + \gamma_k + \delta_l + \epsilon_{ijkl}$$

where $\alpha_i$ is called the direct effect of treatment $i$, $\beta_j$ is called the residual effect of treatment $j$, and $\gamma_k$ and $\delta_l$ are subject and period effects as usual. We make the usual zero-sum assumptions for the block and direct treatment effects. For the $\beta_j$'s we assume that $\beta_0 = 0$ and $\sum_{j=1}^{g} \beta_j = 0$. That is, we assume that there is a zero residual effect when in the first treatment period.

Direct treatment effects are orthogonal to block effects (we have a cross-over design), but residual effects are not orthogonal to direct treatment effects or subjects. Formulae for estimated effects and sums of squares are thus rather opaque, and it seems best just to let your statistical software do its work.

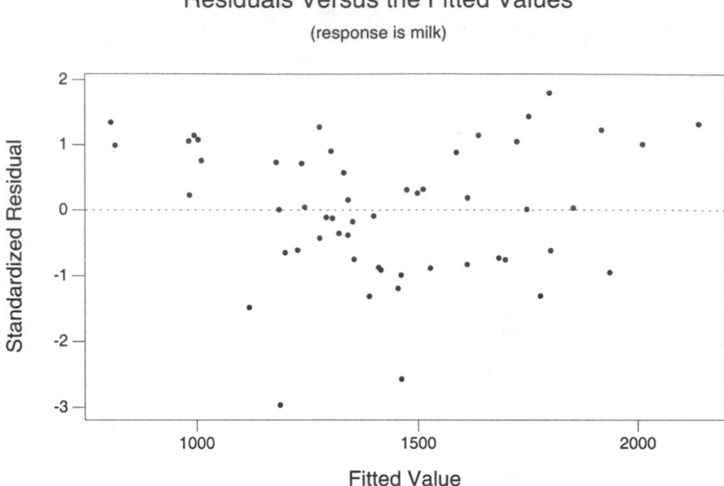


**Figure 13.2:** Residuals versus predicted values for the milk production data on the original scale, using Minitab.

**Example 13.12** **Milk yield**

Milk production in cows may depend on their feed. There is large cow to cow variation in production, so blocking on cow and giving all the treatments to each cow seems appropriate. Milk production for a given cow also tends to decrease during any given lactation, so blocking on period is important. This leads us to a crossover design. The treatments of interest are A—roughage, B—limited grain, and C—full grain. The response will be the milk production during the six week period the cow is on a given feed. There was insufficient time for washout periods, so the design was balanced for residual effects. Table 13.4 gives the data from Cochran, Autrey, and Cannon (1941) via Bellavance and Tardif (1995).

A plot of residuals versus predicted values on the original scale in Figure 13.2 shows problems (I call this shape the flopping fish). The plot seems wider on the right than the left, suggesting a lower power to stabilize the variability. Furthermore, the plot seems bent—low in the middle and high on the ends. This probably means that we are analyzing on the wrong scale, but it can indicate that we have left out important terms. Box-Cox suggests a log transformation, and the new residual plot looks much better (Figure 13.3). There is one potential outlier that should be investigated.

**Listing 13.3:** Minitab output for milk yield data.

```
Analysis of Variance for lmilk, using Sequential SS for Tests

Source      DF     Seq SS     Adj SS     Seq MS        F      P
period       2    0.99807    0.99807    0.49903   123.25  0.000
cow         17    0.90727    0.88620    0.05337    13.18  0.000
trt          2    0.40999    0.42744    0.20500    50.63  0.000
r1           1    0.03374    0.02425    0.03374     8.33  0.007
r2           1    0.00004    0.00004    0.00004     0.01  0.917
Error       30    0.12147    0.12147    0.00405
Total       53    2.47058

Term            Coef      StDev         T       P
Constant     7.23885    0.00866    835.99   0.000
trt
1           -0.12926    0.01369     -9.44   0.000
2            0.01657    0.01369      1.21   0.236
r1          -0.04496    0.01837     -2.45   0.020
r2          -0.00193    0.01837     -0.10   0.917
```

Listing 13.3 gives an ANOVA for the milk production data on the log scale. There is overwhelming evidence of a treatment effect. There is also reasonably strong evidence that residual effects exist.

The direct effects for treatments 1 and 2 are estimated to be $-.129$ and .017; the third must be .113 by the zero sum criterion. These effects are on the log scale, so roughage and full grain correspond to about 12% decreases and increases from the partial grain treatment. The residual effects for treatments 1 and 2 are estimated to be $-.045$ and $-.002$; the third must be .047 by the zero sum criterion. Thus the period after the roughage treatment tends to be about 5% lower than might be expected otherwise, and the period after the full-grain treatment tends to be about 5% higher.

Implementing residual effects

Most statistical software packages are not set up to handle residual effects directly. I implemented residual effects in the last example by including two single-degree-of-freedom terms called `r1` and `r2`. The terms `r1` and `r2` appear in the model as regression variables. The regression coefficients for `r1` and `r2` are the residual effects of treatments 1 and 2; the residual effect of treatment 3 is found by the zero-sum constraint to be minus the sum of the first two residual effects.

To implement residual effects for $g$ treatments, we need $g-1$ terms `ri`, for $i$ running from 1 to $g-1$. Their regression coefficients are the residual effects of the first $g-1$ treatments, and the last residual effect is found by the zero-sum constraint. Begin the construction of term `ri` with a column

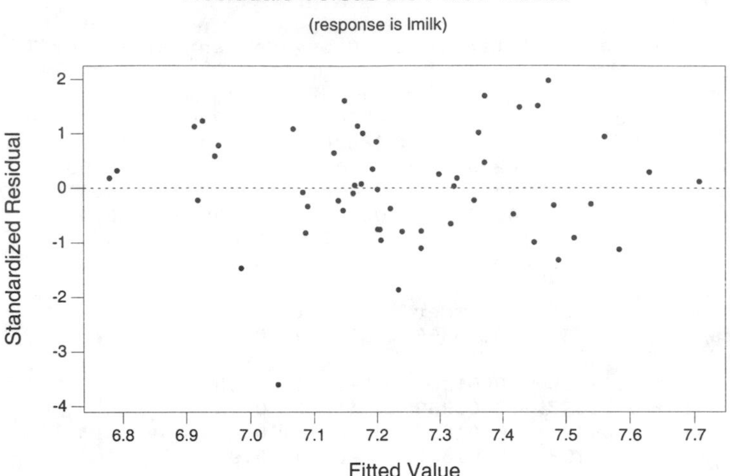


**Figure 13.3:** Residuals versus predicted values for the milk production data on the log scale, using Minitab.

of all zeroes of length $N$, one for each experimental unit. Set to +1 those elements in `ri` corresponding to units that immediately follow treatment $i$, and set to –1 those elements in `ri` corresponding to units that immediately follow treatment $g$. In all these "`r`" terms, an observation has a –1 if it follows treatment $g$; in term `ri`, an observation has a +1 if it follows treatment $i$; all other entries in the "`r`" terms have zeroes. For example, consider just the first two cows in Table 13.4, with treatments A, B, C, and B, C, A. The `r1` term would be (0, 1, 0, 0, 0, -1), and `r2` term would be (0, 0, 1, 0, 1, -1). It is the temporal order in which subjects experience treatments that determines which treatments follow others, not the order in which the units are listed in some display. There are other constructions that give the correct sum of squares in the ANOVA, but their coefficients may be interpreted differently.

Repeat last treatment

When resources permit an additional test period for each subject, considerable gain can be achieved by repeating the last treatment for each subject. For example, if cow 13 received the treatments A, B, and C, then the treatment in the fourth period should also be C. With this structure, every treatment follows every treatment (including itself) an equal number of times, and every residual effect occurs with every subject. These conditions permit more precise estimation of direct and residual treatment effects.

## 13.4 Graeco-Latin Squares

Randomized Complete Blocks allow us to control one extraneous source of variability in our units, and Latin Squares allow us to control two sources. The Latin Square design can be extended to control for three sources of extraneous variability; this is the Graeco-Latin Square. For four or more sources of variability, we use Latin Hyper-Squares. Graeco-Latin Squares allow us to test $g$ treatments using $g^2$ units blocked three different ways. Graeco-Latin Squares don't get used very often, because they require a fairly restricted set of circumstances to be applicable.

Graeco-Latin Squares block three ways

The Graeco-Latin Square is represented as a $g$ by $g$ table or square. Entries in the table correspond to the $g^2$ units. Rows and columns of the square correspond to blocks, as in a Latin Square. Each entry in the table has one Latin letter and one Greek letter. Latin letters correspond to treatments, as in a Latin Square, and Greek letters correspond to the third blocking factor. The Latin letters occur once in each row and column (they form a Latin Square), and the Greek letters occur once in each row and column (they also form a Latin Square). In addition, each Latin letter occurs once with each Greek letter. Here is a four by four Graeco-Latin Square:

Treatments occur once in each blocking factor

| | | | |
|---|---|---|---|
| A $\alpha$ | B $\gamma$ | C $\delta$ | D $\beta$ |
| B $\beta$ | A $\delta$ | D $\gamma$ | C $\alpha$ |
| C $\gamma$ | D $\alpha$ | A $\beta$ | B $\delta$ |
| D $\delta$ | C $\beta$ | B $\alpha$ | A $\gamma$ |

Each treatment occurs once in each row block, once in each column block, and once in each Greek letter block. Similarly, each kind of block occurs once in each other kind of block.

If two Latin Squares are superimposed and all $g^2$ combinations of letters from the two squares once, the Latin Squares are called *orthogonal*. A Graeco-Latin Square is the superposition of two orthogonal Latin Squares.

Orthogonal Latin Squares

Graeco-Latin Squares do not exist for all values of $g$. For example, there are Graeco-Latin Squares for $g$ of 3, 4, 5, 7, 8, 9, and 10, but *not* for $g$ of 6. Appendix C lists orthogonal Latin Squares for $g$ = 3, 4, 5, 7, from which a Graeco-Latin Square can be built.

No GLS for $g = 6$

The usual model for a Graeco-Latin Square has terms for treatments and row, column, and Greek letter blocks and assumes that all these terms are additive. The balance built into these designs allows us to use our standard methods for estimating effects and computing sums of squares, contrasts, and so on, just as for a Latin Square.

Additive blocks plus treatments

Hyper Squares

The Latin Square/Graeco-Latin Square family of designs can be extended to have more blocking factors. These designs, called Hyper-Latin Squares, are rare in practice.

## 13.5 Further Reading and Extensions

Our discussion of the RCB has focused on its standard form, where we have $g$ treatments and blocks of size $g$. There are several other possibilities. For example, we may be able to block our units, but there may not be enough units in each block for each treatment. This leads us to incomplete block designs, which we will consider in Chapter 14.

Alternatively, we may have more than $g$ units in each block. What should we do now? This depends on several issues. If units are very inexpensive, one possibility is to use only $g$ units from each block. This preserves the simplicity of the RCB, without costing too much. If units are expensive, such waste is not tolerable. If there is some multiple of $g$ units per block, say $2g$ or $3g$, then we can randomly assign each treatment to two or three units in each block. This design, sometimes called a Generalized Randomized Complete Block, still has a simple structure and analysis. The standard model has treatments fixed, blocks random, and the treatment by blocks interaction as the denominator for treatments. Figure 13.4 shows a Hasse diagram for a GRCB with $g$ treatments, $r$ blocks of size $kg$ units, and $n$ measurement units per unit.

A third possibility is that units are expensive, but the block sizes are not a nice multiple of the number of treatments. Here, we can combine an RCB (or GRCB) with one of the incomplete block designs from Chapter 14. For example, with three treatments (A, B, and C) and three blocks of size 5, we could use (A, B, C, A, B) in block 1, (A, B, C, A, C) in block 2, and (A, B, C, B, C) in block 3. So each block has one full complement of the treatments, plus two more according to an incomplete block design.

The final possibility that we mention is that we can have blocks with different numbers of units; that is, some blocks have more units than others. Standard designs assume that all blocks have the same number of units, so we must do something special. The most promising approach is probably *optimal design* via special design software. Optimal design allocates treatments to units in such a way as to optimize some criterion; for example, we may wish to minimize the average variance of the estimated treatment effects. See Silvey (1980). The algorithms that do the optimization are complicated, but software exists that will do what is needed (though most statistical analysis packages do not). See Cook and Nachtsheim (1989). Oh yes, in case

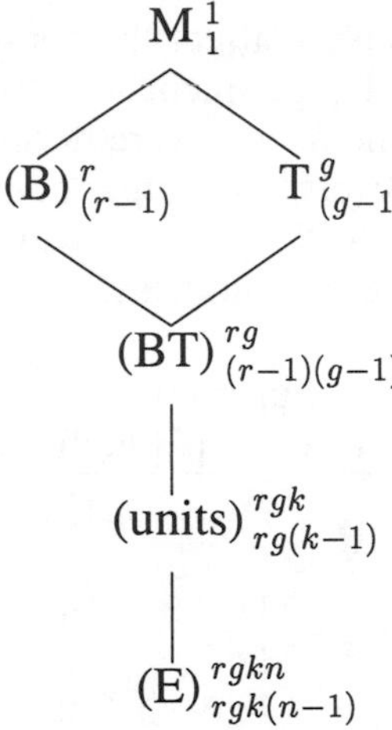


**Figure 13.4:** Hasse diagram for a Generalized Randomized Complete Block with $g$ treatments, $r$ blocks of size $kg$ units, and $n$ measurement units per unit; blocks are assumed random.

you were worried, most standard designs such as RCB's are also "optimal" designs; we just don't need the fancy software in the standard situations.

## 13.6 Problems

**Exercise 13.1**

Winter road treatments to clear snow and ice can lead to cracking in the pavement. An experiment was conducted comparing four treatments: sodium chloride, calcium chloride, a proprietary organic compound, and sand. Traffic level was used as a blocking factor and a randomized complete block experiment was conducted. One observation is missing, because the spreader in that district was not operating properly. The response is new cracks per mile of treated roadway.

| | A | B | C | D |
|---|---|---|---|---|
| Block 1 | | 32 | 27 | 36 |
| Block 2 | 38 | 40 | 43 | 33 |
| Block 3 | 40 | 63 | 14 | 27 |

Our interest is in the following comparisons: chemical versus physical (A,B,C versus D), inorganic versus organic (A,B versus C), and sodium versus calcium (A versus B). Which of these comparisons seem large?

**Exercise 13.2** Grains or crystals adversely affect the sensory qualities of foods using dried fruit pulp. A factorial experiment was conducted to determine which factors affect graininess. The factors were drying temperature (three levels), acidity (pH) of pulp (two levels), and sugar content (two levels). The experiment has two replications, with each replication using a different batch of pulp. Response is a measure of graininess.

| | | Sugar low | | Sugar high | |
|---|---|---|---|---|---|
| Temp. | Rep. | pH low | pH high | pH low | pH high |
| 1 | 1 | 21 | 12 | 13 | 1 |
| | 2 | 21 | 18 | 14 | 8 |
| 2 | 1 | 23 | 14 | 13 | 1 |
| | 2 | 23 | 17 | 16 | 11 |
| 3 | 1 | 17 | 20 | 16 | 14 |
| | 2 | 23 | 17 | 17 | 5 |

Analyze these data to determine which factors effect graininess, and which combination of factors leads to the least graininess.

**Exercise 13.3** The data below are from a replicated Latin Square with four treatments; row blocks were reused, but column blocks were not. Test for treatment differences and use Tukey HSD with level .01 to analyze the pairwise treatment differences.

| | | | | | | | |
|---|---|---|---|---|---|---|---|
| D 44 | B 26 | C 67 | A 77 | B 51 | D 62 | A 71 | C 49 |
| C 39 | A 45 | D 71 | B 74 | C 63 | A 74 | D 67 | B 47 |
| B 52 | D 49 | A 81 | C 88 | A 74 | C 75 | B 60 | D 58 |
| A 73 | C 58 | B 76 | D 100 | D 82 | B 79 | C 74 | A 68 |

**Exercise 13.4** Consider replicating a six by six Latin Square three times, where we use the same row blocks but different column blocks in the three replicates. The six treatments are the factorial combinations of factor A at three levels and factor B at two levels. Give the sources and degrees of freedom for the Analysis of Variance of this design.

**Exercise 13.5** Disk drive substrates may affect the amplitude of the signal obtained during readback. A manufacturer compares four substrates: aluminum (A), nickel-plated aluminum (B), and two types of glass (C and D). Sixteen disk drives will be made, four using each of the substrates. It is felt that operator, machine, and day of production may have an effect on the drives, so these three effects were blocked. The design and responses (in microvolts $\times 10^{-2}$) are given in the following table (data from Nelson 1993, Greek letters indicate day):

| | Operator | | | | | | | |
|---|---|---|---|---|---|---|---|---|
| Machine | 1 | | 2 | | 3 | | 4 | |
| 1 | A$\alpha$ | 8 | C$\gamma$ | 11 | D$\delta$ | 2 | B$\beta$ | 8 |
| 2 | C$\delta$ | 7 | A$\beta$ | 5 | B$\alpha$ | 2 | D$\gamma$ | 4 |
| 3 | D$\beta$ | 3 | B$\delta$ | 9 | A$\gamma$ | 7 | C$\alpha$ | 9 |
| 4 | B$\gamma$ | 4 | D$\alpha$ | 5 | C$\beta$ | 9 | A$\delta$ | 3 |

Analyze these data and report your findings, including a description of the design.

**Problem 13.1**

Ruminant animals, such as sheep, may not be able to quickly utilize protein in their diets, because the bacteria in their stomachs absorb the protein before it reaches the ruminant's intestine. Eventually the bacteria will die and the protein will be available for the ruminant, but we are interested in dietary changes that will help the protein get past the bacteria and to the intestine of the ruminant sooner.

We can vary the cereal source (oats or hay) and the protein source (soy or fish meal) in the diets. There are twelve lambs available for the experiment, and we expect fairly large animal to animal differences. Each diet must be fed to a lamb for at least 1 week before the protein uptake measurement is made. The measurement technique is safe and benign, so we may use each lamb more than once. We do not expect any carryover (residual) effects from one diet to the next, but there may be effects due to the aging of the lambs.

Describe an appropriate designed experiment and its randomization. Give a skeleton ANOVA (source and degrees of freedom only).

**Problem 13.2**

Briefly describe the experimental design you would choose for each of the following situations.

(a) We wish to study the effects of three factors on corn yields: nitrogen added, planting depth, and planting date. The nitrogen and depth factors have two levels, and the date factor has three levels. There are 24 plots available: twelve are in St. Paul, MN, and twelve are in Rosemount, MN.

(b) You manage a french fry booth at the state fair and wish to compare four brands of french fry cutters for amount of potato wasted. You sell a lot of fries and keep four fry cutters and their operators going constantly. Each day you get a new load of potatoes, and you expect some day to day variation in waste due to size and shape of that day's load. Different operators may also produce different amounts of waste. A full day's usage is needed to get a reasonable measure of waste, and you would like to finish in under a week.

(c) A Health Maintenance Organization wishes to test the effect of substituting generic drugs for name brand drugs on patient satisfaction. Satisfaction will be measured by questionnaire after the study. They decide to start small, using only one drug (a decongestant for which they have an analogous generic) and twenty patients at each of their five clinics. The patients at the different clinics are from rather different socioeconomic backgrounds, so some clinic to clinic variation is expected. Drugs may be assigned on an individual basis.

**Problem 13.3** For each of the following, describe the design that was used, give a skeleton ANOVA, and indicate how you would test the various terms in the model.

(a) Birds will often respond to other birds that invade their territory. We are interested in the time it takes nesting red-shouldered hawks to respond to invading calls, and want to know if that time varies according to the type of intruder. We have two state forests that have red-shouldered hawks nesting. In each forest, we choose ten nests at random from the known nesting sites. At each nest, we play two prerecorded calls over a loudspeaker (several days apart). One call is a red-shouldered hawk call; the other call is a great horned owl call. The response we measure is the time until the nesting hawks leave the nest to drive off the intruder.

(b) The food science department conducts an experiment to determine if the level of fiber in a muffin affects how hungry subjects perceive themselves to be. There are twenty subjects—ten randomly selected males and ten randomly selected females—from a large food science class. Each subject attends four sessions lasting 15 minutes. At the beginning of the session, they rate their hunger on a 1 to 100 scale. They then eat the muffin. Fifteen minutes later they again rate their hunger. The response for a given session is the decrease in hunger. At the four sessions they receive two low-fiber muffins and two high-fiber muffins in random order.

**Problem 13.4** Many professions have board certification exams. Part of the certification process for bank examiners involves a "work basket" of tasks that the examinee must complete in a satisfactory fashion in a fixed time period. New work baskets must be constructed for each round of examinations, and much effort is expended to make the workbaskets comparable (in terms of average score) from exam to exam. This year, two new work baskets (A and B) are being evaluated. We have three old work baskets (C, D, and E) to form a basis for comparison. We have ten paid examinees (1 through 6 are certified bank examiners, 7 through 9 are noncertified bank examiners nearing the end of their training, and 10 is a public accountant with no bank examining experience or training) who will each take all five tests. There are five graders who will each grade ten exams. We anticipate differences between the examinees and the graders; our interest is in the exams, which were randomized so that each examinee took each exam and each grader grades two of each exam.

The data follow. The letter indicates exam. Scores are out of 100, and 60 is passing. We want to know if either or both of the new exams are equivalent to the old exams.

| Student | Grader | | | | |
|---|---|---|---|---|---|
| | 1 | 2 | 3 | 4 | 5 |
| 1 | 68 D | 65 A | 76 E | 74 C | 76 B |
| 2 | 68 A | 77 E | 84 B | 65 D | 75 C |
| 3 | 73 C | 85 B | 72 D | 68 E | 62 A |
| 4 | 74 E | 76 C | 57 A | 79 B | 64 D |
| 5 | 80 B | 71 D | 76 C | 59 A | 68 E |
| 6 | 69 D | 75 E | 81 B | 68 A | 68 C |
| 7 | 60 C | 62 D | 62 E | 66 B | 40 A |
| 8 | 70 B | 55 A | 62 C | 57 E | 40 D |
| 9 | 61 E | 67 C | 53 A | 63 D | 69 B |
| 10 | 37 A | 53 B | 31 D | 48 C | 33 E |

**Problem 13.5**

An experiment was conducted to see how variety of soybean and crop rotation practices affect soybean productivity. There are two varieties used, Hodgson 78 and BSR191. These varieties are each used in four different 5-year rotation patterns with corn. The rotation patterns are (1) four years of corn and then soybeans (C-C-C-C-S), (2) three years of corn and then two years of soybeans (C-C-C-S-S), (3) soybean and corn alternation (S-C-S-C-S), and (4) five years of soybeans (S-S-S-S-S). Here we only analyze data from the fifth year.

This experiment was conducted twice in Waseca, MN, and twice in Lamberton, MN. Two groups of eight plots were chosen at each location. The first group of eight plots at each location was randomly assigned to the variety-rotation treatments in 1983. The second group was then assigned in 1984. Responses were measured in 1987 and 1988 (the fifth years) for the two groups.

The response of interest is the weight (g) of 100 random seeds from soybean plants (data from Whiting 1990). Analyze these data and report your findings.

| Location-Year | Variety | Rotation pattern 1 | 2 | 3 | 4 |
|---|---|---|---|---|---|
| W87 | 1 | 155 | 151 | 147 | 146 |
| | 2 | 153 | 156 | 159 | 155 |
| W88 | 1 | 170 | 159 | 157 | 168 |
| | 2 | 164 | 170 | 162 | 169 |
| L87 | 1 | 142 | 135 | 139 | 136 |
| | 2 | 146 | 138 | 135 | 133 |
| L88 | 1 | 170 | 155 | 159 | 173 |
| | 2 | 167 | 162 | 153 | 162 |

**Problem 13.6** An experiment was conducted to determine how different soybean varieties compete against weeds. There were sixteen varieties of soybeans and three weed treatments: no herbicide, apply herbicide 2 weeks after planting the soybeans, and apply herbicide 4 weeks after planting the soybeans. The measured response is weed biomass in kg/ha. There were two replications of the experiment—one in St. Paul, MN, and one in Rosemount, MN—for a total of 96 observations (data from Bussan 1995):

| | Herb. 2 weeks | | Herb. 4 weeks | | No herb. | |
|---|---|---|---|---|---|---|
| Variety | R | StP | R | StP | R | StP |
| Parker | 750 | 1440 | 1630 | 890 | 3590 | 740 |
| Lambert | 870 | 550 | 3430 | 2520 | 6850 | 1620 |
| M89-792 | 1090 | 130 | 2930 | 570 | 3710 | 3600 |
| Sturdy | 1110 | 400 | 1310 | 2060 | 2680 | 1510 |
| Ozzie | 1150 | 370 | 1730 | 2420 | 4870 | 1700 |
| M89-1743 | 1210 | 430 | 6070 | 2790 | 4480 | 5070 |
| M89-794 | 1330 | 190 | 1700 | 1370 | 3740 | 610 |
| M90-1682 | 1630 | 200 | 2000 | 880 | 3330 | 3030 |
| M89-1946 | 1660 | 230 | 2290 | 2210 | 3180 | 2640 |
| Archer | 2210 | 1110 | 3070 | 2120 | 6980 | 2210 |
| M89-642 | 2290 | 220 | 1530 | 390 | 3750 | 2590 |
| M90-317 | 2320 | 330 | 1760 | 680 | 2320 | 2700 |
| M90-610 | 2480 | 350 | 1360 | 1680 | 5240 | 1510 |
| M88-250 | 2480 | 350 | 1810 | 1020 | 6230 | 2420 |
| M89-1006 | 2430 | 280 | 2420 | 2350 | 5990 | 1590 |
| M89-1926 | 3120 | 260 | 1360 | 1840 | 5980 | 1560 |

Analyze these data for the effects of herbicide and variety.

**Problem 13.7** Plant shoots can be encouraged in tissue culture by exposing the cotyledons of plant embryos to cytokinin, a plant growth hormone. However, some shoots become watery, soft, and unviable; this is vitrification. An experiment was performed to study how the orientation of the embryo during exposure to cytokinin and the type of growth medium after exposure to cytokinin affect the rate of vitrification. There are six treatments, which are the factorial combinations of orientation (standard and experimental) and medium (three kinds). On a given day, the experimenters extract embryos from white pine seeds and randomize them to the six treatments. The embryos are exposed using the selected orientation for 1 week, and then go onto the selected medium. The experiment was repeated 22 times on different starting days. The response is the fraction of shoots that are normal (data from David Zlesak):

| | Medium 1 | | Medium 2 | | Medium 3 | |
|---|---|---|---|---|---|---|
| | Exp. | Std. | Exp. | Std. | Exp. | Std. |
| 1 | .67 | .34 | .46 | .26 | .63 | .40 |
| 2 | .70 | .42 | .69 | .42 | .74 | .17 |
| 3 | .86 | .42 | .89 | .33 | .80 | .17 |
| 4 | .76 | .53 | .74 | .60 | .78 | .53 |
| 5 | .63 | .71 | .50 | .29 | .63 | .29 |
| 6 | .65 | .60 | .95 | 1.00 | .90 | .40 |
| 7 | .73 | .50 | .83 | .88 | .93 | .88 |
| 8 | .94 | .75 | .94 | .75 | .80 | 1.00 |
| 9 | .93 | .70 | .77 | .50 | .90 | .80 |
| 10 | .71 | .30 | .48 | .40 | .65 | .30 |
| 11 | .83 | .20 | .74 | .00 | .69 | .30 |
| 12 | .82 | .50 | .72 | .00 | .63 | .30 |
| 13 | .67 | .67 | .67 | .25 | .90 | .42 |
| 14 | .83 | .50 | .94 | .40 | .83 | .33 |
| 15 | 1.00 | 1.00 | .80 | .33 | .90 | 1.00 |
| 16 | .95 | .75 | .76 | .25 | .96 | .63 |
| 17 | .47 | .50 | .71 | .67 | .67 | .50 |
| 18 | .83 | .50 | .94 | .67 | .83 | .83 |
| 19 | .90 | .33 | .83 | .67 | .97 | .50 |
| 20 | 1.00 | .50 | .69 | .25 | .92 | 1.00 |
| 21 | .80 | .63 | .63 | .00 | .70 | .50 |
| 22 | .82 | .60 | .57 | .40 | 1.00 | .50 |

Analyze these data and report your conclusions on how orientation and medium affect vitrification.

**Problem 13.8**

An army rocket development program was investigating the effects of slant range and propellant temperature on the accuracy of rockets. The overall objective of this phase of the program was to determine how these variables affect azimuth error (that is, side to side as opposed to distance) in the rocket impacts.

Three levels were chosen for each of slant range and temperature. The following procedure was repeated on 3 days. Twenty-seven rockets are grouped into nine sets of three, which are then assigned to the nine factor-level combinations in random order. The three rockets in a group are fired all at once in a single volley, and the azimuth error recorded. (Note that meteorological conditions may change from volley to volley.) The data follow (Bicking 1958):

| | Slant range | | | | | | | | |
|---|---|---|---|---|---|---|---|---|---|
| | 1 | | | 2 | | | 3 | | |
| | Days | | | Days | | | Days | | |
| | 1 | 2 | 3 | 1 | 2 | 3 | 1 | 2 | 3 |
| | -10 | -22 | -9 | -5 | -17 | -4 | 11 | -10 | 1 |
| Temp 1 | -13 | 0 | 7 | -9 | 6 | 13 | -5 | 10 | 20 |
| | 14 | -5 | 12 | 21 | 0 | 20 | 22 | 6 | 24 |
| | -15 | -25 | -15 | -14 | -3 | 14 | -9 | 8 | 14 |
| Temp 2 | -17 | -5 | 2 | 15 | -1 | 5 | -3 | -2 | 18 |
| | 7 | -11 | 5 | -11 | -20 | -10 | 20 | -15 | -2 |
| | -21 | -26 | -15 | -18 | -8 | 0 | 13 | -5 | -8 |
| Temp 3 | -23 | -8 | -5 | 5 | 5 | -13 | -9 | -18 | 3 |
| | 0 | -10 | 0 | -10 | -10 | 3 | -13 | -3 | 12 |

Analyze these data and determine how slant range and temperature affect azimuth error. (Hint: how many experimental units per block?)

**Problem 13.9** An experiment is conducted to study the effect of alfalfa meal in the diet of male turkey poults (chicks). There are nine treatments. Treatment 1 is a control treatment; treatments 2 through 9 contain alfalfa meal. Treatments 2 through 5 contain alfalfa meal type 22; treatments 6 through 9 contain alfalfa meal type 27. Treatments 2 and 6 are 2.5% alfalfa, treatments 3 and 7 are 5% alfalfa, treatments 4 and 8 are 7.5% alfalfa. Treatments 5 and 9 are also 7.5% alfalfa, but they have been modified to have the same calories as the control treatment.

The randomization is conducted as follows. Seventy-two pens of eight birds each are set out. Treatments are separately randomized to pens grouped 1–9, 10–18, 19–27, and so on. We do not have the response for pen 66. The response is average daily weight gain per bird for birds aged 7 to 14 days in g/day (data from Turgay Ergul):

| Trt | 1–9 | 10–18 | 19–27 | 28–36 | 37–45 | 46–54 | 55–63 | 64–72 |
|---|---|---|---|---|---|---|---|---|
| 1 | 23.63 | 19.86 | 24.00 | 22.11 | 25.38 | 24.18 | 23.43 | 18.75 |
| 2 | 20.70 | 20.02 | 23.95 | 19.13 | 21.21 | 20.89 | 23.55 | 22.89 |
| 3 | 19.95 | 18.29 | 17.61 | 19.89 | 23.96 | 20.46 | 22.55 | 17.30 |
| 4 | 21.16 | 19.02 | 19.38 | 19.46 | 20.48 | 19.54 | 19.96 | 20.71 |
| 5 | 23.71 | 16.44 | 20.71 | 20.16 | 21.70 | 21.47 | 20.44 | 22.51 |
| 6 | 20.38 | 18.68 | 20.91 | 23.07 | 22.54 | 21.73 | 25.04 | 23.22 |
| 7 | 21.57 | 17.38 | 19.55 | 19.79 | 20.77 | 18.36 | 20.32 | 21.98 |
| 8 | 18.52 | 18.84 | 22.54 | 19.95 | 21.27 | 20.09 | 19.27 | 20.02 |
| 9 | 23.14 | 20.46 | 18.14 | 21.70 | 22.93 | 21.29 | 22.49 | |

Analyze these data to determine the effects of the treatments on weight gain.

**Problem 13.10**

Implantable pacemakers contain a small circuit board called a substrate. Multiple substrates are made as part of a single "laminate." In this experiment, seven laminates are chosen at random. We choose eight substrate locations and measure the length of the substrates at those eight locations on the seven substrates. Here we give coded responses ($10,000 \times [response - 1.45]$, data from Todd Kerkow).

| | Laminate | | | | | | |
|---|---|---|---|---|---|---|---|
| Location | 1 | 2 | 3 | 4 | 5 | 6 | 7 |
| 1 | 28 | 20 | 23 | 29 | 44 | 45 | 43 |
| 2 | 11 | 20 | 27 | 31 | 33 | 38 | 36 |
| 3 | 26 | 26 | 14 | 17 | 41 | 36 | 36 |
| 4 | 23 | 26 | 18 | 21 | 36 | 36 | 39 |
| 5 | 20 | 21 | 30 | 28 | 45 | 31 | 33 |
| 6 | 16 | 19 | 24 | 23 | 33 | 32 | 39 |
| 7 | 37 | 43 | 49 | 33 | 53 | 49 | 32 |
| 8 | 04 | 09 | 13 | 17 | 39 | 29 | 32 |

Analyze these data to determine the effect of location. (Hint: think carefully about the design.)

**Problem 13.11**

The oleoresin of trees is obtained by cutting a tapping gash in the bark and removing the resin that collects there. Acid treatments can also improve collection. In this experiment, four trees (*Dipterocarpus kerrii*) will be tapped seven times each. Each of the tappings will be treated with a different strength of sulfuric acid (0, 2.5, 5, 10, 15, 25, and 50% strength), and the resin collected from each tapping is the response (in grams, data from Bin Jantan, Bin Ahmad, and Bin Ahmad 1987):

| | Acid strength (%) | | | | | | |
|---|---|---|---|---|---|---|---|
| Tree | 0 | 2.5 | 5 | 10 | 15 | 25 | 50 |
| 1 | 3 | 108 | 219 | 276 | 197 | 171 | 166 |
| 2 | 2 | 100 | 198 | 319 | 202 | 173 | 304 |
| 3 | 1 | 43 | 79 | 182 | 123 | 172 | 194 |
| 4 | .5 | 17 | 33 | 78 | 51 | 41 | 70 |

Determine the effect of acid treatments on resin output; if acid makes a difference, which treatments are best?

**Problem 13.12** Hormones can alter the sexual development of animals. This experiment studies the effects of growth hormone (GH) and follicle-stimulating hormone (FSH) on the length of the seminiferous tubules in pigs. The treatments are control, daily injection of GH, daily injection of FSH, and daily injection of GH and FSH. Twenty-four weanling boars are used, four from each of six litters. The four boars in each litter are randomized to the four treatments. The boars are castrated at 100 days of age, and the length (in meters!) of the seminiferous tubules determined as response (data from Swanlund *et al.* 1995).

| | Litter | | | | | |
|---|---|---|---|---|---|---|
| | 1 | 2 | 3 | 4 | 5 | 6 |
| Control | 1641 | 1290 | 2411 | 2527 | 1930 | 2158 |
| GH | 1829 | 1811 | 1897 | 1506 | 2060 | 1207 |
| FSH | 3395 | 3113 | 2219 | 2667 | 2210 | 2625 |
| GH+FSH | 1537 | 1991 | 3639 | 2246 | 1840 | 2217 |

Analyze these data to determine the effects of the hormones on tubule length.

**Problem 13.13** Shade trees in coffee plantations may increase or decrease the yield of coffee, depending on several environmental and ecological factors. Robusta coffee was planted at three locations in Ghana. Each location was divided into four plots, and trees were planted at densities of 185, 90, 70, and 0 trees per hectare. Data are the yields of coffee (kg of fresh berries per hectare) for the 1994-95 cropping season (data from Amoah, Osei-Bonsu, and Oppong 1997):

| Location | 185 | 90 | 70 | 0 |
|---|---|---|---|---|
| 1 | 3107 | 2092 | 2329 | 2017 |
| 2 | 1531 | 2101 | 1519 | 1766 |
| 3 | 2167 | 2428 | 2160 | 1967 |

Analyze these data to determine the effect of tree density on coffee production.

**Problem 13.14**

A sensory experiment was conducted to determine if consumers have a preference between regular potato chips (A) and reduced-fat potato chips (B). Twenty-four judges will rate both types of chips; twelve judges will rate the chips in the order regular fat, then reduced fat; and the other twelve will have the order reduced fat, then regular fat. We anticipate judge to judge differences and possible differences between the first and second chips tasted. The response is a liking scale, with higher scores indicating greater liking (data from Monica Coulter):

| | 1 | 2 | 3 | 4 | 5 | 6 | 7 | 8 | 9 | 10 | 11 | 12 |
|---|---|---|---|---|---|---|---|---|---|---|---|---|
| A first | 8 | 5 | 7 | 8 | 7 | 7 | 4 | 9 | 8 | 7 | 7 | 7 |
| B second | 6 | 6 | 8 | 8 | 4 | 7 | 8 | 9 | 9 | 7 | 5 | 3 |

| | 13 | 14 | 15 | 16 | 17 | 18 | 19 | 20 | 21 | 22 | 23 | 24 |
|---|---|---|---|---|---|---|---|---|---|---|---|---|
| B first | 4 | 6 | 6 | 7 | 6 | 4 | 8 | 6 | 7 | 6 | 8 | 7 |
| A second | 7 | 8 | 7 | 8 | 4 | 8 | 7 | 7 | 7 | 8 | 8 | 8 |

**Question 13.1**

Find conditions under which the estimated variance for a CRD based on RCB data is less than the naive estimate pooling sums of squares and degrees of freedom for error and blocks. Give a heuristic argument, based on randomization, suggesting why your relationship is true.

**Question 13.2**

The inspector general is coming, and an officer wishes to arrange some soldiers for inspection. In the officer's command are men and women of three different ranks, who come from six different states. The officer is trying to arrange 36 soldiers for inspection in a six by six square with one soldier from each state-rank-gender combination. Furthermore, the idea is to arrange the soldiers so that no matter which rank or file (row or column) is inspected by the general, the general will see someone from each of the six states, one woman of each rank, and one man of each rank. Why is this officer so frustrated?

# Chapter 14

# Incomplete Block Designs

Block designs group similar units into blocks so that variation among units within the blocks is reduced. Complete block designs, such as RCB and LS, have each treatment occurring once in each block. Incomplete block designs also group units into blocks, but the blocks do not have enough units to accommodate all the treatments.

Not all treatments appear in an incomplete block

Incomplete block designs share with complete block designs the advantage of variance reduction due to blocking. The drawback of incomplete block designs is that they do not provide as much information per experimental unit as a complete block design with the same error variance. Thus complete blocks are preferred over incomplete blocks when both can be constructed with the same error variance.

Incomplete blocks less efficient than complete blocks

### Eyedrops

**Example 14.1**

Eye irritation can be reduced with eyedrops, and we wish to compare three brands of eyedrops for their ability to reduce eye irritation. (There are problems here related to measuring eye irritation, but we set them aside for now.) We expect considerable subject to subject variation, so blocking on subject seems appropriate. If each subject can only be used during one treatment period, then we must use one brand of drop in the left eye and another brand in the right eye. We are forced into incomplete blocks of size two, because our subjects have only two eyes.

Suppose that we have three subjects that receive brands (A and B), (A and C), and (B and C) respectively. How can we estimate the expected difference in responses between two treatments, say A and B? We can get some information from subject 1 by taking the difference of the A and B responses; the subject effect will cancel in this difference. This first difference has variance $2\sigma^2$. We can also get an estimate of A-B by subtracting the B-C difference in subject three from the A-C difference in subject two. Again, subject effects cancel out, and this difference has variance $4\sigma^2$. Similar approaches yield estimates of A-C and B-C using data from all subjects.

If we had had two complete blocks (three-eyed subjects?) with the same unit variance, then we would have had two independent estimates of A-B each with variance $2\sigma^2$. Thus the incomplete block design has more variance in its estimates of treatment differences than does the complete block design with the same variance and number of units.

There are many kinds of incomplete block designs. This chapter will cover only some of the more common types. Several of the incomplete block designs given in this chapter have "balanced" in their name. It is important to realize that these designs are not balanced in the sense that all block and factor-level combinations occur equally often. Rather they are balanced using somewhat looser criteria that will be described later.

Resolvable designs split into replications

Two general classes of incomplete block designs are *resolvable* designs and *connected* designs. Suppose that each treatment is used $r$ times in the design. A resolvable design is one in which the blocks can be arranged into $r$ groups, with each group representing a complete set of treatments. Resolvable designs can make management of experiments simpler, because each replication can be run at a different time or a different location, or entire replications can be dropped if the need arises. The eyedrop example is not resolvable.

Connected designs can estimate all treatment differences

A design is *disconnected* if you can separate the treatments into two groups, with no treatment from the first group ever appearing in the same block with a treatment from the second group. A *connected* design is one that is not disconnected. In a connected design you can estimate all treatment differences. You cannot estimate all treatment differences in a disconnected design; in particular, you cannot estimate differences between treatments in different groups. Connectedness is obviously a very desirable property.

## 14.1 Balanced Incomplete Block Designs

BIBD

The Balanced Incomplete Block Design (BIBD) is the simplest incomplete block design. We have $g$ treatments, and each block has $k$ units, with $k < g$. Each treatment will be given to $r$ units, and we will use $b$ blocks. The total number of units $N$ must satisfy $N = kb = rg$. The final requirement for a BIBD is that all pairs of treatments must occur together in the same number of blocks. The BIBD is called "balanced" because the variance of the estimated

**Table 14.1:** Plates washed before foam disappears. Letters indicate treatments.

| Session | | | | | | | | | | | |
|---|---|---|---|---|---|---|---|---|---|---|---|
| 1 | 2 | 3 | 4 | 5 | 6 | 7 | 8 | 9 | 10 | 11 | 12 |
| A 19 | D 6 | G 21 | A 20 | B 17 | C 15 | A 20 | B 16 | C 13 | A 20 | B 17 | C 14 |
| B 17 | E 26 | H 19 | D 7 | E 26 | F 23 | E 26 | F 23 | D 7 | F 24 | D 6 | E 24 |
| C 11 | F 23 | J 28 | G 20 | H 19 | J 31 | J 31 | G 21 | H 20 | H 19 | J 29 | G 21 |

difference of treatment effects $\widehat{\alpha}_i - \widehat{\alpha}_j$ is the same for all pairs of treatments $i, j$.

Example 14.1 is the simplest possible BIBD. There are $g = 3$ treatments, with blocks of size $k = 2$. Each treatment occurs $r = 2$ times in the $b = 3$ blocks. There are $N = 6$ total units, and each pair of treatments occurs together in one block.

We may use the BIBD design for treatments with factorial structure. For example, suppose that we have three factors each with two levels for a total of $g = 8$ treatments. If we have $b = 8$ blocks of size $k = 7$, then we can use a BIBD with $r = 7$, with each treatment left out of one block and each pair of treatments occurring together six times.

**Example 14.2**

## Dish detergent

John (1961) gives an example of a BIBD. Nine different dishwashing solutions are to be compared. The first four consist of base detergent I and 3, 2, 1, and 0 parts of an additive; solutions five through eight consist of base detergent II and 3, 2, 1, and 0 parts of an additive; the last solution is a control. There are three washing basins and one operator for each basin. The three operators wash at the same speed during each test, and the response is the number of plates washed when the foam disappears. The speed of washing is the same for all three detergents used at any one session, but could differ from session to session.

Table 14.1 gives the design and the results. There are $g = 9$ treatments arranged in $b = 12$ incomplete blocks of size $k = 3$. Each treatment appears $r = 4$ times, and each pair of treatments appears together in one block.

Treatment pairs occur together $\lambda$ times

The requirement that all pairs of treatments occur together in an equal number of blocks is a real stickler. Any given treatment occurs in $r$ blocks, and there are $k - 1$ other units in each of these blocks for a total of $r(k - 1)$ units. These must be divided evenly between the $g - 1$ other treatments. Thus $\lambda = r(k - 1)/(g - 1)$ must be a whole number for a BIBD to exist. For the eyedrop example, $\lambda = 2(2 - 1)/(3 - 1) = 1$, and for the dishes example, $\lambda = 4(3 - 1)/(9 - 1) = 1$.

A major impediment to the use of the BIBD is that no BIBD may exist for your combination of $kb = rg$. For example, you may have $g = 5$ treatments and $b = 5$ blocks of size $k = 3$. Then $r = 3$, but $\lambda = 3(3-1)/(5-1) = 3/2$ is not a whole number, so there can be no BIBD for this combination of $r$, $k$, and $g$. Unfortunately, $\lambda$ being a whole number is not sufficient to guarantee that a BIBD exists, though one usually does.

Unreduced BIBD has all combinations

A BIBD always exists for every combination of $k < g$. For example, you can always generate a BIBD by using all combinations of the $g$ treatments taken $k$ at a time. Such a BIBD is called *unreduced*. The problem with this approach is that you may need a lot of blocks for the design. For example, the unreduced design for $g = 8$ treatments in blocks of size $k = 4$ requires $b = 70$ blocks. Appendix C contains a list of some BIBD plans for $g \leq 9$. Fisher and Yates (1963) and Cochran and Cox (1957) contain much more extensive lists.

BIBD tables

Design complement

If you have a plan for a BIBD with $g$, $k$, and $b$ blocks, then you can construct a plan for $g$ treatments in $b$ blocks of $g - k$ units per block simply by using in each block of the second design the treatments *not* used in the corresponding block of the first design. The second design is called the *complement* of the first design. When $b = g$ and $r = k$, a BIBD is said to be *symmetric*. The eyedrop example above is symmetric; the detergent example is not symmetric.

Symmetric BIBD

BIBD randomization

Randomization of a BIBD occurs in three steps. First, randomize the assignment of physical blocks to subgroups of treatment letters (or numbers) given in the design. Second, randomize the assignment of these treatment letters to physical units within blocks. Third, randomize the assignment of treatment letters to treatments.

### 14.1.1 Intrablock analysis of the BIBD

BIBD model

Intrablock analysis sounds exotic, but it is just the standard analysis that you would probably have guessed was appropriate. Let $y_{ij}$ be the response for treatment $i$ in block $j$; we do not observe all $i, j$ combinations. Use the model

$$y_{ij} = \mu + \alpha_i + \beta_j + \epsilon_{ij} \quad .$$

If treatments are fixed, we assume that the treatment effects sum to zero; otherwise we assume that they are a random sample from a $N(0, \sigma_\alpha^2)$ distribution. Block effects may be fixed or random.

Our usual methods for estimating treatment effects *do not* work for the BIBD. In this way, this "balanced" design is more like an unbalanced factorial or an RCB with missing data. For those situations, we relied on statistical

**Listing 14.1:** SAS output for intrablock analysis of detergent data.

```
                                     Sum of              Mean
Source                    DF        Squares            Square    F Value       Pr > F

Model                     19     1499.56481          78.92446      95.77       0.0001

Error                     16       13.18519           0.82407                           ①

Source                    DF    Type III SS       Mean Square    F Value       Pr > F

BLOCK                     11       10.06481           0.91498       1.11       0.4127
DETERG                     8     1086.81481         135.85185     164.85       0.0001   ②

Contrast                  DF    Contrast SS       Mean Square    F Value       Pr > F

control vs test            1     345.041667        345.041667     418.70       0.0001   ③
base I vs base II          1     381.337963        381.337963     462.75       0.0001
linear in additive         1     306.134259        306.134259     371.49       0.0001

                                         T for H0:    Pr > |T|   Std Error of
Parameter                  Estimate    Parameter=0                Estimate

base I vs base II       -7.97222222         -21.51      0.0001    0.37060178          ④
```

software to fit the model, and we do so here as well. Similarly, our usual contrast methods do not work either. An RCB with missing data is a good way to think about the analysis of the BIBD, even though in the BIBD the data were planned to be missing in a very systematic way.

Usual estimates of treatment effects do not work for BIBD

For the RCB with missing data, we computed the sum of squares for treatments adjusted for blocks. That is, we let blocks account for as much variation in the data as they could, and then we determined how much additional variation could be explained by adding treatments to the model. Because we had already removed the variation between blocks, this additional variation explained by treatments must be variation within blocks: hence *intra*block analysis. Intrablock analysis of a BIBD is analysis with treatments adjusted for blocks.

Intrablock analysis is treatments adjusted for blocks

## Dish detergent, continued

**Example 14.3**

The basic intrablock ANOVA consists of treatments adjusted for blocks. Listing 14.1 ② shows SAS output for this model; the Type III sum of squares for detergent is adjusted for blocks. Residual plots show that the variance is fairly stable, but the residuals have somewhat short tails. There is strong evidence against the null hypothesis ($p$-value .0001).

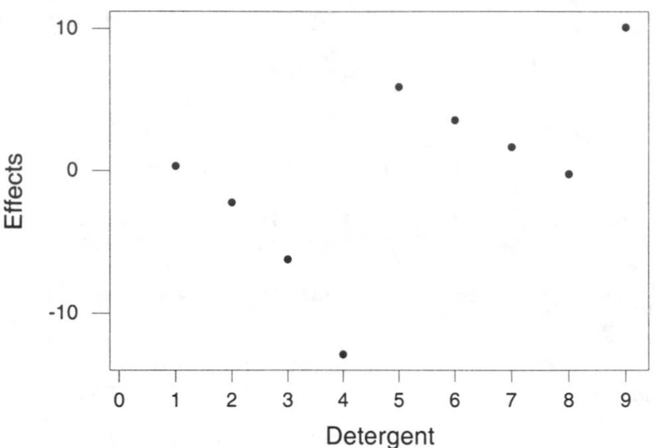


**Figure 14.1:** Treatment effects for intrablock analysis of dish detergent data, using Minitab.

We can examine the treatment effects more closely by comparing the two detergent bases with each other and the control, and by looking at the effects of the additive. Figure 14.1 shows the nine treatment effects. Clearly there is a mostly linear effect due to the amount of additive, with more additive giving a higher response. We also see that detergent base I gives lower responses than detergent base II, and both are lower than the control. For example, the contrast between base I and base II has sum of squares 381.34; the contrast between the control and the other treatments has sum of squares 345.04; and the linear in additive contrast has sum of squares 306.16 (Listing 14.1 ③). These 3 degrees of freedom account for 1032.5 of the total 1086.8 sum of squares between treatments.

Efficiency of BIBD to RCB

There is in fact a fairly simple hand-calculation for treatments adjusted for blocks in the BIBD; the availability of this simple calculation helped make the BIBD attractive before computers. We discuss the calculation not because you will ever be doing the calculations that way, but rather because it helps give some insight into $E_{\text{BIBD:RCB}}$, the efficiency of the BIBD relative to the RCB. Define $E_{\text{BIBD:RCB}}$ to be

$$E_{\text{BIBD:RCB}} = \frac{g(k-1)}{(g-1)k} ,$$

where $g$ is the number of treatments and $k$ is the number of units per block. Observe that $E_{\text{BIBD:RCB}} < 1$, because $k < g$ in the BIBD. For the detergent example, $E_{\text{BIBD:RCB}} = 9 \times 2/(8 \times 3) = 3/4$.

The value $E_{\text{BIBD:RCB}}$ is the relative efficiency of the BIBD to an RCB with the same variance. One way to think about $E_{\text{BIBD:RCB}}$ is that every unit in a BIBD is only worth $E_{\text{BIBD:RCB}}$ units worth of information in an RCB with the same variance. Thus while each treatment is used $r$ times in a BIBD, the effective sample size is only $rE_{\text{BIBD:RCB}}$.

Effective sample size $rE_{\text{BIBD:RCB}}$

The hand-calculation formulae for the BIBD use the effective sample size in place of the actual sample size. Let $\overline{y}_{\bullet j}$ be the mean response in the $j$th block; let $v_{ij} = y_{ij} - \overline{y}_{\bullet j}$ be the data with block means removed; and let $v_{i\bullet}$ be the sum of the $v_{ij}$ values for treatment $i$ (there are $r$ of them). Then we have

Hand formulae for BIBD use effective sample size

$$\widehat{\alpha}_i = \frac{v_{i\bullet}}{rE_{\text{BIBD:RCB}}} ,$$

$$SS_{\text{Trt}} = \sum_{i=1}^{g} (rE_{\text{BIBD:RCB}})\widehat{\alpha}_i^{\,2} ,$$

and

$$Var(\sum_i w_i\widehat{\alpha}_i) = \sigma^2 \sum_i \frac{w_i^2}{rE_{\text{BIBD:RCB}}} .$$

We can also use pairwise comparison procedures with the effective sample size.

In practice, we can often find incomplete blocks with a smaller variance $\sigma^2_{bibd}$ than can be attained using complete blocks $\sigma^2_{rcb}$. We prefer the BIBD design over the RCB if

BIBD beats RCB if variance reduction great enough

$$\frac{\sigma^2_{bibd}}{rE_{\text{BIBD:RCB}}} < \frac{\sigma^2_{rcb}}{r}$$

or

$$\frac{\sigma^2_{bibd}}{\sigma^2_{rcb}} < E_{\text{BIBD:RCB}} ;$$

in words, we prefer the BIBD if the reduction in variance more than compensates for the loss of efficiency. This comparison ignores adjustments for error degrees of freedom.

### 14.1.2 Interblock information

Recovery of interblock information when block effects are random

The first thing we did in the intrablock analysis of the BIBD was to subtract block means from the data to get deviations from the block means. When the block effects in a BIBD are random effects, then these block means also contain information about treatment differences. We can use block means or block totals to produce a second set of estimates for treatment effects, called the *interblock* estimates, independent of the usual intrablock estimates. Combining the interblock and intrablock estimates is called "recovery of interblock information."

First get intrablock estimate

Suppose that we want to estimate a contrast $\zeta = \sum_i w_i \alpha_i$. Recovery of interblock information takes place in three steps. First, compute the *intra*block estimate of the contrast and its variance. Second, compute the *inter*block estimate of the contrast and its variance. Third, combine the two estimates. The intrablock estimate is simply the standard estimate of the last section:

$$\widehat{\zeta} = \sum_{i=1}^{g} w_i \widehat{\alpha}_i$$

with variance

$$Var(\widehat{\zeta}) = \sigma^2 \sum_{i=1}^{g} \frac{w_i^2}{rE_{\text{BIBD:RCB}}} \quad ,$$

using $MS_E$ to estimate $\sigma^2$.

For step 2, start by letting $n_{ij}$ be 1 if treatment $i$ occurs in block $j$, and 0 otherwise. Then the block total $y_{\bullet j}$ can be expressed

$$\begin{aligned} y_{\bullet j} &= k\mu + \sum_{i=1}^{g} n_{ij}\alpha_i + \left\{ k\beta_j + \sum_{i=1}^{g} n_{ij}\epsilon_{ij} \right\} \\ &= k\mu + \sum_{i=1}^{g} n_{ij}\alpha_i + \eta_j \quad . \end{aligned}$$

Interblock estimates from block totals

This has the form of a multiple regression with $g$ predicting variables and an independent and normally distributed error $\eta_j$ having variance $k^2\sigma_\beta^2 + k\sigma^2$. Some tedious algebra shows that the interblock estimates are

$$\begin{aligned} \tilde{\mu} &= \overline{y}_{\bullet\bullet} \\ \tilde{\alpha}_i &= \frac{\sum_{j=1}^{b} n_{ij} y_{\bullet j} - rk\tilde{\mu}}{r - \lambda} \quad , \end{aligned}$$

and the variance of the contrast $\tilde{\zeta} = \sum_{i=1}^{g} w_i \tilde{\alpha}_i$ is

$$Var(\tilde{\zeta}) = (k^2\sigma_\beta^2 + k\sigma^2)\sum_{i=1}^{g} \frac{w_i^2}{r-\lambda} \ .$$

We estimate $\sigma^2$ using the $MS_E$ from the intrablock analysis. Estimating $\sigma_\beta^2$ involves something highly unusual. The expected value of the mean square for *blocks adjusted for treatments* is $\sigma^2 + (N-g)\sigma_\beta^2/(b-1)$. Thus an unbiased estimate of $\sigma_\beta^2$ is

Use blocks adjusted for treatments to get block variance

$$\widehat{\sigma}_\beta^2 = \frac{b-1}{N-g}(MS_{\text{blocks adjusted}} - MS_E) \ .$$

This interblock recovery is the only place we will consider blocks adjusted for treatments.

At this stage, we have the intrablock estimate $\widehat{\zeta}$ and its variance $Var(\widehat{\zeta})$, and we have the interblock estimate $\tilde{\zeta}$ and its variance $Var(\tilde{\zeta})$. If the variances were equal, we would just average the two estimates to get a combined estimate. However, the variance of the intrablock estimate is always less than the interblock estimate, so we want to give the intrablock estimate more weight in the average. The best weight is "inversely proportional to the variance", so the combined estimate for contrast $\zeta$ is

Use weighted average to combine inter- and intrablock estimates

$$\bar{\zeta} = \frac{\dfrac{1}{Var(\widehat{\zeta})}\widehat{\zeta} + \dfrac{1}{Var(\tilde{\zeta})}\tilde{\zeta}}{\dfrac{1}{Var(\widehat{\zeta})} + \dfrac{1}{Var(\tilde{\zeta})}} \ .$$

This combined estimate has variance

$$Var(\bar{\zeta}) = \frac{1}{\dfrac{1}{Var(\widehat{\zeta})} + \dfrac{1}{Var(\tilde{\zeta})}} \ .$$

## Dish detergent, continued

**Example 14.4**

Suppose that we wish to examine the difference between detergent bases I and II. We can do that with a contrast $w$ with coefficients (.25, .25, .25,

.25, -.25, -.25, -.25, -.25, 0). Listing 14.1 ④ shows that this contrast has an estimated value of -7.972 with a standard error of .3706 (variance .1373); this is the intrablock estimate.

We begin the interblock analysis by getting the block totals, the incidence matrix $\{n_{ij}\}$ (shown here with treatments indexing columns), and the sums of the cross products:

| Block total | Treatment incidence 1 | 2 | 3 | 4 | 5 | 6 | 7 | 8 | 9 |
|---|---|---|---|---|---|---|---|---|---|
| 47 | 1 | 1 | 1 | 0 | 0 | 0 | 0 | 0 | 0 |
| 55 | 0 | 0 | 0 | 1 | 1 | 1 | 0 | 0 | 0 |
| 68 | 0 | 0 | 0 | 0 | 0 | 0 | 1 | 1 | 1 |
| 47 | 1 | 0 | 0 | 1 | 0 | 0 | 1 | 0 | 0 |
| 62 | 0 | 1 | 0 | 0 | 1 | 0 | 0 | 1 | 0 |
| 69 | 0 | 0 | 1 | 0 | 0 | 1 | 0 | 0 | 1 |
| 77 | 1 | 0 | 0 | 0 | 1 | 0 | 0 | 0 | 1 |
| 60 | 0 | 1 | 0 | 0 | 0 | 1 | 1 | 0 | 0 |
| 40 | 0 | 0 | 1 | 1 | 0 | 0 | 0 | 1 | 0 |
| 63 | 1 | 0 | 0 | 0 | 0 | 1 | 0 | 1 | 0 |
| 52 | 0 | 1 | 0 | 1 | 0 | 0 | 0 | 0 | 1 |
| 59 | 0 | 0 | 1 | 0 | 1 | 0 | 1 | 0 | 0 |
| $\sum_j n_{ij}\overline{y}_{\bullet j}$ | 234 | 221 | 215 | 194 | 253 | 247 | 234 | 233 | 266 |

Applying the formula, we get that the interblock estimates are .333, -4, -6, -13, 6.667, 4.667, .333, 0, and 11. The interblock estimate $\tilde{\zeta}$ is thus

$$\tilde{\zeta} = (.333 - 4 - 6 - 13)/4 - (6.667 + 4.667 + .333 + 0)/4 = -8.583 \ .$$

The variance of $\tilde{\zeta}$ is

$$\begin{aligned} Var(\tilde{\zeta}) &= (k^2\sigma_\beta^2 + k\sigma^2)\sum_{i=1}^{a} \frac{w_i^2}{r-\lambda} \\ &= (9\sigma_\beta^2 + 3\sigma^2)\frac{8 \times .25^2}{3} \\ &= (3\sigma_\beta^2 + \sigma^2)/2 \end{aligned}$$

The intrablock $MS_E$ of .82407 estimates $\sigma^2$ (Listing 14.1 ①). The mean square for blocks adjusted for treatments is .91498 from Listing 14.1 ②. (We show Type III sums of squares, so blocks are also adjusted for treatments.) The estimate for $\sigma_\beta^2$ is thus

$$\begin{aligned}\widehat{\sigma}_\beta^2 &= \frac{b-1}{N-g}(MS_{\text{blocks adjusted}} - MS_E) \\ &= \frac{11}{27}(.91498 - .82407) \\ &= .0370\end{aligned}$$

Substituting in, we get

$$\begin{aligned}Var(\widetilde{\zeta}) &= (3\sigma_\beta^2 + \sigma^2)/2 \\ &= (3 \times .0370 + .82407)/2 \\ &= .4675\end{aligned}$$

Note that even with an estimated block variance of nearly zero, the intrablock estimate of the contrast is still much more precise than the interblock estimate.

The intrablock estimate and variance are -7.972 and .1374, and the interblock estimate and variance are -8.583 and .4675. The combined estimate is

$$\begin{aligned}\bar{\zeta} &= \frac{\frac{-7.972}{.1374} + \frac{-8.583}{.4675}}{\frac{1}{.1374} + \frac{1}{.4675}} \\ &= -8.111\end{aligned}$$

with variance

$$\begin{aligned}Var(\bar{\zeta}) &= \frac{1}{\frac{1}{.1374} + \frac{1}{.4675}} \\ &= .1062\end{aligned}$$

That was a lot of work. Unfortunately, this effort often provides minimal improvement over the intrablock estimates. When there is no block variance (that is, when $\sigma_\beta^2 = 0$), then the interblock variance for a contrast is $g(k-1)/(g-k)$ times as large as the intrablock variance. When blocking is

Interblock recovery often provides little improvement

successful, the variation between blocks will be large compared to the variation within blocks. Then the variance of intrablock estimates will be much smaller than those of interblock estimates, and the combined estimates are very close to the intrablock estimates.

Weights are variable

Another fact to bear in mind is that the weights used in the weighted average to combine intra- and interblock information are rather variable when $b$ is small. This variation comes from the ratio $MS_{\text{blocks adjusted}}/MS_E$, which appears in the formula for the weights. As we saw when trying to estimate ratios of variance components, we need quite a few degrees of freedom in both the numerator and denominator before the ratio, and thus the weights, are stable.

## 14.2 Row and Column Incomplete Blocks

We use Latin Squares and their variants when we need to block on two sources of variation in complete blocks. We can use *Youden Squares* when we need to block on two sources of variation, but cannot set up the complete blocks for LS designs. I've always been amused by this name, because Youden Squares are not square.

Youden Squares are incomplete Latin Squares

The simplest example of a Youden Square starts with a Latin Square and deletes one of the rows (or columns). The resulting arrangement has $g$ columns and $g - 1$ rows. Each row is a complete block for the treatments, and the columns form an unreduced BIBD for the treatments. Here is a simple Youden Square formed from a four by four Latin Square:

| | | | |
|---|---|---|---|
| A | B | C | D |
| B | A | D | C |
| C | D | A | B |

Youden Square is BIBD on columns and RCB on rows

A more general definition of a Youden Square is a rectangular arrangement of treatments, with the columns forming a BIBD and all treatments occurring and equal number of times in each row. In particular, any symmetric BIBD ($b = g$) can be rearranged into a Youden Square. For example, here is a symmetric BIBD with $g = b = 7$ and $r = k = 3$ arranged as a Youden Square:

| | | | | | | |
|---|---|---|---|---|---|---|
| A | B | C | D | E | F | G |
| B | C | D | E | F | G | A |
| D | E | F | G | A | B | C |

**Table 14.2:** Serum levels of lithium ($\mu$Eq/l) 12 hours after administration. Treatments are 300 mg and 250 mg capsules, 450 mg time delay capsule, and 300 mg solution.

| Week | Subject | | | | | |
|---|---|---|---|---|---|---|
| 1 | A 200 | D 267 | C 156 | B 280 | D 333 | D 233 |
| 2 | B 160 | C 178 | A 200 | C 178 | A 167 | B 200 |
| | | | | | | |
| 1 | B 320 | B 320 | C 111 | A 333 | A 233 | C 244 |
| 2 | A 200 | D 200 | D 133 | D 200 | C 178 | B 160 |

In Appendix C, thoses BIBD's that can be arranged as Youden Squares are so arranged.

Row orthogonal designs

The analysis of a Youden Square is a combination of the Latin Square and BIBD, as might be expected. Because both treatments and columns appear once in each row, row contrasts are orthogonal to treatment and column contrasts, and this makes computation a little easier. Youden Squares are also called *row orthogonal* for this reason. The intrablock ANOVA has terms for rows, columns, treatments (adjusted for columns), and error. Row effects and sums of squares are computed via the standard formulae, ignoring columns and treatments. Column sums of squares (unadjusted) are computed ignoring rows and treatments. Intrablock treatment effects and sums of squares are computed as for a BIBD with columns as blocks. Error sums of squares are found by subtraction. Interblock analysis of the Youden Square and the combination of inter- and intrablock information are exactly like the BIBD.

Intrablock analysis adjusts for rows and columns

Interblock analysis similar to BIBD

**Example 14.5**

## Lithium in blood

We wish to compare the blood concentrations of lithium 12 hours after administering lithium carbonate, using either a 300 mg capsule, 250 mg capsule, 450 mg time delay capsule, or 300 mg solution. There are twelve subjects, each of whom will be used twice, 1 week apart. We anticipate that the responses will be different in the second week, so we block on subject and week. The response is the serum lithium level as shown in Table 14.2 (data from Westlake 1974).

There are $g = 4$ treatments in $b = 12$ blocks of size $k = 2$, so that $r = 6$. We have $\lambda = 2$, $E = 2/3$, and each treatment appears three times in each week for a Youden Square.

The intrablock ANOVA for these data is shown in Listing 14.2. The residual plots (not shown) are passable. There is no evidence for a difference between the treatments 12 hours after administration. However, note that the

**Listing 14.2:** Minitab output for intrablock analysis of lithium data.

```
Source     DF    Seq SS     Adj SS     Seq MS      F      P
week        1  0.031974   0.031974   0.031974   15.79  0.004
subject    11  0.039344   0.029946   0.003577    1.77  0.215
treatmen    3  0.005603   0.005603   0.001868    0.92  0.473
Error       8  0.016203   0.016203   0.002025
```

mean square for the week blocking factor is fairly large. If we had ignored the week effect, we could anticipate an error mean square of

$$\frac{11 \times .0020253 + .031974}{12} = .00452 \ ,$$

more than doubling the error mean square in the Youden Square design.

## 14.3 Partially Balanced Incomplete Blocks

BIBD's are too big for some $g$ and $k$

BIBD's are great, but their balancing requirements may imply that the smallest possible BIBD for a given $g$ and $k$ is too big to be practical. For example, let's look for a BIBD for $g = 12$ treatments in incomplete blocks of size $k = 7$. To be a BIBD, $\lambda = r(k-1)/(g-1) = 6r/11$ must be a whole number; this implies that $r$ is some multiple of 11. In addition, $b = rg/k = (11 \times m) \times 12/7$ must be a whole number, and that implies that $b$ is a multiple of $11 \times 12 = 132$. So the smallest possible BIBD has $r = 77$, $b = 132$, and $N = 924$. This is a bigger experiment that we are likely to run.

PBIBD has $N = gr = bk$; some treatment pairs more frequent

Partially Balanced Incomplete Block Designs (PBIBD) allow us to run incomplete block designs with fewer blocks than may be required for a BIBD. The PBIBD has $g$ treatments and $b$ blocks of $k$ units each; each treatment is used $r$ times, and there is a total of $N = gr = bk$ units. The PBIBD does not have the requirement that each pair of treatments occurs together in the same number of blocks. This in turn implies that not all differences $\widehat{\alpha}_i - \widehat{\alpha}_j$ have the same variance in a PBIBD.

Sample PBIBD

Here is a sample PBIBD with $g = 12$, $k = 7$, $r = 7$, and $b = 12$. In this representation, each row is a block, and the numbers in the row indicate which treatments occur in that block.

| Block | Treatments | | | | | | |
|---|---|---|---|---|---|---|---|
| 1 | 1 | 2 | 3 | 4 | 5 | 8 | 10 |
| 2 | 2 | 3 | 4 | 5 | 6 | 9 | 11 |
| 3 | 3 | 4 | 5 | 6 | 7 | 10 | 12 |
| 4 | 1 | 4 | 5 | 6 | 7 | 8 | 11 |
| 5 | 2 | 5 | 6 | 7 | 8 | 9 | 12 |
| 6 | 1 | 3 | 6 | 7 | 8 | 9 | 10 |
| 7 | 2 | 4 | 7 | 8 | 9 | 10 | 11 |
| 8 | 3 | 5 | 8 | 9 | 10 | 11 | 12 |
| 9 | 1 | 4 | 6 | 9 | 10 | 11 | 12 |
| 10 | 1 | 2 | 5 | 7 | 10 | 11 | 12 |
| 11 | 1 | 2 | 3 | 6 | 8 | 11 | 12 |
| 12 | 1 | 2 | 3 | 4 | 7 | 9 | 12 |

We see, for example, that treatment 1 occurs three times with treatments 5 and 9, and four times with all other treatments.

The design rules for a PBIBD are fairly complicated:

Requirements for PBIBD

1. There are $g$ treatments, each used $r$ times. There are $b$ blocks of size $k < g$. Of course, $bk = gr$. No treatment occurs more than once in a block.
2. There are $m$ *associate classes*. Any pair of treatments that are $i$th associates appears together in $\lambda_i$ blocks. We usually arrange the $\lambda_i$ values in decreasing order, so that first associates appear together most frequently.

Associate classes

3. All treatments have the same number of $i$th associates, namely $\rho_i$.

$\rho_i$ $i$th associates

4. Let A and B be two treatments that are $i$th associates, and let $p^i_{jk}$ be the number of treatments that are $j$th associates of A and $k$th associates of B. This number $p^i_{jk}$ does not depend on the pair of $i$th associates chosen. In particular, $p^i_{jk} = p^i_{kj}$.

The PBIBD is partially balanced, because the variance of $\widehat{\alpha}_i - \widehat{\alpha}_j$ depends upon whether $i, j$ are first, second, or $m$th associates. The randomization of a PBIBD is just like that for a BIBD.

Randomize PBIBD like BIBD

Let's check the design given above and verify that it is a PBIBD. First note that $g = 12$, $k = 7$, $r = 7$, $b = 12$, and no treatment appears twice in a block. Next, there are two associate classes, with first associates appearing together four times and second associates appearing together three times. The pairs (1,5), (1,9), (2,6), (2,10), (3,7), (3,11), (4,8), (4,12), (5,9), (6,10), (7,11), and (8,12) are second associates; all other pairs are first associates. Each treatment has nine first associates and two second associates. For any pair of first associates, there are six other treatments that are first associates of both, four other treatments that are first associates of one and second associates of the other (two each way), and no treatments that are second associates of both. We thus have

$$\{p^1_{ij}\} = \begin{bmatrix} 6 & 2 \\ 2 & 0 \end{bmatrix} .$$

For any pair of second associates, there are nine treatments that are first associates of both, and one treatment that is a second associate of both, so that

$$\{p^2_{ij}\} = \begin{bmatrix} 9 & 0 \\ 0 & 1 \end{bmatrix} .$$

Thus all the design requirements are met, and the example design is a PBIBD.

Intrablock analysis is treatments adjusted for blocks

One historical advantage of the PBIBD was that the analysis could be done by hand. That is, there are relatively simple expressions for the various intra- and interblock analyses. With computers, that particular advantage is no longer very useful. The intrablock analysis of the PBIBD is simply treatments adjusted for blocks, as with the BIBD.

PBIBD less efficient on average than BIBD

The efficiency of a PBIBD is actually an average efficiency. The variance of $\widehat{\alpha}_i - \widehat{\alpha}_j$ depends on whether treatments $i$ and $j$ are first associates, second associates, or whatever. So to compute efficiency $E_{\text{PBIBD:RCB}}$, we divide the variance obtained in an RCB for a pairwise difference $(2\sigma^2/r)$ by the average of the variances of all pairwise differences in the PBIBD. There is an algorithm to determine $E_{\text{PBIBD:RCB}}$, but there is no simple formula. We can say that the efficiency will be less than $g(k-1)/[(g-1)k]$, which is the efficiency of a BIBD with the same block size and number of treatments.

There are several extensive catalogues of PBIBD's, including Bose, Clatworthy, and Shrikhande (1954) (376 separate designs) and Clatworthy (1973).

## 14.4 Cyclic Designs

Cyclic designs are simple

*Cyclic designs* are easily constructed incomplete block designs that permit the study of $g$ treatments in blocks of size $k$. We will only examine the simplest situation, where the replication $r$ for each treatment is a multiple of $k$, the block size. So $r = mk$, and $b = mg$ is the number of blocks. Cyclic designs include some BIBD and PBIBD designs.

Cycles of treatments

A cycle of treatments starts with an initial treatment and then proceeds through the subsequent treatments in order. Once we get to treatment $g$, we go back down to treatment 1 and start increasing again. For example, with seven treatments we might have the cycle (4, 5, 6, 7, 1, 2, 3).

Cyclic construction starts with an initial block and builds $g - 1$ more blocks from the initial block by replacing each treatment in the initial block by its successor in the cycle. Additional sets of $g$ blocks are constructed from new initial blocks. Thus all we need to know to build the design are the initial blocks.

Proceed through cycles from initial block

Write the initial block in a column, and write the cycles for each treatment in the initial block in rows, obtaining a $k$ by $g$ arrangement. The columns of this arrangement are the blocks. For example, suppose we have seven treatments and the initial block [1,4]. The cyclic design has blocks (columns):

| | | | | | | |
|---|---|---|---|---|---|---|
| 1 | 2 | 3 | 4 | 5 | 6 | 7 |
| 4 | 5 | 6 | 7 | 1 | 2 | 3 |

Each row is a cycle started by a treatment in the initial block. Cycles are easy, so cyclic designs are easy, once you have the initial block.

But wait, there's more! Not only do we have an incomplete block design with the columns as blocks, we have a complete block design with the rows as blocks. Thus cyclic designs are row orthogonal designs (and may be Youden Squares if the cyclic design is BIBD).

Cyclic designs are row orthogonal

Appendix C.3 contains a table of initial blocks for cyclic designs for $k$ from 2 through 10 and $g$ from 6 through 15. Several initial blocks are given for the smaller designs, depending on how many replications are required. For example, for $k = 3$ the table shows initial blocks for 3, 6, and 9 replications. Use the first initial block if $r = 3$, use the first and second initial blocks if $r = 6$, and use all three initial blocks if $r = 9$. For $g = 10$, $k = 3$, and $r = 6$, the initial blocks are (1,2,5) and (1,3,8), and the plan is

| | | | | | | | | | |
|---|---|---|---|---|---|---|---|---|---|
| 1 | 2 | 3 | 4 | 5 | 6 | 7 | 8 | 9 | 10 |
| 2 | 3 | 4 | 5 | 6 | 7 | 8 | 9 | 10 | 1 |
| 5 | 6 | 7 | 8 | 9 | 10 | 1 | 2 | 3 | 4 |

| | | | | | | | | | |
|---|---|---|---|---|---|---|---|---|---|
| 1 | 2 | 3 | 4 | 5 | 6 | 7 | 8 | 9 | 10 |
| 3 | 4 | 5 | 6 | 7 | 8 | 9 | 10 | 1 | 2 |
| 8 | 9 | 10 | 1 | 2 | 3 | 4 | 5 | 6 | 7 |

As with the PBIBD, there is an algorithm to compute the (average) efficiency of a cyclic design, but there is no simple formula. The initial blocks given in Appendix C.3 were chosen to make the cyclic designs as efficient as possible.

## 14.5 Square, Cubic, and Rectangular Lattices

Lattice designs for special $g, k$ combinations

Lattice designs work when the number of treatments $g$ and the size of the blocks $k$ follow special patterns. Specifically,

- A Square Lattice can be used when $g = k^2$.
- A Cubic Lattice can be used when $g = k^3$.
- A Rectangular Lattice can be used when $g = k(k + 1)$.

These lattice designs are resolvable and are most useful when we have a large number of treatments to be run in small blocks.

We illustrate the Square Lattice when $g = 9 = 3^2$. Arrange the nine treatments in a square; for example:

$$\begin{matrix} 1 & 2 & 3 \\ 4 & 5 & 6 \\ 7 & 8 & 9 \end{matrix}$$

A simple lattice has two replications made of rows and columns of the square

There is nothing special about this pattern; we could arrange the treatments in any way. The first replicate of the Square Lattice consists of blocks made up of the rows of the square: here (1, 2, 3), (4, 5, 6), and (7, 8, 9). The second replicate consists of blocks made from the columns of the square: (1, 4, 7), (2, 5, 8), and (3, 6, 9). A Square Lattice must have at least these two replicates to be connected, and a Square Lattice with only two replicates is called a *simple lattice*.

Triple lattice uses Latin Square for third replicate

We add a third replication using a Latin Square. A Square Lattice with three replicates is called a *triple lattice*. Here is a three by three Latin Square:

$$\begin{matrix} A & B & C \\ B & C & A \\ C & A & B \end{matrix}$$

Assign treatments to blocks using the letter patterns from the square. The three blocks of the third replicate are (1, 6, 8), (2, 4, 9), and (3, 5, 7).

Additional replicates use orthogonal Latin Squares

You can construct additional replicates for every Latin Square that is orthogonal to those already used. For example, the following square

$$\begin{matrix} A & B & C \\ C & A & B \\ B & C & A \end{matrix}$$

is orthogonal to the first one used. Our fourth replicate is thus (1, 5, 9), (2, 6, 7), and (3, 4, 8). Recall that there are no six by six Graeco-Latin Squares (six by six orthogonal Latin Squares), so only simple and triple lattices are possible for $g = 6^2$.

For $g = k^2$, there are at most $k-1$ orthogonal Latin Squares. The Square Lattice formed when $k-1$ Latin Squares are used has $k+1$ replicates; is called a *balanced lattice*; and is a BIBD with $g = k^2$, $b = k(k+1)$, $r = k+1$, $\lambda = 1$, and $E = k/(k+1)$. The BIBD plan for $g = 9$ treatments in $b = 12$ blocks of size $k = 3$, given in Appendix C, is exactly the balanced lattice constructed above.

Balanced Lattice ($k + 1$ replicates) is a BIBD

The (average) efficiency of a Square Lattice relative to an RCB is

$$E_{\text{SL:RCB}} = \frac{(k+1)(r-1)}{(k+1)(r-1)+r} \quad .$$

This is the best possible efficiency for any resolvable design.

The *Rectangular Lattice* is closely related to the Square Lattice. Arrange the $g = k(k+1)$ treatments in an $(k+1) \times (k+1)$ square with the diagonal blank, for example:

Rectangular Lattice is subset of a square

| | | | |
|---|---|---|---|
| • | 1 | 2 | 3 |
| 4 | • | 5 | 6 |
| 7 | 8 | • | 9 |
| 10 | 11 | 12 | • |

As with the Square Lattice, the first two replicates are formed from the rows and columns of this arrangement, ignoring the diagonal: (1, 2, 3), (4, 5, 6), (7, 8, 9), (10, 11, 12), (4, 7, 10), (1, 8, 11), (2, 5, 12), (3, 6, 9). Additional replicates are formed from the letters of orthogonal Latin Squares that satisfy the extra constraints that all the squares have the same diagonal and all letters appear on the diagonal; for example:

Rows, columns, and Latin Squares for a Rectangular Lattice

| | | | | | | | |
|---|---|---|---|---|---|---|---|
| A | B | C | D | A | C | D | B |
| C | D | A | B | B | D | C | A |
| D | C | B | A | C | A | B | D |
| B | A | D | C | D | B | A | C |

These squares are orthogonal and share the same diagonal containing all treatments. The next two replicates for this Rectangular Lattice design are thus (5, 9, 11), (1, 6, 10), (2, 4, 8), (3, 7, 12) and (6, 8, 12), (3, 4, 11), (1, 5, 7), (2, 9, 10).

The *Cubic Lattice* is a generalization the Square Lattice. In the Square Lattice, each treatment can be indexed by two subscripts $i, j$, with $1 \leq i \leq k$

and $1 \le j \le k$. The subscript $i$ indexes rows, and the subscript $j$ indexes columns. The first row in the Square Lattice is all those treatments with $i = 1$. The second column is all those treatments with $j = 2$. The blocks of the first replicate of a Square Lattice are rows; that is, treatments are the same block if they have the same $i$. The blocks of the second replicate of the Square Lattice are columns; that is, treatments are in the same block if they have the same $j$.

Cubic Lattice for $k^3$ treatments in blocks of $k$

For the Cubic Lattice, we have $g = k^3$ treatments that we index with three subscripts $i, j, l$, with $1 \le i \le k$, $1 \le j \le k$, and $1 \le l \le k$. Each replicate of the Cubic Lattice will be $k^2$ blocks of size $k$. In the first replicate of a Cubic Lattice, treatments are grouped so that all treatments in a block have the same values of $i$ and $j$. In the second replicate, treatments in the same block have the same values of $i$ and $l$, and in the third replicate, treatments in the same block have the same values of $j$ and $l$. For example, when $g = 8 = 2^3$, the cubic lattice will have four blocks of size two in each replicate. These blocks are as follows (using the $ijl$ subscript to represent a treatment):

Form blocks by keeping two subscripts constant

| Replicate 1 | Replicate 2 | Replicate 3 |
|---|---|---|
| (111, 112) | (111, 121) | (111, 211) |
| (121, 122) | (112, 122) | (112, 212) |
| (211, 212) | (211, 221) | (121, 221) |
| (221, 222) | (212, 222) | (122, 222) |

Cubic Lattice designs can have 3, 6, 9, and so forth replicates by repeating this pattern.

Treatments adjusted for blocks

The intrablock Analysis of Variance for a Square, Cubic, or Rectangular Lattice is analogous to that for the BIBD; namely, treatments should be adjusted for blocks.

## 14.6 Alpha Designs

Alpha Designs are resolvable with $g = mk$

Alpha Designs allow us to construct resolvable incomplete block designs when the number of treatments $g$ or block size $k$ does not meet the strict requirements for one of the lattice designs. Alpha Designs require that the number of treatments be a multiple of the block size $g = mk$, so that there are $m$ blocks per replication and $b = rm$ blocks in the complete design.

Three-step construction

We construct an Alpha Design in three steps. First we obtain the "generating array" for $k$, $m$, and $r$. This array has $k$ rows and $r$ columns. Next we expand each column of the generating array to $m$ columns using a cyclic pattern to obtain an "intermediate array" with $k$ rows and $mr$ columns. Finally we add $m$ to the second row of the intermediate array, $2m$ to the third row, and so on. Columns of the final array are blocks.

Finding the generating array

Section C.4 has generating arrays for $m$ from 5 to 15, $k$ at least four but no more than the minimum of $m$ and $100/m$, and $r$ up to four. The major division is by $m$, so first find the full array for your value of $m$. We only need the first $k$ rows and $r$ columns of this full tabulated array.

For example, suppose that we have $g = 20$ treatments and blocks of size $k = 4$, and we desire $r = 2$ replications. Then $m = 5$ and $b = 10$. The full generating array for $m = 5$ from Section C.4 is

$$\begin{array}{cccc} 1 & 1 & 1 & 1 \\ 1 & 2 & 5 & 3 \\ 1 & 3 & 4 & 5 \\ 1 & 4 & 3 & 2 \\ 1 & 5 & 2 & 4 \end{array}$$

We only need the first $k = 4$ rows and $r = 2$ columns, so our generating array is

$$\begin{array}{cc} 1 & 1 \\ 1 & 2 \\ 1 & 3 \\ 1 & 4 \end{array}$$

Construct intermediate array

Step two takes each column of the generating array and does cyclic substitution with $1, 2, \ldots, m$, to get $m$ columns. So, for our array, we get

$$\begin{array}{ccccc@{\qquad}ccccc} 1 & 2 & 3 & 4 & 5 & 1 & 2 & 3 & 4 & 5 \\ 1 & 2 & 3 & 4 & 5 & 2 & 3 & 4 & 5 & 1 \\ 1 & 2 & 3 & 4 & 5 & 3 & 4 & 5 & 1 & 2 \\ 1 & 2 & 3 & 4 & 5 & 4 & 5 & 1 & 2 & 3 \end{array}$$

The first five columns are from the first column of the generating array, and the last five columns are from the last column of the generating array. This is the intermediate array.

Add multiples of $m$ to rows

Finally, we take the intermediate array and add $m = 5$ to the second row, $2m = 10$ to the third row, and $3m = 15$ to the last row, obtaining

$$\begin{array}{ccccc@{\qquad}ccccc} 1 & 2 & 3 & 4 & 5 & 1 & 2 & 3 & 4 & 5 \\ 6 & 7 & 8 & 9 & 10 & 7 & 8 & 9 & 10 & 6 \\ 11 & 12 & 13 & 14 & 15 & 13 & 14 & 15 & 11 & 12 \\ 16 & 17 & 18 & 19 & 20 & 19 & 20 & 16 & 17 & 18 \end{array}$$

This is our final design, with columns being blocks and numbers indicating treatments.

The Alpha Designs constructed from the tables in Section C.4 are with a few exceptions the most efficient Alpha Designs possible. The average efficiencies for these Alpha Designs are very close to the theoretical upper bound for average efficiency of a resolvable design, namely

$$E_{\alpha:\text{RCB}} \leq \frac{(g-1)(r-1)}{(g-1)(r-1)+r(m-1)} .$$

## 14.7 Further Reading and Extensions

Incomplete block designs have been the subject of a great deal of research and theory; we have mentioned almost none of it. Two excellent sources for more theoretical discussions of incomplete blocks are John (1971) and John and Williams (1995). Among the topics relevant to this chapter, John (1971) describes recovery of interblock information for BIBD, PBIBD, and general incomplete block designs; existence and construction of BIBD's; classification, existence, and construction of PBIBD's; and efficiency. John and Williams (1995) is my basic reference for Cyclic Designs, Alpha Designs, and incomplete block efficiencies; and it has a good deal to say about row column designs, interblock information, and other topics as well.

Most of the designs described in this chapter are not recent. Many of these incomplete block designs were introduced by Frank Yates in the late 1930's, including BIBD's (Yates 1936a), Square Lattices (Yates 1936b), and Cubic Lattices (Yates 1939), as well other designs such as Lattice Squares (different from a Square Lattice, Yates 1940). PBIBD's first appear in Bose and Nair (1939). Alpha Designs are the relative newcomers, first appearing in Patterson and Williams (1976).

John and Williams (1995) provide a detailed discussion of the efficiencies of incomplete block designs, including a proof that the BIBD has the highest possible efficiency for equally replicated designs with equal block sizes. Section 3.3 of their book gives an expression for the efficiency of a cyclic design; Sections 2.8 and 4.10 give a variety of upper bounds for the efficiencies of blocked designs and resolvable designs. Chapter 12 of John (1971) and Chapter 1 of Bose, Clatworthy, and Shrikhande (1954) describe efficiency of PBIBD's.

Some experimental situations will not fit into any of the standard design categories. For example, different treatments may have different replication, or blocks may have different sizes. Computer software exists that will search for "optimal" allocations of the treatments to units. *Optimal* can be defined in several ways; for example, you could choose to minimize the average variance for pairwise comparisons. See Silvey (1980) and Cook and Nachtsheim (1989).

## 14.8 Problems

**Exercise 14.1**

Consider the following incomplete block experiment with nine treatments (A-I) in nine blocks of size three.

| Block | | | | | | | | |
|---|---|---|---|---|---|---|---|---|
| 1 | 2 | 3 | 4 | 5 | 6 | 7 | 8 | 9 |
| C 54 | B 35 | A 48 | G 46 | D 61 | C 52 | A 54 | B 45 | A 31 |
| H 56 | G 36 | G 42 | H 56 | E 61 | I 53 | H 59 | I 46 | B 28 |
| D 53 | D 40 | E 43 | I 59 | F 54 | E 48 | F 62 | F 47 | C 25 |

(a) Identify the type of design.
(b) Analyze the data for differences between the treatments.

**Exercise 14.2**

Chemical yield may be influenced by the temperature, pressure, and/or time in the reactor vessel. Each of these factors may be set at a high or a low level. Thus we have a $2^3$ experiment. Unfortunately, the process feedstock is highly variable, so batch to batch differences in feedstock are expected; we must start with new feedstock every day. Furthermore, each batch of feedstock is only big enough for seven runs (experimental units). We have enough money for eight batches of feedstock. We decide to use a BIBD, with each of the eight factor-level combinations missing from one of the blocks.

Give a skeleton ANOVA (source and degrees of freedom only), and describe an appropriate randomization scheme.

**Exercise 14.3**

Briefly describe the following incomplete block designs (BIBD, or PBIBD with what associate classes, and so on).

(a)

| Block | 1 | 2 | 3 | 4 |
|---|---|---|---|---|
| | A | A | B | A |
| | B | C | C | B |
| | C | D | D | D |

(b)

| Block | 1 | 2 | 3 | 4 | 5 |
|---|---|---|---|---|---|
| | A | A | A | B | C |
| | B | B | C | D | D |
| | C | D | E | E | E |

(c)

| Block | 1 | 2 | 3 | 4 |
|---|---|---|---|---|
| | 1 | 3 | 1 | 2 |
| | 2 | 4 | 3 | 4 |

**Exercise 14.4** We wish to compare the average access times of five brands of half-height computer disk drives (denoted A through E). We would like to block on the computer in which they are used, but each computer will only hold four drives. Average access times and the design are given in the following table (data from Nelson 1993):

| Computer | | | | |
|---|---|---|---|---|
| 1 | 2 | 3 | 4 | 5 |
| A 35 | A 41 | B 40 | A 32 | A 40 |
| B 42 | B 45 | C 42 | C 33 | B 38 |
| C 31 | D 32 | D 33 | D 35 | C 35 |
| D 30 | E 40 | E 39 | E 36 | E 37 |

Analyze these data and report your findings, including a description of the design.

**Problem 14.1** Japanese beetles ate the Roma beans in our garden last year, so we ran an experiment this year to learn the best pesticide. We have six garden beds with beans, and the garden store has three different sprays that claim to keep the beetles off the beans. Sprays drift on the wind, so we cannot spray very small areas. We divide each garden bed into two plots and use a different spray on each plot. Below are the numbers of beetles per plot.

| Bed | | | | | |
|---|---|---|---|---|---|
| 1 | 2 | 3 | 4 | 5 | 6 |
| 19 A | 9 A | 25 B | 9 A | 26 A | 13 B |
| 21 B | 16 C | 30 C | 11 B | 33 C | 18 C |

Analyze these data to determine the effects of sprays. Which one should we use?

**Problem 14.2** Milk can be strained through filter disks to remove dirt and debris. Filters are made by surface-bonding fiber webs to both sides of a disk. This experiment is concerned with how the construction of the filter affects the speed of milk flow through the filter.

We have a $2^4$ factorial structure for the filters. The factors are fiber weight (normal or heavy), loft (thickness of the filter, normal or low), bonding solution on bottom surface (A or B), and bonding solution on top surface (A or B). Note the unfortunate fact that the "high" level of the second factor, loft, is low loft. Treatments 1 through 16 are the factor-level combinations in standard order.

These are speed tests, so we pour a measured amount of milk through the disk and record the filtration time as the response. We expect considerable variation from farm to farm, so we block on farm. We also expect variation from milking to milking, so we want all measurements at one farm to be done at a single milking. However, only three filters can be satisfactorily used at a single milking. Thus we must use incomplete blocks of size three.

Sixteen farms were selected. At each farm there will be three strainings at one milking, with the milk strained first with one filter, then a second, then a third. Each treatment will be used three times in the design: once as a first filter, once as second, and once as third. The treatments and responses for the experiment are given below (data from Connor 1958):

Treatments and Responses

| | Filtration time | | | | | |
|---|---|---|---|---|---|---|
| Farm | First | | Second | | Third | |
| 1 | 10 | 451 | 7 | 457 | 16 | 343 |
| 2 | 11 | 260 | 8 | 418 | 13 | 320 |
| 3 | 12 | 464 | 5 | 317 | 14 | 315 |
| 4 | 9 | 306 | 6 | 462 | 15 | 291 |
| 5 | 13 | 381 | 4 | 597 | 6 | 491 |
| 6 | 14 | 362 | 1 | 325 | 7 | 449 |
| 7 | 15 | 292 | 2 | 402 | 8 | 576 |
| 8 | 16 | 431 | 3 | 477 | 5 | 394 |
| 9 | 7 | 329 | 9 | 261 | 4 | 430 |
| 10 | 8 | 389 | 10 | 413 | 1 | 272 |
| 11 | 5 | 368 | 11 | 244 | 2 | 447 |
| 12 | 6 | 398 | 12 | 517 | 3 | 354 |
| 13 | 2 | 490 | 16 | 311 | 9 | 278 |
| 14 | 3 | 467 | 13 | 429 | 10 | 486 |
| 15 | 4 | 735 | 14 | 642 | 11 | 474 |
| 16 | 1 | 402 | 15 | 380 | 12 | 589 |

What type of design is this? Analyze the data and report your findings on the influence of the treatment factors on straining time.

**Problem 14.3**

The State Board of Education has adopted basic skills tests for high school graduation. One of these is a writing test. The student writing samples are graded by professional graders, and the board is taking some care to be sure that the graders are grading to the same standard. We examine grader differences with the following experiment. There are 25 graders. We select 30 writing samples at random; each writing sample will be graded by five graders. Thus each grader will grade six samples, and each pair of graders will have a test in common.

| Exam | Grader | Score | Exam | Grader | Score |
|---|---|---|---|---|---|
| 1 | 1 2 3 4 5 | 60 59 51 64 53 | 16 | 1 9 12 20 23 | 61 67 69 68 65 |
| 2 | 6 7 8 9 10 | 64 69 63 63 71 | 17 | 2 10 13 16 24 | 78 75 76 75 72 |
| 3 | 11 12 13 14 15 | 84 85 86 85 83 | 18 | 3 6 14 17 25 | 67 72 72 75 76 |
| 4 | 16 17 18 19 20 | 72 76 77 74 77 | 19 | 4 7 15 18 21 | 84 81 76 79 77 |
| 5 | 21 22 23 24 25 | 65 73 70 71 70 | 20 | 5 8 11 19 22 | 81 84 85 84 81 |
| 6 | 1 6 11 16 21 | 52 54 62 54 55 | 21 | 1 8 15 17 24 | 70 65 61 66 66 |
| 7 | 2 7 12 17 22 | 56 51 52 57 51 | 22 | 2 9 11 18 25 | 84 82 86 85 86 |
| 8 | 3 8 13 18 23 | 55 60 59 60 61 | 23 | 3 10 12 19 21 | 72 85 77 82 79 |
| 9 | 4 9 14 19 24 | 88 76 77 77 74 | 24 | 4 6 13 20 22 | 85 75 78 82 83 |
| 10 | 5 10 15 20 25 | 65 68 72 74 77 | 25 | 5 7 14 16 23 | 58 64 58 57 58 |
| 11 | 1 10 14 18 22 | 79 77 77 77 79 | 26 | 1 7 13 19 25 | 66 71 73 70 70 |
| 12 | 2 6 15 19 23 | 70 66 63 62 66 | 27 | 2 8 14 20 21 | 73 67 63 70 66 |
| 13 | 3 7 11 20 24 | 48 49 51 48 50 | 28 | 3 9 15 16 22 | 58 70 69 61 71 |
| 14 | 4 8 12 16 25 | 75 64 75 68 65 | 29 | 4 10 11 17 23 | 95 84 88 88 87 |
| 15 | 5 9 13 17 21 | 79 77 81 79 83 | 30 | 5 6 12 18 24 | 47 47 51 49 56 |

Analyze these data to determine if graders differ, and if so, how. Be sure to describe the design.

**Problem 14.4** Thirty consumers are asked to rate the softness of clothes washed by ten different detergents, but each consumer rates only four different detergents. The design and responses are given below:

| | Trts | Softness | | Trts | Softness |
|---|---|---|---|---|---|
| 1 | A B C D | 37 23 37 41 | 16 | A B C D | 52 41 45 48 |
| 2 | A B E F | 35 32 39 37 | 17 | A B E F | 46 42 45 42 |
| 3 | A C G H | 39 45 39 41 | 18 | A C G H | 44 43 41 36 |
| 4 | A D I J | 44 42 46 44 | 19 | A D I J | 32 42 36 29 |
| 5 | A E G I | 44 44 45 50 | 20 | A E G I | 43 42 44 44 |
| 6 | A F H J | 55 45 53 49 | 21 | A F H J | 46 41 43 45 |
| 7 | B C F I | 47 50 48 52 | 22 | B C F I | 43 51 40 42 |
| 8 | B D G J | 37 42 40 37 | 23 | B D G J | 38 37 36 34 |
| 9 | B E H J | 32 34 39 29 | 24 | B E H J | 40 49 43 44 |
| 10 | B G H I | 36 41 39 43 | 25 | B G H I | 23 20 27 29 |
| 11 | C E I J | 45 44 40 36 | 26 | C E I J | 46 49 48 43 |
| 12 | C F G J | 42 38 39 39 | 27 | C F G J | 48 43 48 41 |
| 13 | C D E H | 47 48 46 47 | 28 | C D E H | 35 35 31 26 |
| 14 | D E F G | 43 47 48 41 | 29 | D E F G | 45 47 47 42 |
| 15 | D F H I | 39 32 32 31 | 30 | D F H I | 43 39 38 39 |

Analyze these data for treatment effects and report your findings.

Briefly describe the experimental design you would choose for each of the following situations, and why. **Problem 14.5**

(a) Competition cuts tree growth rate, so we wish to study the effects on tree growth of using four herbicides on the competition. There are many study sites available, but each site is only large enough for three plots. Resources are available for 24 plots (that is, eight sites with three plots per site). Large site differences are expected.

(b) We use 2-inch wide tape to seal moving cartons, and we want to find the brand that seals best. The principal problem is not the tape breaking, but the tape pulling away from the cardboard. Unfortunately, there is considerable variation from carton to carton in the ability of any tape to adhere to the cardboard. There are four brands of tape available. The test is to seal a box bottom with four strips of tape of one or more types, place the carton so that only the edges are supported, drop 50 pounds of old *National Geographics* into the carton from a height of one foot, and then measure the length of tape that pulled away from the cardboard. There is a general tendency for tape to pull away more in the center of the carton than near its ends. Our cheap boss has given us only sixteen boxes to ruin in this destructive fashion before deciding on a tape. Tape placement on the bottom looks like this:

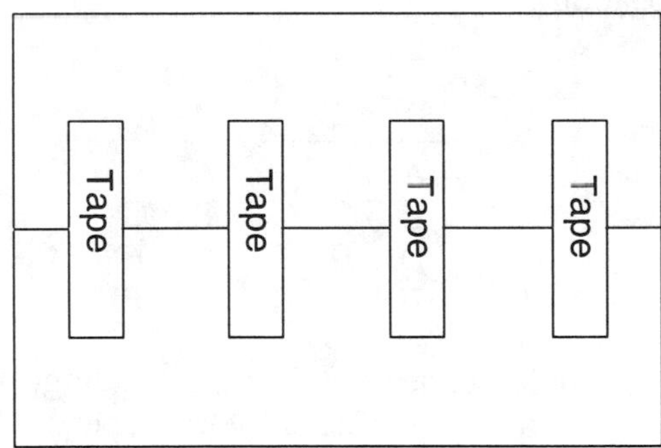


(c) Three treatments are being studied for the rehabilitation of acidified lakes. Unfortunately, there is tremendous lake to lake variability, and we only have six lakes on which we are allowed to experiment. We may treat each lake as a whole, or we may split each lake in two using a plastic "curtain" and treat the halves separately. Sadly, the technology does not allow us to split each lake into three.

(d) A retail bookstore has two checkouts, and thus two checkout advertising displays. These displays are important for enticing impulse purchases, so the bookstore would like to know which of the four types of displays available will lead to the most sales. The displays will be left up for one week, because it is expensive to change displays and you really need a full week to get sufficient volume of sales and overcome day-of-week effects; there are, however, week to week differences in sales. The store wishes to complete the comparison in at most 8 and preferably fewer weeks.

(e) We wish to compare four "dog collars." The thought is that some collars will lead to faster obedience than others. The response we measure will be the time it takes a dog to complete a walking course with lots of potential distractions. We have 24 dogs that can be used, and we expect large dog to dog variability. Dogs can be used more than once, but if they are used more than once there should be at least 1 week between trials. Our experiment should be completed in less than 3 weeks, so no dog could possibly be used more than three times.

**Problem 14.6** For each of the following, describe the experimental design that was used, and give a skeleton ANOVA.

(a) Plant breeders wish to study six varieties of corn. They have 24 plots available, four in each of six locations. The varieties are assigned to location as follows (there is random assignment of varieties to plot within location):

| Locations | | | | | |
|---|---|---|---|---|---|
| 1 | 2 | 3 | 4 | 5 | 6 |
| A | B | A | A | B | A |
| B | C | C | B | C | C |
| D | E | D | D | E | D |
| E | F | F | E | F | F |

(b) We wish to study gender bias in paper grading. We have 12 "lower" level papers and 12 "advanced" level papers. There are four paid graders who do not know the students or their names. Each paper is submitted for grading exactly once (that is, no paper is graded by more than one grader). We examine gender bias by the name put on the paper: either a male first name, a female first name, or just initials. The twelve lower-level papers are assigned at random to the combinations of grader and name gender, as are the advanced-level papers. The response we measure is the grade given (on a 0-100 scale).

(c) Song bird abundance can be measured by sending trained observers to a site to listen for the calls of the birds and make counts. Consider an experiment on the effects of three different forest harvesting techniques on bird abundance. There are six forests and two observers, and there will be two harvests in each of the six forests. The harvest techniques were assigned in the following way:

| | Forest | | | | | |
|---|---|---|---|---|---|---|
| Observer | 1 | 2 | 3 | 4 | 5 | 6 |
| 1 | A | C | B | B | A | C |
| 2 | C | A | A | C | B | B |

(d) Wafer board is a manufactured wood product made from wood chips. One potential problem is warping. Consider an experiment where we compare three kinds of glue and two curing methods. All six combinations are used four times, once for each of four different batches of wood chips. The response is the amount of warping.

**Question 14.1**

When recovering interblock information in a BIBD, we take the weighted average of intra- and interblock estimates

$$\bar{\zeta} = \lambda\hat{\zeta} + (1 - \lambda)\tilde{\zeta} \ .$$

Suppose that $\sigma^2 = \sigma_\beta^2 = 1$, $g = 8$, $k = 7$, and $b = 8$. Find the mean and standard deviation of $1/\lambda$. Do you feel that $\lambda$ is well determined?

# Chapter 15

# Factorials in Incomplete Blocks—Confounding

We may use the complete or incomplete block techniques of the last two chapters when treatments have factorial structure; just consider that there are $g = abc$ treatments and proceed as usual. However, there are some incomplete block techniques that are specialized for factorial treatment structure. We consider these factorial-specific methods in this chapter and the next.

This chapter describes *confounding* as a design technique. A design with confounding is unable to distinguish between some treatment comparisons and other sources of variation. For example, if the experimental drug is only given to patients with advanced symptoms, and the standard therapy is given to other patients, then the treatments are confounded with patient population. We usually go to great lengths to avoid confounding, so why would we deliberately introduce confounding into an experiment?

Use confounding in design

Incomplete blocks are less efficient than complete blocks; we always lose some information when we use incomplete blocks instead of complete blocks. Thus the issue with incomplete blocks is not whether we lose information, but how much information we lose, and which particular comparisons lose information. Incomplete block designs like the BIBD and PBIBD spread the inefficiency around every comparison. Confounded factorials allow us to isolate the inefficiency of incomplete blocks in particular contrasts that we specify at design time and retain full efficiency for all other contrasts.

Confounding isolates incomplete block inefficiency

Let's restate that. With factorial treatment structure we are usually more interested in main effects and low-order interactions than we are in multi-factor interactions. Confounding designs will allow us to isolate the inef-

Put inefficiency in interactions

**Table 15.1:** All contrasts and grand mean for a $2^3$ design.

| | I | A | B | C | AB | AC | BC | ABC |
|---|---|---|---|---|---|---|---|---|
| (1) | + | − | − | − | + | + | + | − |
| a | + | + | − | − | − | − | + | + |
| b | + | − | + | − | − | + | − | + |
| ab | + | + | + | − | + | − | − | − |
| c | + | − | − | + | + | − | − | + |
| ac | + | + | − | + | − | + | − | − |
| bc | + | − | + | + | − | − | + | − |
| abc | + | + | + | + | + | + | + | + |

ficiency of incomplete blocks in the multi-factor interactions and have full efficiency for main effects and low-order interactions.

## 15.1 Confounding the Two-Series Factorial

Letter or digit labels for factor-level combinations

Let's begin with a review of some notation and facts from Chapter 10. The $2^k$ factorial has $k$ factors, each at two levels for a total of $g = 2^k$ treatments. There are two common ways to denote factor-level combinations. First is a lettering method. Let (1) denote all factors at their low level. Otherwise, denote a factor-level combination by including (lower-case) letters for all factors at their high levels. Thus *bc* denotes factors B and C at their high levels and all other factors are their low levels. Second, there is a numbering method. Each factor-level combination is denoted by a $k$-tuple, with a 1 for each factor at the high level and a 0 for each factor at the low level. For example, in a $2^3$, *bc* corresponds to 011. To refer to individual factors, let $x_A$ be the level of A, and so on, so that $x_A = 0$, $x_B = 1$, and $x_C = 1$ in 011.

Standard order

Standard order for a two-series design arranges the factor-level combinations in a specific order. Begin with (1). Then proceed through the remainder of the factor-level combinations with factor A varying fastest, then factor B, and so on. In a $2^3$, the standard order is (1), *a*, *b*, *ab*, *c*, *ac*, *bc*, *abc*.

Table of + and −

Each main effect and interaction in a two-series factorial is a single degree of freedom and can be described with a single contrast. It is customary to use contrast coefficients of +1 and −1, and the contrast is often represented as a set of plus and minus signs, one for each factor-level combination. The full table of contrasts for a $2^3$ is shown in Table 15.1, which also includes a column of all + signs corresponding to the grand mean.

1. Choose a factorial effect to confound with blocks and get its contrast.
2. Put all factor-level combinations with a plus sign in the contrast in one block and all the factor-level combinations with a minus sign in the other block.

**Display 15.1:** Steps to confound a $2^k$ design into two blocks.

The $2^k$ factorial can be confounded into two blocks of size $2^{k-1}$ or four blocks of $2^{k-2}$, and so on, to $2^q$ blocks of size $2^{k-q}$ in general. Let's begin with just one replication of the experiment confounded in two blocks of size $2^{k-1}$; we look at smaller blocks and additional replication later.

$2^q$ blocks of size $2^{k-q}$

### 15.1.1 Two blocks

Confounding a $2^k$ design into two blocks of size $2^{k-1}$ is simple; the steps are given in Display 15.1. Every factorial effect corresponds to a contrast with $2^{k-1}$ plus signs and $2^{k-1}$ minus signs. Choose a factorial effect to confound with blocks; this is the *defining contrast*. Put all factor-level combinations with a plus sign on the defining contrast in one block and all the factor-level combinations with a minus sign in the other block. This confounds the block difference with the defining contrast effect, so we have zero information on that effect. However, all factorial effects are orthogonal, so block differences are orthogonal to the unconfounded factorial effects, and we have complete information and full efficiency for all unconfounded factorial effects.

Confound defining contrast with blocks

It makes sense to choose as defining contrast a multifactor interaction, because multifactor interactions are generally of less interest, and we will lose all information about whatever contrast is used as defining contrast. For the $2^k$ factorial in two blocks of size $2^{k-1}$, the obvious defining contrast is the $k$-factor interaction.

Use $k$-factor interaction as defining contrast

**Example 15.1**

#### $2^3$ in two blocks of size four

Suppose that we wish to confound a $2^3$ into two blocks of size four. We use the ABC interaction as the defining contrast, because it is the highest-order interaction. The pattern of plus and minus signs is the last column of Table 15.1. The four factor-level effects with minus signs are (1), $ab$, $ac$, and $bc$; the four factor-level effects with plus signs are $a$, $b$, $c$, and $abc$. Thus the two blocks are

| (1) | $a$ |
|---|---|
| $ab$ | $b$ |
| $ac$ | $c$ |
| $bc$ | $abc$ |

Alternative methods for finding blocks

This idea of finding the contrast pattern for a defining contrast to confound into two blocks works for any two-series design, but finding the pattern becomes tedious for large designs. For example, dividing a $2^6$ into two blocks of 32 with ABCDEF as defining contrast requires finding the ABCDEF contrast, which is the product of the six main-effects contrasts. Here are two equivalent procedures that you may find easier, though which method you like best is entirely a personal matter.

Even/odd rule and 0/1 rule

First is the "even/odd" rule. Examine the letter designation for every factor-level combination. Divide the factor-level combinations into two groups depending on whether the letters of a factor-level combination contain an even or odd number of letters from the defining contrast. The second approach is the "0/1" rule. Now we work with the numerical 0/1 designations for the factor-level combinations. What we do is compute for each factor-level combination the sum of the 0/1 level indicators for the factors that appear in the defining contrast, and then reduce this modulo 2. (Reduction modulo 2 subtracts any multiples of 2; 0 stays 0, 1 stays 1, 2 becomes 0, 3 becomes 1, and so on.) For the defining contrast ABC, we compute

$$L = x_A + x_B + x_C \bmod 2 \;;$$

those factor-level combinations that yield an $L$ value of 0 go in one block, and those that yield a 1 go in the second block. It is not too hard to see that this 0/1 rule is just the even/odd rule in numerical form.

## Example 15.2 $2^4$ in two blocks of eight

Suppose that we have a $2^4$ that we wish to block into two blocks using BCD as the defining contrast. To choose blocks using the even/odd rule, we first find the letters from each factor-level combination that appear in the defining contrast, as shown in Table 15.2. We then count whether there is an even or odd number of these letters and put the factor-level combinations with an even number of letters matching in one block and those with an odd number matching in a second block. For example, the combination $ac$ has one letter in BCD, so $ac$ goes in the odd group; and the combination $bc$ has two letters in BCD, so it goes in the even group. Note that we would not ordinarily use

**Table 15.2:** Confounding a $2^4$ with defining contrast BCD using the even/odd rule.

| | Matches | Even/odd | Block 1 | Block 2 |
|---|---|---|---|---|
| (1) | none | even | (1) | $b$ |
| $a$ | none | even | $a$ | $ab$ |
| $b$ | B | odd | $bc$ | $c$ |
| $ab$ | B | odd | $abc$ | $ac$ |
| $c$ | C | odd | $bd$ | $d$ |
| $ac$ | C | odd | $abd$ | $ad$ |
| $bc$ | BC | even | $cd$ | $bcd$ |
| $abc$ | BC | even | $acd$ | $abcd$ |
| $d$ | D | odd | | |
| $ad$ | D | odd | | |
| $bd$ | BD | even | | |
| $abd$ | BD | even | | |
| $cd$ | CD | even | | |
| $acd$ | CD | even | | |
| $bcd$ | BCD | odd | | |
| $abcd$ | BCD | odd | | |

BCD as the defining contrast; we use it here for illustration to show that even and odd is not simply the number of letters in a factor-level combination, but the number in that combination that occur in the defining contrast.

To use the 0/1 rule, we start by computing $x_B + x_C + x_D$. We then reduce the sum modulo 2, and assign the zeroes to one block and the ones to a second block. For 0111 ($bcd$), this sum is $1+1+1 = 3$, and 3 mod 2 $= 1$; for 1110 ($abc$), the sum is $1+1+0 = 2$, and 2 mod 2 $= 0$. Table 15.3 shows the results of the 0/1 rule for our example.

Principal block and alternate block

The block containing (1) or 0000 is called the *principal block*. The other block is called the *alternate block*. These blocks have some nice mathematical properties that we will find useful in more complicated confounding situations. Consider the following modified multiplication which we will denote by $\odot$. Let (1) act as an identity—anything multiplied by (1) is just itself. So $a \odot (1) = a$ and $bcd \odot (1) = bcd$. For any other pair of factor-level combinations, multiply as usual but then reduce exponents modulo 2. Thus $a \odot ab = a^2b = a^0b = b$, and $a \odot a = a^2 = a^0 = (1)$.

Multiply and reduce exponents mod 2

There is an analogous operation we can perform with the 0/1 representation of the factor-level combinations. Think of the zeroes and ones as exponents; for example, 1101 corresponds to $a^1b^1c^0d^1 = abd$. Exponents add when we multiply, so the corresponding operation is to add the zeroes and ones componentwise and then reduce them mod 2. Thus $abd \odot acd = a^2bcd^2 = bc$ corresponds to $1101 \oplus 1011 = 2112 = 0110$. Personally, I prefer the letters, but some people prefer the numbers.

Get alternate blocks from principal block

Here are the useful mathematical properties. If you multiply any two elements of the principal block together reducing exponents modulo two, you get another element of the principal block. If you multiply all elements of the principal block by an element not in the principal block, you get an alternate block. What this means is that you can find alternate blocks easily once you have the principal block. This is no big deal when there are only two blocks, but can be very useful when we have four, eight, or more blocks.

**Example 15.3** **$2^4$ in two blocks of eight, continued**

In our $2^4$ example with BCD as the defining contrast, $ac$ is not in the principal block. Multiplying every element of the principal block by $ac$, we get the following

$$\begin{aligned}(1) \odot ac &= ac &&= ac\\ a \odot ac &= a^2c &&= c\\ bc \odot ac &= abc^2 &&= ab\\ abc \odot ac &= a^2bc^2 &&= b\\ bd \odot ac &= abcd &&= abcd\\ abd \odot ac &= a^2bcd &&= bcd\\ cd \odot ac &= ac^2d &&= ad\\ acd \odot ac &= a^2c^2d &&= d\end{aligned}$$

This is the alternate block, but in a different order than Table 15.2.

### 15.1.2 Four or more blocks

Use $q$ defining contrasts for $2^q$ blocks

A single replication of a $2^k$ design can be confounded into two blocks, four blocks, eight blocks, and so on. The last subsection showed how to confound into two blocks using one defining contrast. We can confound into four blocks using two defining contrasts, and in general we can confound into $2^q$ blocks using $q$ defining contrasts. Let's begin with four blocks.

Choose defining contrasts carefully

Start by choosing two defining contrasts for confounding a $2^4$ design into four blocks of size four. It turns out that choosing these defining contrasts is very important, and bad choices lead to poor designs. We will use ABC and BCD as defining contrasts; these are good choices. Later on we will see what can happen with bad choices.

**Table 15.3:** Confounding a $2^4$ with defining contrast BCD using the 0/1 rule.

| | $x_B + x_C + x_D$ | Reduced mod 2 | Block 1 | Block 2 |
|---|---|---|---|---|
| 0000 | 0 | 0 | 0000 | 0100 |
| 1000 | 0 | 0 | 1000 | 1100 |
| 0100 | 1 | 1 | 0110 | 0010 |
| 1100 | 1 | 1 | 1110 | 1010 |
| 0010 | 1 | 1 | 0101 | 0001 |
| 1010 | 1 | 1 | 1101 | 1001 |
| 0110 | 2 | 0 | 0011 | 0111 |
| 1110 | 2 | 0 | 1011 | 1111 |
| 0001 | 1 | 1 | | |
| 1001 | 1 | 1 | | |
| 0101 | 2 | 0 | | |
| 1101 | 2 | 0 | | |
| 0011 | 2 | 0 | | |
| 1011 | 2 | 0 | | |
| 0111 | 3 | 1 | | |
| 1111 | 3 | 1 | | |

Each defining contrast divides the factor-level combinations into evens and odds (or ones and zeroes). If we look at those factor-level combinations that are even for BCD, half of them will be even for ABC and the other half will be odd for ABC. Similarly, those combinations that are odd for BCD are evenly split between even and odd for ABC. Our blocks will be formed as those combinations that are even for both ABC and BCD, those that are odd for both ABC and BCD, those that are even for ABC and odd for BCD, and those that are odd for ABC and even for BCD. Table 15.4 shows the results of confounding on ABC and BCD. Alternatively, we compute $L_1$ and $L_2$ for the two defining contrasts, and take as blocks those combinations that are zero on both, one on both, zero on the first and one on the second, and zero on the second and one on the first.

Combinations of defining contrasts form blocks

We have confounded into four blocks, so there are 3 degrees of freedom between blocks. We know that the two defining contrasts are confounded with block differences, but what is the third degree of freedom that is confounded with block differences? The ABC contrast is constant (plus or minus 1) within each block, and the BCD contrast is also constant within each block. Therefore, their product is constant within each block. Recall that each contrast is formed as the product of the corresponding main-effect contrasts, so the product of the ABC and BCD contrasts must be the contrast for

**Table 15.4:** Confounding the $2^4$ into four blocks using ABC and BCD as defining contrasts.

| | ABC | BCD |
|---|---|---|
| (1) | even | even |
| $a$ | odd | even |
| $b$ | odd | odd |
| $ab$ | even | odd |
| $c$ | odd | odd |
| $ac$ | even | odd |
| $bc$ | even | even |
| $abc$ | odd | even |
| $d$ | even | odd |
| $ad$ | odd | odd |
| $bd$ | odd | even |
| $abd$ | even | even |
| $cd$ | odd | even |
| $acd$ | even | even |
| $bcd$ | even | odd |
| $abcd$ | odd | odd |

| | BCD even | BCD odd |
|---|---|---|
| ABC even | (1), $bc$, $abd$, $acd$ | $ab$, $ac$, $d$, $bcd$ |
| ABC odd | $a$, $abc$, $bd$, $cd$ | $b$, $c$, $ad$, $abcd$ |

Generalized interactions of defining contrasts are confounded

$AB^2C^2D = AD$. Squared terms disappear because their elements are all ones. The term AD is called the *generalized interaction* of ABC and BCD. When we confound into four blocks using two defining contrasts, we not only confound the defining contrasts with blocks, we also confound their generalized interaction. If you examine the blocks in Table 15.4, you will see that two of them always have exactly one of $a$ or $d$, and the other two always have both or neither.

Note that if we had chosen AD and ABC as our defining contrasts, we would get the same four blocks, and the generalized interaction BCD would also be confounded with blocks.

Check generalized interactions when choosing defining contrasts

This fact that we also confound the generalized interaction explains why we need to be careful when choosing defining contrasts. It is very tempting to use the intuition that we want to confound interactions with as high an order as possible, so we choose, say, ABCD and BCD as generators. This intuition leads to disaster, because the generalized interaction of ABCD and BCD is A, and we would thus confound a main effect with blocks.

When choosing defining contrasts, we need to look at the full set of effects that are confounded with blocks. We want first to find a set such that the lowest-order term confounded with blocks is as high an order as possible. Among all the sets that meet the first criterion, we want sets that have as few low-order terms as possible. For example, consider the sets (A, BCD, ABCD), (ABC, BCD, AD), and (AB, CD, ABCD). We prefer the second and third sets to the first, because the first confounds a main effect, and the second and third confound two-factor interactions. We prefer the second set to the third, because the second set confounds only one two-factor interaction, while the third set confounds two two-factor interactions.

We want as few lower order interactions confounded as possible

Section C.5 suggests defining contrasts and their generalized interactions for two-series designs with up to eight factors.

Confounding plans

Use three defining contrasts to get eight blocks. These defining contrasts must be independent of each other, in the sense that none of them is the generalized interaction the other two. Thus we cannot use ABC, BCD, and AD as three defining contrasts to get eight blocks, because AD is the generalized interaction of ABC and BCD. Divide the factor-level combinations into eight groups using the even/odd patterns of the three defining contrasts: (even, even, even), (even, even, odd), (even, odd, even), (even, odd, odd), (odd, even, even), (odd, even, odd), (odd, odd, even), and (odd, odd, odd). There are eight blocks, so there must be 7 degrees of freedom between them. The three defining contrasts are confounded with blocks, as are their three two-way generalized interactions and their three-way generalized interaction, for a total of 7 degrees of freedom.

We again note that once you have the principal block, you can find the other blocks by choosing an element not in the principal block and multiplying all the elements of the principal block by the new element and reducing exponents mod 2.

**Example 15.4**

## $2^5$ in eight blocks of four

Suppose that we wish to block a $2^5$ design into eight blocks of four. Section C.5 suggests ABC, BD, and AE for the defining contrasts. The principal block is that block containing (1), or equivalently those factor-level combinations that are even for ABC, BD, and AE. The principal block is (1), *bcd*, *ace*, and *abde*. This principal block was found by inspection, meaning working through the factor-level combinations finding those that are even for all three defining contrasts.

The remaining blocks can be found by multiplying the elements of the principal block by a factor-level combination not already accounted for. For example, *a* is not in the principal block, so we multiply and get *a*, *abcd*, *ce*, and *bde* for a second block. Next, *b* has not been listed, so we multiply by *b* and get *b*, *cd*, *abce*, and *ade* for the third block. Table 15.5 gives the remaining blocks.

**Table 15.5:** $2^5$ in eight blocks of four using ABC, BD, and AE as defining contrasts, found by products with principal block.

| | | | | Multiply by | | | |
|---|---|---|---|---|---|---|---|
| P.B. | $a$ | $b$ | $c$ | $d$ | $e$ | $ab$ | $ad$ |
| (1) | $a$ | $b$ | $c$ | $d$ | $e$ | $ab$ | $ad$ |
| $bcd$ | $abcd$ | $cd$ | $bd$ | $bc$ | $bcde$ | $acd$ | $abc$ |
| $ace$ | $ce$ | $abce$ | $ae$ | $acde$ | $ac$ | $bce$ | $cde$ |
| $abde$ | $bde$ | $ade$ | $abcde$ | $abe$ | $abd$ | $de$ | $be$ |

$q$ defining contrasts for $2^q$ blocks

For $2^q$ blocks, we use $q$ defining contrasts. These $q$ defining contrasts must be independent; no defining contrast can be a generalized interaction of two or more of the others. Form blocks by grouping the factor-level combinations according to the $2^q$ different even-odd combinations for the $q$ defining contrasts. There will be $2^{k-q}$ factor-level combinations in each block. There are $2^q$ blocks, so there are $2^q - 1$ degrees of freedom confounded with blocks. These are the $q$ defining contrasts, their two-way, three-way, and up to $q$-way generalized interactions.

Doing the actual blocking is rather tedious in large designs, so it is helpful to have software that will do confounding. The usual even/odd or 0/1 methods are available if you must do the confounding by hand, but a little thinking first can save a lot of calculation.

**Example 15.5**

### $2^7$ in 16 blocks of eight

Suppose that we are going to confound a $2^7$ design into 16 blocks of size eight using the defining contrasts ABCD, BCE, ACF, and ABG. The effects that are confounded with blocks will be

| | |
|---|---|
| ABCD | ACEG = (BCE)(ABG) |
| BCE | BCFG = (ACF)(ABG) |
| ACF | CDEF = (ABCD)(BCE)(ACF) |
| ABG | BDEG = (ABCD)(BCE)(ABG) |
| ADE = (ABCD)(BCE) | ADFG = (ABCD)(ACF)(ABG) |
| BDF = (ABCD)(ACF) | EFG = (BCE)(ACF)(ABG) |
| CDG = (ABCD)(ABG) | ABCDEFG = (ABCD)(BCE)(ACF)(ABG) |
| ABEF = (BCE)(ACF) | |

We get exactly the same blocks using BCE, ACF, ABG, and ABCDEFG as defining contrasts. Combinations in the principal block always have an even number of letters from every defining contrast. Because the full seven-way interaction including all the letters is one of the defining contrasts, all elements in the principal block must have an even number of letters. Next, no pair of letters occurs an even number of times in BCE, ACF, and ABG, so no two-letter combinations can be in the principal block. Similarly, no six-letter combinations can be in the principal block. This indicates that the principal block will contain (1) and combinations with four letters.

Start going through groups of four letters. We find $abcd$ is a match right at the start. We next find $abef$. We can either get this with a direct search, or by reasoning that if we have $a$ and $b$, then we can't have $g$, so we must have two of $c$, $d$, $e$, and $f$. The combinations with $c$ or $d$ don't work, but $abef$ does work. Similarly, if we start with $bc$, then we can't have $e$, and we must have two of $a$, $d$, $f$, and $g$. The combinations with $a$ and $d$ don't work, but $bcfg$ does work.

We now have (1), $abcd$, $abef$, and $bcfg$ in the principal block. We know that in the principal group we can multiply any two elements together, reduce the exponent mod 2, and get another element of the block. Thus we find that $abcd \odot abef = cdef$, $abcd \odot bcfg = adfg$, $abef \odot bcfg = aceg$, and $abcd \odot abef \odot bcfg = bdeg$ are also in the principal block.

Now that we have the principal block, we can find alternate blocks by finding a factor-level combination not already accounted for and multiplying the elements of the principal block by this new element. For example, $a$ is not in the principal block, so we can find a second block as $a = (1) \odot a$, $bcd = abcd \odot a$, $bef = abef \odot a$, $abcfg = bcfg \odot a$, $acdef = cdef \odot a$, $dfg = adfg \odot a$, $ceg = aceg \odot a$, and $abdeg = bdeg \odot a$. Next, $b$ is not in these first two blocks, so $b = (1) \odot b$, $acd = abcd \odot b$, $aef = abef \odot b$, $cfg = bcfg \odot b$, $bcdef = cdef \odot b$, $abdfg = adfg \odot b$, $abceg = aceg \odot b$, and $deg = bdeg \odot b$ are the next block.

### 15.1.3 Analysis of an unreplicated confounded two-series

Remember that the trick to the analysis of any unreplicated factorial is obtaining an estimate of error. The additional complication with confounding is that some of the treatment degrees of freedom are confounded with blocks. The approach we take is to compute the sum of squares or total effect for each main effect and interaction, remove from consideration those that are confounded with blocks, and then analyze the remaining nonconfounded effects with standard methods.

Use standard methods with nonblock effects

**Example 15.6**

#### Visual perception

We wish to study how image properties affect visual perception. In this experiment we will have a subject look at a white computer screen. At random

**Table 15.6:** Fraction of images identified in vision experiment. Data in standard order reading down columns.

| | | | | | | | |
|---|---|---|---|---|---|---|---|
| .27 | .47 | .20 | .73 | .40 | .73 | .20 | .33 |
| .40 | .87 | .20 | .33 | .33 | .53 | .27 | .60 |
| .40 | .60 | .53 | .47 | .27 | .60 | .53 | .67 |
| .40 | .87 | .20 | .67 | .27 | .40 | .80 | .93 |
| .47 | .53 | .53 | .53 | .47 | .73 | .47 | .47 |
| .47 | .60 | .13 | .73 | .27 | .87 | .47 | .47 |
| .40 | .33 | .47 | .80 | .53 | .73 | .33 | .80 |
| .33 | .60 | .47 | .47 | .33 | .73 | .33 | .60 |
| .20 | .67 | .20 | .67 | .27 | .53 | .40 | .73 |
| .27 | .33 | .60 | .73 | .33 | .87 | .40 | .53 |
| .60 | .60 | .20 | .53 | .33 | .47 | .27 | .67 |
| .40 | .67 | .47 | .73 | .60 | .40 | .20 | .33 |
| .60 | .27 | .13 | .67 | .07 | .47 | .47 | .73 |
| .27 | .60 | .73 | .60 | .47 | .60 | .33 | .73 |
| .27 | .67 | .27 | .47 | .33 | .67 | .27 | .60 |
| .53 | .80 | .20 | .60 | .27 | .93 | .20 | .47 |

intervals averaging about 5 seconds, we will put a small image on the screen for a very short time. The subject is supposed to click the mouse button when she sees an image on the screen. The experiment takes place in sixteen ten-minute sessions to prevent tiring; during each session we present 120 images. In fact, these are eight images repeated fifteen times each and presented in random order. We record as the response the fraction of times that the mouse is clicked for a given image type.

We wish to study 128 different images, the factorial combinations of seven factors each at two levels: size of image, shape of image, color of image, orientation of image, duration of image, vertical location of image, and horizontal location of image. Because we anticipate session to session variability, we should design the experiment to account for that. A confounded factorial with sixteen blocks of size eight will work. We use the defining contrasts of Example 15.5, and Table 15.6 gives the responses in standard order.

There are fifteen factorial effects confounded with blocks, seven three-way interactions, seven four-way interactions, and the seven-way interaction. The remaining $127 - 15 = 112$ are not confounded with blocks. We could pool the five- and six-way interaction degrees of freedom for a 28-degree-of-freedom estimate of error, and then use this surrogate error in testing the lower-order terms that are not confounded with blocks. Alternatively, we

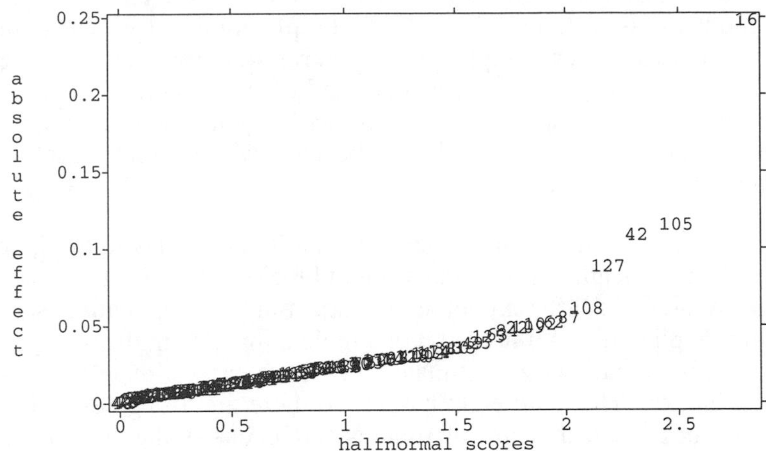


**Figure 15.1:** Halfnormal plot of factorial effects for transformed vision data, including those confounded with blocks. Number indicates effect.

could make a rankit plot or half normal plot of the total effects. It would be best to make these plots using only the 112 nonconfounded terms, but it is usually tedious to remove the confounded terms. Outliers in a plot of all terms will need to be interpreted with blocks in mind.

We begin the analysis by noting that the responses are binomial proportions ranging from .07 to .93; for such data we anticipate nonconstant variance, so we transform using arcsine-square roots at the start. Next we make the half-normal plot of effects shown in Figure 15.1. This plot has all 127 effects in standard order, including those confounded with blocks. Effect 16 (the E main effect) is a clear outlier. Other outliers are effects 105, 42, and 127; these are ADFG, BDF, and ABCDEFG. All three are confounded with blocks, so we regard this as block rather than treatment effects.

We conclude that of the treatments we chose, only factor E (duration) has an effect; images that are on the screen longer are easier to see.

### 15.1.4 Replicating a confounded two-series

We replicate confounded two-series designs for the same reasons that we replicate any design—replication gives us more power, shorter confidence intervals, and better estimates of error. We must choose defining contrasts

Complete versus partial confounding

for the confounding in each replication, and here we have an option. We can confound the same defining contrasts in all replications, or we can confound different contrasts in each replication. Contrasts confounded in all replications are called *completely confounded*, and contrasts confounded in some but not all replications are called *partially confounded*. Partial confounding generally seems like the better choice, because we will have at least some information on every effect.

Suppose that we have four replications of a $2^3$ factorial with two blocks of size four per replication, for a total of eight blocks. One partial confounding scheme would use a different defining contrast in each replication, say ABC in the first replication, AB in the second replication, AC in the third, and BC in the fourth. What can we estimate? First, we can estimate the variation between blocks. There are eight blocks, so there are 7 degrees of freedom between blocks, and the sum of squares for blocks is the sum of squares between the eight groups formed by the blocks. Second, the effects and sums of squares for A, B, and C can be computed in the usual way. This is true for any effect that is never confounded. Next, we can compute the sums of squares and estimated effects for AB, AC, BC, and ABC. Here we must be careful, because all these effects are partially confounded.

Partially confounded effects can be estimated in replications where they are not confounded

Consider first ABC, which is confounded with blocks in the first replication but not in the other replications. The degree of freedom that the ABC effect would estimate in the first replication has already been accounted for as block variation (it is one of the 7 block degrees of freedom), so the first replication tells us nothing about ABC. The ABC effect is not confounded with blocks in replications two through four, so compute the ABC sum of squares and estimated effects from replications two through four. Similarly, we compute the AB effect from replications one, three, and four. In general, estimate an effect and compute its sum of squares from those replication where the effect is not confounded. All that remains after blocks and treatments is error or residual variation. In summary, there are 7 degrees of freedom between blocks, 1 degree of freedom each for A, B, C, AB, AC, BC, and ABC, and $31 - 14 = 17$ degrees of freedom for error.

Treatments adjusted for blocks

Let's repeat the pattern one more time. First remove block to block variation. Compute sums of squares and estimated effects for any main effect or interaction by using the standard formulae applied to those replications in which the main effect or interaction is not confounded. Any effect confounded in every replication cannot be estimated. Error variation is the remainder. This pattern works for complete or partial confounding, and when using statistical software for analysis is most easily expressed as treatments adjusted for blocks.

**Table 15.7:** Milk chiller sensory ratings, by blocks

| Block 1 | | Block 2 | | Block 3 | | Block 4 | |
|---|---|---|---|---|---|---|---|
| (1) | 86 | $a$ | 88 | (1) | 82 | $b$ | 93 |
| $ab$ | 87 | $b$ | 97 | $a$ | 74 | $ab$ | 91 |
| $ac$ | 84 | $c$ | 82 | $bc$ | 84 | $c$ | 79 |
| $bc$ | 91 | $abc$ | 85 | $abc$ | 83 | $ac$ | 81 |

We can estimate all effects in a partially confounded factorial, but we do not have full information on the partially confounded effects. The effective sample size for any effect is the number of replications in which the effect is not confounded. In the example, the effective sample size is four for A, B, and C, but only three for AB, AC, BC, and ABC. Each of these loses one replication due to confounding. The fraction of information available for an effect is the effective sample size divided by the number of replications. Thus in the example we have full or 100% information for the main effects and 3/4 information for the interactions.

Partial information on partially confounded effects

**Example 15.7**

## Milk chiller

Milk is chilled immediately after Pasteurization, and we need to design a chiller. The goal is to get high flow at low capital and operating costs while still chilling the milk quickly enough to maintain sensory qualities. Basic chiller design is a set of refrigerated plates over which the hot milk is pumped. We are investigating the effect of the spacing between the plates (two levels), the temperature of the plates (two levels), and the flow rate of the milk (two levels) on the perceived quality of the resulting milk. There is a fresh batch of raw milk each day, and we expect batch to batch differences in quality. Because of the time involved in modifying the chiller, we can use at most four factor-level combinations in a day.

This constraint of at most four observations a day suggests a confounded design. We use two replicates, confounding ABC and BC in the two replicates. The processed milk is judged daily by a trained expert who is blinded to the treatments used; the design and results are in Table 15.7. Listing 15.1 shows an ANOVA for these data. All effects can be estimated because of the partial confounding. There is evidence for an effect of plate temperature, with lower temperatures giving better sensory results. There is very slight evidence for a rate effect.

By way of illustration, the sum of squares for the three-factor interaction in the second replicate is 10.12, what Listing 15.1 shows for the three-factor interaction after adjusting for blocks. The block sum of squares is the sum of the between replicates, ABC in replicate one, and BC in replicate two sums of squares (68.06, 2.00, and 55.13 respectively).

**Listing 15.1:** Minitab output for chiller data.

```
Source             DF    Seq SS     Adj SS      Seq MS        F       P
block               3    125.19     106.19       41.73     4.07   0.083
space               1     27.56      27.56       27.56     2.69   0.162
temp                1    189.06     189.06      189.06    18.42   0.008
rate                1     52.56      52.56       52.56     5.12   0.073
space*temp          1     18.06      18.06       18.06     1.76   0.242
space*rate          1     14.06      14.06       14.06     1.37   0.295
temp*rate           1      0.00       0.00        0.00     0.00   1.000
space*temp*rate     1     10.12      10.12       10.12     0.99   0.366
Error               5     51.31      51.31       10.26
Total              15    487.94

Term              Coef      StDev         T        P
space
1               1.3125     0.8009      1.64    0.162
temp
1              -3.4375     0.8009     -4.29    0.008
rate
1               1.8125     0.8009      2.26    0.073
```

### 15.1.5 Double confounding

Double confounding blocks on two sources of variation

Latin Squares, Youden Squares, and related designs allow us to block on two sources of variation at once; *double confounding* allows us to block on two sources of variation in a confounding design. Suppose that we have a $2^k$ treatment structure and that we have two sources of variation on which to block; there are $2^q$ levels of blocking on one source and $2^{k-q}$ levels of blocking on the other source. Arrange the treatments in a rectangle with $2^q$ rows and $2^{k-q}$ columns. The rows and columns form the blocks for the two sources of variation.

Products of principal blocks

In double confounding, we choose $q$ defining contrasts to generate row blocking, and $k - q$ defining contrasts to generate column blocking. To produce the design, we find the principal blocks for rows and columns and put these in the first row and column of the rectangular arrangement. The remainder of the arrangement is filled by taking products and reducing exponents modulo 2.

Confound rows and columns separately

For example, in a $2^4$ factorial we could block on two sources of variation with four levels each. Put the treatments in a four by four arrangement, using AB and BCD to generate the row blocking, and ABC and CD to generate the column blocking. The generalized interactions ACD and ABD are also confounded. The column principal block is (1), *ab*, *bcd*, and *acd*; the row principal block is (1), *abc*, *cd*, and *abd*; and the full design is

| | | | |
|---|---|---|---|
| (1) | $ab$ | $acd$ | $bcd$ |
| $abd$ | $d$ | $bc$ | $ac$ |
| $cd$ | $abcd$ | $a$ | $b$ |
| $abc$ | $c$ | $bd$ | $ad$ |

For example, we take the third row element *cd* times the fourth column element *bcd* to get $b$ for the $3,4$ element of the table. Each row of the treatment arrangement contains a block from the row-defining contrasts, and each column of the arrangement contains a block from the column-defining contrasts.

## 15.2 Confounding the Three-Series Factorial

Confounding in the three-series factorial is analogous to confounding in the two-series, but threes keep popping up instead of twos. The $2^k$ is confounded into $2^q$ blocks each with $2^{k-q}$ units. The $3^k$ is confounded into $3^q$ blocks, each with $3^{k-q}$ units. When we replicate a three-series design with confounding, we can use complete or partial confounding, just as for the two-series design.

$3^p$ blocks of $3^{k-p}$ units; partial or complete confounding

The levels of a factor in a three-series design are denoted 0, 1, or 2; for example, the factor-level combinations of a $3^2$ design are 00, 10, 20, 01, 11, 21, 02, 12, and 22. The level for factor A is denoted by $x_A$, just as for the two-series design.

Main effects in a three-series design have 2 degrees of freedom, two-factor interactions have 4 degrees of freedom, and $q$-factor interactions have $2^q$ degrees of freedom. We can partition all three-series effects into two-degree-of-freedom bundles. Each main effect contains one of these bundles, each two-factor interaction contains two of these bundles, each three-factor interaction contains four of these bundles, and so on. Each two-degree-of-freedom bundle arises by, in effect, splitting the factor-level combinations into three groups and assessing the variation in the 2 degrees of freedom between these three groups. These two-degree-of-freedom splits provide the basis for confounding the three series, just as one-degree-of-freedom contrasts are the basis for confounding the two series.

Partition three-series effects into two-degree-of-freedom bundles

Each two-degree-of-freedom split has a label, and the labels can be confused with the ordinary interactions, so let's explain them carefully at the beginning. The label for an interaction effect is the letters in the interaction, for example, BCD. The label for a two-degree-of-freedom split is the letters from the factors, each with an exponent of either 0, 1, or 2. By convention, we drop the letters with exponent 0, and by further convention, the first nonzero exponent is always a 1. Thus $A^1C^2$ and $B^1C^1D^2$ are examples of two-degree-of-freedom splits. The two-degree-of-freedom splits that make up an interaction are those splits that have nonzero exponents for the

Label two-degree-of-freedom splits with exponents

same set of factors as the interaction. Thus the splits in BCD are $B^1C^1D^1$, $B^1C^1D^2$, $B^1C^2D^1$, and $B^1C^2D^2$.

We use these two-degree-of-freedom splits to generate confounding in the three-series in the same way that defining contrasts generate confounding in a two-series, so these splits are often called *defining contrasts*, even though they are not really contrasts (which have just 1 degree of freedom).

### 15.2.1 Building the design

Sums of factor levels mod 3 determine splits

Each two-degree-of-freedom portion corresponds to a different way to split the factor-level combinations into three groups. For concreteness, consider the $B^1C^2D^1$ split in a $3^4$ design. Compute for each factor-level combination

$$L = x_B + 2x_C + x_D \bmod 3 \ .$$

The $L$ values will be 0, 1, or 2, and we split the factor-level combinations into three groups according to their values of $L$. In general, for the split $A^{r_A}B^{r_B}C^{r_C}D^{r_D}$, we compute for each factor-level combination

$$L = r_A x_A + r_B x_B + r_C x_C + r_D x_D \bmod 3 \ .$$

Principal block

These $L$ values will again be 0, 1, or 2, determining three groups. The block containing the combination with all factors low is the principal block.

**Example 15.8**

**A $3^2$ with $A^1B^2$ confounded**

Suppose that we want to confound a $3^2$ design into three blocks of size three using $A^1B^2$ as the defining split. We need to compute the defining split $L$ values, and then group the factor-level combinations into blocks, as shown here:

| $x_Ax_B$ | $x_A + 2x_B$ | $L$ |
|---|---|---|
| 00 | 0 | 0 |
| 10 | 1 | 1 |
| 20 | 2 | 2 |
| 01 | 2 | 2 |
| 11 | 3 | 0 |
| 21 | 4 | 1 |
| 02 | 4 | 1 |
| 12 | 5 | 2 |
| 22 | 6 | 0 |

| $L=0$ | $L=1$ | $L=2$ |
|---|---|---|
| 00 | 10 | 20 |
| 11 | 21 | 01 |
| 22 | 02 | 12 |

This particular arrangement into blocks forms a Latin Square, as can be seen when the block numbers are superimposed on the three by three pattern below:

| | | $x_B$ 0 | 1 | 2 |
|---|---|---|---|---|
| | 0 | 0 | 2 | 1 |
| $x_A$ | 1 | 1 | 0 | 2 |
| | 2 | 2 | 1 | 0 |

If we had used $A^1B^1$ as the defining split, we would again get a Latin Square arrangement, but that Latin Square would be orthogonal to this one.

To block a three-series into nine blocks, we must use two defining splits $P_1$ and $P_2$ with corresponding $L$ values $L_1$ and $L_2$. Each $L$ can take the values 0, 1, or 2, so there are nine combinations of $L_1$ and $L_2$ values, and these form the nine blocks. To get 27 blocks, we use three defining splits and look at all combinations of 0, 1, or 2 from the $L_1$, $L_2$, and $L_3$ values, and so on for more blocks.

Use $q$ defining splits for $3^q$ blocks

For $3^q$ blocks, we follow the same pattern but use $q$ defining splits. The only restriction on these splits is that none can be a generalized interaction of any of the others (see the next section). Thus we cannot use $A^1C^2$, $B^1D^1$, and $A^1B^1C^2D^1$ as our defining splits. As with two-series confounded designs, we try to find defining splits that confound interactions of as high an order as possible.

## Confounding a $3^3$ in nine blocks

**Example 15.9**

Suppose that we wish to confound a $3^3$ design into nine blocks using defining splits $A^1B^1$ and $A^1C^2$. The $L$ equations are

$$L_1 = x_A + x_B \bmod 3$$

and

$$L_2 = x_A + 2x_C \bmod 3$$

We need to go through all 27 factor-level combinations and compute the $L_1$ and $L_2$ values. Once we have the L-values, we can make the split into nine blocks. For example, the 110 treatment has an $L_1$ value of $1 + 1 = 2$ and an $L_2$ value of $1 + 2 \times 0 = 1$, so it belongs in the 2/1 block; the 102 treatment has an $L_1$ value of $1 + 0 = 1$ and an $L_2$ value of $1 + 2 \times 2 \bmod 3 = 2$, so it belongs in the 1/2 block. The full design follows:

| Treatment | $L_1$ | $L_2$ |
|---|---|---|
| 000 | 0 | 0 |
| 100 | 1 | 1 |
| 200 | 2 | 2 |
| 010 | 1 | 0 |
| 110 | 2 | 1 |
| 210 | 0 | 2 |
| 020 | 2 | 0 |
| 120 | 0 | 1 |
| 220 | 1 | 2 |
| 001 | 0 | 2 |
| 101 | 1 | 0 |
| 201 | 2 | 1 |
| 011 | 1 | 2 |
| 111 | 2 | 0 |
| 211 | 0 | 1 |
| 021 | 2 | 2 |
| 121 | 0 | 0 |
| 221 | 1 | 1 |
| 002 | 0 | 1 |
| 102 | 1 | 2 |
| 202 | 2 | 0 |
| 012 | 1 | 1 |
| 112 | 2 | 2 |
| 212 | 0 | 0 |
| 022 | 2 | 1 |
| 122 | 0 | 2 |
| 222 | 1 | 0 |

| 0/0 | 0/1 | 0/2 |
|---|---|---|
| 000<br>121<br>212 | 120<br>211<br>022 | 210<br>001<br>122 |

| 1/0 | 1/1 | 1/2 |
|---|---|---|
| 010<br>101<br>222 | 100<br>221<br>012 | 220<br>011<br>102 |

| 2/0 | 2/1 | 2/2 |
|---|---|---|
| 020<br>111<br>202 | 110<br>201<br>022 | 200<br>021<br>112 |

Combine factor levels mod 3

In the two-series using the 0/1 labels, any two elements of the principal block could be combined using the operation $\oplus$ with the result being an element of the principal block. Furthermore, if you combine the principal block with any element not in the principal block, you get another block. These properties also hold for the three-series design, provided you interpret the operation $\oplus$ as "add the factor levels individually and reduce modulo three."

For example, the principal block in Example 15.9 was 000, 121, and 212. We see that $121 \oplus 121 = 242 = 212$, which is in the principal block. Also, the combination 210 is not in the principal block, so $000 \oplus 210 = 210$, $121 \oplus 210 = 331 = 001$, and $212 \oplus 210 = 422 = 122$ form a block (the one labeled 0/2).

### 15.2.2 Confounded effects

Confounding a three-series design into three blocks uses one defining split with 2 degrees of freedom. There are 2 degrees of freedom between the three blocks, and these 2 degrees of freedom are exactly those of the defining split.

Confounding a three-series design into nine blocks uses two defining splits, each with 2 degrees of freedom. The 4 degrees of freedom for these two defining splits are confounded with block differences. There are 8 degrees of freedom between the nine blocks, so 4 more degrees of freedom must be confounded along with the two defining splits. These additional degrees of freedom are from the generalized interactions of the defining splits. If $P_1$ and $P_2$ are the defining splits, then the generalized interactions are $P_1P_2$ and $P_1P_2^2$.

Confounded effects are $P_1$, $P_2$, $P_1P_2$ and $P_1P_2^2$

Recall that we always write these two-degree-of-freedom splits in a three series with exponents of 0, 1, or 2, with the first nonzero exponent always being a 1. Products like $P_1P_2$ won't always be in that form, so how can we convert? First, reduce exponents modulo three. Second, if the leading nonzero exponent is not a 1, then square the term and reduce exponents modulo three again. The net effect of this second step is to leave zero exponents as zero and swap ones and twos.

Rearrange to get a leading exponent of 1

#### Confounding a $3^3$ in nine blocks, continued

**Example 15.10**

The defining splits in Example 15.9 were $A^1B^1$ and $A^1C^2$, so the generalized interactions are

$$\begin{aligned} P_1P_2 &= A^1B^1 \times A^1C^2 \\ &= A^2B^1C^2 \\ &= (A^2B^1C^2)^2 \quad \text{leading exponent was 2, so square} \\ &= A^4B^2C^4 \\ &= A^1B^2C^1 \quad \text{reduce exponents modulo 3} \end{aligned}$$

$$\begin{aligned} P_1P_2^2 &= A^1B^1\,(A^1C^2)^2 \\ &= A^3B^1C^4 \\ &= B^1C^1 \quad \text{reduce exponents modulo 3} \end{aligned}$$

Thus the full set of confounded effects is $A^1B^1$, $A^1C^2$, $A^1B^2C^1$, $B^1C^1$.

When we confound into 27 blocks using defining splits $P_1$, $P_2$, and $P_3$, there are 26 degrees of freedom between blocks, comprising thirteen two-degree-of-freedom splits. Now it makes sense to give the general rule. Suppose that there are $q$ defining contrasts, $P_1$, $P_2$, $\ldots P_q$. The confounded degrees of freedom will be $P_1^{v_1}P_2^{v_2}\cdots, P_q^{v_q}$, for all exponent sets that use exponents 0, 1, or 2, and with the leading nonzero exponent being a 1. Applying this to $q = 3$, we get the following confounded terms: $P_1$, $P_2$, $P_3$, $P_1P_2$, $P_1P_2^2$, $P_1P_3$, $P_1P_3^2$, $P_2P_3$, $P_1P_3^2$, $P_1P_2P_3$, $P_1P_2P_3^2$, $P_1P_2^2P_3$, and $P_1P_2^2P_3^2$.

**Example 15.11** **Confounding a $3^5$ in 27 blocks**

Suppose that we wish to confound a $3^5$ into 27 blocks using $A^1C^1$, $A^1B^1D^1$, and $A^1B^2E^2$ as defining splits. The the complete list of confounded effects will be

$$
\begin{aligned}
P_1 = A^1C^1 &= A^1C^1 \\
P_2 = A^1B^1D^1 &= A^1B^1D^1 \\
P_3 = A^1B^2E^2 &= A^1B^2E^2 \\
P_1P_2 = A^2B^1C^1D^1 &= A^1B^2C^2D^2 \\
P_1P_2^2 = A^3B^2C^1D^2 = B^2C^1D^2 &= B^1C^2D^1 \\
P_1P_3 = A^2B^2C^1E^2 &= A^1B^1C^2E^1 \\
P_1P_3^2 = A^3B^4C^1E^4 &= B^1C^1E^1 \\
P_2P_3 = A^2B^3D^1E^2 = A^2D^1E^2 &= A^1D^2E^1 \\
P_2P_3^2 = A^3B^5D^1E^4 = B^2D^1E^1 &= B^1D^2E^2 \\
P_1P_2P_3 = A^3B^3C^1D^1E^2 &= C^1D^1E^2 \\
P_1P_2P_3^2 = A^4B^5C^1D^1E^4 &= A^1B^2C^1D^1E^1 \\
P_1P_2^2P_3 = A^4B^4C^1D^2E^2 &= A^1B^1C^1D^2E^2 \\
P_1P_2^2P_3^2 = A^5B^6C^1D^2E^4 = A^2C^1D^2E^1 &= A^1C^2D^1E^2
\end{aligned}
$$

This design confounds 2 degrees of freedom in the AC interaction, but otherwise confounds three-way interactions and higher.

### 15.2.3 Analysis of confounded three-series

Treatments adjusted for blocks

Analysis of a confounded three-series is analogous to analysis of a confounded two-series. First remove variation between blocks, then remove any treatment variation that can be estimated; any remaining variation is used as error. When there is only one replication, the highest-order interaction is typically used as an estimate of error. With most statistical software, you can get this analysis by requesting an ANOVA with treatment sums of squares adjusted for blocks.

The accounting is a little more complicated in a confounded three-series than it was in the two-series, because confounding is done via two-degree-of-freedom splits, whereas the ANOVA is usually tabulated by interaction terms. For example, consider two replications of a $3^2$ with $A^1B^1$ completely confounded. There are eighteen experimental units, with 17 degrees of freedom between them. There are 5 degrees of freedom between the blocks, 2 degrees of freedom for each main effect, 2 degrees of freedom for the AB interaction, and 6 degrees of freedom for error. The 2 degrees of freedom for AB are the $A^1B^2$ degrees of freedom, which are not confounded with blocks.

Interactions containing completely confounded splits have fewer than nominal degrees of freedom

When we use partial confounding, we can estimate all treatment effects, but we will only have partial information on those effects that are partially confounded. Again consider two replications of a $3^2$, but confound $A^1B^1$ in the first replication and $A^1B^2$ in the second. We can estimate $A^1B^1$ in the second replication and $A^1B^2$ in the first, so we have 4 degrees of freedom for interaction. However, the effective sample size for each of these interaction effects is nine, rather than eighteen.

## 15.3 Further Reading and Extensions

Two- and three-series are the easiest factorials to confound, but we can use confounding for other factorials too. John (1971) is a good place to get started with these other designs. Kempthorne (1952) also has a good discussion. Derivation and methods for some of these other designs takes some (abstract) algebra. In fact, this algebra is present in the two- and three-series designs; we've just been ignoring it. For example, we have stated that multiplying two elements of the principal block together gives another element in the principal block, and that multiplying the principal block by any element not in the principal block yields an alternate block. These are a consequence of the facts that the factor-level combinations form an (algebraic) group, the principal block is a subgroup, and the alternate blocks are cosets.

Confounding $s^k$ designs when $s$ is prime is the straightforward generalization of the 0/1 and 0/1/2 methods we used for $2^k$ and $3^k$ designs. For example, when $s = 5$ and $k = 4$, represent the factor levels by 0, 1, 2, 3, and 4. Block into five blocks of size 125 using the defining split $A^{r_A}B^{r_B}C^{r_C}D^{r_D}$ by computing

$$L = r_A x_A + r_B x_B + r_C x_C + r_D x_D \bmod 5$$

and splitting into groups based on $L$. If you have two defining splits $P_1$ and $P_2$, the confounded effects are $P_1$, $P_2$, $P_1P_2$, $P_1P_2^2$, $P_1P_2^3$, and $P_1P_2^4$. More generally, use powers up to $s-1$.

To confound $s^k$ designs when $s$ is the $m$th power of a prime, reexpress the design as a $p^{mk}$ design, where $p$ is the prime factor of $s$. Now use standard methods for confounding a $p^{mk}$, but take care that none of the generalized interactions that get confounded are actually main effects. For example, confound a $4^2$ design into four blocks of four. A $4^2$ design can be reexpressed as a $2^4$ design, with the AB combinations indexing the first four-level factor, and the BC combinations indexing the second four-level factor. We could confound ABC and AD (and their generalized interaction BCD). All three of these degrees of freedom are in the 9-degree-of-freedom interaction for the four-series design. We would not want to confound AB, BCD, and ACD, because AB is a degree of freedom in the main effect of the first four-level factor.

Mixed-base factorials are more limited. Suppose we have a $s_1^{k_1}s_2^{k_2}$ factorial, where $s_1$ and $s_2$ are different primes. It is straightforward to choose $s_1^q$ blocks of size $s_1^{k_1-q}s_2^{k_2}$ or $s_2^q$ blocks of size $s_1^{k_1}s_2^{k_2-q}$. Just use methods for the factors in play and carry the other factors along. Getting $s_1s_2$ blocks of size $s_1^{k_1-1}s_2^{k_2-1}$ is considerably more difficult.

## 15.4 Problems

**Exercise 15.1** Confound a $2^5$ factorial into four blocks of eight, confounding BCD and ACE with blocks. Write out the factor-level combinations that go into each block.

**Exercise 15.2** We want to confound a $2^4$ factorial into four blocks of size four using ACD and ABD as defining contrasts. Find the factor-level combinations that go into each block.

**Exercise 15.3** Suppose that we confound a $2^8$ into sixteen blocks of size 16 using ABCF, ABDE, ACDE, and BCDH as defining contrasts. Find the all the confounded effects.

**Exercise 15.4** Divide the factor-level combinations in a $3^3$ factorial into three groups of nine according to the $ABC^2$ interaction term.

**Exercise 15.5** Suppose that we have a partially confounded $3^3$ factorial design run in four replicates, with $ABC$, $ABC^2$, $AB^2C$, and $AB^2C^2$ confounded in the four replicates. Give a skeletal ANOVA for such an experiment (sources and degrees of freedom only).

**Problem 15.1** Briefly describe the experimental design you would choose for each of the following situations, and why.

(a) Untrained consumer judges cannot reliably rate their liking of more than about fifteen to twenty similar foods at one sitting. However, you have been asked to design an experiment to compare the liking of cookies made with 64 recipes, which are the factorial combinations of six recipe factors, each at two levels. The judges are paid, and you are allowed to use up to 50 judges.

(b) Seed germination is sensitive to environmental conditions, so many experiments are performed in laboratory growth chambers that seek to provide a uniform environment. Even so, we know that the environment is not constant: temperatures vary from the front to the back with the front being a bit cooler. We wish to determine if there is any effect on germination due to soil type. We have resources for 64 units (pots with a given soil type). There are eight soil types of interest, and the growth chamber is big enough for 64 pots in an eight by eight arrangement.

(c) Acid rain seems to kill fish in lakes, and we would like to study the mechanism more closely. We would like to know about effects due to the kind of acid (nitric versus sulfuric), amount of acid exposure (as measured by two levels of pH in the water), amount of aluminum present (two levels of aluminum; acids leach aluminum from soils, so it could be the aluminum that is killing the fish instead of the acid), and time of exposure (that is, a single peak acute exposure versus a chronic exposure over 3 months). We have 32 aquariums to use, and a large supply of homogeneous brook trout.

**Problem 15.2**

Briefly describe the experimental design used in each of the following and give a skeleton ANOVA.

(a) Neurologists use functional Magnetic Resonance Imaging (fMRI) to determine the amount of the brain that is "activated" (in use) during certain activities. We have twelve right-handed subjects. Each subject will lie in the magnet. On a visual signal, the subject will perform an action (tapping of fingers in a certain order) using either the left or the right hand (depending on the signal). The measured response is the number of "pixels" on the left side of the brain that are activated. We expect substantial subject to subject variation in the response, and there may be a consistent difference between the first trial and the second trial. Six subjects are chosen at random for the left-right order, and the other six get right-left. We obtain responses for each subject under both right- and left-hand tapping.

(b) We wish to study the winter hardiness of four new varieties of rosebushes compared with the standard variety. An experimental unit will consist of a plot of land suitable for 4 bushes, and we have 25 plots available in a five by five arrangement (a total of 100 bushes). The plots are located on the side of a hill, so the rows have different drainage. Furthermore, one side of the garden is sheltered by a clump of trees, so that we expect differences in wind exposure from column to column. The five varieties are randomly arranged subject to the constraint that each variety occurs once in each row and each column. The response of interest is the number of blooms produced after the first winter.

(c) Nisin is a naturally occurring antimicrobial substance, and *Listeria* is a microbe we'd like to control. Consider an experiment where we examine the effects of the two factors "amount of nisin" (factor A, three levels, 0, 100, and 200 IU) and "heat" (factor B, three levels, 0, 5, and 10 second scalds) on the number of live *Listeria* bacteria on poultry skin. We use six chicken thighs. The skin of each thigh is divided into three sections, and each section receives a different A-B combination. We expect large thigh to thigh variability in bacteria counts. The factor-level combinations used for each skin section follow (using 0,1,2 type notation for the three levels of each factor):

| | Thigh | | | | | |
|---|---|---|---|---|---|---|
| Section | 1 | 2 | 3 | 4 | 5 | 6 |
| 1 | 00 | 10 | 20 | 00 | 10 | 02 |
| 2 | 11 | 21 | 01 | 21 | 01 | 20 |
| 3 | 22 | 02 | 12 | 12 | 22 | 11 |

(d) Semen potency is measured by counting the number of fertilized eggs produced when the semen is used. Consider a study on the influence of four treatments on the potency of thawed boar semen. The factors are cryoprotector used (factor A, two levels) and temperature regime (factor B, two levels). We expect large sow to sow differences in fertility, so we block on sow by using one factor-level combination in each of the two horns (halves) of the uterus. Eight sows were used, with the following treatment assignment.

| Sow | | | | | | | |
|---|---|---|---|---|---|---|---|
| 1 | 2 | 3 | 4 | 5 | 6 | 7 | 8 |
| a | ab | (1) | b | b | (1) | (1) | a |
| b | (1) | ab | a | a | ab | ab | b |

**Problem 15.3** Choose an experimental design appropriate for the following conditions. Describe treatments, blocks, and so on.

(a) "Habitat improvement" (HI) is the term used to describe the modification of a segment of a stream to increase the numbers of trout in the stream. HI has been used for decades, but there is little experimental evidence on whether it works. We have eight streams in southeastern Minnesota to work with, and we can make up to eight habitat improvements (that is, modify eight stream segments). Each stream flows through both agricultural and forested landscapes, and for each stream we have identified two segments for potential HI, one in the forested area and one in the agricultural area. We anticipate large differences between streams in trout numbers; there may be differences between forested and agricultural areas. We can count the trout in all sixteen segments.

(b) We wish to study how the fracturability of potato chips is affected by the recipe for the chip. (Fracturability is related to crispness.) We are going to study five factors, each at two levels. Thus there are 32 recipes to consider. We can only bake and measure eight recipes a day, and we expect considerable day to day variation due to environmental conditions (primarily temperature and humidity). We have resources for eight days.

(c) One of the issues in understanding the effects of increasing atmospheric $CO_2$ is the degree to which trees will increase their uptake of $CO_2$ as the atmospheric concentration of $CO_2$ increases. We can manipulate the $CO_2$ concentration in a forest by using Free-Air $CO_2$ Enrichment (FACE) rings. Each ring is a collection of sixteen towers (and other equipment) 14 m tall and 30 m in diameter that can be placed around a plot in a forest. A ring can be set to enrich $CO_2$ inside the ring by 0, 100, or 200 ppm. We have money for six rings and can work at two research stations, one in North Carolina and one in South Carolina. Both research stations have plantations of 10-year-old loblolly pine. The response we measure will be the growth of the trees over 3 years.

(d) We wish to study the effects of soil density, pH, and moisture on snapdragon seed germination, with each factor at two levels. Twenty-four pots are prepared with appropriate combinations of the factors, and then seeds are added to each pot. The 24 pots are put on trays that are scattered around the greenhouse, but only 4 pots fit on a tray.

**Problem 15.4**

Individuals perceive odors at different intensities. We have a procedure that allows us to determine the concentration of a solution at which an individual first senses the odor (the threshold concentration). We would like to determine how the threshold concentrations vary over sixteen solutions.

However, the threshold-determining procedure is time consuming and any individual judge can only be used to find threshold concentrations for four solutions.

Each solution is a combination of five compounds in various ratios. The sixteen solutions are formed by manipulating four factors, each at two levels. Factor 1 is the ratio of the concentration of compound 1 to the concentration of compound 5. Factors 2 through 4 are are similar.

We have eight judges. Two judges are assigned at random to each of the solution sets [(1), $bc$, $abd$, $acd$], [$a$, $abc$, $bd$, $cd$], [$ab$, $ac$, $d$, $bcd$], and [$b$, $c$, $ad$, $abcd$]. We then determine the threshold concentration for the solutions for each judge. The threshold concentrations are normalized by dividing by a reference concentration. The ratios are given below:

| Judge 1 | | 2 | | 3 | | 4 | |
|---|---|---|---|---|---|---|---|
| (1) | 8389 | $a$ | 4351 | $ab$ | 6 | $b$ | 375 |
| $bc$ | 816 | $abc$ | 78 | $ac$ | 262 | $c$ | 33551 |
| $abd$ | 4 | $bd$ | 5941 | $d$ | 1230 | $ad$ | 246 |
| $acd$ | 46 | $cd$ | 27138 | $bcd$ | 98 | $abcd$ | 10 |

| Judge 5 | | 6 | | 7 | | 8 | |
|---|---|---|---|---|---|---|---|
| (1) | 56034 | $a$ | 2346 | $ab$ | 67 | $b$ | 40581 |
| $bc$ | 25046 | $abc$ | 35 | $ac$ | 3081 | $c$ | 90293 |
| $abd$ | 109 | $bd$ | 228 | $d$ | 50991 | $ad$ | 19103 |
| $acd$ | 490 | $cd$ | 6842 | $bcd$ | 784 | $abcd$ | 61 |

Analyze these data to determine how the compounds affect the threshold concentration. Are there any deficiencies in the design?

**Problem 15.5** Eurasian water milfoil is a nonnative plant that is taking over many lakes in Minnesota and driving out the native northern milfoil. However, there is a native weevil (an insect) that eats milfoil and may be useful as a control. We wish to investigate how eight treatments affect the damage the weevils do to Eurasian milfoil. The treatments are the combinations of whether a weevil's parents were raised on Eurasian or northern, whether the weevil was hatched on Eurasian or northern, and whether the weevil grew to maturity on Eurasian or northern.

We have eight tanks (big aquariums), each of which is subdivided into four sections. The subdivision is accomplished with a fine mesh that lets water through, but not weevils. The tanks are planted with equal amounts of Eurasian milfoil. We try to maintain uniformity between tanks, but there will be some tank to tank variation due to differences in light and temperature.

The tanks are planted in May, then weevils are introduced. In September, milfoil biomass is measured as response and is shown here:

Tank

| 1 | | 2 | | 3 | | 4 | |
|---|---|---|---|---|---|---|---|
| (1) | 10.4 | $a$ | 4.8 | (1) | 16.8 | $a$ | 12.3 |
| $ab$ | 17.5 | $b$ | 8.9 | $ab$ | 19.6 | $b$ | 17.1 |
| $ac$ | 22.2 | $c$ | 6.8 | $c$ | 16.4 | $ac$ | 13.3 |
| $bc$ | 27.7 | $abc$ | 17.6 | $abc$ | 35.6 | $bc$ | 19.5 |

| 5 | | 6 | | 7 | | 8 | |
|---|---|---|---|---|---|---|---|
| (1) | 7.7 | $a$ | 6.3 | (1) | 14.9 | $b$ | 7.1 |
| $ac$ | 13.3 | $c$ | 7.3 | $bc$ | 34.0 | $c$ | 8.3 |
| $b$ | 12.4 | $ab$ | 11.2 | $a$ | 16.9 | $ab$ | 15.3 |
| $abc$ | 17.7 | $bc$ | 25.0 | $abc$ | 36.8 | $ac$ | 7.0 |

Analyze these data to determine how the treatments affect milfoil biomass.

**Problem 15.6**

Scientists wish to understand how the amount of sugar (two levels), culture strain (two levels), type of fruit (blueberry or strawberry), and pH (two levels) influence shelf life of refrigerated yogurt. In a preliminary experiment, they produce one batch of each of the sixteen kinds of yogurt. The yogurt is then placed in two coolers, eight batches in each cooler. The response is the number of days till an off odor is detected from the batch.

Cooler

| 1 | | 2 | |
|---|---|---|---|
| (1) | 34 | $a$ | 35 |
| $ab$ | 34 | $b$ | 36 |
| $ac$ | 32 | $c$ | 39 |
| $ad$ | 34 | $d$ | 41 |
| $bc$ | 34 | $abc$ | 39 |
| $bd$ | 39 | $abd$ | 44 |
| $cd$ | 38 | $acd$ | 44 |
| $abcd$ | 37 | $bcd$ | 42 |

Analyze these data to determine how the treatments affect time till off odor.

**Question 15.1**

Consider a defining split in a three-series design, say $A^{r_A}B^{r_B}C^{r_C}D^{r_D}$. Now double the exponents and reduce them modulo 3 to generate a new defining split. Show that the two splits lead to the same three sets of factor-level combinations.

**Question 15.2** Show that in a three-series design, any defining split with leading nonzero exponent 2 is equivalent to a a defining split with leading nonzero exponent 1.

**Question 15.3** Show that in a three-series design with defining splits $P_1$ and $P_2$, the generalized interactions $P_1 P_2^2$ and $P_1^2 P_2$ are equivalent.

# Chapter 16

# Split-Plot Designs

Split plots are another class of experimental designs for factorial treatment structure. We generally choose a split-plot design when some of the factors are more difficult or expensive to vary than the others, but split plots can arise for other reasons. Split plots can be described in several ways, including incomplete blocks and restrictions on the randomization, but the key features to recognize are that split plots have more than one randomization and more than one idea of experimental unit.

Use split plots when some factors more difficult to vary

## 16.1 What Is a Split Plot?

The terminology of split plots comes from agricultural experimentation, so let's begin with an agricultural example. Suppose that we wish to determine the effects of four corn varieties and three levels of irrigation on yield. Irrigation is accomplished by using sprinklers, and these sprinklers irrigate a large area. Thus it is logistically difficult to use a design with smallish experimental units, with adjacent units having different levels of irrigation. At the same time, we might want to have small units, because there may be a limit on the total amount of land available for the experiment, or there may be variation in the soils leading us to desire small units grouped in blocks. Split plots give us something of a compromise.

Divide the land into six *whole plots*. These whole plots should be sized so that we can set the irrigation on one whole plot without affecting its neighbors. Randomly assign each irrigation level to two of the whole plots. Irrigation is the *whole-plot factor*, sometimes called the *whole-plot treatment*. Divide each whole plot into four *split plots*. Randomly assign the four corn

Whole plots and whole-plot factor

Split plots and split-plot factor

varieties to the four split plots, with a separate, independent randomization in each whole plot. Variety is the *split-plot factor*. One possible arrangement is as follows, with the six columns representing whole plots with four split plots within each:

| | | | | | |
|---|---|---|---|---|---|
| I2 V1 | I3 V4 | I3 V1 | I1 V3 | I2 V3 | I1 V2 |
| I2 V3 | I3 V3 | I3 V3 | I1 V2 | I2 V1 | I1 V1 |
| I2 V2 | I3 V1 | I3 V4 | I1 V1 | I2 V2 | I1 V4 |
| I2 V4 | I3 V2 | I3 V2 | I1 V4 | I2 V4 | I1 V3 |

Split plots have two sizes of units and two randomizations

What makes a split-plot design different from other designs with factorial treatment structure? Here are three ways to think about what makes the split plot different. First, the split plot has two sizes of units and two separate randomizations. Whole plots act as experimental units for one randomization, which assigns levels of the whole-plot factor irrigation to the whole plots. The other randomization assigns levels of the split-plot factor variety to split plots. In this randomization, split plots act as experimental units, and whole plots act as blocks for the split plots. There are two separate randomizations, with two different kinds of units that can be identified before randomization starts. This is the way I usually think about split plots.

Split plots restrict randomization

Second, a split-plot randomization can be done in one stage, assigning factor-level combinations to split plots, provided that we restrict the randomization so that all split plots in any whole plot get the same level of the whole-plot factor and no two split plots in the same whole plot get the same level of the split-plot factor. Thus a split-plot design is a restricted randomization. We have seen other restrictions on randomization; for example, RCB designs can be considered a restriction on randomization.

Split plots confound whole-plot factor with incomplete blocks

Third, a split plot is a factorial design in incomplete blocks with one main effect confounded with blocks. The whole plots are the incomplete blocks, and the whole-plot factor is confounded with blocks. We will still be able to make inference about the whole-plot factor, because we have randomized the assignment of whole plots to levels of the whole-plot factor. This is analogous to recovering interblock information in a BIBD, but is fortunately much simpler.

Here is another split-plot example to help fix ideas. A statistically oriented music student performs the following experiment. Eight pianos are obtained, a baby grand and a concert grand from each of four manufacturers. Forty music majors are divided at random into eight panels of five students each. Two panels are assigned at random to each manufacturer, and will hear and rate the sound of the baby and concert grand pianos from that manufacturer. Logistically, each panel goes to the concert hall for a 30-minute time period. The panelists are seated and blindfolded. The curtain opens to reveal the two pianos of the appropriate brand, and the same piece of music is played on the two pianos in random order (the pianos are randomized, not the music!). Each panelist rates the sound on a 1–100 scale after each piece.

The whole plots are the eight panels, and the whole-plot factor is manufacturer. The split plots are the two listening sessions for each panel, and the split-plot factor is baby versus concert grand. How can we tell? We have to follow the randomization and see how treatments were assigned to units. Manufacturer was randomized to panel, and piano type was randomized to session within each panel. The randomization was restricted in such a way that both sessions for a panel had to have the same level of manufacturer. Thus panel was the unit for manufacturer, and session was the unit for type. Individual panelist is a measurement unit in this experiment, not an experimental unit. The response for any session must be some summary of the five panelist ratings.

Follow the randomization to identify a split plot

You cannot distinguish a split-plot design from some other design simply by looking at a table of factor levels and responses. You *must* know how the randomization was done. We also have been speaking as if the whole plot randomization was done first; this is often true, but is not required.

Before moving on, we should state that the flexibility that split plots provide for dealing with factors that are difficult to vary comes at a price: comparisons involving the split-plot factor are more precise than those involving the whole-plot factor. This will be more explicit in the Hasse diagrams below, where we will see two separate error terms, the one for whole plots having a larger expectation.

Split-plot comparisons more precise than whole-plot comparisons

## 16.2 Fancier Split Plots

The two examples given in the last section were the simplest possible split-plot design: the treatments have a factorial structure with two factors, levels of the whole-plot factor are assigned to whole plots in a completely randomized fashion; and levels of the split-plot factor are assigned to split plots in randomized complete block fashion with whole plots as blocks. The key to a split plot is two sizes of units and two randomizations; we can increase the number of factors and/or change the whole-plot randomization and still have a split plot.

Begin with the number of factors. The treatments assigned to whole plots need not be just the levels of a single factor: they can be the factor-level combinations of two or more factors. For example, the four piano manufacturers could actually be the two by two factorial combinations of the factors source

Can have more than one whole-plot factor

(levels domestic and imported) and cost (levels expensive and very expensive). Here there would be two whole-plot factors. Other experiments could have more.

Can have more than one split plot factor

Similarly, the treatments assigned to split plots at the split-plot level can be the factor-level combinations of two or more factors. The four varieties of corn could be from the combinations of the two factors insect resistant/not insect resistant, and fungus resistant/not fungus resistant. This would have two split-plot factors, and more are possible.

Randomization is key

Of course, these can be combined to have two or more factors at the whole-plot level and two or more factors at the split-plot level. The key feature of the split plot is not the number of factors, but the kind of randomization.

Whole plots blocked in RCB

Next consider the way that whole-plot treatments are assigned to whole plots. Our first examples used completely randomized design; this is not necessary. It is very common to have the whole plots grouped together into blocks, and assign whole-plot treatments to whole plots in RCB design. For example, the six whole plots in the irrigation experiment could be grouped into two blocks of three whole plots each. Then we randomly assign the three levels of irrigation to the whole plots in the first block, and then perform an independent randomization in the second block of whole plots. In this kind of design, there are two kinds of blocks: blocks of whole plots for the whole-plot treatment randomization, and whole plots acting as blocks for split plots in the split-plot treatment randomization.

Other block designs for whole plots

We can use other designs at the whole-plot level, arranging the whole plots in Balanced Incomplete Blocks, Latin Squares, or other blocking designs. These are not common, but there is no reason that they cannot be used if the experimental situation requires it.

Additional split plot blocking

Whole plots always act as blocks for split plots. Additional blocking at the split-plot level is possible, but fairly rare. For example, we might expect a consistent difference between the first and second pianos rated by a panel. The two panels for a given manufacturer could then be run as a Latin Square, with panel as column-blocking factor and first or second session as the row-blocking factor. This would block on the additional factor time.

## 16.3 Analysis of a Split Plot

Random effect for every randomization

Analysis of a split-plot design is fairly straightforward, once we figure out what the model should be. We assume that there is a random effect for every randomization. Thus we get a random value for each whole plot; if we ignore the split plots, we have a design with whole plot as experimental unit, and this random value is the experimental error. We also get a random value for each split plot to go with the split-plot randomization; this is experimental error at the split-plot level. Here are several examples of split plots and models for them.

### Split plot with one whole-plot factor, one split-plot factor, and CRD at the whole-plot level

**Example 16.1**

Suppose that there is one whole-plot factor A, with $a$ levels, one split-plot factor B, with $b$ levels, and $n$ whole plots for each level of A. The model is

$$\begin{aligned} y_{ijk} &= \mu + \alpha_i + \eta_{k(i)} \\ &\quad + \beta_j + \alpha\beta_{ij} + \epsilon_{k(ij)} \ , \end{aligned}$$

with $\eta_{k(i)}$ as the whole-plot level random error, and $\epsilon_{k(ij)}$ as the split-plot level random error. Note that there is an $\eta_{k(i)}$ value for each whole plot (some whole plots have bigger responses than others), and an $\epsilon_{k(ij)}$ for each split plot. The whole-plot error term nests within whole-plot treatments in the same way that an ordinary error term nests within treatments in a CRD. In fact, if you just look at whole-plot effects (those not involving $j$) and ignore the split-plot effects in the second line, this model is a simple CRD on the whole plots with the whole-plot factor as treatment. Similarly, if you lump together all the whole-plot effects in the first line and think of them as blocks, then we have a model for an RCB with the first line as block, some treatment effects, and an error.

Below are two Hasse diagrams. The first is generic and the second is for a split plot with $an = 10$ whole plots, whole-plot factor A with $a = 2$ levels, and split-plot factor B with $b = 3$ levels. The denominator for the whole-plot factor A is whole-plot error (WPE); the denominator for the split-plot factor B and the AB interaction is split-plot error (SPE).

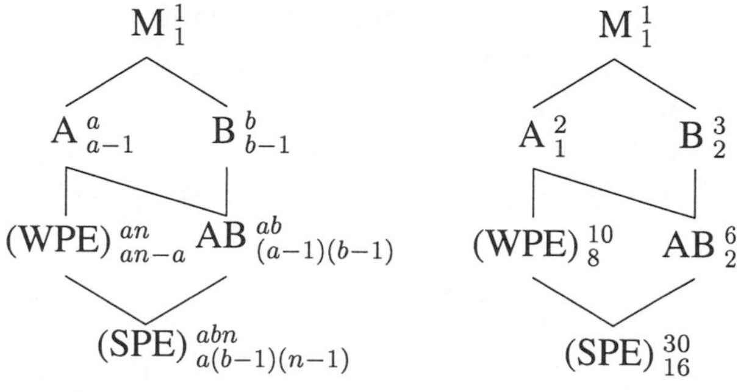

Example 16.2

### Split plot with two whole-plot factors, one split-plot factor, and CRD at the whole-plot level

Now consider a split-plot design with three factors, two at the whole-plot level and one at the split-plot level. We still assume a completely randomized design for whole plots. An appropriate model for this design would be

$$\begin{aligned} y_{ijk} &= \mu + \alpha_i + \beta_j + \alpha\beta_{ij} + \eta_{l(ij)} \\ &\quad + \gamma_k + \alpha\gamma_{ik} + \beta\gamma_{jk} + \alpha\beta\gamma_{ijk} + \epsilon_{l(ijk)} \ , \end{aligned}$$

where we have again arranged the model into a first line with whole-plot effects (those without $k$) and a second line with split-plot effects. The indices $i$, $j$, and $k$ run up to $a$, $b$, and $c$, the number of levels of factors A, B, and C; and the index $l$ runs up to $n$, the replication at the whole-plot level.

Here are two Hasse diagrams. The first is generic for this setup, and the second is for such a split plot with $n = 5$ and whole-plot factors A and B with $a = 2$ and $b = 3$ levels, and split-plot factor C with $c = 5$ levels. The denominator for the whole-plot effects A, B, and AB is whole-plot error; the denominator for the split-plot effects C, AC, BC, and ABC is split-plot error.

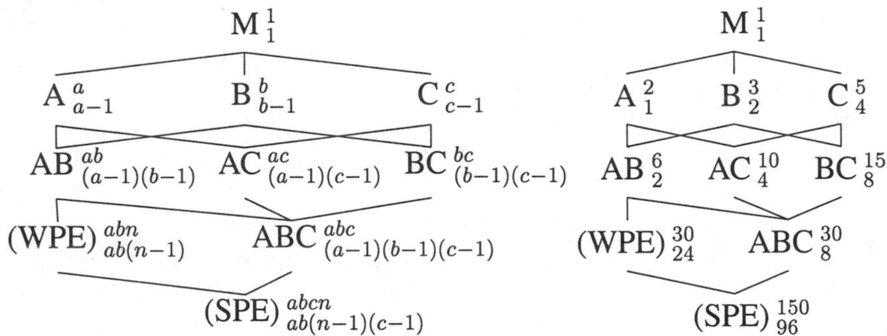


Example 16.3

### Split plot with one whole-plot factor, two split-plot factors, and CRD at the whole-plot level

This split plot again has three factors, but now only one is at the whole-plot level and two are at the split-plot level. We keep a completely randomized design for whole plots. An appropriate model for this design would be

$$\begin{aligned} y_{ijk} &= \mu + \alpha_i + \eta_{l(i)} \\ &\quad + \beta_j + \alpha\beta_{ij} + \gamma_k + \alpha\gamma_{ik} + \beta\gamma_{jk} + \alpha\beta\gamma_{ijk} + \epsilon_{l(ijk)} \ , \end{aligned}$$

where we have arranged the model into a first line with whole-plot effects (those without $j$ or $k$) and a second line with split-plot effects. The indices $i$,

$j$, and $k$ run up to $a$, $b$, and $c$, the number of levels of factors A, B, and C; and the index $l$ runs up to $n$, the amount of replication at the whole-plot level.

Below is the generic Hasse diagram for such a split plot. The denominator for the whole-plot effect A is whole-plot error; the denominator for the split plot effects B, AB, C, AC, BC, and ABC is split-plot error.

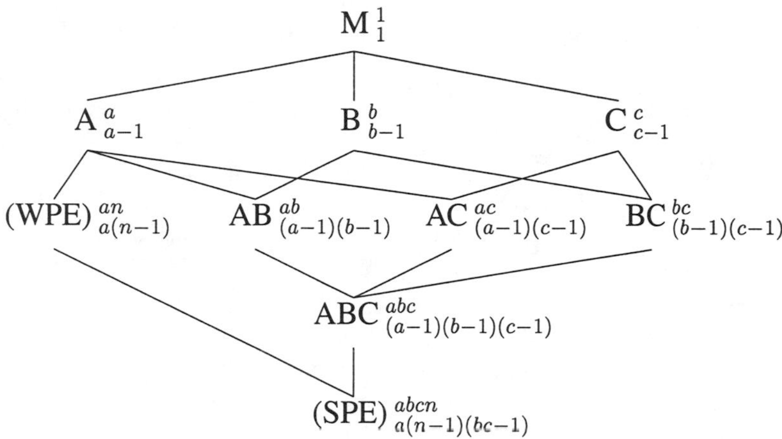


### Split plot with one whole-plot factor, one split-plot factor, and RCB at the whole-plot level

**Example 16.4**

Now consider a split-plot design with two factors, one at the whole-plot level and one at the split-plot level, but use a block design for the whole plots. An appropriate model for this design would be

$$\begin{aligned} y_{ijk} &= \mu + \alpha_i + \gamma_k + \eta_{l(ik)} \\ &\quad + \beta_j + \alpha\beta_{ij} + \epsilon_{l(ijk)} \ , \end{aligned}$$

where we have again arranged the model into a first line with whole-plot effects (those without $j$) and a second line with split-plot effects. The indices $i$ and $j$ run up to $a$ and $b$, the number of levels of factors A and B; the index $k$ runs up to $n$, the number of blocks at the whole-plot level; and the index $l$ runs up to 1, the number of whole plots in each block getting a given whole-plot treatment or the number of split plots in each whole plot getting a given split-plot treatment. Thus the model assumes that block effects are fixed and additive with whole-plot treatments, and there is a random error for each whole plot. This is just the standard RCB model applied to the whole plots.

Below is a generic Hasse diagram for a blocked split plot and a sample Hasse diagram for a split plot with $n = 5$ blocks and whole-plot factor A with

$a = 2$ levels, and split-plot factor B with $b = 3$ levels. The denominator for the whole-plot effect A is whole-plot error; the denominator for the split-plot effects B and AB is split-plot error.

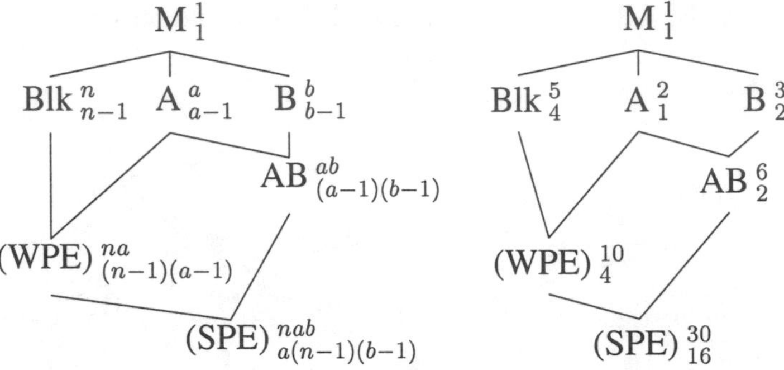


This model assumes that blocks are additive. If we allow a block by whole-plot factor interaction, then there will be no degrees of freedom for whole-plot error, and we will need to use the block by whole-plot factor interaction as surrogate error for whole-plot factor.

Partition variation into between and within whole plots

We can use our standard methods for mixed-effects factorials from Chapter 12 to analyze split-plot designs using these split-plot models. Alternatively, we can achieve the same results using the following heuristic approach. A split plot has two sizes of units and two randomizations, so first split the variation in the data into two bundles, the variation between whole plots and the variation within whole plots (between split plots). Using a simple split-plot design with just two factors, there are $an$ whole plots and $N - 1 = abn - 1$ degrees of freedom between all the responses. We can get the variation between whole plots by considering the whole plots to be $an$ "treatment groups" of $b$ units each and doing an ordinary one-way ANOVA. There are thus $an - 1$ degrees of freedom between the whole plots and $(abn - 1) - (an - 1) = an(b - 1)$ degrees of freedom within whole plots, between split plots. Visualize this decomposition as:

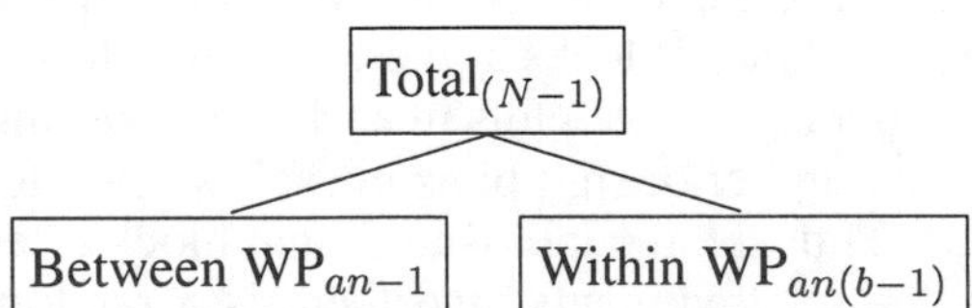


The between whole plots variation is made up of effects that affect complete whole plots. These include the whole-plot treatment factor(s), whole-plot error, and any blocking that might have been done at the whole-plot level. This variation yields the following decomposition, assuming the whole plots were blocked.

Whole-plot variation includes blocks, whole-plot factor, and whole-plot error

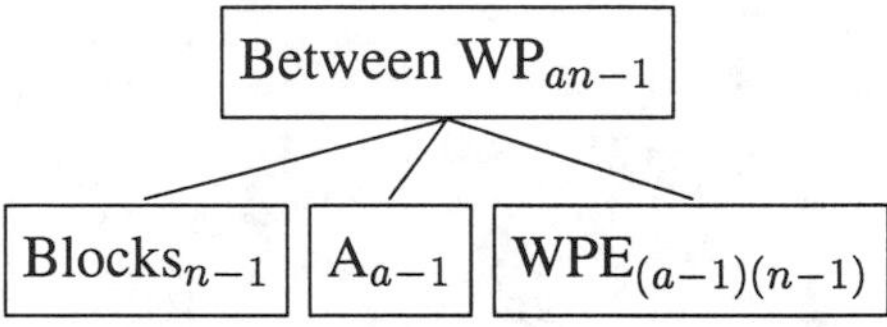


The variation between split plots (within whole plots) is variation in the responses that depends on effects that affect individual split plots, including the split-plot treatment factor(s), interaction between whole-plot and split-plot treatment factors, and split-plot error. The variation is decomposed as

Split-plot variation includes split-plot factor, whole-by-split-factor interaction, and split-plot error

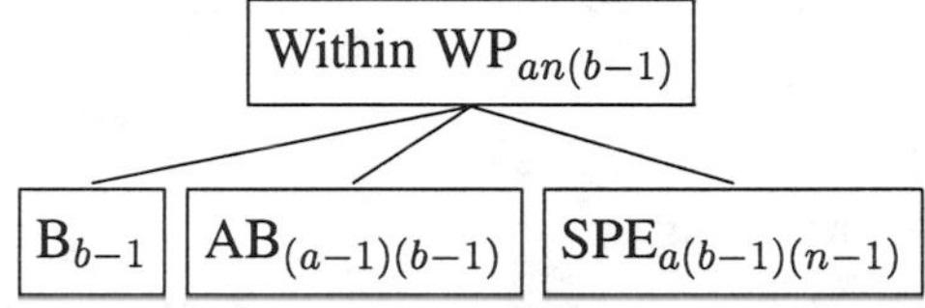


The easiest way to get the degrees of freedom for split-plot error is by subtraction. There are $an(b-1)$ degrees of freedom between split plots within whole plots; $b-1$ of these go to B, $(a-1)(b-1)$ go to AB, and the remainder must be split-plot error.

Get df by subtraction

It may not be obvious why the interaction between the whole- and split-plot factors should be a split-plot level effect. Recall that one way to describe this interaction is how the split-plot treatment effects change as we vary the whole-plot treatment. Because this is dealing with changing split-plot treatment levels, this effect cannot be at the whole-plot level; it must be lower.

Interaction at split-plot level

Assembling the pieces, we get the overall decomposition:

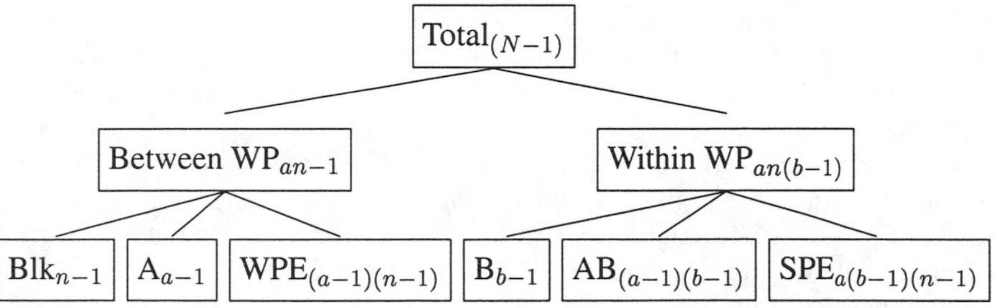


I find that this decomposition gives me a little more understanding about what is going on in the split-plot analysis than just looking at the Hasse diagram.

**Table 16.1:** Number of memory errors by type, tension, and anxiety level; subjects are columns.

| Anxiety | 1 | 1 | 1 | 1 | 1 | 1 | 2 | 2 | 2 | 2 | 2 | 2 |
|---|---|---|---|---|---|---|---|---|---|---|---|---|
| Tension | 1 | 1 | 1 | 2 | 2 | 2 | 1 | 1 | 1 | 2 | 2 | 2 |
| Type 1 | 18 | 19 | 14 | 16 | 12 | 18 | 16 | 18 | 16 | 19 | 16 | 16 |
| Type 2 | 14 | 12 | 10 | 12 | 8 | 10 | 10 | 8 | 12 | 16 | 14 | 12 |
| Type 3 | 12 | 8 | 6 | 10 | 6 | 5 | 8 | 4 | 6 | 10 | 10 | 8 |
| Type 4 | 6 | 4 | 2 | 4 | 2 | 1 | 4 | 1 | 2 | 8 | 9 | 8 |

We compute sums of squares and estimates of treatment effects in the usual way. When it is time for testing or computing standard errors for contrasts, effects at the split-plot level use the split-plot error with its degrees of freedom, and effects at the whole-plot level use the whole-plot error with its degrees of freedom.

**Example 16.5**

### Anxiety, tension, and memory

We wish to study the effects of anxiety and muscular tension on four different types of memory. Twelve subjects are assigned to one of four anxiety-tension combinations at random. The low-anxiety group is told that they will be awarded \$5 for participation and \$10 if they remember sufficiently accurately, and the high-anxiety group is told that they will be awarded \$5 for participation and \$100 if they remember sufficiently accurately. Everyone must squeeze a spring-loaded grip to keep a buzzer from sounding during the testing period. The high-tension group must squeeze against a stronger spring than the low-tension group. All subjects then perform four memory trials in random order, testing four different types of memory. The response is the number of errors on each memory trial, as shown in Table 16.1.

This is a split-plot design. There are two separate randomizations. We first randomly assign the anxiety-tension combinations to each subject. Even though we will have four responses from each subject, the randomization is restricted so that all four of those responses will be at the same anxiety-tension combination. Anxiety and tension are thus whole-plot treatment factors. Each subject will do four memory trials. The trial type is randomized to the four trials for a given subject. Thus the four trials for a subject are the split plots, and the trial type is the split-plot treatment. At the whole-plot level, the anxiety-tension combinations are assigned according to a CRD, so there is no blocking.

Listing 16.1 shows some Minitab output from an analysis of these data. The ANOVA table has been arranged so that the whole-plot analysis is on

**Listing 16.1:** Minitab output for memory errors data.

```
Source                       DF     Seq SS     Adj SS     Adj MS       F      P
anxiety                       1     10.083     10.083     10.083    0.98  0.352
tension                       1      8.333      8.333      8.333    0.81  0.395
anxiety*tension               1     80.083     80.083     80.083    7.77  0.024
subject(anxiety tension)      8     82.500     82.500     10.312    4.74  0.001
type                          3    991.500    991.500    330.500  152.05  0.000
anxiety*type                  3      8.417      8.417      2.806    1.29  0.300
tension*type                  3     12.167     12.167      4.056    1.87  0.162
anxiety*tension*type          3     12.750     12.750      4.250    1.96  0.148
Error                        24     52.167     52.167      2.174
```

top and the split-plot analysis below, as is customary. The whole-plot error is shown as *subject* nested in *anxiety* and *tension,* and the split-plot error is just denoted *Error.* Note that the split-plot error is smaller than the whole-plot error by a factor of nearly 5. Subject to subject variation is not negligible, and split-plot comparisons, which are made with subjects as blocks, are much more precise than whole-plot comparisons, where subjects are units.

At the split-plot level, the effect of type is highly significant. All the type effects $\gamma_k$ differ from each other by more than 3, and the standard error of the difference of two type means is $\sqrt{2.174(1/12+1/12)} = .602$. Thus all type means are at least 5 standard errors apart and can be distinguished from each other. No interactions with type appear to be significant.

Analysis at the whole-plot level is more ambiguous. The main effects of anxiety and tension are both nonsignificant, but their interaction is moderately significant. Figure 16.1 shows an interaction plot for anxiety and tension. We see that more errors occur when anxiety and tension are both low or both high. With such strong interaction, it makes sense to examine the treatment means themselves. The greatest difference between the four whole plot treatment means is 3.5, and the standard error for a difference of two means is $\sqrt{10.312(1/12+1/12)} = 1.311$. This is only a bit more than 2.5 standard errors and is not significant after adjusting for multiple comparisons; for example, the Bonferroni $p$-value is .17. This is in accordance with the result we obtain by considering the four whole-plot treatments to be a single factor with four levels. Pooling sums of squares and degrees of freedom for anxiety, tension, and their interaction, we get a mean square of 32.83 with 3 degrees of freedom and a $p$-value of .08.

The residuals-versus-predicted plot shows slight nonconstant variance; no transformation makes much improvement, so the data have been analyzed on the original scale.

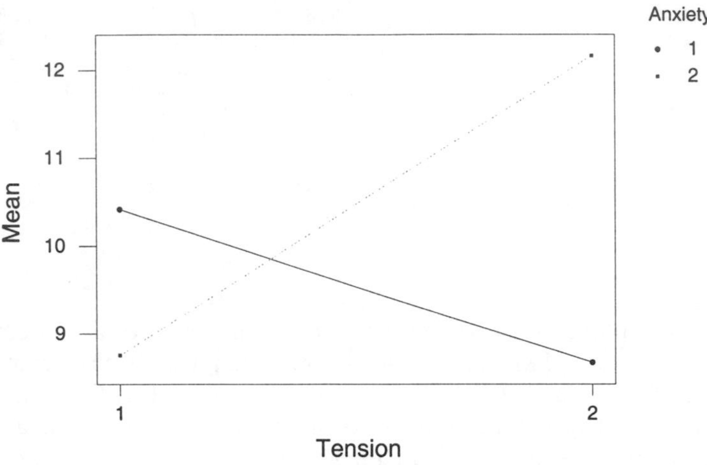


**Figure 16.1:** Anxiety by tension interaction plot for memory errors data, using Minitab.

In conclusion, there is strong evidence that the number of errors differs between memory type. There is no evidence that this difference depends on anxiety or tension individually. There is mild evidence that there are more errors when anxiety and tension are both high or both low, but none of the actual anxiety-tension combinations can be distinguished.

Alternate model has blocks random and interacting

Let me note here that some authors prefer an alternate model for the split plot with one whole-plot factor, one split-plot factor, and RCB structure on the whole plots. This model assumes that blocks are a random effect that interact with all other factors; effectively this is a three-way factorial model with one random factor.

## 16.4 Split-Split Plots

Split the split plots

What we have split once, we can split again. Consider an experiment with three factors. The levels of factor A are assigned at random to $n$ whole plots each (total of $an$ whole plots). Each whole plot is split into $b$ split plots.

The levels of factor B are assigned at random to split plots, using whole plots as blocks. So far, this is just like a split-plot design. Now each split plot is divided into $c$ split-split plots, and the levels of factor C are randomly assigned to split-split plots using split plots as blocks. Obviously, once we get used to splitting, we can split again for a fourth factor, and keep on going.

Split-split plots arise for the same reasons as ordinary split plots: some factors are easier to vary than others. For example, consider a chemical experiment where we study the effects of the type of feedstock, the temperature of the reaction, and the duration of the reaction on yield. Some experimental setups require extensive cleaning between different feedstocks, so we might wish to vary the feedstock as infrequently as possible. Similarly, there may be some delay that must occur when the temperature is changed to allow the equipment to equilibrate at the new temperature. In such a situation, we might choose type of feedstock as the whole-plot factor, temperature of reaction as the split-plot factor, and duration of reaction as the split-split-plot factor. This makes our experiment more feasible logistically, because we have fewer cleanups and temperature delays; comparisons involving time will be more precise than those for temperature, which are themselves more precise than those for feedstock.

Use split-split plots with three levels of difficulty for varying factors

Split-split plots have three sizes of units. Whole plots act as unit for the whole-plot treatments. Whole plots act as blocks for split plots, and split plots act as unit for the split-plot treatments. Split plots act as blocks for split-split plots, and split-split plots act as unit for the split-split-plot treatments. The whole plots can be blocked, just as in the split plot.

**Example 16.6**

## Split-split plot with one whole-plot factor, one split-plot factor, one split-split-plot factor and CRD at the whole plot level

Now consider a split-split-plot design with three factors, one at the whole-plot level, one at the split-plot level, and one at the split-split-plot level, with a completely randomized design for whole plots. An appropriate model for this design would be

$$\begin{aligned} y_{ijk} &= \mu + \alpha_i + \eta_{l(i)} \\ &\quad + \beta_j + \alpha\beta_{ij} + \zeta_{l(ij)} \\ &\quad + \gamma_k + \alpha\gamma_{ik} + \beta\gamma_{jk} + \alpha\beta\gamma_{ijk} + \epsilon_{l(ijk)} \ , \end{aligned}$$

where we have arranged the model into a first line with whole-plot effects (those without $j$ or $k$), a second line with split-plot effects (those with $j$ but not $k$), and the last line with split-split-plot effects. The indices $i$, $j$, and $k$ run up to $a$, $b$, and $c$, the number of levels of factors A, B, and C; and the index $l$ runs up to $n$, the amount of replication at the whole plot level.

Below is a Hasse diagram for this generic split-split plot with three factors and a CRD at the whole-plot level. The denominator for the whole-plot effect A is whole-plot error; the denominator for the split-plot effects B and AB is the split-plot error; and the denominator for the split-split-plot effects C, AC, BC, and ABC is split-split-plot error (SSPE).

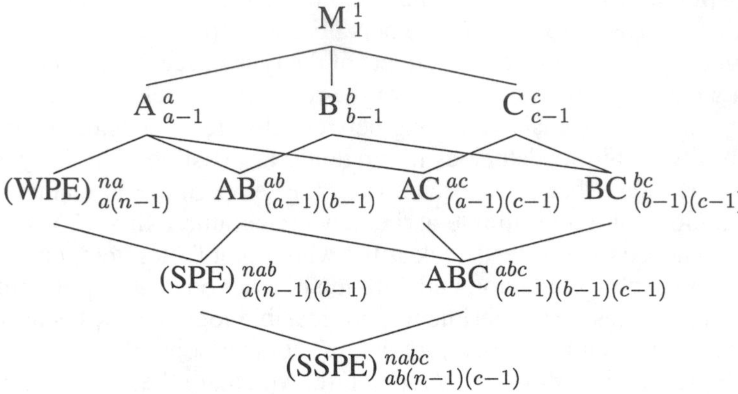


Randomization, not number of factors, determines design

A split-split plot has at least three treatment factors, but it can have more than three. Any of whole-, split-, or split-split-plot treatments can have factorial structure. Thus you cannot distinguish a split plot from a split-split plot or other design solely on the basis of the number of factors; the units and randomization determine the design.

Partition variation between levels of the design

Analysis of a split-split plot can be conducted using standard methods for mixed-effects factorials, but I find that a graphical partitioning of degrees of freedom and their associated sums of squares helps me understand what is going on. Consider three factors with $a$, $b$, and $c$ levels, in a split-split-plot design with $n$ replications. Begin the decomposition just as for a split plot:

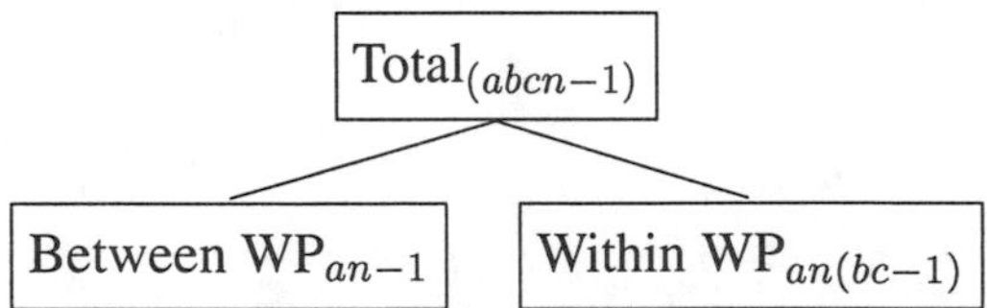


The only difference between this and a split-plot design is that we have $bc-1$ degrees of freedom within each whole plot, because each whole plot is a bundle of $bc$ split-split-plot values instead of just $b$ split-plot values.

The between whole plots variation partitions in the same way as for a split-plot design. For example, with blocking we get:

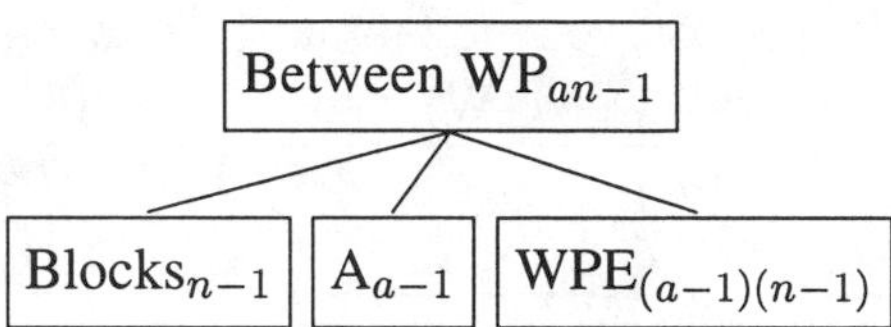


Between and within split plots

Variation within whole plots can be divided into variation between split plots and variation between split-split plots within the split plots. This is like split plots as block variation, and split-split plots as unit to unit within block variation. This partition is:

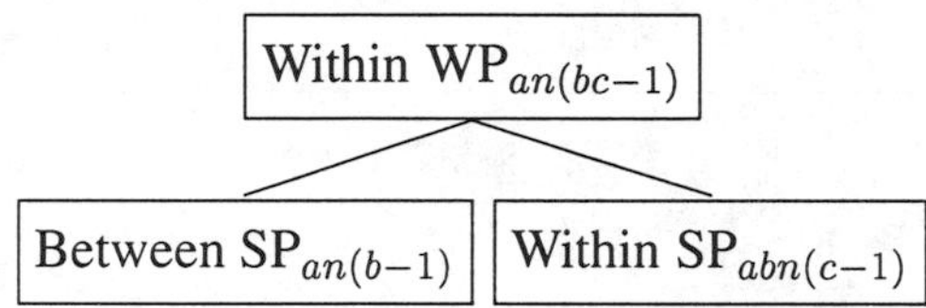


There are $b$ split plots in each whole plot, so $b-1$ degrees of freedom between split plots in a single whole plot, and $an(b-1)$ total degrees of freedom between split plots within whole plots. There are $c$ split-split plots in each split plot, so $c-1$ degrees of freedom between split-split plots in a single split plot, and $abn(c-1)$ total degrees of freedom between split-split plots within a split plot.

Between split plots

The variation between split plots within whole plots is partitioned just as for a split-plot design:

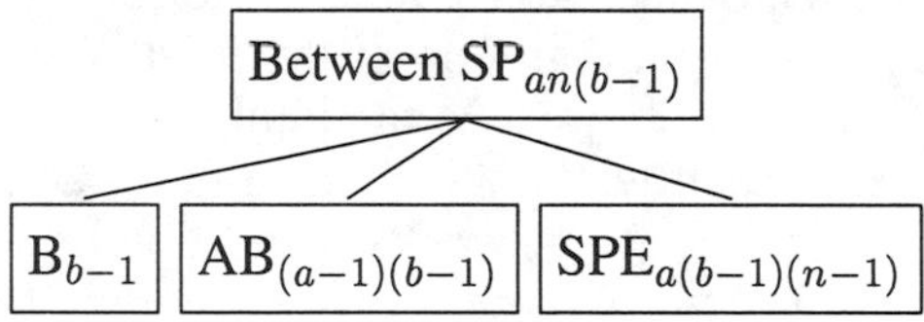


Between split-split plots

Finally, we come to the variation between split-split plots within split plots. This is variation due to factor C and its interactions, and split-split-plot error:

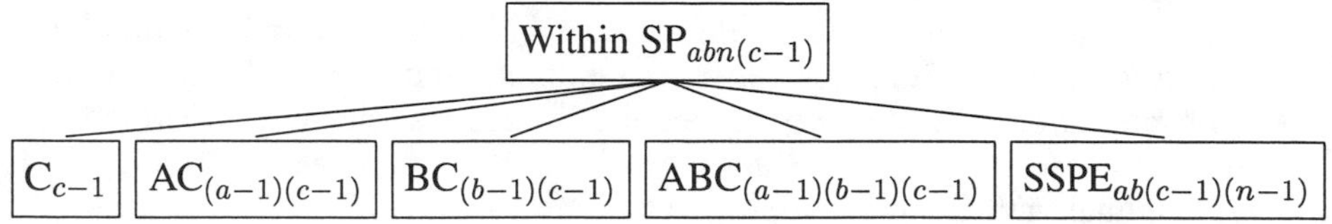

**Table 16.2:** Percent of wetland biomass that is nonweed, by table (T), nitrogen (N), weed (W), and clipping (C).

| | | W 1 | | W 2 | | W 3 | |
|---|---|---|---|---|---|---|---|
| T | N | C 1 | C 2 | C 1 | C 2 | C 1 | C 2 |
| 1 | 1 | 87.2 | 88.8 | 70.4 | 75.7 | 75.9 | 80.6 |
| | 2 | 80.5 | 83.8 | 59.2 | 61.5 | 59.5 | 62.5 |
| | 3 | 76.8 | 80.8 | 47.8 | 49.5 | 48.4 | 52.9 |
| | 4 | 77.7 | 81.5 | 35.7 | 37.3 | 38.3 | 42.4 |
| 2 | 1 | 78.2 | 80.5 | 65.1 | 68.3 | 65.3 | 66.6 |
| | 2 | 79.8 | 85.2 | 57.6 | 61.4 | 58.5 | 61.6 |
| | 3 | 82.4 | 83.1 | 50.5 | 54.0 | 51.6 | 54.7 |
| | 4 | 75.5 | 78.7 | 39.0 | 43.9 | 41.9 | 45.1 |

**Example 16.7**

### Weed biomass in wetlands

An experiment studies the effect of nitrogen and weeds on plant growth in wetlands. We investigate four levels of nitrogen, three weed treatments (no additional weeds, addition of weed species 1, addition of weed species 2), and two herbivory treatments (clipping and no clipping). We have eight trays; each tray holds three artificial wetlands consisting of rectangular wire baskets containing wetland soil. The trays are full of water, so the artificial wetlands stay wet. All of the artificial wetlands receive a standard set of seeds to start growth.

Four of the trays are placed on a table near the door of the greenhouse, and the other four trays are placed on a table in the center of the greenhouse. On each table, we randomly assign one of the trays to each of the four nitrogen treatments. Within each tray, we randomly assign the wetlands to the three weed treatments. Each wetland is split in half. One half is chosen at random and will be clipped after 4 weeks, with the clippings removed; the other half is not clipped. After 8 weeks, we measure the fraction of biomass in each wetland that is nonweed as our response. Responses are given in Table 16.2.

This is a split-split-plot design. Everything in a given tray has the same level of nitrogen, so the trays are whole plots, and nitrogen is the whole-plot factor. The whole plots are arranged in two blocks, with table as block accounting for any differences between the door and center of the greenhouse. Both measurements for a given wetland have the same weed treatment, so the wetlands are split plots, and weed is the split-plot factor. Finally each wetland half gets its own clipping treatment, so wetland halves are split-split plots, and clipping is the split-split-plot factor.

**Listing 16.2:** SAS output for wetland weeds data.

```
                                     Sum of            Mean
Source                  DF          Squares          Square    F Value      Pr > F

Model                   35       11602.7467        331.5070     310.30      0.0001

Error                   12          12.8200          1.0683

Source                  DF        Type I SS     Mean Square    F Value      Pr > F

TABLE                    1         14.30083        14.30083      13.39      0.0033
N                        3       3197.05500      1065.68500     997.52      0.0001
TRAY                     3        278.95083        92.98361      87.04      0.0001
W                        2       7001.25542      3500.62771    3276.72      0.0001
N*W                      6        929.51625       154.91938     145.01      0.0001
WET                      8         50.41833         6.30229       5.90      0.0033
C                        1        125.45333       125.45333     117.43      0.0001
N*C                      3          0.73500         0.24500       0.23      0.8742
W*C                      2          0.24542         0.12271       0.11      0.8925
N*W*C                    6          4.81625         0.80271       0.75      0.6203

Tests of Hypotheses using the Type I MS for TRAY as an error term

Source                  DF        Type I SS     Mean Square    F Value      Pr > F

N                        3       3197.05500      1065.68500      11.46      0.0377

Tests of Hypotheses using the Type I MS for WET as an error term

Source                  DF        Type I SS     Mean Square    F Value      Pr > F

W                        2       7001.25542      3500.62771     555.45      0.0001
N*W                      6        929.51625       154.91938      24.58      0.0001
```

Listing 16.2 shows SAS output for these data. Notice that F-ratios and $p$-values in the ANOVA table use the 12-degree-of-freedom error term as denominator. This is correct for split-split-plot terms (those including clipping), but is incorrect for whole-plot and split-plot terms. Those must be tested separately in SAS by specifying the appropriate denominators. This is important, because the whole-plot error mean square is about 15 times as big as the split-plot error mean square, which is about 6 times as big as the split-split-plot mean square.

All main effects and the nitrogen by weed interaction are significant. An interaction plot for nitrogen and weed shows the nature of the interaction, Figure 16.2. Weeds do better as nitrogen is introduced, but the effect is much larger when the weeds have been seeded. Clipping slightly increases the fraction of nonweed biomass.

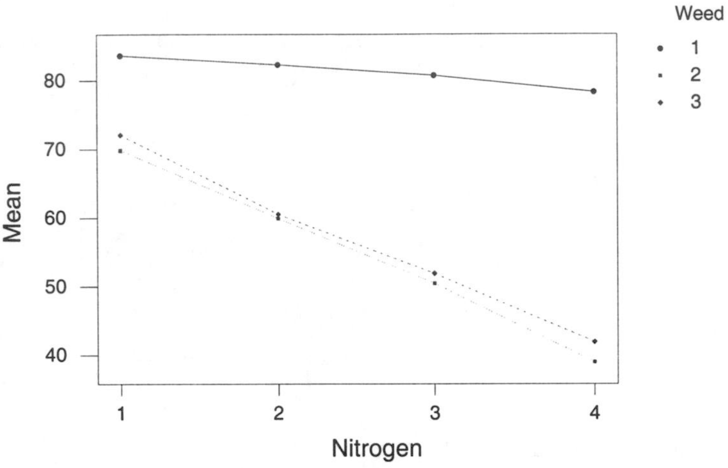


**Figure 16.2:** Nitrogen by weed interaction plot for for wetland weeds data, using Minitab.

Residual plots show that the variance increases somewhat with the mean, but no reasonable transformation fixes the problem.

## 16.5 Other Generalizations of Split Plots

Other unit structures besides nesting are possible

One way to think about split plots is that the units have a structure somewhat like that of nested factorial treatments. In a split plot, the split plots are nested in whole plots; in a split-split plot, the split-split plots are nested in split plots, which are themselves nested in whole plots. In the split-plot design, levels of different factors are assigned to the different kinds of units. This section deals with some other unit structures that are possible.

**Example 16.8** **Machine shop**

Consider a machine shop that is producing parts cut from metal blanks. The quality of the parts is determined by their strength and fidelity to the desired shape. The shop wishes to determine how brand of cutting tool and supplier of metal blank affect the quality. An experiment will be performed one week, and then repeated the next week. Four brands of cutting tools will be obtained, and brand of tool will be randomly assigned to four lathes. A different supplier of metal blank will be randomly selected for each of the 5 work days during the week. That way, all brand-supplier combinations are observed.

A schematic for the design might look like this:

| | Day 1 | Day 2 | Day 3 | Day 4 | Day 5 |
|---|---|---|---|---|---|
| Lathe 1 | Br 3 Sp 5 | Br 3 Sp 1 | Br 3 Sp 2 | Br 3 Sp 4 | Br 3 Sp 3 |
| Lathe 2 | Br 2 Sp 5 | Br 2 Sp 1 | Br 2 Sp 2 | Br 2 Sp 4 | Br 2 Sp 3 |
| Lathe 3 | Br 1 Sp 5 | Br 1 Sp 1 | Br 1 Sp 2 | Br 1 Sp 4 | Br 1 Sp 3 |
| Lathe 4 | Br 4 Sp 5 | Br 4 Sp 1 | Br 4 Sp 2 | Br 4 Sp 4 | Br 4 Sp 3 |

The table shows the combinations of the four lathes and 5 days. Brand is assigned to lathe, or row of the table. Thus the unit for brand is lathe. Supplier of blanks is assigned to day, or column of the table. Thus the unit for supplier is day. There are two separate randomizations done in this design to two different kinds of units, but this is not a split plot, because here the units do not nest as they would in a split plot.

Strip plot or split block, with units that cross

The design used in the machine shop example has been given a couple of different names, including *strip plot* and *split block*. What we have in a strip plot is two different kinds of units, with levels of factors assigned to each unit, but the units *cross* each other. This is in contrast to the split plot, where the units nest.

Strip plot easy to use

Like the split plot, the strip plot arises through ease-of-use considerations. It is easier to use one brand of tool on each lathe than it is to change. Similarly, it is easier to use one supplier all day than to change suppliers during the day. When units are large and treatments difficult to change, but the units and treatments can cross, a strip plot can be the design of choice.

Random term for every unit and every cross of units

The usual assumptions in model building for split plots and related designs such as strip plots are that there is a random term for each kind of unit, or kind of randomization if you prefer, and there is a random term whenever two units cross. For the split plot, there is a random term for whole plots that we call whole-plot error, and a random term for split plots that we call split-plot error. There are no further random terms because the unit structure in a whole plot does not cross; it nests.

For the strip plot, there is a random term for rows and a random term for columns, because these are the two basic units. There is also a random term for each row-column combination, because this is where two units cross. For the machine tool example, we have the model

$$\begin{aligned} y_{ijkl} &= \mu + \gamma_k + \alpha_i + \eta_{l(ik)} + \\ &\quad \beta_j + \zeta_{l(jk)} + \\ &\quad \alpha\beta_{ij} + \epsilon_{l(ijk)} \ , \end{aligned}$$

Strip plot has row, column, and unit errors

where $i$ and $j$ index the levels of brand and supplier, $k$ indexes the week (weeks are acting as blocks), and $l$ is always 1 and indicates a particular unit for a block-treatment-unit size combination. The term $\eta_{l(ik)}$ is the random effect for machine to machine (row to row) differences within a week; the term $\zeta_{l(jk)}$ is the random effect for day to day (column to column) differences within a week; $\epsilon_{l(ijk)}$ is unit experimental error.

Here is a Hasse diagram for the machine shop example. We denote brand and supplier by B and S; R and C denote the row and column random effects.

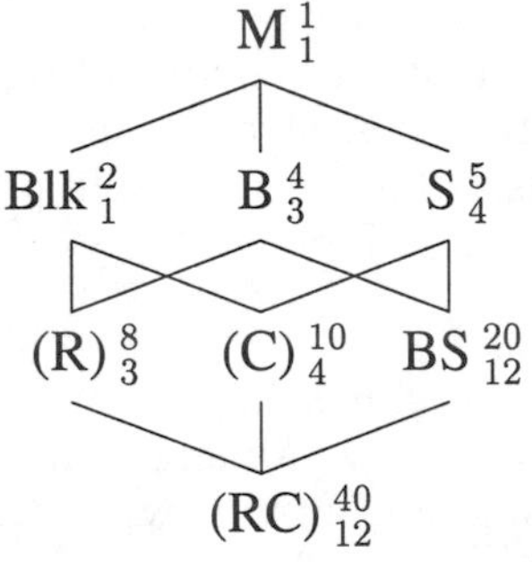


Interaction error smaller

We can see from the Hasse diagram that row and column mean squares tend to be larger than the error for individual cells. This means that a strip plot experiment has less precise comparisons and lower power for main effects, and more precision and power for interactions.

Units can nest and/or cross

When we saw that treatment factors could cross or nest, a whole world of new treatment structures opened to us. Many combinations of crossing and nesting were useful in different situations. The same is true for unit structures—we can construct more diverse designs by combining nesting and crossing of units. Just as with the split plot and strip plot, these unit structures usually arise through ease-of-use requirements.

Three kinds of units crossing

Now extend the machine tool example by supposing that in addition to four brands of tool, there are also two types. Brands of tool are assigned to each lathe at random as before, but we now assign at random the first or second tool type to morning or afternoon use. If all the lathes use the same type of tool in the morning and the other type in the afternoon, then our units have a three-way crossing structure, with lathe, day, and hour being rows, columns, and layers in a three-way table. There will be separate random terms for each unit type (lathe, day, and hour) and for each crossing of unit types (lathe by day, lathe by hour, day by hour, and lathe by day by hour).

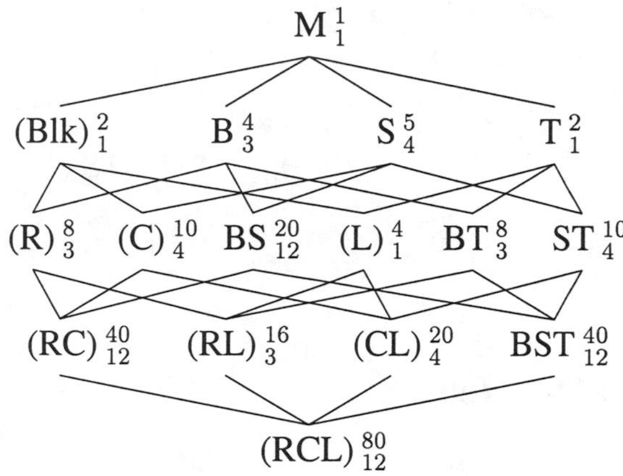


In the Hasse diagram, R, C, and L are the random effects for rows, columns, and layers (lathes, days, and hours). The interaction RCL cannot be distinguished from the usual experimental error E. The appropriate test denominators are

| Term | B | S | T | BS | BT | ST | BST |
|---|---|---|---|---|---|---|---|
| Denominator | R | C | L | RC | RL | CL | RCL |

Units nested and crossed

Alternatively, suppose that instead of using the same type of tool for all lathes in the mornings and afternoons, we instead randomize types to morning or afternoon separately for each lathe. Then ignoring supplier and day, we have hour units nested in lathe units, so that the experiment is a split plot in brand and type. Overall we have three treatment factors, all crossed, and unit structure hour nested in lathe and crossed with day. This is a split plot (in brand and type, with lathe as whole plot, time as split plot, and week as block) crossed with an RCB (in supplier, with day as unit and week as block).

The Hasse diagram for this setup is on the next page. In the Hasse diagram, R, C, and L are the random effects for rows, columns, and layers (lathes, days, and hours). The layer effects L (hours) are nested in rows (lathes). Again, the interaction CL cannot be distinguished from the usual experimental error E. The appropriate test denominators are

| Term | B | T | BT | S | BS | TS | BTS |
|---|---|---|---|---|---|---|---|
| Denominator | R | L | L | C | RC | CL | CL |

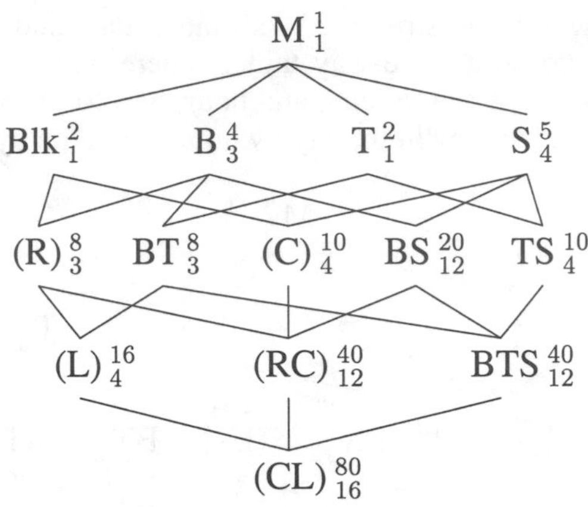


## 16.6 Repeated Measures

Consider the following experiment, which looks similar to a split-plot design but lacks an important ingredient. We wish to study the effects of different infant formulas and time on infant growth. Thirty newborns are assigned at random to three different infant formulas. (All the formulas are believed to provide adequate nutrition, and informed consent of the parents is obtained.) The weights of the infants are measured at birth, 1 week, 4 weeks, 2 months, and 6 months. The main effect of time is expected; the research questions relate to the main effect of formula and interaction between time and formula.

Split plot needs two randomizations

This looks a little like a split-plot design, with infant as whole plot and formula as whole-plot treatment, and infant time periods as split plot and age as split-plot treatment. However, this is not a split-plot design, because age was not randomized; indeed, age cannot be randomized. A split-plot design has two sizes of units and two randomizations. This experiment has two sizes of units, but only one randomization.

Repeated measures have only one randomization

This is the prototypical *repeated-measures* design. The jargon used in repeated measures is a bit different from split plots. Whole plots are usually called "subjects," whole-plot treatment factors are called "grouping factors" or "between subjects factors," and split-plot treatment factors are called "repeated measures" or "within subjects factors" or "trial factors." In a repeated-measures design, the grouping factors are randomized to the subjects, but the repeated measures are not randomized. The example has a single grouping factor applied to subjects in a completely randomized fashion, but there could be multiple grouping factors, and the subject level design could include blocking.

What we really have with a repeated-measures design is that subjects are units, and every unit has a *multivariate* response. That is, instead of a single response, every subject has a whole vector of responses, with one element for each repeated measure. Thus, each infant in the example above has a response that is a vector of length 5, giving weights at the five ages.

Repeated measures have multivariate response

The challenge presented by repeated measures is that the components in a vector of responses tend to be correlated, not independent, and every pair of repeated measures could have a different correlation. This correlation is both a blessing and a curse. It is a blessing because within-subject correlation makes comparisons between repeated measures more precise, in the same way that blocking makes treatment comparisons more precise. It is a curse because correlation complicates the analysis.

Correlated responses can improve precision but complicate analysis

There are three basic choices for the analysis of repeated-measures designs. First, you can do a full multivariate analysis, though such an analysis is beyond the scope of this text. Second, you can make a suitable univariate summary of the data for each subject, and then use these summaries as the response in a standard analysis. For the infant formula example, we could calculate the average growth rate for each infant and then analyze these as responses in a CRD with three treatments, or we could simply use the 6 month weight as response to see if the formulas have any effect on weight after 6 months. In fact, most experiments have more than one response, which we usually analyze separately; the trick comes in analyzing more than one response at a time.

Multivariate analysis

Univariate summaries

The third method is to analyze the data with a suitable ANOVA model. The applicability of the third method depends on whether nature has been kind to us: if the correlation structure of the responses meets certain requirements, then we can ignore the correlation and get a proper analysis using univariate mixed-effects models and ANOVA. For example, if all the repeated measures have the same variance, and all pairs of repeated measures have the same correlation (a condition called *compound symmetry*), then we can get an appropriate analysis by treating the repeated-measures design as if it were a split-plot design. Another important case is when there are only two repeated measures; then the requirements are always met. Thus you can always use the standard split-plot type analysis when there are only two repeated measures. When the ANOVA model is appropriate, it provides more powerful tests than the multivariate procedures.

Univariate ANOVA works in some cases, such as compound symmetry, or two repeated measures

The mysterious "certain requirements" mentioned above are called the Huynh-Feldt condition or circularity, and it states that all differences of repeated measures have the same variance. For example, compound symmetry implies the Huynh-Feldt condition. There is a test for the Huynh-Feldt condition, called the Mauchly test for sphericity, but it is very dependent on

Huynh-Feldt condition and Mauchly test

normality in the same way that most classical tests of equal variance are dependent on normality.

Random subject effect interacts with trial factors

The standard model in a univariate analysis of repeated measures assumes that there is a random effect for each subject, and that this random effect interacts with all repeated-measures effects and their interactions, but not with the grouping by repeated interactions. For example, consider a model for the infant weights:

$$y_{ijk} = \mu + \alpha_i + \epsilon_{k(i)} + \beta_j + \alpha\beta_{ij} + \epsilon\beta_{jk(i)} .$$

The term $\alpha_i$ is the formula effect (F), and $\epsilon_{k(i)}$ is the subject random effect (S); effect $\beta_j$ is age (A), and $\epsilon\beta_{jk(i)}$ is the interaction of age and subject.

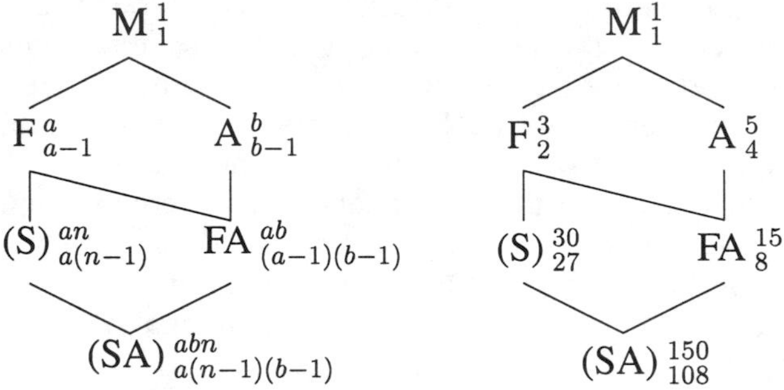


One trial factor is like split plot

We see that formula is tested against subject, and age and the formula by age interaction are tested against the subject by age interaction. This analysis is just like a split-plot design.

Suppose now that the infants are weighed twice at each age, using two different techniques. Now the model looks like

$$\begin{aligned} y_{ijkl} = \; & \mu + \alpha_i + \epsilon_{l(i)} + \\ & \beta_j + \alpha\beta_{ij} + \epsilon\beta_{jl(i)} + \\ & \gamma_k + \alpha\gamma_{ik} + \epsilon\gamma_{kl(i)} + \\ & \beta\gamma_{jk} + \alpha\beta\gamma_{ijk} + \epsilon\beta\gamma_{jkl(i)} . \end{aligned}$$

Two trial factors unlike split plot

The repeated measures effects are $\beta_j$ for age, $\gamma_k$ for measurement technique (T), and $\beta\gamma_{jk}$ for their interaction. Each of these is assumed to interact with the subject effect $\epsilon_{l(i)}$. This leads to the error structure shown in the Hasse diagram below, which is unlike either a split-plot design with two factors at the split-plot level or a split-split plot.

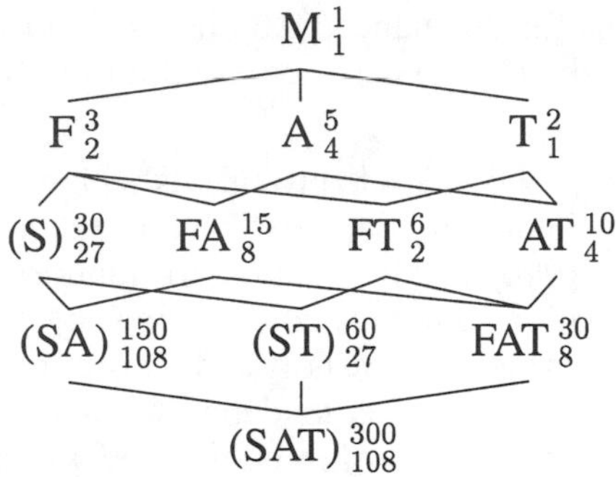


The test denominators are

| Term | F | A | FA | T | FT | AT | FAT |
|---|---|---|---|---|---|---|---|
| Denominator | S | SA | SA | ST | ST | SAT | SAT |

## 16.7 Crossover Designs

In this section we make a brief return to crossover designs, which in Chapter 13 we described as replicated Latin Squares with blocking on subjects and periods. For concreteness suppose that we have three treatments, three periods, and twelve subjects.

Crossover as Latin Square

The three treatments can be given to the subjects in any of six orders. Assign the orders at random to the subject, two subjects per order, and observe the responses to the treatments in the three periods. From this point of view, the crossover design is a repeated measures design. Order is the grouping factor, *period* is the trial factor, and treatment lies in the order by period interaction. Any carryover effects are also in the order by period interaction. It is customary not to fit the entire order by period interaction, but instead to fit only treatment and carryover effects as needed. With this reduced model, the only difference between the repeated measures and Latin Square approaches to a crossover design is that the Latin Square pools all between subjects variation into a single block term, and the repeated measure splits this into between orders and between subjects within order, allowing the estimation and testing of the overall order effect.

Crossover as repeated measure

Fit order effects

## 16.8 Further Reading and Extensions

Unbalanced mixed-effects designs are generally difficult to analyze, and split plots are no different. Software that can compute Type I and III mean squares and their expectations for unbalanced data helps find reasonable test statistics. Mathew and Sinha (1992) describe exact and optimal tests for unbalanced split plots.

Nature is not always so kind as to provide us with repeated-measures data that meet the Huynh-Feldt condition (Huynh and Feldt 1970), and as noted above, the Mauchly (1940) test is sensitive to nonnormality. The result of nonconforming correlations is to make the within subjects procedures liberal; that is, confidence intervals are too short and tests reject the null hypothesis more often than they should. This tendency for tests to be liberal can be reduced by modifying the degrees of freedom used when assessing $p$-values. For example, the within subjects tests for B and AB have $b-1, a(b-1)(n-1)$ and $(a-1)(b-1), a(b-1)(n-1)$ degrees of freedom; these degrees of freedom are adjusted by rescaling to $\lambda(b-1), \lambda a(b-1)(n-1)$ and $\lambda(a-1)(b-1), \lambda a(b-1)(n-1)$, where $1/(b-1) \leq \lambda \leq 1$.

There are two fairly common methods for computing this adjustment $\lambda$. The first is from Greenhouse and Geisser (1959); Huynh and Feldt (1976) provide a slightly less conservative correction. Both adjustments are too tedious for hand computation but are available in many software packages. Greenhouse and Geisser (1959) also provide a simple conservative test that uses the minimum possible value of $\lambda$, namely $1/(b-1)$. For this conservative approach, the tests for B and AB have $1, a(n-1)$ and $(a-1), a(n-1)$ degrees of freedom.

## 16.9 Problems

**Problem 16.1** Briefly describe the experimental design you would choose for each of the following situations, and explain why.

(a) A plant breeder wishes to study the effects of soil drainage and variety of tulip bulbs on flower production. Twelve 3 m by 10 m experimental sites are available in a garden. Each site is a .5 m–deep trench. Soil drainage is changed by adding varying amounts of sand to a clay soil (more sand improves drainage), mixing the two well, and placing the mixture in the trench. The bulbs are then planted in the soils, and flower production is measured the following spring. It is felt that four different levels of soil drainage would suffice, and there are fifteen tulip varieties that need to be studied.

(b) It's Girl Scout cookie time, and the Girl Scout leaders want to find out how to sell even more cookies (make more dough?) in the future. The variables they have to work with are type of sales (two levels: door-to-door sales or table sales at grocery stores, malls, etc.) and cookie selection (four levels comprising four different "menus" of cookies offered to customers). Administratively, the Girl Scouts are organized into "councils" consisting of many "troops" of 30-or-so girls each. Each Troop in the experiment will be assigned a menu and a sales type for the year, and for logistical reasons, all the troops in a given council should have the same cookie selection. Sixteen councils have agreed to participate in the experiment.

(c) Rodent activity may be affected by photoperiod patterns. We wish to test this possibility by treating newly-weaned mouse pups with three different treatments. Treatment 1 is a control with the mice getting 14 hours of light and 10 hours of dark per day. Treatment 2 also has 14 hours of light, but the 10 hours of dark are replaced by 10 hours of a low light level. Treatment 3 has 24 hours of full light.

Mice will be housed in individual cages, and motion detectors connected to computers will record activity. We can use 24 cages, but the computer equipment must be shared and is only available to us for 1 month.

Mice should be on a treatment for 3 days—one day to adjust and then 2 days to take measurements. We may use each mouse for more than one treatment, but if we do, there should be 7 days of standard photoperiod between treatments. We expect large subject-to-subject variation. There may or may not be a change in activity as the rat pups age; we don't know.

**Problem 16.2**

A food scientist is interested in the production of ice cream. He has two different recipes (A and B). Additional factors that may affect the ice cream are the temperature at which the process is run and the pressure used. We wish to investigate the effects of recipe, temperature, and pressure on ice cream viscosity. The production machinery is available for 8 days, and two batches of ice cream can be made each day. A fresh supply of milk will be used each day, and there is probably some day to day variability in the quality of the milk.

The production machinery is such that temperature and pressure have to be set at the start of each day and cannot be changed during the day. Both temperature and pressure can be set at one of two levels (low and high). Each batch of ice cream will be measured for viscosity.

(a) Describe an appropriate experiment. Give a skeleton ANOVA (source and degrees of freedom only), and describe an appropriate randomization scheme.

(b) Explain how to construct simultaneous 95% confidence intervals for the differences in mean viscosity between the various combinations of temperature and pressure.

**Problem 16.3** An experiment was conducted to study the effects of irrigation, crop variety, and aerially sprayed pesticide on grain yield. There were two replicates. Within each replicate, three fields were chosen and randomly assigned to be sprayed with one of the pesticides. Each field was then divided into two east-west strips; one of these strips was chosen at random to be irrigated, and the other was left unirrigated. Each east-west strip was split into north-south plots, and the two varieties were randomly assigned to plots.

| | Rep 1 | | | Rep 2 | | | |
|---|---|---|---|---|---|---|---|
| P1 | P2 | P3 | P1 | P2 | P3 | Irrig | Var |
| 53.4 | 54.3 | 55.9 | 46.5 | 57.2 | 57.4 | yes | 1 |
| 53.8 | 56.3 | 58.6 | 51.1 | 56.9 | 60.2 | yes | 2 |
| 58.2 | 60.4 | 62.4 | 49.2 | 61.6 | 57.2 | no | 1 |
| 59.5 | 64.5 | 64.5 | 51.3 | 66.8 | 62.7 | no | 2 |

What is the design of this experiment? Analyze the data and report your conclusions. What is the standard error of the estimated difference in average yield between pesticide 1 and pesticide 2? irrigation and no irrigation? variety 1 and variety 2?

**Problem 16.4** Most universities teach many sections of introductory calculus, and faculty are constantly looking for a method to evaluate students consistently across sections. Generally, all sections of intro-calculus take the final exam at the same time, so a single exam is used for all sections. An exam service claims that it can supply different exams that consistently evaluate students. Some faculty doubt this claim, in part because they believe that there may be an interaction between the text used and the exam used.

Three math departments (one each at Minnesota, Washington, and Berkeley) propose the following experiment. Three random final exams are obtained from the service: E1, E2, and E3. At Minnesota, the three exams will be used in random order in the fall, winter, and spring quarters. Randomization will also be done at Washington and Berkeley. The three schools all use the same two intro calculus texts. Sections of intro calculus at each school will be divided at random into two groups, with half of the sections using text A and the other half using text B. At the end of the year, the mean test scores are tallied with the following results.

| | | Text | |
|---|---|---|---|
| School | Exam | A | B |
| Wash | 1 | 81 | 87 |
| | 2 | 79 | 85 |
| | 3 | 70 | 78 |
| Minn | 1 | 84 | 82 |
| | 2 | 81 | 81 |
| | 3 | 83 | 84 |
| Berk | 1 | 87 | 98 |
| | 2 | 82 | 93 |
| | 3 | 86 | 90 |

Analyze these data to determine if there is any evidence of variation between exams, text effect, or exam by text interaction. Be sure to include an explicit description of the model you used.

**Problem 16.5**

Companies A, M, and S are three long-distance carriers. All claim to give quality service, but S has been advertising its network as being incredibly clear. A consumer testing agency wishes to determine if S really is any better. A complicating factor in this determination is that you don't hook directly to a long-distance company. Your call must first go through your personal phone, through local lines, and through the local switch before it even gets to the long-distance company equipment, and then the call must go through local switch, local lines, and a local phone on the receiving end. Thus while one long-distance carrier might, in fact, have clearer transmissions than the others, you might not be able to detect the difference due to noise generated by local phones, lines, and switches. Furthermore, the quality may depend on the load on the long-distance system. Load varies during the day and between days, but is fairly constant over periods up to about 15 or 20 minutes.

The consumer agency performs the following experiment. All calls will originate from one of two phones, one in New York and the other in New Haven, CT. Calls will be placed by a computer which will put a very precise 2-minute sequence of tones on the line. All calls will terminate at one of three cities: Washington, DC; Los Angeles; or Ely, MN. All calls will be answered by an answering machine with a high-quality tape recorder. The quality of the transmission is judged by comparing the tape recording of the tones with the known original tones, producing a distortion score D. Calls are placed in the following way. Twenty-four time slots were chosen at random over a period of 7 days. These 24 time slots were randomly assigned to the six originating/terminating city pairs, four time slots per pair. Three calls will be made from the originating city to the terminating city during the time slot, using each of the long-distance companies in a random order. The data follow (and are completely fictitious).

| | | Time slots | | | |
|---|---|---|---|---|---|
| City pair | LD | 1 | 2 | 3 | 4 |
| NY/DC | A | 4.3 | 4.7 | 5.6 | 7.7 |
| | M | 6.8 | 7.6 | 9.2 | 10.7 |
| | S | 2.3 | 2.2 | 2.6 | 4.3 |
| NY/LA | A | 3.2 | 5.4 | 4.9 | 10.7 |
| | M | 5.6 | 8.0 | 7.7 | 13.1 |
| | S | 0.3 | 2.3 | 3.0 | 8.2 |
| NY/Ely | A | 13.7 | 13.5 | 12.3 | 10.6 |
| | M | 16.1 | 16.5 | 15.6 | 13.2 |
| | S | 13.2 | 13.1 | 13.3 | 10.8 |
| NH/DC | A | 7.9 | 6.3 | 8.9 | 6.1 |
| | M | 10.8 | 8.7 | 10.7 | 9.0 |
| | S | 6.2 | 4.6 | 6.4 | 4.4 |
| NH/LA | A | 9.0 | 11.4 | 10.6 | 9.3 |
| | M | 11.1 | 14.5 | 13.2 | 11.6 |
| | S | 6.7 | 9.9 | 8.4 | 6.2 |
| NH/Ely | A | 13.9 | 12.1 | 14.2 | 17.1 |
| | M | 16.1 | 15.9 | 17.8 | 19.8 |
| | S | 14.2 | 11.2 | 14.4 | 16.7 |

We are mostly interested in differences in long-distance carriers, but we are also interested in city pair effects. Analyze these data. What conclusions would you draw, and what implications does the experiment have for people living in Ely?

**Problem 16.6** For each of the following, describe the experimental design used and give a skeleton ANOVA (sources and degrees of freedom only).

(a) A grocery store chain is experimenting with its weekly advertising, trying to decide among cents-off coupons, regular merchandise sales, and special-purchase merchandise sales. There are two cities about 100 km apart in which the chain operates, and the chain will always run one advertisement in each city on Wednesday, with the offer good for 1 week. The response of interest is total sales in each city, and large city to city differences in total sales are expected due to population differences. Furthermore, week to week differences are expected. The chain runs the experiment on 12 consecutive weeks, randomizing the assignment of advertising method to each city, subject to the restrictions that each of the three methods is used eight times, four times in each city, and each of the three pairs of methods is used an equal number of times.

(b) A forest products company conducts a study on twenty sites of 1 hectare each to determine good forestry practice. Their goal is to maximize the production of wood biomass (used for paper) on a given site over 20 years. All sites in the study have been cut recently, and the factors of interest are species to plant (alder or birch) and the thinning regime (thin once at 10 years, or twice at 10 and 15 years). The species is assigned at random to each site. The sites are then split into east-west halves. The thinning regimes are assigned at random to east-west halves independently for each site.

(c) We wish to study the acidity of orange juice available at our grocery store. We choose two national brands. We then choose 3 days at random (from the next month) for each brand; cartons of brand A will be purchased only on the days for brand A, and similarly for brand B. On a purchase day for brand A, we choose five cartons of brand A orange juice at random from the shelf, and similarly for brand B. Each carton is sampled twice and the samples are measured for acidity.

(d) We wish to determine the number of warblers that will respond to three recorded calls. We will get eighteen counts, nine from each of two forest clearings. We expect variation in the counts from early to mid to late morning, and we expect variation in the counts from early to mid to late in the breeding season. Each recorded call is used three times at each clearing, arranged in such a way that each call is used once in each phase of the breeding season and once in each morning hour.

**Problem 16.7**

Artificial insemination is widely used in the beef industry, but there are still many questions about how fresh semen should be frozen for later use. The motility of the thawed semen is the usual laboratory measure of semen quality, and this varies from bull to bull and ejaculate to ejaculate even without the freeze/thaw cycle. We wish to evaluate five freeze/thaw methods for their effects on motility.

Four bulls are selected at random from a population of potential donors; three ejaculates are collected from each of the four bulls (these may be considered a random sample). Each ejaculate is split into five parts, with the parts being randomly assigned to the five freeze/thaw methods. After each part is frozen and thawed, two small subsamples are taken and observed under the microscope for motility.

Give a skeleton ANOVA for this design and indicate how you would test the various effects. (Hint: is this a split plot or not?)

**Problem 16.8**

Traffic engineers are experimenting with two ideas. The first is that erecting signs that say "Accident Reduction Project Area" along freeways will raise awareness and thus reduce accidents. Such signs may have an effect on traffic speed. The second idea is that metering the flow of vehicles onto on-ramps will spread out the entering traffic and lead to an average increase in speed on the freeway. The engineers conduct an experiment to determine how these two ideas affect average traffic speed.

First, twenty more-or-less equivalent freeway interchanges are chosen, spread well around a single metropolitan area and not too close to each other. Ten of these interchanges are chosen at random to get "Accident Reduction Project Area" signs (in both directions); the other ten receive no signs. Traffic lights are installed on all on-ramps to meter traffic. The traffic lights can be turned off (that is, no minimum spacing between entering vehicles) or be adjusted to require 3 or 6 seconds between entering vehicles. Average traffic speed 6:30–8:30 A.M. and 4:30–6:30 P.M. will be measured at each interchange on three consecutive Tuesdays, with our response being the average of morning and evening speeds. At each interchange, the three settings of the traffic lights are assigned at random to the three Tuesdays.

The results of the experiment follow. Analyze the results and report your conclusions.

| | | Timing | | |
|---|---|---|---|---|
| Interchange | Sign? | 0 | 3 | 6 |
| 1 | n | 13 | 25 | 26 |
| 2 | n | 24 | 35 | 37 |
| 3 | n | 22 | 38 | 41 |
| 4 | n | 24 | 32 | 37 |
| 5 | n | 23 | 35 | 38 |
| 6 | n | 23 | 33 | 35 |
| 7 | n | 24 | 35 | 41 |
| 8 | n | 19 | 34 | 35 |
| 9 | n | 21 | 33 | 37 |
| 10 | n | 15 | 30 | 30 |
| 11 | y | 19 | 31 | 33 |
| 12 | y | 12 | 28 | 27 |
| 13 | y | 10 | 24 | 29 |
| 14 | y | 12 | 23 | 28 |
| 15 | y | 26 | 41 | 41 |
| 16 | y | 17 | 31 | 30 |
| 17 | y | 17 | 27 | 31 |
| 18 | y | 18 | 32 | 33 |
| 19 | y | 16 | 29 | 30 |
| 20 | y | 24 | 37 | 37 |

**Problem 16.9** A consumer testing agency wishes to test the ability of laundry detergents, bleaches, and prewash treatments to remove soils and stains from fab-

ric. Three detergents are selected (a liquid, an all-temperature powder, and a hot-water powder). The two bleach treatments are no bleach or chlorine bleach. The three prewash treatments are none, brand A, and brand B. The three stain treatments are mud, grass, and gravy. There are thus 54 factor-level combinations.

Each of 108 white-cotton handkerchiefs is numbered with a random code. Nine are selected at random, and these nine are assigned at random to the nine factor-level combinations of stain and prewash. These nine handkerchiefs along with four single sheets make a "tub" of wash. This is repeated twelve times to get twelve tubs. Each tub of wash is assigned at random to one of the six factor-level combinations of detergent and bleach. After washing and drying, the handkerchiefs are graded (in random order) for whiteness by a single evaluator using a 1 to 100 scale, with 1 being whitest (cleanest).

Analyze these data and report your findings.

| | | | Stain 1 | | | Stain 2 | | | Stain 3 | | |
|---|---|---|---|---|---|---|---|---|---|---|---|
| Tub | Det. | Bl. | P1 | P2 | P3 | P1 | P2 | P3 | P1 | P2 | P3 |
| 1 | 1 | 1 | 1 | 3 | 3 | 3 | 3 | 5 | 10 | 3 | 2 |
| 2 | 1 | 2 | 5 | 3 | 3 | 3 | 5 | 3 | 7 | 3 | 2 |
| 3 | 2 | 1 | 3 | 2 | 2 | 4 | 6 | 1 | 5 | 1 | 2 |
| 4 | 2 | 2 | 3 | 1 | 2 | 2 | 4 | 3 | 8 | 1 | 2 |
| 5 | 3 | 1 | 34 | 29 | 35 | 35 | 34 | 41 | 49 | 25 | 26 |
| 6 | 3 | 2 | 7 | 5 | 6 | 6 | 6 | 7 | 10 | 5 | 4 |
| 7 | 1 | 1 | 4 | 4 | 4 | 5 | 7 | 10 | 11 | 5 | 4 |
| 8 | 1 | 2 | 4 | 6 | 3 | 4 | 7 | 6 | 9 | 7 | 5 |
| 9 | 2 | 1 | 6 | 8 | 7 | 5 | 6 | 7 | 11 | 6 | 4 |
| 10 | 2 | 2 | 6 | 6 | 7 | 8 | 7 | 9 | 12 | 5 | 5 |
| 11 | 3 | 1 | 26 | 28 | 31 | 38 | 30 | 34 | 41 | 27 | 27 |
| 12 | 3 | 2 | 2 | 4 | 2 | 2 | 5 | 3 | 8 | 3 | 2 |

**Problem 16.10**

We wish to study the effect of drought stress on height growth of red maple seedlings. The factors of interest are the amount of stress and variety of tree. Stress is at two levels: no stress (that is, always well watered) and drought-stressed after 6 weeks of being well watered. There are four varieties available, and all individuals within a given variety are clones, that is, genetically identical.

This will be a greenhouse experiment so that we can control the watering. Plants will be grown in six deep sandboxes. There is space in each sandbox for 36 plants in a 6 by 6 arrangement. However, the plants in the outer row have a dissimilar environment and are used as a "guard row," so responses are observed on only the inner 16 plants (in 4 by 4 arrangement).

The six sandboxes are in a three by two arrangement, with three boxes north to south and two boxes east to west. We anticipate considerable differences in light (and perhaps temperature and other related factors) on the north to south axis. No differences are anticipated on the east to west axis.

Only one watering level can be given to each sandbox. Variety can be varied within sandbox. The response is measured after 6 months.

(a) Describe an experimental design appropriate for this setup.

(b) Give a skeleton ANOVA (sources and df only) for this design.

(c) Suppose now that the heights of the seedlings are measured ten times over the course of the experiment. Describe how your analysis would change and any assumptions that you might need to make.

**Problem 16.11** Consider the following experimental design. This design was randomized independently on each of ten fields. First, each field is split into northern and southern halves, and we randomly assign herbicide/no herbicide treatments to the two halves. Next, each field is split into eastern and western halves, and we randomly assign tillage method 1 or tillage method 2 to the two halves. Finally, each tillage half is again split into east and west halves (a quarter of the whole field), and we randomly assign two different insecticides to the two different quarters, independently in the two tillage halves. Thus, within each field we have the following setup:

| 1 | 2 | 3 | 4 |
|---|---|---|---|
| 5 | 6 | 7 | 8 |

Plots 1, 2, 3, and 4 all receive the same herbicide treatment, as do plots 5, 6, 7, and 8. Plots 1, 2, 5, and 6, all receive the same tillage treatment, as do plots 3, 4, 7, and 8. Insecticide A is given to plot pair (1, 5) or plot pair (2, 6); the other pair gets insecticide B. Similarly, one of the plot pairs (3, 7) and (4, 8) gets insecticide A and the other gets B.

Construct a Hasse diagram for this experiment. Indicate how you would test the null hypotheses that the various terms in the model are zero.

**Problem 16.12** Consider the following situation. We have four varieties of wheat to test, and three levels of nitrogen fertilizer to use, for twelve factor-level combinations. We have chosen eight blocks of land at random on an experimental study area; each block of land will be split into twelve plots in a four by three rectangular pattern. We are considering two different experimental designs. In the first design, the twelve factor-level combinations are assigned at random to the twelve plots in each block, and this randomization is redone from block to block. In the second design, a variety of wheat is assigned at random to each row of the four by three pattern, and a level of nitrogen fertilizer is assigned at random to each column of the four by three pattern; this randomization is redone from block to block.

(a) What are the types of the two designs (for example, CRD, RCB, and so on)?

(b) Give Hasse diagrams for these designs, and indicate how you would test the null hypotheses that the various terms in the model are zero.

(c) Which design provides more power for testing main effects? Which design is easier to implement?

**Problem 16.13**

Yellow perch and ruffe are two fish species that compete. An experiment is run to determine the effects of fish density and competition with ruffe on the weight change in yellow perch. There are two levels of fish density (low and high) and two levels of competition (ruffe absent and ruffe present). Sixteen tanks are arranged in four enclosures of four tanks each. Within each enclosure, the four tanks are randomly assigned to the four factor-level combinations of density and competition. The response is the change in the weight of perch after 5 weeks (in grams, data from Julia Frost).

| | | Enclosure | | | |
|---|---|---|---|---|---|
| Ruffe | Density | 1 | 2 | 3 | 4 |
| Absent | Low | .0 | .4 | .9 | -.4 |
| | High | .9 | -.4 | -.6 | -1.2 |
| Present | Low | .0 | -.4 | -.9 | -.9 |
| | High | -1.2 | -1.5 | -1.1 | -.7 |

Analyze these data for the effects of density and competition.

# Chapter 17

# Designs with Covariates

Covariates are predictive responses, meaning that covariates are responses measured for an experimental unit in anticipation that the covariates will be associated with, and thus predictors for, the primary response. The use of covariates is not design in the sense of treatment structure, unit structure, or the way treatments are assigned to units. Instead, a covariate is an additional response that we exploit by modifying our models to include. Nearly any model can be modified to include covariates.

Covariates are predictive responses

**Example 17.1**

**Keyboarding pain**

A company wishes to choose an ergonomic keyboard for its computers to reduce the severity of repetitive motion disorders (RMD) among its staff. Twelve staff known to have mild RMD problems are randomly assigned to three keyboard types. The staff keep daily logs of the amount of time spent keyboarding and their subjective assessment of the RMD pain. After 2 weeks, we get the total number of hours spent keyboarding and the total number of hours in RMD pain.

The primary response here is pain; we wish to choose a keyboard that reduces the pain. However, we know that the amount of pain depends on the amount of time spent keyboarding—more keyboarding usually leads to more pain. If we knew at the outset the amount of keyboarding to be done, we could block on time spent keyboarding. However, we don't know that at the outset of the experiment, we can only measure it along with the primary response. Keyboarding time is a covariate.

## 17.1 The Basic Covariate Model

Covariates make treatment comparisons more precise

Before we show how to use covariates, let's describe what they can do for us. First, we can make comparisons between treatments more precise by including covariates in our model. Thus we get a form of variance reduction through modeling the response-covariate relationship, rather than through blocking. The responses we observe are just as variable as without covariates, but we can account for some of that variability using covariates in our model and obtain many of the benefits of variance reduction via modeling instead of blocking.

Treatment comparisons adjusted to common covariate value

Second—and this is not completely separate from the first advantage—covariate models allow us to compare predicted treatment responses at a common value of the covariate for all treatments. Thus treatments which by chance received above or below average covariate values can be compared in the center.

Treatments should not affect covariates

One potential pitfall of covariate models is that they assume that the covariate is not affected by the treatment. When treatments affect covariates, the comparison of responses at equal covariate values (our second advantage) may, in fact, obscure treatment differences. For example, one of the keyboards may be so awkward that the users avoid typing; trying to compare it to the others at an average amount of typing hides part of the effect of the keyboard.

Treatment and covariate effects

The key to using covariates is building a model that is appropriate for the design and the data. Covariate models have two parts: a usual treatment effect part and a covariate effect part. The treatment effect part is essentially determined by the design, as usual; but there are several possibilities for the covariate effect part, and our model will be appropriate for the data only when we have accurately modeled the relationship between the covariates and the response.

Analysis of covariance

Let's begin with the simplest sort of covariance modeling—in fact, the sort usually called *Analysis of Covariance*. We will generalize to more complicated models later. Consider a completely randomized design with a single covariate $x$; let $x_{ij}$ be the covariate for $y_{ij}$. For the CRD, the model ignoring the covariate is

$$y_{ij} = \mu + \alpha_i + \epsilon_{ij} \quad .$$

We can estimate the $i$th treatment mean $\widehat{\mu} + \widehat{\alpha}_i$ or a contrast between treatments $\sum w_i \widehat{\alpha}_i$, and we can test the null hypothesis that all the $\alpha_i$ values are zero with the usual F-test by comparing the fit of this model to the fit of a model without the $\alpha_i$'s.

Now consider a model that uses the covariate. We augment the previous model to include a regression-like term for the covariate:

Include covariate via regression

$$y_{ij} = \mu^\star + \alpha_i^\star + \beta x_{ij} + \epsilon_{ij}^\star \ .$$

As usual, the treatment effects $\alpha_i^\star$ add to zero. The $\star$'s in this model are shown just this once to indicate that the $\mu$, $\alpha_i$, and $\epsilon_{ij}$ values in this model are different from those in the model without covariates. The $\star$'s will be dropped now for ease of notation.

The difference between the covariate and no-covariate models is the term $\beta x_{ij}$. This term models the response as a linear function of the covariate $x$. The assumption of a linear relationship between $x$ and $y$ is a big one, and writing a model with a linear relationship doesn't make the actual relationship linear. As with any regression, we may need to transform the $x$ or $y$ to improve linearity. Plots of the response versus the covariate are essential for assessing this relationship.

Model assumes linear relationship between response and covariate

Also note that the slope $\beta$ is assumed to be the same for every treatment. The covariate model for treatment $i$ is a linear regression with slope $\beta$ and intercept $\mu + \alpha_i$. Because the $\alpha_i$'s can all differ, this is a set of parallel lines, one for each treatment. Thus this covariate model is called the *parallel-lines* model or the *separate-intercepts* model.

Common slope creates parallel lines

We need to be able to test the same hypotheses and estimate the same quantities as in noncovariate models. To test the null hypothesis of no treatment effects (all the $\alpha_i$'s equal to zero) when covariate effects are present, compare the model with treatment and covariate effects to the reduced model with only covariate effects:

Test via model comparison

$$y_{ij} = \mu + \beta x_{ij} + \epsilon_{ij} \ .$$

This simpler model is called the *single-line* model, because it is a simple linear regression of the response on the covariate. The reduction in error sum of squares going from the single-line model to the parallel-lines model has $g-1$ degrees of freedom. The mean square for this reduction is divided by the mean square for error from the larger parallel-lines model to form an F-test of the null hypothesis of no treatment effects. These treatment effects are said to be covariate-adjusted, because the covariate is present in the model. There are formulae for these sums of squares, but I don't think you'll find them enlightening; just let your software do the computations.

Single-line model

F-test for covariate-adjusted treatment effects

The underlying philosophy of the test is that the covariate relationship with the response is real and exists with or without treatment effects. The test is only to determine if adding treatment effects to a model that already includes a covariate makes any significant improvement in explanatory power.

Analysis of Covariance

**Table 17.1:** Hours keyboarding (x) and hours of repetitive-motion pain (y) during 2 weeks for three styles of keyboards.

| 1 | | 2 | | 3 | |
|---|---|---|---|---|---|
| x | y | x | y | x | y |
| 60 | 85 | 54 | 41 | 56 | 41 |
| 72 | 95 | 68 | 74 | 56 | 34 |
| 61 | 69 | 66 | 71 | 55 | 50 |
| 50 | 58 | 59 | 52 | 51 | 40 |

That is, does the parallel-lines model explain significantly more than the single-line model. This test is the classical Analysis of Covariance.

Computer software can supply estimates of the effects in our models. The estimated treatment effects $\widehat{\alpha}_i$ describe how far apart the parallel lines are, $\widehat{\mu}$ gives an average intercept, $\widehat{\mu} + \widehat{\alpha}_i$ gives the intercept for treatment $i$, and $\widehat{\beta}$ is the estimated slope.

Means depend on covariate

How should we answer the question, "What is the mean response in treatment $i$?" This is a little tricky, because the response depends on the covariate. We need to choose some standard covariate value $\dot{x}$ and evaluate the treatment means there.

Covariate adjusted means at grand mean of covariate

*Covariate-adjusted means* are the estimated values in each treatment group when the covariate is set to $\overline{x}_{\bullet\bullet}$, the grand mean of the covariates, or

$$\widehat{\mu} + \widehat{\alpha}_i + \widehat{\beta}\overline{x}_{\bullet\bullet} \quad .$$

Covariate-adjusted means give us a common basis for comparison, because all treatments are evaluated at the same covariate level. Note that the difference between two covariate-adjusted means is just the difference between the treatment effects; we would get the same differences if we compare the means at the common covariate value $\dot{x} = 0$.

**Example 17.2**

## Keyboarding pain, continued

Table 17.1 shows hours of keyboarding and hours of pain for the twelve subjects, and Figure 17.1 shows a plot of the response versus the covariate, with keyboard type indicated by the plotting symbol. The plot clearly shows a strong, reasonably linear relationship between the response and the covariate. The figure also shows that the keyboard 1 responses tend to be above the keyboard 2 responses for similar covariate values, and keyboard 2 and 3 responses are somewhat mixed at the low end of the covariate. We can further see that keyboard 3 covariates tend to be a bit smaller than the other two

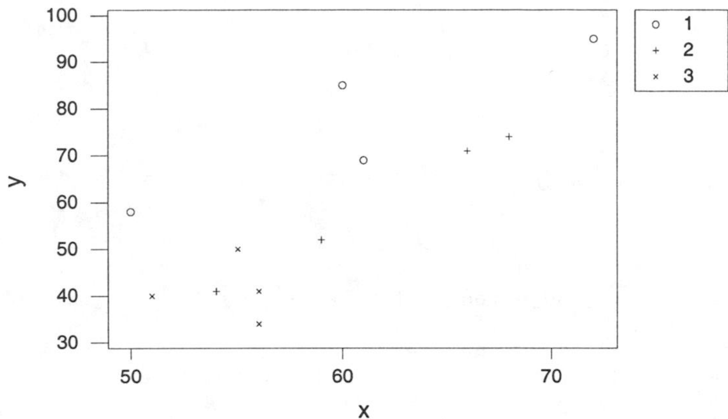


**Figure 17.1:** Hours of pain versus hours of keyboarding for twelve subjects and three keyboard types, using Minitab.

keyboards, so presumably at least some of the explanation for the low responses for keyboard 3 is the low covariate values.

Listing 17.1 shows Minitab output analyzing these data. We first check to see if treatments affect the covariate keyboarding time. The ANOVA ① provides no evidence against the null hypothesis that the treatments have the same average covariate values ($p$-value .29). In these data, keyboard 3 averages about 6 to 7 hours less than the other two keyboards ②, but the difference is within sampling variability.

Next we do the Analysis of Covariance ③. The model includes the covariate and then the treatment. Minitab produces both sequential and Type III sums of squares; in either case, the sum of squares for treatments is treatments adjusted for covariates, which is what we need. The $p$-value is .004, indicating strong evidence against the null hypothesis of no treatment effects.

The covariate-adjusted means and their standard errors are given at ⑤. Note that the standard errors are not all equal. We can also construct the covariate adjusted means from the effects ④. For example, the covariate-adjusted mean for keyboard 1 is

$$-48.21 + 14.399 + 1.8199 \times 59 = 73.57 \ .$$

**Listing 17.1:** Minitab output for keyboarding pain.

```
Analysis of Variance for x

Source      DF         SS          MS        F       P
type         2     123.50       61.75     1.45   0.286            ①
Error        9     384.50       42.72

Means

type     N         x                                              ②
1        4    60.750
2        4    61.750
3        4    54.500

Analysis of Variance for y, using Adjusted SS for Tests

Source      DF     Seq SS      Adj SS      Adj MS       F       P
x            1     2598.8      1273.5      1273.5   24.79   0.001  ③
type         2     1195.8      1195.8       597.9   11.64   0.004
Error        8      411.0       411.0        51.4

Term            Coef      StDev          T        P
Constant      -48.21      21.67      -2.22    0.057               ④
x             1.8199     0.3655       4.98    0.001
type
1             14.399      2.995       4.81    0.001
2             -4.671      3.094      -1.51    0.170

Means for Covariates

Covariate        Mean      StDev
x               59.00      6.796

Least Squares Means for y

type       Mean      StDev                                        ⑤
1         73.57      3.641
2         54.50      3.722
3         49.44      3.943

Tukey 95.0% Simultaneous Confidence Intervals                     ⑥
Response Variable y
All Pairwise Comparisons among Levels of type

type = 1 subtracted from:

type      Lower     Center      Upper    -------+---------+---------+---------
2        -33.59     -19.07     -4.553         (--------*---------)
3        -40.01     -24.13     -8.244    (----------*----------)
                                         -------+---------+---------+---------
                                              -30       -15         0
```

*Listing 17.1, continued*

```
type = 2 subtracted from:

type     Lower    Center    Upper    -------+---------+---------+---------
3       -21.39    -5.056    11.28              (----------*----------)
                                     -------+---------+---------+---------
                                          -30       -15         0

Analysis of Variance for y, using Adjusted SS for Tests

Source     DF     Seq SS     Adj SS     Adj MS        F       P
type        2     2521.2     2521.2     1260.6     6.74   0.016    ⑦
Error       9     1684.5     1684.5      187.2

Term          Coef      StDev         T       P
Constant    59.167      3.949     14.98   0.000                    ⑧
type
1           17.583      5.585      3.15   0.012
2            0.333      5.585      0.06   0.954

Least Squares Means for y
type        Mean      StDev                                        ⑨
1          76.75      6.840
2          59.50      6.840
3          41.25      6.840
```

It appears that keyboards 2 and 3 are about the same, and keyboard 1 is worse (leads to a greater response). This is confirmed by doing a pairwise comparison of the three treatment effects using Tukey HSD ⑥.

We conclude that there are differences between the three keyboards, with keyboard 1 leading to about 21 more hours of pain in the 2-week period for an average number of hours keyboarding. The coefficient of keyboard hours was estimated to be 1.82, so an additional hour of keyboarding is associated with about 1.82 hours of additional pain.

Before leaving the example, a few observations are in order. First, the linear model is only reliable for the range of data over which it was fit. In these data, the hours of keyboarding ranged from about 50 to 70, so it makes no sense to think that doing no keyboarding with keyboard 1 will lead to -34 hours of pain (34 hours of pleasure?).

Next, it is instructive to compare the results of this Analysis of Covariance with those that would be obtained if the covariate had been ignored. You would not ordinarily do this as part of your analysis, but it helps us see what the covariate has done for us. Two things are noteworthy. First, the error mean square for the analysis without the covariate ⑦ is about 3.6 times larger than that with the covariate. Regression on the covariate has explained much of the variation within treatment groups, so that residual variation is reduced. Second, the covariate-adjusted treatment effects ④ are not the same as the unadjusted treatment effects ⑧; likewise, the covariate-adjusted means 73.565, 54.495, and 49.44 ⑤ differ from the raw treatment means 76.75, 59.5, and 41.25 ⑨. This shows the effect of comparing the treatments at a common value of the covariate. For these data, the covariate-adjusted means are more tightly clustered than the raw means; other data sets may show other patterns.

Some authors prefer to write the covariate model

$$y_{ij} = \mu + \alpha_i + \beta x_{ij} + \epsilon_{ij}$$

in the slightly different form

$$y_{ij} = \tilde{\mu} + \alpha_i + \beta(x_{ij} - \overline{x}_{\bullet\bullet}) + \epsilon_{ij} \quad .$$

Centered covariates

The difference is that the covariate $x$ is centered to have mean zero, so that the covariate-adjusted means in the revised model are just $\tilde{\mu} + \alpha_i$. We can see that there is no essential difference between these two models once we realize that $\tilde{\mu} = \mu + \beta\overline{x}_{\bullet\bullet}$.

## 17.2 When Treatments Change Covariates

Covariate adjustment can obscure the treatment effect

The usual Analysis of Covariance assumes that treatments do not affect the covariates. When this is true, it makes sense to compare treatments via covariate-adjusted means—that is, to compare treatments at a common value of the covariate—because any differences between covariates are just random variation. When treatments do affect covariates, differences between covariates are partly treatment effect and partly random variation. Forcing treatment comparisons to be at a common value of the covariate obscures the true treatment differences.

We can make this more precise by reexpressing the covariate in our model. Expand the covariate into a grand mean, deviations of treatment means from the grand mean, and deviations from treatment means to obtain $x_{ij} = \overline{x}_{\bullet\bullet} + (\overline{x}_{i\bullet} - \overline{x}_{\bullet\bullet}) + (x_{ij} - \overline{x}_{i\bullet})$, and substitute it into the model:

$$\begin{aligned} y_{ij} &= \mu + \alpha_i + \beta x_{ij} + \epsilon_{ij} \\ &= \mu + \alpha_i + \beta(\overline{x}_{\bullet\bullet} + (\overline{x}_{i\bullet} - \overline{x}_{\bullet\bullet}) + (x_{ij} - \overline{x}_{i\bullet})) + \epsilon_{ij} \\ &= (\mu + \beta\overline{x}_{\bullet\bullet}) + (\alpha_i + \beta(\overline{x}_{i\bullet} - \overline{x}_{\bullet\bullet})) + \beta(x_{ij} - \overline{x}_{i\bullet}) + \epsilon_{ij} \\ &= \tilde{\mu} \qquad\qquad + \tilde{\alpha}_i \qquad\qquad\qquad + \beta\tilde{x}_{ij} \qquad\qquad + \epsilon_{ij} \end{aligned}$$

**Listing 17.2:** Minitab analysis of keyboarding pain when treatments affect covariates.

```
Analysis of Variance for y, using Adjusted SS for Tests

Source     DF     Seq SS     Adj SS     Adj MS       F      P
xtilde      1     1273.5     1273.5     1273.5   24.79  0.001
type        2     2521.2     2521.2     1260.6   24.54  0.000
Error       8      411.0      411.0       51.4

Least Squares Means for y

type       Mean     StDev
1         76.75     3.584
2         59.50     3.584
3         41.25     3.584
```

We have seen that covariate-adjusted treatment effects may not equal covariate-unadjusted treatment effects. In the preceding equations, $\alpha_i$ is the covariate-adjusted treatment effect, and $\tilde{\alpha}_i$ is the unadjusted effect (see Question 17.1). These differ by $\beta(\overline{x}_{i\bullet} - \overline{x}_{\bullet\bullet})$, so adjusted and unadjusted effects are the same if all treatments have the same average covariate. If the treatments are affecting the covariate, these adjustments should not be made.

Covariate adjustment to means is $\beta(\overline{x}_{i\bullet} - \overline{x}_{\bullet\bullet})$

We can obtain the variance reduction property of covariance analysis without also doing covariate adjustment by using the covariate $\tilde{x}$ instead of $x$. Compute $\tilde{x}$ by treating the covariate $x$ as a response with the treatments as explanatory variables; the residuals from this model are $\tilde{x}$.

Using $\tilde{x}$ gives variance reduction only

Note that the two analyses described here are extremes: ordinary analysis of covariance assumes that treatments cause no variation in the covariate, and the analysis with the altered covariate $\tilde{x}$ assumes that all between treatment variation in the covariates is due to treatment.

### Keyboarding pain, continued

**Example 17.3**

An analysis of variance on the keyboarding times in Table 17.1 showed no evidence that the different keyboards affected keyboarding times. Nonetheless, we use those data here to illustrate the analysis that uses covariates only for variance reduction, and not for covariate adjustment.

The first step is to get the modified covariate as the residuals from a model with treatments and the covariate as the response. The ANOVA for this model is at ① of Listing 17.1; the residuals have been saved as $\tilde{x}$, which we next use in a standard Analysis of Covariance.

Listing 17.2 shows Minitab output using this modified covariate. We can see in the ANOVA table that the error mean square is the same in this analysis as it was in the standard Analysis of Covariance in Listing 17.1 ③. The mean square for treatments adjusted for this modified covariate is the same as the mean square for treatments alone; in fact, we constructed the modified covariate to make this so. For these data, the treatment mean square adjusted for the modified covariate (same as the unadjusted treatment mean square) is over twice the size of the treatments adjusted for covariate mean square; the $p$-value in the modified analysis is thus much smaller.

Finally, we see that the covariate-adjusted treatment means using the modified covariate are the same as the simple treatment means in Listing 17.1 ⑨. The standard errors for these adjusted means are much smaller than the standard errors for the unadjusted means, however, because the modified covariate accounts for a large amount of response variation within each treatment group. Also, the standard errors for the covariate-adjusted means using $\tilde{x}$ are equal, unlike those using $x$.

The covariate-adjusted treatment effects can be larger or smaller than the unadjusted effects (depending on the sign of $\beta$ and the pattern of covariates). Similarly, the covariate-adjusted effects may have a larger or smaller $p$-value than the treatment effects in a model with the modified covariate. We must not choose between the original and modified covariates based on the results of the analysis; we must choose based on whether we wish to ascribe covariate differences to treatments.

## 17.3 Other Covariate Models

More than one covariate

We have been discussing the simplest possible covariate model: a single covariate with the same slope in all treatment groups. It is certainly possible to have two or more covariates. The standard analysis is still treatments adjusted for covariates, and covariate-adjusted means are evaluated with each covariate at its overall average. If one or more covariates are affected by treatments and we wish to identify the variation associated with treatment differences in those covariates as treatment variation, then each of those covariates should be individually modified as described in the preceding section.

Covariates with blocks or factorials

Covariates can also be used in other designs beyond the CRD with a single treatment factor. Blocking designs and fixed-effects factorials can easily accommodate covariates; simply look at treatments adjusted for any blocks and covariates. Note that treatment factors adjusted for covariates will not usually be orthogonal, even for balanced designs, so you will need to do Type II or Type III analyses for factorials.

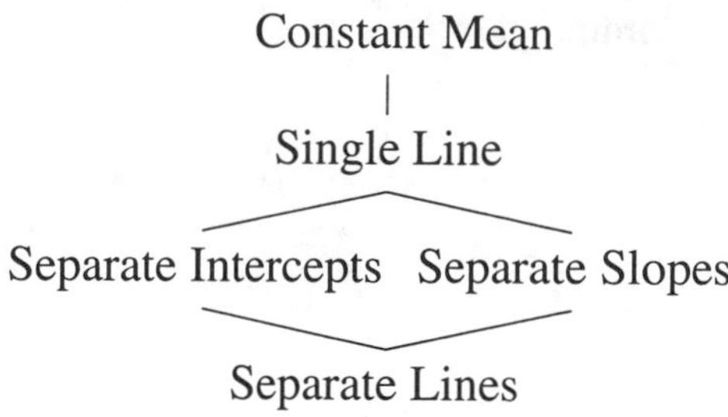


**Figure 17.2:** Lattice of covariate models.

Treatments could change the covariate slope

Our covariate models have assumed that treatments affect the response by an additive constant that is the same for all values of the covariate. This is the parallel-lines model, and it is the standard model for covariates. It is by no means the only possibility for treatment effects. For example, treatments could change the slope of the response-covariate relationship, or treatments could change both the slope and the intercept.

Lattice of covariate models

We can put covariate models into an overall framework as shown in Figure 17.2. Models are simplest on top and add complexity as you move down an edge. Any two models that can be connected by going down one or more edges can be compared using an Analysis of Variance. The lower model is the full model and the upper model is the reduced model, and the change in error sum of squares between the two models is the sum of squares used to compare the two models. The degrees of freedom for any model comparison is the number of additional parameters that must be fit for the larger model.

Constant mean

Single line

The top model is a constant mean; this is a model with no treatment effects and no covariate effect. We only use this model if we are interested in determining whether there is any covariate effect at all (by comparing it to the single-line model). The single line model is the model where the covariate affects the response, but there are no treatment effects. This model has one more parameter than the constant mean model, so there is 1 degree of freedom in the comparison of the constant-mean and single-line models (and that degree of freedom is the slope parameter).

Separate intercepts

Moving down the figure, we have two choices. On the left is the separate-intercepts model. This is the model with a common covariate slope and a different intercept for each treatment. The comparison between the single-line model and the separate-intercepts model is the standard Analysis of Covariance, and it has $g - 1$ degrees of freedom for the $g - 1$ additional intercepts that must be fit.

**Listing 17.3:** MacAnova output for keyboarding pain.

```
Model used is y=x+type+x.type
              DF          SS          MS            F       P-value
x              1      2598.8      2598.8     53.62884    0.00033117  ①
type           2      1195.8      597.91     12.33835     0.0074822
x.type         2      120.27      60.136      1.24095       0.35398
ERROR1         6      290.76      48.459

Model used is y=x+x.type+type
              DF          SS          MS            F       P-value
x              1      2598.8      2598.8     57.62884    0.00033117  ②
x.type         2      1168.4      584.22     12.05596     0.0079111
type           2      147.65      73.826      1.52345       0.29171
ERROR1         6      290.76      48.459

Model used is y=x59+x59.type
              DF          SS          MS            F       P-value
x59            1      2598.8      2598.8     14.66486     0.0050217  ③
x59.type       2      189.13      94.566      0.53363       0.60598
ERROR1         8      1417.7      177.21
```

If instead we move down to the right, we get the separate-slopes model:

$$y_{ij} = \mu + \beta_i(x_{ij} - x_0) + \epsilon_{ij}$$

Separate slopes

In this model, the relationship between response and covariate has a different slope $\beta_i$ for each treatment, but all the lines intersect at the covariate value $x_0$. If you set $x_0 = 0$, then all the lines have the same intercept. Different values of $x_0$ are like different covariates. This model has $g - 1$ more degrees of freedom than the single-line model.

At the bottom, we have the separate-lines model:

$$y_{ij} = \mu + \alpha_i + \beta_i x_{ij} + \epsilon_{ij}$$

Separate lines

This model has $g - 1$ more degrees of freedom than either the separate-intercepts or separate-slopes models. If we move down the left side of the figure, we add intercepts then slopes, while moving down the right side we add the slopes first, then the intercepts.

**Example 17.4** **Keyboarding pain, continued**

Let's fit the full lattice of covariate models to the keyboarding pain data. Listing 17.3 shows MacAnova output for these models; all sums of squares are sequential. ANOVA ① descends the left-hand side of the lattice, start-

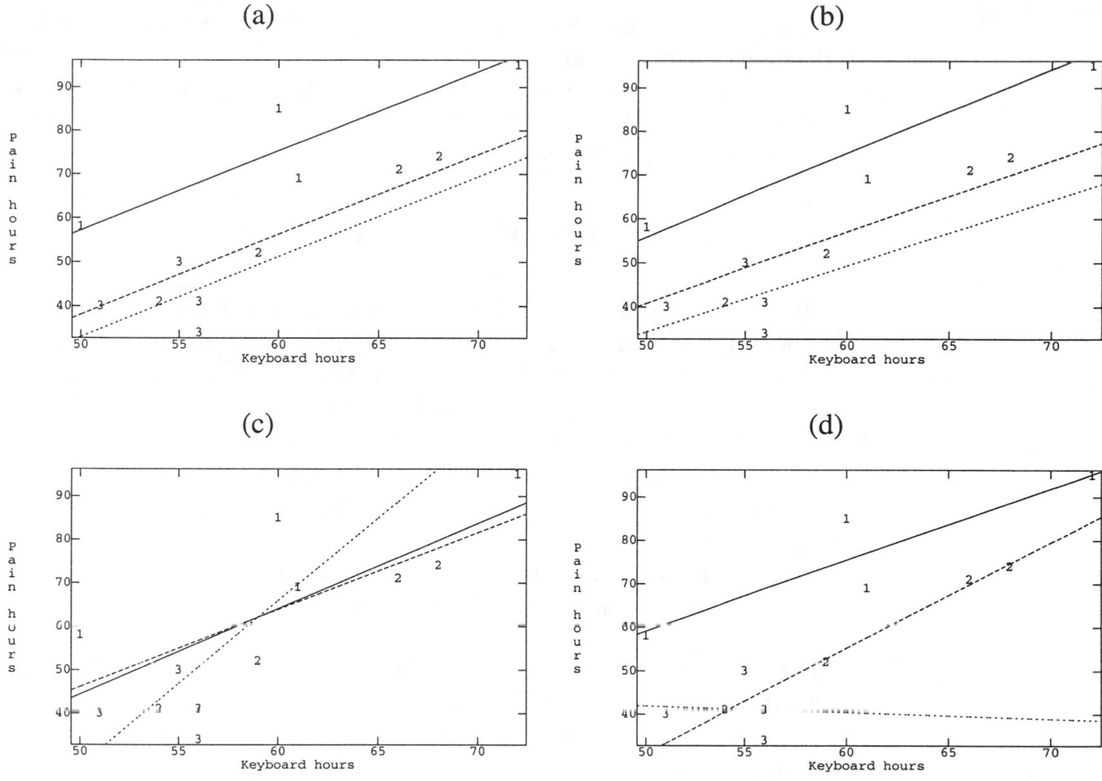


**Figure 17.3:** Covariate model fits for the keyboarding pain data, using MacAnova: (a) separate intercepts, (b) separate slopes $x_0 = 0$, (c) separate slopes $x_0 = 59$, (d) separate lines.

ing with the covariate x (time), adding keyboard type adjusted for covariate (separate intercepts), and finally adding separate slopes to get separate lines. The type mean square of 597.91 is the usual Analysis of Covariance mean square. ANOVA ② descends the right-hand side of the lattice, starting with the covariate x, adding separate slopes, and finally adding separate intercepts to get separate lines. Adding separate slopes makes a significant improvement over a single line ($p$-value of .0079), but adding separate lines is not a significant improvement over separate slopes. The separate slopes model ② uses $x_0 = 0$, so the fitted lines intersect at 0. ANOVA ③ fits a separate slopes model with $x_0 = 59$. In this case, there is no significant improvement going to separate slopes. Figure 17.3 shows the fits for four models.

The single-line and separate-intercepts models are the most commonly used models of this family. They are analogues of treatment models with blocking. However, not all experimental data will fit nicely into this view of the world, and we need to be ready to consider the less common covariate models if the data require it.

## 17.4 Further Reading and Extensions

Federer and Meredith (1992) discuss the use of covariates in split-plot and split-block designs. Consider two situations. First, all split plots in a whole plot have the same covariate, so that the covariate only depends on the whole plot. In this case, covariate is a whole-plot effect, and its 1 degree of freedom and sum of squares are computed at the whole-plot level.

Second, consider when each split plot has its own covariate value $x_{ijk}$. Construct two new covariates from $x$. The first is a covariate at the whole-plot level formed by taking the average covariate for each whole plot: $\overline{x}_{i\bullet k}$. This covariate acts at the whole-plot level, and its 1 degree of freedom and sum of squares are computed at the whole-plot level. The second is a split-plot covariate: $\tilde{x}_{ijk} = x_{ijk} - \overline{x}_{i\bullet k}$. This split-plot covariate is the deviation of the original covariate $x$ from the whole-plot average value for $x$. The 1 degree of freedom and sum of squares for this covariate are at the split-plot level. Note that there may be different coefficients (slopes) for the covariates at the whole- and split-plot levels.

Analysis of Covariance for general random- and mixed-effects models is considerably more difficult. Henderson and Henderson (1979) and Henderson (1982) discuss the problems and possible approaches. In fact, the whole September 1982 issue of *Biometrics* that includes Henderson (1982) is devoted to Analysis of Covariance.

## 17.5 Problems

**Exercise 17.1** What is the difference in randomization between a completely randomized design in which a covariate is measured and a completely randomized design in which no covariate is measured?

**Exercise 17.2** Briefly discuss the difference in *design* between a randomized complete block design with four treatments and five blocks, and a two-way factorial design with factor A having four levels and factor B having five levels.

**Problem 17.1** Pollutants may reduce the strength of bird bones. We believe that the strength reduction, if present, is due to a change in the bone itself, and not a change in the size of the bone. One measure of bone strength is calcium content. We have an instrument which can measure the total amount of calcium in a 1cm length of bone. Bird bones are essentially thin tubes in shape, so the total amount of calcium will also depend on the diameter of the bone.

Thirty-two chicks are divided at random into four groups. Group 1 is a control group and receives a normal diet. Each other group receives a diet including a different toxin (pesticides related to DDT). At 6 weeks, the chicks are sacrificed and the calcium content (in mg) and diameter (in mm) of the right femur is measured for each chick.

| Control | | P #1 | | P #2 | | P #3 | |
|---|---|---|---|---|---|---|---|
| C | Dia | C | Dia | C | Dia | C | Dia |
| 10.41 | 2.48 | 12.10 | 3.10 | 10.33 | 2.57 | 10.46 | 2.6 |
| 11.82 | 2.81 | 10.38 | 2.61 | 10.03 | 2.48 | 8.64 | 2.17 |
| 11.58 | 2.73 | 10.08 | 2.49 | 11.13 | 2.77 | 10.48 | 2.64 |
| 11.14 | 2.67 | 10.71 | 2.69 | 8.99 | 2.30 | 9.32 | 2.35 |
| 12.05 | 2.90 | 9.82 | 2.43 | 10.06 | 2.56 | 11.54 | 2.89 |
| 10.45 | 2.45 | 10.12 | 2.52 | 8.73 | 2.18 | 9.48 | 2.38 |
| 11.39 | 2.69 | 10.16 | 2.54 | 10.66 | 2.65 | 10.08 | 2.55 |
| 12.5 | 2.94 | 10.14 | 2.55 | 11.03 | 2.73 | 9.12 | 2.29 |

Analyze these data with respect to the effect of pesticide on calcium in bones.

**Problem 17.2**

Briefly describe the experimental design you would choose for each of the following situations, and why.

(a) We wish to determine the amount of salt to put in a microwave popcorn so that it has the best overall acceptability. We will test three levels of salt: low, medium, and high. We have recruited 25 volunteers to taste popcorn, and while we expect the individuals to be reasonably consistent in their own personal ratings, we expect large volunteer to volunteer differences in overall ratings.

(b) Some brands of golf balls claim to fly farther. To test this claim, you devise a mechanical golf ball whacker which will strike the golf balls with the same power and stroke time after time. Ten balls of each of six brands will be struck once by the device and measured for distance traveled. Wind speed, which will affect the distance traveled, is variable and unpredictable, but can be measured.

(c) We wish to study the effects of two food additives (plus a control treatment for a total of three treatments) on the milk productivity of cows. We have three large herds available, each of a different breed, and we expect breed to breed differences in the response. Furthermore, we expect an age effect, which we make explicit by dividing cows into three groups: those which have had 0, 1, and 2 or more previous calves. We have enough resources to study 27 animals through one breeding cycle.

**Problem 17.3** For each of the following, describe the experimental design used and give a skeleton ANOVA (sources and degrees of freedom only).

(a) We wish to study the effects of air pressure (low or high) and tire type (radial versus all season radial) on gas mileage. We do this by fitting tires of the appropriate type and pressure on a car, driving the car 150 miles around a closed circuit, then changing the tire settings and driving again. We have obtained eight cars for this purpose and can use each car for one day. Unfortunately, we can only do three of the four tire combinations on one day, so we have each factor-level combination missing for two cars.

(b) Metribuzin is an agricultural chemical that may accumulate in soils. We wish to determine whether the amount of metribuzin retained in the soil depends on the amount applied to the soil. To test the accumulation, we select 24 plots. Each plot is treated with one of three levels of metribuzin, with plots assigned to levels at random. After one growing season, we take a sample of the top three cm of soil from each plot and determine the amount of metribuzin in the soil. We also measure the pH of the soil, as pH may affect the ability of the soil to retain metribuzin.

(c) We wish to test the efficacy of dental sealants for reducing tooth decay on molars in children. There are five treatments (sealants A or B applied at either 6 or 8 years of age, and a control of no sealant). We have 40 children, and the five treatments are assigned at random to the 40 children. As a response, we measure the number of cavities on the molars by age 10. In addition, we measure the number of cavities on the nonmolar teeth (this may be a general measure of quality of brushing or resistance to decay).

(d) A national travel agency is considering new computer hardware and software. There are two hardware setups and three competing software setups. All three software setups will run on both hardware setups, but the different setups have different strengths and weaknesses. Twenty branches of the agency are chosen to take part in an experiment. Ten are high sales volume; ten are low sales volume. Five of the high-sales branches are chosen at random for hardware A; the other five get hardware B. The same is done in the low-sales branches. All three software setups are tried at each branch. One of the three software systems is randomly assigned to each of the first 3 weeks of May (this is done separately at each branch). The measured response for each hardware-software combination is a rating score based on the satisfaction of the sales personnel.

**Problem 17.4**

Advertisers wish to determine if program content affects the success of their ads on those programs. They produce two videos, one containing a depressing drama and some ads, the second containing an upbeat comedy and the same ads. Twenty-two subjects are split at random into two groups of eleven, with the first group watching the drama and the second group watching the comedy. After the videos, the subjects are asked several questions, including "How do you feel?" and "How likely are you to buy?" one of the products mentioned in the ads. "How do you feel" was on a 1 to 6 scale, with 1 being happy and 6 being sad. "How likely are you to buy?" was also on a 1 to 6 scale, with 6 being most likely.

| Drama | | Comedy | |
|---|---|---|---|
| Feel | Buy | Feel | Buy |
| 5 | 1 | 3 | 1 |
| 1 | 3 | 2 | 2 |
| 5 | 1 | 3 | 1 |
| 5 | 3 | 2 | 3 |
| 4 | 5 | 4 | 1 |
| 4 | 3 | 1 | 3 |
| 5 | 2 | 1 | 4 |
| 6 | 1 | 2 | 4 |
| 5 | 5 | 3 | 1 |
| 3 | 4 | 4 | 1 |
| 4 | 1 | 2 | 2 |

Analyze these data to determine if program type affects the likelihood of product purchase.

**Problem 17.5**

A study has been conducted on the environmental impact of an industrial incinerator. One of the concerns is the emission of heavy metals from the stack, and one way to measure the impact is by looking at metal accumulations in soil and seeing if nearby sites have more metals than distant sites (presumably due to deposition of metals from the incinerator).

Eleven sites of one hectare each (100 m by 100 m) were selected around the incinerator. Five sites are on agricultural soils, while the other six are on forested soils. Five of the sites were located near the incinerator (on their respective soil types), while the other sites were located far from the incinerator. At each site, nine locations are randomly selected within the site and mineral soil sampled at each location. We then measure the mercury content in each sample (mg/kg).

Complicating any comparison is the fact that heavy metals are generally held in the organic portion of the soil, so that a soil sample with more carbon will tend to have more heavy metals than a sample with less carbon, regardless of the deposition histories of the samples, soil type, etc. For this reason, we also measure the carbon fraction of each sample (literally the fraction of the soil sample that was carbon).

The data given below are site averages for carbon and mercury. Analyze these data to determine if there is any evidence of an incinerator effect on soil mercury.

| Soil | Distance | Carbon | Mercury |
|---|---|---|---|
| Agricultural | Near | .0084 | .0128 |
| Agricultural | Near | .0120 | .0146 |
| Agricultural | Near | .0075 | .0130 |
| Agricultural | Far | .0087 | .0133 |
| Agricultural | Far | .0105 | .0090 |
| Forest | Near | .0486 | .0507 |
| Forest | Near | .0410 | .0477 |
| Forest | Far | .0370 | .0410 |
| Forest | Far | .0711 | .0613 |
| Forest | Far | .0358 | .0388 |
| Forest | Far | .0459 | .0466 |

**Question 17.1** Show that the covariate-adjusted means using the covariate $\tilde{x}$ equal the unadjusted treatment means.

# Chapter 18

# Fractional Factorials

This chapter and the next deal with *treatment design*. We have been using treatments that are the factor-level combinations of two or more factors. These factors may be fixed or random or nested or crossed, but we have a regular array of factor combinations as treatments. Treatment design investigates other ways for choosing treatments. This chapter investigates fractional factorials, that is, use of a subset of the factor-level combinations in a factorial treatment structure.

Treatment design

## 18.1 Why Fraction?

Factorial treatment structure has the benefits that it is efficient and allows us to study main effects and interactions, but factorials can become really big. For seven factors, the smallest factorial has $2^7 = 128$ treatments and units. There are 127 degrees of freedom in such an experiment, with 7 degrees of freedom for main effects, 21 degrees of freedom for two-factor interactions, 35 degrees of freedom for three-factor interactions, and 64 degrees of freedom for four-, five-, six-, and seven-factor interactions. In many experiments, we either don't expect high-order interactions or we are willing to ignore them at the current stage of experimentation, so we construct a surrogate error by pooling high-order interactions. For example, pooling fourth- and higher-order interactions into error in the $2^7$ gives us 64 degrees of freedom for error.

Factorials have many degrees of freedom in multi-factor interactions

What does a big factorial such as a $2^7$ give us? First, it gives us a large sample size for estimating main effects and interactions; this is a very good thing. Second, it allows us to estimate many-way interactions; this may or may not be useful, depending on the experimental situation. Third, the abundant high-order interactions give us many degrees of freedom for constructing a surrogate error.

High-order interactions and many error df may not be worth the expense

Larger sample sizes always give us more precise estimates, but there are diminishing returns for the second and third advantages. In some experiments we either do not expect high-order interactions, or we are willing to ignore them in the current problem. For such an experiment, being able to estimate high-order interactions is not a major advantage. Similarly, more degrees of freedom for error are always better, but the improvement in power and confidence interval length is modest after 15 degrees of freedom for error and very slight after 30.

Thus the full factorial may be wasteful or infeasible if

- We believe there are no high-order interactions or that they are ignorably small, or
- We are just screening a large number of treatments to determine which affect the response and will study interactions in subsequent experiments on the active factors, or
- We have limited resources.

We need a design that retains as many of the advantages of factorials as possible, but does not use all the factor-level combinations.

Fractional factorial looks at main effects and low-order interactions

A *fractional-factorial* design is a modification of a standard factorial that allows us to get information on main effects and low-order interactions without having to run the full factorial design. Fractional factorials are closely related to the confounding designs of Chapter 15, which you may wish to review. In fact, the simplest way to describe a fractional factorial is to confound the factorial into blocks, but only run one of the blocks.

## 18.2 Fractioning the Two-Series

A fraction is one block of a confounded design

A $2^k$ factorial can be confounded into two blocks of size $2^{k-1}$, four blocks of size $2^{k-2}$, and in general $2^q$ blocks of size $2^{k-q}$. A $2^{k-1}$ fractional factorial is a design with $k$ factors each at two levels that uses $2^{k-1}$ experimental units and factor-level combinations. We essentially block the $2^k$ into two blocks but only run one of the blocks. In general, a $2^{k-q}$ fractional factorial is a design with $k$ factors each at two levels that uses $2^{k-q}$ experimental units and factor-level combinations. Again, this design is one block of a confounded $2^k$

factorial. The principal block of a confounded design becomes the *principal fraction*, and alternate blocks become *alternate fractions*.

Principal and alternate fractions

We confound a $2^k$ factorial by choosing one or more defining contrasts. These defining contrasts are factorial effects that will be confounded with block differences. We construct blocks by partitioning the factor-level combinations into $2^q$ groups according to whether they are $\pm 1$ on the defining contrasts, or equivalently by whether an even or odd number of factors from the defining contrasts are at the high level in the factor-level combination or by whether the $L$ values are 0 or 1.

Review of confounding

In the confounded $2^k$, all possible plus/minus, even/odd, or 0/1 combinations for the defining contrasts occur somewhere in the design, though in different blocks. For example, with two defining contrasts, we will have plus and plus, minus and plus, plus and minus, and minus and minus blocks. A fractional factorial is a single block of this design, so only a single plus/minus combination of the defining contrasts occurs: for example, the plus and plus combination. Thus a fractional factorial is a subset of factor-level combinations that has a particular pattern of plus and minus signs on the defining contrasts, or equivalently a particular pattern of even/odd or 0/1 values.

$q$ defining contrasts constant in a fraction

The jargon and notation of fractional factorials are slightly different from confounding. Recall the tables of plus and minus signs such as Table 15.1 that we used in two-series design. Augment such tables with a column of all plus signs labeled I. Defining contrasts are the effects that we confound to produce confounded factorials; we call these contrasts *generators* or *words* when we work with just a fraction of the design. In a fraction of a two-series, each generator for the design will always be plus or always be minus; thus for each generating word W, either $I = W$ or $I = -W$ will be true on the fraction. The statement $I = W$ is called a *defining relation.* Note that if $I = W_1$ and $I = -W_2$, then $I = -W_1W_2$; that is, generalized interactions of the generators also have constant sign that can be determined from the defining relations.

Fractional factorials have generators and defining relations

## Quarter fraction of a $2^5$ design

**Example 18.1**

Construct a $2^{5-2}$ fractional factorial using ABC and –CDE as generators; I = ABC = –CDE = –ABDE is the full set of defining relations. This is the same as confounding into four blocks using the generators ABC and CDE, but then only using the block where ABC is plus and CDE is minus. Using the even/odd rule, ABC is plus when a factor-level combination has an odd number of factors A, B, or C high, and CDE is minus when a factor-level combination has an even number of C, D, or E high.

**Table 18.1:** Table of pluses and minuses for a $2^{5-2}$ with I = ABC = –CDE.

| | A | B | C | D | E | AB | ⋯ | ABCDE |
|---|---|---|---|---|---|---|---|---|
| $ce$ | − | − | + | − | + | + | ⋯ | − |
| $a$ | + | − | − | − | − | − | | + |
| $b$ | − | + | − | − | − | − | | + |
| $abce$ | + | + | + | − | + | + | | − |
| $cd$ | − | − | + | + | − | + | | − |
| $ade$ | + | − | − | + | + | − | | + |
| $bde$ | − | + | − | + | + | − | | + |
| $abcd$ | + | + | + | + | − | + | ⋯ | − |

The eight factor-level combinations in our fraction are

$$a,\ b,\ ade,\ bde,\ ce,\ abce,\ cd,\ abcd\ .$$

In principle we find the fraction by confounding the full factorial and choosing the correct block. However, we know that we can find alternate blocks from the principal block, so we can find alternate fractions from principal fractions. I found our fraction by first finding the principal fraction,

$$(1),\ ab,\ de,\ abde,\ ace,\ bce,\ acd,\ bcd$$

then finding a factor-level combination in the fraction of interest ($a$), and multiplying everything in the principal fraction by $a$ to get the alternate fraction.

The natural way to estimate the total effect of factor A in a fractional factorial is to subtract the average response where A is low from the average response where A is high. For the $2^{5-2}$ of Example 18.1, this is the contrast

$$\frac{\overline{y}_a + \overline{y}_{abce} + \overline{y}_{ade} + \overline{y}_{abcd}}{4} - \frac{\overline{y}_{ce} + \overline{y}_b + \overline{y}_{cd} + \overline{y}_{bde}}{4}\ .$$

Total effect contrasts as before

This amounts to taking the pattern of pluses and minuses for the A contrast from the complete factorial and just using the elements in it that correspond to the factor-level combinations that we have in our fraction. Part of this reduced table of pluses and minuses is shown in Table 18.1. Using this table, we can compute contrasts for all the factorial effects.

This sounds as if we've just gotten something for nothing. We only have eight observations, but we've (apparently) just extracted estimates of 31 effects and interactions. The laws of physics and economics argue that you don't get something for nothing, and indeed there is a catch here. To see the catch, look at the patterns of signs we use for the C main effect and the AB interaction. These patterns are the same, so our estimate of the C main effect is the same as our estimate of the AB interaction. If we look further, we will also find that the C contrast is the negative of the DE and ABCDE contrasts.

Same contrast for several effects

We say that C, AB, –DE, and –ABCDE are *aliases*, or aliased to each other. Another way of writing this is C = AB = –DE = –ABCDE, meaning that these contrasts have equal coefficients on this fraction. When we apply that contrast, we are estimating the total effect of C, plus the total effect of AB, minus the total effect of DE, minus the total effect of ABCDE, or C + AB – DE – ABCDE. In a $2^{k-q}$ design, every degree of freedom is associated with $2^q$ effects that are aliased to each other. So aliases come in pairs for half-fractions, sets of four for quarter-fractions, and so on.

Fractional factorials have aliased effects

There is a simple rule for determining which effects are aliased. Begin with the defining relations, I = ABC = –CDE = –ABDE in our example. Treat I as an identity, multiply all elements of the defining relations by an effect, and reduce exponents mod 2. For example,

Multiply defining relation to get aliases

$$\begin{array}{lclclcl} C \times I & = & C \times ABC & = & C \times -CDE & = & C \times -ABDE \\ C & = & ABC^2 & = & -C^2DE & = & -ABCDE \\ C & = & AB & = & -DE & = & -ABCDE \end{array}$$

We can continue this to find the complete set of aliases:

$$\begin{array}{lclclcl} I & = & ABC & = & -CDE & = & -ABDE \\ A & = & BC & = & -ACDE & = & -BDE \\ B & = & AC & = & -BCDE & = & -ADE \\ C & = & AB & = & -DE & = & -ABCDE \\ D & = & ABCD & = & -CE & = & -ABE \\ E & = & ABCE & = & -CD & = & -ABD \\ AD & = & BCD & = & -ACE & = & -BE \\ BD & = & ACD & = & -BCE & = & -AE \end{array}$$

It is very important to check the aliasing during the design phase of a fractional factorial. In particular, we do not want to have a two-factor interaction as a generator (or generalized interaction of generators), because that would imply that two main effects will be aliased. The more letters in the generators and their interactions the better.

Check to be sure no important effects are aliased to each other

Aliases for more complicated designs follow the same pattern. The defining relation for the fraction will include I and all $2^q - 1$ of the generators and their interactions. For example, consider a $2^{8-4}$ with generators BCDE, ACDF, ABDG, and –ABCH; the defining relation is I = BCDE = ACDF = ABEF = ABDG = ACEG = BCFG = DEFG = –ABCH = –ADEH = –BDFH = –CEFH = –CDGH = –BEGH = –AFGH = –ABCDEFGH, which is found as the generators, their 6 two-way interactions, their 4 three-way interactions,

All effects have $2^q - 1$ aliases in $2^{k-q}$ design

**Table 18.2:** Aliases for $2^{8-4}$ with generators BCDE, ACDF, ABDG, and –ABCH.

| |
|---|
| I = BCDE = ACDF = ABEF = ABDG = ACEG = BCFG = DEFG = -ABCH = -ADEH = -BDFH = -CEFH = -CDGH = -BEGH = -AFGH = -ABCDEFGH |
| A = ABCDE = CDF = BEF = BDG = CEG = ABCFG = ADEFG = -BCH = -DEH = -ABDFH = -ACEFH = -ACDGH = -ABEGH = -FGH = -BCDEFGH |
| B = CDE = ABCDF = AEF = ADG = ABCEG = CFG = BDEFG = -ACH = -ABDEH = -DFH = -BCEFH = -BCDGH = -EGH = -ABFGH = -ACDEFGH |
| AB = ACDE = BCDF = EF = DG = BCEG = ACFG = ABDEFG = -CH = -BDEH = -ADFH = -ABCEFH = -ABCDGH = -AEGH = -BFGH = -CDEFGH |
| C = BDE = ADF = ABCEF = ABCDG = AEG = BFG = CDEFG = -ABH = -ACDEH = -BCDFH = -EFH = -DGH = -BCEGH = -ACFGH = -ABDEFGH |
| AC = ABDE = DF = BCEF = BCDG = EG = ABFG = ACDEFG = -BH = -CDEH = -ABCDFH = -AEFH = -ADGH = -ABCEGH = -CFGH = -BDEFGH |
| BC = DE = ABDF = ACEF = ACDG = ABEG = FG = BCDEFG = -AH = -ABCDEH = -CDFH = -BEFH = -BDGH = -CEGH = -ABCFGH = -ADEFGH |
| ABC = ADE = BDF = CEF = CDG = BEG = AFG = ABCDEFG = -H = -BCDEH = -ACDFH = -ABEFH = -ABDGH = -ACEGH = -BCFGH = -DEFGH |
| D = BCE = ACF = ABDEF = ABG = ACDEG = BCDFG = EFG = -ABCDH = -AEH = -BFH = -CDEFH = -CGH = -BDEGH = -ADFGH = -ABCEFGH |
| AD = ABCE = CF = BDEF = BG = CDEG = ABCDFG = AEFG = -BCDH = -EH = -ABFH = -ACDEFH = -ACGH = -ABDEGH = -DFGH = -BCEFGH |
| BD = CE = ABCF = ADEF = AG = ABCDEG = CDFG = BEFG = -ACDH = -ABEH = -FH = -BCDEFH = -BCGH = -DEGH = -ABDFGH = -ACEFGH |
| ABD = ACE = BCF = DEF = G = BCDEG = ACDFG = ABEFG = -CDH = -BEH = -AFH = -ABCDEFH = -ABCGH = -ADEGH = -BDFGH = -CEFGH |
| CD = BE = AF = ABCDEF = ABCG = ADEG = BDFG = CEFG = -ABDH = -ACEH = -BCFH = -DEFH = -GH = -BCDEGH = -ACDFGH = -ABEFGH |
| ACD = ABE = F = BCDEF = BCG = DEG = ABDFG = ACEFG = -BDH = -CEH = -ABCFH = -ADEFH = -AGH = -ABCDEGH = -CDFGH = -BEFGH |
| BCD = E = ABF = ACDEF = ACG = ABDEG = DFG = BCEFG = -ADH = -ABCEH = -CFH = -BDEFH = -BGH = -CDEGH = -ABCDFGH = -AEFGH |
| ABCD = AE = BF = CDEF = CG = BDEG = ADFG = ABCEFG = -DH = -BCEH = -ACFH = -ABDEFH = -ABGH = -ACDEGH = -BCDFGH = -EFGH |

and their four-way interaction. Thus every degree of freedom has sixteen names and every effect is aliased to fifteen other effects. The full set of aliases for this design is shown in Table 18.2. We see that no main effect is aliased with a two-factor interaction—only three-way or higher. Thus if we could assume that three-factor and higher interactions are negligible, all main effects would be estimated without aliasing to nonnegligible effects.

Full factorial in $k - q$ factors embedded in $2^{k-q}$

Every $2^{k-q}$ fractional factorial contains a complete factorial in some set of $k - q$ factors (possibly many sets), meaning that if you ignore the letters for the other $q$ factors, all $2^{k-q}$ factor-level combinations of the chosen $k - q$ factors appear in the design. You can use any set of $k - q$ factors that does not contain an alias of I as a subset. For example, the $2^{5-2}$ in Example 18.1 has an embedded complete factorial with three factors. This design has defining relation I = ABC = –CDE = ABDE; there are ten sets of three factors, and any triple except ABC or CDE will provide a complete factorial. Consider A, B, and D. Rearranging the treatments in the fraction, we get

$$ce,\ a,\ b,\ abce,\ cd,\ ade,\ bde,\ abcd;$$

ignoring C and E, we get

$$(1),\ a,\ b,\ ab,\ d,\ ad,\ bd,\ abd,$$

which are in standard order for A, B, and D. We cannot do this with A, B, and C; ignoring D and E, we get

$$c,\ a,\ b,\ abc,\ c,\ a,\ b,\ abc;$$

which is not a complete factorial.

As a second example, the factor-level combinations of the $2^{8-4}$ in Table 18.2 are

$$\begin{array}{l} h,\ afg,\ beg,\ abefh,\ cef,\ acegh,\ bcfgh,\ abc, \\ defgh,\ ade,\ bdf,\ abdgh,\ cdg,\ acdfh,\ bcdeh,\ abcdefg\ \ , \end{array}$$

which are in standard order for A, B, C, and D.

Use embedded factorial to build fractions

The embedded complete factorial is a tool for constructing fractional factorials. Display 18.1 gives the steps. Essentially we start with the factor-level combinations of the embedded factorial. Each additional factor is aliased to an interaction of the embedded factorial, so we can determine the pattern of high and low of the additional factors from the interactions of the embedded factors. Add letters to factor-level combinations of the embedded factorial when the additional factors are at the high level.

1. Choose $q$ generators and get the aliases of I.
2. Find a set of $k - q$ base factors that has an embedded complete factorial.
3. Write the factor-level combinations of the base factors in standard order.
4. Find the aliases of the remaining $q$ factors in terms of interactions of the $k - q$ base factors.
5. Determine the plus/minus pattern for the $q$ remaining factors from their aliased interactions.
6. Add letters to the factor-level combinations of the base factors to indicate when the remaining factors are at their high levels (plus).

**Display 18.1:** Constructing fractional factorials

**Example 18.2**

## Treatments in a $2^{8-4}$ design

Consider the $2^{8-4}$ of Table 18.2 with generators BCDE, ACDF, ABDG, and –ABCH. We can see from the aliases of I that this design has an embedded factorial in A, B, C, and D. The remaining factors E, F, G, and H can be expressed in terms of interactions of the base factors as E = BCD, F = ACD, G = ABC, and H = –ABD.

| Embedded design | E = BCD | F = ACD | G = ABD | H = –ABC | Final design |
|---|---|---|---|---|---|
| (1) | -1 | -1 | -1 | 1 | *h* |
| *a* | -1 | 1 | 1 | -1 | *afg* |
| *b* | 1 | -1 | 1 | -1 | *beg* |
| *ab* | 1 | 1 | -1 | 1 | *abefh* |
| *c* | 1 | 1 | -1 | -1 | *cef* |
| *ac* | 1 | -1 | 1 | 1 | *acegh* |
| *bc* | -1 | 1 | 1 | 1 | *bcfgh* |
| *abc* | -1 | -1 | -1 | -1 | *abc* |
| *d* | 1 | 1 | 1 | 1 | *defgh* |
| *ad* | 1 | -1 | -1 | -1 | *ade* |
| *bd* | -1 | 1 | -1 | -1 | *bdf* |
| *abd* | -1 | -1 | 1 | 1 | *abdgh* |

| Embedded design | E = BCD | F = ACD | G = ABD | H = –ABC | Final design |
|---|---|---|---|---|---|
| *cd* | -1 | -1 | 1 | -1 | *cdg* |
| *acd* | -1 | 1 | -1 | 1 | *acdfh* |
| *bcd* | 1 | -1 | -1 | 1 | *bcdeh* |
| *abcd* | 1 | 1 | 1 | -1 | *abcdefg* |

We can see that each factor-level combination has an even number of letters from the sets BCDE, ACDF, and ABDG, and an odd number of letters from ABCH.

## 18.3 Analyzing a $2^{k-q}$

Analysis of a $2^{k-q}$ is really much like any $2^k$ except that we must always keep the alias structure in mind. Most fractional factorials have only a single replication, so there will be no estimate of pure error. We must either compute a surrogate error by pooling interaction terms, use a graphical approach such as the half-normal plot, or use Lenth's PSE. Keep in mind that if we pool interaction terms, we must look at all the aliases for a given degree of freedom; some interaction terms are aliased to main effects! Similarly, a normal plot of effects may show that an interaction appears to be large. Check the aliases for that degree of freedom, because it could be aliased to a main effect.

Analyze like $2^k$ but remember aliasing

Notice that there is some subjectivity in the analysis of a fractional factorial. For example, we could find that only the degree of freedom D = ABC appears to be significant in a $2^{4-1}$ design with I = ABCD as a defining relation. The most reasonable interpretation is that we are seeing the main effect of D, not an ABC interaction in the absence of any lower-order effects. It is possible that the ABC interaction is large when the A, B, C, AB, AC, and BC effects are null, so we could be making a mistake ascribing this effect to D; but lower-order aliases are usually the safer bet.

Some subjectivity in interpreting aliases

### Welding strength

**Example 18.3**

Taguchi and Wu (1980) describe an experiment carried out to determine factors affecting the strength of welds. There were nine factors at two levels each to be explored. The full experiment was much too large, so a $2^{9-5}$ fractional factorial with sixteen units was used. The factors are coded A though J (skipping I); the generators are –ACE, –ADF, –ACDG, BCDH, ABCDJ. The full defining relation is I = –ACE = –ADF = CDEF = –ACDG = DEG = CFG = –AEFG = BCDH = –ABDEH = –ABCFH = BEFH = –ABGH = BCEGH = BDFGH = –ABCDEFGH = ABCDJ = –BDEJ = –BCFJ = ABEFJ = –BGJ =

**Table 18.3:** Design and responses for welding strength data.

| | A | B | C | D | E | F | G | H | J | y |
|---|---|---|---|---|---|---|---|---|---|---|
| $gj$ | − | − | − | − | − | − | + | − | + | 40.2 |
| $aef$ | + | − | − | − | + | + | − | − | − | 43.7 |
| $bgh$ | − | + | − | − | − | − | + | + | − | 44.7 |
| $abefhj$ | + | + | − | − | + | + | − | + | + | 42.4 |
| $ceh$ | − | − | + | − | + | − | − | + | − | 45.9 |
| $acfghj$ | + | − | + | − | − | + | + | + | + | 42.4 |
| $bcej$ | − | + | + | − | + | − | − | − | + | 40.6 |
| $abcfg$ | + | + | + | − | − | + | + | − | − | 42.2 |
| $dfh$ | − | − | − | + | − | + | − | + | − | 45.5 |
| $adeghj$ | + | − | − | + | + | − | + | + | + | 42.4 |
| $bdfj$ | − | + | − | + | − | + | − | − | + | 40.6 |
| $abdeg$ | + | + | − | + | + | − | + | − | − | 43.6 |
| $cdefgj$ | − | − | + | + | + | + | + | − | + | 40.2 |
| $acd$ | + | − | + | + | − | − | − | − | − | 44.0 |
| $bcdefgh$ | − | + | + | + | + | + | + | + | − | 46.5 |
| $abcdhj$ | + | + | + | + | − | − | − | + | + | 42.5 |

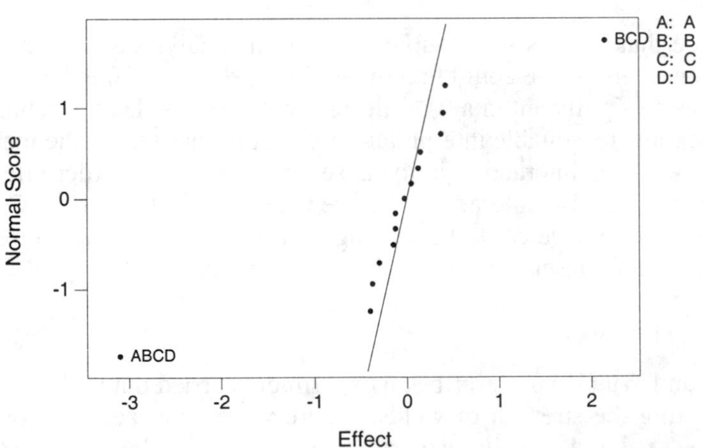


**Figure 18.1:** Normal plot of effects in welding strength data, using Minitab.

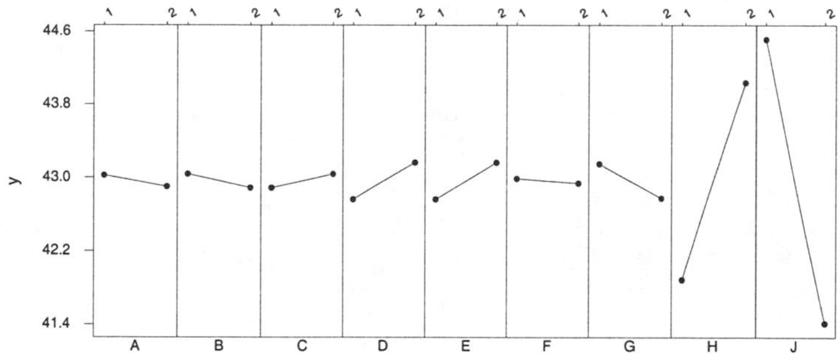


**Figure 18.2:** Main effects in welding strength data, using Minitab.

ABCEGJ = ABDFGJ = –BCDEFGJ = AHJ = –CEHJ = –DFHJ = ACDEFHJ = –CDGHJ = ADEGHJ = ACFGHJ = –EFGHJ; every effect is aliased to 31 other effects. The design and responses are given in Table 18.3.

First note that this design has an embedded $2^4$ design. A check of the defining relation reveals that ABCD is not aliased to I (nor is any subset of ABCD), so we have a complete embedded factorial in those four factors. The data in Table 18.3 are in standard order for A, B, C, and D, so we may compute the main effects and interactions for A, B, C, and D using Yates' algorithm on the responses in the order presented. Figure 18.1 shows a normal plot of these effects. Only the BCD and ABCD interactions are large. Before we interpret these, we must look at their aliases. We find that BCD is aliased to H, and ABCD is aliased to J, so we are probably seeing main effects of H and J.

Alternatively, we may decide to fit just main effects in an Analysis of Variance and pool all remaining degrees of freedom into error. This gives us 9 main-effects degrees of freedom and 6 error degrees of freedom. Listing 18.1 ① shows the estimated effects, their standard errors, and $p$-values. Again, only H and J are significant, which can be seen visually in Figure 18.2. Note that Minitab also computes the low-order aliases of any terms in the model ②.

**Listing 18.1:** Minitab output for welding strength data.

```
Fractional Factorial Fit

Estimated Effects and Coefficients for y (coded units)

Term        Effect      Coef  StDev Coef       T      P
Constant              42.963      0.1359  316.18  0.000
A           -0.125    -0.063      0.1359   -0.46  0.662     ①
B           -0.150    -0.075      0.1359   -0.55  0.601
C            0.150     0.075      0.1359    0.55  0.601
D            0.400     0.200      0.1359    1.47  0.191
E            0.400     0.200      0.1359    1.47  0.191
F           -0.050    -0.025      0.1359   -0.18  0.860
G           -0.375    -0.187      0.1359   -1.38  0.217
H            2.150     1.075      0.1359    7.91  0.000
J           -3.100    -1.550      0.1359  -11.41  0.000

Analysis of Variance for y (coded units)

Source            DF      Seq SS     Adj SS     Adj MS      F      P
Main Effects       9      59.025     59.025     6.5583  22.20  0.001
Residual Error     6       1.772      1.772     0.2954
Total             15      60.797

Alias Structure (up to order 3)

I - A*C*E - A*D*F + A*H*J - B*G*J + C*F*G + D*E*G               ②
A - C*E - D*F + H*J - B*G*H - C*D*G - E*F*G
B - G*J - A*G*H + C*D*H - C*F*J - D*E*J + E*F*H
C - A*E + F*G - A*D*G + B*D*H - B*F*J + D*E*F - E*H*J
D - A*F + E*G - A*C*G + B*C*H - B*E*J + C*E*F - F*H*J
E - A*C + D*G - A*F*G - B*D*J + B*F*H + C*D*F - C*H*J
F - A*D + C*G - A*E*G - B*C*J + B*E*H + C*D*E - D*H*J
G - B*J + C*F + D*E - A*B*H - A*C*D - A*E*F
H + A*J - A*B*G + B*C*D + B*E*F - C*E*J - D*F*J
J + A*H - B*G - B*C*F - B*D*E - C*E*H - D*F*H
```

## 18.4 Resolution and Projection

Resolution determines how short aliases can be

Fractional factorials are classified according to their *resolution,* which tells us which types of effects are aliased. A resolution $R$ design is one in which no interaction of $j$-factors is aliased to an interaction with fewer than $R - j$ factors. For example, in a resolution three design, no main effect ($j = 1$) is aliased with any other main effect, but main effects can be aliased with two-factor interactions ($R - j = 2$). In a resolution four design, no main effect ($j = 1$) is aliased with any main effect or two-factor interaction, but main effects can be aliased with three-factor interactions ($R - j = 3$), and two-factor interactions ($j = 2$) can be aliased with two-factor interactions ($R - j = 2$). In a resolution five design, no main effect is aliased with any main effect, two-factor interaction, or three-factor interaction, but main effects can be aliased with four-factor interactions. Two-factor interactions are not aliased with main effects or two-factor interactions, but they may be aliased with three-factor interactions.

A fractional factorial of resolution $R$ has $R$ letters in the shortest alias of I, so we call these $R$-letter designs. In fact, this is the easy way to remember what resolution means. Resolution is usually written as a Roman numeral subscript for the design. The $2^{8-4}$ design in Table 18.2 has 14 four-letter aliases of I and an eight-letter alias, so it is resolution IV and is written $2_{IV}^{8-4}$.

Resolution equals minimum number of letters in aliases of I

We never want a resolution II design, because such a design would alias two main effects. Thus the minimum acceptable resolution is III. When choosing generators for a $2^{k-p}$ factorial, we want to obtain as high a resolution as possible so that the aliases of main effects will be interactions with as high an order as possible.

Maximize resolution

Resolution isn't the complete picture. Consider three $2^{7-2}$ designs, with defining relations I = ABCF = BCDG = ADCF, I = ABCF = ADEG = BCDEFG, and I = ABCDF = ABCEG = DEFG. All four designs are resolution IV, but we prefer the last design because it has only one 4-letter alias, while the others have two or three. Designs that have the minimum possible number of short aliases are called *minimum-aberration* designs. Thus we want maximum resolution and minimum aberration.

Minimize aberration

Resolution III designs have some main effects aliased to two-factor interactions. If we believe that only main effects are present and all interactions are negligible, then a resolution III design is sufficient for estimating main effects. Resolution III designs are called *main-effects designs* for this reason. If we believe that some two-factor interactions may be nonnegligible but all three-way and higher interactions are negligible, then a resolution IV design is sufficient for main effects.

Main-effects designs

Low-resolution fractional factorials are often used as screening designs, where we are trying to screen many factors to see if any of them has an effect. This is usually an early stage of investigation, so we do not usually require information about interactions, though we would not throw away such information if we can get it.

Screening experiments

We have constructed fractional factorials by augmenting an embedded complete factorial. *Projection* of factorials is somewhat the reverse process, in that we collapse a fractional factorial onto a complete factorial in a subset of factors. A $2^{k-q}$ fractional factorial of resolution $R$ contains a complete factorial in any set of at most $R - 1$ factors. If $R$ is less than $k - q$, then this

Projection onto embedded factorial

**Listing 18.2:** SAS output for welding strength data.

```
Dependent Variable: Y
                                  Sum of          Mean
Source                   DF      Squares        Square    F Value      Pr > F

Model                     3   56.9925000    18.9975000      59.91      0.0001

Error                    12    3.8050000     0.3170833

Source                   DF   Type I SS    Mean Square    F Value      Pr > F

H                         1   18.4900000    18.4900000      58.31      0.0001
J                         1   38.4400000    38.4400000     121.23      0.0001
H*J                       1    0.0625000     0.0625000       0.20      0.6650
```

embedded factorial is replicated. There may also be *some* sets of $R$ or more factors that form a complete factorial, but you are guaranteed a complete factorial for *any* set of $R-1$ factors.

For example, consider the $2^{7-2}_{IV}$ design with defining relation I = ABCDF = ABCEG = DEFG. This design contains a replicated complete factorial in any set of three factors. It also contains a complete factorial in all sets of four factors except D, E, F, and G, which cannot form a complete factorial because their four-factor interaction is aliased to I.

Project onto significant factors

Fractional factorials can be projected onto an embedded factorial during analysis. For example, a half-normal plot of effects in a resolution IV design might indicate that factors A, D, and E look significant. Projection then treats the data as if they were a full factorial in the factors A, D, and E and proceeds with the analysis. Notice that the $p$-values obtained in this way are somewhat suspect. We have put "big" effects into the model and "small" effects wind up in error, so F-statistics and other tests tend to be too big, and $p$-values tend to be too small.

**Example 18.4** **Welding strength, continued**

We found in Example 18.3 that factors H and J were significant. This was a resolution III design, so we can project it onto a factorial in H and J. Listing 18.2 shows an ANOVA for H, J, and their interaction. The main effects are highly significant, as we saw in the earlier analysis. Here we also see that there is no evidence of interaction.

## 18.5 Confounding a Fractional Factorial

We can run a $2^{k-q}$ design in incomplete blocks by confounding one or more degrees of freedom with block differences, just as we did for complete two-series factorials. The only difference is that each defining contrast we confound is aliased with $2^q - 1$ other effects. Similarly, the generalized interactions of the defining contrasts and their aliases are also confounded.

Confound fractions using defining contrasts

**Example 18.5**

### $2^{8-4}$ in two blocks of eight

Example 18.2 has generators BCDE, ACDF, ABDG, and –ABCH, and the factor-level combinations of this fraction are

$$h,\ afg,\ beg,\ abefh,\ cef,\ acegh,\ bcfgh,\ abc,$$
$$defgh,\ ade,\ bdf,\ abdgh,\ cdg,\ acdfh,\ bcdeh,\ abcdefg\ .$$

We must choose a degree of freedom to confound, and Table 18.2 shows that all degrees of freedom have either main-effect or two-factor interaction aliases. We don't want to confound a main effect, so we will confound a two-factor interaction, say AB and its aliases ACDE = BCDF = EF = DG = BCEG = ACFG = ABDEFG = –CH = –BDEH = –ADFH = –ABCEFH = –ABCDGH = –AEGH = –BFGH = –CDEFGH.

To do the confounding, we put all the factor-level combinations with an even number of the letters A and B in one block, and those with an odd number in the other block. These blocks are

$$h,\ abefh,\ cef,\ abc,\ defgh,\ abdgh,\ cdg,\ abcdefg$$

and

$$afg,\ beg,\ acegh,\ bcfgh,\ ade,\ bdf,\ acdfh,\ bcdeh\ .$$

We could have used any of the aliases of AB to get the same blocks. For example, the first block has an even number of B, C, D, and F, and the second block has an odd number.

## 18.6 De-aliasing

Aliasing is the price that we pay for using fractional factorials. Sometimes, aliasing is just a nuisance and it doesn't really affect our analysis. Other times aliasing is crucial. Consider the $2^{5-2}$ design with defining relation I = ABC = –CDE = –ABDE. This design has eight units and 7 degrees of freedom. Suppose that 3 of these degrees of freedom look significant, namely those as-

Check aliases to interpret results

sociated with the main effects of A, C, and E. We cannot interpret the results until we look at the alias structure, and when we do, we find that A = BC = –ACDE = –BDE, C = AB = –DE = –ABCDE, and E = ABCE = –CD = –ABD. The most reasonable explanation of our results is that the main effects of A, C, and E are significant, because other possibilities such as A, C, and the CD interaction seem less plausible. Here aliasing was a nuisance but didn't hurt much.

Aliasing can leave unresolved ambiguity

Suppose instead that the 3 significant degrees of freedom are associated with the main effects of A, B, and C. Now the aliases are A = BC = –ACDE = –BDE, B = AC = –BCDE = –ADE, and C = AB = –DE = –ABCDE. There are four plausible scenarios for significant effects: A, B, and C; A, B, and AB; B, C and BC; or A, C, and AC. All of these interpretations fit the results, and we cannot decide between these interpretations with just these data. We either need additional data or external information that certain interactions are unlikely to choose among the four.

> Fractional factorials can help us immensely by letting us reduce the number of units needed, but they can leave many questions unanswered.

De-aliasing breaks aliases by running an additional fraction

The problem, of course, is that our fractional designs have aliasing. We can *de-alias* by obtaining additional data. Consider the four possible fractions of a $2^5$ using ABC and CDE as generators:

| ABC | CDE | ABDE | Treatments | | | | | | | |
|---|---|---|---|---|---|---|---|---|---|---|
| – | – | + | (1) | *ab* | *acd* | *bcd* | *ace* | *bce* | *de* | *abde* |
| + | – | – | *a* | *b* | *cd* | *abcd* | *ce* | *abce* | *ade* | *bde* |
| – | + | – | *ac* | *bc* | *d* | *abd* | *e* | *abe* | *acde* | *bcde* |
| + | + | + | *c* | *abc* | *ad* | *bd* | *ae* | *be* | *cde* | *abcde* |

Aliasing in common to all fractions is aliasing for full design

Our original fraction is the second one in this table, where ABC is plus and CDE is minus. If we run an additional fraction, then we will have a half-fraction of a $2^5$ run in two blocks of size eight. The aliasing for the half-fraction is the aliasing that is in common to the two quarter-fractions that we use. The defining contrast for blocking is the aliasing that differs between the two fractions.

Aliases that change between fractions are confounded

Suppose that we run the third fraction as an additional fraction. The only aliasing in common to the two fractions is I = –ABDE, so this is the defining relation for the half-fraction. The aliasing that changes between the two fractions is ABC = –CDE, so this is the defining contrast for the confounding.

Note that if we knew ahead of time that we were going to run a second quarter-fraction, we could have designed a resolution V fraction at the start.

By proceeding in two steps, we wound up with resolution IV. The advantage of the two-step procedure is that we might have been able to stop at eight units if the three active factors had been any three other than ABC or CDE; we were just unlucky.

## 18.7 Fold-Over

Use fold-over to construct resolution IV designs

Resolution III fractions are easy to construct, but resolution IV designs are more complicated. *Fold-over* is a technique related to de-aliasing for producing resolution IV designs from resolution III designs. In particular, fold-over produces a $2_{IV}^{k-q}$ design from a $2_{III}^{(k-1)-q}$ design.

Resolution III is easy

Resolution III fractions are easy to produce. Choose a set of base factors for an embedded factorial, and alias every additional factor to an interaction of the base factors. This will always be resolution III or higher.

Fold-over by reversing all signs

To use fold-over, start with a $2_{III}^{(k-1)-q}$ design in the first $k-1$ factors, and produce the table of plus and minus signs for these $k-1$ factors. Augment this table with an additional column of all minuses, labeled for factor $k$. Now double the number of runs by adding the inverse of every row. That is, switch all plus signs to minus, and all minus signs to plus, including the column for factor $k$ that was all minus signs. The result is a $2_{IV}^{k-q}$. The generators for the full design are the generators from the $2_{III}^{(k-1)-q}$, with factor $k$ appended to any generator with an odd number of letters. Note that even though we have constructed this with two fractions, the design is run in one randomization.

Odd-length generators gain last factor

**Example 18.6**

### Fold-over for a $2_{IV}^{15-10}$

A $2_{IV}^{15-10}$ design is too big for most tables, and you will need to work hard to find one by trial and error, but fold-over will do the job easily. Begin with a $2^{14-10}$ design. We will use the generators AB = E, AC = F, AD = G, BC = H, BD = J, CD = K, ABC = L, ABD = M, ACD = N, BCD = O. This just aliases ten additional factors to interactions of the first four. The factor-level combinations and columns of pluses and minuses for the main effects are in the top half of Table 18.4. This includes a column of all minuses for the fifteenth factor P.

In the bottom half, we reverse all the signs from above to produce the second half of the design. In this half, P is always plus. The generators for the full design are ABEP, ACFP, ADGP, BCHP, BDJP, CDKP, ABCL, ABDM, ACDN, BCDO; the odd-length generators for the resolution III design (ABE, ACF, ADG, BCH, BDJ, CDK, and ABC) gain a P in the fold-over design.

**Table 18.4:** Folding over to produce a $2^{15-10}_{IV}$.

| | A | B | C | D | E | F | G | H | J | K | L | M | N | O | P |
|---|---|---|---|---|---|---|---|---|---|---|---|---|---|---|---|
| $efghjk$ | − | − | − | − | + | + | + | + | + | + | − | − | − | − | − |
| $ahjklmn$ | + | − | − | − | − | − | − | + | + | + | + | + | + | − | − |
| $bfgklmo$ | − | + | − | − | − | + | + | − | − | + | + | + | − | + | − |
| $abekno$ | + | + | − | − | + | − | − | − | − | + | − | − | + | + | − |
| $cegjlno$ | − | − | + | − | + | − | + | − | + | − | + | − | + | + | − |
| $acfjmo$ | + | − | + | − | − | + | − | − | + | − | − | + | − | + | − |
| $bcghmn$ | − | + | + | − | − | − | + | + | − | − | − | + | + | − | − |
| $abcefhl$ | + | + | + | − | + | + | − | + | − | − | + | − | − | − | − |
| $defhmno$ | − | − | − | + | + | + | − | + | − | − | − | + | + | + | − |
| $adghlo$ | + | − | − | + | − | − | + | + | − | − | + | − | − | + | − |
| $bdfjln$ | − | + | − | + | − | + | − | − | + | − | + | − | + | − | − |
| $abdegjm$ | + | + | − | + | + | − | + | − | + | − | − | + | − | − | − |
| $cdeklm$ | − | − | + | + | + | − | − | − | − | + | + | + | − | − | − |
| $acdfgkn$ | + | − | + | + | − | + | + | − | − | + | − | − | + | − | − |
| $bcdhjko$ | − | + | + | + | − | − | − | + | + | + | − | − | − | + | − |
| $abcdefghjklmno$ | + | + | + | + | + | + | + | + | + | + | + | + | + | + | − |
| $abcdlmnop$ | + | + | + | + | − | − | − | − | − | − | + | + | + | + | + |
| $bcdefgop$ | − | + | + | + | + | + | + | − | − | − | − | − | − | + | + |
| $acdehjnp$ | + | − | + | + | + | − | − | + | + | − | − | − | + | − | + |
| $cdfghjlmp$ | − | − | + | + | − | + | + | + | + | − | + | + | − | − | + |
| $abdfhkmp$ | + | + | − | + | − | + | − | + | − | + | − | + | − | − | + |
| $bdeghklnp$ | − | + | − | + | + | − | + | + | − | + | + | − | + | − | + |
| $adefjklop$ | + | − | − | + | + | + | − | − | + | + | + | − | − | + | + |
| $dgjkmnop$ | − | − | − | + | − | − | + | − | + | + | − | + | + | + | + |
| $abcgjklp$ | + | + | + | − | − | − | + | − | + | + | + | − | − | − | + |
| $bcefjkmnp$ | − | + | + | − | + | + | − | − | + | + | − | + | + | − | + |
| $aceghkmop$ | + | − | + | − | + | − | + | + | − | + | − | + | − | + | + |
| $cfhklnop$ | − | − | + | − | − | + | − | + | − | + | + | − | + | + | + |
| $abfghjnop$ | + | + | − | − | − | + | + | + | + | − | − | − | + | + | + |
| $behjlmop$ | − | + | − | − | + | − | − | + | + | − | + | + | − | + | + |
| $aefglmnp$ | + | − | − | − | + | + | + | − | − | − | + | + | + | − | + |
| $p$ | − | − | − | − | − | − | − | − | − | − | − | − | − | − | + |

There are 105 four-factor, 280 six-factor, 435 eight-factor, 168 ten-factor, and 35 twelve-factor aliases of I in this fold-over design, a complete enumeration of which you will be spared.

## 18.8 Sequences of Fractions

De-aliasing makes routine use of fractional factorials possible, because we can always use additional fractions to break any aliases that are giving us trouble. In particular, one thing that makes fractional factorials attractive is the ability to run fractions in sequence.

For example, suppose you have six factors that you wish to explore, and money for 32 experimental units. You could use those 32 units to run a $2_{VI}^{6-1}$ design. Or you could use 16 of those units and run a $2_{IV}^{6-2}$ design with ABCE and BCDF as generators and save the remaining 16. Why is the second approach often better? If three or fewer factors are active, then you have a replicated complete factorial in those three factors (projection of a fraction). In this case, these first 16 units may be enough to answer our questions. If more factors are active—in particular if A, B, C, and E or B, C, D, and F are active—we can always use the remaining 16 units to run an additional fraction, and we can choose that fraction to break aliases that appear troublesome in the first fraction. The combined quarter-fractions are as good as the original half-fraction (except for a single degree of freedom between the two blocks), because we can choose our second quarter-fraction after seeing the first.

Sequences of fractions can save money

Use results of first fraction to select later fractions

Thus by using a sequence of fractions, you can often learn everything you need to learn with fewer units; and if you cannot, you can use the first fraction to guide your choice of subsequent fraction for remaining units.

Sequences of fractions make sense when each experiment is of short duration so that running experiments in sequence is feasible. If each experiment takes months to complete (for example, many agronomy experiments), then a sequence of fractions is a poor choice of design.

Sequences need quick turnaround

## 18.9 Fractioning the Three-Series

Fractional factorials for the three-series are constructed in the same way as the two-series: confound the full factorial into blocks and then run just one block. Three-series factorials are confounded into 3, 9, 27, and other powers of three blocks, so three-series can be fractioned into fractions of one third, one ninth, and so on.

Recall that the factor levels in a three-series are represented by the digits 0, 1, or 2, and that all degrees of freedom are partitioned into two-degree-of-freedom bundles. The bundles are obtained by splitting the factor-level combinations according to their values on a defining split $L$. For example,

A fraction is a single block from a confounded three-series

the defining split $A^1B^1C^2$ separates the factor-level combinations into three groups according to

$$L = 1 \times x_A + 1 \times x_B + 2 \times x_C \bmod 3 \ ,$$

where $x_A$, $x_B$, and $x_C$ are the the levels of factors A, B, and C; $L$ takes the values 0, 1, or 2. The factor-level combinations that have value 0 for the defining split(s) form the principal block, and all others are alternate blocks. These become principal and alternate fractions. The defining splits are the generators for the fraction.

$3^{k-1}$ aliases come in threes

In a $2^{k-q}$ factorial, every degree of freedom has $2^q$ names, and every effect is aliased to $2^q - 1$ other effects. It's just a little more complicated for three-series fractions. In a $3^{k-1}$, the constant is aliased to a two-degree-of-freedom split (the generator); all other two-degree-of-freedom bundles have three names, and all other splits are aliased to two other splits. If $W$ is the generator, then the aliases of a split $P$ are $PW$ and $PW^2$. (Recall that exponents of these products are reduced modulo 3, and if the leading nonzero exponent is a 2, double the exponents and reduce modulo 3 again.) For example, the aliases in a $3^{3-1}$ with $W = A^1B^2C^2$ as generator are

| | $W$ | $W^2$ |
|---|---|---|
| $I$ | $A^1B^2C^2$ | |
| $A$ | $A^1B^1C^1 = A(A^1B^2C^2)$ | $B^1C^1 = A(A^1B^2C^2)^2$ |
| $B$ | $A^1C^2 = B(A^1B^2C^2)$ | $A^1B^1C^2 = B(A^1B^2C^2)^2$ |
| $C$ | $A^1B^2 = C(A^1B^2C^2)$ | $A^1B^2C^1 = C(A^1B^2C^2)^2$ |
| $A^1B^1$ | $A^1C^1 = A^1B^1(A^1B^2C^2)$ | $B^1C^2 = A^1B^1(A^1B^2C^2)^2$ |

$3^{k-2}$ aliases come in nines

In a $3^{k-2}$, the constant is aliased to four two-degree-of-freedom splits; all other two-degree-of-freedom bundles have nine names, and all other splits are aliased to eight other splits. Using two generators $W_1$ and $W_2$, the aliases of I are $W_1$, $W_2$, $W_1W_2$, and $W_1W_2^2$. Which generator is labeled one or two does not matter, because $W_1W_2^2 = W_1^2W_2$ after reducing exponents modulo 3 and making the leading nonzero exponent a 1. The aliases of any other split $P$ are $PW_1$, $PW_2$, $PW_1W_2$, $PW_1W_2^2$, $PW_1^2$, $PW_2^2$, $PW_1^2W_2^2$, and $PW_1^2W_2$. (Again, reduce exponents modulo 3; double and reduce modulo 3 again if the leading nonzero exponent is not a 1.) For a $3^{4-2}$ factorial with generators $A^1B^1C^1$ and $B^1C^2D^1$, the complete alias structure is

| | $W_1$ | $W_2$ | $W_1W_2$ | $W_1W_2^2$ |
|---|---|---|---|---|
| $I$ | $A^1B^1C^1$ | $B^1C^2D^1$ | $A^1B^2D^1$ | $A^1C^2D^2$ |
| $A$ | $A^1B^2C^2$ | $A^1B^1C^2D^1$ | $A^1B^1D^2$ | $A^1C^1D^1$ |
| $B$ | $A^1B^2C^1$ | $B^1C^1D^2$ | $A^1D^1$ | $A^1B^1C^2D^2$ |
| $C$ | $A^1B^1C^2$ | $B^1D^1$ | $A^1B^2C^1D^1$ | $A^1D^2$ |
| $D$ | $A^1B^1C^1D^1$ | $A^1C^2D^2$ | $A^1B^2D^2$ | $A^1C^2$ |

| | $W_1^2$ | $W_2^2$ | $W_1^2W_2^2$ | $W_1^2W_2$ |
|---|---|---|---|---|
| $I$ | | | | |
| $A$ | $B^1C^1$ | $A^1B^2C^1D^2$ | $B^1D^2$ | $C^1D^1$ |
| $B$ | $A^1C^1$ | $C^1D^2$ | $A^1B^1D^1$ | $A^1B^2C^2D^2$ |
| $C$ | $A^1B^1$ | $A^1C^1D^1$ | $A^1B^2C^2D^1$ | $A^1C^1D^2$ |
| $D$ | $A^1B^1C^1D^2$ | $B^1C^2$ | $A^1B^2$ | $A^1C^2D^1$ |

General $3^{k-q}$ aliasing

Further fractions require more generators. A $3^{k-q}$ has $q$ generators $W_1$ through $W_q$. The constant is aliased to $1 + 3 + \cdots + 3^{q-1}$ two-degree-of-freedom splits; these splits aliased to I are of the form $W_1^{i_1}W_2^{i_2}\cdots W_q^{i_q}$ where the exponents are 0, 1, or 2, and the first nonzero exponent is a 1. All other two-degree-of-freedom bundles have $3^q$ names, and all other splits are aliased to $3^q - 1$ other splits. The aliases of a split $P$ are products of the form $PW_1^{i_1}W_2^{i_2}\cdots W_q^{i_q}$, where the exponents $i_j$ are allowed to range over all $3^q$ combinations of 0, 1, and 2. There are $1+3+\cdots+3^{k-q-1}$ sets of aliases in addition to the aliases of I.

Design resolution

Resolution in the $3^{k-q}$ is the same as in the two-series: a fractional factorial has resolution $R$ indexFractional factorials!resolution if no interaction of $j$ factors is aliased to an interaction of fewer than $R-j$ factors. And again like the two-series, the resolution of a $3^{k-q}$ is the number of letters in the shortest alias of I.

Add levels of aliased factors to embedded factorial

We can construct a $3^{k-q}$ using embedded factorials as we did for two-series. In the $3^{3-1}$ described above, recall the aliasing $C = A^1B^2$. Construct a full factorial in A and B, and then set the levels of C according to the $A^1B^2$ interaction; this will generate the fraction. Consider the following table:

| | | |
|---|---|---|
| 00 0 | 01 2 | 02 1 |
| 10 1 | 11 0 | 12 2 |
| 20 2 | 21 1 | 22 0 |

The pairs of digits form a complete $3^2$ design, and the single digits are the values of

$$1 \times x_A + 2 \times x_B \bmod 3 \ ,$$

the $A^1B^2$ interaction. These are also the levels of C for the principal fraction. Group the triples together, and we have the principal fraction of a $3^{3-1}$ with generator $A^1B^2C^2$. If we want an alternate fraction, use

Add 1 or 2 to get alternate fraction

$$1 \times x_A + 2 \times x_B + 1 \bmod 3$$

or

$$1 \times x_A + 2 \times x_B + 2 \bmod 3$$

to generate the levels of C.

## 18.10 Problems with Fractional Factorials

Fractions offer many chances for mistakes

Fractional factorials can be extremely advantageous in situations where we want to screen factors, can ignore interactions, or have restricted resources. However, the sophistication of the fractional factorial gives us many ways in which to err, and fractional factorials are a bit more brittle than complete factorials in the face of real-world data. Daniel (1976) discusses these problems in detail.

Choose fraction size carefully

Here are some common pitfalls that you must try to avoid when using fractional factorials. During the design stage, you can make your fractional factorial too large or too small. A design that is too small tries to estimate too many effects for the number of experimental units used; this is called oversaturation. Designs that are too small tend to be limited in how you can estimate error, because all the degrees of freedom are tied up in interesting effects, and resolution tends to be small. Designs that are too large are being wasteful of resources; you may be able to estimate all terms of interest with a smaller design. This ties in with power. Fractional designs have smaller sample sizes and thus less power for a given set of effects and error variance. When planning the size of the design, we need to keep power in mind. All of these design issues depend on having at least some prior knowledge or belief of how the system works. This will allow us to decide what resolution and replication is needed.

Check aliasing and watch for bad data

In the analysis stage, the most obvious problem is dealing incorrectly with aliasing. You thus wind up with a misinterpretation of which effects are important. You may also miss a need to de-alias. Finally, outliers and missing data tend to cause more problems for fractional factorials than complete factorials. For example, consider an outlier in a $2^{k-q}$. In the complete two-series, an outlier can sometimes be detected by a pattern of smallish effects of about the same size, usually high-order interactions. In the fraction, many degrees of freedom have a main effect or low-order interaction in their aliases, so there are few opportunities to see the flat pattern in effects that we expect to be null.

## 18.11 Using Fractional Factorials in Off-Line Quality Control

One of the areas in which fractional factorials and related designs have been used with much success, profit, and acclaim is off-line quality control. Quality control has on-line and off-line aspects. On-line means "on the production line"; on-line quality control includes inspection of manufactured parts to make sure that they meet specifications. Off-line quality control is off the production line; this includes designing the product and manufacturing process so that the product will meet specifications when manufactured. The explicit goal is to have the product on target, with minimum variation around the target.

Goal of off-line quality control is to make products on target with minimum variation

Suppose that you manufacture exhaust tubing for the automotive industry. Your client orders a tubing part that should be 2.1 meters long and bent into a specific shape; parts from 2.09 to 2.11 meters in length are acceptable. One step of the manufacturing process is cutting the tubing to length. On-line quality control will include inspection of the cut tubing and rejection of those tubes out of specification. Off-line quality control designs the tube cutting process so that the average tube length is 2.1 meters and the variation around that average is as small as possible.

Off-line quality control has become quite the rage under the banner of "Taguchi methods," named for Genechi Taguchi, the Japanese statistician who developed and advocated the methods. The principle of off-line quality control is to put a product on target with minimum variation. This principle is absolutely golden, but the exact methods Taguchi recommended for achieving this have flaws and inefficiencies in both design and analysis (see Box, Bisgaard, and Fung 1988 or Pignatiello and Ramberg 1991). What we discuss here is very much in the spirit of Taguchi, but the analysis approach is closer to Box (1988).

Taguchi methods

Most manufacturing processes have many controllable design parameters. For the exhaust tubes, design parameters include the speed at which tubing moves down the line, the air pressure for tubing clamps, cutting saw speed, the type of sensor for recognizing the end of a tube, and so on. These parameters might influence product quality, but we generally don't know which ones are important. Manufacturing processes also have uncontrollable aspects, including variation in raw materials and environmental varia-

Inner noise controllable, outer noise uncontrollable

tion such as temperature and humidity. Some of these "uncontrollables" can actually be controlled under laboratory or testing conditions. Taguchi uses the term "inner noise" for variation that arises from changes in the controllable parameters and the term "outer noise" for variation due to the uncontrollable parameters.

### 18.11.1 Designing an off-line quality experiment

Study means and variances

We want to find settings for the controllable variables so that the product is on target and the variation due to the outer noise is as small as possible. This implies that we need experiments that can study both means *and* variances. We are also explicitly considering the possibility that the variance will not be constant, so we will need some form of replication at all design points to allow us to estimate the variances separately.

Use replicated fractional factorials

Replicated two- and three-series factorials are the basic designs for off-line quality control. From these we can estimate mean responses as usual, and replication allows us to estimate the variance at each factor-level combination as well. There are often ten to fifteen or more factors identified as potentially important. A complete factorial with this many factors would be prohibitively large, so off-line quality control designs are frequently highly-fractioned factorials, but with replication.

Is outer noise micro or macro scale?

Two situations present themselves. In the first situation, the outer noise is at something of a micro scale, meaning that you tend to experience the full range of outer noise whenever you experiment. One of Taguchi's early successes was at the Ina Tile Company, where there was temperature variation in the kilns. This noise was always present, as tiles in different parts of the kiln experienced different temperatures. In the second situation, the outer noise is at a more macro scale, meaning that you tend to experience only part of the range of outer noise in one experiment. In the exhaust tubing, for example, temperature and humidity in the factory may affect the machinery, but you tend not to get hot and cold, dry and humid conditions scattered randomly among your experimental runs. It is hot and humid in the summer and cold and dry in the winter.

Design plan should include macro-level outer noise

These two situations require different experimental approaches. When you have outer noise at the micro level, it is generally enough to plan an experiment using the controllable variables and let the outer noise appear naturally during replication. When the outer noise is at the macro level, you must take steps to make sure that the range of outer noise is included in your experiment. If the outer-noise factors can be controlled under experimental conditions, then these factors should also be included in the design to ensure their full range.

Let's return to the exhaust tube problem to make things explicit. Our controllable factors are tube speed, air pressure, saw speed, and sensor type; the outer-noise factors are temperature and humidity. Assume for simplicity that we can choose two levels for all factors, so that there are sixteen combinations for the controllable factors and four combinations for the outer-noise factors. We need to include the outer-noise factors in our design, because we are unlikely to see the full range of outer-noise variation if we do not.

There are several possibilities for this experiment. For example, we could run the full $2^6$ design. This gives four "replications" at each combination of the controllable factors, and these replications span the range of the noise factors. Or we could run a $2^{6-1}$ fraction with 32 points. This is smaller (and possibly quicker and cheaper), but with a smaller sample size we have less power for detecting effects and only 1 degree of freedom for estimating variation at each of the sixteen combinations of controllable factors.

### 18.11.2 Analysis of off-line quality experiments

Design variables affect mean and variation, adjustment variables affect only mean

Analysis is based on the following idea. Some of the controllable factors affect the variance and the mean, and an additional set of controllable factors affects only the mean. The factors that affect the variance and mean are called *design* variables; those that affect only the mean are called *adjustment* variables. The idea is to use the design variables to minimize the variance, and then use the adjustment variables to bring the mean on target.

This approach is complicated by the fact that mean and variance are often linked in the usual nonconstant-variance sense that we check with residual plots and remove using a transformation. If we have this kind of nonconstant variance, then every variable that affects the mean also affects the variance, and we will have no adjustment variables. Therefore we need to accommodate this kind of nonconstant variance before dealing with variation that depends on controllable variables but not directly through the mean.

Transform to "constant" variance

First, find a transformation of the responses that removes the dependence of variance on mean as much as possible. This is essentially a Box-Cox transformation analysis. On this transformed scale, we hope that there are variables that affect the mean but not the variance.

Analyze log variances to determine design variables

Next, compute the log of the variance of the transformed data at every factor-level combination of the controllable factors. Treat these log variances as responses, and analyze them via ANOVA to see which, if any, controllable factors affect the variance; these are the design variables. Find the factor-level combination that minimizes the variance. For highly-fractioned designs we may only be able to do this by looking at main effects and hoping that

**Table 18.5:** Variance of natural-log sample variances from normal data for 1 through 10 degrees of freedom.

| 1 | 2 | 3 | 4 | 5 | 6 | 7 | 8 | 9 | 10 |
|---|---|---|---|---|---|---|---|---|---|
| 4.93 | 1.64 | .93 | .64 | .49 | .39 | .33 | .28 | .25 | .22 |

there are no interactions. One complication that arises in this step is that once we have log variance as a response, there is no replication. Thus we must use a method for unreplicated factorials to assess whether treatments affect variances.

Variance of log sample variance is known for normally distributed data

If we can assume that the (transformed) responses that go into each of these variances are independent and normally distributed, then we can calculate an approximate $MS_E$ for the ANOVA with log variances as the responses. Suppose that there are $n$ experimental units at each factor-level combination of the controllable factors; then each of these sample variances has $n-1$ degrees of freedom. The variance of the (natural) log of a sample variance depends only on the degrees of freedom. Table 18.5 lists the variance of the log of a sample variance for up to 10 degrees of freedom. Note that the variances in that table are *very* sensitive to the normality assumption.

Put response on target using adjustment variables with design variables set to minimum variance

Finally, return to the original scale. Analyze the response to determine which factors affect the mean response, and find settings for the adjustment variables that put the response on target when the design variables are at their variance-minimizing settings. This step generally makes the assumptions that the adjustment factors can be varied continuously and that the response is linear between the two observed levels of a factor. Please note that adjusting a transformation of $y$ to a target $T$, say $\sqrt{y}$ to $\sqrt{T}$, will result in a bias on the original scale and thus a deviation from the target.

**Example 18.7**

**Free height of leaf springs**

Pignatiello and Ramberg (1985) present a set of data from a quality experiment on the manufacture of leaf springs for trucks. The free height should be as close to 8 inches as possible, with minimum variation. There are four inner noise factors, each at two levels: furnace temperature (B), heating time (C), transfer time (D), and hold-down time (E). There was one outer noise factor: quench oil temperature (O). A $2^{5-1}$ design with three replications was conducted. We will analyze this as a $2^{4-1}$ design in the inner noise factors with six replications, because quench-oil temperature is not easily controlled in factory conditions. Table 18.6 shows the results.

**Table 18.6:** Free height of leaf springs.

| B | C | D | E | O low | | | O high | | | $\overline{y}$ | $s^2$ |
|---|---|---|---|---|---|---|---|---|---|---|---|
| − | − | − | − | 7.78 | 7.78 | 7.81 | 7.50 | 7.25 | 7.12 | 7.54 | .0900 |
| + | − | − | + | 8.15 | 8.18 | 7.88 | 7.88 | 7.88 | 7.44 | 7.90 | .0707 |
| − | + | − | + | 7.50 | 7.56 | 7.50 | 7.50 | 7.56 | 7.50 | 7.52 | .0010 |
| + | + | − | − | 7.59 | 7.56 | 7.75 | 7.63 | 7.75 | 7.56 | 7.64 | .0079 |
| − | − | + | + | 7.94 | 8.00 | 7.88 | 7.32 | 7.44 | 7.44 | 7.67 | .0908 |
| + | − | + | − | 7.69 | 8.09 | 8.06 | 7.56 | 7.69 | 7.62 | 7.79 | .0529 |
| − | + | + | − | 7.56 | 7.62 | 7.44 | 7.18 | 7.18 | 7.25 | 7.37 | .0380 |
| + | + | + | + | 7.56 | 7.81 | 7.69 | 7.81 | 7.50 | 7.59 | 7.66 | .0173 |

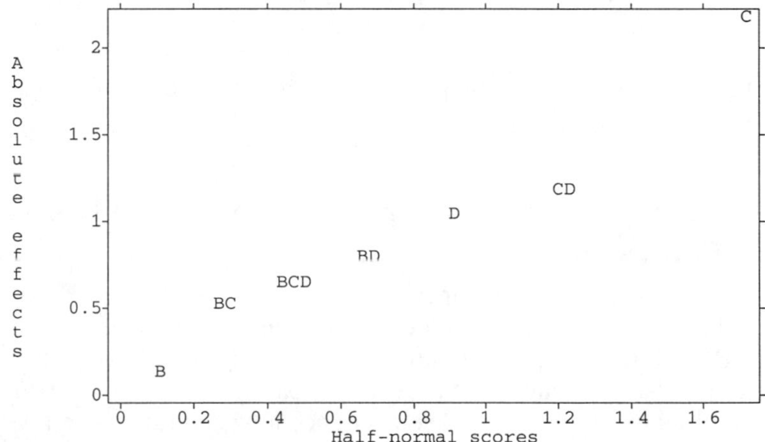


**Figure 18.3:** Half-normal plot of dispersion effects for leaf spring data, using MacAnova.

We first examine whether the data should be transformed. A plot of log treatment variance against log treatment mean shows no pattern, and Box-Cox does not indicate the need for a transformation, so we use the data on the original scale.

We now do a factorial analysis using log treatment variance as response. (If we had transformed the data, the response would be the log of the variance of the transformed data.) Figure 18.3 shows a half-normal plot of the dispersion effects, that is, the factorial effects with log variance as response. Only

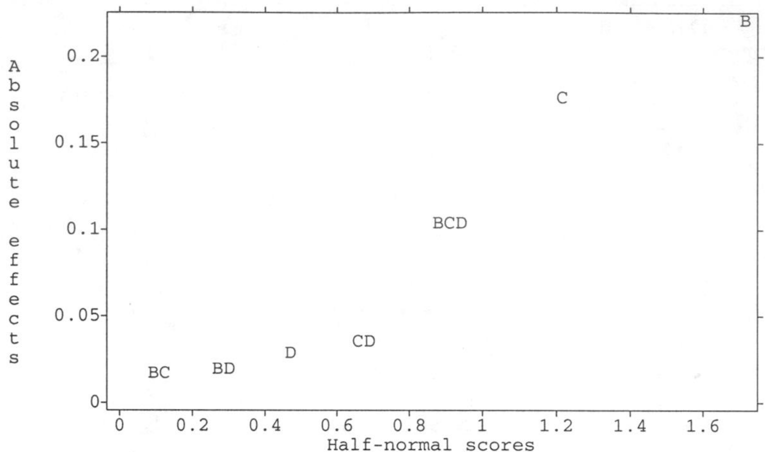


**Figure 18.4:** Half-normal plot of location effects for leaf spring data, using MacAnova.

factor C appears to affect dispersion, and inspection of Table 18.6 shows that the high level of C has lower variance.

Now examine how the treatments affect average response. Figure 18.4 shows a half-normal plot of the location effects. Here we see that B, C, and the BCD interaction are significant. Recalling the aliasing, the BCD interaction is also the main effect of E. Thus heating time is a design variable that we will set to a high level to keep variance low, and furnace temperature and hold-down time are adjustment variables.

Listing 18.3 shows the location effects for these variables. We have set C to the high level to get a small variance. To get the mean close to the target of 8, we need B and E to be at their high levels as well; this gives us 7.636 + .111 − .088 + .052, or 7.711, as our estimated response. This is still a little low, so we may need to explore the possibility of expanding the ranges for factors B and E to get the response closer to target.

## 18.12 Further Reading and Extensions

Orthogonal-main-effects plans are resolution III designs constructed so that the main effects are orthogonal. Resolution III two- and three-series fraction factorials are orthogonal-main-effects plans, but there are several addi-

**Listing 18.3:** Location effects for the leaf spring data, using MacAnova.

```
component: CONSTANT
(1)         7.636
component: b
(1)      -0.11062        0.11063
component: c
(1)      0.088125      -0.088125
component: e
(1)     -0.051875       0.051875
```

tional families of designs that have these properties as well. Plackett-Burman designs (Plackett and Burman 1946) are orthogonal-main-effects plans for $N - 1$ factors at two levels each using $N$ experimental units when N is an integer multiple of 4. When N is a power of 2, these are resolution III fractions of the kind discussed in this chapter. Addelman (1962) constructs orthogonal-main-effects plans for mixed factorials by collapsing factors. For example, start with a $3^{4-2}$ fraction. Replace factor A by a two level factor E, using the low level of E when A is 0 or 2, and the high level of E when A is 1. This produces a fraction of a $2^1 3^3$ design in nine units. John (1971) discusses these two classes, as well as some other mixed factorial fractions. The aliasing structure of these designs can be quite complex.

Orthogonal arrays are a third class of orthogonal-main-effects plans that are often used in quality experiments. An orthogonal array for $k$ factors in $N$ units is described by an $N$ by $k$ matrix of integers; rows for units, columns for factors, and integers giving factor levels. To be an orthogonal array, all possible pairs of factor levels must occur together an equal number of times for any pair of factors. Standard two- and three-series fractional factorials of resolution III meet this criterion, but so do many additional designs. Hedayat and Wallis (1978) review some of the theory and applications of these arrays.

Fractional factorials can also be run using split-plot and related unit structures. See Miller (1997).

## 18.13 Problems

**Exercise 18.1**

Food scientists are trying to determine what chemical compounds make heated butter smell like heated butter. If they could figure that out, then they could make foods that smell like butter without having all the fat of butter. There are eight compounds that they wish to investigate, with each compound at either a high or low level. They use a $2^{8-4}$ fractional factorial design with I = ABDE = ABCF = -ACDG = -BCDH.

(a) Find the factor-level combinations used in this design.

(b) Find the aliases of I and A.

(c) If A, B, D, E, and AB look big, are there any unresolved ambiguities? If so, which further fraction would you run to resolve the ambiguity?

**Exercise 18.2** Consider a $2^{6-2}$ fractional factorial using I=ABDF = -BCDE.

(a) Find the aliases of the main effects.

(b) Find the factor-level combinations used.

(c) Show how you would block these combinations into two blocks of size eight.

**Exercise 18.3** Consider the $2^{8-4}$ fractional factorial with generator I = BCDE = ACDF = ABCG = ABDH. Find the aliases of C.

**Exercise 18.4** Design a $2^{7-2}$ resolution IV fractional factorial. Give the factor-level combinations used in the principal fraction and show how you would block these combinations into two blocks of size sixteen.

**Exercise 18.5** Design an experiment. There are eight factors, each at two levels. However, we can only afford 64 experimental units. Furthermore, there is considerable unit to unit variability, so blocking will be required, and the maximum block size possible is 16 units. You may assume that three-way and higher-order interactions are negligible, but two-factor interactions may be present.

**Exercise 18.6** Find the factor-level combinations used in the principal fraction of a $3^{4-1}$ with the generator $A^1B^1C^1D^1$. Report the alias structure, and show how you would block the design into blocks of size nine.

**Problem 18.1** Briefly describe the experimental design used in each of the following situations (list units, blocks, covariates, factors, whole/split plots, and so forth). Give a skeleton ANOVA (sources and degrees of freedom only).

(a) We wish to study the effects of stress and activity on the production of a hormone present in the saliva of children. The high-stress treatment is participation in a play group containing children with whom the subject child in unacquainted; the low-stress treatment is participation in a play group with other children already known to the subject child. The activities are a group activity, where all children play together, and an individual activity, where each child plays separately. Thirty-two children are split at random into two groups of sixteen. The first group is assigned to high stress, the other to low stress. For each child the order of group or individual activity is randomized, and a saliva sample is taken during each activity.

(b) Neighbors near the municipal incinerator are concerned about mercury emitted in stack gasses. They want a measure of the accumulation rate of mercury in soil at various distances and directions from the incinerator. They collect a bunch of soil, mix it as well as they can, divide it into 30 buckets, and have a lab measure the mercury concentration in each bucket. The buckets are then randomly divided into fifteen sets of two; the pairs are placed in fifteen locations around the incinerator, left for 2 years, and then analyzed again for mercury. The response is the increase in mercury. The lab informed the activists that the amount of increase will be related to the amount of carbon in the soil, because mercury is held in the organic fraction; so they also take a carbon measurement.

(c) We wish to discover the effects of food availability on the reproductive success of anole lizards as measured by the number of new adults appearing after the breeding season. There are twelve very small islands with anole populations available for the study. The islands are man-made and more or less equally spaced along a north-south line. The treatments will be manipulation of the food supply on the islands during peak breeding season. There are three treatments: control (leave natural), add supplemental food, and reduced food (set out traps to deplete the population of insects the anoles eat). One potential source of variation is that the lizards are eaten by birds, and there is a wildlife refuge with a large bird population near the northern extreme of the study area. To control for this, we group the islands into the northern three, the next three, and so on, and randomize the treatments within these groups.

(d) A fast-food restaurant offers both smoking and non-smoking sections for its customers. However, there is considerable smoke "leakage" from the smoking section to the non-smoking section. The manager wants to minimize this leakage by finding a good division of the restaurant into the two sections. She has three possible divisions of the tables, and conducts an experiment by assigning divisions at random to days for 3 weeks (7 days per division) and surveying non-smoking patrons about the amount of smoke. In addition, she monitors the number of smokers per day, as that has an obvious effect on the amount of leakage.

**Problem 18.2** Briefly describe the experimental design you would choose for each of the following situations, and why.

(a) Asbestos fiber concentrations in air are measured by drawing a fixed volume of air through a disk-shaped filter, taking a wedge of the filter (generally 1/4 of the filter), preparing it for microscopic analysis, and then counting the number of asbestos fibers found on the prepared wedge when looking through an optical microscope. (Actually, we only count on a random subsample of the area of the prepared wedge, but for the purposes of the question, consider the wedge counted.) We wish to compare four methods of preparing the wedges for their effects on the subsequent fiber counts. We have available 24 filters from a broad range of asbestos air concentrations; we can use each filter entirely, so that we can get four wedges from each filter. We can also use four trained microscopists. Despite the training, we anticipate considerable microscopist to microscopist variation in the counts (that is, some tend to count high, and some tend to count low).

(b) A food scientist wishes to study the effect that eating a given food will have on the ratings given to a similar food (sensory-specific satiety). There is a pool of 24 volunteers to work with. Each volunteer must eat a "load food" (a large portion of hamburger or potato), and then eat and rate two "test foods" (small portions of roast beef and rice). After eating, the volunteer will rate the appeal of the roast and rice.

(c) Scientists studying the formation of tropospheric ozone believe that five factors might be important: amount of hydrocarbon present, amount of $NO_X$ present, humidity, temperature, and level of ultraviolet light. They propose to set up a "model atmosphere" with the appropriate ingredients, "let it cook" for 6 hours, and then measure the ozone produced. They only have funding sufficient for sixteen experimental units, and their ozone-measuring device can only be used eight times before it needs to be cleaned and recalibrated.

(d) A school wishes to evaluate four reading texts for use in the sixth grade. One of the factors in the evaluation is a student rating of the stories in the texts. The principal of the school decides to use four sixth-grade rooms in the study, and she expects large room to room differences in ratings. Due to the length of the reading texts and the organization of the school year into trimesters, each room can evaluate three texts. The faculty do not expect systematic differences in ratings between the trimesters.

(e) The sensory quality of prepared frozen pizza can vary dramatically. Before the quality control department begins remedial action to reduce the variability, they first attempt to learn where the variability arises. Three broad sources are production (variation in quality from batch to batch at the factory), transportation (freeze/thaw cycles degrade the product, and our five shipping/warehouse companies might not keep the product fully frozen), and stores (grocery store display freezers may not keep the product frozen). Design an experiment to estimate the various sources of variability from measurements made on pizzas taken from grocery freezers. All batches of pizza are shipped by all shipping companies, but each grocery store is served by only one shipping company. You should buy no more than 500 pizzas.

(f) Food scientists are trying to figure out what makes cheddar cheese smell like cheddar cheese. To this end, they have been able to identify fifteen compounds in the "odor" of the cheese, and they wish to make a preliminary screen of these compounds to see if consumers identify any of these compounds or combinations of compounds as "cheddary." At this preliminary stage, the scientists are willing to ignore interactions. They can construct test samples in which the compounds are present or absent in any combination. They have resources to test sixteen consumers, each of whom should sample at most sixteen combinations.

(g) The time until germination for seeds can be affected by several variables. In our current experiment, a batch of seeds is pretreated with one of three chemicals and stored for one of three time periods in one of two container types. After storage time is complete, the average time to germination is measured for the batch. We have 54 essentially uniform batches of seeds, and wish to understand the relationships between the chemicals, storage times, and storage containers.

(h) The U.S. Department of Transportation needs to compare five new types of pavement for durability. They do this by selecting "stretches" of highway, installing an experimental pavement in the stretch, and then measuring the condition of the stretch after 3 years. There are resources allocated for 25 stretches of highway. From past experience, the department knows that traffic level and weather patterns affect the durability of pavement. The department is organized into five regional districts, and within each district the weather patterns are reasonably uniform. Also within each district are highways from each of the five traffic level groups.

Avocado oil may be extracted from avocado paste using the following steps: (1) dilute the paste with water, (2) adjust the pH of the paste, (3) heat the paste at 98°C for 5 minutes, (4) let the paste settle, (5) centrifuge the **Problem 18.3** paste. We may vary the dilution rate (3:1 water or 5:1 water), pH (4.0 or 5.5), settling (9 days at 23°C or 4 days at 37°C), and centrifugation (6000g or 12000g). Briefly describe experimental designs for each of the following situations. You may assume that the paste (prior to any of the five steps mentioned) may be used any time up to a week after its preparation. You may also assume that the primary cost is the analysis; the cost of the paste is trivial.

(a) We wish to study effects of the four factors mentioned on the extraction efficiency. Avocado paste is rather uniform, and we have enough money for 48 experimental units.

(b) We wish to study effects of the four factors mentioned on the extraction efficiency. Avocado paste is not uniform but varies from individual fruit to fruit. Each fruit produces enough paste for about 20 experimental units, and we have enough money for 48 experimental units.

(c) We wish to study effects of the four factors mentioned on the extraction efficiency. Avocado paste is not uniform but varies from individual fruit to fruit. Each fruit produces enough paste for about 10 experimental units, and we have enough money for 48 experimental units.

(d) We wish to determine the effects of the pH, settling, and centrifugation treatments on the concentration of $\alpha$-tocopherol (vitamin E) in the oil. Each fruit produces enough paste for about six experimental units, and we have enough money for 32 experimental units. Furthermore, we can only use four experimental units per day and the instruments need to be recalibrated each day.

**Problem 18.4** An experiment was conducted to determine the factors that affect the amount of shrinkage in speedometer cable casings. There were fifteen factors, each at two levels, but the design used only sixteen factor-level combinations ($2^{15-11}_{III}$). The generators were I = –DHM = –BHK = BDF = BDHO = –AHJ = –ADE = –ADHN = –ABC = ABHL = ABDG = –ABDHP, and the factors were: liner OD (A); liner die (B); liner material (C); liner line speed (D); wire braid type (E); braiding tension (F); wire diameter (G); liner tension (H); liner temperature (J); coating material (K); coating die type (L); melt temperature (M); screen pack (N); cooling method (O); and line speed (P). The response is the average of four shrinkage measurements (data from Quinlan 1985).

| A | B | C | D | E | F | G | H | J | K | L | M | N | O | P | $y$ |
|---|---|---|---|---|---|---|---|---|---|---|---|---|---|---|---|
| − | − | − | − | − | + | − | − | − | − | − | − | − | − | − | .4850 |
| − | − | − | − | − | + | − | + | + | + | + | + | + | + | + | .5750 |
| − | − | − | + | + | − | + | − | − | − | − | + | + | + | + | .0875 |
| − | − | − | + | + | − | + | + | + | + | + | − | − | − | − | .1750 |
| − | + | + | − | − | − | + | − | − | + | + | − | − | + | + | .1950 |
| − | + | + | − | − | − | + | + | + | − | − | + | + | − | − | .1450 |
| − | + | + | + | + | + | − | − | − | + | + | + | + | − | − | .2250 |
| − | + | + | + | + | + | − | + | + | − | − | − | − | + | + | .1750 |
| + | − | + | − | + | + | + | − | + | − | + | − | + | − | + | .1250 |
| + | − | + | − | + | + | + | + | − | + | − | + | − | + | − | .1200 |
| + | − | + | + | − | − | − | − | + | − | + | + | − | + | − | .4550 |
| + | − | + | + | − | − | − | + | − | + | − | − | + | − | + | .5350 |
| + | + | − | − | + | − | − | − | + | + | − | − | + | + | − | .1700 |
| + | + | − | − | + | − | − | + | − | − | + | + | − | − | + | .2750 |
| + | + | − | + | − | + | + | − | + | + | − | + | − | − | + | .3425 |
| + | + | − | + | − | + | + | + | − | − | + | − | + | + | − | .5825 |

Analyze these data to determine which factors affect shrinkage, and how they affect shrinkage.

**Problem 18.5**

Seven factors are believed to control the softness of cold-foamed car seats, and an experiment was conducted to determine how these factors influence the softness. A $2^{7-4}_{III}$ design was run with generators I = ABD = ACE = BDF = ABCG. The response is the average softness of the seats (data from Bergman and Hynén 1997)

| A | B | C | D | E | F | G | $y$ |
|---|---|---|---|---|---|---|---|
| − | − | − | + | + | − | − | 25.3 |
| + | − | − | − | − | + | + | 20.6 |
| − | + | − | − | + | − | + | 26.7 |
| + | + | − | + | − | + | − | 23.8 |
| − | − | + | + | − | − | + | 23.5 |
| + | − | + | − | + | + | − | 24.0 |
| − | + | + | − | − | − | − | 23.5 |
| + | + | + | + | + | + | + | 24.2 |

Analyze these data to determine how the factors affect softness.

**Problem 18.6**

Silicon wafers for integrated circuits are grown in a device called a susceptor, and a response of interest is the thickness of the silicon. Eight factors, each at two levels, were believed to contribute: rotation method (A), wafer code (B), deposition temperature (C), deposition time (D), arsenic flow rate (E), HCl etch temperature (F), HCl flow rate (G), and nozzle position (H). A

$2^{8-4}_{IV}$ design was run with generators I = ABCD = BCEF = ACEG = BCEH. The average thickness of the silicon follows (data from Shoemaker, Tsui, and Wu 1991)

| A | B | C | D | E | F | G | H | $y$ |
|---|---|---|---|---|---|---|---|---|
| – | – | – | – | – | – | – | – | 14.80 |
| – | – | – | – | + | + | + | + | 14.86 |
| – | – | + | + | – | + | + | + | 14.00 |
| – | – | + | + | + | – | – | – | 13.91 |
| – | + | – | + | – | + | – | + | 14.14 |
| – | + | – | + | + | – | + | – | 13.80 |
| – | + | + | – | – | – | + | – | 14.73 |
| – | + | + | – | + | + | – | + | 14.89 |
| + | – | – | + | – | – | + | – | 13.93 |
| + | – | – | + | + | + | – | + | 14.09 |
| + | – | + | – | – | + | – | + | 14.79 |
| + | – | + | – | + | – | + | – | 14.33 |
| + | + | – | – | – | + | + | + | 14.77 |
| + | + | – | – | + | – | – | – | 14.88 |
| + | + | + | + | – | – | – | – | 13.76 |
| + | + | + | + | + | + | + | + | 13.97 |

Analyze these data to determine how silicon thickness depends on the factors.

**Problem 18.7** The responses shown in Problem 18.5 are the averages of sixteen individual units. The variances among those units were: 3.24, .64, 1.00, 2.56, 1.96, 1.00, 1.00, and 2.56 for the eight factor-level combinations used in the design. Which factor-levels should we use to reduce variation?

**Problem 18.8** We have a replicated $2^3$ design with data (in standard order, first replicate then second replicate) 6, 10, 32, 60, 4, 15, 26, 60, 8, 12, 34, 60, 16, 5, 37, 52. We would like the mean response to be about 30, with minimum variability. How should we choose our factor levels?

**Problem 18.9** A product is produced that should have a score as close to 2 as possible. Eight factors are believed to influence the score, and a completely randomized experiment is conducted using 64 units and sixteen treatments in a $2^{8-4}_{IV}$ fractional-factorial treatment structure. Analyze these data and report how you would achieve the score of 2. You may assume that the treatments are continuous and can take any level between -1 (low) and 1 (high). Increasing any factor costs more money, and factors are named in order of increasing expense.

| | y1 | y2 | y3 | y4 |
|---|---|---|---|---|
| (1) | 2.50 | 2.85 | 2.80 | 2.92 |
| $aefg$ | 1.83 | 1.87 | 1.87 | 1.70 |
| $befh$ | 1.55 | 1.56 | 1.64 | 1.56 |
| $abgh$ | 1.12 | 1.14 | 1.23 | 1.18 |
| $cegh$ | 1.67 | 1.65 | 1.83 | 1.89 |
| $acfh$ | 2.79 | 2.75 | 2.95 | 3.18 |
| $bcfg$ | 1.15 | 1.19 | 1.18 | 1.16 |
| $abce$ | 1.55 | 1.52 | 1.62 | 1.66 |
| $dfgh$ | 2.95 | 4.05 | 2.73 | 2.13 |
| $adeh$ | 9.41 | 4.37 | 5.06 | 4.20 |
| $bdeg$ | 1.38 | 1.88 | 2.05 | 1.54 |
| $abdf$ | 2.14 | 2.79 | 2.65 | 1.85 |
| $cdef$ | 7.48 | 5.79 | 3.55 | 13.63 |
| $acdg$ | 3.13 | 1.98 | 2.24 | 3.14 |
| $bcdh$ | 2.48 | 1.87 | 2.92 | 2.21 |
| $abcdefgh$ | 2.00 | 1.42 | 1.36 | 1.23 |

**Problem 18.10**

Suppose you have seven factors to study, each at two levels, but that you can only afford 32 runs. Further assume that at most four of the factors are active, and the rest inert. You may safely assume that all three-factor or higher-order interactions are negligible, but many or all of the two-factor interactions in the active factors are present.

(a) Design a single-stage experiment that uses all 32 runs. Show that this experiment may not be able to estimate all effects of interest.

(b) Design a two-stage experiment, where you use 16 runs in the first stage, and then use an additional 16 runs if needed. Show that you can always estimate the effects of interest with the two-stage design.

(c) Suppose that we had assigned the seven labels A, B, C, D, E, F, and G to the seven factors at random. There are 35 (seven choose four) ways of assigning the four active factors to labels, ignoring the order. What is the probability that you can estimate main effects and all two-factor interactions in the active factors with your design from part (a)? What is the probability that you can estimate main effects and all two factor interactions in the active factors with your first 16-point design from (b) and your full two-stage design from part (b)?

(d) What is the main lesson you draw from (a), (b), and (c)?

**Problem 18.11**

We wish to determine the tolerance of icings to ingredient changes and variation in the preparation. Ingredient changes are represented by factors C, D, E, F, G, and H. All are at two levels. C and D are two types of sugars;

E, F, and G are three stabilizers; and H is a setting agent. The levels of these factors represent changes in the amounts of these constituents in the mix. Variation in preparation is modeled as the amount of water added to the product. This has four levels and is represented as the combinations of factors A and B. The response we measure is (coded) viscosity of the icing. A quarter-fraction with 64 observations was run; data follow (Carroll and Dykstra 1958):

| | | | | | | | |
|---|---|---|---|---|---|---|---|
| (1) | 26 | *agh* | 6 | *bh* | 43 | *abg* | -3 |
| *cg* | 16 | *ach* | 10 | *bcgh* | 69 | *abc* | -5 |
| *dgh* | 12 | *ad* | 13 | *bdg* | 45 | *abdh* | -13 |
| *cdh* | 22 | *acdg* | 17 | *bcd* | 45 | *abcdgh* | -4 |
| *eh* | 29 | *aeg* | 13 | *be* | 54 | *abegh* | 4 |
| *cegh* | 30 | *ace* | 17 | *bceg* | 54 | *abceh* | 5 |
| *deg* | 29 | *adeh* | 16 | *bdegh* | 43 | *abde* | -2 |
| *cde* | 34 | *acdegh* | 16 | *bcdeh* | 67 | *abcdeg* | -3 |
| *fgh* | 32 | *af* | 19 | *bfg* | 64 | *abfh* | 6 |
| *cfh* | 30 | *acfg* | 18 | *bcf* | 57 | *abcfgh* | 6 |
| *df* | 27 | *adfgh* | 29 | *bdfh* | 50 | *abdfg* | 6 |
| *cdfg* | 35 | *acdfh* | 22 | *bcdfgh* | 53 | *abcdf* | 7 |
| *efg* | 53 | *aefh* | 29 | *befgh* | 74 | *abef* | 8 |
| *cef* | 46 | *acefgh* | 21 | *bcefh* | 73 | *abcefg* | 13 |
| *defh* | 35 | *adefg* | 23 | *bdef* | 69 | *abdefgh* | 20 |
| *cdefgh* | 42 | *acdef* | 27 | *bcdefg* | 69 | *abcdefh* | 10 |

Determine which factors affect the viscosity of the icing, and in what ways. The response should lie between 25 and 30; what does the experiment tell us about the icing's tolerance to changes in ingredients?

**Question 18.1** Use the fact that the shortest alias of I in a resolution $R$ design has $R$ letters to show that a $2^{k-p}$ design of resolution $R$ contains a complete factorial in any $R-1$ factors.

**Question 18.2** Show that fold-over breaks all aliases of odd length.

**Question 18.3** Show that (1) there are $1+3+3^2+\cdots+3^{k-1}$ two-degree-of-freedom splits in a $3^k$ factorial; (2) there are $1+3+3^2+\cdots+3^{k-q-1}$ two-degree-of-freedom splits in a $3^{k-q}$ fractional factorial, each with $3^q$ labels; and (3) there are $1+3+\cdots+3^{q-1}$ two-degree-of-freedom splits aliased to I in a $3^{k-q}$ fractional factorial.

# Chapter 19

# Response Surface Designs

Many experiments have the goals of describing how the response varies as a function of the treatments and determining treatments that give optimal responses, perhaps maxima or minima. Factorial-treatment structures can be used for these kinds of experiments, but when treatment factors can be varied across a continuous range of values, other treatment designs may be more efficient. *Response surface methods* are designs and models for working with continuous treatments when finding optima or describing the response is the goal.

Response surface methods

## 19.1 Visualizing the Response

In some experiments, the treatment factors can vary continuously. When we bake a cake, we bake for a certain time $x_1$ at a certain temperature $x_2$; time and temperature can vary continuously. We could, in principle, bake cakes for any time and temperature combination. Assuming that all the cake batters are the same, the quality of the cakes $y$ will depend on the time and temperature of baking. We express this as

Response is a function of continuous design variables

$$y_{ij} = f(x_{1i}, x_{2i}) + \epsilon_{ij} \quad ,$$

meaning that the response $y$ is some function $f$ of the design variables $x_1$ and $x_2$, plus experimental error. Here $j$ indexes the replication at the $i$th unique set of design variables.

One common goal when working with response surface data is to find the settings for the design variables that optimize (maximize or minimize)

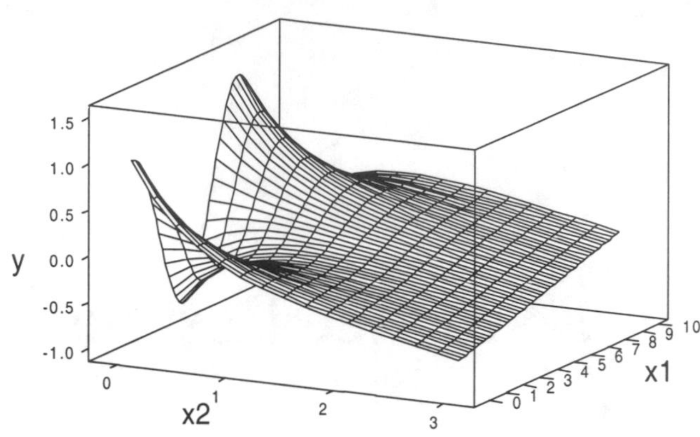


**Figure 19.1:** Sample perspective plot, using Minitab.

the response. Often there are complications. For example, there may be several responses, and we must seek some kind of compromise optimum that makes all responses good but does not exactly optimize any single response. Alternatively, there may be constraints on the design variables, so that the goal is to optimize a response, subject to the design variables meeting some constraints.

Compromise or constrained optimum

A second goal for response surfaces is to understand "the lie of the land." Where are the hills, valleys, ridge lines, and so on that make up the topography of the response surface? At any give design point, how will the response change if we alter the design variables in a given direction?

Describe the shape of the response

We can visualize the function $f$ as a surface of heights over the $x_1, x_2$ plane, like a relief map showing mountains and valleys. A perspective plot shows the surface when viewed from the side; Figure 19.1 is a perspective plot of a fairly complicated surface that is wiggly for low values of $x_2$, and flat for higher values of $x_2$. A contour plot shows the contours of the surface, that is, curves of $x_1, x_2$ pairs that have the same response value. Figure 19.2 is a contour plot for the same surface as Figure 19.1.

Perspective plots and contour plots

Graphics and visualization techniques are some of our best tools for understanding response surfaces. Unfortunately, response surfaces are difficult to visualize when there are three design variables, and become almost impossible for more than three. We thus work with models for the response

Use models for $f$

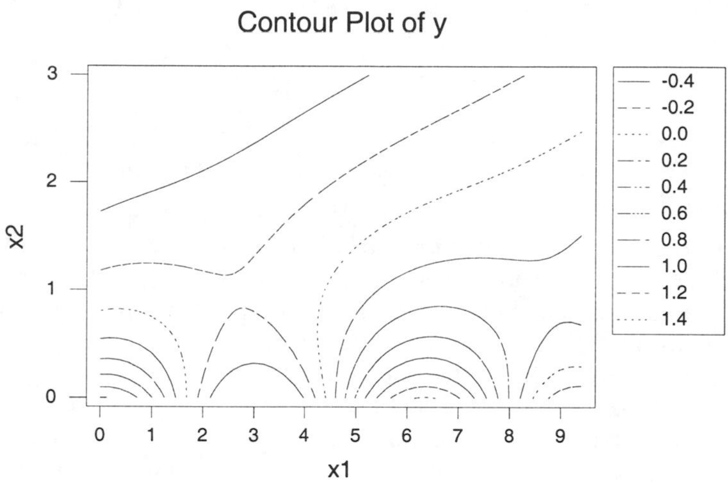


**Figure 19.2:** Sample contour plot, using Minitab.

function $f$.

## 19.2 First-Order Models

All models are wrong; some models are useful. *George Box*

We often don't know anything about the shape or form of the function $f$, so any mathematical model that we assume for $f$ is surely wrong. On the other hand, experience has shown that simple models using low-order polynomial terms in the design variables are generally sufficient to describe sections of a response surface. In other words, we know that the polynomial models described below are almost surely incorrect, in the sense that the response surface $f$ is unlikely to be a true polynomial; but in a "small" region, polynomial models are usually a close enough approximation to the response surface that we can make useful inferences using polynomial models.

Polynomials are often adequate models

We will consider *first-order models* and *second-order models* for response surfaces. A first-order model with $q$ variables takes the form

First-order model has linear terms

$$y_{ij} = \beta_0 + \beta_1 x_{1i} + \beta_2 x_{2i} + \cdots + \beta_q x_{qi} + \epsilon_{ij}$$

$$= \beta_0 + \sum_{k=1}^{q} \beta_k x_{ki} + \epsilon_{ij}$$

$$= \beta_0 + \mathbf{x}_i'\boldsymbol{\beta} + \epsilon_{ij} ,$$

where $\mathbf{x}_i = (x_{1i}, x_{2i}, \ldots, x_{qi})'$ and $\boldsymbol{\beta} = (\beta_1, \beta_2, \ldots, \beta_q)'$. The first-order model is an ordinary multiple-regression model, with design variables as predictors and $\beta_k$'s as regression coefficients.

First-order models describe flat, but tilted, surfaces

First-order models describe inclined planes: flat surfaces, possibly tilted. These models are appropriate for describing portions of a response surface that are separated from maxima, minima, ridge lines, and other strongly curved regions. For example, the side slopes of a hill might be reasonably approximated as inclined planes. These approximations are local, in the sense that you need different inclined planes to describe different parts of the mountain. First-order models can approximate $f$ reasonably well as long as the region of approximation is not too big and $f$ is not too curved in that region. A first-order model would be a reasonable approximation for the part of the surface in Figures 19.1 or 19.2 where $x_2$ is large; a first-order model would work poorly where $x_2$ is small.

First-order models show direction of steepest ascent

Bearing in mind that these models are only approximations to the true response, what can these models tell us about the surface? First-order models can tell us which way is up (or down). Suppose that we are at the design variables $\mathbf{x}$, and we want to know in which direction to move to increase the response the most. This is the direction of *steepest ascent*. It turns out that we should take a step proportional to $\boldsymbol{\beta}$, so that our new design variables are $\mathbf{x} + r\boldsymbol{\beta}$, for some $r > 0$. If we want the direction of *steepest descent,* then we move to $\mathbf{x} - r\boldsymbol{\beta}$, for some $r > 0$. Note that this direction of steepest ascent is only approximately correct, even in the region where we have fit the first-order model. As we move outside that region, the surface may change and a new direction may be needed.

Contours are flat for first-order models

Contours or level curves are sets of design variables that have the same expected response. For a first-order surface, design points $\mathbf{x}$ and $\mathbf{x} + \boldsymbol{\delta}$ are on the same contour if $\sum \beta_k \delta_k = 0$. First-order model contours are straight lines for $q = 2$, planes for $q = 3$, and so on. Note that directions of steepest ascent are perpendicular to contours.

## 19.3 First-Order Designs

We have three basic needs from a response surface design. First, we must be able to estimate the parameters of the model. Second, we must be able to estimate *pure error* and *lack of fit*. As described below, pure error and lack of fit are our tools for determining if the first-order model is an adequate approximation to the true mean structure of the data. And third, we need the design to be efficient, both from a variance of estimation point of view and a use of resources point of view.

Get parameters, pure error, and LoF efficiently

The concept of pure error needs a little explanation. Data might not fit a model because of random error (the $\epsilon_{ij}$ sort of error); this is pure error. Data also might not fit a model because the model is misspecified and does not truly describe the mean structure; this is lack of fit. Our models are approximations, so we need to know when the lack of fit becomes large relative to pure error. This is particularly true for first-order models, which we will then replace with second-order models. It is also true for second-order models, though we are more likely to reduce our region of modeling rather than move to higher orders.

Large lack of fit implies model does not describe mean structure adequately

We do not have lack of fit for factorial models when the full factorial model is fit. In that situation, we have fit a degree of freedom for every factor-level combination—in effect, a mean for each combination. There can be no lack of fit in that case because all means have been fit exactly. We can get lack of fit when our models contain fewer degrees of freedom than the number of distinct design points used; in particular, first- and second-order models may not fit the data.

Response surface designs are usually given in terms of *coded variables*. Coding simply means that the design variables are rescaled so that 0 is in the center of the design, and $\pm 1$ are reasonable steps up and down from the center. For example, if cake baking time should be about 35 minutes, give or take a couple of minutes, we might rescale time by $(x_1 - 35)/2$, so that 33 minutes is a –1, 35 minutes is a 0, and 37 minutes is a 1.

Coded variables simply design

First-order designs collect data to fit first-order models. The standard first-order design is a $2^q$ factorial with *center points*. The (coded) low and high values for each variable are $\pm 1$; the center points are $m$ observations taken with all variables at 0. This design has $2^q + m$ points. We may also use any $2^{q-k}$ fraction with resolution III or greater.

Two-series with center points for first order

The replicated center points serve two uses. First, the variation among the responses at the center point provides an estimate of pure error. Second, the contrast between the mean of the center points and the mean of the factorial points provides a test for lack of fit. When the data follow a first-order model, this contrast has expected value zero; when the data follow a second-order model, this contrast has an expectation that depends on the pure quadratic terms.

Center points for pure error and lack of fit

**Example 19.1** **Cake baking**

Our cake mix recommends 35 minutes at 350°, but we are going to try to find a time and temperature that suit our palate better. We begin with a first-order design in baking time and temperature, so we use a $2^2$ factorial with three center points. Use the coded values –1, 0, 1 for 33, 35, and 37 minutes for time, and the coded values –1, 0, 1 for 340, 350, and 360 degrees for temperature. We will thus have three cakes baked at the package-recommended time and temperature (our center point), and four cakes with time and temperature spread around the center. Our response is an average palatability score, with higher values being desirable:

| $x_1$ | $x_2$ | $y$ |
|---|---|---|
| -1 | -1 | 3.89 |
| 1 | -1 | 6.36 |
| -1 | 1 | 7.65 |
| 1 | 1 | 6.79 |
| 0 | 0 | 8.36 |
| 0 | 0 | 7.63 |
| 0 | 0 | 8.12 |

## 19.4 Analyzing First-Order Data

Here are three possible goals when analyzing data from a first-order design:

- Determine which design variables affect the response.
- Determine whether there is lack of fit.
- Determine the direction of steepest ascent.

Multiple regression to estimate $\beta_k$'s

Some experimental situations can involve a sequence of designs and all these goals. In all cases, model fitting for response surfaces is done using multiple linear regression. The model variables ($x_1$ through $x_q$ for the first-order model) are the "independent" or "predictor" variables of the regression. The estimated regression coefficients are estimates of the model parameters $\beta_k$. For first-order models using data from $2^q$ factorials with or without center points, the estimated regression slopes using coded variables are equal to the ordinary main effects for the factorial model. Let **b** be the vector of estimated coefficients for first-order terms (an estimate of $\boldsymbol{\beta}$).

Model testing is done with F-tests on mean squares from the ANOVA of the regression; each term has its own line in the ANOVA table. Predictor variables are orthogonal to each other in many designs and models, but not in all cases, and certainly not when there is missing data; so it seems easiest just to treat all testing situations as if the model variables were nonorthogonal.

To test the null hypothesis that the coefficients for a set of model terms are all zero, get the error sum of squares for the full model and the error sum of squares for the reduced model that does not contain the model terms being tested. The difference in these error sums of squares is the improvement sum of squares for the model terms under test. The improvement mean square is the improvement sum of squares divided by its degrees of freedom (the number of model terms in the multiple regression being tested). This improvement mean square is divided by the error mean square from the full model to obtain an F-test of the null hypothesis. The sum of squares for improvement can also be computed from a sequential (Type I) ANOVA for the model, provided that the terms being tested are the last terms entered into the model. The F-test of $\beta_k = 0$ (with one numerator degree of freedom) is equivalent to the $t$-test for $\beta_k$ that is printed by most regression software.

Test terms of interest adjusted for other terms in model

In many response surface experiments, all variables are important, as there has been preliminary screening to find important variables prior to exploring the surface. However, inclusion of noise variables into models can alter subsequent analysis. It is worth noting that variables can look inert in some parts of a response surface, and active in other parts.

Test to exclude noise variables from model

The direction of steepest ascent in a first-order model is proportional to the coefficients $\boldsymbol{\beta}$. Our estimated direction of steepest ascent is then proportional to $\mathbf{b}$. Inclusion of inert variables in the computation of this direction increases the error in the direction of the active variables. This effect is worst when the active variables have relatively small effects. The net effect is that our response will not increase as quickly as possible per unit change in the design variables, because the direction could have a nonnegligible component on the inert axes.

Direction of steepest ascent proportional to estimated $\beta$'s

Residual variation can be divided into two parts: pure error and lack of fit. Pure error is variation among responses that have the same explanatory variables (and are in the same blocks, if there is blocking). We use replicated points, usually center points, to get an estimate of pure error. All the rest of residual variation that is not pure error is lack of fit. Thus we can make the decompositions

Divide residual into pure error and lack of fit

$$\begin{aligned} SS_{\text{Tot}} &= SS_{\text{Model}} + SS_{LoF} + SS_{PE} \\ N-1 &= df_{\text{Model}} + df_{LoF} + df_{PE} \ . \end{aligned}$$

The mean square for pure error estimates $\sigma^2$, the variance of $\epsilon$. If the model we have fit has the correct mean structure, then the mean square for lack of fit also estimates $\sigma^2$, and the F-ratio $MS_{LoF}/MS_{PE}$ will have an F-distribution with $df_{LoF}$ and $df_{PE}$ degrees of freedom. If the model we have fit has the wrong mean structure—for example, if we fit a first-order model and a second-order model is correct—then the expected value of $MS_{LoF}$ is larger than $\sigma^2$. Thus we can test for lack of fit by comparing the F-ratio $MS_{LoF}/MS_{PE}$ to an F-distribution with $df_{LoF}$ and $df_{PE}$ degrees of freedom.

Pure error estimates $\sigma^2$; lack of fit measures deviation of model from true mean structure

For a $2^q$ factorial design with $m$ center points, there are $2^q + m - 1$ degrees of freedom, with $q$ for the model, $m - 1$ for pure error, and all the rest for lack of fit.

Quantities in the analysis of a first-order model are not very reliable when there is significant lack of fit. Because the model is not tracking the actual mean structure of the data, the importance of a variable in the first-order model may not relate to the variable's importance in the mean structure of the data. Likewise, the direction of steepest ascent from a first-order model may be meaningless if the the model is not describing the true mean structure.

All bets off when lack of fit present

**Example 19.2**

## Cake baking, continued

Example 19.1 was a $2^2$ design with three center points. Our first-order model includes a constant and linear terms for time and temperature. With seven data points, there will be 4 residual degrees of freedom. The only replication in the design is at the three center points, so we have 2 degrees of freedom for pure error. The remaining 2 residual degrees of freedom are lack of fit.

Listing 19.1 shows results for this analysis. Using the 4-degree-of-freedom residual mean square, neither time nor temperature has an F-ratio much bigger than one, so neither appears to affect the response ①. However, look at the test for lack of fit ②. This test has an F-ratio of 31.5 and $p$-value of .03, indicating that the first-order model is missing some of the mean structure.

The 2 degrees of freedom for lack of fit are the interaction in the factorial points and the contrast between the factorial points and the center points. The sums of squares for these contrasts are 2.77 and 5.96, so most of the lack of fit is due to the center points not lying on the plane fit from the factorial points. In fact, the center points are about 1.86 higher on average than what the first-order model predicts.

The direction of steepest ascent in this model is proportional to (.40, 1.05), the estimated $\beta_1$ and $\beta_2$. That is, the model says that a maximal increase in response can be obtained by increasing $x_1$ by .38 (coded) units for every increase of 1 (coded) unit in $x_2$. However, we have already seen that

**Listing 19.1:** Minitab output for first-order model of cake baking data.

```
Estimated Regression Coefficients for y

Term           Coef       StDev         T       P
Constant     6.9714      0.5671    12.292   0.000
x1           0.4025      0.7503     0.536   0.620          ①
x2           1.0475      0.7503     1.396   0.235

S = 1.501        R-Sq = 35.9%      R-Sq(adj) = 3.8%

Analysis of Variance for y

Source            DF     Seq SS      Adj SS      Adj MS       F       P
Regression         2     5.0370      5.0370      2.5185    1.12   0.411
  Linear           2     5.0370      5.0370      2.5185    1.12   0.411
Residual Error     4     9.0064      9.0064      2.2516
  Lack-of-Fit      2     8.7296      8.7296      4.3648   31.53   0.031   ②
  Pure Error       2     0.2769      0.2769      0.1384
Total              6    14.0435
```

there is significant lack of fit using the first-order model with these data, so this direction of steepest ascent is not reliable.

## 19.5 Second-Order Models

We use second-order models when the portion of the response surface that we are describing has curvature. A second-order model contains all the terms in the first-order model, plus all quadratic terms like $\beta_{11}x_{1i}^2$ and all cross product terms like $\beta_{12}x_{1i}x_{2i}$. Specifically, it takes the form

Second-order models include quadratic and cross product terms

$$
\begin{aligned}
y_{ij} &= \beta_0 + \beta_1 x_{1i} + \beta_2 x_{2i} + \cdots + \beta_q x_{qi} + \\
&\quad \beta_{11}x_{1i}^2 + \beta_{22}x_{2i}^2 + \cdots + \beta_{qq}x_{qi}^2 + \\
&\quad \beta_{12}x_{1i}x_{2i} + \beta_{13}x_{1i}x_{3i} + \cdots + \beta_{1q}x_{1i}x_{qi} + \\
&\quad \beta_{23}x_{2i}x_{3i} + \beta_{24}x_{2i}x_{4i} + \cdots + \beta_{2q}x_{2i}x_{qi} + \\
&\quad \cdots + \beta_{(q-1)q}x_{(q-1)i}x_{qi} + \epsilon_{ij} \\
&= \beta_0 + \sum_{k=1}^{q}\beta_k x_{ki} + \sum_{k=1}^{q}\beta_{kk}x_{ki}^2 + \sum_{k=1}^{q-1}\sum_{l=k+1}^{q}\beta_{kl}x_{ki}x_{li} + \epsilon_{ij} \\
&= \beta_0 + \mathbf{x}_i'\boldsymbol{\beta} + \mathbf{x}_i'\mathcal{B}\mathbf{x}_i + \epsilon_{ij} \ ,
\end{aligned}
$$

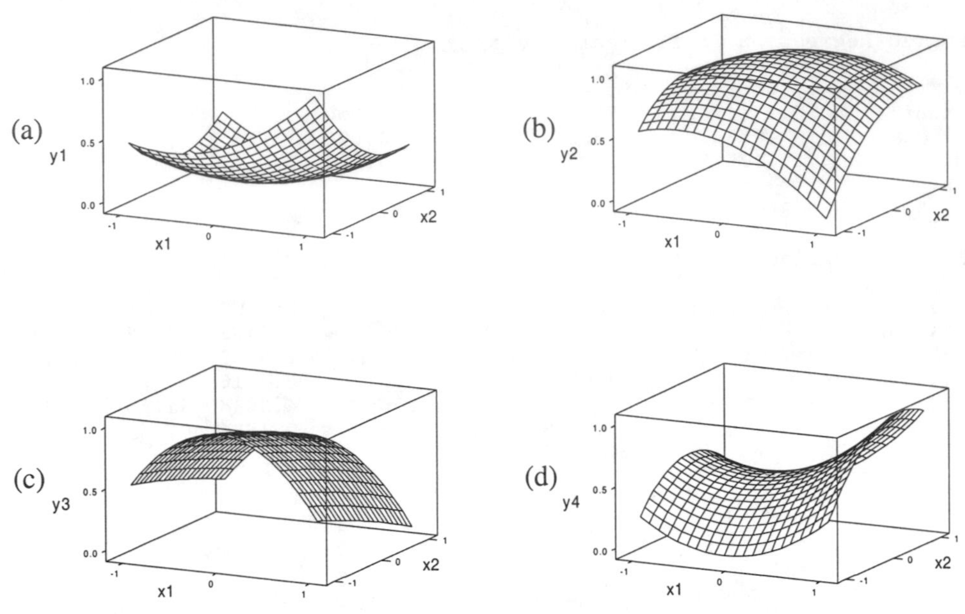


**Figure 19.3:** Sample second-order surfaces: (a) minimum, (b) maximum, (c) ridge, and (d) saddle, using Minitab.

where once again $\mathbf{x}_i = (x_{1i}, x_{2i}, \ldots, x_{qi})'$, $\boldsymbol{\beta} = (\beta_1, \beta_2, \ldots, \beta_q)'$, and $\mathcal{B}$ is a $q \times q$ matrix with $\mathcal{B}_{kk} = \beta_{kk}$ and $\mathcal{B}_{kl} = \mathcal{B}_{lk} = \beta_{kl}/2$ for $k < l$. Note that the model only includes the $kl$ cross product for $k < l$; the matrix form with $\mathcal{B}$ includes both $kl$ and $lk$, so the coefficients are halved to take this into account.

Quadratic surfaces take many shapes

Second-order models describe quadratic surfaces, and quadratic surfaces can take several shapes. Figure 19.3 shows four of the shapes that a quadratic surface can take. First, we have a simple minimum and maximum. Then we have a ridge; the surface is curved (here a maximum) in one direction, but is fairly constant in another direction. Finally, we see a saddle point; the surface curves up in one direction and curves down in another.

Second-order models are easier to understand if we change from the original design variables $x_1$ and $x_2$ to *canonical variables* $v_1$ and $v_2$. Canonical variables will be defined shortly, but for now consider that they shift the origin (the zero point) and rotate the coordinate axes to match the second-order surface; the second-order model is very simple when expressed in canonical variables:

Use canonical variables

$$f_v(\mathbf{v}) = f_v(0) + \sum_{k=1}^{q} \lambda_k {v_k}^2 \ ,$$

where $\mathbf{v} = (v_1, v_2, \ldots, v_q)'$ is the design variables expressed in canonical coordinates; $f_v$ is the response as a function of the canonical variables; and $\lambda_k$'s are numbers computed from the $\mathcal{B}$ matrix. The $\mathbf{x}$ value that maps to 0 in the canonical variables is called the *stationary point* and is denoted by $\mathbf{x}_0$; thus $f_v(0) = f(\mathbf{x}_0)$.

The key to understanding canonical variables is the stationary point of the second-order surface. The stationary point is that combination of design variables where the surface is at either a maximum or a minimum in all directions. If the stationary point is a maximum in all directions, then the stationary point is the maximum response on the whole modeled surface. If the stationary point is a minimum in all directions, then it is the minimum response on the whole modeled surface. If the stationary point is a maximum in some directions and a minimum in other directions, then the stationary point is a saddle point, and the modeled surface has no overall maximum or minimum. If a ridge surface is absolutely level in some direction, then it does not have a unique stationary point; this rarely happens in practice.

Stationary point is maximum, minimum, or saddle point

The stationary point will be the origin (0 point) for our canonical variables. Now imagine yourself situated at the stationary point of a second-order surface. The first canonical axis is the direction in which you would move so that a step of unit length yields a response as large as possible (either increase the response as much as possible or decrease it as little as possible). The second canonical axis is the direction, among all those directions perpendicular to the first canonical axis, that yields a response as large as possible. There are as many canonical axes as there are design variables. Each additional canonical axis that we find must be perpendicular to all those we have already found.

From stationary point, response increases as quickly as possible in first canonical direction (axis)

Figure 19.4 shows contours, stationary points, and canonical axes for the four sample second-order surfaces. As shown in this figure, contours for surfaces with maxima or minima are ellipses. The stationary point $\mathbf{x}_0$ is the center of these ellipses, and the canonical axes are the major and minor axes of the elliptical contours. For the ridge system, we still have elliptical contours, but they are very long and skinny, and the stationary point is outside the region where we have fit the model. If the ridge is absolutely flat, then the contours are parallel lines. For the saddle point, contours are hyperbolic instead of elliptical. The stationary point is in the center of the hyperbolas, and the canonical axes are the axes of the hyperbolas.

Second-order contours are ellipses or hyperbolas centered at stationary point

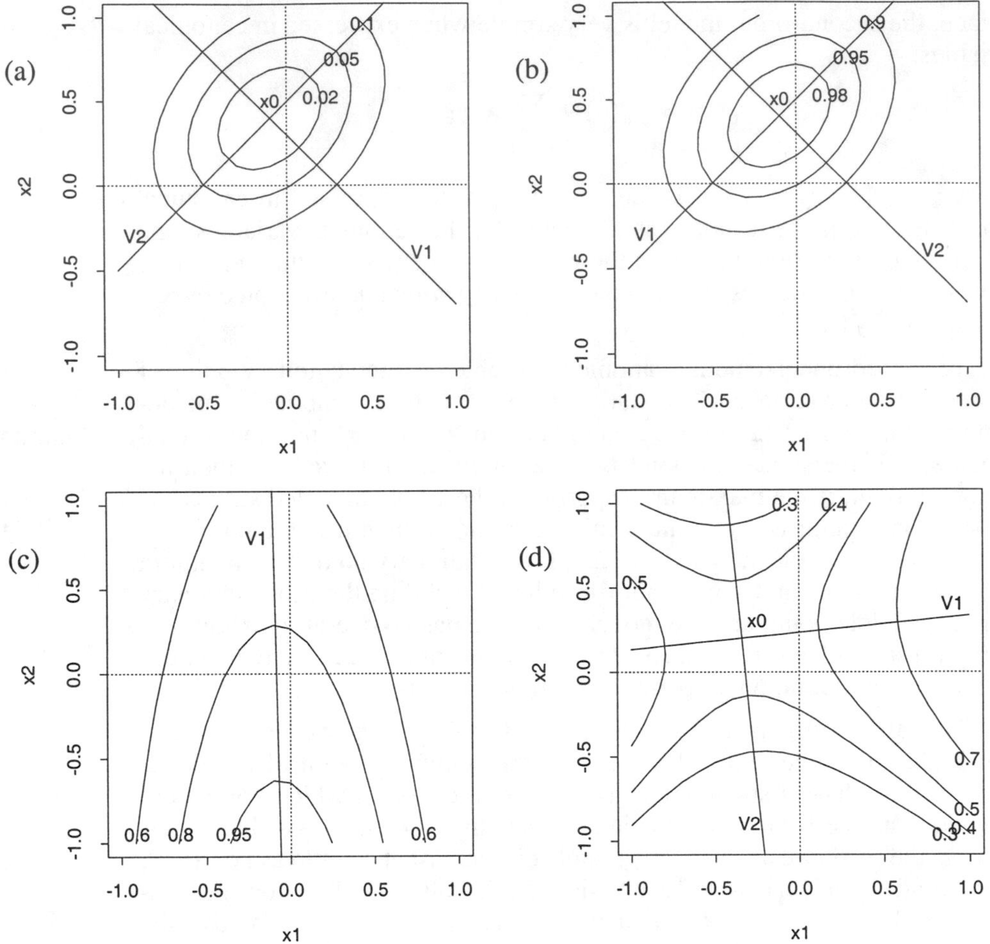


**Figure 19.4:** Contours, stationary points, and canonical axes for sample second-order surfaces: (a) minimum, (b) maximum, (c) ridge, and (d) saddle, using S-Plus.

This description of second-order surfaces has been geometric; pictures are an easy way to understand these surfaces. It is difficult to calculate with pictures, though, so we also have an algebraic description of the second-order surface. Recall that the matrix form of the response surface is written

$$f(\mathbf{x}) = \beta_0 + \mathbf{x}'\boldsymbol{\beta} + \mathbf{x}'\mathcal{B}\mathbf{x} \quad .$$

Our algebraic description of the surface depends on the following facts:

Two results from linear algebra

1. The stationary point for this quadratic surface is at

$$\mathbf{x}_0 = -\frac{1}{2}\mathcal{B}^{-1}\boldsymbol{\beta} \ ,$$

   where $\mathcal{B}^{-1}$ is the matrix inverse of $\mathcal{B}$.

2. For the $q \times q$ symmetric matrix $\mathcal{B}$, we can find a $q \times q$ matrix $H$ such that $H'H = HH' = I_q$ and $H'\mathcal{B}H = \Lambda$, where $I_q$ is the $q \times q$ identity matrix and $\Lambda$ is a matrix with elements $\lambda_1, \ldots, \lambda_q$ on the diagonal and zeroes off the diagonal.

The numbers $\lambda_k$ are the *eigenvalues* of $\mathcal{B}$, and the columns of $H$ are the corresponding *eigenvectors*.

Get canonical coordinates

We saw in Figure 19.4 that the stationary point and canonical axes give us a new coordinate system for the design variables. We get the new coordinates $\mathbf{v}' = (v_1, v_2, \ldots, v_q)$ via

$$\mathbf{v} = H'(\mathbf{x} - \mathbf{x}_0) \ .$$

Subtracting $\mathbf{x}_0$ shifts the origin, and multiplying by $H'$ rotates to the canonical axes.

Response in canonical coordinates

Finally, the payoff: in the canonical coordinates, we can express the response surface as

$$f_v(\mathbf{v}) = f_v(0) + \sum_{k=1}^{q} \lambda_k v_k^2 \ ,$$

where

$$f_v(0) = f(\mathbf{x}_0) = \beta_0 + \frac{1}{2}\mathbf{x}_0'\boldsymbol{\beta} \ .$$

Signs of $\lambda_k$'s determine maximum, minimum, or saddle

That is, when looked at in the canonical coordinates, the response surface is a constant plus a simple squared term from each of the canonical variables $v_i$. If all of the $\lambda_k$'s are positive, $\mathbf{x}_0$ is a minimum. If all of the $\lambda_k$'s are negative, $\mathbf{x}_0$ is a maximum. If some are negative and some are positive, $\mathbf{x}_0$ is a saddle point. If all of the $\lambda_k$'s are of the same sign, but some are near zero in value, we have a ridge system. The $\lambda_k$'s for our four examples in Figure 19.4 are (.31771, .15886) for the surface with a minimum, (-.31771, -.15886) for the surface with a maximum, (-.021377, -.54561) for the surface with a ridge, and (.30822, -.29613) for the surface with a saddle point.

In principal, we could also use third- or higher-order models. This is rarely done, as second-order models are generally sufficient.

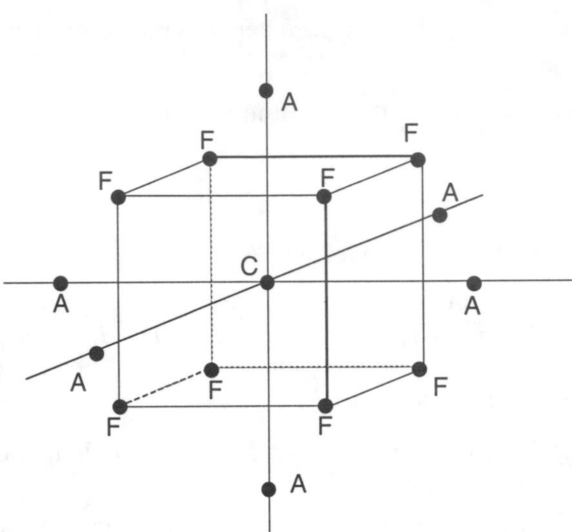


**Figure 19.5:** A central composite design in three dimensions, showing center (C), factorial (F), and axial (A) points.

## 19.6 Second-Order Designs

Central composite (CCD) has factorial points, axial points, and center points

There are several choices for second-order designs. One of the most popular is the *central composite design* (CCD). A CCD is composed of factorial points, *axial* points, and center points. Factorial points are the points from a $2^q$ design with levels coded as $\pm 1$ or the points in a $2^{q-k}$ fraction with resolution V or greater; center points are again $m$ points at the origin. The axial points have one design variable at $\pm\alpha$ and all other design variables at 0; there are $2q$ axial points. Figure 19.5 shows a CCD for $q = 3$.

Augment first-order design to CCD

One of the reasons that CCD's are so popular is that you can start with a first-order design using a $2^q$ factorial and then augment it with axial points and perhaps more center points to get a second-order design. For example, we may find lack of fit for a first-order model fit to data from a first-order design. Augment the first-order design by adding axial points and center points to get a CCD, which is a second-order design and can be used to fit a second-order model. We consider such a CCD to have been run in two incomplete blocks.

We get to choose $\alpha$ and the number of center points $m$. Suppose that we run our CCD in incomplete blocks, with the first block having the factorial points and center points, and the second block having axial points and cen-

**Table 19.1:** Design parameters for Central Composite Designs with orthogonal blocking.

| p<br>rep | 2<br>1 | 3<br>1 | 4<br>1 | 5<br>1 | 5<br>$\frac{1}{2}$ | 6<br>1 | 6<br>$\frac{1}{2}$ | 7<br>1 | 7<br>$\frac{1}{2}$ |
|---|---|---|---|---|---|---|---|---|---|
| Number of blocks in factorial | 1 | 2 | 2 | 4 | 1 | 8 | 2 | 16 | 8 |
| Center points per factorial block | 3 | 2 | 2 | 2 | 6 | 1 | 4 | 1 | 1 |
| $\alpha$ for axial points | 1.414 | 1.633 | 2.000 | 2.366 | 2.000 | 2.828 | 2.366 | 3.364 | 2.828 |
| Center points for axial block | 3 | 2 | 2 | 4 | 1 | 6 | 2 | 11 | 4 |
| Total points in design | 14 | 20 | 30 | 54 | 33 | 90 | 54 | 169 | 80 |

ter points. Block effects should be orthogonal to treatment effects, so that blocking does not affect the shape of our estimated response surface. We can achieve this orthogonality by choosing $\alpha$ and the number of center points in the factorial and axial blocks as shown in Table 19.1 (Box and Hunter 1957).

Choose $\alpha$ and $m$ so that effects are orthogonal to blocks

Table 19.1 deserves some explanation. When blocking the CCD, factorial points and axial points will be in different blocks. The factorial points may also be blocked using the confounding schemes of Chapter 15. The table gives the maximum number of blocks into which the factorial portion can be confounded, while main effects and two-way interactions are confounded only with three-way and higher-order interactions. The table also gives the number of center points for *each* of these blocks. If fewer blocks are desired, the center points are added to the combined blocks. For example, the $2^5$ can be run in four blocks, with two center points per block. If we instead use two blocks, then each should have four center points; with only one block, use all eight center points. The final block consists of all axial points and additional center points.

There are a couple of heuristics for choosing $\alpha$ and the number of center points when the CCD is not blocked, but these are just guidelines and not overly compelling. If the precision of the estimated response surface at some point $\mathbf{x}$ depends only on the distance from $\mathbf{x}$ to the origin, not on the direction, then the design is said to be *rotatable.* Thus rotatable designs do not favor one direction over another when we explore the surface. This is reasonable when we know little about the surface before experimentation. We get a rotatable design by choosing $\alpha = 2^{q/4}$ for the full factorial or $\alpha = 2^{(q-k)/4}$ for a fractional factorial. Some of the blocked CCD's given in Table 19.1 are exactly rotatable, and all are nearly rotatable.

$\alpha$ for rotatable design

**Table 19.2:** Parameters for rotatable, uniform precision Central Composite Designs.

| p | 2 | 3 | 4 | 5 | 5 | 6 | 6 | 7 | 7 |
|---|---|---|---|---|---|---|---|---|---|
| Replication | 1 | 1 | 1 | 1 | $\frac{1}{2}$ | 1 | $\frac{1}{2}$ | 1 | $\frac{1}{2}$ |
| Number of center points | 5 | 6 | 7 | 10 | 6 | 15 | 9 | 21 | 14 |

Rotatable designs need five levels of every factor and depend on coding

Rotatable designs are nice, and I would probably choose one as a default. However, I don't obsess on rotatability, for a couple of reasons. First, rotatability depends on the coding we choose. The property that the precision of the estimated surface does not depend on direction disappears when we go back to the original, uncoded variables. It also disappears if we keep the same design points in the original variables but then express them with a different coding. Second, rotatable designs use five levels of every variable, and this may be logistically awkward. Thus choosing $\alpha = 1$ so that all variables have only three levels may make a more practical design. Third, using $\alpha = \sqrt{q}$ so that all the noncenter points are on the surface of a sphere (only rotatable for $q = 2$) gives a better design when we are only interested in the response surface within that sphere.

$\alpha$ for uniform precision

A second-order design has *uniform precision* if the precision of the fitted surface is the same at the origin and at a distance of 1 from the origin. Uniform precision is a reasonable criterion, because we are unlikely to know just how close to the origin a maximum or other surface feature may be; (relatively) too many center points give us much better precision near the origin, and too few give us better precision away from the origin. It is impossible to achieve this exactly; Table 19.2 shows the number of center points to get as close as possible to uniform precision for rotatable CCD's.

**Example 19.3**

## Cake baking, continued

We saw in Example 19.2 that the first-order model was a poor fit; in particular, the contrast between the factorial points and the center points indicated curvature of the response surface. We will need a second-order model to fit the curved surface, so we will need a second-order design to collect the data for the fit.

We already have factorial points and three center points. Looking in Table 19.1, we see that adding three more center points and axial points at $\alpha = 1.414$ will give us a design with two blocks with blocks orthogonal to treatments. This design is also rotatable, but not uniform precision.

Here is the complete design, including responses for the seven additional cakes we bake to complete the CCD:

| Block | $x_1$ | $x_2$ | $y$ |
|---|---|---|---|
| 1 | –1 | –1 | 3.89 |
| 1 | 1 | –1 | 6.36 |
| 1 | –1 | 1 | 7.65 |
| 1 | 1 | 1 | 6.79 |
| 1 | 0 | 0 | 8.36 |
| 1 | 0 | 0 | 7.63 |
| 1 | 0 | 0 | 8.12 |
| 2 | 1.414 | 0 | 8.40 |
| 2 | –1.414 | 0 | 5.38 |
| 2 | 0 | 1.414 | 7.00 |
| 2 | 0 | –1.414 | 4.51 |
| 2 | 0 | 0 | 7.81 |
| 2 | 0 | 0 | 8.44 |
| 2 | 0 | 0 | 8.06 |

There are several other second-order designs in addition to central composite designs. The simplest are $3^q$ factorials and fractions with resolution V or greater. These designs are not much used for $q \geq 3$, as they require large numbers of design points.

$3^q$ designs

Box-Behnken designs are rotatable, second-order designs that are incomplete $3^q$ factorials, but not ordinary fractions. Box-Behnken designs are formed by combining incomplete block designs with factorials. For $q$ factors, find an incomplete block design for $q$ treatments in blocks of size two. (Blocks of other sizes may be used, we merely illustrate with two.) Associate the "treatment" letters A, B, C, and so on with "factor" letters A, B, C, and so on. When two factor letters appear together in a block, use all combinations where those factors are at the $\pm 1$ levels, and all other factors are at 0. The combinations from all blocks are then joined with some center points to form the Box-Behnken design.

Box-Behnken designs

For example, for $q = 3$, we can use the BIBD with three blocks and (A,B), (A,C), and (B,C) as assignment of treatments to blocks. From the three blocks, we get the combinations:

| A | B | C |
|---|---|---|
| $x_1$ | $x_2$ | $x_3$ |
| –1 | –1 | 0 |
| –1 | 1 | 0 |
| 1 | –1 | 0 |
| 1 | 1 | 0 |

| A | B | C |
|---|---|---|
| $x_1$ | $x_2$ | $x_3$ |
| –1 | 0 | –1 |
| –1 | 0 | 1 |
| 1 | 0 | –1 |
| 1 | 0 | 1 |

| A | B | C |
|---|---|---|
| $x_1$ | $x_2$ | $x_3$ |
| 0 | –1 | –1 |
| 0 | –1 | 1 |
| 0 | 1 | –1 |
| 0 | 1 | 1 |

To this we add some center points, say five, to form the complete design.

This design takes only 17 points, instead of the 27 (plus some for replication) needed in the full factorial.

## 19.7 Second-Order Analysis

Here are three possible goals for the analysis of second-order models:

- Determine which design variables affect the response.
- Determine whether there is lack of fit.
- Determine the stationary point and surface type.

Use regression and F-tests

As with first-order models, fitting is done with multiple linear regression, and testing is done with F-tests. Let $\mathbf{b}$ be the estimated coefficients for first-order terms, and let $\mathbf{B}$ be the estimate of the second-order terms.

Test all coefficients to exclude a variable

The goal of determining which variables affect the response is a bit more complex for second-order models. To test that a variable—say variable 1—has no effect on the response, we must test that its linear, quadratic, and cross product coefficients are all zero: $\beta_1 = \beta_{11} = \cdots = \beta_{1q} = 0$. This is a $q + 1$-degree-of-freedom null hypothesis which we must test using an F-test.

Testing for lack of fit in the second-order model is completely analogous to the first-order model. Compute an estimate of pure error variability from the replicated points; all other residual variability is lack of fit. Significant lack of fit indicates that our model is not capturing the mean structure in our region of experimentation. When we have significant lack of fit, we should first consider whether a transformation of the response will improve the quality of the fit. For example, a second-order model may be a good fit for the log of the response. Alternatively, we can investigate higher-order models for the mean or obtain data to fit the second-order model in a smaller region.

Canonical analysis for shape of surface

*Canonical analysis* is the determination of the type of second-order surface, the location of its stationary point, and the canonical directions. These quantites are functions of the estimated coefficients $\mathbf{b}$ and $\mathbf{B}$ computed in the multiple regression. We estimate the stationary point as $\widehat{x}_0 = -\mathbf{B}^{-1}\mathbf{b}/2$, and the eigenvectors and eigenvalues of $\mathcal{B}$ are estimated by the eigenvectors and eigenvalues of $\mathbf{B}$ using special software.

**Example 19.4** **Cake baking, continued**

We now fit a second-order model to the data from the blocked central composite design of Example 19.3. This model will have linear terms, quadratic

**Listing 19.2:** Minitab output for second-order model of cake baking data.

```
Estimated Regression Coefficients for y

Term          Coef      StDev         T      P
Constant     8.070     0.1842    43.809  0.000          ①
Block       -0.057     0.1206    -0.473  0.651
x1           0.735     0.1595     4.608  0.002
x2           0.964     0.1595     6.042  0.001
x1*x1       -0.628     0.1661    -3.779  0.007
x2*x2       -1.195     0.1661    -7.197  0.000
x1*x2       -0.832     0.2256    -3.690  0.008

S = 0.4512       R-Sq = 95.0%      R-Sq(adj) = 90.8%

Analysis of Variance for y

Source            DF     Seq SS     Adj SS     Adj MS      F      P
Blocks             1     0.0457     0.0455    0.04546   0.22  0.651
Regression         5    27.2047    27.2047    5.44094  26.72  0.000
  Linear           2    11.7562    11.7562    5.87808  28.87  0.000
  Square           2    12.6763    12.6763    6.33816  31.13  0.000
  Interaction      1     2.7722     2.7722    2.77223  13.62  0.008
Residual Error     7     1.4252     1.4252    0.20359
  Lack-of-Fit      3     0.9470     0.9470    0.31567   2.64  0.186  ②
  Pure Error       4     0.4781     0.4781    0.11953
Total             13    28.6756
```

terms, a cross product term, and a block term. Listing 19.2 shows the results. At ① we see that all first- and second-order terms are significant, so that no variables need to be deleted from the model. We also see that lack of fit is not significant ②, so the second-order model should be a reasonable approximation to the mean structure in the region of experimentation.

Figure 19.6 shows a contour plot of the fitted second-order model. We see that the optimum is at about .4 coded time units above 0, and .2 coded temperature units above zero, corresponding to 35.8 minutes and 352°. We also see that the ellipse slopes northwest to southeast, meaning that we can trade time for temperature and still get a cake that we like.

Listing 19.3 shows a canonical analysis for this surface. The estimated coefficients are at ① ($\widehat{\beta}_0$), ② (**b**), and ③ (**B**). The estimated stationary point and its response are at ④ and ⑤; I guessed (.4, .2) for the stationary point from Figure 19.6—it was actually (.42, .26). The estimated eigenvectors and eigenvalues are at ⑥ and ⑦. Both eigenvalues are negative, indicating a maximum. The smallest decrease is associated with the first eigenvector (-.884, .467), so increasing the temperature by .53 coded units for every decrease in 1 coded unit of time keeps the response as close to maximum as possible.

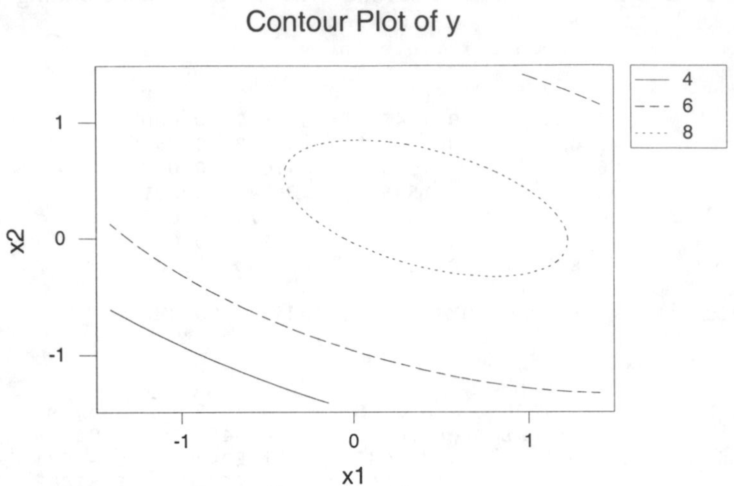


**Figure 19.6:** Contour plot of fitted second-order model for cake baking data, using Minitab.

**Listing 19.3:** MacAnova output for canonical analysis of cake baking data.

```
component: b0                               ①
(1)          8.07
component: b                                ②
(1)       0.73515          0.964
component: B                                ③
(1,1)      -0.62756       -0.41625
(2,1)      -0.41625        -1.1952
component: x0                               ④
(1,1)       0.41383
(2,1)       0.25915
component: y0                               ⑤
(1,1)         8.347
component: H                                ⑥
(1,1)      -0.88413       -0.46724
(2,1)       0.46724       -0.88413
component: lambda                           ⑦
(1)      -0.40758         -1.4152
```

The results of a canonical analysis have an aura of precision that is often not justified. Many software packages can compute and print the estimated stationary point, but few give a standard error for this estimate. In fact, the standard error is difficult to compute and tends to be rather large. Likewise, there can be considerable error in the estimated canonical directions.

## 19.8 Mixture Experiments

Mixture experiments are a special case of response surface experiments in which the response depends on the proportions of the various components, but not on absolute amounts. For example, the taste of a punch depends on the proportion of ingredients, not on the amount of punch that is mixed, and the strength of an alloy may depend on the proportions of the various metals in the alloy, but not on the total amount of alloy produced.

Mixtures depend on proportions

The design variables $x_1, x_2, \ldots, x_q$ in a mixture experiment are proportions, so they must be nonnegative and add to one:

$$x_k \geq 0, \qquad k = 1, 2, \cdots, q$$

and

$$x_1 + x_2 + \cdots + x_q = 1 \ .$$

This design space is called a *simplex* in $q$ dimensions. In two dimensions, the design space is the segment from (1,0) to (0,1); in three dimensions, it is bounded by the equilateral triangle (0,0,1), (0,1,0), and (1,0,0); and so on. Note that a point in the simplex in $q$ dimensions is determined by any $q-1$ of the coordinates, with the remaining coordinate determined by the constraint that the coordinates add to one.

Mixtures have a simplex design space

### Fruit punch

**Example 19.5**

Cornell (1985) gave an example of a three-component fruit punch mixture experiment, where the goal is to find the most appealing mixture of watermelon juice ($x_1$), pineapple juice ($x_2$), and orange juice ($x_3$). Appeal depends on the recipe, not on the quantity of punch produced, so it is the proportions of the constituents that matter. Six different punches are produced, and eighteen judges are assigned at random to the punches, three to a punch. The recipes and results are given in Table 19.3.

As in ordinary response surfaces, we have some response $y$ that we wish to model as a function of the explanatory variables:

$$y_{ij} = f(x_{1i}, x_{2i}, \cdots, x_{qi}) + \epsilon_{ij} \ .$$

**Table 19.3:** Blends of fruit punch.

| $x_1$ | $x_2$ | $x_3$ | Appeal | | |
|---|---|---|---|---|---|
| 1 | 0 | 0 | 4.3 | 4.7 | 4.8 |
| 0 | 1 | 0 | 6.2 | 6.5 | 6.3 |
| .5 | .5 | 0 | 6.3 | 6.1 | 5.8 |
| 0 | 0 | 1 | 7.0 | 6.9 | 7.4 |
| .5 | 0 | .5 | 6.1 | 6.5 | 5.9 |
| 0 | .5 | .5 | 6.2 | 6.1 | 6.2 |

Model response with low-order polynomial

We use a low-order polynomial for this model, not because we believe that the function really is polynomial, but rather because we usually don't know what the correct model form is; we are willing to settle for a reasonable approximation to the underlying function. We can use this model for various purposes:

- To predict the response at any combination of design variables,
- To find combinations of design variables that give best response, and
- To measure the effects of various factors on the response.

### 19.8.1 Designs for mixtures

Simplex lattice design

A $\{q,m\}$ *simplex lattice* design for $q$ components consists of all design points on the simplex where each component is of the form $r/m$, for some integer $r = 0, 1, 2, \ldots, m$. For example, the $\{3,2\}$ simplex lattice consists of the six combinations (1, 0, 0), (0, 1, 0), (0, 0, 1), (1/2, 1/2, 0), (1/2, 0, 1/2), and (0, 1/2, 1/2). The fruit punch experiment in Example 19.5 is a $\{3,2\}$ simplex lattice. The $\{3,3\}$ simplex lattice has the ten combinations (1, 0, 0), (0, 1, 0), (0, 0, 1), (2/3, 1/3, 0), (2/3, 0, 1/3), (1/3, 2/3, 0), (0, 2/3, 1/3), (1/3, 0, 2/3), (0, 1/3, 2/3), and (1/3, 1/3, 1/3). In general, $m$ needs to be at least as large as $q$ to get any points in the interior of the simplex, and $m$ needs to be larger still to get more points into the interior of the simplex. Figure 19.7(a) illustrates a $\{3,4\}$ simplex lattice.

Simplex centroid design

The second class of models is the *simplex centroid* designs. These designs have $2^q - 1$ design points for $q$ factors. The design points are the pure mixtures, all the 1/2-1/2 two-component mixtures, all the 1/3-1/3-1/3 three-component mixtures, and so on, through the equal mixture of all $q$ components. Alternatively, we may describe this design as all the permutations of $(1, 0, \ldots, 0)$, all the permutations of $(1/2, 1/2, \ldots, 0)$, all the permutations of

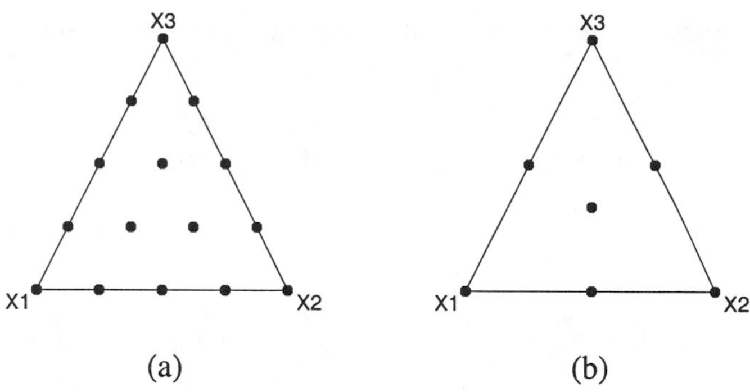


**Figure 19.7:** (a) {3,4} simplex lattice and (b) three variable simplex centroid designs.

$(1/3, 1/3, 1/3, \ldots, 0)$, and so on to the point $(1/q, 1/q, \ldots, 1/q)$. A simplex centroid design only has one point in the interior of the simplex; all the rest are on the boundary. Figure 19.7(b) illustrates a simplex centroid in three factors.

Mixtures in the interior of the simplex—that is, mixtures which include at least some of each component—are called *complete* mixtures. We sometimes need to do our experiments with complete mixtures. This may arise for several reasons, for example, all components may need to be present for a chemical reaction to take place.

Complete mixtures have all $x_k > 0$

*Factorial ratios* provide one class of designs for complete mixtures. This design is a factorial in the ratios of the first $q - 1$ components to the last component. We may want to reorder our components to obtain a convenient "last" component. The design points will have ratios $x_k/x_q$ that take a few fixed values (the factorial levels) for each $k$, and we then solve for the actual proportions of the components. For example, if $x_1/x_3 = 4$ and $x_2/x_3 = 2$, then $x_1 = 4/7$, $x_2 = 2/7$, and $x_3 = 1/7$. Only complete mixtures occur in a factorial ratios design with all ratios greater than 0.

Factorial ratios vary $x_k/x_q$

## Harvey Wallbangers

**Example 19.6**

Sahrmann, Piepel, and Cornell (1987) ran an experiment to find the best proportions for orange juice (O), vodka (V), and Galliano (G) in a mixed drink called a Harvey Wallbanger. Only complete mixtures are considered, because it is the mixture of these three ingredients that defines a Wallbanger (as opposed to say, orange juice and vodka, which is a drink called a screwdriver). Furthermore, preliminary screening established some approximate limits for the various components.

**Table 19.4:** Harvey Wallbanger mixture experiment.

| O/G | V/G | G | V | O | Rating |
|---|---|---|---|---|---|
| 4.0 | 1.2 | .161 | .194 | .645 | 3.6 |
| 9.0 | 1.2 | .089 | .107 | .804 | 5.1 |
| 4.0 | 2.8 | .128 | .359 | .513 | 3.8 |
| 9.0 | 2.8 | .078 | .219 | .703 | 3.8 |
| 6.5 | 2.0 | .105 | .211 | .684 | 4.7 |
| 4.0 | 2.0 | .143 | .286 | .571 | 2.4 |
| 9.0 | 2.0 | .083 | .167 | .750 | 4.0 |

The authors used a factorial ratios model, with three levels of the ratio V/G (1.2, 2.0, and 2.8) and two levels of the ratio O/G (4 and 9). They also ran a center point at V/G = 2 and O/G = 6.5. Their actual design included incomplete blocks (so that no evaluator consumed more than a small number of drinks). However, there were no apparent evaluator differences, so the average score was used as response for each mixture, and blocks were ignored. Evaluators rated the drinks on a 1 to 7 scale. The data are given in Table 19.4, which also shows the actual proportions of the three components.

Pseudocomponents

A second class of complete-mixture designs arises when we have lower bounds for each component: $x_k \geq d_k > 0$, where $\sum d_k = D < 1$. Here, we define *pseudocomponents*

$$x'_k = \frac{x_k - d_k}{1 - D}$$

and do a simplex lattice or simplex centroid design in the pseudocomponents. The pseudocomponents map back to the original components via

$$x_k = d_k + (1 - D)x'_k \quad .$$

Many mixture problems have constrained design spaces

Many realistic mixture problems are constrained in some way so that the available design space is not the full simplex or even a simplex of pseudocomponents. A regulatory constraint might say that ice cream must contain at least a certain percent fat, so we are constrained to use mixtures that contain at least the required amount of fat; and an economic constraint requires that our recipe cost less than a fixed amount. Mixture designs can be adapted to such situations, but we often need special software to determine a good design for a specific model over a constrained space.

### 19.8.2 Models for mixture designs

Polynomial models for a mixture response have fewer parameters than the general polynomial model found in ordinary response surfaces for the same number of design variables. This reduction in parameters arises from the simplex constraints on the mixture components—some terms disappear due to the linear restrictions among the mixture components. For example, consider a first-order model for a mixture with three components. In such a mixture, we have $x_1 + x_2 + x_3 = 1$. Thus,

Mixture constraints reduce parameter count

$$
\begin{aligned}
f(x_1, x_2, x_3) &= \beta_0 + \beta_1 x_1 + \beta_2 x_2 + \beta_3 x_3 \\
&= \beta_0(x_1 + x_2 + x_3) + \beta_1 x_1 + \beta_2 x_2 + \beta_3 x_3 \\
&= (\beta_1 + \beta_0)x_1 + (\beta_2 + \beta_0)x_2 + (\beta_3 + \beta_0)x_3 \\
&= \tilde{\beta}_1 x_1 \qquad + \tilde{\beta}_2 x_2 \qquad + \tilde{\beta}_3 x_3
\end{aligned}
$$

In this model, the linear constraint on the mixture components has allowed us to eliminate the constant from the model. This reducted model is called the *canonical form* of the mixture polynomial. We will simply use $\beta$ in place of $\tilde{\beta}$ in the sequel.

Canonical form of first-order model

Mixture constraints also permit simplifications in second-order models. Not only can we eliminate the constant, but we can also eliminate the pure quadratic terms! For example:

$$
\begin{aligned}
x_1^2 &= x_1 x_1 \\
&= x_1(1 - x_2 - x_3 - \cdots - x_q) \\
&= x_1 - x_1 x_2 - x_1 x_3 - \cdots - x_1 x_q \ .
\end{aligned}
$$

By making similar substitutions for all pure quadratic terms, we get the canonical form:

Canonical form of second-order model

$$
f(x_1, x_2, \cdots, x_q) = \sum_{k=1}^{q} \beta_k x_k + \sum_{k<l}^{q} \beta_{kl} x_k x_l \ .
$$

Third-order models are sometimes fit for mixtures; the canonical form for the full third-order model is:

Canonical form of third-order model

$$
\begin{aligned}
f(x_1, x_2, \cdots, x_q) &= \sum_{k=1}^{q} \beta_k x_k + \sum_{k<l}^{q} \beta_{kl} x_k x_l \\
&\quad + \sum_{k<l}^{q} \delta_{kl} x_k x_l (x_k - x_l) + \sum_{k<l<n}^{q} \beta_{klm} x_k x_l x_n \ .
\end{aligned}
$$

Special cubic model

A subset of the full cubic model called the *special cubic* model sometimes appears:

$$f(x_1, x_2, \cdots, x_q) = \sum_{k=1}^{q} \beta_k x_k + \sum_{k<l}^{q} \beta_{kl} x_k x_l + \sum_{k<l<n}^{q} \beta_{kln} x_k x_l x_n \ .$$

Mixture coefficients have special interpretations

Coefficients in mixture canonical polynomials have interpretations that are somewhat different from standard polynomials. If the mixture is pure (that is, contains only a single component, say component $k$), then $x_k$ is 1 and the other components are 0. The predicted response is $\beta_k$. Thus the "linear" coefficients give the predicted response when the mixture is simply a single component. If the mixture is a 50-50 mix of components $k$ and $l$, then the predicted response is $\beta_k/2 + \beta_l/2 + \beta_{kl}/4$. Thus the bivariate interaction terms correspond to deviations from a simple additive fit, and in particular show how the response for pairwise blends varies from additive. The three-way interaction term $\beta_{klm}$ has a similar interpretation for triples. The cubic interaction term $\delta_{kl}$ provides some asymmetry in the response to two-way blends.

Fewer factors as an alternative to reduced models

We may use ordinary polynomial models in $q - 1$ factors instead of reduced polynomial models in $q$ factors. For example, the canonical quadratic model in $q = 3$ factors is

$$y = \beta_1 x_1 + \beta_2 x_2 + \beta_3 x_3 + \beta_{12} x_1 x_2 + \beta_{13} x_1 x_3 + \beta_{23} x_2 x_3 \ .$$

We can instead use the model

$$y = \tilde{\beta}_0 + \tilde{\beta}_1 x_1 + \tilde{\beta}_2 x_2 + \tilde{\beta}_{12} x_1 x_2 + \tilde{\beta}_{11} x_1^2 + \tilde{\beta}_{22} x_2^2 \ ,$$

which is the usual quadratic model for $q = 2$ factors. The models are equivalent mathematically, and which model you choose is personal preference. There are linear relations between the models that allow you to transfer between the representations. For example, $\tilde{\beta}_0 = \beta_3$ ($x_3 = 1$, $x_1 = x_2 = 0$), and $\tilde{\beta}_0 + \tilde{\beta}_1 + \tilde{\beta}_{11} = \beta_1$ ($x_1 = 1$, $x_2 = x_3 = 0$).

Factorial ratios experiments also have the option of using polynomials in the components, polynomials in the ratios, or a combination of the two. The choice of model can sometimes be determined *a priori* but will frequently be determined by choosing the model that best fits the data.

**Example 19.7** **Harvey Wallbangers, continued**

Example 19.6 introduced the Harvey Wallbanger data. Listing 19.4 shows the results from fitting the canonical second-order model. All terms are signifi-

**Listing 19.4:** MacAnova output for second-order model of Harvey Wallbanger data.

```
                Coef        StdErr           t
g            -518.14        41.143     -12.594
o            -12.625        1.1111     -11.363
v             100.56        5.8373      17.226
og            812.73        55.472      14.651
vg            126.64        56.449      2.2435
ov           -101.53        5.8706     -17.294

N: 7, MSE: 0.0042851, DF: 1, R^2: 0.99996
Regression F(6,1): 4344.4, Durbin-Watson: 2.1195
```

cant with the exception of the vodka by Galliano interaction (though there is only 1 degree of freedom for error, so significance testing is rather dubious).

It is difficult to interpret the coefficients directly. The usual interpretations for coefficients are for pure mixtures and two-component mixtures, but this experiment was conducted on a small region in the interior of the design space. Thus using the model for pure mixtures or two-component mixtures would be an unwarranted extrapolation. The best approach is to plot the contours of the fitted response surface, as shown in Figure 19.8. We see that there is a saddle point near the fifth design point (the center point), and the highest estimated responses are on the boundary between the first two design points. This has the V/G ratio at 1.2 and the O/G ratio between 4.0 and 9.0, but somewhat closer to 9.

## 19.9 Further Reading and Extensions

As might be expected, there is much more to the subjects discussed in this chapter. Box and Draper (1987) and Cornell (1990) provide excellent book-length coverage of response surfaces and mixture experiments respectively.

Earlier we alluded to the issue of constraints on the design space. These constraints can make it difficult to run standard response surface or mixture designs. Special-purpose computer software (for example, Design-Expert) can construct good designs for constrained situations. These designs are generally chosen to be optimal in the sense of minimizing the estimation variance. See Cook and Nachtsheim (1980) or Cook and Nachtsheim (1989). A second interesting area is trying to optimize when there is more than one response. Multiple responses are common in the real world, and methods have been proposed to compromise among the competing criteria. See Myers, Khuri, and Carter (1989) and the references cited there.

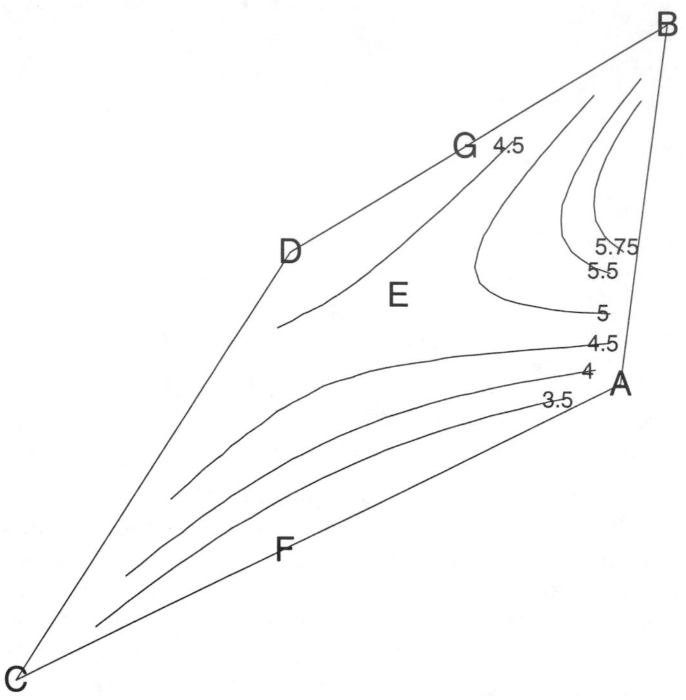


**Figure 19.8:** Contour plot for Harvey Wallbanger data, using S-Plus. Letters indicate the points of Table 19.4 in the table order.

## 19.10 Problems

**Exercise 19.1** We run a central composite design and fit a second-order model. The fitted coefficients are:

$$y = 86 + 9.2x_1 + 7.3x_2 - 7.8x_1^2 - 3.9x_2^2 - 6.0x_1x_2 \ .$$

Perform the canonical analysis on this response surface.

**Exercise 19.2** Fit the second-order model to the fruit punch data of Example 19.5. Which mixture gives the highest appeal?

**Exercise 19.3** The whiteness of acrylic fabrics after being washed at different detergent concentrations (.09 to .21 percent) and temperatures (29 to 41°C) was measured and the following model was obtained (Prato and Morris 1984):

$$y = -116.27 + 819.58x_1 + 1.77x_2 - 1145.34x_1^2 - .01x_2^2 - 3.48x_1x_2 \ .$$

Perform the canonical analysis on this response surface.

**Problem 19.1**

Three components of a rocket propellant are the binder ($x_1$), the oxidizer ($x_2$), and the fuel ($x_3$). We want to find the mixtures that yield coefficients of elasticity ($y$) less than 3000. All components must be present and there are minimum proportions, so the investigators used a pseudocomponents design, with the following pseudocomponent values and results (data from Kurotori 1966 via Park 1978):

| $x_1$ | $x_2$ | $x_3$ | $y$ |
|---|---|---|---|
| 1 | 0 | 0 | 2350 |
| 0 | 1 | 0 | 2450 |
| 0 | 0 | 1 | 2650 |
| 1/2 | 1/2 | 0 | 2400 |
| 1/2 | 0 | 1/2 | 2750 |
| 0 | 1/2 | 1/2 | 2950 |
| 1/3 | 1/3 | 1/3 | 3000 |
| 2/3 | 1/6 | 1/6 | 2690 |
| 1/6 | 2/3 | 1/6 | 2770 |
| 1/6 | 1/6 | 2/3 | 2980 |

Does this design correspond to any of our standard mixture designs? Does it have an estimate of pure error? Fit the second-order mixture model. Is the estimated maximum above 3000? Where is the estimated maximum, and where is the region that has elasticity less than 3000?

**Problem 19.2**

Millers want to make bread flours that bake into large loaves. They need to mix flours from four varieties of wheat, so they run an experiment with different mixtures and measure the volume of the resulting loaves (ml/100 g dough). The experiment was performed on 2 separate days, obtaining the following results (data from Draper et al. 1993):

| Day 1 | | | | | Day 2 | | | | |
|---|---|---|---|---|---|---|---|---|---|
| $x_1$ | $x_2$ | $x_3$ | $x_4$ | Volume | $x_1$ | $x_2$ | $x_3$ | $x_4$ | Volume |
| 0 | .25 | 0 | .75 | 403 | 0 | .75 | 0 | .25 | 423 |
| .25 | 0 | .75 | 0 | 425 | .25 | 0 | .75 | 0 | 417 |
| 0 | .75 | 0 | .25 | 442 | 0 | .25 | 0 | .75 | 388 |
| .75 | 0 | .25 | 0 | 433 | .75 | 0 | .25 | 0 | 407 |
| 0 | .75 | .25 | 0 | 445 | 0 | 0 | .25 | .75 | 338 |
| .25 | 0 | 0 | .75 | 435 | .25 | .75 | 0 | 0 | 435 |
| 0 | 0 | .75 | .25 | 385 | 0 | .25 | .75 | 0 | 379 |
| .75 | .25 | 0 | 0 | 425 | .75 | 0 | 0 | .25 | 406 |
| .25 | .25 | .25 | .25 | 433 | .25 | .25 | .25 | .25 | 439 |

Analyze these data to determine which mixture of flours yields the largest loaves.

**Problem 19.3** An experiment is performed to determine how a gasoline engine responds to various factors. The response of interest is CO emissions in grams per hour. The design factors are engine load, in Newton meters, range (30,70); engine speed, in rpm, range (1000, 4000); spark advance, in degrees, range (10, 30); air-to-fuel ratio, dimensionless, range (13, 16.4); and exhaust gas recycle, in percent, range (0, 10). The experimental design has 46 observations in two blocks of 23 each. The design factors have been coded to the range (-1, 1) in the table below (data from Draper et al. 1994). Analyze these data and describe how CO emissions depend on engine settings.

| Load | Speed | Advance | Ratio | Recycle | Block | Response |
|---|---|---|---|---|---|---|
| –1 | –1 | 0 | 0 | 0 | 1 | 81 |
| 1 | –1 | 0 | 0 | 0 | 1 | 148 |
| –1 | 1 | 0 | 0 | 0 | 1 | 348 |
| 1 | 1 | 0 | 0 | 0 | 1 | 530 |
| 0 | 0 | –1 | –1 | 0 | 1 | 1906 |
| 0 | 0 | 1 | –1 | 0 | 1 | 1717 |
| 0 | 0 | –1 | 1 | 0 | 1 | 91 |
| 0 | 0 | 1 | 1 | 0 | 1 | 42 |
| 0 | –1 | 0 | 0 | –1 | 1 | 86 |
| 0 | 1 | 0 | 0 | –1 | 1 | 435 |
| 0 | –1 | 0 | 0 | 1 | 1 | 93 |
| 0 | 1 | 0 | 0 | 1 | 1 | 474 |
| –1 | 0 | –1 | 0 | 0 | 1 | 224 |
| 1 | 0 | –1 | 0 | 0 | 1 | 346 |
| –1 | 0 | 1 | 0 | 0 | 1 | 147 |
| 1 | 0 | 1 | 0 | 0 | 1 | 287 |
| 0 | 0 | 0 | –1 | –1 | 1 | 1743 |
| 0 | 0 | 0 | 1 | –1 | 1 | 46 |
| 0 | 0 | 0 | –1 | 1 | 1 | 1767 |
| 0 | 0 | 0 | 1 | 1 | 1 | 73 |
| 0 | 0 | 0 | 0 | 0 | 1 | 195 |
| 0 | 0 | 0 | 0 | 0 | 1 | 233 |
| 0 | 0 | 0 | 0 | 0 | 1 | 236 |
| 0 | –1 | –1 | 0 | 0 | 2 | 100 |
| 0 | 1 | –1 | 0 | 0 | 2 | 559 |
| 0 | –1 | 1 | 0 | 0 | 2 | 118 |
| 0 | 1 | 1 | 0 | 0 | 2 | 406 |
| –1 | 0 | 0 | –1 | 0 | 2 | 1255 |
| 1 | 0 | 0 | –1 | 0 | 2 | 2513 |
| –1 | 0 | 0 | 1 | 0 | 2 | 53 |

| Load | Speed | Advance | Ratio | Recycle | Block | Response |
|---|---|---|---|---|---|---|
| 1 | 0 | 0 | 1 | 0 | 2 | 54 |
| 0 | 0 | –1 | 0 | –1 | 2 | 270 |
| 0 | 0 | 1 | 0 | –1 | 2 | 277 |
| 0 | 0 | –1 | 0 | 1 | 2 | 303 |
| 0 | 0 | 1 | 0 | 1 | 2 | 213 |
| –1 | 0 | 0 | 0 | –1 | 2 | 171 |
| 1 | 0 | 0 | 0 | –1 | 2 | 344 |
| –1 | 0 | 0 | 0 | 1 | 2 | 180 |
| 1 | 0 | 0 | 0 | 1 | 2 | 280 |
| 0 | –1 | 0 | –1 | 0 | 2 | 548 |
| 0 | 1 | 0 | –1 | 0 | 2 | 3046 |
| 0 | –1 | 0 | 1 | 0 | 2 | 13 |
| 0 | 1 | 0 | 1 | 0 | 2 | 123 |
| 0 | 0 | 0 | 0 | 0 | 2 | 228 |
| 0 | 0 | 0 | 0 | 0 | 2 | 201 |
| 0 | 0 | 0 | 0 | 0 | 2 | 238 |

**Problem 19.4**

Briefly describe an experimental design appropriate for each of the following situations.

(a) Whole house air exchangers have become important as houses become more tightly sealed and the dangers of indoor air pollution become known. Exchangers are used primarily in winter, when they draw in fresh air from the outside and exhaust an equal volume of indoor air. In the process, heat from the exhausted indoor air is used to warm the incoming air. The design problem is to construct an exchanger that maximizes energy efficiency while maintaining air flow volume within tolerances. Energy efficiency is energy saved by heating the incoming air minus energy used to power the fan. There are two design variables: the pore size of the exchanger and the fan speed. In general, as the pore size decreases the energy saved through heat exchange increases, but for smaller pores the fan must be run faster to maintain air flow, thus using more energy.

We have a current guess as to the best settings for maximum energy efficiency (pore size P and fan speed S). Any settings with 15% of P and S will provide acceptable air flow, and we feel that the optimum is probably within about 5% of these current settings.

(b) Neuropeptide Y (NPY) is believed to be involved in the regulation of feeding and basal metabolism. When rat brains are perfused with NPY, the rats dramatically increase their food intake over the next 24 hours. Naloxone (NLX) may potentially block the effects of NPY. If so, it could be an important line of research in obesity studies. We wish to test the effect of four treatments, the factorial combinations of brain perfusion by either NPY or saline (as a control), and the subcutaneous injection of either NLX or saline (as a control) on 24-hour post-treatment food intake. We have available 32 male inbred, essentially similar rats.

(c) We are trying to produce a new cleaning solvent for circuit boards. We anticipate that a combination of three standard solvents will work as well as the specialty solvent currently in use, but beyond knowing that we want each of the three to be at least 10% of the combination, we don't know how much of each to use.

(d) Child development specialists are interested in factors affecting the ability of children to solve "ten questions" puzzles. In these puzzles the child is given a set of pictures, one of which has been chosen by the researcher. The child gets to ask questions that the researcher answers either yes or no; on the basis of these answers the child tries to determine which of the pictures has been chosen. The response the researchers are looking at is the number of questions (ten maximum) that the child asks before determining the chosen picture. Two factors are under study: the number of pictures to choose from (either fifteen or twenty), and the familiarity of the objects in the pictures (either dinosaurs or birds, and oddly enough, I think the dinosaurs are the familiar objects!). The researchers have funds to study twelve children, and they expect substantial child to child variation. All children will do four puzzles, one of each type. They expect learning to take place, so that the later puzzles will generally be solved more quickly.

(e) A fertilizer company is developing a rose fertilizer which consists of a nitrogen compound N, a phosphorus compound P, a potassium compound K, and an inert binder to hold it all together. (The binder can be disregarded in the experiment.) The company believes that there are optimum levels of N, P, and K to give best rose yield, and they believe that their current settings $N_0 = 6$, $P_0 = 6$, and $K_0 = 4$ (kg per 100 kg of fertilizer) are pretty close to optimal; probably each is within 10% of the optimal values. They want to find the optimal values.

**Problem 19.5** Curing time and temperature affect the shear strength of an adhesive that bonds galvanized steel bars. The following experiment was repeated on 2 separate days. Twenty-four pieces of steel are obtained by random sampling from warehouse stock. These are grouped into twelve pairs; the twelve pairs are glued and then cured with one of nine curing treatments assigned at random. The treatments are the three by three factorial combinations of temperature (375°, 400°, and 450°F, coded -1, 0, 2) and time (30, 35, or 40 seconds, coded -1, 0, 1). Four pairs were assigned to the center point, and one pair to all other conditions. The response is shear strength (in psi, data from Khuri 1992):

| Temp. | Time | Day 1 | Day 2 |
|---|---|---|---|
| -1 | -1 | 1226 | 1213 |
| 0 | -1 | 1898 | 1961 |
| 2 | -1 | 2142 | 2184 |
| -1 | 0 | 1472 | 1606 |
| 0 | 0 | 2010 | 2450 |
| 0 | 0 | 1882 | 2355 |
| 0 | 0 | 1915 | 2420 |
| 0 | 0 | 2106 | 2240 |
| 2 | 0 | 2352 | 2298 |
| -1 | 1 | 1491 | 2298 |
| 0 | 1 | 2078 | 2531 |
| 2 | 1 | 2531 | 2609 |

Determine the temperature and time settings that give strong bonds.

**Problem 19.6**

For each of the following, briefly describe the design used and give a skeleton ANOVA.

(a) National forests are managed for multiple uses, including wildlife habitat. Suppose that we are managing our multiple-use forest, and we want to know how snowmobiling and timber harvest method affect timber wolf reproductive success (as measured by number of pups surviving to 1 year of age over a 5-year interval). We may permit or ban snowmobiles; snowmobiles cover a lot of area when present, so we can only change the snowmobile factor over large areas. We have three timber harvest methods, and they are fairly easy to change over small areas. We have six large, widely dispersed forest sections that we may use for the experiment. We choose three sections at random and ban snowmobiles there. The other three sections allow snowmobiles. Each of these sections is divided into three zones, and we randomly assign one of the three harvest methods to each zone within each section. (Note that we do not harvest the entire zone; we merely use that harvest method when we do harvest within the zone.) We observe timber wolf success in each zone.

(b) Some aircraft have in-flight deicing systems that are designed to prevent or remove ice buildup from the wings. A manufacturer wishes to compare three different deicing systems. This is done by installing the system on a test aircraft and flying the test aircraft behind a second plane that sprays a fine mist into the path of the test aircraft. The wings are photographed, and the ice buildup is estimated from interpretation of the photographs. They make five test flights for each of the three systems. The amount of buildup is influenced by temperature and humidity at flight altitude. The flights will be made at constant temperature (achieved by slightly varying the altitude); relative humidity cannot be controlled, but will be measured at the time of the flight.

(c) We wish to study new varieties of corn for disease resistance. We start by taking four varieties (A, B, C, D) and cross them (pollen from type A, B, C or D fertilizing flowers from type A, B, C, or D), getting sixteen crosses. (This is called a diallel cross experiment, and yes, four of the sixteen "crosses" are actually pure varieties.) The sixteen crosses produce seed, and we now treat the crosses as varieties for our experiment. We have 48 plots available, 16 plots in each of St. Paul, Crookston, and Waseca. We randomly assign each of the crosses to one of the sixteen plots at each location.

(d) A political scientist wishes to study how polling methods affect results. Two candidates (A and B) are seeking endorsement at their party convention. A random sample of 3600 voters has been taken and divided at random into nine sets of 400. All voters were asked if they support candidate A. However, before the question was asked, they were either told (a) that the poll is funded by candidate A, (b) that the poll is funded by candidate B, or (c) nothing. Due to logistical constraints, all voters in a given set (of 400) were given the same information; the response for a set of 400 is the number supporting candidate A. The three versions of information were randomly assigned to the nine sets.

**Question 19.1** Suppose we are fitting a first-order model using data from a $2^q$ design with $m$ center points, but a second-order model is actually correct. Show that the contrast formed by taking the average response at the factorial points minus the average at the center points estimates the sum of the quadratic coefficients of the second-order model. Show that the two-factor interaction effects in the factorial points estimate the cross product terms in the second-order model.

# Chapter 20

# On Your Own

Adult birds push their babies out of the nest to force them to learn to fly. As I write this, I have a 16-year-old daughter learning to drive. And you, our statistical children, must leave the cozy confines of textbook problems and graduate to the real world of designing and analyzing your own experiments for your own goals. This final chapter is an attempt at a framework for the experimental design process, to help you on your way to designing real-world experiments.

## 20.1 Experimental Context

An individual experiment is usually part of a larger research enterprise; thus planning an experiment takes place within this larger context. One way to frame this larger context is hierarchically, with goals, objectives, and hypotheses. The (overall) goals are for the large research enterprise. For example, we might have the goal of developing artificial heated-butter aromas for the food industry. The (immediate) objective is a refinement of the goals to narrow the scope of investigation. Continuing the butter aroma example, we might have the objective of determining which naturally occurring odorants in heated butter influence the perceived butter aroma. Finally, hypotheses are specific, answerable questions regarding an objective that can be addressed in an experiment. We might ask, can human subjects detect the difference in aroma between heated butter and this particular mixture of compounds?

Goals, objectives, and hypotheses

| We design experiments to answer the questions raised in our hypotheses. |
|---|

## 20.2 Experiments by the Numbers

Many authors have presented guidelines for designing experiments. Noteworthy among these are Kempthorne (1952), Cochran and Cox (1957), Cox (1958), Daniel (1976), and Box, Hunter, and Hunter (1978). I have tried to synthesize a number of these recommendations into a sequence of steps for designing an experiment, which are presented below. Experimentation, like all science, is not one-size-fits-all, but these steps will work for many investigations.

Information and simplicity

I have two basic rules when planning an experiment. The first is "Use all the information you have available to you." Most of this information is subject matter information (what you know about treatments, units, and so on) rather than statistical tactics. The second is "Use the simplest possible design that gets the job done." Thus when designing an experiment I consider the fancy tricks of the trade only when they are needed.

1. Do background research. At a minimum, you should

    - Determine what is already *known* about your problem. Researchers know things that have been discovered by experiment and verified by repeated experiments. You may wish to repeat a "known" experiment if you are trying to verify it, extend it to a new population, or learn an experimental technique, but more often you will be looking at new hypotheses.

    - Determine what other researchers *suspect* about your problem. Many experiments are follow-up experiments on vague indications from earlier research. For example, a preliminary experiment may have indicated the possibility that a particular drug was effective against breast cancer, but the sample size was too small to be conclusive.

    - Determine what background or extraneous factors (for example, environmental factors) might affect the outcome of your experiment. Here we are looking ahead to the possibility that blocking might be needed, so we identify the sources of extraneous variation on which we may need to block.

    - Find out what related experiments have been done, what types of designs were used, and what kinds of problems were encountered. There is always room for innovation, particularly if earlier experiments encountered problems, but experimental designs that work well are worth imitating.

- Determine the cost or availability of experimental material such as animals, equipment, and chemical stocks; determine your time and monetary budgets. Time and money are major constraints on experimentation. Determine these constraints early.

This research takes time, but it will save you time later.

2. Decide which question to address next, and clearly state your question. This process should include:

- A list of hypotheses to be tested or effects to be estimated.
- An ordering of these hypotheses or effects by importance.
- An ordering of these hypotheses or effects by logical or time sequence if some should be examined before others.

Your experiment is part of the research enterprise, so choose your hypotheses to address your current objectives. Knowing if some hypotheses are more important than others will matter for designs such as split plots, which are more precise for split-plot factors than for whole-plot factors.

Remember, science is sequential, with new results building on old results. Unless you have an overwhelming argument to the contrary, plan for a sequence of hypotheses and experiments and *don't try to do everything in a single experiment!*

3. Determine the treatments to be studied, experimental units to be used, and responses to be measured. These depend on the hypotheses being addressed and the population about which you wish to make inferences. Choice of treatments includes the consideration of controls (probably needed) and/or placebo treatments.

The type of experimental units you use will determine the population about which you can make inferences and usually the size of your experimental errors. Homogeneous units generally lead to smaller experimental errors and thus shorter confidence intervals and more powerful tests. On the other hand, homogeneous units often represent a narrow subset of all potential units, and it can be difficult to argue that conclusions reached about a homogeneous subset of a population hold for the entire population. If you need to work with a heterogeneous population of units, you will probably need to consider blocking the experiment.

The response or responses to be measured are usually determined by the hypotheses, but you must still determine how they will be measured, what the measurement units are, and whether blinding will be needed.

4. Design the current experiment. Try simple designs first; if upon inspection the simple design won't do the job for some reason, you can design a fancier experiment. But at least contemplate the simple experiment first. Keep the qualities of a good design in mind—design to avoid systematic error, to be precise, to allow esimation of error, and to have broad validity.

5. Inspect the design for scientific adequacy and practicality.

   - Are there any systematic problems that would invalidate your results or reduce their range of generalization? For example, does your design have confounding that biases your comparisons?
   - Are there treatments or factor-level combinations that are impractical or simply cannot be used? For example, you may have several factors that involve time, and the overall time may be impractical when all factors are at the high level; or perhaps some treatments are "a little too exothermic" (as my chemistry T.A. described one of our proposed experiments).
   - Do you have the time and resources to carry out the experiment?

If there are problems in any of these areas, you will need to go back to step 4 and revise your design. For example, the simple design was a full factorial, but it was too big, so we could move to a fancier design such as a fractional factorial.

6. Inspect the design for statistical adequacy and practicality.

   - Do you know how to analyze the results?
   - Will your experiment satisfy the statistical or model assumptions implicit in the statistical analysis?
   - Do you have enough degrees of freedom for error for all terms of interest?
   - Will you have adequate power or precision?
   - Will the analysis be easy to interpret?
   - Can you account for aliasing?

If you answer any of these in the negative, you will need to go back to step 4 and revise your design. For example, you might need to add blocking to reduce variability, or you might decide that a design with an unbalanced mixed-effects model was simply too difficult to analyze. Study the design carefully for oversights or mistakes. For example, I have seen split-plot designs with no degrees of freedom for error at the whole-plot level. (The investigator had intended to use an interaction for a surrogate error, but all interactions were at the split-plot level.)

7. Run the experiment.

8. Analyze the results. Pay close attention to where model or distributional assumptions might fail, and take corrective action if necessary. For example,

- Do factors assumed to be additive actually interact, or do treatments act differently in different blocks?
- Is the error variance nonconstant?
- Are there outliers in the data?
- Do the random errors follow the normal distribution?
- Are there unmodeled dependencies in the data (for example, time dependencies)?

Consider whether the experiment as run answers the questions, or if some further observations are needed. For example, you might want to rerun suspected outlier points, or you might need another fraction of a factorial to disentangle some aliases.

9. Draw conclusions, giving estimates of error or reliability. Assess this experiment in relation to similar experiments. Reporting is crucial, and it is only a slight exaggeration to say that an experiment not reported is an experiment not conducted. I like to begin reports with a short "executive summary" giving the conclusions, and then add sections on the experimental design and analysis (many journals call such sections "Materials and Methods" and "Results").

10. Consider what needs to be studied next. Research is ongoing and sequential, and one completed experiment leads to the design of the next.

It is clear that a carefully planned experiment requires a great deal of effort. Many of the steps in planning an experiment are nonstatistical and require considerable background knowledge in the subject being studied, while other steps require substantial statistical knowledge. Thus experimental design is often a team effort, with subject matter experts and statistical experts working together. One goal of this book has been to make the statistical part of the planning a little easier.

## 20.3 Final Project

Design an experiment, run the experiment, analyze the results, and report your findings.

This is not an overnight homework problem, but a project with several stages. Stage one is the project proposal, which should include a description of your hypotheses and proposed experimental design. This proposal should be sufficiently complete that anyone could replicate your experiment given just your proposal. Submit your proposal to your instructor for approval before conducting the experiment.

Stage two is running the experiment. Here you are on your own.

Stage three is analysis and reporting. Your report will typically be in the five to ten page range and should include a summary giving the conclusions, an introduction to the problem stating the background and hypothesis to be tested, a description of the experimental design (similar to stage one), and a description of the analysis. The description of the analysis should not be a batch of unannotated computer output. It should say what you are doing, why you are doing it, and what it tells you. Output and figures can be intermixed or appended separately.

The subject of the experiment is up to you and your instructor. Those of you in graduate school or at work in a research area may be able to adapt your own ongoing work to this project. Or just try something fun—food experiments (particularly desserts!) are always attractive, as are the experiments of youth such as rolling balls down inclined planes.

# Bibliography

Addelman, S. (1962). Orthogonal main-effects plans for asymmetrical factorial experiments. *Technometrics 4*, 21–46.

Ali, M. M. (1984). An approximation to the null distribution and power of the Durbin-Watson statistic. *Biometrika 71*, 253–261.

Alley, M. C., C. B. Uhl, and M. M. Lieber (1982). Improved detection of drug cytotoxicity in the soft agar colony formation assay through use of a metabolizable tetrazolium salt. *Life Sciences 31*, 3071–3078.

Amoah, F. M., K. Osei-Bonsu, and F. K. Oppong (1997). Response of improved robusta coffee to location and management practices in Ghana. *Experimental Agriculture 33*, 103–111.

Anderson, R. L. (1954). The problem of autocorrelation in regression analysis. *Journal of the American Statistical Association 49*, 113–129.

Andrews, D. F., P. J. Bickel, F. R. Hampel, P. J. Huber, W. H. Rogers, and J. W. Tukey (1972). *Robust Estimates of Location: Survey and Advances*. Princeton, NJ: Princeton University Press.

Aniche, N. G. and N. Okafor (1989). Studies on the effect of germination time and temperature on malting of rice. *Journal of the Institute of Brewing 95*, 165–167.

Atkinson, A. C. (1985). *Plots, Transformations, and Regression*. Oxford, U.K.: Oxford University Press.

Barnett, V. and T. Lewis (1994). *Outliers in Statistical Data* (Third ed.). New York: Wiley.

Beckman, R. J. and R. D. Cook (1983). Outlier. . . . . . . . . .s. *Technometrics 25*, 119–149.

Bellavance, F. and S. Tardif (1995). A nonparametric approach to the analysis of three-treatment three-period crossover designs. *Biometrika 82*, 865–875.

Benjamini, Y. and Y. Hochberg (1995). Controlling the false discovery rate: a practical and powerful approach to multiple testing. *Journal of the Royal Statistical Society, Series B 57*, 289–300.

Bergman, B. and A. Hynén (1997). Dispersion effects from unreplicated designs in the $2^{k-p}$ series. *Technometrics 39*, 191–198.

Bernhardson, C. S. (1975). Type I error rates when multiple comparison procedures follow a significant F test of ANOVA. *Biometrics 31*, 229–332.

Berry, D. A. (1989). Comments on "Investigating therapies of potentially great benefit: ECMO". *Statistical Science 4*, 306–310.

Bezjak, Z. and B. Knez (1995). Workplace design and loadings in the process of sewing garments. *International Journal of Clothing Science and Technology 7*, 89–101.

Bicking, C. A. (1958). Experiences and needs for design in ordnance experimentation. In V. Chew (Ed.), *Experimental Designs in Industry*, pp. 247–252. New York: Wiley.

Bin Jantan, I., A. S. Bin Ahmad, and A. R. Bin Ahmad (1987). Tapping of oleo-resin from *Dipterocarpus kerrii*. *The Malaysian Forester 50*, 343–353.

Bishop, Y. M. M., S. E. Fienberg, and P. W. Holland (1975). *Discrete Multivariate Analyses: Theory and Practice*. Cambridge, MA: MIT Press.

Boehner, A. W. (1975). *The Effect of Three Species of Logging Slash on the Properties of Aspen Planer Shavings Particleboard*. Ph. D. thesis, University of Minnesota, St. Paul, MN.

Bose, R. C., W. H. Clatworthy, and S. S. Shrikhande (1954). Tables of partially balanced designs with two associate classes. Technical Bulletin 107, North Carolina Agricultural Experiment Station.

Bose, R. C. and K. R. Nair (1939). Partially balanced incomplete block designs. *Sankhya 4*, 337–372.

Box, G. E. P. (1954). Some theorems on quadratic forms applied in the study of analysis of variance problems, I. Effect of inequality of variance in the one-way classification. *Annals of Mathematical Statistics 25*, 290–302.

Box, G. E. P. (1988). Signal-to-noise ratios, performance criteria, and transformations. *Technometrics 30*, 1–17.

Box, G. E. P. and S. L. Andersen (1955). Permutation theory in the derivation of robust criteria and the study of departures from assumptions. *Journal of the Royal Statistical Society, Series B 17*, 1–34.

Box, G. E. P., S. Bisgaard, and C. A. Fung (1988). An explanation and critique of Taguchi's contributions to quality engineering. *Quality and Reliability Engineering International 4*, 123–131.

Box, G. E. P. and D. R. Cox (1964). An analysis of transformations. *Journal of the Royal Statistical Society, Series B 26*, 211–243.

Box, G. E. P. and N. R. Draper (1987). *Empirical Model-Building with Response Surfaces*. New York: Wiley.

Box, G. E. P. and J. S. Hunter (1957). Multi-factor experimental designs for exploring response surfaces. *Annals of Mathematical Statistics 28*, 195–241.

Box, G. E. P., W. G. Hunter, and J. S. Hunter (1978). *Statistics for Experimenters*. New York: Wiley.

Bréfort, H., J. X. Guinard, and M. J. Lewis (1989). The contribution of dextrins of beer sensory properties, part II. Aftertaste. *Journal of the Institute of Brewing 95*, 431–435.

Brown, M. B. (1975). Exploring interaction effects in the ANOVA. *Applied Statistics 24*, 288–298.

Bussan, A. J. (1995). Selection for weed competitiveness among soybean genotypes. Master's thesis, University of Minnesota, St. Paul, MN.

Carmer, S. G. and W. M. Walker (1982). Baby bear's dilemma: A statistical tale. *Agronomy Journal 74*, 122–124.

Caro, M. R., E. Zamora, L. Leon, F. Cuello, J. Salinas, D. Megias, M. J. Cubero, and A. Contreras (1990). Isolation and identification of *Listeria monocytogenes* in vegetable byproduct silages containing preser vative additives and destined for animal feeding. *Animal Feed Science and Technology 31*, 285–291.

Carroll, M. B. and O. Dykstra (1958). Application of fractional factorials in a food research laboratory. In V. Chew (Ed.), *Experimental Designs in Industry*, pp. 224–234. New York: Wiley.

CAST Investigators (1989). Effect of encainide and flecanide on mortality in a randomized trial of arrhythmia. *New England Journal of Medicine 312*, 406–412.

Christensen, R. and L. G. Blackwood (1993). Tests for precision and accuracy of multiple measuring devices. *Technometrics 35*, 411–420.

Chu, Y. C. (1970). Comparison of *in vivo* and *vitro* inhibition of ATPases by the insecticide Chlordane. Master's thesis, University of Minnesota, St. Paul, MN.

Clatworthy, W. H. (1973). *Tables of Two-Associate Class Partially Balanced Designs*. National Bureau of Standards, Applied Mathematics Series, No. 63.

Cochran, W. G., K. M. Autrey, and C. Y. Cannon (1941). A double changeover design for dairy cattle feeding experiments. *Journal of Dairy Science 24*, 937–951.

Cochran, W. G. and G. M. Cox (1957). *Experimental Designs*. New York: Wiley.

Cole, D. N. (1993). Trampling effects on mountain vegetation in Washington, Colorado, New Hampshire, and North Carolina. Research Paper INT-464, U.S. Forest Service, Intermountain Research Station.

Connolloy, H. M., J. L. Crary, M. D. McGoon, D. D. Hensrud, B. S. Edwards, W. D. Edwards, and H. V. Schaff (1997). Valvular heart disease associated with fenfluramine-phentermine. *New England Journal of Medicine 337*, 581–588.

Connor, W. S. (1958). Experiences with incomplete block designs. In V. Chew (Ed.), *Experimental Designs in Industry*, pp. 193–206. New York: Wiley.

Conover, W. J. (1980). *Practical Nonparametric Statistics* (Second ed.). New York: Wiley.

Conover, W. J., M. E. Johnson, and M. M. Johnson (1981). A comparative study of tests for homogeneity of variances, with applications to the outer continental shelf bidding data. *Technometrics 23*, 351–361.

Cook, R. D. and C. J. Nachtsheim (1980). A comparison of algorithms for constructing exact $D$-optimal designs. *Technometrics 22*, 315–324.

Cook, R. D. and C. J. Nachtsheim (1989). Computer-aided blocking of factorial and response-surface designs. *Technometrics 31*, 339–346.

Cook, R. D. and S. Weisberg (1982). *Residuals and Influence in Regression*. London: Chapman and Hall.

Cornell, J. A. (1985). Mixture experiments. In S. Kotz and N. Johnson (Eds.), *Encyclopedia of Statistical Sciences*, Volume 5, pp. 569–579. New York: Wiley.

Cornell, J. A. (1990). *Experiments with Mixtures* (Second ed.). New York: Wiley.

Cox, D. R. (1958). *Planning of Experiments*. New York: Wiley.

Cressie, N. A. C. (1991). *Statistics for Spatial Data*. New York: Wiley.

Dale, T. B. (1992). Biological and chemical effects of acidified snowmelt on seasonal wetlands in Minnesota. Master's thesis, University of Minnesota.

Daniel, C. (1959). Use of half-normal plots in interpreting factorial two level experiments. *Technometrics 1*, 311–341.

Daniel, C. (1976). *Applications of Statistics to Industrial Experimentation*. New York: Wiley.

de Belleroche, J., A. Dick, and A. Wyrley-Birch (1982). Anticonvulsants and trifluoperazine inhibit the evoked release of GABA from cerebral cortex of rat at different sites. *Life Sciences 31*, 2875–2882.

Draper, N. R., T. P. Davis, L. Pozueta, and D. M. Grove (1994). Isolation of degrees of freedom for Box-Behnken designs. *Technometrics 36*, 283–291.

Draper, N. R., S. M. Lewis, P. W. M. John, P. Prescott, A. M. Dean, and M. G. Tuck (1993). Mixture designs for four components in orthogonal blocks. *Technometrics 35*, 268–276.

Duncan, D. B. (1955). Multiple range and multiple *F* tests. *Biometrics 11*, 1–42.

Dunnett, C. W. (1955). A multiple comparisons procedure for comparing several treatments with a control. *Journal of the American Statistical Association 50*, 1096–1121.

Dunnett, C. W. (1989). Algorithm AS 251: Multivariate normal probability integrals with product correlation structure. *Applied Statistics 38*, 564–579.

Durbin, J. (1960). Estimation of parameters in time-series regression. *Journal of the Royal Statistical Society, Series B 22*, 139–153.

Durbin, J. and G. S. Watson (1950). Testing for serial correlation in least squares regression I. *Biometrika 37*, 409–428.

Durbin, J. and G. S. Watson (1951). Testing for serial correlation in least squares regression II. *Biometrika 38*, 159–178.

Durbin, J. and G. S. Watson (1971). Testing for serial correlation in least squares regression III. *Biometrika 58*, 1–19.

Efron, B. (1979). Bootstrap methods: another look at the jackknife. *Annals of Statistics 7*, 1–26.

Efron, B. and R. J. Tibshirani (1993). *An Introduction to the Bootstrap*. London: Chapman and Hall.

Einot, I. and K. R. Gabriel (1975). A study of the powers of several methods of multiple comparisons. *Journal of the American Statistical Association 70*, 574–583.

Fairfield Smith, H. (1938). An empirical law describing heterogeneity in the yields of agricultural crops. *Journal of Agricultural Science 26*, 1–29.

Federer, W. T. and M. P. Meredith (1992). Covariance analysis for split-plot and split-block designs. *The American Statistician 46*, 155–162.

Fisher, R. A. (1918). The correlation between relatives on the supposition of Mendelian inheritance. *Transactions of the Royal Society of Edinburgh 52*, 399–433.

Fisher, R. A. (1925a). Applications of "student's" distribution. *Metron 5*, 90–104.

Fisher, R. A. (1925b). *Statistical Methods for Research Workers*. London: Oliver & Boyd, Ltd.

Fisher, R. A. (1935). *The Design of Experiments*. London: Oliver & Boyd, Ltd.

Fisher, R. A. and F. Yates (1963). *Statistical Tables for Biological, Agricultural, and Medical Research* (Sixth ed.). Edinburgh: Oliver and Boyd.

Gauss, K. F. (1809). *Theoria motus corporum coelestrium in sectionibus conicis solem ambientium*. Hamburg: Perthes and Besser.

Gayen, A. K. (1950). The distribution of the variance ratio in random samples of any size drawn from non-normal universes. *Biometrika 37*, 236–255.

Geary, R. C. (1970). Relative efficiency of count of sign changes for assessing residual autoregression in least squares regression. *Biometrika 57*, 123–127.

Giguere, V., G. Lefevre, and F. Labrie (1982). Site of calcium requirement for stimulation of ACTH release in rat anterior pituitary cells in culture by synthetic ovine corticotropin-releasing factor. *Life Sciences 31*, 3057–3062.

Greenhouse, S. W. and S. Geisser (1959). On methods in the analysis of profile data. *Psychometrika 24*, 95–112.

Grondona, M. O. and N. Cressie (1991). Using spatial considerations in the analysis of experiments. *Technometrics 33*, 381–392.

Hampel, F. R., E. M. Ronchetti, P. J. Rousseeuw, and W. A. Stahel (1986). *Robust Statistics: The Approach Based on Influence Functions*. New York: Wiley.

Hanley, J. A. and S. H. Shapiro (1994). Sexual activity and the lifespan of male fruitflies: a dataset that gets attention. *Journal of Statistics Education 2*.

Hareland, G. A. and M. A. Madson (1989). Barley dormancy and fatty acid composition of lipids isolated from freshly harvested and stored kernels. *Journal of the Institute of Brewing 95*, 437–442.

Hawkins, D. M. (1980). *Identification of Outliers*. London: Chapman and Hall.

Hayter, A. J. (1984). A proof of the conjecture that the Tukey-Kramer multiple comparisons procedure is conservative. *Annals of Statistics 12*, 61–75.

Hedayat, A. and W. D. Wallis (1978). Hadamard matrices and their applications. *Annals of Statistics 6*, 1184–1238.

Henderson, Jr., C. R. (1982). Analysis of covariance in the mixed model: Higher-level, nonhomogeneous, and random regressions. *Biometrics 38*, 623–640.

Henderson, Jr., C. R. and C. R. Henderson (1979). Analysis of covariance in mixed models with unequal subclass numbers. *Communications in Statistics A 8*, 751–788.

Herr, D. G. (1986). On the history of ANOVA in unbalanced, factorial designs: The first 30 years. *The American Statistician 40*, 265–270.

Hoaglin, D. C., F. Mosteller, and J. W. Tukey (Eds.) (1983). *Understanding Robust and Exploratory Data Analysis*. New York: Wiley.

Hoaglin, D. C., F. Mosteller, and J. W. Tukey (Eds.) (1991). *Fundamentals of Exploratory Analysis of Variance*. New York: Wiley.

Hochberg, Y. (1988). A sharper Bonferroni procedure for multiple tests of significance. *Biometrika 75*, 800–802.

Hochberg, Y. and A. C. Tamhane (1987). *Multiple Comparison Procedures*. New York: Wiley.

Hocking, R. R. (1985). *The Analysis of Linear Models*. Monterey, CA: Brooks/Cole.

Holm, S. (1979). A simple sequentially rejective multiple test procedure. *Scandanavian Journal of Statistics 6*, 65–70.

Huber, P. J. (1981). *Robust Statistics*. New York: Wiley.

Hunt, J. R. and B. J. Larson (1990). Meal protein and zinc levels interact to influence zinc retention by the rat. *Nutrition Research 10*, 697–705.

Hunt, R. (1973). A method for estimating root efficiency. *Journal of Applied Ecology 10*, 157–164.

Huynh, H. and L. S. Feldt (1970). Conditions under which mean square ratios in repeated measurements designs have exact $F$-distributions. *Journal of the American Statistical Association 65*, 1582–1589.

Huynh, H. and L. S. Feldt (1976). Estimation of the Box correction for degrees of freedom from sample data in randomized block and split-plot designs. *Journal of Educational Statistics 1*, 69–82.

Ito, P. K. (1980). Robustness of ANOVA and MANOVA test procedures. In P. R. Krishnaiah (Ed.), *Handbook of Statistics*, Volume 1, pp. 199–236. Amsterdam: North Holland.

John, J. A. and E. R. Williams (1995). *Cyclic and Computer Generated Designs*. London: Chapman and Hall.

John, P. W. M. (1961). An application of a balanced incomplete block design. *Technometrics 3*, 51–54.

John, P. W. M. (1971). *Statistical Design and Analysis of Experiments*. New York: Macmillan.

Johnson, D. E. and F. A. Graybill (1972). Analysis of two-way model with interaction and no replication. *Journal of the American Statistical Association 67*, 862–868.

Johnson, N. L. and S. Kotz (1970). *Continuous Univariate Distributions*, Volume 1. New York: Wiley.

Kempthorne, O. (1952). *The Design and Analysis of Experiments*. New York: Wiley.

Kempthorne, O. (1955). The randomization theory of experimental inference. *Journal of the American Statistical Association 50*, 946–967.

Kempthorne, O. (1966). Some aspects of experimental inference. *Journal of the American Statistical Association 61*, 11–34.

Keuls, M. (1952). The use of the studentized range in connection with an analysis of variance. *Euphytica 1*, 112–122.

Khuri, A. I. (1992). Response surface models with random block effects. *Technometrics 34*, 26–37.

Kurotori, I. S. (1966). Experiments with mixtures of components having lower bounds. *Industrial Quality Control*, May, 592–596.

Land, C. E. (1972). An evaluation of approximate confidence interval estimation methods for lognormal means. *Technometrics 14*, 145–158.

Legendre, A. M. (1806). *Nouvelles méthodes pour la détermination des orbites des comètes; avec un suppleément contenant divers perfectionnements de ces méthodes et leur application aux deux comètes de 1805*. Paris: Courcier.

Lehmann, E. L. (1959). *Testing Statistical Hypotheses*. New York: Wiley.

Lenth, R. V. (1989). Quick and easy analysis of unreplicated factorials. *Technometrics 31*, 469–473.

Lohr, S. L. (1995). Hasse diagrams in statistical consulting and teaching. *American Statistician 49*, 376–381.

Low, C. K. and A. R. Bin Mohd. Ali (1985). Experimental tapping of pine oleoresin. *The Malaysian Forester 48*, 248–253.

Lund, R. E. and J. R. Lund (1983). Algorithm AS 190: Probabilities and upper quantiles for the studentized range. *Applied Statistics 32*, 204–210.

Lynch, S. M. and J. J. Strain (1990). Effects of skimmed milk powder, whey or casein on tissue trace elements status and antioxidant enzyme activites in rats fed control and copper-deficient diets. *Nutrition Research 10*, 449–460.

Mallows, C. L. (1973). Some comments on $C_p$. *Technometrics 15*, 661–675.

Mandel, J. (1961). Non-additivity in two-way analysis of variance. *Journal of the American Statistical Association 56*, 878–888.

Mathew, T. and B. K. Sinha (1992). Exact and optimum tests in unbalanced split-plot designs under mixed and random models. *Journal of the American Statistical Association 87*, 192–200.

Mauchly, J. W. (1940). Significance test for sphericity of a normal n-variate distribution. *Annals of Mathematical Statistics 11*, 204–209.

McCall, A. L., W. R. Millington, and R. J. Wurtman (1982). Blood-brain barrier transport of caffeine: dose-related restriction of adenine transport. *Life Sciences 31*, 2709–2715.

McCullagh, P. and J. A. Nelder (1989). *Generalized Linear Models* (Second ed.). London: Chapman and Hall.

McElhoe, H. B. and W. D. Conner (1986). Remote measurement of sulfur dioxide emissions using an ultraviolet light sensitive video system. *Journal of the Air Pollution Control Association 36*, 42–47.

Michicich, M. K. (1995). Sensory characteristics, consumer acceptance, and consumption of dairy products made from designer fats. Master's thesis, University of Minnesota.

Miller, A. (1997). Strip-plot configurations of fractional factorials. *Technometrics 39*, 153–161.

Miller, Jr., R. G. (1981). *Simultaneous Statistical Inference* (Second ed.). New York: Springer-Verlag.

Moore, D. and G. McCabe (1999). *Introduction to the Practice of Statistics* (Third ed.). New York: Freeman.

Moses, L. E. (1987). Analysis of Bernoulli data from a $2^5$ design done in blocks of size four. In C. Mallows (Ed.), *Design, Data, and Analysis*, pp. 275–290. New York: Wiley.

Mosteller, F., R. E. K. Rourke, and G. B. Thomas (1970). *Probability with Statistical Applications*. Reading, MA: Addison-Wesley.

Mosteller, F. and J. W. Tukey (1977). *Data Analysis and Regression: A Second Course in Statistics*. Reading, MA: Addison-Wesley.

Myers, R. H., A. I. Khuri, and W. H. Carter, Jr. (1989). Response surface methodology: 1966-1988. *Technometrics 31*, 137–157.

Nelson, J. W. (1961). *The Nature of Wheat Lipids and Their Role in Flour Deterioration*. Ph. D. thesis, University of Minnesota, St. Paul, MN.

Nelson, P. R. (1993). Additional uses for the analysis of means and extended tables of critical values. *Technometrics 35*, 61–71.

Nelson, T. S., L. K. Kriby, and Z. B. Johnson (1990). Effect of minerals on the incidence of leg abnormalities in growing broiler chicks. *Nutrition Research 10*, 525–533.

Nelson, W. (1990). *Accelerated Testing*. New York: Wiley.

Newman, D. (1939). The distribution of the range in samples from a normal population, expressed in terms of an independent estimate of the standard deviation. *Biometrika 31*, 20–30.

Oehlert, G. W. (1992). A note on the delta method. *The American Statistician 46*, 27–29.

Oehlert, G. W. (1994). Isolating one-cell interactions. *Technometrics 36*, 403–408.

Orman, B. A. (1986). Maize germination and seedling growth at suboptimal temperatures. Master's thesis, University of Minnesota, St. Paul, MN.

Park, S. H. (1978). Selecting contrasts among parameters in Scheffé's mixture models: Screening components and model reduction. *Technometrics 20*, 273–279.

Paskova, T. and C. Meyer (1997). Low-cycle fatigue of plain and fiber-reinforced concrete. *ACI Materials Journal 94*, 273–285.

Patterson, H. D. and E. R. Williams (1976). A new class of resolvable incomplete block designs. *Biometrika 63*, 83–92.

Patterson, H. D., E. R. Williams, and E. A. Hunter (1978). Block designs for variety trials. *Journal of Agricultural Science 90*, 395–400.

Pearson, E. S. (1931). The analysis of variance in cases of non-normal variation. *Biometrika 23*, 114–133.

Pierce, D. A. (1971). Least squares estimation in the regression model with autoregressive-moving average errors. *Biometrika 58*, 299–312.

Pignatiello, J. J. and J. S. Ramberg (1985). Comments on "Off-line quality control, parameter design, and the Taguchi method," by R. N. Kackar. *Journal of Quality Technology 17*, 198–206.

Pignatiello, J. J. and J. S. Ramberg (1991). Top ten triumphs and tragedies of Genichi Taguchi. *Quality Engineering 4*, 211–225.

Pitman, E. J. G. (1937). Significance tests which may be applied to samples from any populations: I and II. *Journal of the Royal Statistical Society, Series B 4*, 119–130, 225–237.

Pitman, E. J. G. (1938). Significance tests which may be applied to samples from any populations: III. *Biometrika 29*, 322–335.

Plackett, R. L. and J. P. Burman (1946). The design of optimum multifactorial experiments. *Biometrika 33*, 305–325.

Prato, H. H. and M. A. Morris (1984). Soil remaining on fabric after laundering as evaluated by response surface methodology. *Textile Research Journal*, 637–644.

Quinlan, J. (1985). Product improvement by application of Taguchi methods. In *American Supplier Institute News* (special symposium ed.), pp. 11–16. Dearborn, MI: American Supplier Institute.

Rey, D. K. (1981). *Characterization of the Effect of Solutes on the Water-Binding and Gel Strength Properties of Carrageenan*. Ph. D. thesis, University of Minnesota, St. Paul, MN.

Rey, W. J. J. (1983). *Introduction to Robust and Quasi-Robust Statistical Methods*. New York: Sprinter-Verlag.

Richards, J. A. (1965). Effects of fertilizers and management on three promising tropical grasses in Jamaica. *Expl. Agric. 1*, 281–288.

Ripley, B. D. (1987). *Stochastic Simulation*. New York: Wiley.

Rollag, M. D. (1982). Ability of tryptophan derivatives to mimic melatonin's action upon the Syrian hamster reproductive system. *Life Sciences 31*, 2699–2707.

Ryan, T. A. (1960). Significance tests for multiple comparison of proportions, variances and other statistics. *Psychological Bulletin 57*, 318–328.

Sahrmann, H. F., G. F. Piepel, and J. A. Cornell (1987). In search of the optimum Harvey Wallbanger recipe via mixture experiment techniques. *The American Statistician 41*, 190–194.

Satterthwaite, F. E. (1946). An approximate distribution of estimates of variance components. *Biometrics 2*, 110–114.

Scheffé, H. (1953). A method for judging all contrasts in the analysis of variance. *Biometrika 40*, 87–104.

Scheffé, H. (1956). Alternative models for the analysis of variance. *Annals of Mathematical Statistics 27*, 251–271.

Scheffé, H. (1959). *The Analysis of Variance*. New York: Wiley.

Searle, S. R. (1971). Topics in variance component estimation. *Biometrics 27*, 1–76.

Searle, S. R., G. Casella, and C. E. McCulloch (1992). *Variance Components*. New York: Wiley.

Sellke, T., M. J. Bayarri, and J. O. Berger (1999). Calibration of p-values for testing precise null hypotheses. Technical report, Institute of Statistics and Decision Sciences, Duke University, Durham, NC.

Selwyn, M. R. and N. R. Hall (1984). On Bayesian methods for bioequivalence. *Biometrics 40*, 1103–1108.

Shoemaker, A. C., K.-L. Tsui, and C. F. J. Wu (1991). Economical experimentation methods for robust design. *Technometrics 33*, 415–427.

Silvey, S. D. (1980). *Optimal Design: An Introduction to the Theory for Parameter Estimation*. London: Chapman and Hall.

Simes, R. J. (1986). An improved Bonferroni procedure for multiple tests of significance. *Biometrika 73*, 751–754.

Simpson, J., A. Olsen, and J. C. Eden (1975). A Bayesian analysis of a multiplicative treatment effect in weather modification. *Technometrics 17*, 161–166.

Smith, J. R. and J. M. Beverly (1981). The use and analysis of staggered nested factorial designs. *Journal of Quality Technology 13*, 166–173.

Straus, M. A., D. B. Sugarman, and J. Giles-Sims (1997). Spanking by parents and subsequent antisocial behavior of children. *Archives of Pediatrics and Adolescents Medicine 151*, 761–767.

Student (1908). On the probable error of the mean. *Biometrika 6*, 1–25.

Sutheerawattananonda, M. (1994). Variation in physical properties and microstructure of extruded wheat flours. Master's thesis, University of Minnesota, St. Paul, MN.

Swallow, W. H. and S. R. Searle (1978). Minimum variance quadratic unbiased estimation (MIVQUE) of variance components. *Technometrics 20*, 265–272.

Swanlund, D. J., M. R. N'Diaye, K. J. Loseth, J. L. Pryor, and B. G. Crabo (1995). Diverse testicular responses to exogenous growth hormone and follicle-stimulating hormone in prepubertal boars. *Biology of Reproduction 53*, 749–757.

Taam, W. and M. Hamada (1993). Detecting spatial effects from factorial experiments: An application from integrated-circuit manufacturing. *Technometrics 35*, 149–160.

Taguchi, G. and Y. Wu (1980). *Introduction to Off-Line Quality Engineering*. Nagoya, Japan: Central Japan Quality Control Association.

Tajima, A. (1987). *Some Aspects of Preserving Chicken Semen: Glycerol Effect, Assay Method, and Application*. Ph. D. thesis, University of Minnesota, St. Paul, MN.

Thomas, L. and C. J. Krebs (1997). A review of statistical power analysis software. *Bulletin of the Ecological Society of America 78*, 128–139.

Tjahjadi, C. (1983). *Isolation and Characterization of Adzuki Bean (Vigna angularis) Protein and Starch*. Ph. D. thesis, University of Minnesota, St. Paul, MN.

Tsay, R. (1984). Regression models with time series errors. *Journal of the American Statistical Association 79*, 118–124.

Tukey, J. W. (1952). Allowances for various types of error rates. Unpublished IMS address.

Tukey, J. W. (1956). Variances of variance components: I. Balanced designs. *Annals of Mathematical Statistics 27*, 722–736.

Tukey, J. W. (1957a). On the comparative anatomy of transformations. *Annals of Mathematical Statistics 28*, 602–632.

Tukey, J. W. (1957b). Variances of variance components: II. The unbalanced single classification. *Annals of Mathematical Statistics 28*, 43–56.

Tukey, J. W. (1991). The philosophy of multiple comparisons. *Statistical Science 6*, 100–116.

US FDA (1997). FDA announces withdrawl of fenfluramine and dexfenfluramine. Press release P97-32.

Vangel, M. G. (1992). New methods for one-sided tolerance limits for a one-way balanced random-effects ANOVA model. *Technometrics 34*, 176–185.

Wedin, D. A. (1990). *Nitrogen Cycling and Competition among Grass Species*. Ph. D. thesis, University of Minnesota, Minneapolis, MN.

Weisberg, S. (1985). *Applied linear regression* (Second ed.). New York: Wiley.

Welch, B. (1996). Effects of humidity on storing big sagebrush seed. Technical Report Research Paper INT-RP-493, USDA Forest Service, Intermountain Research Station.

Welch, W. J. (1990). Construction of permutation tests. *Journal of the American Statistical Association 85*, 693–698.

Welsch, R. E. (1977). Stepwise multiple comparison procedures. *Journal of the American Statistical Association 72*, 566–575.

Westlake, W. J. (1974). The use of balanced incomplete block designs in comparative bioavailability trials. *Biometrics 30*, 319–327.

Whiting, K. R. (1990). *Host-Specific Pathogens and the Corn/Soybean Rotation Effect*. Ph. D. thesis, University of Minnesota, St. Paul, MN.

Williams, J. S. (1962). A confidence interval for variance components. *Biometrika 49*, 278–281.

Windels, H. F. (1964). *The Influence of Diet and of Duration of Fast upon Plasma Levels of Free Leucine, Isoleucine, and Valine in the Growing Pig*. Ph. D. thesis, University of Minnesota, St. Paul, MN.

Wood, T. and F. H. Bormann (1974). Effects of an artificial acid mist upon the growth of *Betula alleghaninsis* britt. *Environmental Pollution 7*, 259–268.

Xhonga, R. (1971). Direct gold alloys—part II. *Journal of the American Academy of Gold Foil Operators 14*, 5–15.

Yates, F. (1936a). Incomplete randomized blocks. *Annals of Eugenics 7*, 121–140.

Yates, F. (1936b). A new method of arranging variety trials involving a large number of varieties. *Journal of Agricultural Science 26*, 424–455.

Yates, F. (1939). The recovery of inter-block information in variety trials arranged in three dimensional lattices. *Annals of Eugenics 9*, 136–156.

Yates, F. (1940). Lattice squares. *Journal of Agricultural Science 30*, 672–687.

# Appendix A

# Linear Models for Fixed Effects

Much of our analysis has used the Analysis of Variance, and we have approached ANOVA in a classical way, with lots of sums over indices $i$, $j$, and $k$. This approach is valid, but does not give insight into why ANOVA works or where the formulae come from. This appendix is meant as a *brief* introduction and survey of the theory of linear models for fixed effects. We can achieve a great deal of simplification and unity in our analysis approach through the use of linear models. Hocking (1985) is a good book-length reference for this material.

## A.1 Models

Let $\boldsymbol{y} \in \mathcal{R}^N$ be a vector of length $N$; $\boldsymbol{y}$ contains the responses in an experiment. A *model* $\boldsymbol{M}$ is a linear subspace of $\mathcal{R}^N$. For example, in a one-factor ANOVA the hypothesis of zero treatment effects corresponds to a model in $\mathcal{R}^N$ where all the vectors in $\boldsymbol{M}$ are constant vectors: $x \in \boldsymbol{M} \leftrightarrow x = \mathbf{1}\beta$, where $\mathbf{1} = (1, 1, \ldots, 1)'$ is a vector of all ones. In a one-factor ANOVA, the hypothesis of $k$ separate treatment means corresponds to a model in $\mathcal{R}^N$ where for any $x \in \boldsymbol{M}$, the elements of $x$ corresponding to the same treatment must all be the same, but the elements corresponding to different treatments can be different. Such a model can also be described as the range of a matrix $X_{N\times k}$, where $X_{i,j}$ is 1 if the $i$th response was in the $j$th treatment group, and zero otherwise. This means that $Y \in \boldsymbol{M}$ can be written as $Y = X\beta$ for a $k$-vector $\beta$ with elements interpreted $\mu_1, \mu_2, \ldots, \mu_k$. If $k = 3$; the treatment sample sizes were 2, 3, and 5; and the units were in treatment order; then $X$ could be written

$$X = \begin{bmatrix} 1 & 0 & 0 \\ 1 & 0 & 0 \\ 0 & 1 & 0 \\ 0 & 1 & 0 \\ 0 & 1 & 0 \\ 0 & 0 & 1 \\ 0 & 0 & 1 \\ 0 & 0 & 1 \\ 0 & 0 & 1 \\ 0 & 0 & 1 \end{bmatrix}.$$

There are many other matrices that span the same space, including:

$$\text{(a)} \begin{bmatrix} 1 & 1 & 0 & 0 \\ 1 & 1 & 0 & 0 \\ 1 & 0 & 1 & 0 \\ 1 & 0 & 1 & 0 \\ 1 & 0 & 1 & 0 \\ 1 & 0 & 0 & 1 \\ 1 & 0 & 0 & 1 \\ 1 & 0 & 0 & 1 \\ 1 & 0 & 0 & 1 \\ 1 & 0 & 0 & 1 \end{bmatrix}, \quad \text{(b)} \begin{bmatrix} 1 & 0 & 0 \\ 1 & 0 & 0 \\ 1 & 1 & 0 \\ 1 & 1 & 0 \\ 1 & 1 & 0 \\ 1 & 0 & 1 \\ 1 & 0 & 1 \\ 1 & 0 & 1 \\ 1 & 0 & 1 \\ 1 & 0 & 1 \end{bmatrix},$$

$$\text{(c)} \begin{bmatrix} 1 & 1 & 0 \\ 1 & 1 & 0 \\ 1 & 0 & 1 \\ 1 & 0 & 1 \\ 1 & 0 & 1 \\ 1 & -1 & -1 \\ 1 & -1 & -1 \\ 1 & -1 & -1 \\ 1 & -1 & -1 \\ 1 & -1 & -1 \end{bmatrix}, \quad \text{and (d)} \begin{bmatrix} 1 & 1 & 0 \\ 1 & 1 & 0 \\ 1 & 0 & 1 \\ 1 & 0 & 1 \\ 1 & 0 & 1 \\ 1 & -0.4 & -0.6 \\ 1 & -0.4 & -0.6 \\ 1 & -0.4 & -0.6 \\ 1 & -0.4 & -0.6 \\ 1 & -0.4 & -0.6 \end{bmatrix}.$$

These matrices are shown because they illustrate the use of restrictions. For matrix (a), $Y \in \boldsymbol{M}$ if $Y = X\beta$, where $\beta$ is a 4-vector with elements interpreted $(\mu, \alpha_1, \alpha_2, \alpha_3)$. Recall that the separate means model is overparameterized if we don't put some kind restrictions on the $\alpha_i$'s. This is what happens with matrix (a); if we add 100 to $\mu$ and subtract 100 from the $\alpha_i$'s, we get the same $Y$. Note that matrix (a) has 4 columns but only spans a subspace of dimension 3; matrix (a) is rank deficient.

To make the parameters unique, we need some restrictions. Some statistics programs assume that $\alpha_1$ is zero and use $\mu$, $\mu + \alpha_2$, and $\mu + \alpha_3$ as the treatment means. Thus $\alpha_2$ is the difference in means between groups 2 and 1. Matrix (b) reflects this parameterization if we interpret the coefficients $\beta$ as $(\mu, \alpha_2, \alpha_3)$.

One standard set of restrictions is that the treatment effects sum to 0, or equivalently, that $\alpha_g = -\sum_{i=1}^{g-1} \alpha_i$. Thus we may replace the last $\alpha_g$ with minus the sum of the others. Matrix (c) reflects this parameterization. For matrix (c), $Y \in \boldsymbol{M}$ if $Y = X\beta$, where $\beta$ is a 3-vector with elements interpreted $(\mu, \alpha_1, \alpha_2)$. The mean in the last treatment is $\mu - \alpha_1 - \alpha_2 = \mu + \alpha_3$.

Finally, a fourth possible set of restrictions is that the weighted sum of the treatment effects is 0, or equivalently, that $\alpha_g = -\sum_{i=1}^{g-1} n_i\alpha_i/n_g$. Matrix (d) reflects this parameterization. For matrix (d), $Y \in \boldsymbol{M}$ if $Y = X\beta$, where $\beta$ is a 3-vector with elements interpreted $(\mu, \alpha_1, \alpha_2)$. The mean in the last treatment is $\mu - n_1\alpha_1/n_3 - n_2\alpha_2/n_3 = \mu + \alpha_3$. Notice that the last two columns of matrix (d) are orthogonal to the first. This orthogonality is what makes the weighted-sum restrictions easier for hand work.

We arrange models in a lattice. A *lattice* is a partially ordered set in which every pair has a union and an intersection. For a lattice of models, the intersection is the largest submodel contained in both models (the intersection of the two model subspaces), and the union is the smallest (or simplest) model containing both submodels (the subspace spanned by the two models). The role of lattices in linear models is that it is easy to compare models up and down a lattice, but difficult to compare models if one model is not a subset of the other. Here is a sample lattice for a two-factor factorial:

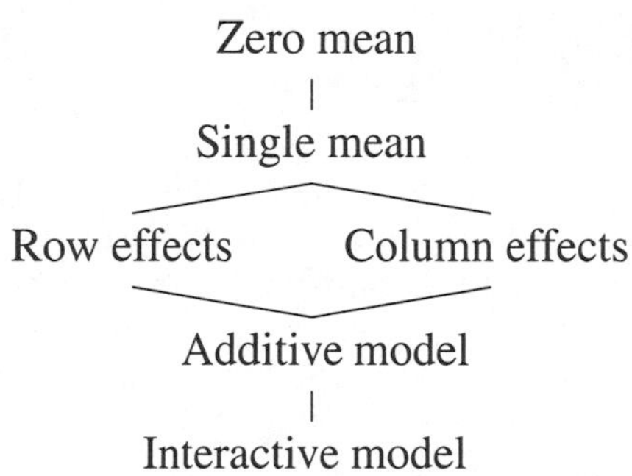


We can easily compare the "no row effects" model with the "interactive model," but it is more difficult to compare the "no row effects" model with the "no column effects" model. It should also be rather clear that lattice representations of several models and Hasse diagrams are related.

## A.2 Least Squares

Suppose that we have a model $M$ which is spanned by a matrix $X_{N\times r}$; thus $M = \mathcal{C}(X)$, where $\mathcal{C}(X)$ is the column space of $X$. We want to fit the model $M$ to the data $y \in \mathcal{R}^N$. This means we want to find the $Y \in M$ that is closest to $y$. We measure closeness by the sum of squared errors: $(y - Y)'(y - Y)$. This is the same as finding the least squares regression of $y$ on the $r$ independent variables given by the columns of $X$. The minimum occurs when

$$X'\,Xb = X'y \ ,$$

(the normal equations), or when

$$X'(y - Xb) = 0 \ .$$

The latter says that the residuals $(y - Xb)$ are orthogonal to $X$, or equivalently, to $\mathcal{C}(X)$. The observations are then decomposed into the sum of fitted values $Y$ and residuals $y - Y$. This may be formalized as a theorem.

***Theorem A.1*** *For any $y \in \mathcal{R}^N$ and any model $M = \mathcal{C}(X_{N\times r})$, there exists a unique $Y \in \mathcal{C}(X)$ such that $y - Y \perp \mathcal{C}(X)$. This $Y$ is the least squares fit of the model $M$ to $y$. $Y$ may be written as $Xb$ for* any *$b$ that solves the normal equations. If $X$ has full rank, then $b$ is unique and $b = (X'\,X)^{-1}X'y$. If $M$ is reparameterized to $M = \mathcal{C}(X^\star)$ where $\mathcal{C}(X) = \mathcal{C}(X^\star)$, then $Y$ remains the same, though the parameter estimates $b$ may change.*

Look at Figure A.1; the triangle formed by $Y_0$, $Y$, and $y$ will be a right triangle for any $Y_0$ in $\mathcal{C}(X)$, so the Pythagorean Theorem gives us the following for any $Y_0 \in \mathcal{C}(X)$:

$$(y - Y_0)'(y - Y_0) = (Y - Y_0)'(Y - Y_0) + (y - Y)'(y - Y) \ .$$

In particular, if we take $Y_0$ to be zero, this tells us that we may decompose the (uncorrected) total sum of squares in $y$ into a model sum of squares $(Y - Y_0)'(Y - Y_0)$ and a residual sum of squares $(y - Y)'(y - Y)$. If the vector $\mathbf{1}$ lies in $M$, then we may decompose the corrected total sum of squares in $y$ into a model sum of squares around the overall mean $(Y - \overline{y}\mathbf{1})'(Y - \overline{y}\mathbf{1})$ and a residual sum of squares $(y - Y)'(y - Y)$.

We may revise the usual ANOVA terminology to reflect this geometric perspective. A source of variation is a model subspace. Variation of a certain type is variation that lies in a particular subspace. The degrees of freedom for a source or model is merely the dimension of the subspace. The sum of squares for a model (source) is the squared length of the part of $y$ that

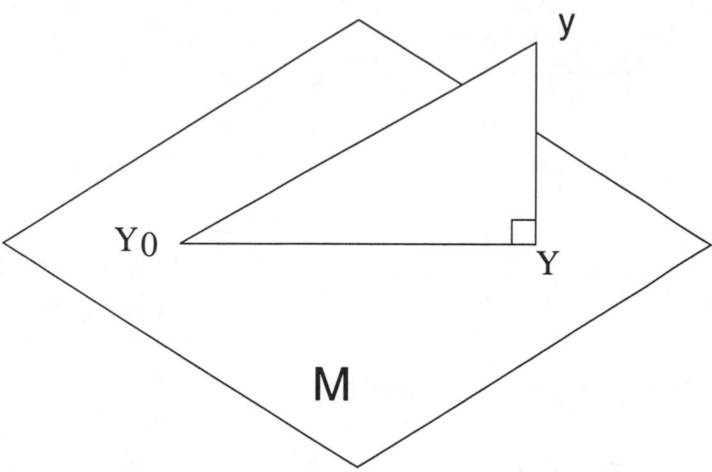


**Figure A.1:** Fitting a model.

lies in that subspace. The ANOVA table becomes (assuming that the model subspace has dimension $r$)

| Source<br>Model subspace | DF<br>Dimension of subspace | SS<br>Squared length in subspace |
|---|---|---|
| Model<br>$\boldsymbol{M}$ | $r$ | $Y'Y$ |
| Deviations<br>$\boldsymbol{M}^{\perp}$ | $N-r$ | $(\boldsymbol{y}-Y)'(\boldsymbol{y}-Y)$ |
| Total<br>$\mathcal{R}^N$ | $N$ | $\boldsymbol{y}'\boldsymbol{y}$ |

We can also construct an ANOVA table for observations corrected for the grand mean, assuming that $\mathbf{1} \in \boldsymbol{M}$, as is usually the case.

| Source Subspace | DF Dimension | SS Squared length |
|---|---|---|
| Model corrected for grand mean $\boldsymbol{M} \cap \mathbf{1}^\perp$ | $r-1$ | $(Y-\overline{y}\mathbf{1})'(Y-\overline{y}\mathbf{1})$ |
| Deviations $\boldsymbol{M}^\perp$ | $N-r$ | $(\boldsymbol{y}-Y)'(\boldsymbol{y}-Y)$ |
| Corrected total $\mathcal{R}^N \cap \mathbf{1}^\perp$ | $N-1$ | $(\boldsymbol{y}-\overline{y}\mathbf{1})'(\boldsymbol{y}-\overline{y}\mathbf{1})$ |

## A.3 Comparison of Models

Suppose that we have two models with $\boldsymbol{M}_1 \cap \boldsymbol{M}_2 = \boldsymbol{M}_1$. Thus $\boldsymbol{M}_1$ is above $\boldsymbol{M}_2$ in the model lattice. If we have $\boldsymbol{M}_1 = \mathcal{C}(X_1)$ and $\boldsymbol{M}_2 = \mathcal{C}(X_2)$, then $\boldsymbol{M}_1 \cap \boldsymbol{M}_2 = \boldsymbol{M}_1$ is equivalent to $\mathcal{C}(X_1) \subset \mathcal{C}(X_2)$. Let $\mathcal{C}(X_1)$ have dimension $r_1$, and let $\mathcal{C}(X_2)$ have dimension $r_2$. $Y_1$ is the fit of $\boldsymbol{M}_1$ to $\boldsymbol{y}$, and $Y_2$ is the fit of $\boldsymbol{M}_2$ to $\boldsymbol{y}$.

Look at Figure A.2. Not only is $Y_1$ the fit of $\boldsymbol{M}_1$ to $\boldsymbol{y}$, $Y_1$ is the fit of $\boldsymbol{M}_1$ to $Y_2$. We have right triangles everywhere we look.

| Right angle | Right triangle |
|---|---|
| $(\boldsymbol{y}-Y_2) \perp \boldsymbol{M}_2$ | $(0, Y_2, \boldsymbol{y})$ |
| $(\boldsymbol{y}-Y_1) \perp \boldsymbol{M}_1$ | $(0, Y_1, \boldsymbol{y})$ |
| $(Y_2-Y_1) \perp \boldsymbol{M}_1$ | $(0, Y_1, Y_2)$ |

Using these right triangles and the Pythagorean Theorem, we can make a variety of squared-length decompositions.

$$\boldsymbol{y}'\boldsymbol{y} = Y_2'Y_2 + (\boldsymbol{y}-Y_2)'(\boldsymbol{y}-Y_2)$$

$$\boldsymbol{y}'\boldsymbol{y} = Y_1'Y_1 + (\boldsymbol{y}-Y_1)'(\boldsymbol{y}-Y_1)$$

$$Y_2'Y_2 = Y_1'Y_1 + (Y_2-Y_1)'(Y_2-Y_1)$$

$$\boldsymbol{y}'\boldsymbol{y} = Y_1'Y_1 + (Y_2-Y_1)'(Y_2-Y_1) + (\boldsymbol{y}-Y_2)'(\boldsymbol{y}-Y_2)$$

$$(\boldsymbol{y}-Y_1)'(\boldsymbol{y}-Y_1) = (Y_2-Y_1)'(Y_2-Y_1) + (\boldsymbol{y}-Y_2)'(\boldsymbol{y}-Y_2)$$

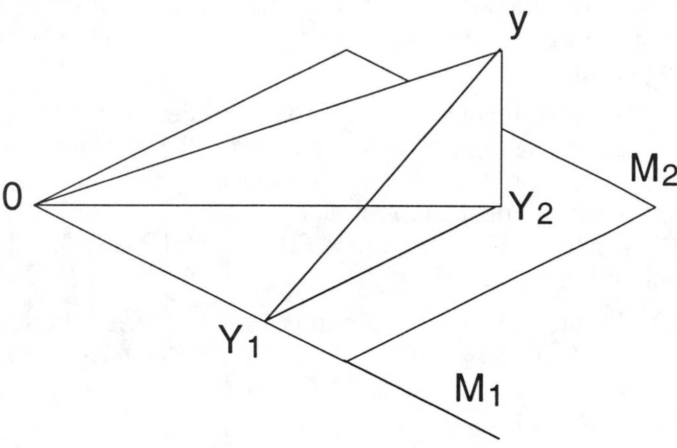


**Figure A.2:** Comparing two model fits.

In an Analysis of Variance, these squared-length decompositions are usually arranged as follows:

| Source<br>Subspace | DF<br>Dimension | SS<br>Squared length |
|---|---|---|
| Model 1<br>$\boldsymbol{M}_1$ | $r_1$ | $Y_1' Y_1$ |
| Improvement of model 2 over model 1<br>$\boldsymbol{M}_2 \cap \boldsymbol{M}_1^{\perp}$ | $r_2 - r_1$ | $(Y_2 - Y_1)'(Y_2 - Y_1)$ |
| Deviations<br>$\boldsymbol{M}_2^{\perp}$ | $N - r_2$ | $(\boldsymbol{y} - Y_2)'(\boldsymbol{y} - Y_2)$ |
| Total<br>$\mathcal{R}^N$ | N | $\boldsymbol{y}'\boldsymbol{y}$ |

For example, consider model 1 to be the model of common means, $\boldsymbol{M}_1 = \mathcal{C}(\mathbf{1})$, and model 2 to be the model of separate treatment means in a one-factor ANOVA. Then $\boldsymbol{M}_1 \subset \boldsymbol{M}_2$, because the separate treatment means could all be equal. We have $r_1 = 1$, and $r_2 = g$; thus the improvement in going from model 1 to model 2 is a $g-1$ dimensional improvement. In the ANOVA, model 1 is usually called the constant or grand mean, and the improvement sum of squares going from model 1 to model 2 is called the between treatments sum of squares.

The parameterization in matrix (d) above is easier for hand work. It arises when we want to compute the sum of squares for the improvement of model 2 ($g$ group means) over model 1 (common mean). This is the sum of squares for the orthogonal complement of model 1 in model 2. However, for matrix (d), the orthogonal complement of model 1 in model 2 is spanned by the last two columns of matrix (d). The orthogonality is built in.

We can, of course, extend model comparison to a series of three (or more) nested models: $\boldsymbol{M}_1 \subset \boldsymbol{M}_2 \subset \boldsymbol{M}_3$. This gives an ANOVA table as follows:

| Source<br>Subspace | DF<br>Dimension | SS<br>Squared length |
|---|---|---|
| Model 1<br>$\boldsymbol{M}_1$ | $r_1$ | $Y_1'Y_1$ |
| Improvement of model 2<br>over model 1<br>$\boldsymbol{M}_2 \cap \boldsymbol{M}_1^\perp$ | $r_2 - r_1$ | $(Y_2 - Y_1)'(Y_2 - Y_1)$ |
| Improvement of model 3<br>over model 2<br>$\boldsymbol{M}_3 \cap \boldsymbol{M}_2^\perp$ | $r_3 - r_2$ | $(Y_3 - Y_2)'(Y_3 - Y_2)$ |
| Deviations<br>$\boldsymbol{M}_3^\perp$ | $N - r_3$ | $(y - Y_3)'(y - Y_3)$ |
| Total<br>$\mathcal{R}^N$ | $N$ | $y'y$ |

## A.4 Projections

The *sum* of two subspaces $U_1$ and $U_2$ of a vector space $V$ is $U_1 + U_2 = \{u_1 + u_2 : u_1 \in U_1, u_2 \in U_2\}$; $U_1 + U_2$ is also a subspace of $V$. If $U_1 \cap U_2 = \{0\}$, the sum is called *direct* and is written $U_1 \dot{+} U_2$. If $V$ is the direct sum of $U_1$ and $U_2$, then $v \in V$ may be written uniquely as $v = u_1 + u_2$, where $u_1 \in U_1$ and $u_2 \in U_2$.

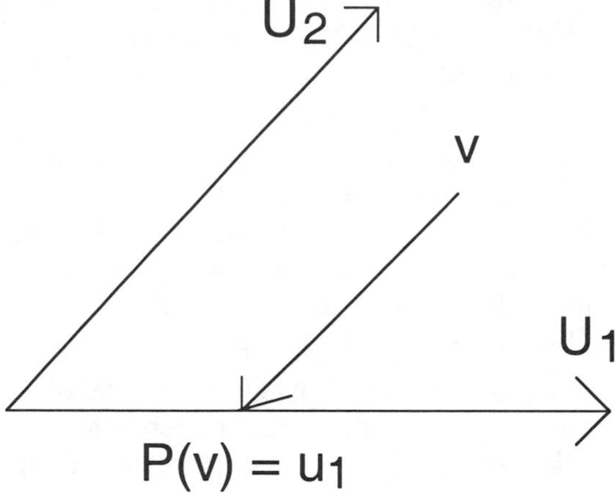


**Figure A.3:** Projection onto $U_1$ parallel to $U_2$.

If $V$ is the direct sum of $U_1$ and $U_2$ with $v \in V$ written as $v = u_1 + u_2$ ($u_1 \in U_1$, $u_2 \in U_2$), then the *projection of* $V$ *onto* $U_1$ *parallel to* $U_2$ is the linear map $P : V \rightarrow U_1$ given by $P(v) = u_1$. See Figure A.3. A linear mapping is a projection if and only if $P^2 = P$.

If two subspaces are orthogonal ($U_1 \perp U_2$), we write their direct sum as $U_1 \oplus U_2$ to emphasize their orthogonality. If $V = U_1 \oplus U_2$, then the projection of $V$ onto $U_1$ is called an *orthogonal projection.*

Suppose we have a space $V = U_1 \oplus U_2$, with $P_i$ being the orthogonal projection onto $U_i$. Then $P_1 P_2 = 0$. (Figure out why!) Furthermore, we have that since $v = u_1 + u_2$, then $v = P_1 v + P_2 v$, so that $(I - P_1) = P_2$.

Linear maps from $\mathcal{R}^N$ to $\mathcal{R}^N$ can be written as $N$ by $N$ matrices. Thus, we can express projections in $\mathcal{R}^N$ as matrices. The $N$ by $N$ matrix $P$ is an orthogonal projection onto $U \in \mathcal{R}^N$ if and only if $P$ is symmetric, idempotent (that is, $P^2 = P$), and $\mathcal{C}(P) = U$. If $U = \mathcal{C}(X)$ and $X$ has full rank, then $P = X(X'X)^{-1}X'$.

What does all this have to do with linear models? If $\boldsymbol{M}$ is a model and $P$ is the orthogonal projection onto $\boldsymbol{M}$, then the fitted values for fitting $\boldsymbol{M}$ to $\boldsymbol{y}$ are $P\boldsymbol{y}$. Least-squares fitting of models to data is simply the use of the orthogonal projection onto the model subspace.

Suppose we have two models $\boldsymbol{M}_1$ and $\boldsymbol{M}_2$, along with their union $\boldsymbol{M}_{12} = \boldsymbol{M}_1 \dot{+} \boldsymbol{M}_2$. When does the sum of squares for $\boldsymbol{M}_{12}$ equal the sum of squares for $\boldsymbol{M}_1$ plus the sum of squares for $\boldsymbol{M}_2$? By Pythagorean Theorem, the sum of squares for $\boldsymbol{M}_{12}$ is the sum of the sum of squares for $\boldsymbol{M}_1$ and the sum of squares for $\boldsymbol{M}_{12} \cap \boldsymbol{M}_1^{\perp}$. This second model is $\boldsymbol{M}_2$ if and only if model 2 is orthogonal to model 1, so the sums of squares add up if and only if the two original models are orthogonal.

How do we use this in ANOVA? We will have sums of squares that add up properly if we break $\mathcal{R}^N$ up into orthogonal subspaces. Our model lattices are hierarchical, with higher models including lower models. Thus to get orthogonal subspaces, we must look at the orthogonal complement of the smaller subspace in the larger subspace. This is the improvement in going from the smaller subspace to the larger subspace.

In the usual two-factor balanced ANOVA, the model of separate column means ($\boldsymbol{M}_C$) is not orthogonal to the model of separate row means ($\boldsymbol{M}_R$); these models have the constant-mean model as intersection. However, the model "improvement going from constant mean to separate column means" ($\boldsymbol{M}_C \cap \mathbf{1}^{\perp}$) is orthogonal to the model "improvement going from constant mean to separate row means" ($\boldsymbol{M}_R \cap \mathbf{1}^{\perp}$). This orthogonality is not present in the general unbalanced case.

When we have two nonorthogonal models, we will get different sums of squares if we decompose $\boldsymbol{M}_{12}$ as $\boldsymbol{M}_1 \oplus \boldsymbol{M}_{12} \cap \boldsymbol{M}_1^{\perp}$ or $\boldsymbol{M}_2 \oplus \boldsymbol{M}_{12} \cap \boldsymbol{M}_2^{\perp}$. The first corresponds to fitting model 1, and then getting the improvement going to $\boldsymbol{M}_{12}$, and the second corresponds to fitting model 2, and then getting the improvement going to $\boldsymbol{M}_{12}$. These have different projections in different orders. See Figure A.4. These changing subspaces are why sequential sums of squares (Type I) depend on order. Thus the sum of squares for B will not equal the sum of squares for B after A unless B and A represent orthogonal subspaces. The same applies for A and A after B.

## A.5 Random Variation

So far, the linear models computations have not included any random variation, but we add that in. Our observations $\boldsymbol{y} \in \mathcal{R}^N$ will have a normal distribution with mean $\boldsymbol{\mu}$ and variance matrix $\Sigma$. The mean $\boldsymbol{\mu}$ will lie in some model $\boldsymbol{M}$. We usually assume that $\Sigma = \sigma^2 I$, where $I$ is the $N$ by $N$ identity matrix. If $\boldsymbol{y}$ has the above distribution, then $C\boldsymbol{y}$ (where $C$ is a $p$ by $N$ matrix of constants) has a normal distribution with mean $C\mu$ and variance matrix $C\Sigma C'$.

Let's assume that $\boldsymbol{y} \sim N(\boldsymbol{\mu}, \sigma^2 I)$, where $\boldsymbol{\mu} \in \boldsymbol{M}$, and $\boldsymbol{M} = \mathcal{C}(X)$ has dimension $r$. We can thus find a $\beta$ (possibly infinitely many $\beta$'s) such that $\boldsymbol{\mu} = X\beta$. Let $P$ be the orthogonal projection onto $\boldsymbol{M}$; $(I - P)$ is thus the

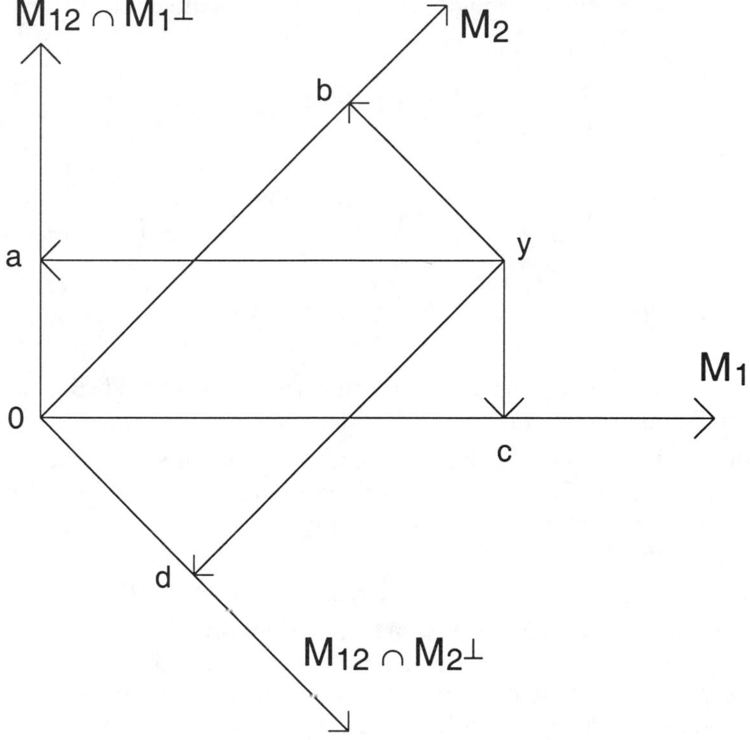


**Figure A.4:** Projecting in different orders.

orthogonal projection onto $\boldsymbol{M}^{\perp}$. The fitted values Y have the distribution

$$\begin{aligned} Y &= P\boldsymbol{y} \sim N(P\boldsymbol{\mu},\, \sigma^2 PP') \\ &= N(\boldsymbol{\mu},\, \sigma^2 P) \\ &= N(X\beta,\, \sigma^2 P) \ . \end{aligned}$$

The residuals have the distribution

$$\begin{aligned} \boldsymbol{y} - Y &= (I-P)\boldsymbol{y} \sim N((I-P)\mu,\, \sigma^2 (I-P)(I-P)') \\ &= N(0,\, \sigma^2 (I-P)) \ . \end{aligned}$$

These derivations give us the distributions of the fitted values and the residuals: they are both normal. However, we need to know their joint distribution. To discover this, we use a little trick and look at two copies of $\boldsymbol{y}$

just stacked into a vector of length $2N$, and we do separate projections on the two copies.

$$\begin{pmatrix} y \\ y \end{pmatrix} \sim N\left(\begin{pmatrix} \mu \\ \mu \end{pmatrix}, \sigma^2\begin{pmatrix} I & I \\ I & I \end{pmatrix}\right)$$

$$\begin{aligned}\begin{pmatrix} P & 0 \\ 0 & (I-P) \end{pmatrix}\begin{pmatrix} y \\ y \end{pmatrix} &\sim N\left(\begin{pmatrix} \mu \\ 0 \end{pmatrix}, \sigma^2\begin{pmatrix} P & P-P^2 \\ P-P^2 & I-P \end{pmatrix}\right) \\ &\sim N\left(\begin{pmatrix} \mu \\ 0 \end{pmatrix}, \sigma^2\begin{pmatrix} P & 0 \\ 0 & I-P \end{pmatrix}\right)\end{aligned}$$

This shows that the residuals and fitted values are uncorrelated. Because they are normally distributed, they are also independent.

How are the sums of squares distributed? Sums of squares are squared lengths, or quadratic forms, of normally distributed vectors. Normal vectors are easier to work with if they have a diagonal variance matrix, so let's work towards a diagonal variance matrix.

Let $H_1$ ($N$ by $r$) be an orthonormal basis for $\boldsymbol{M}$; then $H_1'H_1$ is the $r$ by $r$ identity matrix. Let $H_2$ ($N$ by $N-r$) be an orthonormal basis for $\boldsymbol{M}^{\perp}$; then $H_2'H_2$ is the $N-r$ by $N-r$ identity matrix. Furthermore, both $H_1'H_2$ and $H_2'H_1$ are 0. (The two matrices have columns that are bases for orthogonal subspaces; their columns must be orthogonal.) Now let $H$ be the $N$ by $N$ matrix formed by joining $H_1$ and $H_2$ by $H = (H_1 : H_2)$. $H$ is an orthogonal matrix, meaning that $H'H = HH' = I$.

The squared length of $z$ and $H'z$ is the same for any $z \in \mathcal{R}^N$, because

$$z'z = z'Iz = zHH'z = (H'z)'(H'z)$$

So for sums of squares calculations, we may premultiply by $H'$ before taking the squared length without changing the value or distribution.

Let's look at the residual sum of squares by looking at $H'(I-P)\boldsymbol{y}$.

$$\begin{aligned}H'(I-P)\boldsymbol{y} &\sim N\left(\begin{pmatrix} H_1' \\ H_2' \end{pmatrix}(I-P)\boldsymbol{\mu}, \sigma^2\begin{pmatrix} H_1' \\ H_2' \end{pmatrix}(I-P)(H_1, H_2)\right) \\ &\sim N\left(\begin{pmatrix} 0 \\ 0 \end{pmatrix}, \sigma^2\begin{pmatrix} H_1' \\ H_2' \end{pmatrix}(0, H_2)\right) \\ &\sim N\left(\begin{pmatrix} 0 \\ 0 \end{pmatrix}, \sigma^2\begin{pmatrix} 0 & 0 \\ 0 & I_{N-r} \end{pmatrix}\right)\end{aligned}$$

Thus the distribution of the sum of squared residuals is the same as the distribution of the sum of $N-r$ independent normals with mean 0 and variance $\sigma^2$. This is, of course, $\sigma^2$ times a chi-square distribution with $N-r$ degrees of freedom. The expected sum of squared errors is just $(N-r)\sigma^2$.

What about the model sum of squares? Look at $H'P\boldsymbol{y}$.

$$
\begin{aligned}
H'P\boldsymbol{y} &\sim N\left(\begin{pmatrix} H_1' \\ H_2' \end{pmatrix} P\boldsymbol{\mu},\ \sigma^2 \begin{pmatrix} H_1' \\ H_2' \end{pmatrix} P(H_1, H_2)\right) \\
&\sim N\left(\begin{pmatrix} H_1'\boldsymbol{\mu} \\ 0 \end{pmatrix},\ \sigma^2 \begin{pmatrix} H_1' \\ H_2' \end{pmatrix} (H_1, 0)\right) \\
&\sim N\left(\begin{pmatrix} H_1'\boldsymbol{\mu} \\ 0 \end{pmatrix},\ \sigma^2 \begin{pmatrix} I_r & 0 \\ 0 & 0 \end{pmatrix}\right)
\end{aligned}
$$

Thus the distribution of the model sum of squares is $\sigma^2$ times a noncentral chi-square with noncentrality parameter $\boldsymbol{\mu}'H_1H_1'\boldsymbol{\mu}/\sigma^2$ and $r$ degrees of freedom. The noncentrality parameter $\boldsymbol{\mu}'H_1H_1'\boldsymbol{\mu}/\sigma^2$ also equals $\boldsymbol{\mu}'\boldsymbol{\mu}/\sigma^2$, so the expected model sum of squares is $\boldsymbol{\mu}'\boldsymbol{\mu}+r\sigma^2$. We may test the null hypothesis $H_0 : \boldsymbol{\mu} = 0$ against the alternative $H_a : \boldsymbol{\mu} \neq 0$ by taking the ratio of the model mean square to the error mean square; this ratio has an F-distribution under the null hypothesis and a noncentral F-distribution under the alternative.

We can generalize these distributional results to a sequence of models. Consider models $\boldsymbol{M}_1 = \mathcal{C}(X_1)$ and $\boldsymbol{M}_2 = \mathcal{C}(X_2)$ with $\boldsymbol{M}_1 \subset \boldsymbol{M}_2$. Let $P_1$ and $P_2$ be the orthogonal projections onto $\boldsymbol{M}_1$ and $\boldsymbol{M}_2$. As usual, $\boldsymbol{\mu} \in \boldsymbol{M}_2$ is the expected value of $\boldsymbol{y}$; decompose $\boldsymbol{\mu}$ into $P_1\boldsymbol{\mu}$ and $(P_2 - P_1)\boldsymbol{\mu}$. These are the parts of the mean that lie in $\boldsymbol{M}_1$ and that are orthogonal to $\boldsymbol{M}_1$. Work with a pile of three copies of $\boldsymbol{y}$.

$$
\begin{bmatrix} \boldsymbol{y} \\ \boldsymbol{y} \\ \boldsymbol{y} \end{bmatrix} \sim N\left(\begin{bmatrix} \boldsymbol{\mu} \\ \boldsymbol{\mu} \\ \boldsymbol{\mu} \end{bmatrix},\ \sigma^2 \begin{bmatrix} I & I & I \\ I & I & I \\ I & I & I \end{bmatrix}\right)
$$

$$
\begin{bmatrix} P_1 & 0 & 0 \\ 0 & P_2 - P_1 & 0 \\ 0 & 0 & I - P_2 \end{bmatrix} \begin{bmatrix} \boldsymbol{y} \\ \boldsymbol{y} \\ \boldsymbol{y} \end{bmatrix} \sim N\left(\begin{bmatrix} P_1\boldsymbol{\mu} \\ (P_2 - P_1)\boldsymbol{\mu} \\ 0 \end{bmatrix},\right.
$$

$$
\left.\sigma^2 \begin{bmatrix} P_1 & 0 & 0 \\ 0 & P_2 - P_1 & 0 \\ 0 & 0 & I - P_2 \end{bmatrix}\right)
$$

Thus the fitted values $Y_1$, the difference in fitted values between the two models $Y_2 - Y_1$, and the residuals are all independent. The sum of squares for error is a multiple of chi-square with $N - r_2$ degrees of freedom. The improvement sum of squares going from the smaller to the larger model is a multiple of a chi-square with $r_2 - r_1$ degrees of freedom if the null is true ($(P_2 - P_1)\boldsymbol{\mu} = 0$); otherwise it is a multiple of a noncentral chi-square.

## A.6 Estimable Functions

Assume that $\boldsymbol{y} = \boldsymbol{\mu} + \epsilon$, where $\boldsymbol{\mu} \in \boldsymbol{M} = \mathcal{C}(X)$ and $\epsilon \sim N(0, \sigma^2 I)$. Since $\boldsymbol{\mu} \in \mathcal{C}(X)$, we have that $\boldsymbol{\mu} = X\beta$ for some $\beta$. Let $Y = Xb$ be the projection of $\boldsymbol{y}$ onto $\boldsymbol{M}$.

A linear combination of the $\beta$'s given by $h'\beta$ is *estimable* if there exists a vector $\boldsymbol{t} \in \mathcal{R}^N$ such that

$$E(\boldsymbol{t}'\boldsymbol{y}) = h'\beta,$$

for all values of $\beta$. Note that estimability is defined in terms of a particular set of parameters, so estimability depends on the basis $X$, not just the model space $\boldsymbol{M}$. For $h'\beta$ to be estimable, we must have

$$h'\beta = E(\boldsymbol{t}'\boldsymbol{y}) = \boldsymbol{t}'E(\boldsymbol{y}) = \boldsymbol{t}'X\beta$$

for all $\beta$, so that

$$h = X'\boldsymbol{t} \ .$$

Thus $h'\beta$ is estimable if and only if $h = X'\boldsymbol{t}$, or in other words, if $h$ is a linear combination of the rows of $X$.

We estimate $h'\beta$ by $h'b$, where $b$ is any solution of the normal equations. There may be many solutions to the normal equations; is $h'b$ unique? Yes, it is unique because

$$h'b = \boldsymbol{t}'Xb = \boldsymbol{t}'Y \ ,$$

so the estimable function only depends on the fitted value $Y$. Note that $\boldsymbol{t}'y$ has the same expectation as $h'b$, but we will see below that $\boldsymbol{t}'y$ can have a larger variance.

What are the mean and variance of an estimable function? Let $\boldsymbol{t}^\star$ be the projection of $\boldsymbol{t}$ onto $\boldsymbol{M}$, and let $\boldsymbol{t} = \boldsymbol{t}^\star + \boldsymbol{t}_r$. Then

$$\begin{aligned} E(h'b) &= E(\boldsymbol{t}'\boldsymbol{y}) \\ &= E(\boldsymbol{t}^{\star\prime}\boldsymbol{y} + \boldsymbol{t}_r'\boldsymbol{y}) \end{aligned}$$

$$
\begin{aligned}
&= \boldsymbol{t}^{\star\prime} X\beta + \boldsymbol{t}_r' X\beta \\
&= \boldsymbol{t}^{\star\prime} X\beta + 0\beta \\
&= \boldsymbol{t}^{\star\prime} X\beta
\end{aligned}
$$

So the expected value of $\boldsymbol{t}'\boldsymbol{y}$ only depends on the part of $\boldsymbol{t}$ that lies in $\boldsymbol{M}$. Variance is a bit trickier. If we directly attack $h'b$ we get

$$\text{Var}(h'b) = \text{Var}(\boldsymbol{t}'Y) = \sigma^2 \boldsymbol{t}' P \boldsymbol{t} = \sigma^2 \boldsymbol{t}^{\star\prime} \boldsymbol{t}^\star \ .$$

On the other hand, if we look at $\boldsymbol{t}'\boldsymbol{y}$, we find

$$\text{Var}(\boldsymbol{t}'\boldsymbol{y}) = \sigma^2 \boldsymbol{t}'\boldsymbol{t} = \sigma^2 (\boldsymbol{t}^{\star\prime}\boldsymbol{t}^\star + \boldsymbol{t}_r'\boldsymbol{t}_r) \ .$$

In the second version we only get minimum variance if $\boldsymbol{t}_r$ is 0. Because $\boldsymbol{t}_r$ does not affect expected value, we may restrict our attention to $\boldsymbol{t}$'s that lie entirely in $\boldsymbol{M}$; these will give us minimum variance no matter which way we use them.

Consider a one-factor model with $g$ treatments, parameterized by $\mu$ and $\alpha_i$, for $i = 1, 2, \ldots, g$. The $i$th treatment group has $n_i$ observations and mean $\mu + \alpha_i$. The $X$ matrix looks like

$$
\begin{array}{ccccccc}
1 & 1 & & 0 & & & 0 \\
1 & 1 & & 0 & & & 0 \\
\vdots & \vdots & \Bigg\} n_1 & \vdots & & \cdots & \vdots \\
1 & 1 & & 0 & & & 0 \\
1 & 0 & & 1 & & & 0 \\
1 & 0 & & 1 & & & 0 \\
\vdots & \vdots & & \vdots & \Bigg\} n_2 & \cdots & \vdots \\
1 & 0 & & 1 & & & 0 \\
\vdots & \vdots & & \vdots & & \cdots & \vdots \\
1 & 0 & & 0 & & & 1 \\
1 & 0 & & 0 & & & 1 \\
\vdots & \vdots & & \vdots & & \cdots & \vdots \ \Bigg\} n_g \\
1 & 0 & & 0 & & & 1
\end{array}
$$

For an estimable function given by a vector $\boldsymbol{t} \in \boldsymbol{M}$, the first $n_1$ elements of $\boldsymbol{t}$ are the same, the next $n_2$ are the same, and so on. Call these $g$ unique values $s_1, s_2, \cdots, s_a$. An estimable $h$ is of the form $h = X'\boldsymbol{t}$, and with this $X$, $X'\boldsymbol{t}$ leads to

$$\begin{aligned} h_\mu &= \sum_{i=1}^{g} n_i s_i \\ h_{\alpha_1} &= n_1 s_1 \\ h_{\alpha_2} &= n_2 s_2 \\ &\vdots \\ h_{\alpha_g} &= n_g s_g \end{aligned}$$

Thus for $h'\beta$ to be estimable, we only need to have that

$$h_\mu = h_{\alpha_1} + h_{\alpha_2} + \cdots + h_{\alpha_g} \quad .$$

## A.7 Contrasts

An estimable function $h'\beta$ for which the associated $\boldsymbol{t} \in \boldsymbol{M}$ satisfies $\boldsymbol{t}'\mathbf{1} = 0$ is called a *contrast*. A contrast thus describes a direction $\boldsymbol{t} \in \boldsymbol{M}$ that is orthogonal to the grand mean. For the one-factor ANOVA problem, an estimable function is a contrast if

$$0 = h_\mu = \sum_{i=0}^{a} n_i s_i = \sum_{i=0}^{a} h_{\alpha_i} \quad .$$

For contrasts, the overall mean must have a 0 coefficient, so we usually don't bother with a coefficient for $\mu$ at all, and denote the $h_{\alpha_i}$ by $w_i$.

Two contrasts are *orthogonal* if their corresponding $\boldsymbol{t}$ vectors are orthogonal:

$$\boldsymbol{t} \perp \boldsymbol{t}^\star \Leftrightarrow 0 = \sum_{i=0}^{n} \boldsymbol{t}_i \boldsymbol{t}_i^\star = \sum_{i=0}^{a} n_i s_i s_i^\star = \sum_{i=0}^{a} \frac{w_i w_i^\star}{n_i} \quad .$$

$\boldsymbol{M}$ has $r$ dimensions, so $\boldsymbol{M} \cap \mathbf{1}^\perp$ has $r - 1$ dimensions. All contrasts lie in $\boldsymbol{M} \cap \mathbf{1}^\perp$, so we can have at most $r - 1$ mutually orthogonal contrasts in a collection. These contrasts form an orthogonal basis for $\boldsymbol{M} \cap \mathbf{1}^\perp$, and of course there are many such bases.

Every contrast determines a model $\mathcal{C}(\boldsymbol{t})$, and we may compute a sum of squares for this model via

$$SS(\boldsymbol{t}) = \frac{(\boldsymbol{t}'Y)^2}{\boldsymbol{t}'\boldsymbol{t}} \quad .$$

We may do F-tests on this sum of squares exactly as we would on any model sum of squares. For a complete set of orthogonal contrasts $\boldsymbol{t}_{(k)}$, we have

$$\boldsymbol{M} \cap \mathbf{1}^{\perp} = \mathcal{C}(\boldsymbol{t}_{(1)}) \oplus \mathcal{C}(\boldsymbol{t}_{(2)}) \oplus \cdots \oplus \mathcal{C}(\boldsymbol{t}_{(r-1)})$$

so that

$$SS(\boldsymbol{M} \cap \mathbf{1}^{\perp}) = SS(\boldsymbol{t}_{(1)}) + SS(\boldsymbol{t}_{(2)}) + \cdots + SS(\boldsymbol{t}_{(r-1)}) \ .$$

Alternatively, $\boldsymbol{t}'\boldsymbol{y} = h'b \sim N(h'\beta, \sigma^2 \boldsymbol{t}'\boldsymbol{t})$, so we may use $t$-style inference with the error mean square estimating $\sigma^2$. If $\boldsymbol{t}'\boldsymbol{t}^\star = 0$, then $\boldsymbol{t}'\boldsymbol{y}$ and $\boldsymbol{t}^{\star\prime}\boldsymbol{y}$ are independent.

## A.8 The Scheffé Method

How large can the sum of squares for a contrast be? The sum of squares for a contrast is the sum of squares for $\mathcal{C}(\boldsymbol{t})$, the model subspace spanned by the contrast. All contrast subspaces lie in $\boldsymbol{M} \cap \mathbf{1}^{\perp}$, so we can make the decomposition

$$SS(\boldsymbol{M} \cap \mathbf{1}^{\perp}) = SS(\boldsymbol{t}) + SS(\boldsymbol{M} \cap \mathbf{1}^{\perp} \cap \boldsymbol{t}^{\perp}) \ .$$

Thus the maximum that $SS(\boldsymbol{t})$ could possibly be is $SS(\boldsymbol{M} \cap \mathbf{1}^{\perp})$, which equals $(Y - \overline{Y}\mathbf{1})'(Y - \overline{Y}\mathbf{1})$. We can achieve this maximum by taking $\boldsymbol{t} = (Y - \overline{Y}\mathbf{1})$:

$$\begin{aligned} \frac{(\boldsymbol{t}'Y)^2}{\boldsymbol{t}'\boldsymbol{t}} &= \frac{((Y - \overline{Y}\mathbf{1})'Y)^2}{(Y - \overline{Y}\mathbf{1})'(Y - \overline{Y}\mathbf{1})} \\ &= \frac{((Y - \overline{Y}\mathbf{1})'(Y - \overline{Y}\mathbf{1}))^2}{(Y - \overline{Y}\mathbf{1})'(Y - \overline{Y}\mathbf{1})} \\ &= (Y - \overline{Y}\mathbf{1})'(Y - \overline{Y}\mathbf{1}) \ . \end{aligned}$$

In a one-factor ANOVA, the maximum sum of squares for a contrast is the between groups sum of squares. Under the null hypothesis of no treatment differences, this sum of squares is distributed as $\sigma^2$ times a chi-square with $g - 1$ degrees of freedom. We do inference by comparing the F-ratio to the F distribution. Notice, however, that the maximal contrast sum of squares is equal to the treatment sum of squares. Thus we can do inference on arbitrarily many contrasts by treating them as if they were the maximal contrast. This is the basis for the Scheffé method of multiple comparisons.

## A.9 Problems

**Question A.1** Let $y$ be an $N$ by 1 random vector with $E\,y = X\beta$, and $Var(y) = \sigma^2 I_N$, where $X$ is $N$ by $p$ and $\beta$ is $p$ by 1. Let $Y = Py$, where $P$ is a projection (not necessarily orthogonal) onto the range of $X$. (a) Find the mean and (co)variance of $Y$ and $y - Y$. (b) Prove that Cov($Y, y - Y$) is 0 if and only if $P$ is an orthogonal projection.

**Question A.2** Let $y = X\beta + \epsilon$, where $\epsilon$ is $iid$ $N(0, \sigma^2)$; $y$ is $N$ by 1, $X$ is $N$ by $p$, and $\beta$ is $p$ by 1. Let $g$ be any $N$ by 1 vector. What is the distribution of $(g'y)^2$? What, if anything, changes when $g'X$ is zero?

**Question A.3** Consider a linear model $M = C(X)$ with parameters $\mu$, $\beta_1$, $\beta_2$, and $\beta_3$, where X is as follows:

$$\begin{matrix} 1 & 1 & 0 & 0 \\ 1 & 1 & 0 & 0 \\ 1 & 0 & 1 & 0 \\ 1 & 0 & 1 & 0 \\ 1 & 0 & 0 & 1 \\ 1 & 0 & 0 & 1 \end{matrix}$$

Which of the following are estimable (give a brief reason): (a) $\mu$, (b) $\beta_1$, (c) $\beta_2 - \beta_3$, (d) $\mu + (\beta_1 + \beta_2 + \beta_3)/3$, (e) $\beta_1 + \beta_2 - \beta_3$.

**Question A.4** Consider a two by three factorial with proportional balance: $n_{ij} = n_{i\bullet} n_{\bullet j} / n_{\bullet\bullet}$. Show that contrasts in factor A are orthogonal to contrasts in factor B.

**Question A.5** Consider the following $X$ matrices parameterizing models 1 and 2.

| $X1$ | | | $X2$ | |
|---|---|---|---|---|
| 1 | 0 | | 1 | 0 |
| 1 | 0 | | 0 | 1 |
| 1 | 0 | | -1 | -1 |
| 0 | 1 | | 1 | 0 |
| 0 | 1 | | 0 | 1 |
| 0 | 1 | | -1 | -1 |
| -1 | -1 | | 1 | 0 |
| -1 | -1 | | 0 | 1 |
| -1 | -1 | | -1 | -1 |

Let model 3 be the union of the models spanned by these two matrices. Will the sum of squares for model 3 be the sum of the sums of squares for models 1 and 2? Why or why not?

In the one-way ANOVA problem, show that the three restrictions $\sum \alpha_i = 0$, $\sum n_i\alpha_i = 0$, and $\alpha_1 = 0$ lead to the same values of $\alpha_1 - \alpha_2$. Interpret this result in terms of estimable functions. **Question A.6**

Consider a one-factor model parameterized by the following matrix: **Question A.7**

$$\begin{matrix} 1 & 1 & 0 \\ 1 & 1 & 0 \\ 1 & 0 & 1 \\ 1 & 0 & 1 \\ 1 & -1 & -1 \\ 1 & -1 & -1 \end{matrix}$$

The parameters are $\mu$, $\alpha_1$, and $\alpha_2$. Which of the following are estimable: (a) $\mu$ , (b) $\mu + \alpha_1$, (c) $\alpha_1 + \alpha_2$, (d) $\mu - \alpha_1$, and (e) $\alpha_1 - \alpha_2$?

Consider a completely randomized design with twelve treatments and 24 units (all $n_i = 2$). The twelve treatments have a three by four factorial structure. **Question A.8**

(a) Find the variance/covariance matrix for the estimated factor A effects.

(b) Find the variance/covariance matrix for the estimated interaction effects.

(c) Show that the $t$-test for testing the equality of two factor A main effects can be found by treating the two estimated main effects as means of independent samples of size eight.

(d) Show that the $t$-test for testing the equality of two interaction effects can *not* be found by treating the two estimated interaction effects as means of independent samples of size two.

Consider the one-way ANOVA model with $g$ groups. The sample sizes are $n_i$ are not all equal. The treatments correspond to the levels of a quantitative factor; the level for treatment $i$ is $z_i$, and the $z_i$ are not equally spaced. We may compute linear, quadratic (adjusted for linear), and cubic (adjusted for linear and quadratic) sums of squares by linear regression. We may also compute these sums of squares via contrasts in the treatment means, but we need to find the contrast coefficients. Describe how to find the contrast coefficients for linear and quadratic (adjusted for linear). (Hint: use the $t$ and $s_i$ formulation in Sections A.6 and A.7, and remember your linear regression.) **Question A.9**

**Question A.10** Suppose that $Y_{N\times 1}$ is multivariate normal with mean $\mu$ and variance $\sigma^2 I$, and that we have models $M_1$ and $M_2$ with $M_1$ contained in $M_2$; $M_1$ has dimension $r_1$, $M_2$ has dimension $r_2$, and $P_1$ and $P_2$ are the orthogonal projections onto $M_1$ and $M_2$.

(a) Find the distribution of $(P_2 - P_1)Y$.

(b) What can you say in addition about the distribution of $(P_2 - P_1)Y$ when $\mu$ lies in $M_1$?

**Question A.11** Consider a proportionally balanced two-factor model with $n_{ij}$ units in the $ij$th factor-level combination. Let $M_A$ be the model of factor A effects $(Ey_{ijk} = \mu + \alpha_i)$ and let $M_B$ be the model of factor B effects $(Ey_{ijk} = \mu + \beta_j)$. Show that $M_A \cap 1^\perp$ is orthogonal to $M_B \cap 1^\perp$.

**Question A.12** If $X$ and $X^\star$ are $n$ by $p$ matrices and $X$ has rank $p$, show that the range of $X$ equals the range of $X^\star$ if and only if there exists a $p$ by $p$ nonsingular matrix $Q$ such that $X^\star$ = XQ.

# Appendix B

# Notation

| Symbol | Page | Meaning |
|---|---|---|
| $\Lambda$ | 521 | A diagonal matrix of eigenvalues |
| $\boldsymbol{\Sigma}$ | 572 | Variance (matrix) of $\boldsymbol{y}$ |
| $\alpha$ | 343 | First level of third blocking factor in a Graeco-Latin Square |
| $\alpha$ | 522 | Distance from origin for axial points in central composite design |
| $\alpha_i$ | 38 | $i$th treatment effect, or main effect of factor A |
| $\alpha_i$ | 254 | A random treatment effect |
| $\alpha_i$ | 339 | Direct effect of treatment $i$ in a residual-effects model |
| $\alpha_{[i]}$ | 92 | Effect for the treatment with $i$th smallest observed effect |
| $\widehat{\alpha}_{(i)}$ | 92 | $i$th smallest treatment effect |
| $\widehat{\alpha}_i$ | 39 | An estimator of $\alpha_i$ |
| $\alpha_i^\star$ | 455 | Effect of $i$th treatment in a covariate model |
| $\tilde{\alpha}_i$ | 364 | Interblock estimate of $\alpha_i$ in BIBD |
| $\tilde{\alpha}_i$ | 460 | Treatment effect, not covariate adjusted |

| Symbol | Page | Meaning |
|---|---|---|
| $\alpha\beta_{ij}$ | 175 | AB interaction |
| $\widehat{\alpha\beta}_{ij}$ | 177 | Estimate of AB interaction |
| $\alpha\beta\gamma_{ijk}$ | 183 | ABC interaction |
| $\widehat{\alpha\beta\gamma}_{ijk}$ | 184 | Estimated ABC interaction |
| $\alpha\beta\gamma\delta_{ijkl}$ | 183 | ABCD interaction |
| $\alpha\beta\delta_{ijl}$ | 183 | ABD interaction |
| $\alpha\gamma_{ik}$ | 183 | AC interaction |
| $\widehat{\alpha\gamma}_{ik}$ | 184 | Estimate of AC interaction |
| $\alpha\gamma\delta_{ikl}$ | 183 | ACD interaction |
| $\alpha\delta_{il}$ | 183 | AD interaction |
| $\beta$ | 343 | Second level of third blocking factor in a Graeco-Latin Square |
| $\beta$ | 563 | A vector of coefficients for the columns of a matrix $X$ which spans a model |
| $\beta$ | 455 | Coefficient of the covariate in a covariate model |
| $\boldsymbol{\beta}$ | 512 | Vector form of first-order model coefficients |
| $\beta_j$ | 328 | Effect of $j$th row block in a Latin Square |
| $\beta_j$ | 339 | Residual effect of treatment $j$ in a residual effects model |
| $\beta_j$ | 319 | Effect of $j$th block |
| $\beta_j$ | 168 | Main effect of factor B |
| $\widehat{\beta}_j$ | 177 | Estimate of main effect of factor B |
| $\beta_{j(i)}$ | 280 | Effect of B nested in A |
| $\widehat{\beta}_{j(i)}$ | 283 | Estimated effect for B nested in A |
| $\beta_{j(l)}$ | 331 | Effect of $j$th row in $l$th Latin Square |
| $\beta_k$ | 512 | A first-order parameter in a response surface model |

| Symbol | Page | Meaning |
|---|---|---|
| $\beta_{kk}$ | 517 | A pure quadratic parameter in a response surface model |
| $\beta_{kl}$ | 517 | A cross product parameter in a response surface model |
| $\beta_{klm}$ | 533 | Coefficient of pure third order term for mixture model |
| $\hat{\beta}$ | 128 | Slope of log variances regressed on log means |
| $\widehat{\beta}$ | 456 | Estimated coefficient (slope) of the co-variate |
| $\beta\gamma_{jk}$ | 183 | BC interaction |
| $\widehat{\beta\gamma}_{jk}$ | 184 | Estimate of BC interaction |
| $\beta\gamma_{jk(i)}$ | 284 | BC interaction nested in A effect |
| $\beta\gamma\delta_{jkl}$ | 183 | BCD interaction |
| $\beta\delta_{jl}$ | 183 | BD interaction |
| $\gamma$ | 343 | Third level of third blocking factor in a Graeco-Latin Square |
| $\gamma_1$ | 134 | Skewness |
| $\gamma_2$ | 134 | Kurtosis |
| $\gamma_k$ | 328 | Effect of $k$th column block in a Latin Square |
| $\gamma_k$ | 339 | Effect of subject $k$ in a residual-effects model |
| $\gamma_k$ | 183 | Main effect of factor C |
| $\widehat{\gamma}_k$ | 183 | Estimated main effect of factor C |
| $\gamma_{k(l)}$ | 331 | Effect of $k$th column in $l$th Latin Square |
| $\gamma\delta_{kl}$ | 183 | CD interaction |
| $\delta$ | 343 | Fourth level of third blocking factor in a Graeco-Latin Square |
| $\delta$ | 222 | Coefficient in a Johnson and Graybill interaction |

| Symbol | Page | Meaning |
|---|---|---|
| $\boldsymbol{\delta}$ | 512 | A vector of offsets to the current design variables in a response surface |
| $\delta_i$ | 161 | Mean for normal used in computing noncentrality parameter |
| $\delta_k$ | 512 | Offset for the $k$th design variable in a response surface |
| $\delta_{kl}$ | 533 | Coefficient of asymmetric term for third-order mixture model |
| $\delta_l$ | 331 | Effect of $l$th square in replicated Latin Square design |
| $\delta_l$ | 339 | Effect of period $l$ in a residual-effects model |
| $\delta_l$ | 183 | Main effect of factor D |
| $\epsilon_{ij}$ | 37 | Experimental error for $y_{ij}$ |
| $\epsilon_{ijk}$ | 175 | Experimental error for $y_{ijk}$ |
| $\epsilon_{ijklm}$ | 183 | Random error for $y_{ijklm}$ |
| $\epsilon_{k(i)}$ | 440 | Subject effect in a repeated-measures design |
| $\epsilon\beta_{jk(i)}$ | 440 | Subject by trial-factor interaction in a repeated measures design |
| $\eta$ | 217 | Coefficient of a Tukey interaction |
| $\eta_j$ | 364 | Random error for the total of block $j$ responses in interblock analysis of BIBD |
| $\eta_{k(i)}$ | 421 | Random error for $k$th whole plot at $i$th level of the whole-plot factor |
| $\hat{\eta}$ | 114 | An intermediate quantity used in Land's confidence intervals for log-normal data |
| $\lambda$ | 442 | Degrees of freedom adjustment for repeated measures designs that do not meet the Huynh-Feldt conditions |
| $\lambda$ | 359 | Number of blocks in which any pair of treatments occurs in a BIBD |

| Symbol | Page | Meaning |
|---|---|---|
| $\lambda$ | 128 | Power in a power family transformation |
| $\lambda$ | 220 | A transformation power in a Tukey one-degree-of-freedom interaction |
| $\lambda_i$ | 371 | Number of blocks in which two treatments in associate class $i$ of a PBIBD occur together |
| $\lambda_k$ | 519 | Eigenvalue for $i$th canonical variable |
| $\lambda^\star$ | 129 | Optimum Box-Cox transformation power |
| $\mu$ | 21 | Mean of a normal distribution |
| $\mu$ | 37 | Common or overall mean |
| $\mu$ | 572 | Expected value of $\boldsymbol{y}$ |
| $\mu_0$ | 21 | A null hypothesis mean |
| $\mu_1$ | 25 | Expected value of responses in first treatment |
| $\mu_2$ | 25 | Expected value of responses in second treatment |
| $\mu_{\bullet j}$ | 231 | Equally weighted average of treatment expectations for column $j$ |
| $\mu_i$ | 37 | Expected value for responses in $i$th treatment |
| $\mu_{i\bullet}$ | 231 | Equally weighted average of treatment expectations for row $i$ |
| $\widehat{\mu}_i$ | 39 | An estimator of $\mu_i$ |
| $\mu_{ij}$ | 167 | Treatment expected value in a two-factor factorial design |
| $\mu_{i\star}$ | 244 | A weighted average of treatment expected values for the $i$th row |

| Symbol | Page | Meaning |
|---|---|---|
| $(\mu_{i\star})_{\star j}$ | 244 | Column-weighted averages of row-weighted means |
| $\mu_{\star j}$ | 244 | A weighted average of treatment expected values for the $j$th column |
| $(\mu_{\star j})_{i\star}$ | 244 | Row-weighted averages of column-weighted means |
| $\widehat{\mu}$ | 39 | An estimator of $\mu$ |
| $\widehat{\mu}$ | 177 | Estimate of overall mean in a factorial |
| $\mu^\star$ | 38 | An overall expected value |
| $\mu^\star$ | 455 | Overall mean or intercept in a covariate model |
| $\widehat{\mu}^\star$ | 39 | An estimator of $\mu^\star$ |
| $\tilde{\mu}$ | 364 | Interblock estimate of $\mu$ in BIBD |
| $\tilde{\mu}$ | 460 | Estimate of average intercept for centered covariate |
| $\nu$ | 85 | Degrees of freedom, typically for error |
| $\nu_1$ | 262 | Degrees of freedom for a mean square |
| $\nu_2$ | 262 | Degrees of freedom for a mean square |
| $\nu_3$ | 262 | Degrees of freedom for a mean square |
| $\nu_{crd}$ | 323 | Error degrees of freedom for a CRD |
| $\nu_{ls}$ | 336 | Error degrees of freedom for a Latin Square |
| $\nu_{rcb}$ | 323 | Error degrees of freedom for an RCB |
| $\nu^\star$ | 262 | Approximate degrees of freedom |
| $\rho$ | 127 | Correlation coefficient |
| $\rho$ | 138 | Serial correlation |
| $\rho_i$ | 371 | Number of $i$th associates of a given treatment in a PBIBD |
| $\hat{\rho}$ | 127 | Sample correlation |
| $\sigma^2$ | 21 | Variance, often of experimental error |

| Symbol | Page | Meaning |
|---|---|---|
| $\sigma_\alpha^2$ | 254 | Variance of the random effect $\alpha_i$ |
| $\widehat{\sigma}_\alpha^2$ | 264 | An estimated variance component |
| $\sigma_{\alpha\beta}^2$ | 255 | Variance of the random effect $\alpha\beta_{ij}$ |
| $\sigma_{\alpha\beta\gamma}^2$ | 256 | Variance of the random effect $\alpha\beta\gamma_{ijk}$ |
| $\widehat{\sigma}_{\alpha\beta\gamma}^2$ | 265 | An estimated variance component |
| $\widehat{\sigma}_{\alpha\beta}^2$ | 265 | An estimated variance component |
| $\sigma_{\alpha\gamma}^2$ | 256 | Variance of the random effect $\alpha\gamma_{ik}$ |
| $\widehat{\sigma}_{\alpha\gamma}^2$ | 265 | An estimated variance component |
| $\sigma_\beta^2$ | 255 | Variance of the random effect $\beta_j$ |
| $\sigma_\beta^2$ | 364 | Variance of block effects in interblock analysis of BIBD |
| $\widehat{\sigma}_\beta^2$ | 365 | Estimate of block variance in BIBD |
| $\widehat{\sigma}_\beta^2$ | 265 | An estimated variance component |
| $\sigma_{\beta\gamma}^2$ | 256 | Variance of the random effect $\beta\gamma_{jk}$ |
| $\widehat{\sigma}_{\beta\gamma}^2$ | 265 | An estimated variance component |
| $\widehat{\sigma}_\gamma^2$ | 265 | An estimated variance component |
| $\sigma_\gamma^2$ | 256 | Variance of the random effect $\gamma_k$ |
| $\sigma_\eta^2$ | 269 | A variance component |
| $\sigma_{bibd}^2$ | 363 | Error variance in a BIBD |
| $\sigma_{crd}^2$ | 323 | Error variance for a CRD |
| $\widehat{\sigma}_{crd}^2$ | 323 | Estimate of $\sigma_{crd}^2$ based on results of an RCB |
| $\widehat{\sigma}_{crd}^2$ | 337 | Estimate of error variance in CRD based on data from LS |
| $\sigma_{ls}^2$ | 336 | Error variance in a Latin Square |

| Symbol | Page | Meaning |
|---|---|---|
| $\sigma^2_{rcb}$ | 323 | Error variance for an RCB |
| $\widehat{\sigma}^2_{rcb}$ | 336 | Estimate of error variance in RCB based on data from LS |
| $\widehat{\sigma}^2$ | 41 | An estimator of $\sigma^2$ |
| $\tau$ | 269 | An expected mean square |
| $\tau_d$ | 260 | A denominator expected mean square |
| $\tau_n$ | 260 | A numerator expected mean square |
| $\theta_0$ | 212 | Intercept in a dose-response relationship |
| $\theta_1$ | 55 | Linear coefficient in polynomial dose-response model |
| $\theta_2$ | 55 | Quadratic coefficient in polynomial dose-response model |
| $\theta_{Ar}$ | 212 | Coefficient of $z^r_{Ai}$ in a dose-response relationship |
| $\theta_{Ar0}$ | 215 | Coefficient for $z^r_{Ai}$ averaged across all levels of factor B |
| $\theta_{ArBs}$ | 212 | Coefficient of $z^r_{Ai}z^s_{Bj}$ in a dose-response relationship |
| $\theta_{Arj}$ | 215 | Coefficient for $z^r_{Ai}$ at the $j$th level of factor B |
| $\theta_{Bs}$ | 212 | Coefficient of $z^s_{Bj}$ in a dose-response relationship |
| $\theta\beta_{Arj}$ | 215 | Deviation from overall coefficient of $z^r_{Ai}$ for level $j$ of factor B |
| $\chi^2_{\mathcal{E},\nu}$ | 268 | Upper $\mathcal{E}$ percent point of a chi-square distribution with $\nu$ degrees of freedom |
| $\xi_i$ | 221 | Coefficient in a column-model of interaction |
| $\chi^2_n$ | 161 | Chi-square distribution with $n$ degrees of freedom |

| Symbol | Page | Meaning |
|---|---|---|
| $\chi_n^2(\zeta)$ | 161 | A noncentral chi-square with $n$ degrees of freedom and noncentrality parameter $\zeta$ |
| $\zeta$ | 154 | A noncentrality parameter |
| $\zeta$ | 364 | A contrast in treatment effects |
| $\zeta_j$ | 221 | Coefficient in a row-model of interaction |
| $\zeta_{l(ij)}$ | 429 | A split-plot error |
| $\bar{\zeta}$ | 365 | Estimate of a contrast in BIBD after recovery of interblock information |
| $\widehat{\zeta}$ | 364 | Intrablock estimate of a contrast in a BIBD |
| $\widetilde{\zeta}$ | 365 | Interblock estimate of a contrast in a BIBD |
| (1) | 236 | The treatment in a two-series design with all factors at their low levels |
| $\mathbf{1}$ | 563 | An $N$-vector of all ones |
| $2^{k-q}$ | 472 | A $1/2^q$ fraction of a $2^k$ factorial |
| $A^1C^2$ | 403 | A two-degree-of-freedom split in a three-series design |
| $A^{r_A}B^{r_B}C^{r_C}D^{r_D}$ | 404 | A generic two-degree-of-freedom split in a three-series design |
| $\mathcal{B}$ | 518 | Matrix form of the second-order coefficients in a second-order model |
| $\mathbf{B}$ | 526 | An estimate of $\mathcal{B}$ |
| $B^1C^1D^2$ | 403 | A two-degree-of-freedom split in a three-series design |
| B(A) | 281 | Factor B nested in A |
| $BF$ | 133 | Brown-Forsythe modified F-test |
| BIBD | 358 | A balanced incomplete block design |
| BSD | 91 | Bonferroni significant difference |
| $BSD_{ij}$ | 91 | Unequal sample size form of BSD |

| Symbol | Page | Meaning |
|---|---|---|
| $C_1$ | 159 | A fixed cost |
| $C_2$ | 159 | A cost per experimental unit |
| $C_3$ | 159 | A cost per measurement unit |
| $\mathrm{C}_p$ | 100 | Mallows' criterion for minimizing prediction error |
| C(AB) | 282 | C nested in A and B |
| CCD | 522 | A central composite design |
| CRD | 31 | A completely randomized design |
| $\mathcal{C}(X)$ | 566 | Column space of matrix $X$ |
| $D$ | 88 | A significant difference for all pairwise comparisons |
| $D$ | 532 | Total of component lower bounds in a mixture design |
| $D_{ij}$ | 87 | A significant difference for a pairwise test |
| $D_{max}$ | 122 | Maximum distance between units, used in binning a variogram |
| DSD | 102 | Dunnett significant difference |
| $DW$ | 121 | The Durbin-Watson statistic |
| $E$ | 140 | Factor for determing effective sample size with correlated data |
| $\mathcal{E}$ | 43 | Generic error rate for a test or confidence interval |
| $\mathcal{E}_\chi$ | 270 | Error rate for the chi-square portion of a Williams' confidence interval of a variance component |
| $E_{\text{BIBD:RCB}}$ | 362 | Efficiency of the BIBD relative to the RCB |
| $\mathcal{E}_F$ | 270 | Error rate for the F portion of a Williams' confidence interval of a variance component |
| $\mathcal{E}_I$ | 150 | A Type I error rate |

| Symbol | Page | Meaning |
|---|---|---|
| $\mathcal{E}_{II}$ | 150 | A Type II error rate |
| $E_{\text{LS:CRD}}$ | 336 | Relative efficiency of a Latin Square to a CRD |
| $E_{\text{LS:RCB}}$ | 336 | Relative efficiency of a Latin Square to an RCB |
| $E_{\text{PBIBD:RCB}}$ | 372 | Average efficiency of PBIBD to RCB |
| $E_{\text{RCB:CRD}}$ | 323 | Relative efficiency of RCB to CRD |
| $\widehat{E}_{\text{RCB:CRD}}$ | 323 | Estimated relative efficiency of RCB to CRD |
| $E_{\text{SL:RCB}}$ | 375 | Average efficiency of a Square Lattice to an RCB |
| $\mathcal{E}_i$ | 77 | Type I error rate for hypothesis $i$ |
| EMS | 257 | An expected mean square |
| $EMS_1$ | 269 | Expected value of $MS_1$ |
| $EMS_2$ | 269 | Expected value of $MS_2$ |
| $EMS_{\text{Trt}}$ | 52 | Expected mean square for treatments |
| $F_{\mathcal{E},g-1,\nu}$ | 85 | Upper $\mathcal{E}$ percent point of an F distribution with $g-1$ and $\nu$ degrees of freedom |
| $H$ | 521 | An orthogonal matrix of eigenvectors of $\mathcal{B}$ |
| $H$ | 574 | Orthogonal matrix $(H_1 : H_2)$ |
| $H_0$ | 21 | A null hypothesis |
| $H_{01}$ | 78 | First null hypothesis of a family |
| $H_{0K}$ | 78 | Last null hypothesis of a family |
| $H_{0(i)}$ | 82 | Null hypothesis corresponding to $i$th smallest $p$-value |
| $H_1$ | 574 | An orthonormal basis for $\boldsymbol{M}$ |
| $H_1$ | 21 | An alternative hypothesis |
| $H_2$ | 574 | An orthonormal basis for $\boldsymbol{M}^{\perp}$ |
| $H_{ij}$ | 114 | Leverage for $y_{ij}$ |

| Symbol | Page | Meaning |
|---|---|---|
| HSD | 90 | Tukey's honest significant difference |
| $HSD_{ij}$ | 91 | Tukey-Kramer form of HSD |
| I | 473 | A column of all ones in the analysis of a two-series |
| $I_q$ | 521 | $q$ by $q$ identity matrix |
| $K$ | 78 | Number of null hypotheses in a family |
| $K$ | 122 | Number of bins in a variogram |
| $L$ | 390 | Numerically evaluated defining contrast |
| $L_1$ | 393 | Numerically evaluated first defining contrast |
| $L_2$ | 393 | Numerically evaluated second defining contrast |
| LS | 325 | A Latin Square design |
| LSD | 97 | Least significant difference |
| $\boldsymbol{M}$ | 563 | A model, that is, a linear subspace of $\mathcal{R}^N$ |
| $\boldsymbol{M}_1$ | 568 | A model subspace |
| $\boldsymbol{M}_{12}$ | 571 | Union of models $\boldsymbol{M}_1$ and $\boldsymbol{M}_2$ |
| $\boldsymbol{M}_2$ | 568 | A model subspace |
| $\boldsymbol{M}_3$ | 570 | A model subspace |
| $\boldsymbol{M}_C$ | 572 | Model of separate column means |
| $\boldsymbol{M}_R$ | 572 | Model of separate row means |
| $\boldsymbol{M}^{\perp}$ | 567 | Orthogonal complement of the subspace $\boldsymbol{M}$ in $\mathcal{R}^N$ |
| MCB | 104 | Multiple comparisons with the best procedure |
| $MS_1$ | 262 | A mean square |
| $MS_2$ | 262 | A mean square |
| $MS_3$ | 262 | A mean square |
| $MS_A$ | 181 | Mean square for factor A |

| Symbol | Page | Meaning |
|---|---|---|
| $MS_{AB}$ | 181 | Mean square for the AB interaction |
| $MS_B$ | 181 | Mean square for factor B |
| $MS_{\text{Cols}}$ | 336 | Mean square for columns in a Latin Square |
| $MS_E$ | 41 | Mean square for error |
| $MS_{LoF}$ | 516 | Mean square for lack of fit |
| $MS_{PE}$ | 516 | Mean square for pure error |
| $MS_{\text{Rows}}$ | 336 | Mean square for rows in a Latin Square |
| $MS_{\text{Trt}}$ | 48 | Mean square for treatments |
| $MS_w$ | 69 | Mean square for a contrast |
| $N$ | 18 | Total number of units |
| NPP | 115 | Normal probablity plot |
| $P$ | 490 | A two-degree-of-freedom split in a three-series design |
| $P$ | 571 | A projection mapping |
| $P_1$ | 405 | A defining split for a confounded three-series |
| $P_2$ | 405 | A defining split for a confounded three-series |
| PBIBD | 370 | A partially balanced incomplete block design |
| PSE | 241 | Lenth's pseudostandard error for unreplicated two-series |
| $\mathcal{P}(p)$ | 49 | "Calibrated" $p$-value (lower bound on Type I error probability) |
| Q | 293 | Representative element for a fixed term in an EMS |
| $R$ | 482 | Resolution of a fractional factorial |
| $\mathcal{R}^N$ | 563 | $N$-dimensional Euclidean space |
| RCB | 316 | A randomized complete block design |

| Symbol | Page | Meaning |
|---|---|---|
| REGWR | 94 | Ryan-Einot-Gabriel-Welsch Range test |
| SNK | 96 | Student-Newman-Keuls pairwise comparisons procedure |
| SPE | 421 | Split-plot error |
| $SS_1$ | 56 | Linear sum of squares |
| $SS_2$ | 56 | Quadratic sum of squares |
| $SS_A$ | 167 | Sum of squares for factor A |
| $SS_{AB}$ | 168 | Sum of squares for the AB interaction |
| $SS_{ABD}$ | 184 | Sum of squares for ABD interaction |
| $SS_B$ | 168 | Sum of squares for factor B |
| $SS_E$ | 40 | Sum of squared errors |
| $SS_E(\lambda)$ | 129 | Sum of squared errors as a function of Box-Cox transformation power $\lambda$ |
| $SS_{LoF}$ | 515 | Sum of squares for lack of fit |
| $SS_{PE}$ | 515 | Sum of squares for pure error |
| $SS_{\text{Rows}}$ | 328 | Sum of squares for rows in a Latin Square |
| $SS_T$ | 46 | Corrected total sum of squares |
| $SS_{\text{Trt}}$ | 46 | Treatment sum of squares |
| $SS_{\text{Cols}}$ | 328 | Sum of squares for columns in a Latin Square |
| $SS_{\text{linear}}$ | 56 | Linear sum of squares |
| $SS_{\text{quadratic}}$ | 56 | Quadratic sum of squares |
| $SS_w$ | 69 | Sum of squares for a contrast |
| $SS(B\|1, A, C)$ | 226 | Sum of squares for B adjusted for 1, A, and C |
| SSPE | 430 | Split-split-plot error |
| $SSR$ | 45 | Sum of squared residuals |
| $SSR_0$ | 53 | Sum of squared residuals for a reduced (null) model |

| Symbol | Page | Meaning |
|---|---|---|
| $SSR_A$ | 53 | Sum of squared residuals for a full (alternate) model |
| $SSR_k$ | 56 | Residual sum of squares for a polynomial model including powers up to $k$ |
| $SS(\boldsymbol{t})$ | 578 | Sum of squares for the model spanned by $\boldsymbol{t}$ |
| Type I | 227 | Sequential sums of squares |
| Type I | 77 | (Error) where the null is falsely rejected |
| Type II | 228 | A sum of squares with a term adjusted for the largest hierarchical model that does not include the term |
| Type II | 150 | A Type II error, failing to reject a false null hypothesis |
| $U_1$ | 570 | A subspace of the vector space $V$ |
| $U_2$ | 570 | A subspace of the vector space $V$ |
| $V$ | 570 | A vector space |
| W | 473 | A generating word in a fractional factorial |
| $W_1$ | 473 | A generating word in a fractional factorial |
| $W_2$ | 473 | A generating word in a fractional factorial |
| WPE | 421 | Whole-plot error |
| $X$ | 563 | A matrix, the columns of which span a model |
| $X_1$ | 568 | A matrix which spans model $\boldsymbol{M}_1$ |
| $X_2$ | 568 | A matrix which spans model $\boldsymbol{M}_2$ |
| $Y$ | 563 | Fitted values when fitting a model $\boldsymbol{M}$ to data $\boldsymbol{y}$ |
| $Y_0$ | 566 | A point in the model space $\boldsymbol{M}$ |
| $Y_1$ | 568 | Fit of $\boldsymbol{M}_1$ to $\boldsymbol{y}$ |
| $Y_2$ | 568 | Fit of $\boldsymbol{M}_2$ to $\boldsymbol{y}$ |

| Symbol | Page | Meaning |
|---|---|---|
| $Y_3$ | 570 | Fit of $\boldsymbol{M}_3$ to $\boldsymbol{y}$ |
| $\overline{Y}$ | 579 | Mean of $Y$ |
| $Z_1$ | 161 | A standard normal random variable |
| $a$ | 175 | Number of levels of factor A |
| $b$ | 358 | Number of blocks in a BIBD or PBIBD |
| $b$ | 175 | Number of levels of factor B |
| $\mathbf{b}$ | 514 | Estimated first-order coefficients in a response surface |
| $b$ | 566 | Least squares estimates of the parameters $\beta$ |
| $bcd$ | 236 | A factor-level combination in a two-series design |
| $c$ | 183 | Number of levels for factor C |
| $c_j$ | 220 | A column effect in the derivation of Tukey one-degree-of-freedom for interaction |
| $d$ | 183 | Number of levels for factor D |
| $d_1$ | 21 | A difference in a paired $t$-test |
| $d_{\mathcal{E}}(g-1,\nu)$ | 102 | Upper $\mathcal{E}$ percent point of the two-sided Dunnett distribution for comparing $g-1$ treatments to a control |
| $d'_{\mathcal{E}}(g-1,\nu)$ | 102 | Upper $\mathcal{E}$ percent point of the one-sided Dunnett distribution for comparing $g-1$ treatments to a control |
| $d_i$ | 133 | A scaled sample variance used in the Brown-Forsythe modified F-test |
| $d_{ij}$ | 119 | Absolute deviation of response from treatment mean, as used in the Levene test |
| $d_k$ | 532 | Lower bound for a component in a mixture design |
| $\bar{d}$ | 21 | Mean of differences in a paired $t$-test |

| Symbol | Page | Meaning |
|---|---|---|
| $df_{LoF}$ | 515 | Degrees of freedom for lack of fit |
| $df_{PE}$ | 515 | Degrees of freedom for pure error |
| $f_v$ | 519 | Response surface as a function of canonical variables |
| $f(x_{1i}, x_{2i})$ | 509 | A response function of variables $x_1$ and $x_2$ |
| $f(z_i; \theta)$ | 55 | A dose-response function |
| $g$ | 18 | Number of treatments or groups |
| $g_k$ | 263 | A coefficient in a linear combination of mean squares |
| $h$ | 576 | A vector defining a linear combination $h'\beta$ |
| $i$ | 37 | An index, usually the treatment number or the level of the first factor |
| $j$ | 37 | An index, usually the level of the second factor or an indicator of replication |
| $k$ | 358 | Number of units per block in a BIBD or PBIBD |
| $k$ | 166 | Index denoting level of replication in a two-factor factorial or level of third factor |
| $l$ | 183 | Replication in a three-factor factorial, or level of factor D |
| $m$ | 331 | Number of squares in a design with replicated LS |
| $m$ | 371 | Number of associate classes in a PBIBD |
| $m$ | 513 | Number of center points in a response surface design |
| $n$ | 421 | Replication in a split plot, the number of whole plots for each whole-plot factor level |
| $n$ | 21 | A sample size, for example, of a $t$-test |
| $n_1$ | 18 | Number of units in first treatment |

| Symbol | Page | Meaning |
|---|---|---|
| $n_c$ | 160 | Sample size for control treatment |
| $n_g$ | 18 | Number of units in $g$th treatment |
| $n_i$ | 37 | Sample size for $i$th treatment |
| $n_{ij}$ | 364 | Number of times treatment $i$ occurs in block $j$ (0 or 1 in a BIBD) |
| $n_t$ | 160 | Sample size for noncontrol treatments |
| $p$ | 21 | $p$-value of a test |
| $p$ | 100 | Number of classes into which the $g$ treatments are partitioned for prediction |
| $p_{(1)}$ | 82 | Smallest $p$-value in a family |
| $p_{(K)}$ | 82 | Largest $p$-value in a family |
| $p_i$ | 81 | $p$-value for testing $H_{0i}$ |
| $p^i_{jk}$ | 371 | Number of treatments that are $j$th associates of A and $k$th associates of B when A and B are $i$th associates in a PBIBD |
| $q$ | 511 | Number of variables in a response surface model |
| $q_{\mathcal{E}}(g, \nu)$ | 90 | Upper $\mathcal{E}$ percent point of the Studentized range distribution for $g$ groups and $\nu$ error degrees of freedom |
| $r$ | 316 | Number of blocks in an RCB |
| $r$ | 358 | Number of times each treatment is used in a BIBD or PBIBD |
| $r$ | 567 | Dimension of a model space |
| $r$ | 512 | A positive multiplier for steps in steepest ascent |
| $r_1$ | 288 | Product of the number of levels of fixed factors in a mixed term |
| $r_1$ | 568 | Dimension of model $\boldsymbol{M}_1$ |
| $r_2$ | 288 | Product across fixed factors of the number of levels minus one in a mixed term |
| $r_2$ | 568 | Dimension of model $\boldsymbol{M}_2$ |

| Symbol | Page | Meaning |
|---|---|---|
| $r_3$ | 570 | Dimension of $M_3$ |
| $r_{ab}$ | 288 | Scaling factor for the variance of a mixed effect in the restricted model |
| $r_i$ | 220 | A row effect in the derivation of Tukey one-degree-of-freedom for interaction |
| $r_{ij}$ | 45 | A raw residual |
| $r_k$ | 121 | The $k$th residual in time order, used in the Durbin-Watson statistic |
| $s$ | 21 | A sample standard deviation, for example, as used in a $t$-test |
| $s$ | 41 | Alternate notation for $\widehat{\sigma}$ |
| $s_0$ | 241 | In a PSE computation, 1.5 times the median of the absolute values of the contrast results |
| $s_1$ | 577 | An element of an estimable function $t$ |
| $s_i^2$ | 132 | Sample variance for treatment $i$ |
| $s_{ij}$ | 114 | Internally Studentized residual for $y_{ij}$ |
| $s_p^2$ | 25 | Pooled estimate of variance |
| $t$ | 21 | A $t$ test statistic |
| $\boldsymbol{t}$ | 576 | A vector in $\mathcal{R}^N$ |
| $t_{\mathcal{E}/2,N-g}$ | 43 | Upper $\mathcal{E}/2$ percent point of a $t$-distribution with $N - g$ degrees of freedom |
| $t_{ij}$ | 115 | Externally Studentized residual for $y_{ij}$ |
| $t_{ij}$ | 132 | $t$-test comparing treatments $i$ and $j$ |
| $\boldsymbol{t}_{(k)}$ | 579 | One of a set of orthogonal contrasts |
| $\boldsymbol{t}_r$ | 576 | Projection of $\boldsymbol{t}$ onto $\boldsymbol{M}^{\perp}$ |
| $\boldsymbol{t}^{\star}$ | 576 | Projection of $\boldsymbol{t}$ onto $\boldsymbol{M}$ |
| $u$ | 87 | A critical value in a pairwise comparison test |

| Symbol | Page | Meaning |
|---|---|---|
| $u_1$ | 570 | An element of $U_1$ |
| $u_2$ | 570 | An element of $U_2$ |
| $u(\mathcal{E}, \nu)$ | 97 | A pairwise comparison critical value depending on $\mathcal{E}$ and the error degrees of freedom |
| $u(\mathcal{E}, \nu, K)$ | 91 | A pairwise comparison critical value depending on $\mathcal{E}$, the error degrees of freedom, and the number of pairwise comparisons |
| $u(\mathcal{E}, \nu, g)$ | 90 | A pairwise comparison critical value depending on $\mathcal{E}$, the error degrees of freedom, and number of treatments |
| $u(\mathcal{E}, \nu, k, g)$ | 94 | A pairwise comparison critical value depending on $\mathcal{E}$, the error degrees of freedom, the length of the stretch, and the number of treatments |
| $u_j$ | 222 | Column singular vector in a Johnson and Graybill interaction |
| $\mathbf{v}$ | 519 | Vector form of canonical variables in a second-order model |
| $v$ | 570 | An element of vector space $V$ |
| $v_1$ | 518 | A canonical variable in a second-order model |
| $v_2$ | 518 | A canonical variable in a second-order model |
| $v_i$ | 222 | Row singular vector in a Johnson and Graybill interaction |
| $v_{i\bullet}$ | 363 | Total for treatment $i$ of block-adjusted responses in a BIBD |
| $v_{ij}$ | 363 | Data with block means subtracted in BIBD |
| $v_k$ | 519 | A design variable in canonical coordinates |
| $w_A$ | 238 | The contrast for factor A in a two-series design |

| Symbol | Page | Meaning |
|---|---|---|
| $w_{Aijk}$ | 238 | The $ijk$ element of the $w_A$ contrast in a two-series design |
| $\{w_i\}$ | 66 | A set of contrast coefficients |
| $w_i$ | 66 | A contrast coefficient |
| $w_i^\star$ | 71 | A contrast coefficient |
| $w_{ij}$ | 167 | A two-factor arrangment of contrast coefficients |
| $w_{ijk}$ | 204 | Contrast coefficients for a three-factor factorial |
| $w_{jk}$ | 208 | Contrast coefficients for a BC interaction contrast |
| $\{w^\star\}$ | 71 | A set of contrast coefficients |
| $w^\star(\{\overline{y}_{i\bullet\bullet}\})$ | 169 | An observed contrast in the factor A average responses |
| $w(\{\alpha_i\})$ | 66 | A contrast in treatment effects |
| $w(\{\widehat{\alpha}_i\})$ | 66 | A contrast in observed treatment effects |
| $w(\{\mu_i\})$ | 66 | A contrast in treatment expected values |
| $w(\{\overline{y}_{i\bullet}\})$ | 66 | A contrast in observed treatment means |
| $x$ | 563 | A vector in a model $M$ |
| $\mathbf{x}_0$ | 519 | Stationary point of a rcsponsc surfacc |
| $x_0$ | 464 | An intersection point in a separate slopes model |
| $x_1$ | 509 | A continuously variable treatment factor |
| $x_2$ | 509 | A continuously variable treatment factor |
| $x_A$ | 388 | Level of factor A |
| $x_B$ | 388 | Level of factor B |
| $x_C$ | 388 | Level of factor C |
| $\overline{x}_{\bullet\bullet}$ | 456 | The grand mean of the covariates |
| $\mathbf{x}_i$ | 512 | Vector form of design variables for $i$th data point |

| Symbol | Page | Meaning |
|---|---|---|
| $x_{ij}$ | 454 | Covariate corresponding to $y_{ij}$ |
| $\tilde{x}_{ij}$ | 460 | Covariate with treatment mean subtracted |
| $x'_k$ | 532 | A pseudocomponent in a mixture design |
| $\overline{x}_{i\bullet}$ | 460 | Average covariate in treatment $i$ |
| $\dot{x}$ | 456 | A standard covariate value |
| $\boldsymbol{y}$ | 563 | An $N$-dimensional vector of responses |
| $y_{14}$ | 25 | A response, here the fourth response in the first treatment group |
| $y_{\bullet\bullet}$ | 40 | Total of all responses |
| $y_{\bullet j}$ | 364 | Total of responses for block $j$ |
| $y_{i\bullet}$ | 40 | Total of responses in the $i$th treatment |
| $\overline{y}_{i\bullet}$ | 40 | Average of responses in $i$th treatment |
| $y_{ij}$ | 319 | Response for the $i$th treatment in the $j$th block |
| $y_{ij}$ | 37 | $j$th response in $i$th treatment |
| $y_{ijk}$ | 166 | A response in a two-factor factorial experiment |
| $y_{ijkl}$ | 339 | In a design balanced for residual effects, the response for the $k$th subject in the $l$th time period; the subject received treatment $i$ in period $l$ and treatment $j$ in period $l-1$ |
| $y_{ijklm}$ | 183 | Response in a four-factor factorial |
| $\tilde{y}_i$ | 119 | Median response in treatment $i$ |
| $y^{(\lambda)}$ | 129 | A Box-Cox transformation |
| $\overline{y}$ | 566 | Mean of $\boldsymbol{y}$ |
| $\overline{y}_{1\bullet}$ | 25 | Mean of responses in the first treatment |
| $\overline{y}_{(1)\bullet}$ | 92 | Smallest treatment mean |

| Symbol | Page | Meaning |
|---|---|---|
| $\overline{y}_{2\bullet}$ | 25 | Mean of responses in the second treatment |
| $\overline{y}_{abc}$ | 238 | The average response for treatment $abc$ in a two-series design |
| $\overline{y}_{\bullet\bullet}$ | 40 | Grand mean of the responses |
| $\overline{y}_{\bullet\bullet\bullet}$ | 177 | Grand mean in a two-factor factorial |
| $\overline{y}_{\bullet\bullet\bullet\bullet\bullet}$ | 183 | Grand mean in a four-factor factorial |
| $\overline{y}_{\bullet\bullet k\bullet\bullet}$ | 183 | Mean response at level $k$ of factor C |
| $\overline{y}_{\bullet j\bullet}$ | 167 | Observed mean at level $j$ of factor B |
| $\overline{y}_{(g)\bullet}$ | 92 | Largest treatment mean |
| $\overline{y}_{i\bullet\bullet}$ | 167 | Observed mean at level $i$ of factor A |
| $\overline{y}_{ij\bullet}$ | 167 | Observed mean in the $ij$ treatment |
| $\overline{y}_{ijk\bullet\bullet}$ | 184 | Marginal mean at level $i$ of factor A, level $j$ of factor B, and level $k$ of factor C |
| $\dot{y}$ | 128 | Geometric mean of the data |
| $z$ | 574 | A vector in $\mathcal{R}^N$ |
| $z_{Ai}$ | 212 | Dose for level $i$ of factor A |
| $z_{Bj}$ | 212 | Dose for level $j$ of factor B |
| $z_{\mathcal{E}/2}$ | 114 | Upper $\mathcal{E}/2$ percent point of the standard normal |
| $z_i$ | 55 | Dose for treatment $i$ |

# Appendix C

# Experimental Design Plans

## C.1 Latin Squares

The plans are presented in two groups. First we present sets of standard squares for several values of $g$. These sets are complete for $g = 3, 4$ and are incomplete for larger $g$. Next we present sets of up to four orthogonal Latin Squares (there are at most $g - 1$ orthogonal squares for any $g$). Graeco-Latin squares (and hyper-Latin squares) may be constructed by combining two (or more) orthogonal Latin Squares. All plans come from Fisher and Yates (1963).

### C.1.1 Standard Latin Squares

**3 × 3**

```
A B C
B C A
C A B
```

**4 × 4**

```
A B C D    A B C D    A B C D    A B C D
B A D C    B C D A    B D A C    B A D C
C D B A    C D A B    C A D B    C D A B
D C A B    D A B C    D C B A    D C B A
```

**5 × 5**

A B C D E    A B C D E    A B C D E    A B C D E
B A E C D    B C E A D    B D A E C    B E A C D
C D A E B    C D B E A    C E D B A    C A D E B
D E B A C    D E A C B    D C E A B    D C E B A
E C D B A    E A D B C    E A B C D    E D B A C

**6 × 6**

A B C D E F    A B C D E F    A B C D E F
B C A F D E    B A E F C D    B A E C F D
C A B E F D    C F A B D E    C F B A D E
D B E B A C    D E B A F C    D E F B C A
E D F A C B    E D F C B A    E D A F B C
F E D C B A    F C D E A B    F C D E A B

**7 × 7**

A B C D E F G    A B C D E F G    A B C D E F G
B E A G F D C    B F E G C A D    B C D E F G A
C F G B D A E    C D A E B G F    C D E F G A B
D G E F B C A    D C G A F E B    D E F G A B C
E D B C A G F    E G B F A D C    E F G A B C D
F C D A G E B    F A D C G B E    F G A B C D E
G A F E C B D    G E F B D C A    G A B C D E F

### C.1.2 Orthogonal Latin Squares

**3 × 3**

A B C    A B C
B C A    C A B
C A B    B C A

**4 × 4**

A B C D    A B C D    A B C D
B A D C    C D A B    D C B A
C D A B    D C B A    B A D C
D C B A    B A D C    C D A B

**5 × 5**

| | | | |
|---|---|---|---|
| A B C D E | A B C D E | A B C D E | A B C D E |
| B C D E A | C D E A B | D E A B C | E A B C D |
| C D E A B | E A B C D | B C D E A | D E A B C |
| D E A B C | B C D E A | E A B C D | C D E A B |
| E A B C D | D E A B C | C D E A B | B C D E A |

**7 × 7**

| | | |
|---|---|---|
| A B C D E F G | A B C D E F G | A B C D E F G |
| E F G A B C D | F G A B C D E | G A B C D E F |
| B C D E F G A | D E F G A B C | F G A B C D E |
| F G A B C D E | B C D E F G A | E F G A B C D |
| C D E F G A B | G A B C D E F | D E F G A B C |
| G A B C D E F | E F G A B C D | C D E F G A B |
| D E F G A B C | C D E F G A B | B C D E F G A |

## C.2 Balanced Incomplete Block Designs

The plans are sorted first by number of treatments $g$, then by size of block $k$. The number of blocks is $b$; the replication for any treatment is $r$; any pair of treatments occurs together in $\lambda = r(k-1)/(g-1)$ blocks; and the efficiency is $E = g(k-1)/[(g-1)k]$. Designs that can be arranged as Youden Squares are marked with YS and shown as Youden Squares. Designs involving all combinations of $g$ treatments taken $k$ at a time that cannot be arranged as Youden Squares are simply labeled *unreduced.* Some designs are generated as complements of other designs, that is, by including in one block all those treatments not appearing in the corresponding block of the other design. Additional plans can be found in Cochran and Cox (1957), who even include some plans with 91 treatments. Fisher and Yates (1963) describe methods for generating BIBD designs. BIBD plans given here were generated using the instructions in Fisher and Yates or de novo and then arranged in Youden Squares when feasible.

**BIBD 1 g = 3, k = 2, b = 3, r = 2, $\lambda$ = 1, E = .75, YS**

| | | |
|---|---|---|
| 1 | 2 | 3 |
| 2 | 3 | 1 |

**BIBD 2 g = 4, k = 2, b = 6, r = 3, $\lambda$ = 1, E = .67**

Unreduced

**BIBD 3** **g = 4, k = 3, b = 4, r = 3, $\lambda$ = 2, E = .89, YS**

| | | | |
|---|---|---|---|
| 1 | 2 | 3 | 4 |
| 2 | 3 | 4 | 1 |
| 3 | 4 | 1 | 2 |

**BIBD 4** **g = 5, k = 2, b = 10, r = 4, $\lambda$ = 1, E = .63, YS**

| | | | | | | | | | |
|---|---|---|---|---|---|---|---|---|---|
| 1 | 1 | 4 | 5 | 2 | 5 | 3 | 3 | 4 | 2 |
| 2 | 3 | 1 | 1 | 4 | 2 | 4 | 5 | 5 | 3 |

**BIBD 5** **g = 5, k = 3, b = 10, r = 6, $\lambda$ = 3, E = .83, YS**

| | | | | | | | | | |
|---|---|---|---|---|---|---|---|---|---|
| 1 | 2 | 5 | 1 | 3 | 4 | 2 | 5 | 4 | 3 |
| 2 | 4 | 1 | 3 | 1 | 5 | 3 | 2 | 5 | 4 |
| 3 | 1 | 2 | 4 | 5 | 1 | 4 | 3 | 2 | 5 |

**BIBD 6** **g = 5, k = 4, b = 5, r = 4, $\lambda$ = 3, E = .94, YS**

| | | | | |
|---|---|---|---|---|
| 1 | 2 | 3 | 4 | 5 |
| 2 | 3 | 4 | 5 | 1 |
| 3 | 4 | 5 | 1 | 2 |
| 4 | 5 | 1 | 2 | 3 |

**BIBD 7** **g = 6, k = 2, b = 15, r = 5, $\lambda$ = 1, E = .6**

Unreduced

**BIBD 8** **g = 6, k = 3, b = 10, r = 5, $\lambda$ = 2, E = .8**

| | | | | | | | | | |
|---|---|---|---|---|---|---|---|---|---|
| 1 | 2 | 3 | 5 | 5 | 6 | 4 | 1 | 5 | 6 |
| 4 | 4 | 4 | 6 | 6 | 1 | 1 | 2 | 2 | 3 |
| 5 | 6 | 5 | 1 | 2 | 3 | 2 | 3 | 3 | 4 |

**BIBD 9** **g = 6, k = 4, b = 15, r = 10, $\lambda$ = 6, E = .9**

Unreduced

**BIBD 10** **g = 6, k = 5, b = 6, r = 5, $\lambda$ = 4, E = .96, YS**

| | | | | | |
|---|---|---|---|---|---|
| 1 | 2 | 3 | 4 | 5 | 6 |
| 2 | 3 | 4 | 5 | 6 | 1 |
| 3 | 4 | 5 | 6 | 1 | 2 |
| 4 | 5 | 6 | 1 | 2 | 3 |
| 5 | 6 | 1 | 2 | 3 | 4 |

**BIBD 11 g = 7, k = 2, b = 21, r = 6, $\lambda$ = 1, E = .58, YS**

| 1 | 1 | 1 | 5 | 6 | 7 | 3 | 4 | 2 | 2 | 2 |
|---|---|---|---|---|---|---|---|---|---|---|
| 2 | 3 | 4 | 1 | 1 | 1 | 2 | 2 | 5 | 6 | 7 |

| 3 | 3 | 6 | 7 | 5 | 4 | 4 | 5 | 7 | 6 |
|---|---|---|---|---|---|---|---|---|---|
| 4 | 5 | 3 | 3 | 4 | 6 | 7 | 6 | 5 | 7 |

**BIBD 12 g = 7, k = 3, b = 7, r = 3, $\lambda$ = 1, E = .78, YS**

| 1 | 3 | 7 | 5 | 4 | 2 | 6 |
|---|---|---|---|---|---|---|
| 2 | 1 | 4 | 3 | 6 | 7 | 5 |
| 5 | 6 | 1 | 4 | 2 | 3 | 7 |

**BIBD 13 g = 7, k = 4, b = 7, r = 4, $\lambda$ = 2, E = .88, YS**

| 3 | 1 | 2 | 7 | 6 | 5 | 4 |
|---|---|---|---|---|---|---|
| 4 | 2 | 7 | 1 | 5 | 6 | 3 |
| 6 | 7 | 4 | 5 | 3 | 1 | 2 |
| 7 | 6 | 5 | 3 | 2 | 4 | 1 |

**BIBD 14 g = 7, k = 5, b = 21, r = 15, $\lambda$ = 10, E = .93, YS**

| 1 | 6 | 4 | 3 | 2 | 1 | 5 | 7 | 2 | 6 | 1 | 4 | 7 | 3 | 5 |
|---|---|---|---|---|---|---|---|---|---|---|---|---|---|---|
| 2 | 1 | 7 | 5 | 3 | 2 | 1 | 4 | 6 | 5 | 6 | 1 | 3 | 7 | 4 |
| 3 | 2 | 1 | 6 | 5 | 3 | 2 | 1 | 4 | 7 | 4 | 7 | 1 | 5 | 6 |
| 4 | 3 | 2 | 1 | 7 | 6 | 4 | 5 | 1 | 2 | 3 | 5 | 6 | 1 | 7 |
| 5 | 4 | 3 | 2 | 1 | 7 | 6 | 2 | 7 | 1 | 5 | 3 | 4 | 6 | 1 |

| 2 | 7 | 6 | 5 | 4 | 3 |
|---|---|---|---|---|---|
| 3 | 2 | 7 | 6 | 5 | 4 |
| 4 | 3 | 2 | 7 | 6 | 5 |
| 5 | 4 | 3 | 2 | 7 | 6 |
| 6 | 5 | 4 | 3 | 2 | 7 |

**BIBD 15 g = 7, k = 6, b = 7, r = 6, $\lambda$ = 5, E = .97, YS**

| 1 | 2 | 3 | 4 | 5 | 6 | 7 |
|---|---|---|---|---|---|---|
| 2 | 3 | 4 | 5 | 6 | 7 | 1 |
| 3 | 4 | 5 | 6 | 7 | 1 | 2 |
| 4 | 5 | 6 | 7 | 1 | 2 | 3 |
| 5 | 6 | 7 | 1 | 2 | 3 | 4 |
| 6 | 7 | 1 | 2 | 3 | 4 | 5 |

**BIBD 16 g = 8, k = 2, b = 28, r = 7, λ = 1, E = .57**

Unreduced

**BIBD 17 g = 8, k = 3, b = 56, r = 21, λ = 6, E = .76, YS**

| 1 | 4 | 2 | 1 | 7 | 2 | 3 | 5 | 1 | 3 | 8 | 1 | 6 | 4 | 1 |
|---|---|---|---|---|---|---|---|---|---|---|---|---|---|---|
| 2 | 1 | 5 | 2 | 1 | 8 | 1 | 3 | 6 | 1 | 3 | 4 | 1 | 7 | 4 |
| 3 | 2 | 1 | 6 | 2 | 1 | 4 | 1 | 3 | 7 | 1 | 5 | 4 | 1 | 8 |

| 6 | 5 | 1 | 1 | 8 | 7 | 2 | 3 | 4 | 5 | 6 | 7 | 8 | 2 | 3 |
|---|---|---|---|---|---|---|---|---|---|---|---|---|---|---|
| 1 | 7 | 5 | 8 | 1 | 6 | 3 | 4 | 5 | 6 | 7 | 8 | 2 | 4 | 5 |
| 5 | 1 | 8 | 6 | 6 | 1 | 7 | 8 | 2 | 3 | 4 | 5 | 6 | 8 | 2 |

| 4 | 5 | 6 | 7 | 8 | 2 | 3 | 4 | 5 | 6 | 7 | 8 | 2 | 3 | 4 |
|---|---|---|---|---|---|---|---|---|---|---|---|---|---|---|
| 6 | 7 | 8 | 2 | 3 | 3 | 4 | 5 | 6 | 7 | 8 | 2 | 5 | 6 | 8 |
| 3 | 4 | 5 | 6 | 7 | 5 | 6 | 7 | 8 | 2 | 3 | 4 | 7 | 8 | 2 |

| 5 | 6 | 7 | 8 | 2 | 3 | 4 | 5 | 6 | 7 | 8 |
|---|---|---|---|---|---|---|---|---|---|---|
| 2 | 3 | 4 | 5 | 6 | 7 | 8 | 2 | 3 | 4 | 5 |
| 3 | 4 | 5 | 6 | 8 | 2 | 3 | 4 | 5 | 6 | 7 |

**BIBD 18 g = 8, k = 4, b = 14, r = 7, λ = 3, E = .86**

| 1 | 5 | 1 | 3 | 1 | 2 | 1 | 2 | 1 | 3 | 1 | 2 | 1 | 2 |
|---|---|---|---|---|---|---|---|---|---|---|---|---|---|
| 2 | 6 | 2 | 4 | 3 | 4 | 4 | 3 | 2 | 4 | 3 | 4 | 4 | 3 |
| 3 | 7 | 7 | 5 | 6 | 5 | 6 | 5 | 5 | 7 | 5 | 6 | 5 | 6 |
| 4 | 8 | 8 | 6 | 8 | 7 | 7 | 8 | 6 | 8 | 7 | 8 | 8 | 7 |

**BIBD 19 g = 8, k = 5, b = 56, r = 35, λ = 20, E = .91, YS**

| 1 | 6 | 4 | 3 | 2 | 1 | 5 | 7 | 2 | 6 | 1 | 4 | 7 | 3 | 5 |
|---|---|---|---|---|---|---|---|---|---|---|---|---|---|---|
| 2 | 1 | 7 | 5 | 3 | 2 | 1 | 4 | 6 | 5 | 6 | 1 | 3 | 7 | 4 |
| 3 | 2 | 1 | 6 | 5 | 3 | 2 | 1 | 4 | 7 | 4 | 7 | 1 | 5 | 6 |
| 4 | 3 | 2 | 1 | 7 | 6 | 4 | 5 | 1 | 2 | 3 | 5 | 6 | 1 | 7 |
| 5 | 4 | 3 | 2 | 1 | 7 | 6 | 2 | 7 | 1 | 5 | 3 | 4 | 6 | 1 |

| 2 | 7 | 6 | 5 | 4 | 3 | 8 | 8 | 8 | 8 | 8 | 8 | 8 | 1 | 2 |
|---|---|---|---|---|---|---|---|---|---|---|---|---|---|---|
| 3 | 2 | 7 | 6 | 5 | 4 | 1 | 2 | 3 | 4 | 5 | 6 | 7 | 8 | 8 |
| 4 | 3 | 2 | 7 | 6 | 5 | 2 | 3 | 4 | 5 | 6 | 7 | 1 | 2 | 3 |
| 5 | 4 | 3 | 2 | 7 | 6 | 3 | 4 | 5 | 6 | 7 | 1 | 2 | 3 | 4 |
| 6 | 5 | 4 | 3 | 2 | 7 | 4 | 5 | 6 | 7 | 1 | 2 | 3 | 5 | 6 |

| 3 | 4 | 5 | 6 | 7 | 1 | 2 | 3 | 4 | 5 | 6 | 7 | 1 | 3 | 3 |
|---|---|---|---|---|---|---|---|---|---|---|---|---|---|---|
| 8 | 8 | 8 | 8 | 8 | 2 | 3 | 4 | 5 | 6 | 7 | 1 | 2 | 3 | 4 |
| 4 | 5 | 6 | 7 | 1 | 8 | 8 | 8 | 8 | 8 | 8 | 8 | 4 | 5 | 6 |
| 5 | 6 | 7 | 1 | 2 | 3 | 4 | 5 | 6 | 7 | 1 | 2 | 8 | 8 | 8 |
| 7 | 1 | 2 | 3 | 4 | 6 | 7 | 1 | 2 | 3 | 4 | 5 | 5 | 6 | 7 |

| 4 | 5 | 6 | 7 | 1 | 2 | 3 | 4 | 5 | 6 | 7 |
|---|---|---|---|---|---|---|---|---|---|---|
| 5 | 6 | 7 | 1 | 2 | 3 | 4 | 5 | 6 | 7 | 1 |
| 7 | 1 | 2 | 3 | 4 | 5 | 6 | 7 | 1 | 2 | 3 |
| 8 | 8 | 8 | 8 | 6 | 7 | 1 | 2 | 3 | 4 | 5 |
| 1 | 2 | 3 | 4 | 8 | 8 | 8 | 8 | 8 | 8 | 8 |

**BIBD 20 g = 8, k = 6, b = 28, r = 21, $\lambda$ = 15, E = .95**

Unreduced

**BIBD 21 g = 8, k = 7, b = 8, r = 7, $\lambda$ = 6, E = .98, YS**

| 1 | 2 | 3 | 4 | 5 | 6 | 7 | 8 |
|---|---|---|---|---|---|---|---|
| 2 | 3 | 4 | 5 | 6 | 7 | 8 | 1 |
| 3 | 4 | 5 | 6 | 7 | 8 | 1 | 2 |
| 4 | 5 | 6 | 7 | 8 | 1 | 2 | 3 |
| 5 | 6 | 7 | 8 | 1 | 2 | 3 | 4 |
| 6 | 7 | 8 | 1 | 2 | 3 | 4 | 5 |
| 7 | 8 | 1 | 2 | 3 | 4 | 5 | 6 |

**BIBD 22 g = 9, k = 2, b = 36, r = 8, $\lambda$ = 1, E = .56, YS**

| 1 | 1 | 1 | 1 | 6 | 7 | 8 | 9 | 3 | 4 | 5 | 2 | 2 | 2 | 2 | 7 | 8 | 9 |
|---|---|---|---|---|---|---|---|---|---|---|---|---|---|---|---|---|---|
| 2 | 3 | 4 | 5 | 1 | 1 | 1 | 1 | 2 | 2 | 2 | 6 | 7 | 8 | 9 | 8 | 7 | 9 |

| 3 | 3 | 3 | 7 | 8 | 9 | 5 | 6 | 4 | 4 | 4 | 5 | 5 | 8 | 9 | 7 | 6 | 6 |
|---|---|---|---|---|---|---|---|---|---|---|---|---|---|---|---|---|---|
| 4 | 5 | 6 | 3 | 3 | 3 | 4 | 4 | 7 | 8 | 9 | 6 | 7 | 5 | 5 | 6 | 8 | 9 |

**BIBD 23 g = 9, k = 3, b = 12, r = 4, $\lambda$ = 1, E = .75**

| | | | | | | | | | | | |
|---|---|---|---|---|---|---|---|---|---|---|---|
| 1 | 4 | 7 | 1 | 2 | 3 | 1 | 2 | 3 | 1 | 2 | 3 |
| 2 | 5 | 8 | 4 | 5 | 6 | 6 | 4 | 5 | 5 | 6 | 4 |
| 3 | 6 | 9 | 7 | 8 | 9 | 8 | 9 | 7 | 9 | 7 | 8 |

**BIBD 24 g = 9, k = 4, b = 18, r = 8, $\lambda$ = 3, E = .84, YS**

| | | | | | | | | | | | | | | | | | |
|---|---|---|---|---|---|---|---|---|---|---|---|---|---|---|---|---|---|
| 1 | 2 | 3 | 4 | 5 | 6 | 7 | 8 | 9 | 1 | 2 | 3 | 4 | 5 | 6 | 7 | 8 | 9 |
| 2 | 3 | 4 | 5 | 6 | 7 | 8 | 9 | 1 | 4 | 5 | 6 | 7 | 8 | 9 | 1 | 2 | 3 |
| 3 | 4 | 5 | 6 | 7 | 8 | 9 | 1 | 2 | 6 | 7 | 8 | 9 | 1 | 2 | 3 | 4 | 5 |
| 5 | 6 | 7 | 8 | 9 | 1 | 2 | 3 | 4 | 9 | 1 | 2 | 3 | 4 | 5 | 6 | 7 | 8 |

**BIBD 25 g = 9, k = 5, b = 18, r = 10, $\lambda$ = 5, E = .9, YS**

| | | | | | | | | | | | | | | | | | |
|---|---|---|---|---|---|---|---|---|---|---|---|---|---|---|---|---|---|
| 4 | 5 | 6 | 7 | 8 | 9 | 1 | 2 | 3 | 2 | 3 | 4 | 5 | 6 | 7 | 8 | 9 | 1 |
| 6 | 7 | 8 | 9 | 1 | 2 | 3 | 4 | 5 | 3 | 4 | 5 | 6 | 7 | 8 | 9 | 1 | 2 |
| 7 | 8 | 9 | 1 | 2 | 3 | 4 | 5 | 6 | 5 | 6 | 7 | 8 | 9 | 1 | 2 | 3 | 4 |
| 8 | 9 | 1 | 2 | 3 | 4 | 5 | 6 | 7 | 7 | 8 | 9 | 1 | 2 | 3 | 4 | 5 | 6 |
| 9 | 1 | 2 | 3 | 4 | 5 | 6 | 7 | 8 | 8 | 9 | 1 | 2 | 3 | 4 | 5 | 6 | 7 |

**BIBD 26 g = 9, k = 6, b = 12, r = 8, $\lambda$ = 5, E = .94**

| | | | | | | | | | | | |
|---|---|---|---|---|---|---|---|---|---|---|---|
| 4 | 1 | 1 | 2 | 1 | 1 | 2 | 1 | 1 | 2 | 1 | 1 |
| 5 | 2 | 2 | 3 | 3 | 2 | 3 | 3 | 2 | 3 | 3 | 2 |
| 6 | 3 | 3 | 5 | 4 | 4 | 4 | 5 | 4 | 4 | 4 | 5 |
| 7 | 7 | 4 | 6 | 6 | 5 | 5 | 6 | 6 | 6 | 5 | 6 |
| 8 | 8 | 5 | 8 | 7 | 7 | 7 | 7 | 8 | 7 | 8 | 7 |
| 9 | 9 | 6 | 9 | 9 | 8 | 9 | 8 | 9 | 8 | 9 | 9 |

**BIBD 27 g = 9, k = 7, b = 36, r = 28, $\lambda$ = 21, E = .96, YS**

| | | | | | | | | | | | | | | | | | |
|---|---|---|---|---|---|---|---|---|---|---|---|---|---|---|---|---|---|
| 3 | 4 | 5 | 6 | 7 | 8 | 9 | 1 | 2 | 2 | 3 | 4 | 5 | 6 | 7 | 8 | 9 | 1 |
| 4 | 5 | 6 | 7 | 8 | 9 | 1 | 2 | 3 | 4 | 5 | 6 | 7 | 8 | 9 | 1 | 2 | 3 |
| 5 | 6 | 7 | 8 | 9 | 1 | 2 | 3 | 4 | 5 | 6 | 7 | 8 | 9 | 1 | 2 | 3 | 4 |
| 6 | 7 | 8 | 9 | 1 | 2 | 3 | 4 | 5 | 6 | 7 | 8 | 9 | 1 | 2 | 3 | 4 | 5 |
| 7 | 8 | 9 | 1 | 2 | 3 | 4 | 5 | 6 | 7 | 8 | 9 | 1 | 2 | 3 | 4 | 5 | 6 |
| 8 | 9 | 1 | 2 | 3 | 4 | 5 | 6 | 7 | 8 | 9 | 1 | 2 | 3 | 4 | 5 | 6 | 7 |
| 9 | 1 | 2 | 3 | 4 | 5 | 6 | 7 | 8 | 9 | 1 | 2 | 3 | 4 | 5 | 6 | 7 | 8 |

| 2 | 3 | 4 | 5 | 6 | 7 | 8 | 9 | 1 | 2 | 3 | 4 | 5 | 6 | 7 | 8 | 9 | 1 |
|---|---|---|---|---|---|---|---|---|---|---|---|---|---|---|---|---|---|
| 3 | 4 | 5 | 6 | 7 | 8 | 9 | 1 | 2 | 3 | 4 | 5 | 6 | 7 | 8 | 9 | 1 | 2 |
| 5 | 6 | 7 | 8 | 9 | 1 | 2 | 3 | 4 | 4 | 5 | 6 | 7 | 8 | 9 | 1 | 2 | 3 |
| 6 | 7 | 8 | 9 | 1 | 2 | 3 | 4 | 5 | 6 | 7 | 8 | 9 | 1 | 2 | 3 | 4 | 5 |
| 7 | 8 | 9 | 1 | 2 | 3 | 4 | 5 | 6 | 7 | 8 | 9 | 1 | 2 | 3 | 4 | 5 | 6 |
| 8 | 9 | 1 | 2 | 3 | 4 | 5 | 6 | 7 | 8 | 9 | 1 | 2 | 3 | 4 | 5 | 6 | 7 |
| 9 | 1 | 2 | 3 | 4 | 5 | 6 | 7 | 8 | 9 | 1 | 2 | 3 | 4 | 5 | 6 | 7 | 8 |

**BIBD 28 g = 9, k = 8, b = 9, r = 8, $\lambda$ = 7, E = .98, YS**

| 1 | 2 | 3 | 4 | 5 | 6 | 7 | 8 | 9 |
|---|---|---|---|---|---|---|---|---|
| 2 | 3 | 4 | 5 | 6 | 7 | 8 | 9 | 1 |
| 3 | 4 | 5 | 6 | 7 | 8 | 9 | 1 | 2 |
| 4 | 5 | 6 | 7 | 8 | 9 | 1 | 2 | 3 |
| 5 | 6 | 7 | 8 | 9 | 1 | 2 | 3 | 4 |
| 6 | 7 | 8 | 9 | 1 | 2 | 3 | 4 | 5 |
| 7 | 8 | 9 | 1 | 2 | 3 | 4 | 5 | 6 |
| 8 | 9 | 1 | 2 | 3 | 4 | 5 | 6 | 7 |

## C.3 Efficient Cyclic Designs

Using this table you can generate an incomplete block design for $g$ treatments in $b = mg$ blocks of size $k$ with each treatment appearing $r = mk$ times. The design will be the union of $m$ individual cyclic patterns, with these $m$ patterns determined by the first $m$ rows of this table for a given $k$. See John and Williams (1995).

| | | | $k$th treatment, $g =$ | | | | | | | | | |
|---|---|---|---|---|---|---|---|---|---|---|---|---|
| $k$ | $r$ | First $k-1$ treatments | 6 | 7 | 8 | 9 | 10 | 11 | 12 | 13 | 14 | 15 |
| 2 | 2 | 1 | 2 | 2 | 2 | 2 | 2 | 2 | 2 | 2 | 2 | 2 |
| | 4 | 1 | 3 | 4 | 4 | 4 | 4 | 4 | 4 | 6 | 5 | 5 |
| | 6 | 1 | 4 | 3 | 3 | 3 | 3 | 6 | 6 | 3 | 7 | 3 |
| | 8 | 1 | 6 | 5 | 5 | 5 | 5 | 3 | 3 | 5 | 4 | 8 |
| | 10 | 1 | 5 | 6 | 6 | 6 | 6 | 5 | 5 | 4 | 6 | 6 |
| 3 | 3 | 1 2 | 4 | 4 | 4 | 4 | 5 | 5 | 5 | 5 | 5 | 5 |
| | 6 | 1 3 | 2 | 4 | 8 | 7 | 8 | 8 | 6 | 8 | 8 | 9 |
| | 9 | 1 2 | 4 | 4 | 5 | 6 | 4 | 4 | 7 | 5 | 7 | 6 |
| 4 | 4 | 1 2 4 | 3 | 7 | 8 | 8 | 7 | 8 | 8 | 10 | 8 | 8 |
| | 8 | 1 2 5 | 3 | 7 | 8 | 9 | 3 | 7 | 7 | 7 | 7 | 7 |

| $k$ | $r$ | First $k-1$ treatments | $k$th treatment, $g =$ | | | | | | | | | |
|---|---|---|---|---|---|---|---|---|---|---|---|---|
| | | | 6 | 7 | 8 | 9 | 10 | 11 | 12 | 13 | 14 | 15 |
| 5 | 5 | 1 2 3 5 | 6 | 6 | 8 | 8 | 8 | 8 | 8 | 8 | 10 | 11 |
| | 10 | 1 3 4 5 | 6 | 6 | 8 | 9 | 10 | 9 | | | | |
| | 10 | 1 3 4 7 | | | | | | | 8 | 12 | 13 | 11 |
| 6 | 6 | 1 2 3 4 7 | | 6 | 6 | 6 | 6 | 11 | 11 | 11 | 11 | 11 |
| 7 | 7 | 1 2 3 4 5 8 | | | 6 | 6 | 10 | 10 | 10 | 10 | 10 | 11 |
| 8 | 8 | 1 2 3 4 5 7 9 | | | | 6 | 10 | 10 | 10 | 10 | 12 | 12 |
| 9 | 9 | 1 2 3 4 5 6 8 10 | | | | | 9 | 9 | 9 | 11 | 11 | 11 |
| 10 | 10 | 1 2 3 4 5 6 7 10 11 | | | | | | 8 | 8 | 8 | 13 | 13 |

## C.4 Alpha Designs

Alpha Designs are resolvable block designs for $g = mk$ treatments in $b = mr$ blocks of size $k$. These tables give the initial alpha arrays for $5 \le m \le 15$, block sizes from 4 up to the minimum of $m$ and $100/m$, and up to four replications. These tables are adapted from Table 2 of Patterson, Williams, and Hunter (1978).

$m = 5$, $4 \le k \le 5$

| | | | |
|---|---|---|---|
| 1 | 1 | 1 | 1 |
| 1 | 2 | 5 | 3 |
| 1 | 3 | 4 | 5 |
| 1 | 4 | 3 | 2 |
| 1 | 5 | 2 | 4 |

$m = 6$, $4 \le k \le 6$

| | | | |
|---|---|---|---|
| 1 | 1 | 1 | 1 |
| 1 | 2 | 6 | 5 |
| 1 | 4 | 3 | 6 |
| 1 | 3 | 4 | 2 |
| 1 | 5 | 2 | 3 |
| 1 | 6 | 2 | 4 |

$m = 7$, $4 \le k \le 7$

| | | | |
|---|---|---|---|
| 1 | 1 | 1 | 1 |
| 1 | 2 | 4 | 3 |
| 1 | 3 | 7 | 5 |
| 1 | 5 | 6 | 2 |
| 1 | 4 | 3 | 7 |
| 1 | 6 | 2 | 4 |
| 1 | 7 | 5 | 6 |

$m = 8$, $4 \le k \le 8$

| | | | |
|---|---|---|---|
| 1 | 1 | 1 | 1 |
| 1 | 2 | 3 | 7 |
| 1 | 4 | 8 | 2 |
| 1 | 6 | 4 | 5 |
| 1 | 3 | 6 | 4 |
| 1 | 5 | 2 | 7 |
| 1 | 7 | 1 | 3 |
| 1 | 8 | 7 | 6 |

$m = 9$, $4 \le k \le 9$

| | | | |
|---|---|---|---|
| 1 | 1 | 1 | 1 |
| 1 | 2 | 9 | 8 |
| 1 | 4 | 7 | 5 |
| 1 | 8 | 3 | 4 |
| 1 | 3 | 4 | 6 |
| 1 | 5 | 2 | 7 |
| 1 | 6 | 8 | 3 |
| 1 | 7 | 6 | 2 |
| 1 | 9 | 5 | 8 |

$m = 10$, $4 \le k \le 10$

| | | | |
|---|---|---|---|
| 1 | 1 | 1 | 1 |
| 1 | 2 | 10 | 6 |
| 1 | 4 | 7 | 10 |
| 1 | 6 | 8 | 3 |
| 1 | 5 | 6 | 7 |
| 1 | 7 | 4 | 2 |
| 1 | 8 | 3 | 5 |
| 1 | 9 | 5 | 8 |
| 1 | 10 | 9 | 3 |
| 1 | 3 | 7 | 4 |

| $m = 11$, $4 \leq k \leq 9$ | | | |
|---|---|---|---|
| 1 | 1 | 1 | 1 |
| 1 | 2 | 7 | 8 |
| 1 | 5 | 9 | 2 |
| 1 | 10 | 8 | 6 |
| 1 | 3 | 4 | 7 |
| 1 | 6 | 2 | 4 |
| 1 | 7 | 6 | 11 |
| 1 | 4 | 10 | 5 |
| 1 | 8 | 5 | 2 |

| $m = 12$, $4 \leq k \leq 8$ | | | |
|---|---|---|---|
| 1 | 1 | 1 | 1 |
| 1 | 2 | 3 | 4 |
| 1 | 8 | 6 | 2 |
| 1 | 10 | 7 | 5 |
| 1 | 5 | 12 | 9 |
| 1 | 12 | 4 | 11 |
| 1 | 11 | 5 | 8 |
| 1 | 6 | 2 | 7 |

| $m = 13$, $4 \leq k \leq 7$ | | | |
|---|---|---|---|
| 1 | 1 | 1 | 1 |
| 1 | 2 | 5 | 11 |
| 1 | 4 | 9 | 12 |
| 1 | 10 | 3 | 2 |
| 1 | 13 | 11 | 7 |
| 1 | 9 | 6 | 13 |
| 1 | 7 | 8 | 9 |

| $m = 14$, $4 \leq k \leq 7$ | | | |
|---|---|---|---|
| 1 | 1 | 1 | 1 |
| 1 | 2 | 9 | 11 |
| 1 | 10 | 11 | 8 |
| 1 | 12 | 14 | 3 |
| 1 | 3 | 7 | 2 |
| 1 | 6 | 12 | 13 |
| 1 | 4 | 2 | 12 |

| $m = 15$, $4 \leq k \leq 6$ | | | |
|---|---|---|---|
| 1 | 1 | 1 | 1 |
| 1 | 2 | 9 | 8 |
| 1 | 4 | 13 | 15 |
| 1 | 8 | 3 | 6 |
| 1 | 11 | 14 | 12 |
| 1 | 15 | 4 | 9 |

## C.5 Two-Series Confounding and Fractioning Plans

The table gives suggested defining contrasts for confounding a $2^k$ design into $2^p$ blocks. It also gives the generalized interactions that are confounded. When only a particular block of the design is run, the resulting $2^{k-p}$ fractional factorial has aliases of $I$ the same as the defining contrasts and their interactions. Other fractions have the same basic aliases, though the signs differ.

| $k$ | $2^p$ | Defining contrasts | Generalized interactions |
|---|---|---|---|
| 3 | 2 | ABC | |
| | 4 | AB, BC | AC |
| 4 | 2 | ABCD | |
| | 4 | ABC, AD | BCD |
| | 8 | AB, BC, CD | AC, AD, BD, ABCD |

| $k$ | $2^p$ | Defining contrasts | Generalized interactions |
|---|---|---|---|
| 5 | 2 | ABCDE | |
| | 4 | ABCD, BCE | ADE |
| | 8 | ABC, BD, AE | ACD, BCE, ABDE, CDE |
| | 16 | AB, BC, CD, DE | AC, ABCD, BD, AD, ABDE, BCDE, ACDE, CE, ABCE, BE, AE |
| 6 | 2 | ABCDEF | |
| | 4 | BCDE, ABDF | ACEF |
| | 8 | ABCD, BCE, ACF | ADE, BDF, ABEF, CDEF |
| | 16 | CD, ACE, BCF, ABC | ADE, BDF, ABEF, ABCDEF, ABD, BE, BCDE, AF, ACDF, CEF, DEF |
| | 32 | AB, BC, CD, DE, EF | All other two-factor interactions, plus all four-factor and six-factor interactions |
| 7 | 2 | ABCDEFG | |
| | 4 | ADEF, ABCDG | BCEFG |
| | 8 | BCDE, ACDF, ABCG | ABEF, ADEG, BDFG, CEFG |
| | 16 | ABCD, BCE, ACF, ABG | ADE, BDF, ABEF, CDEF, CDG, ACEG, BDEG, BCFG, ADFG, EFG, ABCDEFG |
| | 32 | ADG, ACG, ABG, ABF, CEF | CD, BD, BC, ABCDG, BDFG, BCFG, ABCDF, FG, ADF, ACF, CDFG, ACDEFG, AEFG, DEF, ABCEFG, BCDEF, BEF, ABDEFG, ABCE, BCDEG, BEG, ABDE, CEG, ACDE, AE, DEG |
| | 64 | AB, BC, CD, DE, EF, FG | All other two-factor interactions, plus all four-factor and six-factor interactions |
| 8 | 2 | ABCDEFGH | |
| | 4 | ABDFG, BCDEH | ACEFGH |
| | 8 | BCEG, BCDH, ACDEF | DEGH, ABDFG, ABEFH, ACFGH |
| | 16 | BCDE, ACDF, ABDG, ABCH | ABEF, ACEG, BCFG, DEFG, ADEH, BDFH, CEFH, CDGH, BEGH, AFGH, ABCDEFGH |

| $k$ | $2^p$ | Defining contrasts | Generalized interactions |
|---|---|---|---|
| 8 | 32 | ABD, ACE, BCF, ABCG, ABCH | BCDE, ACDF, ABEF, DEF, CDG, BEG, ADEG, AFG, BDFG, CEFG, ABCDEFG, CDH, BEH, ADEH, AFH, BDFH, CEFH, ABCDEFH, GH, ABDGH, ACEGH, BCDEGH, BCFGH, ACDFGH, ABEFGH, DEFGH |
| | 64 | AG, BF, BCE, AEF, BDG, ADH | ABFG, ABCEG, CEF, ACEFG, EFG, ABE, BEG, ABCF, BCFG, AC, CG, ABD, DFG, ADF, CDEG, ACDE, BCDEFG, ABCDEF, ABDEFG, BDEF, ADEG, DE, ACDFG, CDF, ABCDG, BCD, DGH, ABDFH, BDFGH, ABCDEH, BCDEGH, ACDEFH, CDEFGH, DEFH, ADEFGH, BDEH, ABDEGH, BCDFH, ABCDFGH, CDH, ACDGH, ABGH, BH, AFGH, FH, ACEGH, CEH, ABCEFGH, BCEFH, BEFGH, ABEFH, EGH, AEH, CFGH, ACFH, BCGH, ABCH |

# Appendix D

# Tables

| | |
|---|---|
| **Table D.1** | Random digits. |
| **Table D.2** | Tail areas for the standard normal distribution. |
| **Table D.3** | Percent points for the Student's $t$ distribution. |
| **Table D.4** | Percent points for the chi-square distribution. |
| **Table D.5** | Percent points for the F distribution. You may use the relation $F_{1-\mathcal{E},\nu_1,\nu_2} = 1/F_{\mathcal{E},\nu_2,\nu_1}$ to determine lower percent points of $F$. |
| **Table D.6** | Coefficients of orthogonal polynomial contrasts. |
| **Table D.7** | Critical values for Bonferroni $t$. |
| **Table D.8** | Percent points for the Studentized range. |
| **Table D.9** | Critical values for Dunnett's $t$. |
| **Table D.10** | Power curves for fixed-effects ANOVA. For each numerator degrees of freedom, thin and thick lines indicate power at the .05 and .01 levels respectively. Within a significance level, the lines indicate 8, 9, 10, 12, 15, 20, 30, and 60 denominator degrees of freedom (8 df on the bottom, 60 on top of each group). The vertical axis is power, and the horizontal axis is the noncentrality parameter $\sum_{i=1}^{g} \alpha_i^2/\sigma^2$. The curves for the .01 level are shifted to the right by 40 units. |
| **Table D.11** | Power curves for random-effects ANOVA. For each numerator degrees of freedom, thin and thick lines indicate power at the .05 and .01 levels respectively. Within a significance level, the lines indicate 2, 3, 4, 6, 8, 16, 32, and 256 denominator degrees of freedom (2 df on the bottom, 256 on top of each group). The vertical axis is power, and the horizontal axis is the ratio of the numerator and denominator EMS's. The curves for the .01 level are shifted to the right a factor of 10 on the horizontal axis. |

All table values were computed in MacAnova.

**Table D.1:** Random digits.

| | | | | | | | | | |
|---|---|---|---|---|---|---|---|---|---|
| 68094 | 23539 | 18913 | 86955 | 39327 | 02225 | 69423 | 06689 | 99791 | 76722 |
| 01909 | 10889 | 72439 | 61293 | 21529 | 36388 | 14555 | 95914 | 25254 | 38422 |
| 81253 | 33731 | 00873 | 30545 | 50227 | 94749 | 07761 | 77740 | 19743 | 21724 |
| 20501 | 57876 | 10081 | 07431 | 91817 | 25296 | 52198 | 75278 | 45922 | 19728 |
| 30557 | 32116 | 68368 | 18292 | 37433 | 27636 | 92360 | 74374 | 00155 | 19623 |
| 91740 | 24671 | 12987 | 73192 | 97251 | 12516 | 38695 | 12790 | 63529 | 58111 |
| 08388 | 48988 | 91806 | 24777 | 61809 | 84551 | 29619 | 26471 | 87362 | 05818 |
| 76006 | 06178 | 10765 | 76938 | 42086 | 66950 | 90720 | 88483 | 66611 | 19710 |
| 72600 | 85770 | 88793 | 66291 | 41081 | 61031 | 60104 | 02545 | 86041 | 62345 |
| 32209 | 77328 | 41324 | 68614 | 57322 | 94583 | 07415 | 27313 | 26322 | 93218 |
| 38420 | 57120 | 12268 | 15017 | 44456 | 90919 | 73640 | 69974 | 61200 | 82209 |
| 49690 | 34002 | 11553 | 49387 | 44354 | 92179 | 79960 | 61804 | 70374 | 71782 |
| 85210 | 59681 | 38002 | 41958 | 90125 | 02819 | 78165 | 44800 | 17792 | 96272 |
| 35229 | 78839 | 46776 | 00944 | 67288 | 59471 | 23715 | 05753 | 87214 | 06758 |
| 78568 | 94584 | 71728 | 81741 | 38433 | 59390 | 57344 | 27554 | 90465 | 95245 |
| 00679 | 26121 | 29667 | 83237 | 67154 | 10246 | 33005 | 72851 | 34876 | 29007 |
| 15398 | 98457 | 22406 | 30927 | 90111 | 14065 | 51246 | 18592 | 85397 | 92122 |
| 89014 | 44909 | 62227 | 24503 | 59774 | 69233 | 29556 | 14126 | 26810 | 67044 |
| 84538 | 98456 | 19149 | 54714 | 36332 | 89999 | 02248 | 26089 | 77989 | 98072 |
| 33618 | 91123 | 84227 | 34110 | 74523 | 73244 | 27365 | 89167 | 02035 | 90366 |
| 48194 | 17487 | 33892 | 64522 | 69065 | 98755 | 49765 | 90609 | 57786 | 31991 |
| 54929 | 29666 | 72716 | 59146 | 86232 | 38765 | 33335 | 35127 | 71464 | 69505 |
| 13639 | 16775 | 89564 | 73978 | 73321 | 63868 | 65447 | 15689 | 37789 | 22178 |
| 28420 | 16687 | 25081 | 99131 | 15641 | 59055 | 11472 | 31110 | 58669 | 49621 |
| 57905 | 96871 | 07126 | 01978 | 06563 | 18504 | 80138 | 96710 | 51019 | 13183 |
| 36490 | 13154 | 96356 | 90278 | 47401 | 47783 | 14283 | 47107 | 43874 | 73050 |
| 15852 | 60522 | 54438 | 97802 | 18869 | 06219 | 62244 | 67309 | 21556 | 62034 |
| 28614 | 54310 | 58953 | 24393 | 09880 | 69588 | 34399 | 19114 | 17086 | 19286 |
| 92594 | 10130 | 04030 | 12348 | 62118 | 35368 | 11032 | 28513 | 38832 | 49642 |
| 10119 | 22185 | 14692 | 59461 | 98941 | 51851 | 82728 | 60066 | 75060 | 48027 |
| 27970 | 68214 | 84216 | 82761 | 54280 | 98276 | 48123 | 50611 | 11562 | 44945 |
| 83423 | 24025 | 55539 | 30343 | 44943 | 79061 | 54400 | 09157 | 08448 | 81417 |
| 91821 | 56637 | 02232 | 65331 | 24585 | 58902 | 70981 | 84902 | 30673 | 66372 |
| 56385 | 90995 | 94482 | 90187 | 15461 | 78394 | 38276 | 07567 | 17556 | 42504 |
| 45081 | 92518 | 67475 | 26920 | 36524 | 67476 | 11973 | 65938 | 74470 | 80782 |
| 87655 | 77363 | 79749 | 74171 | 35109 | 51652 | 32671 | 47315 | 50862 | 24683 |
| 77287 | 08196 | 64511 | 04557 | 45941 | 87701 | 00805 | 64707 | 43178 | 32760 |
| 60633 | 66288 | 95791 | 18232 | 14346 | 80974 | 50836 | 21944 | 24407 | 95112 |
| 03089 | 42195 | 14802 | 55732 | 92821 | 48338 | 27293 | 61239 | 70050 | 83121 |
| 10570 | 71691 | 04943 | 33707 | 35118 | 06278 | 28534 | 79418 | 85857 | 52665 |

Table D.1: Random digits, continued.

| | | | | | | | | | |
|---|---|---|---|---|---|---|---|---|---|
| 30263 | 25135 | 17075 | 56131 | 64430 | 43573 | 77506 | 09510 | 65985 | 17159 |
| 13811 | 98464 | 48063 | 98483 | 60748 | 07379 | 89540 | 07699 | 60560 | 93391 |
| 80280 | 46665 | 54480 | 90895 | 94555 | 77376 | 55074 | 69674 | 22124 | 86546 |
| 96302 | 09821 | 31198 | 06423 | 69016 | 71408 | 48673 | 22035 | 92401 | 40242 |
| 34922 | 65539 | 17012 | 69492 | 97661 | 66351 | 94296 | 00451 | 99255 | 98999 |
| 81090 | 48413 | 74876 | 24165 | 42912 | 58517 | 51494 | 80415 | 28758 | 96355 |
| 67224 | 24891 | 38160 | 78489 | 73226 | 95368 | 19123 | 78424 | 47010 | 44371 |
| 63204 | 25405 | 51831 | 00562 | 23640 | 97596 | 73613 | 31668 | 81299 | 13975 |
| 39678 | 79440 | 84900 | 06251 | 93120 | 57470 | 68970 | 82673 | 88484 | 93689 |
| 30374 | 19502 | 99804 | 25596 | 07763 | 02914 | 05334 | 52321 | 74595 | 47068 |
| 06813 | 76019 | 12479 | 03459 | 51078 | 44527 | 02086 | 01367 | 26591 | 69118 |
| 57097 | 14846 | 92151 | 95357 | 73479 | 53708 | 04442 | 30282 | 82320 | 99043 |
| 09521 | 48055 | 19823 | 82346 | 38890 | 31327 | 98995 | 37520 | 73670 | 48277 |
| 77991 | 19227 | 65802 | 92645 | 13378 | 06593 | 52303 | 15173 | 98557 | 43631 |
| 47605 | 33709 | 36996 | 22976 | 78611 | 39221 | 95962 | 06137 | 72056 | 44395 |
| 29969 | 01292 | 47429 | 28477 | 72881 | 83330 | 57842 | 96953 | 66190 | 29761 |
| 26978 | 10916 | 24087 | 68880 | 42657 | 93404 | 74540 | 22069 | 56907 | 53591 |
| 43115 | 41945 | 85148 | 43539 | 19452 | 69583 | 88827 | 22232 | 52494 | 19895 |
| 51493 | 62141 | 57091 | 26829 | 61899 | 03433 | 04983 | 85869 | 31376 | 31307 |
| 57731 | 27002 | 19954 | 12314 | 10234 | 99589 | 59101 | 28150 | 65083 | 85057 |
| 37816 | 75263 | 68459 | 32095 | 15844 | 20352 | 46919 | 82419 | 59487 | 78779 |
| 65009 | 90859 | 76655 | 46234 | 24073 | 93183 | 85770 | 60190 | 69870 | 44997 |
| 89443 | 17030 | 30366 | 18026 | 64815 | 64790 | 24439 | 24153 | 75360 | 85068 |
| 19978 | 11146 | 54195 | 18001 | 39458 | 50082 | 47801 | 79655 | 11199 | 00978 |
| 69137 | 35105 | 62192 | 60958 | 32109 | 00787 | 79202 | 74700 | 27231 | 39559 |
| 00102 | 19753 | 27900 | 16409 | 42548 | 81604 | 16881 | 03009 | 62624 | 94651 |
| 86465 | 06647 | 56974 | 45774 | 38612 | 54604 | 35113 | 14259 | 08609 | 86134 |
| 74692 | 64914 | 61361 | 55581 | 79265 | 85121 | 94402 | 66705 | 02455 | 63518 |
| 25531 | 67924 | 61704 | 95032 | 48824 | 40759 | 83063 | 89562 | 74811 | 42721 |
| 87057 | 63223 | 84910 | 27744 | 36979 | 00578 | 63738 | 47473 | 66356 | 59676 |
| 22723 | 61335 | 89609 | 98968 | 78238 | 94353 | 11790 | 62264 | 78866 | 86637 |
| 61837 | 60095 | 22904 | 83603 | 57362 | 85576 | 24298 | 25868 | 08558 | 17143 |
| 07208 | 30664 | 53006 | 15714 | 92246 | 91157 | 97898 | 43295 | 26162 | 85001 |
| 09265 | 97806 | 06556 | 70909 | 24791 | 81907 | 92463 | 80405 | 32493 | 57985 |
| 60079 | 09778 | 70500 | 69276 | 16192 | 39024 | 42519 | 69661 | 59750 | 15740 |
| 11620 | 30055 | 59498 | 63231 | 90667 | 12729 | 99405 | 17906 | 20684 | 65483 |
| 20210 | 31650 | 23408 | 32631 | 87779 | 62148 | 03322 | 98071 | 41217 | 03952 |
| 91935 | 61772 | 67324 | 44921 | 75176 | 32383 | 21611 | 23145 | 51109 | 13168 |
| 15449 | 91085 | 09246 | 06833 | 93677 | 60567 | 20180 | 59763 | 01650 | 41798 |
| 33759 | 00216 | 03782 | 18185 | 98508 | 07890 | 02365 | 50624 | 55194 | 85954 |
| 59706 | 03210 | 55372 | 71993 | 55247 | 40554 | 12783 | 36287 | 19884 | 58491 |

**Table D.2:** Tail areas for the standard normal distribution.

Table entries are $\mathcal{E} = P(Z > z_{\mathcal{E}}) = 1 - \Phi(z_{\mathcal{E}})$.

| $z_{\mathcal{E}}$ | .00 | .01 | .02 | .03 | .04 | .05 | .06 | .07 | .08 | .09 |
|---|---|---|---|---|---|---|---|---|---|---|
| .0 | .50000 | .49601 | .49202 | .48803 | .48405 | .48006 | .47608 | .47210 | .46812 | .46414 |
| .1 | .46017 | .45620 | .45224 | .44828 | .44433 | .44038 | .43644 | .43251 | .42858 | .42465 |
| .2 | .42074 | .41683 | .41294 | .40905 | .40517 | .40129 | .39743 | .39358 | .38974 | .38591 |
| .3 | .38209 | .37828 | .37448 | .37070 | .36693 | .36317 | .35942 | .35569 | .35197 | .34827 |
| .4 | .34458 | .34090 | .33724 | .33360 | .32997 | .32636 | .32276 | .31918 | .31561 | .31207 |
| .5 | .30854 | .30503 | .30153 | .29806 | .29460 | .29116 | .28774 | .28434 | .28096 | .27760 |
| .6 | .27425 | .27093 | .26763 | .26435 | .26109 | .25785 | .25463 | .25143 | .24825 | .24510 |
| .7 | .24196 | .23885 | .23576 | .23270 | .22965 | .22663 | .22363 | .22065 | .21770 | .21476 |
| .8 | .21186 | .20897 | .20611 | .20327 | .20045 | .19766 | .19489 | .19215 | .18943 | .18673 |
| .9 | .18406 | .18141 | .17879 | .17619 | .17361 | .17106 | .16853 | .16602 | .16354 | .16109 |
| 1.0 | .15866 | .15625 | .15386 | .15151 | .14917 | .14686 | .14457 | .14231 | .14007 | .13786 |
| 1.1 | .13567 | .13350 | .13136 | .12924 | .12714 | .12507 | .12302 | .12100 | .11900 | .11702 |
| 1.2 | .11507 | .11314 | .11123 | .10935 | .10749 | .10565 | .10383 | .10204 | .10027 | .09853 |
| 1.3 | .09680 | .09510 | .09342 | .09176 | .09012 | .08851 | .08691 | .08534 | .08379 | .08226 |
| 1.4 | .08076 | .07927 | .07780 | .07636 | .07493 | .07353 | .07215 | .07078 | .06944 | .06811 |
| 1.5 | .06681 | .06552 | .06426 | .06301 | .06178 | .06057 | .05938 | .05821 | .05705 | .05592 |
| 1.6 | .05480 | .05370 | .05262 | .05155 | .05050 | .04947 | .04846 | .04746 | .04648 | .04551 |
| 1.7 | .04457 | .04363 | .04272 | .04182 | .04093 | .04006 | .03920 | .03836 | .03754 | .03673 |
| 1.8 | .03593 | .03515 | .03438 | .03362 | .03288 | .03216 | .03144 | .03074 | .03005 | .02938 |
| 1.9 | .02872 | .02807 | .02743 | .02680 | .02619 | .02559 | .02500 | .02442 | .02385 | .02330 |
| 2.0 | .02275 | .02222 | .02169 | .02118 | .02068 | .02018 | .01970 | .01923 | .01876 | .01831 |
| 2.1 | .01786 | .01743 | .01700 | .01659 | .01618 | .01578 | .01539 | .01500 | .01463 | .01426 |
| 2.2 | .01390 | .01355 | .01321 | .01287 | .01255 | .01222 | .01191 | .01160 | .01130 | .01101 |
| 2.3 | .01072 | .01044 | .01017 | .00990 | .00964 | .00939 | .00914 | .00889 | .00866 | .00842 |
| 2.4 | .00820 | .00798 | .00776 | .00755 | .00734 | .00714 | .00695 | .00676 | .00657 | .00639 |
| 2.5 | .00621 | .00604 | .00587 | .00570 | .00554 | .00539 | .00523 | .00508 | .00494 | .00480 |
| 2.6 | .00466 | .00453 | .00440 | .00427 | .00415 | .00402 | .00391 | .00379 | .00368 | .00357 |
| 2.7 | .00347 | .00336 | .00326 | .00317 | .00307 | .00298 | .00289 | .00280 | .00272 | .00264 |
| 2.8 | .00256 | .00248 | .00240 | .00233 | .00226 | .00219 | .00212 | .00205 | .00199 | .00193 |
| 2.9 | .00187 | .00181 | .00175 | .00169 | .00164 | .00159 | .00154 | .00149 | .00144 | .00139 |
| 3.0 | .00135 | .00131 | .00126 | .00122 | .00118 | .00114 | .00111 | .00107 | .00104 | .00100 |
| 3.1 | .00097 | .00094 | .00090 | .00087 | .00084 | .00082 | .00079 | .00076 | .00074 | .00071 |
| 3.2 | .00069 | .00066 | .00064 | .00062 | .00060 | .00058 | .00056 | .00054 | .00052 | .00050 |
| 3.3 | .00048 | .00047 | .00045 | .00043 | .00042 | .00040 | .00039 | .00038 | .00036 | .00035 |
| 3.4 | .00034 | .00032 | .00031 | .00030 | .00029 | .00028 | .00027 | .00026 | .00025 | .00024 |

**Table D.3:** Percent points for the Student $t$ distribution.

Table entries are $t_{\mathcal{E},\nu}$ where $P_\nu(t > t_{\mathcal{E},\nu}) = \mathcal{E}$.

| | $\mathcal{E}$ | | | | | | | | |
|---|---|---|---|---|---|---|---|---|---|
| $\nu$ | .2 | .1 | .05 | .025 | .01 | .005 | .001 | .0005 | .0001 |
| 1 | 1.376 | 3.078 | 6.314 | 12.71 | 31.82 | 63.66 | 318.3 | 636.6 | 3183 |
| 2 | 1.061 | 1.886 | 2.920 | 4.303 | 6.965 | 9.925 | 22.33 | 31.60 | 70.70 |
| 3 | .978 | 1.638 | 2.353 | 3.182 | 4.541 | 5.841 | 10.22 | 12.92 | 22.20 |
| 4 | .941 | 1.533 | 2.132 | 2.776 | 3.747 | 4.604 | 7.173 | 8.610 | 13.03 |
| 5 | .920 | 1.476 | 2.015 | 2.571 | 3.365 | 4.032 | 5.893 | 6.869 | 9.678 |
| 6 | .906 | 1.440 | 1.943 | 2.447 | 3.143 | 3.707 | 5.208 | 5.959 | 8.025 |
| 7 | .896 | 1.415 | 1.895 | 2.365 | 2.998 | 3.499 | 4.785 | 5.408 | 7.063 |
| 8 | .889 | 1.397 | 1.860 | 2.306 | 2.896 | 3.355 | 4.501 | 5.041 | 6.442 |
| 9 | .883 | 1.383 | 1.833 | 2.262 | 2.821 | 3.250 | 4.297 | 4.781 | 6.010 |
| 10 | .879 | 1.372 | 1.812 | 2.228 | 2.764 | 3.169 | 4.144 | 4.587 | 5.694 |
| 11 | .876 | 1.363 | 1.796 | 2.201 | 2.718 | 3.106 | 4.025 | 4.437 | 5.453 |
| 12 | .873 | 1.356 | 1.782 | 2.179 | 2.681 | 3.055 | 3.930 | 4.318 | 5.263 |
| 13 | .870 | 1.350 | 1.771 | 2.160 | 2.650 | 3.012 | 3.852 | 4.221 | 5.111 |
| 14 | .868 | 1.345 | 1.761 | 2.145 | 2.624 | 2.977 | 3.787 | 4.140 | 4.985 |
| 15 | .866 | 1.341 | 1.753 | 2.131 | 2.602 | 2.947 | 3.733 | 4.073 | 4.880 |
| 16 | .865 | 1.337 | 1.746 | 2.120 | 2.583 | 2.921 | 3.686 | 4.015 | 4.791 |
| 17 | .863 | 1.333 | 1.740 | 2.110 | 2.567 | 2.898 | 3.646 | 3.965 | 4.714 |
| 18 | .862 | 1.330 | 1.734 | 2.101 | 2.552 | 2.878 | 3.610 | 3.922 | 4.648 |
| 19 | .861 | 1.328 | 1.729 | 2.093 | 2.539 | 2.861 | 3.579 | 3.883 | 4.590 |
| 20 | .860 | 1.325 | 1.725 | 2.086 | 2.528 | 2.845 | 3.552 | 3.850 | 4.539 |
| 21 | .859 | 1.323 | 1.721 | 2.080 | 2.518 | 2.831 | 3.527 | 3.819 | 4.493 |
| 22 | .858 | 1.321 | 1.717 | 2.074 | 2.508 | 2.819 | 3.505 | 3.792 | 4.452 |
| 23 | .858 | 1.319 | 1.714 | 2.069 | 2.500 | 2.807 | 3.485 | 3.768 | 4.415 |
| 24 | .857 | 1.318 | 1.711 | 2.064 | 2.492 | 2.797 | 3.467 | 3.745 | 4.382 |
| 25 | .856 | 1.316 | 1.708 | 2.060 | 2.485 | 2.787 | 3.450 | 3.725 | 4.352 |
| 26 | .856 | 1.315 | 1.706 | 2.056 | 2.479 | 2.779 | 3.435 | 3.707 | 4.324 |
| 27 | .855 | 1.314 | 1.703 | 2.052 | 2.473 | 2.771 | 3.421 | 3.690 | 4.299 |
| 28 | .855 | 1.313 | 1.701 | 2.048 | 2.467 | 2.763 | 3.408 | 3.674 | 4.275 |
| 29 | .854 | 1.311 | 1.699 | 2.045 | 2.462 | 2.756 | 3.396 | 3.659 | 4.254 |
| 30 | .854 | 1.310 | 1.697 | 2.042 | 2.457 | 2.750 | 3.385 | 3.646 | 4.234 |
| 35 | .852 | 1.306 | 1.690 | 2.030 | 2.438 | 2.724 | 3.340 | 3.591 | 4.153 |
| 40 | .851 | 1.303 | 1.684 | 2.021 | 2.423 | 2.704 | 3.307 | 3.551 | 4.094 |
| 45 | .850 | 1.301 | 1.679 | 2.014 | 2.412 | 2.690 | 3.281 | 3.520 | 4.049 |
| 50 | .849 | 1.299 | 1.676 | 2.009 | 2.403 | 2.678 | 3.261 | 3.496 | 4.014 |
| 60 | .848 | 1.296 | 1.671 | 2.000 | 2.390 | 2.660 | 3.232 | 3.460 | 3.962 |

**Table D.4:** Percent points for the chi-square distribution.

Table entries are $\chi^2_{\mathcal{E},\nu}$ where $P_\nu(\chi^2 > \chi^2_{\mathcal{E},\nu}) = \mathcal{E}$.

| | $\mathcal{E}$ | | | | | | | |
|---|---|---|---|---|---|---|---|---|
| $\nu$ | .995 | .99 | .975 | .95 | .05 | .025 | .01 | .005 |
| 1 | .000039 | .00016 | .0010 | .0039 | 3.841 | 5.024 | 6.635 | 7.879 |
| 2 | .0100 | .0201 | .0506 | .1026 | 5.991 | 7.378 | 9.210 | 10.60 |
| 3 | .0717 | .1148 | .2158 | .3518 | 7.815 | 9.348 | 11.34 | 12.84 |
| 4 | .2070 | .2971 | .4844 | .7107 | 9.488 | 11.14 | 13.28 | 14.86 |
| 5 | .4117 | .5543 | .8312 | 1.145 | 11.07 | 12.83 | 15.09 | 16.75 |
| 6 | .6757 | .8721 | 1.237 | 1.635 | 12.59 | 14.45 | 16.81 | 18.55 |
| 7 | .9893 | 1.239 | 1.690 | 2.167 | 14.07 | 16.01 | 18.48 | 20.28 |
| 8 | 1.344 | 1.646 | 2.180 | 2.733 | 15.51 | 17.53 | 20.09 | 21.95 |
| 9 | 1.735 | 2.088 | 2.700 | 3.325 | 16.92 | 19.02 | 21.67 | 23.59 |
| 10 | 2.156 | 2.558 | 3.247 | 3.940 | 18.31 | 20.48 | 23.21 | 25.19 |
| 11 | 2.603 | 3.053 | 3.816 | 4.575 | 19.68 | 21.92 | 24.72 | 26.76 |
| 12 | 3.074 | 3.571 | 4.404 | 5.226 | 21.03 | 23.34 | 26.22 | 28.30 |
| 13 | 3.565 | 4.107 | 5.009 | 5.892 | 22.36 | 24.74 | 27.69 | 29.82 |
| 14 | 4.075 | 4.660 | 5.629 | 6.571 | 23.68 | 26.12 | 29.14 | 31.32 |
| 15 | 4.601 | 5.229 | 6.262 | 7.261 | 25.00 | 27.49 | 30.58 | 32.80 |
| 16 | 5.142 | 5.812 | 6.908 | 7.962 | 26.30 | 28.85 | 32.00 | 34.27 |
| 17 | 5.697 | 6.408 | 7.564 | 8.672 | 27.59 | 30.19 | 33.41 | 35.72 |
| 18 | 6.265 | 7.015 | 8.231 | 9.390 | 28.87 | 31.53 | 34.81 | 37.16 |
| 19 | 6.844 | 7.633 | 8.907 | 10.12 | 30.14 | 32.85 | 36.19 | 38.58 |
| 20 | 7.434 | 8.260 | 9.591 | 10.85 | 31.41 | 34.17 | 37.57 | 40.00 |
| 21 | 8.034 | 8.897 | 10.28 | 11.59 | 32.67 | 35.48 | 38.93 | 41.40 |
| 22 | 8.643 | 9.542 | 10.98 | 12.34 | 33.92 | 36.78 | 40.29 | 42.80 |
| 23 | 9.260 | 10.20 | 11.69 | 13.09 | 35.17 | 38.08 | 41.64 | 44.18 |
| 24 | 9.886 | 10.86 | 12.40 | 13.85 | 36.42 | 39.36 | 42.98 | 45.56 |
| 25 | 10.52 | 11.52 | 13.12 | 14.61 | 37.65 | 40.65 | 44.31 | 46.93 |
| 26 | 11.16 | 12.20 | 13.84 | 15.38 | 38.89 | 41.92 | 45.64 | 48.29 |
| 27 | 11.81 | 12.88 | 14.57 | 16.15 | 40.11 | 43.19 | 46.96 | 49.64 |
| 28 | 12.46 | 13.56 | 15.31 | 16.93 | 41.34 | 44.46 | 48.28 | 50.99 |
| 29 | 13.12 | 14.26 | 16.05 | 17.71 | 42.56 | 45.72 | 49.59 | 52.34 |
| 30 | 13.79 | 14.95 | 16.79 | 18.49 | 43.77 | 46.98 | 50.89 | 53.67 |
| 35 | 17.19 | 18.51 | 20.57 | 22.47 | 49.80 | 53.20 | 57.34 | 60.27 |
| 40 | 20.71 | 22.16 | 24.43 | 26.51 | 55.76 | 59.34 | 63.69 | 66.77 |
| 45 | 24.31 | 25.90 | 28.37 | 30.61 | 61.66 | 65.41 | 69.96 | 73.17 |
| 50 | 27.99 | 29.71 | 32.36 | 34.76 | 67.50 | 71.42 | 76.15 | 79.49 |
| 60 | 35.53 | 37.48 | 40.48 | 43.19 | 79.08 | 83.30 | 88.38 | 91.95 |

**Table D.5:** Percent points for the F distribution.

Table entries are $F_{.05,\nu_1,\nu_2}$ where $P_{\nu_1,\nu_2}(F > F_{.05,\nu_1,\nu_2}) = .05$ .

| $\nu_2$ \ $\nu_1$ | 1 | 2 | 3 | 4 | 5 | 6 | 7 | 8 | 9 | 10 | 12 | 15 | 20 | 25 | 30 | 40 |
|---|---|---|---|---|---|---|---|---|---|---|---|---|---|---|---|---|
| 1 | 161 | 200 | 216 | 225 | 230 | 234 | 237 | 239 | 241 | 242 | 244 | 246 | 248 | 249 | 250 | 251 |
| 2 | 18.5 | 19.0 | 19.2 | 19.2 | 19.3 | 19.3 | 19.4 | 19.4 | 19.4 | 19.4 | 19.4 | 19.4 | 19.4 | 19.5 | 19.5 | 19.5 |
| 3 | 10.1 | 9.55 | 9.28 | 9.12 | 9.01 | 8.94 | 8.89 | 8.85 | 8.81 | 8.79 | 8.74 | 8.70 | 8.66 | 8.63 | 8.62 | 8.59 |
| 4 | 7.71 | 6.94 | 6.59 | 6.39 | 6.26 | 6.16 | 6.09 | 6.04 | 6.00 | 5.96 | 5.91 | 5.86 | 5.80 | 5.77 | 5.75 | 5.72 |
| 5 | 6.61 | 5.79 | 5.41 | 5.19 | 5.05 | 4.95 | 4.88 | 4.82 | 4.77 | 4.74 | 4.68 | 4.62 | 4.56 | 4.52 | 4.50 | 4.46 |
| 6 | 5.99 | 5.14 | 4.76 | 4.53 | 4.39 | 4.28 | 4.21 | 4.15 | 4.10 | 4.06 | 4.00 | 3.94 | 3.87 | 3.83 | 3.81 | 3.77 |
| 7 | 5.59 | 4.74 | 4.35 | 4.12 | 3.97 | 3.87 | 3.79 | 3.73 | 3.68 | 3.64 | 3.57 | 3.51 | 3.44 | 3.40 | 3.38 | 3.34 |
| 8 | 5.32 | 4.46 | 4.07 | 3.84 | 3.69 | 3.58 | 3.50 | 3.44 | 3.39 | 3.35 | 3.28 | 3.22 | 3.15 | 3.11 | 3.08 | 3.04 |
| 9 | 5.12 | 4.26 | 3.86 | 3.63 | 3.48 | 3.37 | 3.29 | 3.23 | 3.18 | 3.14 | 3.07 | 3.01 | 2.94 | 2.89 | 2.86 | 2.83 |
| 10 | 4.96 | 4.10 | 3.71 | 3.48 | 3.33 | 3.22 | 3.14 | 3.07 | 3.02 | 2.98 | 2.91 | 2.85 | 2.77 | 2.73 | 2.70 | 2.66 |
| 11 | 4.84 | 3.98 | 3.59 | 3.36 | 3.20 | 3.09 | 3.01 | 2.95 | 2.90 | 2.85 | 2.79 | 2.72 | 2.65 | 2.60 | 2.57 | 2.53 |
| 12 | 4.75 | 3.89 | 3.49 | 3.26 | 3.11 | 3.00 | 2.91 | 2.85 | 2.80 | 2.75 | 2.69 | 2.62 | 2.54 | 2.50 | 2.47 | 2.43 |
| 13 | 4.67 | 3.81 | 3.41 | 3.18 | 3.03 | 2.92 | 2.83 | 2.77 | 2.71 | 2.67 | 2.60 | 2.53 | 2.46 | 2.41 | 2.38 | 2.34 |
| 14 | 4.60 | 3.74 | 3.34 | 3.11 | 2.96 | 2.85 | 2.76 | 2.70 | 2.65 | 2.60 | 2.53 | 2.46 | 2.39 | 2.34 | 2.31 | 2.27 |
| 15 | 4.54 | 3.68 | 3.29 | 3.06 | 2.90 | 2.79 | 2.71 | 2.64 | 2.59 | 2.54 | 2.48 | 2.40 | 2.33 | 2.28 | 2.25 | 2.20 |
| 16 | 4.49 | 3.63 | 3.24 | 3.01 | 2.85 | 2.74 | 2.66 | 2.59 | 2.54 | 2.49 | 2.42 | 2.35 | 2.28 | 2.23 | 2.19 | 2.15 |
| 17 | 4.45 | 3.59 | 3.20 | 2.96 | 2.81 | 2.70 | 2.61 | 2.55 | 2.49 | 2.45 | 2.38 | 2.31 | 2.23 | 2.18 | 2.15 | 2.10 |
| 18 | 4.41 | 3.55 | 3.16 | 2.93 | 2.77 | 2.66 | 2.58 | 2.51 | 2.46 | 2.41 | 2.34 | 2.27 | 2.19 | 2.14 | 2.11 | 2.06 |
| 19 | 4.38 | 3.52 | 3.13 | 2.90 | 2.74 | 2.63 | 2.54 | 2.48 | 2.42 | 2.38 | 2.31 | 2.23 | 2.16 | 2.11 | 2.07 | 2.03 |
| 20 | 4.35 | 3.49 | 3.10 | 2.87 | 2.71 | 2.60 | 2.51 | 2.45 | 2.39 | 2.35 | 2.28 | 2.20 | 2.12 | 2.07 | 2.04 | 1.99 |
| 21 | 4.32 | 3.47 | 3.07 | 2.84 | 2.68 | 2.57 | 2.49 | 2.42 | 2.37 | 2.32 | 2.25 | 2.18 | 2.10 | 2.05 | 2.01 | 1.96 |
| 22 | 4.30 | 3.44 | 3.05 | 2.82 | 2.66 | 2.55 | 2.46 | 2.40 | 2.34 | 2.30 | 2.23 | 2.15 | 2.07 | 2.02 | 1.98 | 1.94 |
| 23 | 4.28 | 3.42 | 3.03 | 2.80 | 2.64 | 2.53 | 2.44 | 2.37 | 2.32 | 2.27 | 2.20 | 2.13 | 2.05 | 2.00 | 1.96 | 1.91 |
| 24 | 4.26 | 3.40 | 3.01 | 2.78 | 2.62 | 2.51 | 2.42 | 2.36 | 2.30 | 2.25 | 2.18 | 2.11 | 2.03 | 1.97 | 1.94 | 1.89 |
| 25 | 4.24 | 3.39 | 2.99 | 2.76 | 2.60 | 2.49 | 2.40 | 2.34 | 2.28 | 2.24 | 2.16 | 2.09 | 2.01 | 1.96 | 1.92 | 1.87 |
| 30 | 4.17 | 3.32 | 2.92 | 2.69 | 2.53 | 2.42 | 2.33 | 2.27 | 2.21 | 2.16 | 2.09 | 2.01 | 1.93 | 1.88 | 1.84 | 1.79 |
| 40 | 4.08 | 3.23 | 2.84 | 2.61 | 2.45 | 2.34 | 2.25 | 2.18 | 2.12 | 2.08 | 2.00 | 1.92 | 1.84 | 1.78 | 1.74 | 1.69 |
| 50 | 4.03 | 3.18 | 2.79 | 2.56 | 2.40 | 2.29 | 2.20 | 2.13 | 2.07 | 2.03 | 1.95 | 1.87 | 1.78 | 1.73 | 1.69 | 1.63 |
| 75 | 3.97 | 3.12 | 2.73 | 2.49 | 2.34 | 2.22 | 2.13 | 2.06 | 2.01 | 1.96 | 1.88 | 1.80 | 1.71 | 1.65 | 1.61 | 1.55 |
| 100 | 3.94 | 3.09 | 2.70 | 2.46 | 2.31 | 2.19 | 2.10 | 2.03 | 1.97 | 1.93 | 1.85 | 1.77 | 1.68 | 1.62 | 1.57 | 1.52 |
| 200 | 3.89 | 3.04 | 2.65 | 2.42 | 2.26 | 2.14 | 2.06 | 1.98 | 1.93 | 1.88 | 1.80 | 1.72 | 1.62 | 1.56 | 1.52 | 1.46 |
| $\infty$ | 3.84 | 3.00 | 2.61 | 2.37 | 2.21 | 2.10 | 2.01 | 1.94 | 1.88 | 1.83 | 1.75 | 1.67 | 1.57 | 1.51 | 1.46 | 1.40 |

Table D.5: Percent points for the F distribution, continued.

Table entries are $F_{.01,\nu_1,\nu_2}$ where $P_{\nu_1,\nu_2}(F > F_{.01,\nu_1,\nu_2}) = .01$ .

| $\nu_2$ \ $\nu_1$ | 1 | 2 | 3 | 4 | 5 | 6 | 7 | 8 | 9 | 10 | 12 | 15 | 20 | 25 | 30 | 40 |
|---|---|---|---|---|---|---|---|---|---|---|---|---|---|---|---|---|
| 2 | 98.5 | 99.0 | 99.2 | 99.2 | 99.3 | 99.3 | 99.4 | 99.4 | 99.4 | 99.4 | 99.4 | 99.4 | 99.4 | 99.5 | 99.5 | 99.5 |
| 3 | 34.1 | 30.8 | 29.5 | 28.7 | 28.2 | 27.9 | 27.7 | 27.5 | 27.3 | 27.2 | 27.1 | 26.9 | 26.7 | 26.6 | 26.5 | 26.4 |
| 4 | 21.2 | 18.0 | 16.7 | 16.0 | 15.5 | 15.2 | 15.0 | 14.8 | 14.7 | 14.5 | 14.4 | 14.2 | 14.0 | 13.9 | 13.8 | 13.7 |
| 5 | 16.3 | 13.3 | 12.1 | 11.4 | 11.0 | 10.7 | 10.5 | 10.3 | 10.2 | 10.1 | 9.89 | 9.72 | 9.55 | 9.45 | 9.38 | 9.29 |
| 6 | 13.7 | 10.9 | 9.78 | 9.15 | 8.75 | 8.47 | 8.26 | 8.10 | 7.98 | 7.87 | 7.72 | 7.56 | 7.40 | 7.30 | 7.23 | 7.14 |
| 7 | 12.2 | 9.55 | 8.45 | 7.85 | 7.46 | 7.19 | 6.99 | 6.84 | 6.72 | 6.62 | 6.47 | 6.31 | 6.16 | 6.06 | 5.99 | 5.91 |
| 8 | 11.3 | 8.65 | 7.59 | 7.01 | 6.63 | 6.37 | 6.18 | 6.03 | 5.91 | 5.81 | 5.67 | 5.52 | 5.36 | 5.26 | 5.20 | 5.12 |
| 9 | 10.6 | 8.02 | 6.99 | 6.42 | 6.06 | 5.80 | 5.61 | 5.47 | 5.35 | 5.26 | 5.11 | 4.96 | 4.81 | 4.71 | 4.65 | 4.57 |
| 10 | 10.0 | 7.56 | 6.55 | 5.99 | 5.64 | 5.39 | 5.20 | 5.06 | 4.94 | 4.85 | 4.71 | 4.56 | 4.41 | 4.31 | 4.25 | 4.17 |
| 11 | 9.65 | 7.21 | 6.22 | 5.67 | 5.32 | 5.07 | 4.89 | 4.74 | 4.63 | 4.54 | 4.40 | 4.25 | 4.10 | 4.01 | 3.94 | 3.86 |
| 12 | 9.33 | 6.93 | 5.95 | 5.41 | 5.06 | 4.82 | 4.64 | 4.50 | 4.39 | 4.30 | 4.16 | 4.01 | 3.86 | 3.76 | 3.70 | 3.62 |
| 13 | 9.07 | 6.70 | 5.74 | 5.21 | 4.86 | 4.62 | 4.44 | 4.30 | 4.19 | 4.10 | 3.96 | 3.82 | 3.66 | 3.57 | 3.51 | 3.43 |
| 14 | 8.86 | 6.51 | 5.56 | 5.04 | 4.69 | 4.46 | 4.28 | 4.14 | 4.03 | 3.94 | 3.80 | 3.66 | 3.51 | 3.41 | 3.35 | 3.27 |
| 15 | 8.68 | 6.36 | 5.42 | 4.89 | 4.56 | 4.32 | 4.14 | 4.00 | 3.89 | 3.80 | 3.67 | 3.52 | 3.37 | 3.28 | 3.21 | 3.13 |
| 16 | 8.53 | 6.23 | 5.29 | 4.77 | 4.44 | 4.20 | 4.03 | 3.89 | 3.78 | 3.69 | 3.55 | 3.41 | 3.26 | 3.16 | 3.10 | 3.02 |
| 17 | 8.40 | 6.11 | 5.18 | 4.67 | 4.34 | 4.10 | 3.93 | 3.79 | 3.68 | 3.59 | 3.46 | 3.31 | 3.16 | 3.07 | 3.00 | 2.92 |
| 18 | 8.29 | 6.01 | 5.09 | 4.58 | 4.25 | 4.01 | 3.84 | 3.71 | 3.60 | 3.51 | 3.37 | 3.23 | 3.08 | 2.98 | 2.92 | 2.84 |
| 19 | 8.18 | 5.93 | 5.01 | 4.50 | 4.17 | 3.94 | 3.77 | 3.63 | 3.52 | 3.43 | 3.30 | 3.15 | 3.00 | 2.91 | 2.84 | 2.76 |
| 20 | 8.10 | 5.85 | 4.94 | 4.43 | 4.10 | 3.87 | 3.70 | 3.56 | 3.46 | 3.37 | 3.23 | 3.09 | 2.94 | 2.84 | 2.78 | 2.69 |
| 21 | 8.02 | 5.78 | 4.87 | 4.37 | 4.04 | 3.81 | 3.64 | 3.51 | 3.40 | 3.31 | 3.17 | 3.03 | 2.88 | 2.79 | 2.72 | 2.64 |
| 22 | 7.95 | 5.72 | 4.82 | 4.31 | 3.99 | 3.76 | 3.59 | 3.45 | 3.35 | 3.26 | 3.12 | 2.98 | 2.83 | 2.73 | 2.67 | 2.58 |
| 23 | 7.88 | 5.66 | 4.76 | 4.26 | 3.94 | 3.71 | 3.54 | 3.41 | 3.30 | 3.21 | 3.07 | 2.93 | 2.78 | 2.69 | 2.62 | 2.54 |
| 24 | 7.82 | 5.61 | 4.72 | 4.22 | 3.90 | 3.67 | 3.50 | 3.36 | 3.26 | 3.17 | 3.03 | 2.89 | 2.74 | 2.64 | 2.58 | 2.49 |
| 25 | 7.77 | 5.57 | 4.68 | 4.18 | 3.85 | 3.63 | 3.46 | 3.32 | 3.22 | 3.13 | 2.99 | 2.85 | 2.70 | 2.60 | 2.54 | 2.45 |
| 30 | 7.56 | 5.39 | 4.51 | 4.02 | 3.70 | 3.47 | 3.30 | 3.17 | 3.07 | 2.98 | 2.84 | 2.70 | 2.55 | 2.45 | 2.39 | 2.30 |
| 40 | 7.31 | 5.18 | 4.31 | 3.83 | 3.51 | 3.29 | 3.12 | 2.99 | 2.89 | 2.80 | 2.66 | 2.52 | 2.37 | 2.27 | 2.20 | 2.11 |
| 50 | 7.17 | 5.06 | 4.20 | 3.72 | 3.41 | 3.19 | 3.02 | 2.89 | 2.78 | 2.70 | 2.56 | 2.42 | 2.27 | 2.17 | 2.10 | 2.01 |
| 75 | 6.99 | 4.90 | 4.05 | 3.58 | 3.27 | 3.05 | 2.89 | 2.76 | 2.65 | 2.57 | 2.43 | 2.29 | 2.13 | 2.03 | 1.96 | 1.87 |
| 100 | 6.90 | 4.82 | 3.98 | 3.51 | 3.21 | 2.99 | 2.82 | 2.69 | 2.59 | 2.50 | 2.37 | 2.22 | 2.07 | 1.97 | 1.89 | 1.80 |
| 200 | 6.76 | 4.71 | 3.88 | 3.41 | 3.11 | 2.89 | 2.73 | 2.60 | 2.50 | 2.41 | 2.27 | 2.13 | 1.97 | 1.87 | 1.79 | 1.69 |
| $\infty$ | 6.63 | 4.61 | 3.78 | 3.32 | 3.02 | 2.80 | 2.64 | 2.51 | 2.41 | 2.32 | 2.18 | 2.04 | 1.88 | 1.77 | 1.70 | 1.59 |

Table D.5: Percent points for the F distribution, continued.

Table entries are $F_{.001,\nu_1,\nu_2}$ where $P_{\nu_1,\nu_2}(F > F_{.001,\nu_1,\nu_2}) = .001$ .

| | $\nu_1$ | | | | | | | | | | | | | | | |
|---|---|---|---|---|---|---|---|---|---|---|---|---|---|---|---|---|
| $\nu_2$ | 1 | 2 | 3 | 4 | 5 | 6 | 7 | 8 | 9 | 10 | 12 | 15 | 20 | 25 | 30 | 40 |
| 2 | 999 | 999 | 999 | 999 | 999 | 999 | 999 | 999 | 999 | 999 | 999 | 999 | 999 | 999 | 999 | 999 |
| 3 | 167 | 149 | 141 | 137 | 135 | 133 | 132 | 131 | 130 | 129 | 128 | 127 | 126 | 126 | 125 | 125 |
| 4 | 74.1 | 61.2 | 56.2 | 53.4 | 51.7 | 50.5 | 49.7 | 49.0 | 48.5 | 48.1 | 47.4 | 46.8 | 46.1 | 45.7 | 45.4 | 45.1 |
| 5 | 47.2 | 37.1 | 33.2 | 31.1 | 29.8 | 28.8 | 28.2 | 27.6 | 27.2 | 26.9 | 26.4 | 25.9 | 25.4 | 25.1 | 24.9 | 24.6 |
| 6 | 35.5 | 27.0 | 23.7 | 21.9 | 20.8 | 20.0 | 19.5 | 19.0 | 18.7 | 18.4 | 18.0 | 17.6 | 17.1 | 16.9 | 16.7 | 16.4 |
| 7 | 29.2 | 21.7 | 18.8 | 17.2 | 16.2 | 15.5 | 15.0 | 14.6 | 14.3 | 14.1 | 13.7 | 13.3 | 12.9 | 12.7 | 12.5 | 12.3 |
| 8 | 25.4 | 18.5 | 15.8 | 14.4 | 13.5 | 12.9 | 12.4 | 12.0 | 11.8 | 11.5 | 11.2 | 10.8 | 10.5 | 10.3 | 10.1 | 9.92 |
| 9 | 22.9 | 16.4 | 13.9 | 12.6 | 11.7 | 11.1 | 10.7 | 10.4 | 10.1 | 9.89 | 9.57 | 9.24 | 8.90 | 8.69 | 8.55 | 8.37 |
| 10 | 21.0 | 14.9 | 12.6 | 11.3 | 10.5 | 9.93 | 9.52 | 9.20 | 8.96 | 8.75 | 8.45 | 8.13 | 7.80 | 7.60 | 7.47 | 7.30 |
| 11 | 19.7 | 13.8 | 11.6 | 10.3 | 9.58 | 9.05 | 8.66 | 8.35 | 8.12 | 7.92 | 7.63 | 7.32 | 7.01 | 6.81 | 6.68 | 6.52 |
| 12 | 18.6 | 13.0 | 10.8 | 9.63 | 8.89 | 8.38 | 8.00 | 7.71 | 7.48 | 7.29 | 7.00 | 6.71 | 6.40 | 6.22 | 6.09 | 5.93 |
| 13 | 17.8 | 12.3 | 10.2 | 9.07 | 8.35 | 7.86 | 7.49 | 7.21 | 6.98 | 6.80 | 6.52 | 6.23 | 5.93 | 5.75 | 5.63 | 5.47 |
| 14 | 17.1 | 11.8 | 9.73 | 8.62 | 7.92 | 7.44 | 7.08 | 6.80 | 6.58 | 6.40 | 6.13 | 5.85 | 5.56 | 5.38 | 5.25 | 5.10 |
| 15 | 16.6 | 11.3 | 9.34 | 8.25 | 7.57 | 7.09 | 6.74 | 6.47 | 6.26 | 6.08 | 5.81 | 5.54 | 5.25 | 5.07 | 4.95 | 4.80 |
| 16 | 16.1 | 11.0 | 9.01 | 7.94 | 7.27 | 6.80 | 6.46 | 6.19 | 5.98 | 5.81 | 5.55 | 5.27 | 4.99 | 4.82 | 4.70 | 4.54 |
| 17 | 15.7 | 10.7 | 8.73 | 7.68 | 7.02 | 6.56 | 6.22 | 5.96 | 5.75 | 5.58 | 5.32 | 5.05 | 4.78 | 4.60 | 4.48 | 4.33 |
| 18 | 15.4 | 10.4 | 8.49 | 7.46 | 6.81 | 6.35 | 6.02 | 5.76 | 5.56 | 5.39 | 5.13 | 4.87 | 4.59 | 4.42 | 4.30 | 4.15 |
| 19 | 15.1 | 10.2 | 8.28 | 7.27 | 6.62 | 6.18 | 5.85 | 5.59 | 5.39 | 5.22 | 4.97 | 4.70 | 4.43 | 4.26 | 4.14 | 3.99 |
| 20 | 14.8 | 9.95 | 8.10 | 7.10 | 6.46 | 6.02 | 5.69 | 5.44 | 5.24 | 5.08 | 4.82 | 4.56 | 4.29 | 4.12 | 4.00 | 3.86 |
| 21 | 14.6 | 9.77 | 7.94 | 6.95 | 6.32 | 5.88 | 5.56 | 5.31 | 5.11 | 4.95 | 4.70 | 4.44 | 4.17 | 4.00 | 3.88 | 3.74 |
| 22 | 14.4 | 9.61 | 7.80 | 6.81 | 6.19 | 5.76 | 5.44 | 5.19 | 4.99 | 4.83 | 4.58 | 4.33 | 4.06 | 3.89 | 3.78 | 3.63 |
| 23 | 14.2 | 9.47 | 7.67 | 6.70 | 6.08 | 5.65 | 5.33 | 5.09 | 4.89 | 4.73 | 4.48 | 4.23 | 3.96 | 3.79 | 3.68 | 3.53 |
| 24 | 14.0 | 9.34 | 7.55 | 6.59 | 5.98 | 5.55 | 5.23 | 4.99 | 4.80 | 4.64 | 4.39 | 4.14 | 3.87 | 3.71 | 3.59 | 3.45 |
| 25 | 13.9 | 9.22 | 7.45 | 6.49 | 5.89 | 5.46 | 5.15 | 4.91 | 4.71 | 4.56 | 4.31 | 4.06 | 3.79 | 3.63 | 3.52 | 3.37 |
| 30 | 13.3 | 8.77 | 7.05 | 6.12 | 5.53 | 5.12 | 4.82 | 4.58 | 4.39 | 4.24 | 4.00 | 3.75 | 3.49 | 3.33 | 3.22 | 3.07 |
| 40 | 12.6 | 8.25 | 6.59 | 5.70 | 5.13 | 4.73 | 4.44 | 4.21 | 4.02 | 3.87 | 3.64 | 3.40 | 3.14 | 2.98 | 2.87 | 2.73 |
| 50 | 12.2 | 7.96 | 6.34 | 5.46 | 4.90 | 4.51 | 4.22 | 4.00 | 3.82 | 3.67 | 3.44 | 3.20 | 2.95 | 2.79 | 2.68 | 2.53 |
| 75 | 11.7 | 7.58 | 6.01 | 5.16 | 4.62 | 4.24 | 3.96 | 3.74 | 3.56 | 3.42 | 3.19 | 2.96 | 2.71 | 2.55 | 2.44 | 2.29 |
| 100 | 11.5 | 7.41 | 5.86 | 5.02 | 4.48 | 4.11 | 3.83 | 3.61 | 3.44 | 3.30 | 3.07 | 2.84 | 2.59 | 2.43 | 2.32 | 2.17 |
| 200 | 11.2 | 7.15 | 5.63 | 4.81 | 4.29 | 3.92 | 3.65 | 3.43 | 3.26 | 3.12 | 2.90 | 2.67 | 2.42 | 2.26 | 2.15 | 2.00 |
| $\infty$ | 10.8 | 6.91 | 5.42 | 4.62 | 4.10 | 3.74 | 3.47 | 3.27 | 3.10 | 2.96 | 2.74 | 2.51 | 2.27 | 2.10 | 1.99 | 1.84 |

**Table D.6:** Coefficients of orthogonal polynomial contrasts.

| | | Coefficients | | | | | | |
|---|---|---|---|---|---|---|---|---|
| $g$ | Order | 1 | 2 | 3 | 4 | 5 | 6 | 7 |
| 3 | 1 | -1 | 0 | 1 | | | | |
| | 2 | 1 | -2 | 1 | | | | |
| 4 | 1 | -3 | -1 | 1 | 3 | | | |
| | 2 | 1 | -1 | -1 | 1 | | | |
| | 3 | -1 | 3 | -3 | 1 | | | |
| 5 | 1 | -2 | -1 | 0 | 1 | 2 | | |
| | 2 | 2 | -1 | -2 | -1 | 2 | | |
| | 3 | -1 | 2 | 0 | -2 | 1 | | |
| | 4 | 1 | -4 | 6 | -4 | 1 | | |
| 6 | 1 | -5 | -3 | -1 | 1 | 3 | 5 | |
| | 2 | 5 | -1 | -4 | -4 | -1 | 5 | |
| | 3 | -5 | 7 | 4 | -4 | -7 | 5 | |
| | 4 | 1 | -3 | 2 | 2 | -3 | 1 | |
| | 5 | -1 | 5 | -10 | 10 | -5 | 1 | |
| 7 | 1 | -3 | -2 | -1 | 0 | 1 | 2 | 3 |
| | 2 | 5 | 0 | -3 | -4 | -3 | 0 | 5 |
| | 3 | -1 | 1 | 1 | 0 | -1 | -1 | 1 |
| | 4 | 3 | -7 | 1 | 6 | 1 | -7 | 3 |
| | 5 | -1 | 4 | -5 | 0 | 5 | -4 | 1 |
| | 6 | 1 | -6 | 15 | -20 | 15 | -6 | 1 |

**Table D.7:** Critical values for the two-sided Bonferroni $t$ statistic.

Table entries are $t_{\mathcal{E},\nu}$ where $P_\nu(t > t_{\mathcal{E},\nu}) = \mathcal{E}$ and $\mathcal{E} = .05/2/K$ .

| | K | | | | | | | | | | | | |
|---|---|---|---|---|---|---|---|---|---|---|---|---|---|
| $\nu$ | 2 | 3 | 4 | 5 | 6 | 7 | 8 | 9 | 10 | 15 | 20 | 30 | 50 |
| 1 | 25.5 | 38.2 | 50.9 | 63.7 | 76.4 | 89.1 | 102 | 115 | 127 | 191 | 255 | 382 | 637 |
| 2 | 6.21 | 7.65 | 8.86 | 9.92 | 10.9 | 11.8 | 12.6 | 13.4 | 14.1 | 17.3 | 20.0 | 24.5 | 31.6 |
| 3 | 4.18 | 4.86 | 5.39 | 5.84 | 6.23 | 6.58 | 6.90 | 7.18 | 7.45 | 8.58 | 9.46 | 10.9 | 12.9 |
| 4 | 3.50 | 3.96 | 4.31 | 4.60 | 4.85 | 5.07 | 5.26 | 5.44 | 5.60 | 6.25 | 6.76 | 7.53 | 8.61 |
| 5 | 3.16 | 3.53 | 3.81 | 4.03 | 4.22 | 4.38 | 4.53 | 4.66 | 4.77 | 5.25 | 5.60 | 6.14 | 6.87 |
| 6 | 2.97 | 3.29 | 3.52 | 3.71 | 3.86 | 4.00 | 4.12 | 4.22 | 4.32 | 4.70 | 4.98 | 5.40 | 5.96 |
| 7 | 2.84 | 3.13 | 3.34 | 3.50 | 3.64 | 3.75 | 3.86 | 3.95 | 4.03 | 4.36 | 4.59 | 4.94 | 5.41 |
| 8 | 2.75 | 3.02 | 3.21 | 3.36 | 3.48 | 3.58 | 3.68 | 3.76 | 3.83 | 4.12 | 4.33 | 4.64 | 5.04 |
| 9 | 2.69 | 2.93 | 3.11 | 3.25 | 3.36 | 3.46 | 3.55 | 3.62 | 3.69 | 3.95 | 4.15 | 4.42 | 4.78 |
| 10 | 2.63 | 2.87 | 3.04 | 3.17 | 3.28 | 3.37 | 3.45 | 3.52 | 3.58 | 3.83 | 4.00 | 4.26 | 4.59 |
| 11 | 2.59 | 2.82 | 2.98 | 3.11 | 3.21 | 3.29 | 3.37 | 3.44 | 3.50 | 3.73 | 3.89 | 4.13 | 4.44 |
| 12 | 2.56 | 2.78 | 2.93 | 3.05 | 3.15 | 3.24 | 3.31 | 3.37 | 3.43 | 3.65 | 3.81 | 4.03 | 4.32 |
| 13 | 2.53 | 2.75 | 2.90 | 3.01 | 3.11 | 3.19 | 3.26 | 3.32 | 3.37 | 3.58 | 3.73 | 3.95 | 4.22 |
| 14 | 2.51 | 2.72 | 2.86 | 2.98 | 3.07 | 3.15 | 3.21 | 3.27 | 3.33 | 3.53 | 3.67 | 3.88 | 4.14 |
| 15 | 2.49 | 2.69 | 2.84 | 2.95 | 3.04 | 3.11 | 3.18 | 3.23 | 3.29 | 3.48 | 3.62 | 3.82 | 4.07 |
| 16 | 2.47 | 2.67 | 2.81 | 2.92 | 3.01 | 3.08 | 3.15 | 3.20 | 3.25 | 3.44 | 3.58 | 3.77 | 4.01 |
| 17 | 2.46 | 2.65 | 2.79 | 2.90 | 2.98 | 3.06 | 3.12 | 3.17 | 3.22 | 3.41 | 3.54 | 3.73 | 3.97 |
| 18 | 2.45 | 2.64 | 2.77 | 2.88 | 2.96 | 3.03 | 3.09 | 3.15 | 3.20 | 3.38 | 3.51 | 3.69 | 3.92 |
| 19 | 2.43 | 2.63 | 2.76 | 2.86 | 2.94 | 3.01 | 3.07 | 3.13 | 3.17 | 3.35 | 3.48 | 3.66 | 3.88 |
| 20 | 2.42 | 2.61 | 2.74 | 2.85 | 2.93 | 3.00 | 3.06 | 3.11 | 3.15 | 3.33 | 3.46 | 3.63 | 3.85 |
| 21 | 2.41 | 2.60 | 2.73 | 2.83 | 2.91 | 2.98 | 3.04 | 3.09 | 3.14 | 3.31 | 3.43 | 3.60 | 3.82 |
| 22 | 2.41 | 2.59 | 2.72 | 2.82 | 2.90 | 2.97 | 3.02 | 3.07 | 3.12 | 3.29 | 3.41 | 3.58 | 3.79 |
| 23 | 2.40 | 2.58 | 2.71 | 2.81 | 2.89 | 2.95 | 3.01 | 3.06 | 3.10 | 3.27 | 3.39 | 3.56 | 3.77 |
| 24 | 2.39 | 2.57 | 2.70 | 2.80 | 2.88 | 2.94 | 3.00 | 3.05 | 3.09 | 3.26 | 3.38 | 3.54 | 3.75 |
| 25 | 2.38 | 2.57 | 2.69 | 2.79 | 2.86 | 2.93 | 2.99 | 3.03 | 3.08 | 3.24 | 3.36 | 3.52 | 3.73 |
| 26 | 2.38 | 2.56 | 2.68 | 2.78 | 2.86 | 2.92 | 2.98 | 3.02 | 3.07 | 3.23 | 3.35 | 3.51 | 3.71 |
| 27 | 2.37 | 2.55 | 2.68 | 2.77 | 2.85 | 2.91 | 2.97 | 3.01 | 3.06 | 3.22 | 3.33 | 3.49 | 3.69 |
| 28 | 2.37 | 2.55 | 2.67 | 2.76 | 2.84 | 2.90 | 2.96 | 3.00 | 3.05 | 3.21 | 3.32 | 3.48 | 3.67 |
| 29 | 2.36 | 2.54 | 2.66 | 2.76 | 2.83 | 2.89 | 2.95 | 3.00 | 3.04 | 3.20 | 3.31 | 3.47 | 3.66 |
| 30 | 2.36 | 2.54 | 2.66 | 2.75 | 2.82 | 2.89 | 2.94 | 2.99 | 3.03 | 3.19 | 3.30 | 3.45 | 3.65 |
| 35 | 2.34 | 2.51 | 2.63 | 2.72 | 2.80 | 2.86 | 2.91 | 2.96 | 3.00 | 3.15 | 3.26 | 3.41 | 3.59 |
| 40 | 2.33 | 2.50 | 2.62 | 2.70 | 2.78 | 2.84 | 2.89 | 2.93 | 2.97 | 3.12 | 3.23 | 3.37 | 3.55 |
| 45 | 2.32 | 2.49 | 2.60 | 2.69 | 2.76 | 2.82 | 2.87 | 2.91 | 2.95 | 3.10 | 3.20 | 3.35 | 3.52 |
| 50 | 2.31 | 2.48 | 2.59 | 2.68 | 2.75 | 2.81 | 2.85 | 2.90 | 2.94 | 3.08 | 3.18 | 3.32 | 3.50 |
| 100 | 2.28 | 2.43 | 2.54 | 2.63 | 2.69 | 2.75 | 2.79 | 2.83 | 2.87 | 3.01 | 3.10 | 3.23 | 3.39 |
| $\infty$ | 2.24 | 2.39 | 2.50 | 2.58 | 2.64 | 2.69 | 2.73 | 2.77 | 2.81 | 2.94 | 3.02 | 3.14 | 3.29 |

Table D.7: Critical values for the two-sided Bonferroni t statistic, continued.

Table entries are $t_{\mathcal{E},\nu}$ where $P_\nu(t > t_{\mathcal{E},\nu}) = \mathcal{E}$ and $\mathcal{E} = .01/2/K$ .

| | K | | | | | | | | | | | | |
|---|---|---|---|---|---|---|---|---|---|---|---|---|---|
| $\nu$ | 2 | 3 | 4 | 5 | 6 | 7 | 8 | 9 | 10 | 15 | 20 | 30 | 50 |
| 1 | 127 | 191 | 255 | 318 | 382 | 446 | 509 | 573 | 637 | 955 | 1273 | 1910 | 3183 |
| 2 | 14.1 | 17.3 | 20.0 | 22.3 | 24.5 | 26.4 | 28.3 | 30.0 | 31.6 | 38.7 | 44.7 | 54.8 | 70.7 |
| 3 | 7.45 | 8.58 | 9.46 | 10.2 | 10.9 | 11.5 | 12.0 | 12.5 | 12.9 | 14.8 | 16.3 | 18.7 | 22.2 |
| 4 | 5.60 | 6.25 | 6.76 | 7.17 | 7.53 | 7.84 | 8.12 | 8.38 | 8.61 | 9.57 | 10.3 | 11.4 | 13.0 |
| 5 | 4.77 | 5.25 | 5.60 | 5.89 | 6.14 | 6.35 | 6.54 | 6.71 | 6.87 | 7.50 | 7.98 | 8.69 | 9.68 |
| 6 | 4.32 | 4.70 | 4.98 | 5.21 | 5.40 | 5.56 | 5.71 | 5.84 | 5.96 | 6.43 | 6.79 | 7.31 | 8.02 |
| 7 | 4.03 | 4.36 | 4.59 | 4.79 | 4.94 | 5.08 | 5.20 | 5.31 | 5.41 | 5.80 | 6.08 | 6.50 | 7.06 |
| 8 | 3.83 | 4.12 | 4.33 | 4.50 | 4.64 | 4.76 | 4.86 | 4.96 | 5.04 | 5.37 | 5.62 | 5.97 | 6.44 |
| 9 | 3.69 | 3.95 | 4.15 | 4.30 | 4.42 | 4.53 | 4.62 | 4.71 | 4.78 | 5.08 | 5.29 | 5.60 | 6.01 |
| 10 | 3.58 | 3.83 | 4.00 | 4.14 | 4.26 | 4.36 | 4.44 | 4.52 | 4.59 | 4.85 | 5.05 | 5.33 | 5.69 |
| 11 | 3.50 | 3.73 | 3.89 | 4.02 | 4.13 | 4.22 | 4.30 | 4.37 | 4.44 | 4.68 | 4.86 | 5.12 | 5.45 |
| 12 | 3.43 | 3.65 | 3.81 | 3.93 | 4.03 | 4.12 | 4.19 | 4.26 | 4.32 | 4.55 | 4.72 | 4.96 | 5.26 |
| 13 | 3.37 | 3.58 | 3.73 | 3.85 | 3.95 | 4.03 | 4.10 | 4.16 | 4.22 | 4.44 | 4.60 | 4.82 | 5.11 |
| 14 | 3.33 | 3.53 | 3.67 | 3.79 | 3.88 | 3.96 | 4.03 | 4.09 | 4.14 | 4.35 | 4.50 | 4.71 | 4.99 |
| 15 | 3.29 | 3.48 | 3.62 | 3.73 | 3.82 | 3.90 | 3.96 | 4.02 | 4.07 | 4.27 | 4.42 | 4.62 | 4.88 |
| 16 | 3.25 | 3.44 | 3.58 | 3.69 | 3.77 | 3.85 | 3.91 | 3.96 | 4.01 | 4.21 | 4.35 | 4.54 | 4.79 |
| 17 | 3.22 | 3.41 | 3.54 | 3.65 | 3.73 | 3.80 | 3.86 | 3.92 | 3.97 | 4.15 | 4.29 | 4.47 | 4.71 |
| 18 | 3.20 | 3.38 | 3.51 | 3.61 | 3.69 | 3.76 | 3.82 | 3.87 | 3.92 | 4.10 | 4.23 | 4.42 | 4.65 |
| 19 | 3.17 | 3.35 | 3.48 | 3.58 | 3.66 | 3.73 | 3.79 | 3.84 | 3.88 | 4.06 | 4.19 | 4.36 | 4.59 |
| 20 | 3.15 | 3.33 | 3.46 | 3.55 | 3.63 | 3.70 | 3.75 | 3.80 | 3.85 | 4.02 | 4.15 | 4.32 | 4.54 |
| 21 | 3.14 | 3.31 | 3.43 | 3.53 | 3.60 | 3.67 | 3.73 | 3.78 | 3.82 | 3.99 | 4.11 | 4.28 | 4.49 |
| 22 | 3.12 | 3.29 | 3.41 | 3.50 | 3.58 | 3.64 | 3.70 | 3.75 | 3.79 | 3.96 | 4.08 | 4.24 | 4.45 |
| 23 | 3.10 | 3.27 | 3.39 | 3.48 | 3.56 | 3.62 | 3.68 | 3.72 | 3.77 | 3.93 | 4.05 | 4.21 | 4.42 |
| 24 | 3.09 | 3.26 | 3.38 | 3.47 | 3.54 | 3.60 | 3.66 | 3.70 | 3.75 | 3.91 | 4.02 | 4.18 | 4.38 |
| 25 | 3.08 | 3.24 | 3.36 | 3.45 | 3.52 | 3.58 | 3.64 | 3.68 | 3.73 | 3.88 | 4.00 | 4.15 | 4.35 |
| 26 | 3.07 | 3.23 | 3.35 | 3.43 | 3.51 | 3.57 | 3.62 | 3.67 | 3.71 | 3.86 | 3.97 | 4.13 | 4.32 |
| 27 | 3.06 | 3.22 | 3.33 | 3.42 | 3.49 | 3.55 | 3.60 | 3.65 | 3.69 | 3.84 | 3.95 | 4.11 | 4.30 |
| 28 | 3.05 | 3.21 | 3.32 | 3.41 | 3.48 | 3.54 | 3.59 | 3.63 | 3.67 | 3.83 | 3.94 | 4.09 | 4.28 |
| 29 | 3.04 | 3.20 | 3.31 | 3.40 | 3.47 | 3.52 | 3.58 | 3.62 | 3.66 | 3.81 | 3.92 | 4.07 | 4.25 |
| 30 | 3.03 | 3.19 | 3.30 | 3.39 | 3.45 | 3.51 | 3.56 | 3.61 | 3.65 | 3.80 | 3.90 | 4.05 | 4.23 |
| 35 | 3.00 | 3.15 | 3.26 | 3.34 | 3.41 | 3.46 | 3.51 | 3.55 | 3.59 | 3.74 | 3.84 | 3.98 | 4.15 |
| 40 | 2.97 | 3.12 | 3.23 | 3.31 | 3.37 | 3.43 | 3.47 | 3.51 | 3.55 | 3.69 | 3.79 | 3.92 | 4.09 |
| 45 | 2.95 | 3.10 | 3.20 | 3.28 | 3.35 | 3.40 | 3.44 | 3.48 | 3.52 | 3.66 | 3.75 | 3.88 | 4.05 |
| 50 | 2.94 | 3.08 | 3.18 | 3.26 | 3.32 | 3.38 | 3.42 | 3.46 | 3.50 | 3.63 | 3.72 | 3.85 | 4.01 |
| 100 | 2.87 | 3.01 | 3.1 | 3.17 | 3.23 | 3.28 | 3.32 | 3.36 | 3.39 | 3.51 | 3.60 | 3.72 | 3.86 |
| $\infty$ | 2.81 | 2.94 | 3.02 | 3.09 | 3.14 | 3.19 | 3.23 | 3.26 | 3.29 | 3.40 | 3.48 | 3.59 | 3.72 |

**Table D.8:** Percent points for the Studentized range.

Table entries are $q_{.05}(K, \nu)$.

| | K | | | | | | | | | | | | |
|---|---|---|---|---|---|---|---|---|---|---|---|---|---|
| $\nu$ | 2 | 3 | 4 | 5 | 6 | 7 | 8 | 9 | 10 | 15 | 20 | 30 | 50 |
| 1 | 18.0 | 27.0 | 32.8 | 37.1 | 40.4 | 43.1 | 45.4 | 47.4 | 49.1 | 55.4 | 59.6 | 65.1 | 71.7 |
| 2 | 6.09 | 8.33 | 9.80 | 10.9 | 11.7 | 12.4 | 13.0 | 13.5 | 14.0 | 15.7 | 16.8 | 18.3 | 20.0 |
| 3 | 4.50 | 5.91 | 6.82 | 7.50 | 8.04 | 8.48 | 8.85 | 9.18 | 9.46 | 10.5 | 11.2 | 12.2 | 13.4 |
| 4 | 3.93 | 5.04 | 5.76 | 6.29 | 6.71 | 7.05 | 7.35 | 7.60 | 7.83 | 8.66 | 9.23 | 10.0 | 10.9 |
| 5 | 3.64 | 4.60 | 5.22 | 5.67 | 6.03 | 6.33 | 6.58 | 6.80 | 6.99 | 7.72 | 8.21 | 8.87 | 9.67 |
| 6 | 3.46 | 4.34 | 4.90 | 5.30 | 5.63 | 5.90 | 6.12 | 6.32 | 6.49 | 7.14 | 7.59 | 8.19 | 8.91 |
| 7 | 3.34 | 4.16 | 4.68 | 5.06 | 5.36 | 5.61 | 5.82 | 6.00 | 6.16 | 6.76 | 7.17 | 7.73 | 8.40 |
| 8 | 3.26 | 4.04 | 4.53 | 4.89 | 5.17 | 5.40 | 5.60 | 5.77 | 5.92 | 6.48 | 6.87 | 7.40 | 8.03 |
| 9 | 3.20 | 3.95 | 4.41 | 4.76 | 5.02 | 5.24 | 5.43 | 5.59 | 5.74 | 6.28 | 6.64 | 7.14 | 7.75 |
| 10 | 3.15 | 3.88 | 4.33 | 4.65 | 4.91 | 5.12 | 5.30 | 5.46 | 5.60 | 6.11 | 6.47 | 6.95 | 7.53 |
| 11 | 3.11 | 3.82 | 4.26 | 4.57 | 4.82 | 5.03 | 5.20 | 5.35 | 5.49 | 5.98 | 6.33 | 6.79 | 7.35 |
| 12 | 3.08 | 3.77 | 4.20 | 4.51 | 4.75 | 4.95 | 5.12 | 5.27 | 5.39 | 5.88 | 6.21 | 6.66 | 7.21 |
| 13 | 3.06 | 3.73 | 4.15 | 4.45 | 4.69 | 4.88 | 5.05 | 5.19 | 5.32 | 5.79 | 6.11 | 6.55 | 7.08 |
| 14 | 3.03 | 3.70 | 4.11 | 4.41 | 4.64 | 4.83 | 4.99 | 5.13 | 5.25 | 5.71 | 6.03 | 6.46 | 6.98 |
| 15 | 3.01 | 3.67 | 4.08 | 4.37 | 4.59 | 4.78 | 4.94 | 5.08 | 5.20 | 5.65 | 5.96 | 6.38 | 6.89 |
| 16 | 3.00 | 3.65 | 4.05 | 4.33 | 4.56 | 4.74 | 4.90 | 5.03 | 5.15 | 5.59 | 5.90 | 6.31 | 6.81 |
| 17 | 2.98 | 3.63 | 4.02 | 4.30 | 4.52 | 4.70 | 4.86 | 4.99 | 5.11 | 5.54 | 5.84 | 6.25 | 6.74 |
| 18 | 2.97 | 3.61 | 4.00 | 4.28 | 4.49 | 4.67 | 4.82 | 4.96 | 5.07 | 5.50 | 5.79 | 6.20 | 6.68 |
| 19 | 2.96 | 3.59 | 3.98 | 4.25 | 4.47 | 4.65 | 4.79 | 4.92 | 5.04 | 5.46 | 5.75 | 6.15 | 6.63 |
| 20 | 2.95 | 3.58 | 3.96 | 4.23 | 4.45 | 4.62 | 4.77 | 4.90 | 5.01 | 5.43 | 5.71 | 6.10 | 6.58 |
| 21 | 2.94 | 3.56 | 3.94 | 4.21 | 4.42 | 4.60 | 4.74 | 4.87 | 4.98 | 5.40 | 5.68 | 6.07 | 6.53 |
| 22 | 2.93 | 3.55 | 3.93 | 4.20 | 4.41 | 4.58 | 4.72 | 4.85 | 4.96 | 5.37 | 5.65 | 6.03 | 6.49 |
| 23 | 2.93 | 3.54 | 3.91 | 4.18 | 4.39 | 4.56 | 4.70 | 4.83 | 4.94 | 5.34 | 5.62 | 6.00 | 6.45 |
| 24 | 2.92 | 3.53 | 3.90 | 4.17 | 4.37 | 4.54 | 4.68 | 4.81 | 4.92 | 5.32 | 5.59 | 5.97 | 6.42 |
| 25 | 2.91 | 3.52 | 3.89 | 4.15 | 4.36 | 4.53 | 4.67 | 4.79 | 4.90 | 5.30 | 5.57 | 5.94 | 6.39 |
| 26 | 2.91 | 3.51 | 3.88 | 4.14 | 4.35 | 4.51 | 4.65 | 4.77 | 4.88 | 5.28 | 5.55 | 5.92 | 6.36 |
| 27 | 2.90 | 3.51 | 3.87 | 4.13 | 4.33 | 4.50 | 4.64 | 4.76 | 4.86 | 5.26 | 5.53 | 5.89 | 6.34 |
| 28 | 2.90 | 3.50 | 3.86 | 4.12 | 4.32 | 4.49 | 4.62 | 4.74 | 4.85 | 5.24 | 5.51 | 5.87 | 6.31 |
| 29 | 2.89 | 3.49 | 3.85 | 4.11 | 4.31 | 4.47 | 4.61 | 4.73 | 4.84 | 5.23 | 5.49 | 5.85 | 6.29 |
| 30 | 2.89 | 3.49 | 3.85 | 4.10 | 4.30 | 4.46 | 4.60 | 4.72 | 4.82 | 5.21 | 5.47 | 5.83 | 6.27 |
| 35 | 2.87 | 3.46 | 3.81 | 4.07 | 4.26 | 4.42 | 4.56 | 4.67 | 4.77 | 5.15 | 5.41 | 5.76 | 6.18 |
| 40 | 2.86 | 3.44 | 3.79 | 4.04 | 4.23 | 4.39 | 4.52 | 4.63 | 4.73 | 5.11 | 5.36 | 5.70 | 6.11 |
| 45 | 2.85 | 3.43 | 3.77 | 4.02 | 4.21 | 4.36 | 4.49 | 4.61 | 4.70 | 5.07 | 5.32 | 5.66 | 6.06 |
| 50 | 2.84 | 3.42 | 3.76 | 4.00 | 4.19 | 4.34 | 4.47 | 4.58 | 4.68 | 5.04 | 5.29 | 5.62 | 6.02 |
| 100 | 2.81 | 3.36 | 3.70 | 3.93 | 4.11 | 4.26 | 4.38 | 4.48 | 4.58 | 4.92 | 5.15 | 5.46 | 5.83 |
| $\infty$ | 2.77 | 3.31 | 3.63 | 3.86 | 4.03 | 4.17 | 4.29 | 4.39 | 4.47 | 4.80 | 5.01 | 5.30 | 5.65 |

Table D.8: Percent points for the Studentized range, continued.

Table entries are $q_{.01}(K, \nu)$.

| | K | | | | | | | | | | | |
|---|---|---|---|---|---|---|---|---|---|---|---|---|
| $\nu$ | 2 | 3 | 4 | 5 | 6 | 7 | 8 | 9 | 10 | 15 | 20 | 30 | 50 |
| 1 | 90.2 | 135 | 164 | 186 | 202 | 216 | 227 | 237 | 246 | 277 | 298 | 326 | 359 |
| 2 | 14.0 | 19.0 | 22.3 | 24.7 | 26.6 | 28.2 | 29.5 | 30.7 | 31.7 | 35.4 | 38.0 | 41.3 | 45.3 |
| 3 | 8.27 | 10.6 | 12.2 | 13.3 | 14.2 | 15.0 | 15.6 | 16.2 | 16.7 | 18.5 | 19.8 | 21.4 | 23.4 |
| 4 | 6.51 | 8.12 | 9.17 | 9.96 | 10.6 | 11.1 | 11.5 | 11.9 | 12.3 | 13.5 | 14.4 | 15.6 | 17.0 |
| 5 | 5.70 | 6.98 | 7.80 | 8.42 | 8.91 | 9.32 | 9.67 | 9.97 | 10.2 | 11.2 | 11.9 | 12.9 | 14.0 |
| 6 | 5.24 | 6.33 | 7.03 | 7.56 | 7.97 | 8.32 | 8.61 | 8.87 | 9.10 | 9.95 | 10.5 | 11.3 | 12.3 |
| 7 | 4.95 | 5.92 | 6.54 | 7.01 | 7.37 | 7.68 | 7.94 | 8.17 | 8.37 | 9.12 | 9.65 | 10.4 | 11.2 |
| 8 | 4.75 | 5.64 | 6.20 | 6.62 | 6.96 | 7.24 | 7.47 | 7.68 | 7.86 | 8.55 | 9.03 | 9.68 | 10.5 |
| 9 | 4.60 | 5.43 | 5.96 | 6.35 | 6.66 | 6.91 | 7.13 | 7.33 | 7.49 | 8.13 | 8.57 | 9.18 | 9.91 |
| 10 | 4.48 | 5.27 | 5.77 | 6.14 | 6.43 | 6.67 | 6.87 | 7.05 | 7.21 | 7.81 | 8.23 | 8.79 | 9.49 |
| 11 | 4.39 | 5.15 | 5.62 | 5.97 | 6.25 | 6.48 | 6.67 | 6.84 | 6.99 | 7.56 | 7.95 | 8.49 | 9.15 |
| 12 | 4.32 | 5.05 | 5.50 | 5.84 | 6.10 | 6.32 | 6.51 | 6.67 | 6.81 | 7.36 | 7.73 | 8.25 | 8.87 |
| 13 | 4.26 | 4.96 | 5.40 | 5.73 | 5.98 | 6.19 | 6.37 | 6.53 | 6.67 | 7.19 | 7.55 | 8.04 | 8.65 |
| 14 | 4.21 | 4.89 | 5.32 | 5.63 | 5.88 | 6.08 | 6.26 | 6.41 | 6.54 | 7.05 | 7.39 | 7.87 | 8.46 |
| 15 | 4.17 | 4.84 | 5.25 | 5.56 | 5.80 | 5.99 | 6.16 | 6.31 | 6.44 | 6.93 | 7.26 | 7.73 | 8.29 |
| 16 | 4.13 | 4.79 | 5.19 | 5.49 | 5.72 | 5.92 | 6.08 | 6.22 | 6.35 | 6.82 | 7.15 | 7.60 | 8.15 |
| 17 | 4.10 | 4.74 | 5.14 | 5.43 | 5.66 | 5.85 | 6.01 | 6.15 | 6.27 | 6.73 | 7.05 | 7.49 | 8.03 |
| 18 | 4.07 | 4.70 | 5.09 | 5.38 | 5.60 | 5.79 | 5.94 | 6.08 | 6.20 | 6.65 | 6.97 | 7.40 | 7.92 |
| 19 | 4.05 | 4.67 | 5.05 | 5.33 | 5.55 | 5.73 | 5.89 | 6.02 | 6.14 | 6.58 | 6.89 | 7.31 | 7.83 |
| 20 | 4.02 | 4.64 | 5.02 | 5.29 | 5.51 | 5.69 | 5.84 | 5.97 | 6.09 | 6.52 | 6.82 | 7.24 | 7.74 |
| 21 | 4.00 | 4.61 | 4.99 | 5.26 | 5.47 | 5.65 | 5.79 | 5.92 | 6.04 | 6.47 | 6.76 | 7.17 | 7.67 |
| 22 | 3.99 | 4.59 | 4.96 | 5.22 | 5.43 | 5.61 | 5.75 | 5.88 | 5.99 | 6.42 | 6.71 | 7.11 | 7.60 |
| 23 | 3.97 | 4.57 | 4.93 | 5.20 | 5.40 | 5.57 | 5.72 | 5.84 | 5.95 | 6.37 | 6.66 | 7.05 | 7.53 |
| 24 | 3.96 | 4.55 | 4.91 | 5.17 | 5.37 | 5.54 | 5.69 | 5.81 | 5.92 | 6.33 | 6.61 | 7.00 | 7.48 |
| 25 | 3.94 | 4.53 | 4.89 | 5.14 | 5.35 | 5.51 | 5.65 | 5.78 | 5.89 | 6.29 | 6.57 | 6.95 | 7.42 |
| 26 | 3.93 | 4.51 | 4.87 | 5.12 | 5.32 | 5.49 | 5.63 | 5.75 | 5.86 | 6.26 | 6.53 | 6.91 | 7.37 |
| 27 | 3.92 | 4.49 | 4.85 | 5.10 | 5.30 | 5.46 | 5.60 | 5.72 | 5.83 | 6.22 | 6.50 | 6.87 | 7.33 |
| 28 | 3.91 | 4.48 | 4.83 | 5.08 | 5.28 | 5.44 | 5.58 | 5.70 | 5.80 | 6.20 | 6.47 | 6.84 | 7.29 |
| 29 | 3.90 | 4.47 | 4.81 | 5.06 | 5.26 | 5.42 | 5.56 | 5.67 | 5.78 | 6.17 | 6.44 | 6.80 | 7.25 |
| 30 | 3.89 | 4.45 | 4.80 | 5.05 | 5.24 | 5.40 | 5.54 | 5.65 | 5.76 | 6.14 | 6.41 | 6.77 | 7.21 |
| 35 | 3.85 | 4.40 | 4.74 | 4.98 | 5.17 | 5.32 | 5.45 | 5.57 | 5.67 | 6.04 | 6.29 | 6.64 | 7.07 |
| 40 | 3.82 | 4.37 | 4.70 | 4.93 | 5.11 | 5.26 | 5.39 | 5.50 | 5.60 | 5.96 | 6.21 | 6.55 | 6.96 |
| 45 | 3.80 | 4.34 | 4.66 | 4.89 | 5.07 | 5.22 | 5.34 | 5.45 | 5.55 | 5.90 | 6.14 | 6.47 | 6.88 |
| 50 | 3.79 | 4.32 | 4.63 | 4.86 | 5.04 | 5.19 | 5.31 | 5.41 | 5.51 | 5.85 | 6.09 | 6.42 | 6.81 |
| 100 | 3.71 | 4.22 | 4.52 | 4.73 | 4.90 | 5.03 | 5.14 | 5.24 | 5.33 | 5.65 | 5.86 | 6.16 | 6.51 |
| $\infty$ | 3.64 | 4.12 | 4.40 | 4.60 | 4.76 | 4.88 | 4.99 | 5.08 | 5.16 | 5.45 | 5.65 | 5.91 | 6.23 |

**Table D.9:** Critical values for one-sided Dunnett's $t$.

Entries are $d'_{.05}(K,\nu)$ where $P(\max_{j=1}^{K} t_{0j} > d'_{.05}(K,\nu)) = .05$ .

| | K | | | | | | | | | | | | |
|---|---|---|---|---|---|---|---|---|---|---|---|---|---|
| $\nu$ | 2 | 3 | 4 | 5 | 6 | 7 | 8 | 9 | 10 | 15 | 20 | 30 | 40 |
| 1 | 9.51 | 11.6 | 13.1 | 14.3 | 15.2 | 16.0 | 16.7 | 17.3 | 17.9 | 19.9 | 21.3 | 23.2 | 24.5 |
| 2 | 3.80 | 4.34 | 4.71 | 5.00 | 5.24 | 5.43 | 5.60 | 5.75 | 5.88 | 6.38 | 6.72 | 7.18 | 7.50 |
| 3 | 2.94 | 3.28 | 3.52 | 3.70 | 3.85 | 3.97 | 4.08 | 4.17 | 4.25 | 4.56 | 4.78 | 5.07 | 5.27 |
| 4 | 2.61 | 2.88 | 3.08 | 3.22 | 3.34 | 3.44 | 3.52 | 3.59 | 3.66 | 3.90 | 4.07 | 4.30 | 4.46 |
| 5 | 2.44 | 2.68 | 2.85 | 2.98 | 3.08 | 3.16 | 3.24 | 3.30 | 3.36 | 3.57 | 3.71 | 3.92 | 4.05 |
| 6 | 2.34 | 2.56 | 2.71 | 2.83 | 2.92 | 3.00 | 3.06 | 3.12 | 3.17 | 3.37 | 3.50 | 3.68 | 3.81 |
| 7 | 2.27 | 2.48 | 2.62 | 2.73 | 2.81 | 2.89 | 2.95 | 3.00 | 3.05 | 3.23 | 3.36 | 3.53 | 3.64 |
| 8 | 2.22 | 2.42 | 2.55 | 2.66 | 2.74 | 2.81 | 2.87 | 2.92 | 2.96 | 3.14 | 3.25 | 3.41 | 3.52 |
| 9 | 2.18 | 2.37 | 2.50 | 2.60 | 2.68 | 2.75 | 2.81 | 2.86 | 2.90 | 3.06 | 3.18 | 3.33 | 3.44 |
| 10 | 2.15 | 2.34 | 2.47 | 2.56 | 2.64 | 2.70 | 2.76 | 2.81 | 2.85 | 3.01 | 3.12 | 3.27 | 3.37 |
| 11 | 2.13 | 2.31 | 2.43 | 2.53 | 2.60 | 2.67 | 2.72 | 2.77 | 2.81 | 2.96 | 3.07 | 3.21 | 3.31 |
| 12 | 2.11 | 2.29 | 2.41 | 2.50 | 2.58 | 2.64 | 2.69 | 2.74 | 2.78 | 2.93 | 3.03 | 3.17 | 3.27 |
| 13 | 2.09 | 2.27 | 2.39 | 2.48 | 2.55 | 2.61 | 2.66 | 2.71 | 2.75 | 2.90 | 3.00 | 3.14 | 3.23 |
| 14 | 2.08 | 2.25 | 2.37 | 2.46 | 2.53 | 2.59 | 2.64 | 2.69 | 2.73 | 2.87 | 2.97 | 3.11 | 3.20 |
| 15 | 2.07 | 2.24 | 2.36 | 2.44 | 2.51 | 2.57 | 2.62 | 2.67 | 2.71 | 2.85 | 2.95 | 3.08 | 3.17 |
| 16 | 2.06 | 2.23 | 2.34 | 2.43 | 2.50 | 2.56 | 2.61 | 2.65 | 2.69 | 2.83 | 2.93 | 3.06 | 3.15 |
| 17 | 2.05 | 2.22 | 2.33 | 2.42 | 2.49 | 2.54 | 2.59 | 2.64 | 2.67 | 2.81 | 2.91 | 3.04 | 3.13 |
| 18 | 2.04 | 2.21 | 2.32 | 2.41 | 2.48 | 2.53 | 2.58 | 2.62 | 2.66 | 2.80 | 2.89 | 3.02 | 3.11 |
| 19 | 2.03 | 2.20 | 2.31 | 2.40 | 2.47 | 2.52 | 2.57 | 2.61 | 2.65 | 2.79 | 2.88 | 3.01 | 3.10 |
| 20 | 2.03 | 2.19 | 2.30 | 2.39 | 2.46 | 2.51 | 2.56 | 2.60 | 2.64 | 2.77 | 2.87 | 2.99 | 3.08 |
| 21 | 2.02 | 2.19 | 2.30 | 2.38 | 2.45 | 2.50 | 2.55 | 2.59 | 2.63 | 2.76 | 2.86 | 2.98 | 3.07 |
| 22 | 2.02 | 2.18 | 2.29 | 2.37 | 2.44 | 2.50 | 2.54 | 2.58 | 2.62 | 2.75 | 2.85 | 2.97 | 3.06 |
| 23 | 2.01 | 2.17 | 2.28 | 2.37 | 2.43 | 2.49 | 2.54 | 2.58 | 2.61 | 2.75 | 2.84 | 2.96 | 3.05 |
| 24 | 2.01 | 2.17 | 2.28 | 2.36 | 2.43 | 2.48 | 2.53 | 2.57 | 2.60 | 2.74 | 2.83 | 2.95 | 3.04 |
| 25 | 2.00 | 2.17 | 2.27 | 2.36 | 2.42 | 2.48 | 2.52 | 2.56 | 2.60 | 2.73 | 2.82 | 2.94 | 3.03 |
| 26 | 2.00 | 2.16 | 2.27 | 2.35 | 2.42 | 2.47 | 2.52 | 2.56 | 2.59 | 2.72 | 2.81 | 2.94 | 3.02 |
| 27 | 2.00 | 2.16 | 2.27 | 2.35 | 2.41 | 2.47 | 2.51 | 2.55 | 2.59 | 2.72 | 2.81 | 2.93 | 3.01 |
| 28 | 1.99 | 2.15 | 2.26 | 2.34 | 2.41 | 2.46 | 2.51 | 2.55 | 2.58 | 2.71 | 2.80 | 2.92 | 3.01 |
| 29 | 1.99 | 2.15 | 2.26 | 2.34 | 2.40 | 2.46 | 2.50 | 2.54 | 2.58 | 2.71 | 2.80 | 2.92 | 3.00 |
| 30 | 1.99 | 2.15 | 2.25 | 2.34 | 2.40 | 2.45 | 2.50 | 2.54 | 2.57 | 2.70 | 2.79 | 2.91 | 2.99 |
| 35 | 1.98 | 2.13 | 2.24 | 2.32 | 2.38 | 2.44 | 2.48 | 2.52 | 2.55 | 2.68 | 2.77 | 2.89 | 2.97 |
| 40 | 1.97 | 2.13 | 2.23 | 2.31 | 2.37 | 2.42 | 2.47 | 2.51 | 2.54 | 2.67 | 2.75 | 2.87 | 2.95 |
| 45 | 1.96 | 2.12 | 2.22 | 2.30 | 2.36 | 2.41 | 2.46 | 2.50 | 2.53 | 2.66 | 2.74 | 2.86 | 2.94 |
| 50 | 1.96 | 2.11 | 2.22 | 2.29 | 2.36 | 2.41 | 2.45 | 2.49 | 2.52 | 2.65 | 2.73 | 2.85 | 2.93 |
| 100 | 1.94 | 2.09 | 2.19 | 2.26 | 2.32 | 2.37 | 2.42 | 2.45 | 2.48 | 2.61 | 2.69 | 2.80 | 2.88 |
| $\infty$ | 1.92 | 2.06 | 2.16 | 2.23 | 2.29 | 2.34 | 2.38 | 2.42 | 2.45 | 2.57 | 2.65 | 2.75 | 2.83 |

Table D.9: Critical values for one-sided Dunnett's $t$, continued.

Entries are $d'_{.01}(K, \nu)$ where $P(\max_{j=1}^{K} t_{0j} > d'_{.01}(K, \nu)) = .01$ .

| | K | | | | | | | | | | | | |
|---|---|---|---|---|---|---|---|---|---|---|---|---|---|
| $\nu$ | 2 | 3 | 4 | 5 | 6 | 7 | 8 | 9 | 10 | 15 | 20 | 30 | 40 |
| 1 | 47.7 | 58.1 | 65.6 | 71.5 | 76.3 | 80.3 | 83.8 | 86.8 | 89.5 | 99.6 | 107 | 116 | 122 |
| 2 | 8.88 | 10.0 | 10.9 | 11.5 | 12.0 | 12.5 | 12.8 | 13.2 | 13.5 | 14.6 | 15.3 | 16.4 | 17.1 |
| 3 | 5.48 | 6.04 | 6.44 | 6.74 | 6.99 | 7.20 | 7.38 | 7.54 | 7.67 | 8.20 | 8.56 | 9.06 | 9.41 |
| 4 | 4.41 | 4.80 | 5.07 | 5.28 | 5.45 | 5.59 | 5.72 | 5.82 | 5.92 | 6.28 | 6.53 | 6.87 | 7.11 |
| 5 | 3.90 | 4.21 | 4.43 | 4.60 | 4.73 | 4.85 | 4.94 | 5.03 | 5.11 | 5.39 | 5.59 | 5.87 | 6.06 |
| 6 | 3.61 | 3.88 | 4.06 | 4.21 | 4.32 | 4.42 | 4.51 | 4.58 | 4.64 | 4.89 | 5.06 | 5.30 | 5.46 |
| 7 | 3.42 | 3.66 | 3.83 | 3.96 | 4.06 | 4.15 | 4.22 | 4.29 | 4.35 | 4.57 | 4.72 | 4.93 | 5.08 |
| 8 | 3.29 | 3.51 | 3.66 | 3.78 | 3.88 | 3.96 | 4.03 | 4.09 | 4.14 | 4.35 | 4.49 | 4.68 | 4.81 |
| 9 | 3.19 | 3.40 | 3.54 | 3.66 | 3.75 | 3.82 | 3.89 | 3.94 | 3.99 | 4.18 | 4.31 | 4.49 | 4.62 |
| 10 | 3.11 | 3.31 | 3.45 | 3.56 | 3.64 | 3.72 | 3.78 | 3.83 | 3.88 | 4.06 | 4.18 | 4.35 | 4.47 |
| 11 | 3.06 | 3.25 | 3.38 | 3.48 | 3.56 | 3.63 | 3.69 | 3.74 | 3.79 | 3.96 | 4.08 | 4.24 | 4.35 |
| 12 | 3.01 | 3.19 | 3.32 | 3.42 | 3.50 | 3.56 | 3.62 | 3.67 | 3.71 | 3.88 | 3.99 | 4.15 | 4.26 |
| 13 | 2.97 | 3.15 | 3.27 | 3.37 | 3.44 | 3.51 | 3.56 | 3.61 | 3.65 | 3.81 | 3.92 | 4.08 | 4.18 |
| 14 | 2.94 | 3.11 | 3.23 | 3.33 | 3.40 | 3.46 | 3.52 | 3.56 | 3.60 | 3.76 | 3.87 | 4.01 | 4.12 |
| 15 | 2.91 | 3.08 | 3.20 | 3.29 | 3.36 | 3.42 | 3.47 | 3.52 | 3.56 | 3.71 | 3.82 | 3.96 | 4.06 |
| 16 | 2.88 | 3.05 | 3.17 | 3.26 | 3.33 | 3.39 | 3.44 | 3.48 | 3.52 | 3.67 | 3.78 | 3.92 | 4.01 |
| 17 | 2.86 | 3.03 | 3.14 | 3.23 | 3.30 | 3.36 | 3.41 | 3.45 | 3.49 | 3.64 | 3.74 | 3.88 | 3.97 |
| 18 | 2.84 | 3.01 | 3.12 | 3.21 | 3.28 | 3.33 | 3.38 | 3.43 | 3.46 | 3.61 | 3.71 | 3.84 | 3.94 |
| 19 | 2.83 | 2.99 | 3.10 | 3.18 | 3.25 | 3.31 | 3.36 | 3.40 | 3.44 | 3.58 | 3.68 | 3.81 | 3.90 |
| 20 | 2.81 | 2.97 | 3.08 | 3.17 | 3.23 | 3.29 | 3.34 | 3.38 | 3.42 | 3.56 | 3.65 | 3.78 | 3.88 |
| 21 | 2.80 | 2.96 | 3.07 | 3.15 | 3.22 | 3.27 | 3.32 | 3.36 | 3.40 | 3.53 | 3.63 | 3.76 | 3.85 |
| 22 | 2.79 | 2.94 | 3.05 | 3.13 | 3.20 | 3.25 | 3.30 | 3.34 | 3.38 | 3.51 | 3.61 | 3.74 | 3.83 |
| 23 | 2.78 | 2.93 | 3.04 | 3.12 | 3.18 | 3.24 | 3.28 | 3.33 | 3.36 | 3.50 | 3.59 | 3.72 | 3.81 |
| 24 | 2.77 | 2.92 | 3.03 | 3.11 | 3.17 | 3.22 | 3.27 | 3.31 | 3.35 | 3.48 | 3.57 | 3.70 | 3.79 |
| 25 | 2.76 | 2.91 | 3.02 | 3.10 | 3.16 | 3.21 | 3.26 | 3.30 | 3.33 | 3.47 | 3.56 | 3.68 | 3.77 |
| 26 | 2.75 | 2.90 | 3.01 | 3.08 | 3.15 | 3.20 | 3.25 | 3.29 | 3.32 | 3.45 | 3.54 | 3.67 | 3.75 |
| 27 | 2.74 | 2.89 | 3.00 | 3.07 | 3.14 | 3.19 | 3.24 | 3.27 | 3.31 | 3.44 | 3.53 | 3.65 | 3.74 |
| 28 | 2.74 | 2.88 | 2.99 | 3.07 | 3.13 | 3.18 | 3.22 | 3.26 | 3.30 | 3.43 | 3.52 | 3.64 | 3.72 |
| 29 | 2.73 | 2.88 | 2.98 | 3.06 | 3.12 | 3.17 | 3.22 | 3.25 | 3.29 | 3.42 | 3.51 | 3.63 | 3.71 |
| 30 | 2.72 | 2.87 | 2.97 | 3.05 | 3.11 | 3.16 | 3.21 | 3.25 | 3.28 | 3.41 | 3.50 | 3.62 | 3.70 |
| 35 | 2.70 | 2.84 | 2.94 | 3.02 | 3.08 | 3.13 | 3.17 | 3.21 | 3.24 | 3.37 | 3.45 | 3.57 | 3.65 |
| 40 | 2.68 | 2.82 | 2.92 | 2.99 | 3.05 | 3.10 | 3.14 | 3.18 | 3.21 | 3.34 | 3.42 | 3.54 | 3.62 |
| 45 | 2.67 | 2.81 | 2.90 | 2.98 | 3.03 | 3.08 | 3.12 | 3.16 | 3.19 | 3.31 | 3.40 | 3.51 | 3.59 |
| 50 | 2.65 | 2.79 | 2.89 | 2.96 | 3.02 | 3.07 | 3.11 | 3.14 | 3.18 | 3.30 | 3.38 | 3.49 | 3.57 |
| 100 | 2.61 | 2.74 | 2.83 | 2.90 | 2.95 | 3.00 | 3.04 | 3.07 | 3.10 | 3.22 | 3.29 | 3.40 | 3.47 |
| $\infty$ | 2.56 | 2.69 | 2.77 | 2.84 | 2.89 | 2.93 | 2.97 | 3.00 | 3.03 | 3.14 | 3.21 | 3.31 | 3.38 |

Table D.9: Critical values for two-sided Dunnett's $t$, continued.

Entries are $d_{.05}(K,\nu)$ where $P(\max_{j=1}^{K} t_{0j} > d_{.05}(K,\nu)) = .05$ .

| | K | | | | | | | | | | | | |
|---|---|---|---|---|---|---|---|---|---|---|---|---|---|
| $\nu$ | 2 | 3 | 4 | 5 | 6 | 7 | 8 | 9 | 10 | 15 | 20 | 30 | 40 |
| 1 | 17.4 | 20.0 | 21.9 | 23.2 | 24.3 | 25.2 | 25.9 | 26.6 | 27.1 | 29.3 | 30.7 | 32.6 | 33.9 |
| 2 | 5.42 | 6.06 | 6.51 | 6.85 | 7.12 | 7.35 | 7.54 | 7.71 | 7.85 | 8.40 | 8.77 | 9.28 | 9.62 |
| 3 | 3.87 | 4.26 | 4.54 | 4.75 | 4.92 | 5.06 | 5.18 | 5.28 | 5.37 | 5.72 | 5.95 | 6.27 | 6.49 |
| 4 | 3.31 | 3.62 | 3.83 | 3.99 | 4.13 | 4.23 | 4.33 | 4.41 | 4.48 | 4.75 | 4.94 | 5.19 | 5.36 |
| 5 | 3.03 | 3.29 | 3.48 | 3.62 | 3.73 | 3.82 | 3.90 | 3.97 | 4.03 | 4.26 | 4.42 | 4.64 | 4.79 |
| 6 | 2.86 | 3.10 | 3.26 | 3.39 | 3.49 | 3.57 | 3.64 | 3.71 | 3.76 | 3.97 | 4.11 | 4.31 | 4.45 |
| 7 | 2.75 | 2.97 | 3.12 | 3.24 | 3.33 | 3.41 | 3.47 | 3.53 | 3.58 | 3.78 | 3.91 | 4.09 | 4.22 |
| 8 | 2.67 | 2.88 | 3.02 | 3.13 | 3.22 | 3.29 | 3.35 | 3.41 | 3.46 | 3.64 | 3.76 | 3.93 | 4.05 |
| 9 | 2.61 | 2.81 | 2.95 | 3.05 | 3.14 | 3.20 | 3.26 | 3.32 | 3.36 | 3.53 | 3.65 | 3.82 | 3.93 |
| 10 | 2.57 | 2.76 | 2.89 | 2.99 | 3.07 | 3.14 | 3.19 | 3.24 | 3.29 | 3.45 | 3.57 | 3.72 | 3.83 |
| 11 | 2.53 | 2.72 | 2.84 | 2.94 | 3.02 | 3.08 | 3.14 | 3.19 | 3.23 | 3.39 | 3.50 | 3.65 | 3.76 |
| 12 | 2.50 | 2.68 | 2.81 | 2.90 | 2.98 | 3.04 | 3.09 | 3.14 | 3.18 | 3.34 | 3.45 | 3.59 | 3.69 |
| 13 | 2.48 | 2.65 | 2.78 | 2.87 | 2.94 | 3.00 | 3.06 | 3.10 | 3.14 | 3.29 | 3.40 | 3.54 | 3.64 |
| 14 | 2.46 | 2.63 | 2.75 | 2.84 | 2.91 | 2.97 | 3.02 | 3.07 | 3.11 | 3.26 | 3.36 | 3.50 | 3.60 |
| 15 | 2.44 | 2.61 | 2.73 | 2.82 | 2.89 | 2.95 | 3.00 | 3.04 | 3.08 | 3.23 | 3.33 | 3.47 | 3.56 |
| 16 | 2.42 | 2.59 | 2.71 | 2.80 | 2.87 | 2.92 | 2.97 | 3.02 | 3.06 | 3.20 | 3.30 | 3.43 | 3.53 |
| 17 | 2.41 | 2.58 | 2.69 | 2.78 | 2.85 | 2.90 | 2.95 | 3.00 | 3.03 | 3.18 | 3.27 | 3.41 | 3.50 |
| 18 | 2.40 | 2.56 | 2.68 | 2.76 | 2.83 | 2.89 | 2.94 | 2.98 | 3.01 | 3.16 | 3.25 | 3.38 | 3.48 |
| 19 | 2.39 | 2.55 | 2.66 | 2.75 | 2.81 | 2.87 | 2.92 | 2.96 | 3.00 | 3.14 | 3.23 | 3.36 | 3.45 |
| 20 | 2.38 | 2.54 | 2.65 | 2.73 | 2.80 | 2.86 | 2.90 | 2.95 | 2.98 | 3.12 | 3.22 | 3.34 | 3.43 |
| 21 | 2.37 | 2.53 | 2.64 | 2.72 | 2.79 | 2.84 | 2.89 | 2.93 | 2.97 | 3.11 | 3.20 | 3.33 | 3.42 |
| 22 | 2.36 | 2.52 | 2.63 | 2.71 | 2.78 | 2.83 | 2.88 | 2.92 | 2.96 | 3.09 | 3.19 | 3.31 | 3.40 |
| 23 | 2.36 | 2.51 | 2.62 | 2.70 | 2.77 | 2.82 | 2.87 | 2.91 | 2.95 | 3.08 | 3.17 | 3.30 | 3.38 |
| 24 | 2.35 | 2.51 | 2.61 | 2.70 | 2.76 | 2.81 | 2.86 | 2.90 | 2.94 | 3.07 | 3.16 | 3.29 | 3.37 |
| 25 | 2.34 | 2.50 | 2.61 | 2.69 | 2.75 | 2.81 | 2.85 | 2.89 | 2.93 | 3.06 | 3.15 | 3.27 | 3.36 |
| 26 | 2.34 | 2.49 | 2.60 | 2.68 | 2.74 | 2.80 | 2.84 | 2.88 | 2.92 | 3.05 | 3.14 | 3.26 | 3.35 |
| 27 | 2.33 | 2.49 | 2.59 | 2.67 | 2.74 | 2.79 | 2.84 | 2.88 | 2.91 | 3.04 | 3.13 | 3.25 | 3.34 |
| 28 | 2.33 | 2.48 | 2.59 | 2.67 | 2.73 | 2.78 | 2.83 | 2.87 | 2.90 | 3.03 | 3.12 | 3.24 | 3.33 |
| 29 | 2.32 | 2.48 | 2.58 | 2.66 | 2.73 | 2.78 | 2.82 | 2.86 | 2.90 | 3.03 | 3.11 | 3.24 | 3.32 |
| 30 | 2.32 | 2.47 | 2.58 | 2.66 | 2.72 | 2.77 | 2.82 | 2.86 | 2.89 | 3.02 | 3.11 | 3.23 | 3.31 |
| 35 | 2.30 | 2.46 | 2.56 | 2.64 | 2.70 | 2.75 | 2.79 | 2.83 | 2.86 | 2.99 | 3.08 | 3.20 | 3.28 |
| 40 | 2.29 | 2.44 | 2.54 | 2.62 | 2.68 | 2.73 | 2.77 | 2.81 | 2.84 | 2.97 | 3.05 | 3.17 | 3.25 |
| 45 | 2.28 | 2.43 | 2.53 | 2.61 | 2.67 | 2.72 | 2.76 | 2.80 | 2.83 | 2.95 | 3.04 | 3.15 | 3.23 |
| 50 | 2.28 | 2.42 | 2.52 | 2.60 | 2.66 | 2.71 | 2.75 | 2.79 | 2.82 | 2.94 | 3.02 | 3.14 | 3.22 |
| 100 | 2.24 | 2.39 | 2.48 | 2.55 | 2.61 | 2.66 | 2.70 | 2.74 | 2.77 | 2.88 | 2.96 | 3.07 | 3.15 |
| $\infty$ | 2.21 | 2.35 | 2.44 | 2.51 | 2.57 | 2.61 | 2.65 | 2.69 | 2.72 | 2.83 | 2.91 | 3.01 | 3.08 |

Table D.9: Critical values for two-sided Dunnett's $t$, continued.

Entries are $d_{.01}(K, \nu)$ where $P(\max_{j=1}^{K} t_{0j} > d_{.01}(K, \nu)) = .01$ .

| | K | | | | | | | | | | | | |
|---|---|---|---|---|---|---|---|---|---|---|---|---|---|
| $\nu$ | 2 | 3 | 4 | 5 | 6 | 7 | 8 | 9 | 10 | 15 | 20 | 30 | 40 |
| 1 | 87.0 | 100 | 109 | 116 | 122 | 126 | 130 | 133 | 136 | 146 | 154 | 163 | 169 |
| 2 | 12.4 | 13.8 | 14.8 | 15.6 | 16.2 | 16.7 | 17.1 | 17.5 | 17.8 | 19.1 | 19.9 | 21.0 | 21.8 |
| 3 | 6.97 | 7.64 | 8.10 | 8.46 | 8.75 | 8.99 | 9.19 | 9.37 | 9.53 | 10.1 | 10.5 | 11.1 | 11.5 |
| 4 | 5.36 | 5.81 | 6.12 | 6.36 | 6.55 | 6.72 | 6.85 | 6.98 | 7.08 | 7.49 | 7.77 | 8.15 | 8.41 |
| 5 | 4.63 | 4.97 | 5.22 | 5.41 | 5.56 | 5.68 | 5.79 | 5.89 | 5.97 | 6.29 | 6.51 | 6.81 | 7.02 |
| 6 | 4.21 | 4.51 | 4.71 | 4.87 | 5.00 | 5.10 | 5.20 | 5.28 | 5.35 | 5.62 | 5.80 | 6.06 | 6.24 |
| 7 | 3.95 | 4.21 | 4.39 | 4.53 | 4.64 | 4.74 | 4.82 | 4.89 | 4.95 | 5.19 | 5.35 | 5.58 | 5.74 |
| 8 | 3.77 | 4.00 | 4.17 | 4.29 | 4.40 | 4.48 | 4.56 | 4.62 | 4.68 | 4.90 | 5.05 | 5.25 | 5.40 |
| 9 | 3.63 | 3.85 | 4.01 | 4.12 | 4.22 | 4.30 | 4.37 | 4.43 | 4.48 | 4.68 | 4.82 | 5.01 | 5.15 |
| 10 | 3.53 | 3.74 | 3.88 | 3.99 | 4.08 | 4.16 | 4.22 | 4.28 | 4.33 | 4.52 | 4.65 | 4.83 | 4.96 |
| 11 | 3.45 | 3.65 | 3.79 | 3.89 | 3.98 | 4.05 | 4.11 | 4.16 | 4.21 | 4.39 | 4.52 | 4.69 | 4.81 |
| 12 | 3.39 | 3.58 | 3.71 | 3.81 | 3.89 | 3.96 | 4.02 | 4.07 | 4.12 | 4.29 | 4.41 | 4.57 | 4.69 |
| 13 | 3.33 | 3.52 | 3.65 | 3.74 | 3.82 | 3.89 | 3.94 | 3.99 | 4.04 | 4.20 | 4.32 | 4.48 | 4.59 |
| 14 | 3.29 | 3.47 | 3.59 | 3.69 | 3.76 | 3.83 | 3.88 | 3.93 | 3.97 | 4.13 | 4.24 | 4.40 | 4.50 |
| 15 | 3.25 | 3.43 | 3.55 | 3.64 | 3.71 | 3.78 | 3.83 | 3.88 | 3.92 | 4.07 | 4.18 | 4.33 | 4.43 |
| 16 | 3.22 | 3.39 | 3.51 | 3.60 | 3.67 | 3.73 | 3.78 | 3.83 | 3.87 | 4.02 | 4.13 | 4.27 | 4.37 |
| 17 | 3.19 | 3.36 | 3.47 | 3.56 | 3.63 | 3.69 | 3.74 | 3.79 | 3.83 | 3.98 | 4.08 | 4.22 | 4.32 |
| 18 | 3.17 | 3.33 | 3.45 | 3.53 | 3.60 | 3.66 | 3.71 | 3.75 | 3.79 | 3.94 | 4.04 | 4.18 | 4.28 |
| 19 | 3.15 | 3.31 | 3.42 | 3.50 | 3.57 | 3.63 | 3.68 | 3.72 | 3.76 | 3.90 | 4.00 | 4.14 | 4.24 |
| 20 | 3.13 | 3.29 | 3.40 | 3.48 | 3.55 | 3.60 | 3.65 | 3.69 | 3.73 | 3.87 | 3.97 | 4.11 | 4.20 |
| 21 | 3.11 | 3.27 | 3.37 | 3.46 | 3.52 | 3.58 | 3.63 | 3.67 | 3.71 | 3.85 | 3.94 | 4.08 | 4.17 |
| 22 | 3.09 | 3.25 | 3.36 | 3.44 | 3.50 | 3.56 | 3.61 | 3.65 | 3.68 | 3.82 | 3.92 | 4.05 | 4.14 |
| 23 | 3.08 | 3.23 | 3.34 | 3.42 | 3.48 | 3.54 | 3.59 | 3.63 | 3.66 | 3.80 | 3.89 | 4.02 | 4.11 |
| 24 | 3.07 | 3.22 | 3.32 | 3.40 | 3.47 | 3.52 | 3.57 | 3.61 | 3.64 | 3.78 | 3.87 | 4.00 | 4.09 |
| 25 | 3.05 | 3.21 | 3.31 | 3.39 | 3.45 | 3.51 | 3.55 | 3.59 | 3.63 | 3.76 | 3.85 | 3.98 | 4.07 |
| 26 | 3.04 | 3.19 | 3.30 | 3.37 | 3.44 | 3.49 | 3.54 | 3.58 | 3.61 | 3.74 | 3.83 | 3.96 | 4.05 |
| 27 | 3.03 | 3.18 | 3.28 | 3.36 | 3.42 | 3.48 | 3.52 | 3.56 | 3.60 | 3.73 | 3.82 | 3.94 | 4.03 |
| 28 | 3.03 | 3.17 | 3.27 | 3.35 | 3.41 | 3.46 | 3.51 | 3.55 | 3.58 | 3.71 | 3.80 | 3.93 | 4.01 |
| 29 | 3.02 | 3.16 | 3.26 | 3.34 | 3.40 | 3.45 | 3.50 | 3.54 | 3.57 | 3.70 | 3.79 | 3.91 | 3.99 |
| 30 | 3.01 | 3.15 | 3.25 | 3.33 | 3.39 | 3.44 | 3.49 | 3.52 | 3.56 | 3.69 | 3.77 | 3.90 | 3.98 |
| 35 | 2.98 | 3.12 | 3.22 | 3.29 | 3.35 | 3.40 | 3.44 | 3.48 | 3.51 | 3.64 | 3.72 | 3.84 | 3.92 |
| 40 | 2.95 | 3.09 | 3.19 | 3.26 | 3.32 | 3.37 | 3.41 | 3.44 | 3.48 | 3.60 | 3.68 | 3.80 | 3.88 |
| 45 | 2.93 | 3.07 | 3.16 | 3.24 | 3.29 | 3.34 | 3.38 | 3.42 | 3.45 | 3.57 | 3.65 | 3.76 | 3.84 |
| 50 | 2.92 | 3.05 | 3.15 | 3.22 | 3.27 | 3.32 | 3.36 | 3.40 | 3.43 | 3.55 | 3.63 | 3.74 | 3.82 |
| 100 | 2.86 | 2.98 | 3.07 | 3.14 | 3.19 | 3.24 | 3.27 | 3.31 | 3.34 | 3.45 | 3.52 | 3.63 | 3.70 |
| $\infty$ | 2.79 | 2.92 | 3.00 | 3.06 | 3.11 | 3.15 | 3.19 | 3.22 | 3.25 | 3.35 | 3.42 | 3.52 | 3.59 |

**Table D.10:** Power curves for fixed-effects ANOVA.

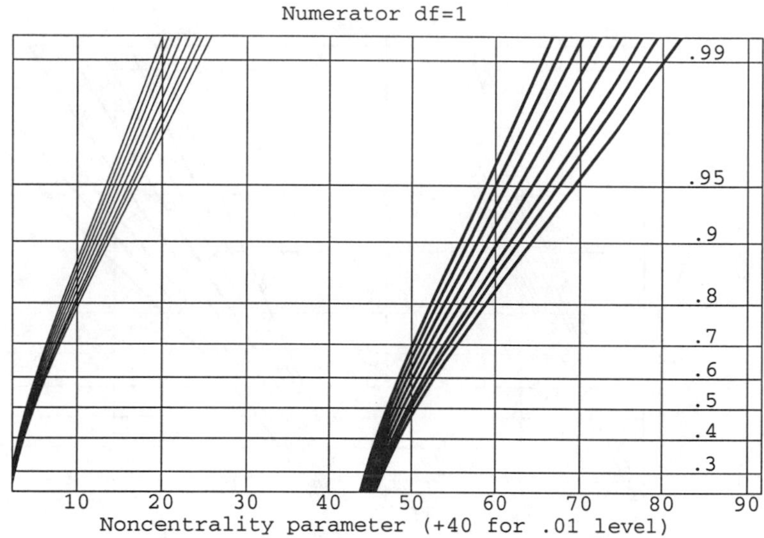


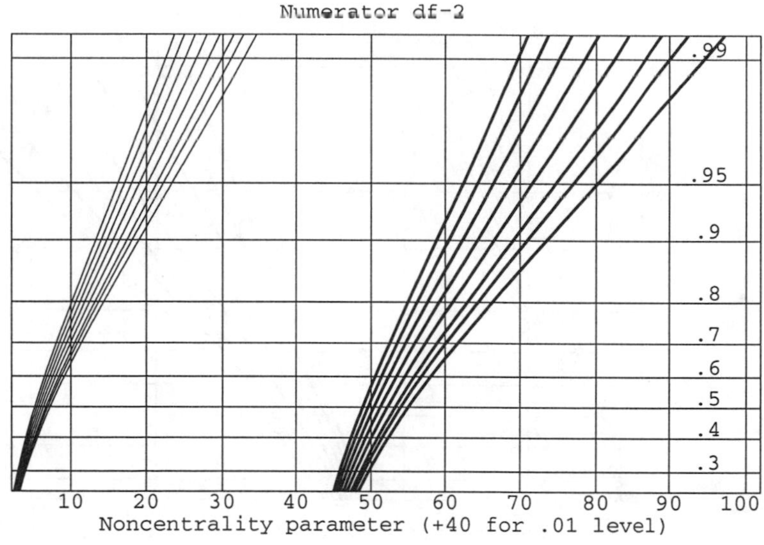

Table D.10: Power curves for fixed-effects ANOVA, continued.

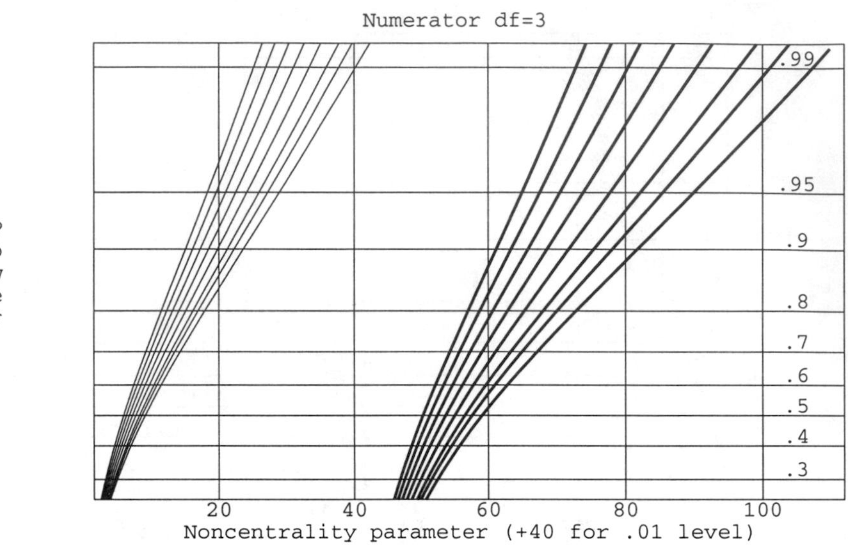


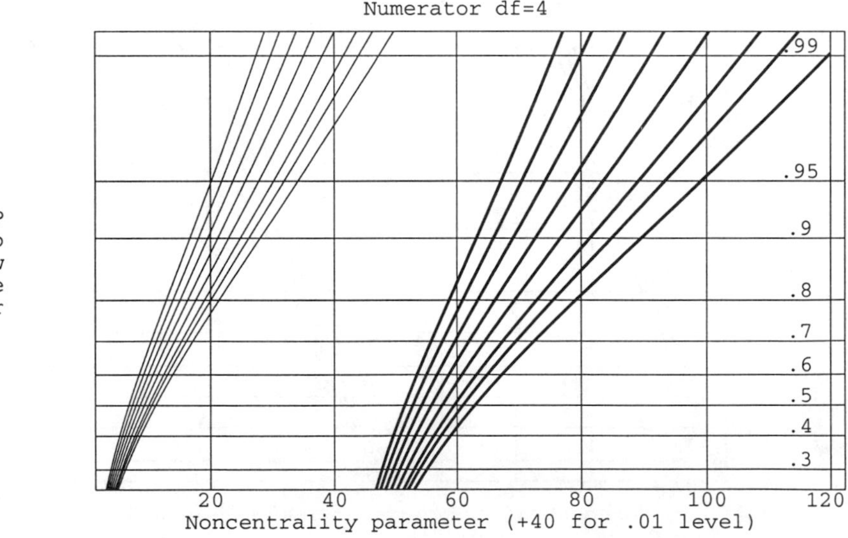

Table D.10: Power curves for fixed-effects ANOVA, continued.

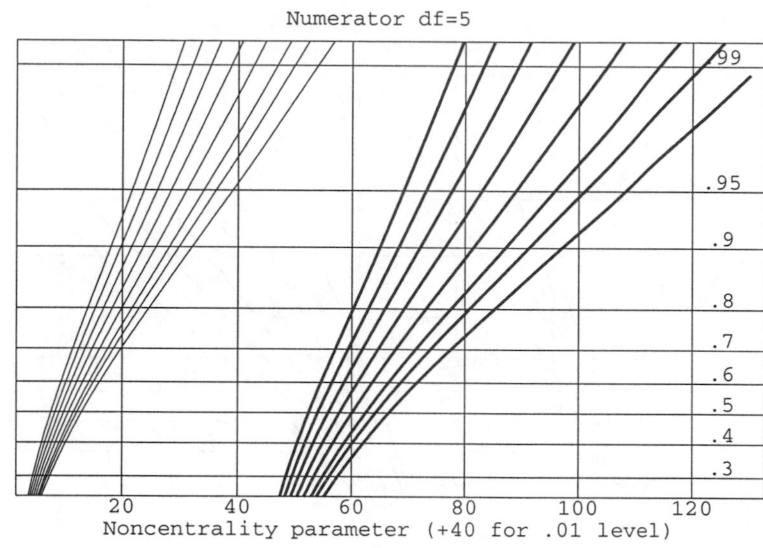


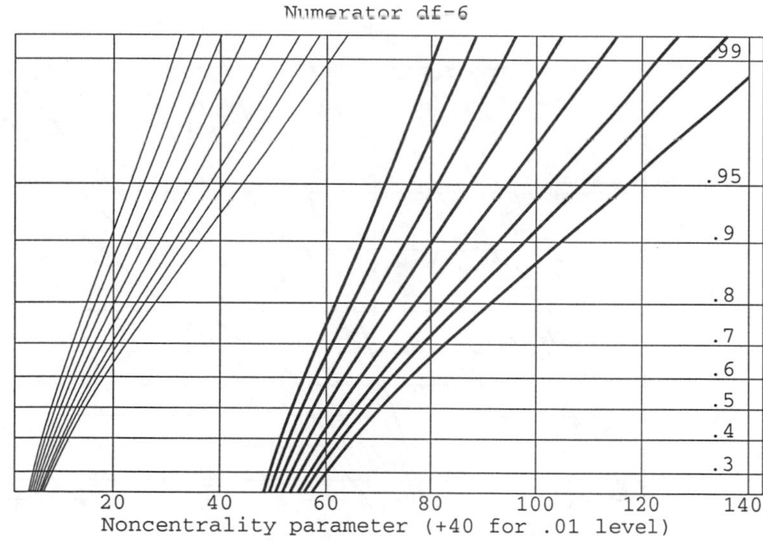

Table D.10: Power curves for fixed-effects ANOVA, continued.

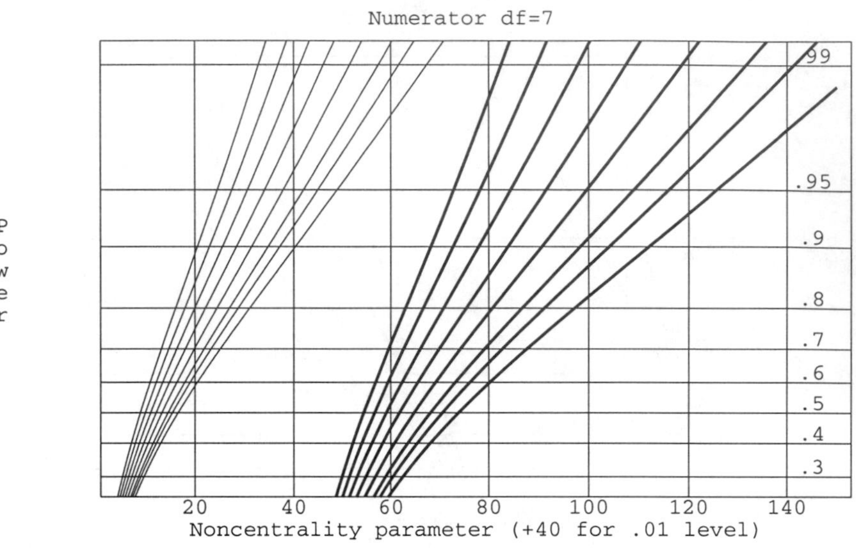


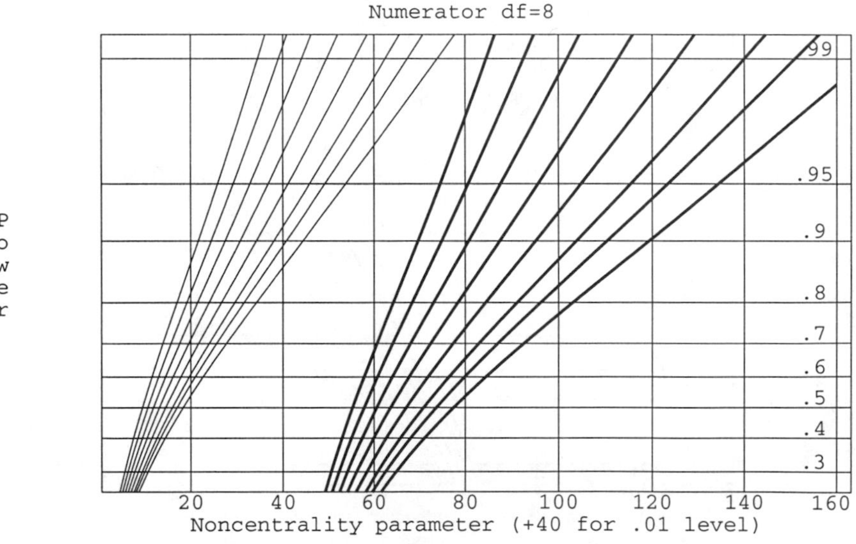

**Table D.11:** Power curves for random-effects ANOVA.

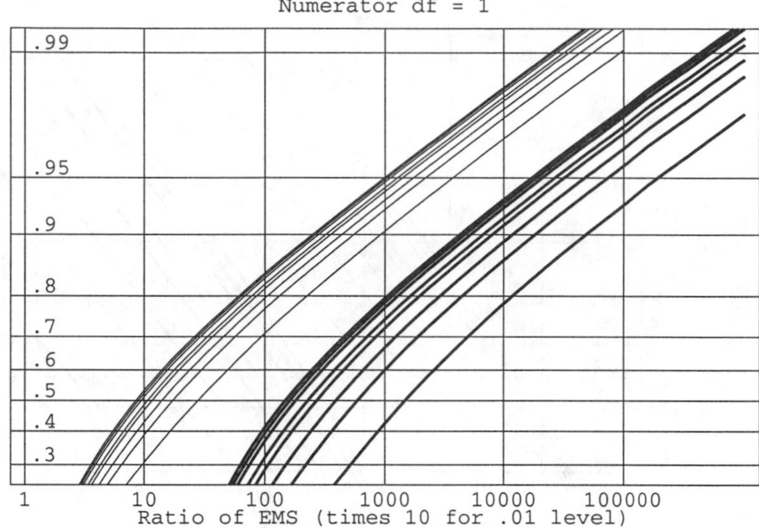

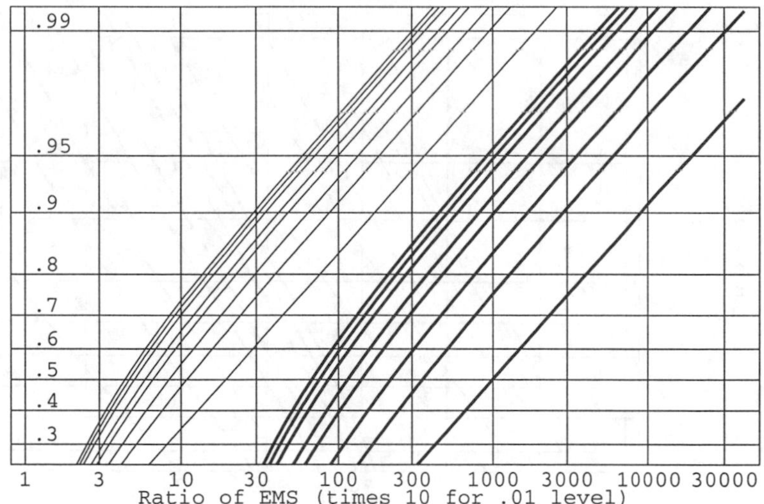

Table D.11: Power curves for random-effects ANOVA, continued.

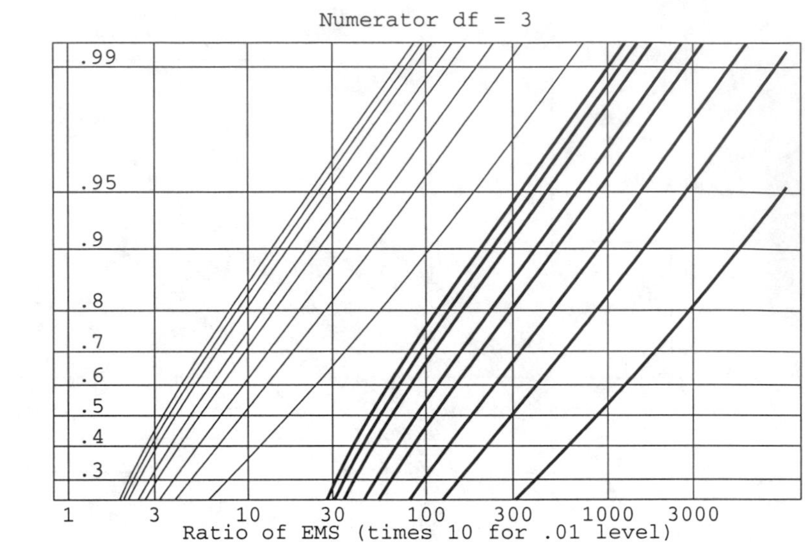


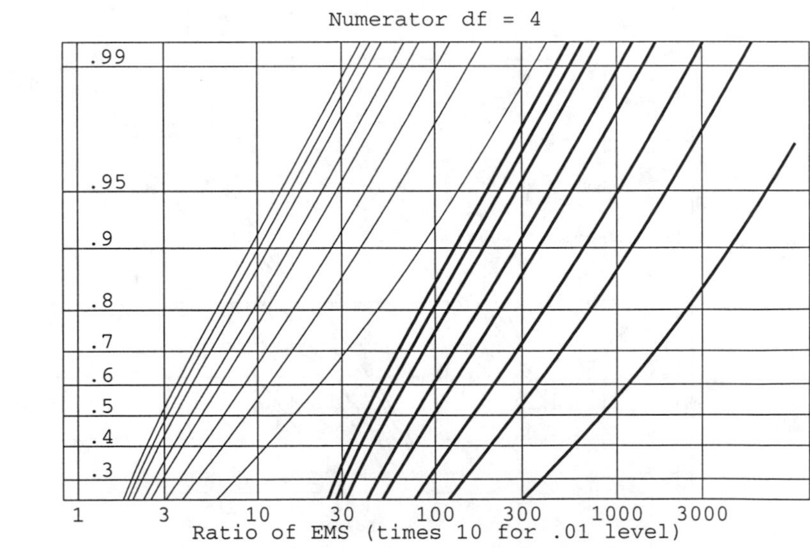

Table D.11: Power curves for random-effects ANOVA, continued.

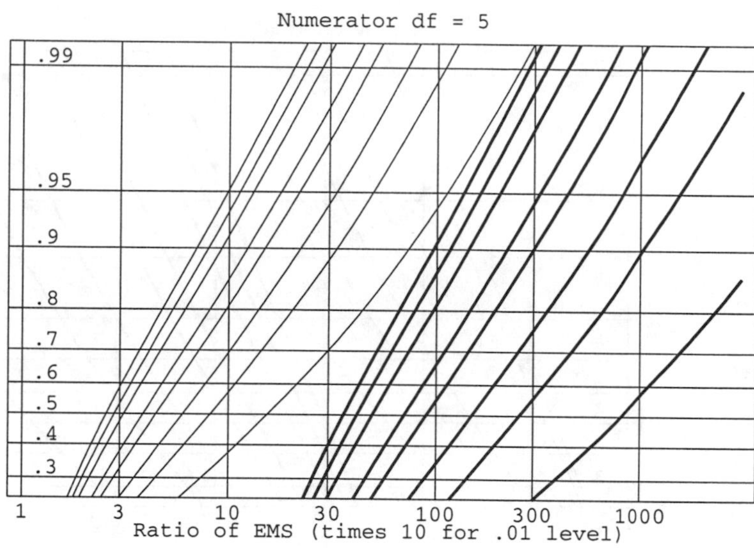


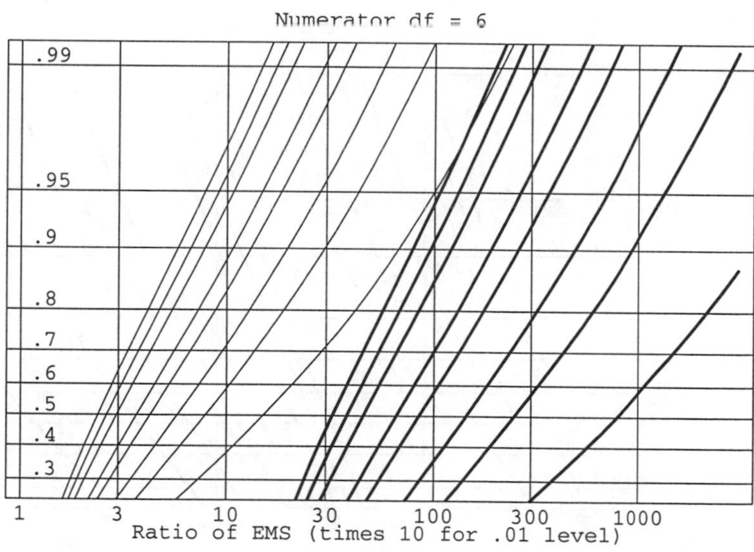

Table D.11: Power curves for random-effects ANOVA, continued.

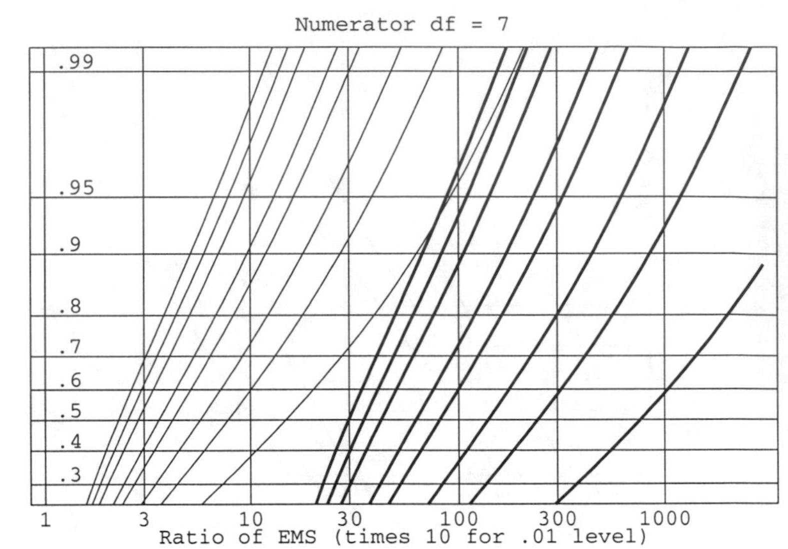


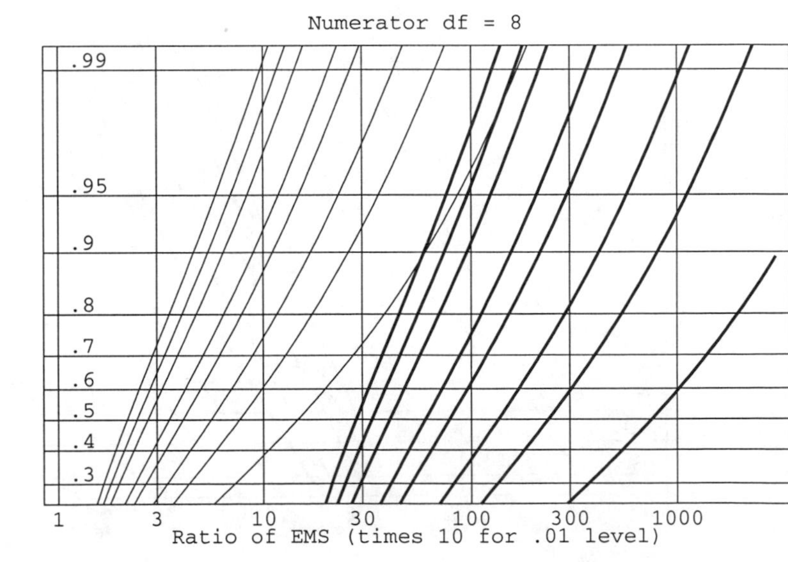

# Index